S_pringer_ **M**_onographs in_ **M**_athematics_

Springer
Berlin
Heidelberg
New York
Barcelona
Budapest
Hong Kong
London
Milan
Paris
Santa Clara
Singapore
Tokyo

Dedicated to my parents

Foreword

Algebraic topology is supported by three pillars: "ordinary" Cohomology, K-theory, and Cobordism. An introduction to the first pillar and some of its applications constitute the curriculum of a typical first year graduate course. There have been all too few books addressed to students who have completed such an introduction, and the present volume is the first such guide in the subject of Cobordism since Robert Stong's encyclopedic and influential notes of a generation ago.

The pioneering work of Pontryagin and Thom forged a deep connection between certain geometric problems (such as the classification of manifolds) and homotopy theory, through the medium of the Thom space. Computations become possible upon stabilization, and this provided some of the first and most compelling examples of "spectra."

Since its inception the subject has thus represented a merger of the Russian and Western mathematical schools. This international tradition was continued with the more or less simultaneous work by Novikov and Milnor on complex cobordism, and later by Quillen. More recently Dennis Sullivan opened the way to the study of "manifolds with singularities," a study taken up most forcefully by the Russian school, notably by Vershinin, Botvinnik, and Rudyak.

Attention to pedagogy is another Russian tradition which you will find amply fulfilled in this book. There is a fine introduction to the stable homotopy category. The subtle and increasingly important issue of phantom maps is addressed here with care. Equally careful is the treatment of orientability, a subject to which the author has contributed greatly. And the various aspects of the theory of Cobordism, especially the central case of complex cobordism, are naturally given a detailed and ample telling.

Professor Rudyak has also performed a service to the history of science in this book, giving detailed and informed attributions. This same care makes the book easy to use by the student, for when proofs are not given here specific references are.

It is to be hoped that this book is the first in a new generation of textbooks, reflecting the current vigor of the subject.

Haynes Miller
Cambridge, MA
January, 1998

Preface

I started to write this book in Moscow and finished in Heidelberg. I am grateful to the Chair of Higher Mathematics of the Moscow Civil Engineering Institute (the Chairman is S.Ja.Havinson) and to the Forschergruppe "Topologie und nichtkommutative Geometrie" (sponsored by Deutsche Forschungsgemeinschaft) at the Mathematical Institute of the University of Heidelberg which partially supported me during the writing of the book.

I express my especial gratitude to my Ph.D. advisor Michael Postnikov, and I am glad to thank all the participants of the Algebraic Topology Seminar (supervised by M.Postnikov) at Moscow State University, where my topological tastes and preferences were formed and developed.

I am also very grateful to Albrecht Dold, Hans–Werner Henn and Dieter Puppe for a lot of fruitful discussions at Heidelberg.

Among the others whose suggestions have served me well, I note Nils Baas, Boris Botvinnik, Victor Buchstaber, Alexei Chernavski, Werner End, Oswald Gschnitzer, Edward Fadell, Andrey Khokhlov, Karl-Heinz Knapp, Matthias Kreck, Peter Landweber, Ran Levi, Mark Mahowald, Peter May, Haynes Miller, Alexandr Mishchenko, Sergey Novikov, Erich Ossa, Andrey Pazhitnov, Yury Solov'ev, Michael Stan'ko, Robert Stong, Paul Turner, and Vladimir Vershinin.

I am grateful to the mathematics editorial of Springer-Verlag (Heidelberg), especially to Ruth Allewelt, Martin Peters and Catriona Byrne, for their kind assistance and help. Thanks are also due to the production editors Leonie Kunz and Karl-Friedrich Koch as well as to the TEX editor Thomas Rudolf.

I would like to thank Samuel Maltby for proof-reading the manuscript for publication.

Finally, Gregor Hoffleit, Waldemar Klemm and Hannes Reinecke helped me with UNIX and TEX.

To all of these and others who have helped me, I express my best thanks.

Yuli B. Rudyak
Heidelberg
February, 1998

Table of Contents

Introduction

> First, tell what you are going to talk about, then tell this, and then tell what you have talked about.
>
> *Manuals of a senior country priest for beginners*

The contents of this book are concentrated around Thom spaces (spectra), orientability theory and (co)bordism theory (including (co)bordism with singularities), framed by (co)homology theories and spectra. These matters have formed one of the main lines of development for the last 50 years in the area of algebraic and geometric topology. In the book I consider some results obtained in this field in the last 20–30 years, settled enough in order to be exposed in a monograph and close to my research interests. As far as I know, there are no books which cover substantial parts of the presented material.

In the book I tried to prove those referenced results which were not proved in any monograph (unfortunately, there are a few exceptions there). Moreover, when I quote a result which I do not prove here, I quote the original paper and a monograph where this result is treated as well. There are also occasional remarks containing historical and bibliographical comments, additional results not included in the text, exercises, etc.

A reference to Theorem III.4.5 is to Theorem 4.5 of Ch. III (which is in §4 of the chapter); if the chapter number is omitted, it is to a theorem of the chapter at hand.

The scheme of interconnections of chapters is very simple:

$$\begin{array}{c} \mathrm{I} \Rightarrow \mathrm{II} \Rightarrow \mathrm{III} \Rightarrow \mathrm{IV} \Rightarrow \mathrm{V} \Rightarrow \mathrm{VII} \Rightarrow \mathrm{VIII} \Rightarrow \mathrm{IX} \\ \Downarrow \\ \mathrm{VI} \end{array}$$

I will not overview the contents, but I will discuss the subject of the book and the place which it occupies in algebraic topology.

Conceptional foundations

From the conceptual point of view, we consider the (inter)connections between geometry and homotopy theory, since Thom spectra and related matters are now the main tools for this interplay. Here I say a few words about this.

Algebraic topology studies topological spaces via their algebraic invariants. Evidently, these algebraic invariants should be simple enough in order to be computable and deep (and so complicated) enough in order to keep some essential information about a space. How does algebraic topology succeed in slipping between these two dangers: non-computable informativity and non-informative computability? The answer is that *homotopy* provides a desired balance between informativity and computability. Therefore, a reasonable way from topology to algebra passes through homotopy theory. (If you like artistic expressions, I can say that homotopy theory works like a camera when we make an algebraic photograph of the topological world.) In other words, one should reduce a geometric problem to a homotopic one and then compute the corresponding homotopy invariants. Thus, interconnections between geometry and homotopy theory play a pivotal role in algebraic topology.

One of the first results in this area was the Gauss–Bonnet formula, relating a geometrical invariant (the curvature) to a homotopical one (the Euler characteristic). Proceeding, we can recall the Riemann–Roch Theorem, the Poincaré integrality theory, relationships between critical points of a smooth function on a smooth manifold and its homotopy type (Lusternik–Schnirelmann, Morse), the de Rham Theorem, etc. Hence, the geometry–homotopy interconnections are very classical things, with a noble genealogy. On the other hand, we'll see below that this instrument works very successfully in the present as well.

The Characters

Here I discuss (briefly and roughly, because the body of the book contains the details) the main concepts which appear in the book.

(Co)homology theories. We reserve the term "classical cohomology" or "ordinary cohomology" for the functors $H^*(-;G)$. The term "cohomology theory" is used for what was previously called "generalized" or "extraordinary" cohomology theory, i.e., for functors which satisfy all the Eilenberg–Steenrod axioms except the dimension axioms. Similarly for homology theories.

Every homology theory $h_*(-)$ yields a so-called dual cohomology theory $h^*(-)$, and vice versa. They are connected via the equality $\widetilde{h}^i(X) = \widetilde{h}_{n-i}(Y)$ where Y is n-dual to X (and tilde denotes the reduced (co)homology).

Thom spaces. The Thom space $T\xi$ of a locally trivial $\mathbb{R}^n$-bundle $\xi = \{p : E \to B\}$ is defined as follows. Let $\xi^\bullet$ be the S^n-bundle obtained from ξ by the fiberwise one-point compactification, and let $E^\bullet$ be the total space of $\xi^\bullet$. Then the "infinities" of the fibers form a section $s : B \to E^\bullet$, and we define $T\xi := E^\bullet/s(B)$. Furthermore, the Thom space of a spherical fibration $\{p : E \to B\}$ is the cone $C(p)$ of the projection p. For example, the Thom space of the $\mathbb{R}^n$-bundle over a point is S^n, the Thom space of the open Möbius band (considered as the $\mathbb{R}^1$-bundle over S^1) is the real projective plane RP^2, the Thom space of the Hopf bundle $S^3 \to S^2$ (with fiber S^1) is the complex projective plane CP^2. We use Thom's notation MO_n for the Thom space $T\gamma^n$ of the universal n-dimensional vector bundle γ^n over the classifying space BO_n, i.e., $MO_n := T\gamma^n$; e.g., $MO_1 = RP^\infty$.

A source of interest in Thom spaces is the unifying role which they play in algebraic topology. Namely, they interlock geometric topology and homotopy theory and, in particular, enable us to apply methods of one of them to problems of the other. Now I discuss some examples.

J.H.C. Whitehead observed the importance of the structure on the normal bundle in classifying structures on manifolds. It turns out that *Thom spaces* establish an adequate context for this. Namely, for every closed smooth manifold M^n, the set of (diffeomorphism classes of) smooth manifolds homotopy equivalent to M is controlled by the group $\pi_{n+N}(T\nu)$, where ν is the normal bundle of an embedding of M in $\mathbb{R}^{n+N}$ with N large enough, see Novikov [2,3], Browder [1,2].

This is closely related to the Milnor–Spanier–Atiyah Duality Theorem, which asserts that $T\nu$ and $M/\partial M$ are stable N-dual for every compact manifold M. This theorem clarifies connections between manifolds and their normal bundles and enables us to transmit properties of bundles to properties of manifolds. For example, we have the Thom isomorphism $\varphi : H^i(X;\mathbb{Z}/2) \to \widetilde{H}^{i+n}(T\xi;\mathbb{Z}/2)$ for every locally trivial $\mathbb{R}^n$-bundle ξ over a space X, and the above theorem transforms it to the Poincaré duality $H^i(M;\mathbb{Z}/2) \cong H_{n-i}(M,\partial M;\mathbb{Z}/2)$ for every compact n-dimensional manifold M.

Turning to another example, I recall the Thom formula

$$w_i(\xi) = \varphi^{-1} Sq^i u_\xi$$

where ξ is an n-dimensional vector bundle over a space X, $w_i(\xi)$ is its i-th Stiefel–Whitney class, $\varphi : H^i(X;\mathbb{Z}/2) \to \widetilde{H}^{i+n}(T\xi;\mathbb{Z}/2)$ is the Thom isomorphism and $u_\xi \in H^n(T\xi;\mathbb{Z}/2)$ is the Thom class of ξ. This formula expands a geometric invariant (the Stiefel–Whitney class) via the Steenrod operation which is a purely homotopic thing. Moreover, we can use the formula in order to *define* the Stiefel–Whitney classes of spherical fibrations. In particular, it becomes clear that the Stiefel–Whitney classes are invariants of the fiber homotopy type of a vector bundle. I note also that, in the book Milnor–Stasheff [1], the authors preferred to define the Stiefel–Whitney classes via the Thom formula and not to use the original geometric definitions.

Generalizing, we can consider an arbitrary natural transformation $\tau : h^* \to k^*$ of cohomology theories instead of Sq^i. Then, under suitable conditions on ξ, there is a generalized Thom class $u_\xi^h \in \widetilde{h}^n(T\xi)$ and a generalized Thom isomorphism $\varphi_k : k^i(X) \to \widetilde{k}^{i+n}(T\xi)$, and so we can form the class

$$K(\xi) = \varphi_k^{-1} \tau u_\xi^h$$

which is an analogue and generalization of the Stiefel–Whitney class. So, we have a large source of invariants of $\mathbb{R}^n$-bundles. For example, the Todd genus and the $\widehat{A}$ genus are particular cases of this construction. Moreover, the well-known integrality theorems which are related to Todd and $\widehat{A}$ genera can be generalized for the class K as above, see Ch. V, §3.

Now I turn to the most impressive example: the relations of Thom spaces to (co)bordism.

Bordism and cobordism. To start with, consider the following problem. Given a closed manifold M, how can one recognize whether it bounds, i.e., when is it the boundary of a compact manifold? This can be developed as follows. One says that two closed manifolds M and N are bordant if the manifold $M \sqcup N$ (the disjoint union of M and N) bounds. Clearly, "to be bordant" is an equivalence relation, and so we have a set $\mathfrak{N}_k$ of bordantness classes of k-dimensional manifolds. It is easy to see that $\mathfrak{N}_k$ is a group with respect to disjoint union; it is called the *bordism group* of k-dimensional manifolds.

Pontrjagin [1] proved that if a manifold bounds then all its characteristic numbers are trivial. In particular, RP^2 does not bound because $w_2(RP^2) \neq 0$. So, $\mathfrak{N}_2 \neq 0$, i.e., some groups $\mathfrak{N}_k$ are non-trivial.

Well, but how to compute $\mathfrak{N}_k$? Clearly, $\mathfrak{N}_0 = \mathbb{Z}/2$, $\mathfrak{N}_1 = 0$. Using the classification of closed surfaces, one can prove that $\mathfrak{N}_2 = \mathbb{Z}/2$: every orientable surface bounds, and every non-orientable surface either bounds or is bordant to RP^2; and RP^2 does not bound. Rokhlin [1] proved that $\mathfrak{N}_3 = 0$, using complicated and tricky geometry. The further computation of $\mathfrak{N}_k$ looked absolutely hopeless; however this was done by Thom [2] via an exciting and successful application of *homotopy theory*. Namely, Thom proved that

$$\mathfrak{N}_k = \pi_{k+N}(M\mathcal{O}_N)$$

for N large enough. Now one can apply all the mighty machinery of homotopy theory and compute the right hand side groups. Thom did it and thus computed the groups $\mathfrak{N}_i$. The answer is

$$\mathfrak{N}_* = \mathbb{Z}/2\,[x_i],\ \dim x_i = i,\ i \in \mathbb{N}, i \neq 2^s - 1$$

where $\mathfrak{N}_* = \oplus \mathfrak{N}_k$ is the graded ring with the multiplication induced by the direct product of manifolds. It is important to remark that, fortunately, powerful computational methods (obtained mainly by the French topological school) came just at the right time, and Thom took advantage of this.

I want to mention here that Pontrjagin (in 1937, the available publication is Pontrjagin [2]) interpreted homotopy groups of spheres in terms of smooth manifolds, and this result anticipated the contemporary research in the area of interconnections between homotopy theory and geometry. In fact, as remarked by Stong [2], "Thom brought the Pontrjagin technique to the study of manifolds, largely reversing the original idea".

The above constructions can be generalized: we can consider oriented manifolds or, more generally, manifolds equipped with some extra structures. As above, there arise certain bordism groups, and they can be interpreted as homotopy groups of certain Thom spaces. Many mathematicians studied and study this zoo of bordism groups. The monograph Stong [2] summed up this level of the development of the theory.

Proceeding, consider two maps $f : M \to X$ and $g : N \to X$ of closed smooth k-dimensional manifolds M, N. We say that these maps are bordant if there is a map $F : W \to X$ with $\partial W = M \sqcup N$ and $F|M \sqcup N = f \sqcup g$. Similarly to the above, "to be bordant" is an equivalence relation, and we have a bordism group $\mathfrak{N}_k(X)$. One can prove that in this way we get a homology theory $\mathfrak{N}_*(-)$ which is called a *bordism theory*. The dual cohomology theory $\mathfrak{N}^k(-)$ is called a *cobordism theory*.

Clearly, $\mathfrak{N}_k = \mathfrak{N}_k(\mathrm{pt})$. Moreover,

$$\mathfrak{N}_k(X) = \pi_{k+N}(X^+ \wedge M\mathcal{O}_N)$$

for N large enough, where X^+ is the disjoint union of the space X and a point.

Spectra. The reader should have noted that we deal with the condition "N large enough", i.e., with the so-called stable situation. However, as remarked by Milnor [4], it is much more pleasant to work in a category where there is, say, a single object $M\mathcal{O}$ rather than the spaces $M\mathcal{O}_n$ which approximate it, i.e., "to put $N = \infty$". This approach has a convenient formalization; its main tool is the conception of a spectrum. There are different categories of spectra proposed by different authors, and for some particular applications one may have an advantage over another.

We use a category of spectra proposed by Adams [5]. So, a spectrum E is a sequence $\{E_n, s_n\}_{n=-\infty}^{\infty}$ of pointed CW-spaces E_n and pointed CW-embeddings $s_n : SE_n \to E_{n+1}$ where S denotes the pointed suspension. There are the following examples.

(1) For every pointed space X we have the spectrum $\Sigma^\infty X = \{S^n X, s_n\}$ where $s_n : SS^n X \to S^{n+1} X$ is the identity map.

(2) For every pointed space X and every spectrum $E = \{E_n, s_n\}$ we have the spectrum $X \wedge E = \{X \wedge E_n, 1 \wedge s_n\}$.

(3) Let θ^1 be the trivial 1-dimensional vector bundle over $B\mathcal{O}_n$, and let the map $B\mathcal{O}_n \to B\mathcal{O}_{n+1}$ (assuming it to be an embedding) classify the vector bundle $\gamma^n \oplus \theta^1$. Then we have a map $s_n : T(\gamma^n \oplus \theta^1) \to T\gamma^{n+1}$. Moreover,

one can prove that $T(\gamma^n \oplus \theta^1) = ST\gamma^n = SMO_n$, and so we have the *Thom spectrum* $MO = \{MO_n, s_n\}$.

Given a spectrum E, we have the homomorphisms

$$h_{k,n} : \pi_k(E_n) \to \pi_{k+1}(SE_n) \xrightarrow{(s_n)_*} \pi_{k+1}(E_{n+1}).$$

We define the homotopy group $\pi_k(E)$ to be the direct limit of the sequence

$$\cdots \to \pi_{k+n}(E_n) \xrightarrow{h_{k+n,n}} \pi_{k+n+1}(E_{n+1}) \to \cdots ,$$

i.e., $\pi_k(E) = \lim_{n\to\infty} \pi_{i+n}(E_n)$. Now we can rewrite the above equalities as

$$\mathfrak{N}_k = \pi_k(MO), \quad \mathfrak{N}_k(X) = \pi_k(X^+ \wedge MO)$$

and so get rid of "N large enough".

More generally, we can define bordism groups for manifolds with a structure, and they can also be interpreted as homotopy groups of certain Thom spectra.

There is a remarkable connection between spectra and (co)homology theories. Every spectrum E yields a homology theory $E_*(-)$ and a cohomology theory $E^*(-)$ by the formulae

$$E_i(X) := \lim_{n\to\infty} \pi_{i+n}(X^+ \wedge E_n), \qquad E^i(X) := \lim_{n\to\infty} [S^n X^+, E_{i+n}].$$

Moreover, $E_*(-)$ and $E^*(-)$ are dual to each other.

Conversely, every (co)homology theory can be represented by a spectrum via the above formulae.

Note that, in particular, the spectrum MO yields the bordism (resp. cobordism) theory $\mathfrak{N}_*(-)$ (resp. $\mathfrak{N}^*(-)$).

Orientability. We consider in this book orientability with respect to arbitrary cohomology theories, but it makes sense to go back to classical things for a moment.

The orientation of $\mathbb{R}^n$ is defined as an equivalence class of its bases, but it can also be defined homologically, as one of the two generators of the group $\mathbb{Z} = H_n(\hat{\mathbb{R}}^n)$ (or $H^n(\hat{\mathbb{R}}^n)$), where $\hat{\mathbb{R}}^n = S^n$ is the one-point compactification of $\mathbb{R}^n$. This approach is very useful from the global point of view, i.e., when we consider the orientability of certain families of $\mathbb{R}^n$'s, like manifolds or $\mathbb{R}^n$-bundles. It is reasonable to treat an orientation of such a family as a family of compatible orientations of its members. For example, if M is a closed connected manifold with $H_n(M) = \mathbb{Z}$ then every generator $[M]$ of $H_n(M)$ can be considered as an orientation of M. Indeed, in this case $[M]$ yields an orientation of every chart: this orientation has the form $\varepsilon_*([M])$ where $\varepsilon : M \to S^n$ collapses the complement of the chart. Moreover, we can *define* orientations of charts to be compatible if there is a fundamental class $[M]$ as above. So, in this way we *define* M to be orientable if $H_n(M) = \mathbb{Z}$, and an orientation of M is defined to be a generator of $H_n(M)$. Similarly, we define a locally trivial $\mathbb{R}^n$-bundle ξ over a connected base to be orientable if $H^n(T\xi) = \mathbb{Z}$, and an orientation of ξ is a generator of the group $H^n(T\xi)$.

The homological approach to orientability enables us to develop orientability theory for arbitrary (co)homology theories h. For example, an h-orientation of an $\mathbb{R}^n$-bundle ξ is a suitable element $u_\xi \in \widetilde{h}^n(T\xi)$, an h-orientation of a closed manifold M^n is an element $[M] \in h_n(M^n)$. Orientable objects have a lot of good properties, and, because of this, "... an orientation is a necessary point of departure for many cohomological constructions" (Adams [9]). For instance, there are a Thom–Dold isomorphism $\varphi : h^i(X) \to \widetilde{h}^{i+n}(T\xi)$ for every h-oriented $\mathbb{R}^n$-bundle ξ over X and a Poincaré (or Poincaré–Milnor–Spanier–Atiyah) duality $h^i(M) \to h_{n-i}(M, \partial M)$ for every h-oriented manifold M^n, and the last one can be deduced from the first one similarly to the classical case as above.

Now we can also tell more about the class $K(\xi) = \varphi^{-1}\tau u_\xi$ considered above: in order to define it, ξ must be h- and k-oriented; so, orientability can tell something about integrality.

Furthermore, one can develop an elegant theory of characteristic classes taking values in $h^*(-)$ provided that all complex vector bundles are h-orientable; these classes generalize the classical Chern classes.

There is not enough space to give all applications of orientability. As the last example, we mention that general orientabilty theory provides a formal group input to algebraic topology; this matter is completely degenerate for classical cohomology, and so this remarkable theory was able to appear only under the general approach.

So, you can see that the orientatibilty theory yields new results as well as makes clear some classical constructions. Summarizing, I cite May [4]: "Orientations of bundles with respect to cohomology theories play a central role in topology."

(Co)bordism with singularities. (Co)bordism with singularities is now a common and convenient notion, being a favorite tool as well as subject of research in algebraic topology. Roughly speaking, we take a class of manifolds and extend it to a class of suitable polyhedra (manifolds with singularities) where a notion of a boundary is reasonably defined. Then, based on these polyhedra, we can define the bordism groups of topological spaces. Under certain circumstances, these bordism groups form a homology theory and, dually, the corresponding cohomology theory. This (co)homology theory is called (co)bordism with singularities. In fact, the passage from (co)bordism to (co)bordism with singularities can be treated as an analogue and far developed generalization of the introduction of coefficients in classical (co)homology.

Varying the classes of manifolds with singularities, we get a big enough stock of (co)homology theories and, in particular, are able to construct ones with prescribed properties. For example, in this way we can construct classical (co)homology and connected complex k-theory. Moreover, the famous Morava k-theories are also constructed as certain cobordism with singularities.

I also want to mention an application of (co)bordism with singularities to the topological quantum field theory: for example, the elliptic (co)homology can be constructed as (co)bordism with singularities.

Finally, (co)bordism with singularities gives a natural geometric flavor to algebraic or homotopical matters. For example, the Adams resolution of certain spectra can be interpreted in terms of (co)bordism with singularities, and this enables us to get useful information about some classical (co)bordism theories, like MSU and MSp, see e.g. Botvinnik [1], Vershinin [1].

Landmarks

The paper Thom [2] made a revolution and formed the contemporary paradigm of algebraic topology, and it freshly demonstrated the power and usefulness of the relations between homotopy theory and geometry. In order to exhibit relatively recent advantages of this matter, I just write down a list (unavoidably incomplete) of certain geometric problems which were (partially or completely) solved via an application of homotopy theory. [1]

(1) When can a manifold M be immersed in a manifold N, and how can one classify these immersions? (Smale [1], Hirsch [1].)

(2) When can a homology class in a space be realized by a map of a closed manifold? (Thom [2].)

(3) When is a closed manifold a boundary of a compact manifold with boundary? (Thom [2].)

(4) Which spaces are homotopy equivalent to closed smooth manifolds? (Browder [1,2], Novikov [2,3].)

(5) How can one classify manifolds up to diffeomorphism (PL isomorphism, homeomorphism)? (Smale [1], Kervaire–Milnor [1], Browder [1,2], Novikov [2,3], Hirsch–Mazur [1], Sullivan [1], Kirby–Siebenmann [1], Freedman [1], Donaldson [1].)

(6) How many pointwise linearly independent tangent vector fields exist on the n-dimensional sphere? (Adams [3].)

(7) Which smooth manifolds admit a Riemannian metric of positive scalar curvature? (Gromov–Lawson [1], Stolz [1].)

This completes my introduction.

---

[1] You can see that the list of authors contains six Fields Medal Award Winners. All of them got this award for the research in question.

Chapter I. Notation, Conventions and Other Preliminaries

The main goal of this chapter is to introduce some notation and terminology. We assume that the reader is more or less familiar with the basic concepts of algebraic topology (homotopy and homology). Typical references are: tom Dieck–Kamps–Puppe [1], tom Dieck [2], Dold [5], Fomenko–Fuchs–Gutenmacher [1], Fritsch–Piccinini [1], Fuks-Rokhlin [1], Gray [1], Hilton–Wiley [1], Hu [1], Ossa [1], Postnikov [2], Spanier [2], Switzer [1], Vick [1].

§1. Generalities

As usual, $\mathbb{N}, \mathbb{Z}, \mathbb{Q}, \mathbb{R}$ and $\mathbb{C}$ will denote the sets of natural, integer, rational, real and complex numbers.

We mark the end of the proof of a lemma, theorem, etc. by the symbol $\square$. If the proof is omitted for some reason (for example, because it is obvious), then we place the symbol $\square$ at the end of the statement. Furthermore, we use the symbol $\boxplus$ to label the end of a proof of a lemma *inside* a theorem. (The lemma is proved, but the theorem is not proved yet.)

The symbol "$:=$" will mean "is defined to be".

We use the abbreviation "iff" for "if and only if".

The symbol "$\cong$" will usually denote an isomorphism of algebraic objects (groups, modules, rings, coalgebras, etc.) or a homeomorphism of topological spaces. For example, given two topological spaces X, Y, the notation $f : X \xrightarrow{\cong} Y$ means that f is a homeomorphism.

For the definitions of category, functor and natural transformation the reader is referred to Mac Lane [2], see also Dold [5] and Switzer [1]. Given a category $\mathscr{K}$, we write $X \in \mathscr{K}$ iff X is an object of $\mathscr{K}$. The identity morphism of any object X is denoted by 1_X. Given two objects X, B of $\mathscr{K}$, the set of all morphisms $X \to B$ is usually denoted by $\mathscr{K}(X, B)$ unless otherwise noted. As usual, a morphism $\alpha : A \to B$ is called an *isomorphism* if there exists a morphism $\beta : B \to A$ such that $\alpha\beta = 1_B$ and $\beta\alpha = 1_A$. In the categorical context we say "product" and "coproduct" rather than "direct product" and "direct sum". (However, in some particular situations we use the standard terminology, e.g., we can say "the direct sum of abelian groups".) Given a family $\{X_k\}$ in $\mathscr{K}$, its product (if the product exists) is denoted by $\prod_k X_k$, and we let $p_i : \prod_k X_k \to X_i$ denote the projection onto the i-th factor. For a finite family $\{X, Y, \dots, Z\}$ we denote the product by $X \times Y \times \cdots \times Z$.

By the definition of the product, for every family $\{f_k : X \to A_k\}$ of morphisms in $\mathscr{K}$ there exists a unique morphism $f : X \to \prod_k A_k$ such that $p_k f = f_k$ for every k. We denote this f by $\{f_k\}$.

The category consisting of sets (as objects) and functions (as morphisms) is denoted by $\mathscr{E}ns$. The category of pointed sets and pointed functions is denoted by $\mathscr{E}ns^\bullet$.

Given a family $\{X_i\}$ of sets, we use the standard notation

$$\bigcup_i X_i,\ \bigcap_i X_i,\ \prod_i X_i,\ \coprod_i X_i$$

for the union, intersection, Cartesian product and disjoint union of sets, respectively. (Note that the Cartesian product of sets yields the product in $\mathscr{E}ns$, and so this notation $\prod_i X_i$ does not lead to confusion.)

Given a function $f : X \to Y$, a *restriction* of f is any function $g : A \to B$ such that the diagram

$$\begin{array}{ccc} A & \xrightarrow{\subset} & X \\ {\scriptstyle g}\downarrow & & \downarrow{\scriptstyle f} \\ B & \xrightarrow{\subset} & Y \end{array}$$

commutes. Given a function $f : X \to Y$ and a subset A of X, we denote the composition $A \subset X \xrightarrow{f} Y$ by $f|A$.

1.1. Definition. (a) A *quasi-ordered set* is a category Λ such that its objects form a set, and there is at most one morphism $\lambda \to \mu$ for every $\lambda, \mu \in \Lambda$. In this case we write $\lambda \leq \mu$. It is clear that

(1) $\lambda \leq \lambda$ for every $\lambda \in \Lambda$;

(2) If $\lambda \leq \mu$ and $\mu \leq \nu$ then $\lambda \leq \nu$.

(b) A quasi-ordered set Λ is called *directed* (with respect to increasing) if for every $\lambda, \mu \in \Lambda$ there exists $\nu \in \Lambda$ such that $\lambda \leq \nu$ and $\mu \leq \nu$.

(c) A *cofinal subset* of a quasi-ordered set Λ is any full subcategory Λ' of Λ such that for every $\lambda \in \Lambda$ there is $\mu \in \Lambda'$ with $\lambda \leq \mu$.

(d) A quasi-rdered set Λ is called *discrete* when $\lambda \leq \mu$ iff $\lambda = \mu$ for every $\lambda, \mu \in \Lambda$.

In fact, $\leq$ can be considered as a relation on the set Λ, and a quasi-ordered set can be defined as a set equipped with a relation satisfying (1) and (2). Such a relation is called a *quasi-ordering*.

1.2. Definition. A *partially ordered set*, or a *poset*, is a quasi-ordered set with the following condition: if $\lambda \leq \mu \leq \lambda$ then $\lambda = \mu$. A *maximal element* of a poset Λ is any $\lambda \in \Lambda$ such that $\lambda \leq \mu$ implies $\lambda = \mu$. A *greatest element* of a poset Λ is an element $\mu \in \Lambda$ such that $\lambda \leq \mu$ for every $\lambda \in \Lambda$. Clearly, the greatest element is a maximal element, but the converse is not true.

A *chain* in a poset is a family $\{a_i\}$ such that, for every pair i, j of indices, either $a_i \leq a_j$ or $a_j \leq a_i$. An *upper bound* of the chain is any a such that $a_i \leq a$ for every i. A poset is called *inductive* if every chain in it has an upper bound.

We use the transfinite induction principle in the following form, see e.g. Kelley [1].

1.3. Zorn's Lemma. *Every inductive set has a maximal element.* □

Let $\mathscr{K}$ be an arbitrary category. Every object B of $\mathscr{K}$ induces a contravariant functor $T_B : \mathscr{K} \to \mathscr{E}ns$ given by $T_B(X) := \mathscr{K}(X, B)$ for every object X of $\mathscr{K}$ and $T_B(f) := \mathscr{K}(f, 1_B) : \mathscr{K}(Y, B) \to \mathscr{K}(X, B)$ for every morphism $f : X \to Y$ of $\mathscr{K}$.

1.4. Definition. We say that a contravariant functor $F : \mathscr{K} \to \mathscr{E}ns$ is *represented* by a certain object B of $\mathscr{K}$ if there exists a natural equivalence $F \cong T_B$. In this case B is called a *classifying* or *representing* object for F. Furthermore, F is called *representable* if it can be represented by some B.

Let $F, G : \mathscr{K} \to \mathscr{E}ns$ be represented by B, C respectively. It is obvious that every morphism $f : B \to C$ yields a natural transformation $T_f : T_B \to T_C$ and hence $F \to G$. The converse is also true.

1.5. Lemma (Yoneda). *Fix natural equivalences $b : F \xrightarrow{\cong} T_B$, $c : G \xrightarrow{\cong} T_C$. For every natural transformation $\varphi : F \to G$ there exists a morphism $f : B \to C$ such that for every object X of $\mathscr{K}$ the diagram*

$$\begin{array}{ccc} F(X) & \xrightarrow{\varphi} & G(X) \\ \downarrow b & & \downarrow c \\ T_B(X) & \xrightarrow{T_f} & T_C(X) \end{array}$$

commutes, and such a morphism f is unique. In particular, the representing object B for F is determined by F uniquely up to isomorphism.

Proof. Consider the function

$$h : \mathscr{K}(B, B) \xrightarrow{b^{-1}} F(B) \xrightarrow{\varphi} G(B) \xrightarrow{c} \mathscr{K}(B, C)$$

and set $f = h(1_B)$. Then the diagram commutes for $X = B$ and hence, by naturality, for arbitrary X. Let $f' : B \to C$ be another morphism satisfying the conditions of the lemma. Then $f' = h(1_B)$, since the diagram commutes for $X = B$. So, φ is unique. To prove the last assertion, put $\varphi = 1_F$. □

§2. Algebra

The category of abelian groups and homomorphisms is denoted by $\mathscr{AG}$. Note that the usual direct product of abelian groups is the categorical product in $\mathscr{AG}$, while the usual direct sum is the categorical coproduct in $\mathscr{AG}$.

We denote the cyclic group of order m by $\mathbb{Z}/m$.

In algebraic context, we reserve the word "unit" for the neutral element of a monoid (group). In particular, the multiplicative identity element of a ring is also called the unit.

We restrict the notion of ring to rings which are associative and unital (i.e. possess a unit), and every ring homomorphism is required to preserve units. Furthermore, every module is required to be unitary (i.e. $1a = a$, where 1 is the unit of the ring and a is any element of the module). Finally, modules over a graded ring are required to be graded.

The degree of a homogeneous element x of a graded object (group, ring, etc.) is denoted by $\deg x$ or $|x|$. If A is a graded object then A_n denotes its component of homogeneous elements of degree n. A graded object A is called *bounded below* if there exists n such that $A_i = 0$ for $i < n$.

Given a commutative ring R, we denote by $R[x, y, \dots, z]$ the polynomial ring of indeterminates $x, y, \dots, z$. The corresponding power series ring is denoted by $R[[x, y, \dots, z]]$. If R is a graded ring, we assume that $x, y, \dots, z$ are homogeneous indeterminates. Furthermore, $\Lambda_R(x, y, \dots, z)$ denotes the free exterior algebra (with a unit) over R of indeterminates $x, y, \dots, z$, and for a graded R we assume that $x, y, \dots, z$ have odd degrees. We use the notation $\Lambda(x, y, \dots, z)$ for the ring $\Lambda_{\mathbb{Z}}(x, y, \dots, z)$.

The set of multiplicatively invertible elements of a commutative ring R is denoted by R^*.

Let $\rho : A \to B$ be a ring homomorphism, and let M be a right A-module. The homomorphism ρ turns B into a left A-module ${}_\rho B$, where $a \cdot b = \rho(a)b$ for $a \in A, b \in B$, cf. Cartan–Eilenberg [1]. We can therefore form the tensor product over A of A-modules M, B. This tensor product is denoted by $M \otimes_\rho B$.

We use the Five Lemma in the following form, see e.g. Mac Lane [2], I.3.3.

2.1. The Five Lemma. *Consider a commutative diagram in* $\mathscr{AG}$

$$\begin{array}{ccccccccc} \bullet & \longrightarrow & \bullet & \longrightarrow & \bullet & \longrightarrow & \bullet & \longrightarrow & \bullet \\ \downarrow{\scriptstyle\alpha_1} & & \downarrow{\scriptstyle\alpha_2} & & \downarrow{\scriptstyle\alpha_3} & & \downarrow{\scriptstyle\alpha_4} & & \downarrow{\scriptstyle\alpha_5} \\ \bullet & \longrightarrow & \bullet & \longrightarrow & \bullet & \longrightarrow & \bullet & \longrightarrow & \bullet \end{array}$$

with exact rows.

(i) *If* α_1 *is an epimorphism and* α_2*,* α_4 *are monomorphisms, then* α_3 *is a monomorphism.*

(ii) *If* α_5 *is a monomorphism and* α_2*,* α_4 *are epimorphisms, then* α_3 *is an epimorphism.*

In particular, if α_1*,* α_2*,* α_4*,* α_5 *are isomorphisms, then* α_3 *is an isomorphism.* □

2.2. Notation. Given a family $\{A_i\}$ of abelian groups, we use the notation $\oplus_i A_i$ for its direct sum. By the definition of the direct sum, for every abelian group G and every family $\varphi_i : A_i \to G$ of homomorphisms there is a unique homomorphism

$$\langle \varphi_i \rangle : \oplus_i A_i \to G$$

such that $\langle \varphi_i \rangle | A_i = \varphi_i$.

2.3. Definition. Let Λ be a quasi-ordered set.

(a) Let $\mathscr{K}$ be a category. A *direct system* over Λ, or briefly, a direct Λ-system, in $\mathscr{K}$ is a covariant functor $\mathscr{M} : \Lambda \to \mathscr{K}$. In other words, $\mathscr{M}$ is a family $\mathscr{M} = \{M_\lambda, j_\lambda^\mu\}_{\lambda,\mu \in \Lambda}$, where $M_\lambda \in \mathscr{K}$ and where $j_\lambda^\mu : M_\mu \to M_\lambda$ for $\mu \leq \lambda$ are morphisms such that $j_\lambda^\mu j_\mu^\nu = j_\lambda^\nu$ for $\nu \leq \mu \leq \lambda$ and $j_\lambda^\lambda = 1_{M_\lambda}$.

(b) A *morphism* $f : \{M_\lambda, j_\lambda^\mu\} \to \{N_\lambda, h_\lambda^\mu\}$ of direct Λ-systems is a natural transformation of functors, i.e., a family $\{f_\lambda : M_\lambda \to N_\lambda\}$ with $h_\lambda^\mu f_\mu = f_\lambda j_\lambda^\mu$.

2.4. Definition. Let Λ be a quasi-ordered set, and let $\{A_\lambda\}_{\lambda \in \Lambda}$ be a direct Λ-system of abelian groups. Let $i_\lambda : A_\lambda \to \oplus_\lambda A_\lambda$ be the inclusion, and let $B \subset \oplus_\lambda A_\lambda$ be the subgroup generated by all elements of the form $(i_\mu a_\mu - i_\lambda j_\lambda^\mu a_\mu)$. The quotient group $(\oplus_\lambda A_\lambda)/B$ is called the *direct limit* of the direct system $\{A_\lambda\}$ and is denoted by $\varinjlim\{A_\lambda\}$.

It is clear that every morphism $f : \{A_\lambda\} \to \{B_\lambda\}$ of direct Λ-systems induces a morphism $\varinjlim f : \varinjlim\{A_\lambda\} \to \varinjlim\{B_\lambda\}$.

Let $q : \oplus A_\lambda \to \varinjlim\{A_\lambda\}$ be the quotient map. Define k_λ to be the composition

$$k_\lambda : A_\lambda \xrightarrow{i_\lambda} \oplus A_\lambda \xrightarrow{q} \varinjlim\{A_\lambda\}.$$

The direct limit has the following universal property.

2.5. Theorem. *Let* G *be an abelian group, and let* $\varphi_\lambda : A_\lambda \to G$ *be a family of homomorphisms such that* $\varphi_\mu = \varphi_\lambda j_\lambda^\mu$ *for every* $\mu \leq \lambda$*. Then there exists a homomorphism* $\varphi : \varinjlim\{A_\lambda\} \to G$ *such that* $\varphi k_\lambda = \varphi_\lambda$ *for every* λ*. Furthermore, such a homomorphism* φ *is unique.*

It is useful to have a notation for φ in terms of φ_λ. I suggest denoting φ by $\langle \varphi_\lambda | \varinjlim \rangle$.[2]

[2] If Λ is a discrete quasi-ordered set, then, clearly, $\langle \varphi_\lambda | \varinjlim \rangle = \langle \varphi_\lambda \rangle$.

Proof. If such φ exists, then $\varphi q = \langle \varphi_\lambda \rangle$, and so φ is unique. To prove the existence, note that $\langle \varphi_\lambda \rangle : \oplus A_\lambda \to G$ passes through $\varinjlim\{A_\lambda\}$, i.e., $\langle \varphi_\lambda \rangle$ can be decomposed as

$$\oplus A_\lambda \xrightarrow{q} \varinjlim\{A_\lambda\} \xrightarrow{\varphi} G.$$

Clearly, $\varphi k_\lambda = \varphi_\lambda$. □

2.6. Definition. Given a quasi-ordered set Λ, let $f : \{A_\lambda\} \to \{B_\lambda\}$ and $g : \{B_\lambda\} \to \{C_\lambda\}$ be two morphisms of direct Λ-systems in $\mathscr{AG}$. We say that the sequence $\{A_\lambda\} \xrightarrow{f} \{B_\lambda\} \xrightarrow{g} \{C_\lambda\}$ is *exact* if the sequence $A_\lambda \xrightarrow{f_\lambda} B_\lambda \xrightarrow{g_\lambda} C_\lambda$ is exact for every $\lambda \in \Lambda$.

2.7. Theorem. *Given a directed quasi-ordered set* Λ, *let*

$$\{A_\lambda\} \xrightarrow{f} \{B_\lambda\} \xrightarrow{g} \{C_\lambda\}$$

be an exact sequence of direct Λ*-systems. Then the sequence*

$$\varinjlim\{A_\lambda\} \xrightarrow{\varinjlim f} \varinjlim\{B_\lambda\} \xrightarrow{\varinjlim g} \varinjlim\{C_\lambda\}$$

is exact.

Proof. See Dold [5], VIII.5.21 or Eilenberg–Steenrod [1], VIII.5.4. □

We discuss the *inverse limit* in Ch. III.

§3. Topology

3.1. Conventions. We reserve the term "map" for a continuous function between two topological spaces.

All neighborhoods and coverings are assumed to be open, unless something else is said explicitly.

When we say "connected space" we mean "path connected space".

Following Bourbaki [2], when we call a space compact we include the Hausdorff property. In particular, every compact space is normal, see loc. cit.

We denote the one-point space by "pt".

A *pair* (of topological spaces) (X, A) is a topological space X with a fixed closed subspace A. A *map* $f : (X, A) \to (Y, B)$ *of pairs* is just a map $f : X \to Y$ such that $f(A) \subset B$. Given a pair (X, A), a *collapse* $c : X \to X/A$ is a quotient map which maps A to a point and induces a homeomorphism of $X \setminus A$ onto its image.

Given two pairs (X, A), (Y, B) of spaces and a map $f : A \to B$, the space $X \cup_f Y$ is defined to be the quotient space $(X \coprod Y)/\sim$, where $\sim$ is the smallest equivalence relation generated by the following relation: $a \sim b$ if $f(a) = b$ for $a \in A, b \in B$. We say that the space $X \cup_f Y$ is obtained from X by *adjoining*, or *gluing*, Y via f.

For instance, if $Y = \text{pt} = B$ then $X \cup_f Y \cong X/A$.

A *triad* $(X; A, B)$ is a topological space and two of its closed subspaces A, B such that $X = A \cup B$.

A *filtration* of a topological space X is a sequence

$$\{\cdots \subset X_0 \subset \cdots \subset X_n \subset \cdots \subset X\}$$

such that:

(1) $X = \cup_n X_n$.
(2) Every X_n is closed in X.
(3) X inherits the direct limit topology, i.e., U is open in X iff $U \cap X_n$ is open in X_n for every n.

A *pointed space* is a pair $(X, \{x_0\})$ where x_0 is a point of X. We use also the notation (X, x_0) and call x_0 the *base point* of X. If there is no reason to indicate the base point, we may write $(X, *)$ (or even X if it is clear that X is pointed). A *pointed map* $f : (X, x_0) \to (Y, y_0)$ is just the map $f : (X, \{x_0\}) \to (Y, \{y_0\})$ of pairs.

Given a space X, we denote by X^+ the disjoint union of X and a point, and the added point is assumed to be the base point.

A *pointed pair* is a triple (X, A, x_0) where (X, x_0) is a pointed space and (X, A) is a pair such that $x_0 \in A$.

A *pointed triad* is a quadruple $(X; A, B; x_0)$ where (X, x_0) is a pointed space and $(X; A, B)$ is a triad such that $x_0 \in A \cap B$.

Algebraic topologists prefer to deal with "nice" spaces, such as *CW*-spaces. However, a class of spaces in which algebraic topologists work should be closed under standard operations which topologists use. In other words, the suitable category of spaces should be large enough to accommodate operations and small enough to rule out pathologies at the same time. One such category was suggested by Steenrod [2] and improved by McCord [1][3], and is known as the category of weak Hausdorff compactly generated spaces. We recall the definitions here, see also Fritsch–Piccinini [1].

3.2. Definition. (a) A topological space X is called *weak Hausdorff* if, for every map $\varphi : C \to X$ of a compact space C, the set $\varphi(C)$ is closed in X.

(b) A subset U of a topological space X is called *compactly open* if $\varphi^{-1}(U)$ is open for every map $\varphi : C \to X$ of a compact space C. A topological space X is called *compactly generated* if each of its compactly open sets is open.

Clearly, every open set is compactly open.

[3] We recommend also the paper of Vogt [1].

Note that every point of a weak Hausdorff space is closed, and that every Hausdorff space is weak Hausdorff. Thus, the weak Hausdorff property lies in between T_1 and T_2.

We denote by $\mathscr{W}$ the category of weak Hausdorff compactly generated spaces and their maps. Similarly, we denote by $\mathscr{W}^\bullet$ the category of weak Hausdorff compactly generated pointed spaces and their pointed maps.

3.3. Proposition. (i) *Let $\varphi : C \to X$ be a map of a compact space C to a weak Hausdorff space X. Then $\varphi(C)$ is compact.*

(ii) *Let X be a weak Hausdorff space. Then a subset U of X is compactly open iff $U \cap C$ is open in C for every compact subspace C of X.*

(iii) *If $X \in \mathscr{W}$ and A is a closed subspace of X then $A \in \mathscr{W}$ and $X/A \in \mathscr{W}$.*

(iv) *Let*

$$X_0 \subset X_1 \subset \cdots \subset X_n \subset \cdots$$

be a sequence in $\mathscr{W}$, and let $X := \bigcup_{n=0}^{\infty} X_n$ have the direct limit topology. Then $X \in \mathscr{W}$.

(v) *Let $X, Y \in \mathscr{W}$, let A be a closed subset of X, and let $f : A \to Y$ be a map. Then $X \cup_f Y \in \mathscr{W}$.*

Proof. See McCord [1, §2] or Fritsch-Piccinini [1, Appendix]. □

Note that X/A is generally non-Hausdorff even when X is Hausdorff. So we cannot restrict our class to that of compactly generated Hausdorff spaces.

3.4. Construction (Steenrod [2]). Given a topological space X, we denote by kX the topological space which coincides with X as a set but has the following topology: a set U is open in kX iff $\varphi^{-1}(U)$ is open in C for every map $\varphi : C \to X$ from a compact space C. We leave it to the reader to check that this family of open sets is a topology.

We define the function $w = w_X : kX \to X, w(x) = x$. Furthermore, given a function $f : X \to Y$, we define the function $kf := w_Y^{-1} f w_X : kX \to kY$.

3.5. Theorem (cf. Steenrod [2]). (i) *For every space X, w_X is a map.*

(ii) *For every space X, kX is a compactly generated space.*

(iii) *If X is compactly generated then w_X is a homeomorphism.*

(iv) *X and kX have the same compact subspaces.*

(v) *If $f : X \to Y$ is a map then so is $kf : kX \to kY$.*

(vi) *If Z is compactly generated and $f : Z \to X$ is a map then so is $w_X^{-1} f : Z \to kX$.*

(vii) *If X is weak Hausdorff then so is kX.*

Proof. (i)–(iv) follow from the definitions.

(v) Let U be open in kY, and let $\varphi : C \to X$ be a map of a compact space C. Then $\varphi^{-1}(f^{-1}(U)) = (f\varphi)^{-1}(U)$ is open in C. Thus, $f^{-1}(U)$ is open in kX.

(vi) By (ii) and (iii), $w_X^{-1} f = (kf)w_Z^{-1}$. Now, by (v) and (iii), $(kf)w_Z^{-1}$ is a map.

(vii) Let $\varphi : C \to kX$ be a map of a compact space. We must prove that $\varphi(C)$ is closed in kX. But this follows, since $w\varphi(C)$ is closed in X. □

Generally speaking, the usual Cartesian product of two spaces from $\mathscr{W}$ is not in $\mathscr{W}$. See Dowker [1], §5. Nevertheless, the category $\mathscr{W}$ admits products.

3.6. Definition. Given a family $\{X_i\}$ of topological spaces, we define their *compactly generated direct product*

$$\prod_i X_i := k\left(\prod_i^c X_i\right)$$

where $\prod^c$ is the usual Cartesian product of topological spaces.

3.7. Lemma. *The compactly generated direct product is the product in* $\mathscr{W}$.

Proof. Firstly, we prove that $\prod X_i \in \mathscr{W}$ if every $X_i \in \mathscr{W}$. In view of 3.5(ii), it suffices to prove that $\prod^c X_i$ is weak Hausdorff. Let

$$p_i^c : \prod^c X_i \to X_i$$

be the projection. Consider a map $\varphi : C \to X$ of a compact space C and set $C_i = p_i^c(\varphi(C))$. Then, by 3.3(i), C_i is a Hausdorff subspace of X_i. Furthermore, $\varphi(C) \subset \prod^c C_i$, and so $\varphi(C)$ is closed in $\prod^c C_i$ since the latter is Hausdorff. Finally, $\prod^c C_i$ is closed in $\prod^c X_i$ since C_i is closed in X_i, see Bourbaki [2].

Now consider the projection $p_i : \prod X_i \xrightarrow{w} \prod^c X_i \xrightarrow{p_i^c} X_i$. We must prove that, for every $Y \in \mathscr{W}$ and every family $f_i : Y \to X_i$ of maps, there is a unique map $f : Y \to \prod X_i$ such that $p_i f = f_i$. Indeed, since $\prod^c$ is the product in the category of all topological spaces, there is a map $f' : Y \to \prod^c X_i$ such that $p_i^c f' = f_i$, and we set $f := w^{-1} f' : Y \to \prod X_i$. By 3.5(vi), f is a map. Now, if there is another map $g : Y \to \prod X_i$ with $p_i g = f$ then $wg = f'$, and so $g = f$. □

3.8. Proposition. *If X is a locally compact Hausdorff space and $Y \in \mathscr{W}$ then $w : X \times Y \to X \times^c Y$ is a homeomorphism.* [4]

Proof. See Steenrod [2], 4.3 or Vogt [1], §3. □

---

[4] Clearly, every locally compact Hausdorff space belongs to $\mathscr{W}$.

Note that the disjoint union yields the coproduct in $\mathscr{W}$.

Define the *compact-open topology* as follows: let $\varphi : C \to X$ be a map of a compact space C, and let U be an open set in Y. We denote by $W(\varphi, U)$ the set of all maps $f : X \to Y$ such that $f\varphi(C) \subset U$. Then the family $\{W(\varphi, U)\}$ for all such pairs (φ, U) forms a subbasis of the compact-open topology on the set of maps from X to Y.

3.9. Definition. (a) Given two spaces X, Y, we let $C(X, Y)$ denote the topological space of all maps $X \to Y$ equipped with the compact-open topology. We let

$$Y^X := kC(X, Y).$$

(b) Given two pairs (X, A) and (Y, B), we define $(Y, B)^{(X,A)}$ to be the subspace of Y^X consisting of maps $f : X \to Y$ such that $f(A) \subset B$. In particular, given two pointed spaces $(X, *)$ and $(Y, *)$, $(Y, *)^{(X,*)}$ is a pointed space, whose base point is given by the constant map $X \to \{*\} \subset Y$.

(c) The *loop space* $\Omega(X, *)$ of a pointed space $(X, *)$ is just the pointed space $(X, *)^{(S^1,*)}$ where S^1 is the circle.

3.10. Theorem. *Let* $X, Y, Z \in \mathscr{W}$.

(i) *The map*

$$u : (Y \times Z)^X \to Y^X \times Z^X,\ u(f) = (p_1 f, p_2 f)$$

is a homeomorphism.

(ii) *The map*

$$e : Z^{Y \times X} \to (Z^Y)^X,\ (e(f)(x))(y) = f(y, x)$$

is a homeomorphism.

(iii) *The function*

$$\mu : Z^Y \times Y^X \to Z^X,\ \mu(f, g) = fg$$

is continuous.

Proof. See Steenrod [2], 5.4, 5.6 and 5.9. □

3.11. Convention. Throughout the book we will assume that all spaces belong to $\mathscr{W}$ unless somthing else is said explicitly, i.e., the word "space" means "weak Hausdorff compactly generated space". Furthermore, all the products and function spaces are taken as in 3.6 and 3.9.

Clearly, the direct product topology on $\mathbb{R}^n = \mathbb{R} \times \cdots \times \mathbb{R}$ coincides with the standard topology on $\mathbb{R}^n$ (defined e.g. by the inner product).

We define the *standard n-dimensional disk*

$$D^n := \left\{ (x^1, \dots, x^n) \in \mathbb{R}^n \,\middle|\, \sum_{i=1}^{n} (x^i)^2 \le 1 \right\}$$

and the *standard n-dimensional sphere*

$$S^n := \left\{ (x^1, \dots, x^{n+1}) \in \mathbb{R}^{n+1} \,\middle|\, \sum_{i=1}^{n+1} (x^i)^2 = 1 \right\}.$$

3.12. Basic homotopy theory. (a) Two maps $f, g : X \to Y$ are called *homotopic* if there is a map (*homotopy*, or *deformation*) $H : X \times I \to Y$ such that $H|X \times \{0\} = f$ and $H|X \times \{1\} = g$. In this case we use the notation $f \simeq g$ or $H : f \simeq g$. The homotopy class of a map f is denoted by $[f]$. The set of all homotopy classes of maps $X \to Y$ is denoted by $[X, Y]$.

(b) A map $f : X \to Y$ is called a *homotopy equivalence* if there is a map $g : Y \to X$ such that $gf \simeq 1_X$ and $fg \simeq 1_Y$. In this case we say that f and g are *homotopy inverse* to each other. Two spaces X, Y are called *homotopy equivalent* if there is a homotopy equivalence $X \to Y$, and we write $X \simeq Y$. The *homotopy type* of a space X is the class of all spaces homotopy equivalent to X.

(c) By saying that two maps $f, g : (X, A) \to (Y, B)$ are homotopic we mean that there exists a homotopy $(X \times I, A \times I) \to (Y, B)$. Furthermore, we say that two maps $f, g : (X, A) \to (Y, B)$ are *homotopic relative to* A, and write $f \simeq g$ rel A, if there is a homotopy $H : f \simeq g$ such that $H(a, t) = f(a)$ for every $a \in A, t \in I$. Similarly, one can define homotopy equivalences of pairs and homotopy equivalences rel A. We leave further such definitions to the reader.

3.13. Definition. We say that a map is *essential* if it is not homotopic to a constant map. Otherwise we say that a map is *inessential.*

3.14. Definition. Let $\mathscr{H}\mathscr{W}$ denote the category whose objects are the same as those of $\mathscr{W}$ but whose morphisms are the homotopy classes of maps. Clearly, every diagram in $\mathscr{W}$ yields a diagram in $\mathscr{H}\mathscr{W}$. We say that a diagram in $\mathscr{W}$ is *homotopy commutative* if the corresponding diagram in $\mathscr{H}\mathscr{W}$ is commutative.

3.15. Definition. We say that two sequences (finite or not)

$$X_1 \xrightarrow{f_1} X_2 \xrightarrow{f_2} \cdots$$

and

$$Y_1 \xrightarrow{g_1} Y_2 \xrightarrow{g_2} \cdots$$

of maps are *homotopy equivalent* if there exists a homotopy commutative diagram

$$\begin{array}{ccccc} X_1 & \xrightarrow{f_1} & X_2 & \xrightarrow{f_2} & \cdots \\ {\scriptstyle h_1}\downarrow & & \downarrow{\scriptstyle h_2} & & \\ Y_1 & \xrightarrow{g_1} & Y_2 & \xrightarrow{g_2} & \cdots \end{array}$$

where every h_i is a homotopy equivalence. In particular, two maps $f : X_1 \to X_2$ and $g : Y_1 \to Y_2$ are homotopy equivalent if there are homotopy equivalences $h_i : X_i \to Y_i, i = 1, 2$ such that $h_2 f = g h_1$.

3.16. Definition. Let $f : X \to Y$ be a map.

(a) The *mapping cylinder*, or just the *cylinder*, of f is the space

$$Mf := X \times [0, 1] \cup_f Y,$$

where f is considered as the map $X \times \{0\} = X \xrightarrow{f} Y$. Recall that there is a standard deformation $F : Mf \times I \to Y$ where

$$\begin{aligned} F((x,t),s) &= (x, st) \text{ if } (x,t) \in X \times (0,1] \text{ and } s > 0 \\ F((x,t),0) &= f(x) \text{ if } (x,t) \in X \times (0,1] \\ F(y,s) &= y \text{ if } y \in Y. \end{aligned}$$

Note that $F \mid Mf \times \{0\} : Mf \to Y$ is a retraction and $F \mid Mf \times \{1\} = 1_{Mf}$, i.e., Y is a deformation retract of Mf.

(b) The *mapping cone*, or the *cofiber*, or just the *cone*, of f is the space $Cf := Mf/(X \times \{1\})$. Recall that the mapping cone has the following universal property: If $h : Y \to Z$ is a map such that hf is inessential then there exists a map $g : Cf \to Z$ such that $g|Y = h$.

(c) We define the *canonical inclusion*

$$k : Y \to Cf \tag{3.17}$$

by setting $k(y) = y$.

3.18. Definition. (a) Given two maps $f : X \to Y$ and $g : X \to Z$, the *double mapping cylinder* of the diagram $Y \xleftarrow{f} X \xrightarrow{g} Z$ is the space

$$D := X \times [0, 2] \cup_\varphi (Y \sqcup Z)$$

where φ is defined to be the composition

$$\varphi : (X \times \{0\}) \sqcup (X \times \{2\}) = X \sqcup X \xrightarrow{f \sqcup g} Y \sqcup Z.$$

Furthermore, we have the inclusions

$$i_{\text{left}} : Y \subset D,\ i_{\text{right}} : Z \subset D,\ i_{\text{mid}} : X = X \times \{1\} \subset D.$$

For instance, Cf is (homeomorphic to) the double mapping cylinder of the diagram $Y \xleftarrow{f} X \to \text{pt}$.

(b) The mapping cone of the constant map $X \to \text{pt}$ is called the *suspension* over a space X and denoted by SX. Thus, the suspension is the double mapping cylinder of the diagram $\text{pt} \leftarrow X \to \text{pt}$. Given a point $(x,t) \in X \times I$, we denote by $[x,t]$ its image under the quotient map $X \times I \to SX = X \times I / X \times \{0,1\}$. Furthermore, given a map $f : X \to Y$, we define a map $Sf : SX \to SY$ by setting $(Sf)[x,t] := [f(x),t]$.

(c) The mapping cylinder of the trivial map $X \to \text{pt}$ is denoted by CX. So, $Cf = CX \cup_f Y$, and $SX = CX/X \times \{1\}$.

(d) The *join* $X * Y$ of the spaces X, Y is defined to be the double mapping cylinder of the diagram

$$X \xleftarrow{p_1} X \times Y \xrightarrow{p_2} Y.$$

For instance, $X * S^0 = SX$. Given a point $(x,t,y) \in X \times [0,2] \times Y$, we denote by $[x,t,y]$ its image under the canonical map $X \times [0,2] \times Y \to X * Y$.

(e) We define the *iterated suspension* $S^n X$ by induction, by setting $S^0 X := X$ and $S^n X := S(S^{n-1}X)$. By induction, every map $f : X \to Y$ yields a map $S^n f : S^n X \to S^n Y$; this turns the suspension into a functor.

3.19. Definition. Given a sequence $\mathfrak{X} = \{\cdots \xrightarrow{f_{n-1}} X_n \xrightarrow{f_n} X_{n+1} \xrightarrow{f_{n+1}} \cdots\}$ of maps, define its *telescope* $T\mathfrak{X}$ to be the space

$$T\mathfrak{X} := \left(\bigcup (X_n \times [n, n+1])\right) / \sim,$$

where under $\sim$, $(x, n+1) \in X_n \times [n, n+1]$ is identified with $(f_n(x), n+1) \in X_{n+1} \times [n+1, n+2]$.

Let $T_{\text{ev}}\mathfrak{X}$ be the subspace

$$\bigcup (X_{2n-1} \times [2n-1, 2n] \cup X_{2n} \times \{2n\}) / \sim$$

of $T\mathfrak{X}$, and let $T_{\text{od}}\mathfrak{X}$ be the subspace

$$\bigcup (X_{2n} \times [2n, 2n+1] \cup X_{2n+1} \times \{2n+1\}) / \sim$$

of $T\mathfrak{X}$. Then

$$T_{\text{ev}}\mathfrak{X} \simeq \coprod_n X_{2n}, \; T_{\text{od}}\mathfrak{X} \simeq \coprod_n X_{2n+1}, \; T_{\text{ev}}\mathfrak{X} \cap T_{\text{od}}\mathfrak{X} \simeq \coprod_n X_n,$$
$$T_{\text{ev}}\mathfrak{X} \cup T_{\text{od}}\mathfrak{X} = T\mathfrak{X}.$$

As an important special case, one can consider a filtration

$$\mathfrak{F} = \{\cdots \subset X_n \subset X_{n+1} \subset \cdots\}$$

of a space X as a sequence of inclusion maps, see 3.51 below.

3.20. Definition. Let $\{(X_i, x_i)\}$ be a family of pointed spaces.

(a) The *pointed direct product* is the pointed space

$$\prod(X_i, x_i) := \left(\prod X_i, *\right)$$

where $*$ is the point $\prod\{x_i\}$.

(b) The *wedge* is the pointed space

$$\bigvee_i (X_i, x_i) := \left(\frac{\coprod_i X_i}{\cup_i\{x_i\}}, *\right).$$

where $*$ is the image of $\cup_i\{x_i\}$.

(c) The obvious injective maps $(X_i, x_i) \to \left(\prod X_i, *\right)$ yield an injective map

$$(\vee_i X_i, *) \to \left(\prod X_i, *\right).$$

Generally speaking, this map is not closed, but it is closed for a finite set of spaces. So, given two pointed spaces $(X,*), (Y,*)$, we define the *smash product*

$$(X,*) \wedge (Y,*) := \frac{(X,*) \times (Y,*)}{(X,*) \vee (Y,*)}.$$

Furthermore, we set

$$(X, x_0) \wedge (Y, y_0) \wedge \cdots \wedge (Z, z_0) := (\cdots((X, x_0) \wedge (Y, y_0)) \wedge \cdots) \wedge (Z, z_0).$$

3.21. Definition. Let $\{(X_i, x_i)\}$ be a family of copies of a pointed space (X, x). We define the *folding map*

$$\pi : \vee(X_i, x_i) \to X,$$

to be the unique map π such that $\pi|X_i = 1_X$.

There are also analogs of constructions 3.16 for pointed spaces.

3.22. Definition. We say that two pointed maps $f, g : (X, x_0) \to (Y, y_0)$ are *pointed homotopic* if $f \simeq g$ rel $\{x_0\}$. In this case we also write $f \simeq^\bullet g$. The set of all pointed homotopy classes of maps $(X,*) \to (Y,*)$ is denoted by $[(X,*),(Y,*)]$ or $[X,Y]^\bullet$. The set $[(S^n,*),(X,*)]$ is denoted by $\pi_n(X,*)$. If $n \geq 1$ then it possesses a natural structure of a group (abelian for $n > 1$) and is called the *n-th homotopy group of* $(X,*)$, see any text book for details.

3.23. Definition. (a) The *reduced mapping cylinder* of a pointed map $f : (X,*) \to (Y,*)$ is the space $Mf = (X \times [0,1] \cup_f Y)/(* \times [0,1])$. Note that the base points of X and Y yield the same point $* \in Mf$; we agree that $*$ is the base point of Mf.

(b) The *reduced mapping cone* of f is defined to be $Cf = Mf/(X \times \{1\})$. It is a pointed space in the obvious way: its base point is the image of the base point of Mf.

(c) The reduced mapping cone of the constant map $(X, *) \to (\mathrm{pt}, *)$ is called the *reduced suspension* over a space X and denoted by SX. Furthermore, we can define the *iterated reduced suspension* $S^n X$, and S^n turns out to be a functor on $\mathscr{W}^\bullet$, see 3.18(e).

(d) The *reduced telescope* of a sequence

$$\mathfrak{X} = \{\cdots \xrightarrow{f_{n-1}} X_n \xrightarrow{f_n} X_{n+1} \xrightarrow{f_{n+1}} \cdots\}$$

of pointed maps is defined to be the pointed space

$$T\mathfrak{X} := \Big(\bigcup (X_n \times [n, n+1])\Big) / \sim,$$

where $(x, n+1) \in X_n \times [n, n+1]$ is identified with $(f_n(x), n+1) \in X_{n+1} \times [n+1, n+2]$ and all the points of the form $(*, t)$ are identified. These points of the form $(*, t)$ yield the base point of $T\mathfrak{X}$.

Let $T_{\mathrm{ev}}\mathfrak{X}$ be the subspace

$$\bigcup (X_{2n-1} \times [2n-1, 2n] \cup X_{2n} \times \{2n\}) / \sim$$

of $T\mathfrak{X}$, and let $T_{\mathrm{od}}\mathfrak{X}$ be the subspace

$$\bigcup (X_{2n} \times [2n, 2n+1] \cup X_{2n+1} \times \{2n+1\}) / \sim$$

of $T\mathfrak{X}$. We have $T\mathfrak{X} = T_{\mathrm{ev}}\mathfrak{X} \vee T_{\mathrm{od}}\mathfrak{X}$, $T_{\mathrm{ev}}\mathfrak{X} \simeq \vee_n X_{2n}$, $T_{\mathrm{od}}\mathfrak{X} \simeq \vee_n X_{2n+1}$, and $T_{\mathrm{ev}}\mathfrak{X} \cap T_{\mathrm{od}}\mathfrak{X} \simeq \vee_n X_n$.

Again, given a pointed filtration $\mathfrak{F} = \{X_n\}$ of a pointed space X, we can introduce the reduced telescope $T\mathfrak{F}$.

You can see that we introduce no special notation for reduced objects. (In fact, the reduced and unreduced cone (cylinder, etc.) of any map(s) of CW-spaces are homotopy equivalent, see 3.26 below.) Moreover, we omit the adjective "reduced" when it is clear that we work with pointed spaces and maps, i.e., we just say "the cone of a pointed map", etc.

Note that, because of 3.3, the categories $\mathscr{W}$ and $\mathscr{W}^\bullet$ are closed under constructions defined in 3.16–3.23.

Prove as an exercise that $SX \cong S^1 \wedge X$ for every $X \in \mathscr{W}^\bullet$.

3.24. Definition. Given a pair (X, A), the inclusion $i : A \to X$ is called a *cofibration* if it satisfies the homotopy extension property, i.e., given maps $g : X \to Y$ and $F : A \times I \to Y$ such that $F|A \times \{0\} = g|A$, there is a map $G : X \times I \to Y$ such that $G|X \times \{0\} = g$ and $G|A \times I = F$. In this case we also say that (X, A) is a *cofibered pair*.

We discuss *fibrations* in Ch. IV.

3.25. Proposition. (i) *(X, A) is a cofibered pair iff every map $h : X \times \{0\} \cup A \times I \to Y$ can be extended to a map $X \times I \to Y$.*

(ii) *(X, A) is a cofibered pair iff $X \times \{0\} \cup A \times I$ is a retract of $X \times I$.*

Proof. (i) Let (X, A) be a cofibered pair, and let $h : X \times \{0\} \cup A \times I \to Y$ be a map. We set $F := h|A \times I$, $g := h|X \times \{0\}$. Then $G : X \times I \to Y$ as in 3.24 is the desired extension of h. Conversely, consider $g : X \to Y$ and $F : A \times I \to Y$ as in 3.24. We define $h : X \times I \to Y$ be setting $h|A \times I := F$, $h|X \times \{0\} := g$. Since A is assumed to be closed, h is continuous. This map h has an extension $G : X \times I \to Y$. Clearly, $G|X \times \{0\} = g$ and $G|A \times I = F$. Thus, (X, A) is a cofibered pair.

(ii) Let (X, A) be a cofibered pair. We put $Y = X \times \{0\} \cup A \times I$, $h = 1_Y$ in (i). Then any extension $G : X \times I \to Y$ of h is a retraction. Conversely, let $r : X \times I \to X \times \{0\} \cup A \times I$ be a retraction. Then every map $h : X \times \{0\} \cup A \times I \to Y$ has the extension $hr : X \times I \to Y$. Thus, by (i), (X, A) is a cofibered pair. □

3.26. Proposition. (i) *For every map $f : X \to Y$, the inclusion*

$$i : X = X \times \{1\} \to Mf$$

is a cofibration. In particular, every map is homotopy equivalent to a cofibration.

(ii) *Let (X, A) be a cofibered pair. Then $Ci \simeq X/A$.*

(iii) *Let (X, A) be a cofibered pair. If A is contractible then the collapsing map $c : X \to X/A$ is a homotopy equivalence.*

Proof. (i) Let I' be a copy of the segment I, and let

$$\rho : I \times I' \to \{0\} \times I' \cup I \times \{0\}$$

be a retraction. We must prove that $X \times I' \cup (X \times I \cup Y)$ is a retract of $(X \times I \times I') \cup (Y \times I') \times \{0\}$. To this end, we define a retraction

$$r : (X \times I \times I') \cup (Y \times I') \to X \times I' \cup (X \times I \cup Y)$$

by setting $r(x, s, t) = (x, \rho(s, t))$ and $r(y, t) = y$.

(ii) Let $q : Ci \to X/A$ be the quotient map which collapses CA. We define $f : X \cup A \times I \to Ci$ to be the quotient map which collapses $A \times \{1\}$. Since (X, A) is a cofibered pair, f can be extended to a map $F : X \times I \to Ci$, and we set $g := F|X \times \{1\} : X \to Ci$. Since $g(A)$ is a point, g passes through a map $j : X/A \to Ci$. We leave it to the reader to prove that q and j are homotopy inverse.

(iii) This can be proved similarly to (i) and (ii), see e.g. Switzer [1], 6.6. □

3.27. Lemma. *If (X, A) and (Y, B) are cofibered pairs then so is the pair $(X \times Y, X \times B \cup A \times Y)$.*

Proof. See Strøm [2], or Spanier [2], Ch. 1, Exercises. □

3.28. Definition. A pointed space (X, x_0) is called *well-pointed* if $(X, \{x_0\})$ is a cofibered pair.

3.29. Lemma (Puppe [1]). *Let $f : (X, x_0) \to (Y, y_0)$ be a pointed map of well-pointed spaces. If $f : X \to Y$ is a homotopy equivalence then $f : (X, x_0) \to (Y, y_0)$ is a pointed homotopy equivalence.*

Proof. Firstly, we prove the following sublemma.

3.30. Sublemma. *Let $\varphi : (X, x_0) \to (X, x_0)$ be a map, and let $H : \varphi \simeq 1_X$ be a free (i.e., unpointed) homotopy. Suppose that the loop $H|(\{x_0\} \times I) : I \to X$ is homotopic to the constant loop. Then there is a pointed homotopy between φ and 1_X.*

Proof. We set $A = X \times \{0\} \cup \{x_0\} \times I \cup X \times \{1\}$ and consider the map

$$F : A \to X, F|X \times \{0\} = \varphi, F|X \times \{1\} = 1_X, F(\{x_0\} \times I) = x_0.$$

Then $F \simeq H|A$,. By 3.27, $(X \times I, A)$ is a cofibered pair, and hence F can be extended to a map $G : X \times I \to X$. Clearly, G is a pointed homotopy between φ and 1_X. ⊞

We continue the proof of the lemma. Let $g' : Y \to X$ be free homotopy inverse to f, and let $F : g'f \simeq 1_X$ be a free homotopy. We define the map $u : \{y_0\} \times I \to X$ by setting $u(y_0, t) = F(x_0, 1 - t)$. Since (Y, y_0) is well-pointed, there is a map $G : Y \times I \to X$ such that $G|Y \times \{1\} = g'$ and $G|\{y_0\} \times I = u$. We set $g := G|Y \times \{0\} : Y \to X$ and prove that $gf \simeq^\bullet 1_X$. We define the free homotopy $H : gf \simeq 1_X$ by setting

$$H(x, t) = \begin{cases} G(f(x), 2t) \text{ if } 0 \leq t \leq 1/2, \\ F(x, 2t - 1) \text{ if } 1/2 \leq t \leq 1. \end{cases}$$

Clearly, the "shrunk" loop $H|(\{x_0\} \times I) : I \to X$ is homotopic to the constant loop. (Indeed, the point x_0 runs along some path until $t = 1/2$ and then runs along the same path but in the opposite direction.) Hence, by the sublemma, $gf \simeq^\bullet 1_X$.

Now we prove that $fg \simeq^\bullet 1_Y$. Indeed, we can apply the arguments as above to g and find $h : (X, x_0) \to (Y, y_0)$ such that $hg \simeq^\bullet 1_Y$. Now,

$$h \simeq^\bullet h(gf) \simeq^\bullet hg(f) \simeq^\bullet f,$$

so that we indeed have $fg \simeq^\bullet 1_Y$. □

3.31. Definition. Let $(X,*), (Y,*)$ be two pointed spaces. We define their *homotopy wedge* (or, briefly, *h-wedge*) $(X,*) \vee^h (Y,*)$ to be the double mapping cylinder (unreduced) of the diagram

$$X \leftarrow \mathrm{pt} \rightarrow Y$$

of pointed maps. We equip $(X,*) \vee^h (Y,*)$ with the base point via the map $i_{\mathrm{mid}} : \mathrm{pt} \rightarrow X \vee^h Y$.

3.32. Convention. For brevity, we write $(\vee X_i, *)$ or just $\vee X_i$ instead of $\vee(X_i, x_i)$, and use similar shorthand for smash products and h-wedges.

3.33. Construction. Let $f : (X,*) \rightarrow (Z,*)$ and $g : (Y,*) \rightarrow (Z,*)$ be two pointed maps. Regarding $X \vee^h Y$ as the quotient space of the space $X \sqcup [0,2] \sqcup Y$, we define a pointed map $f \top g : X \vee^h Y \rightarrow Z$ by setting

$$(f \top g)(x) = f(x),\ (f \top g)(y) = g(y),\ (f \top g)(t) = *,\ \ x \in X, y \in Y, t \in [0,2].$$

Clearly, $f \top g$ is well-defined. Furthermore,

$$(f \top g) i_{\mathrm{left}} = f \text{ and } (f \top g) i_{\mathrm{right}} = g.$$

3.34. Lemma. *Let X, Y be two pointed spaces, and let $a : X \rightarrow X \vee Y, a(x) = x$, and $b : Y \rightarrow X \vee Y, b(y) = y$ be the obvious inclusions. Then $a \top b : X \vee^h Y \rightarrow X \vee Y$ is a homotopy equivalence. Moreover, if X and Y are well-pointed then $a \top b$ is a pointed homotopy equivalence.*

Proof. By 3.26(iii), $a \top b$ is a homotopy equivalence. Hence, by 3.29 and 3.26(i), $a \top b$ is a pointed homotopy equivalence provided that X and Y are well-pointed. □

3.35. Definition. Let $(X,*), (Y,*)$ be two pointed spaces. We consider the maps

$$u : X \rightarrow X \times Y,\ u(x) = (x,*) \ \text{ and } \ v : Y \rightarrow X \times Y,\ v(y) = (*,y)$$

and define the *homotopy smash product* (or, briefly, the *h-smash product*) $(X,*) \wedge^h (Y,*)$ of $(X,*)$ and $(Y,*)$ to be the double mapping cylinder of the diagram

$$\mathrm{pt} \leftarrow X \vee^h Y \xrightarrow{u \top v} X \times Y$$

of pointed maps. We turn $X \wedge^h Y$ into a pointed space by choosing $i_{\mathrm{left}}(\mathrm{pt})$ to be the base point.

Since the composition $X \vee^h Y \xrightarrow{i_{\mathrm{mid}}} X \times Y \xrightarrow{q} X \wedge Y$ is inessential, the quotient map q can be extended to a map $f : (X \wedge^h Y, *) \rightarrow (X \wedge Y, *)$.

3.36. Proposition. *If X, Y are well-pointed then f is a pointed homotopy equivalence.*

Proof. Indeed, we have $X \wedge^h Y = X \times Y \cup C(X \vee^h Y)$. Now, by 3.27, $(X \times Y, X \vee Y)$ is a cofibered pair. We can also see that $(X \wedge^h Y, X \vee^h Y)$ is a cofibered pair. Hence, by 3.26, f is a homotopy equivalence since it collapses a contractible space $C(X \vee^h Y)$. Thus, by 3.29, f is a pointed homotopy equivalence. □

3.37. Lemma. *Let $(X; A, B)$ be a pointed triad such that A and B are well-pointed. Suppose that there are two maps $u, v : X \to [0,1]$ such that $u|X \setminus A = 0 = v|X \setminus B$. Define a map $f : A \vee B \to X, f(a) = a, f(b) = b$, $a \in A, b \in B$. Then $C(f) \simeq S(A \cap B)$.*

Proof. Set $C = A \cap B$ and consider the double mapping cylinder Y of the diagram $A \leftarrow C \to B$ of inclusions. We claim that $Y \simeq X$. Indeed, there is the obvious map

$$g : Y \to X, g(c,t) = c, g(a,0) = a, g(b,2) = b, a \in A, b \in B, c \in C, t \in [0,2].$$

We define $\bar{h} : X \to X \times I$ by setting $\bar{h}(x) = (x, 2v(x))$. Clearly, $\bar{h}(X) \subset Y$, and so we have the map $h : X \to Y, h(x) = \bar{h}(x)$. We leave it to the reader to prove that h is homotopy inverse to g.

Now, the inclusions $A \subset X$, $B \subset X$ induce a map $F : A \vee^h B \to Y$ and, by the above, this map is homotopy equivalent to f. It remains to note that $C(F) \simeq SC$. □

3.38. Definition. (a) A *strict cofiber sequence* is a diagram $A \xrightarrow{u} B \xrightarrow{v} C$ where $u : A \to B$ is a map and v is the canonical inclusion as in (3.17).

(b) A sequence $X \xrightarrow{f} Y \xrightarrow{g} Z$ is called a *cofiber sequence* if there exists a homotopy commutative diagram

$$\begin{array}{ccccc} X & \xrightarrow{f} & Y & \xrightarrow{g} & Z \\ {\scriptstyle a}\downarrow & & {\scriptstyle b}\downarrow & & {\scriptstyle c}\downarrow \\ A & \xrightarrow{u} & B & \xrightarrow{v} & C \end{array}$$

such that all the vertical arrows are homotopy equivalences and the bottom row is a strict cofiber sequence.

(c) A *long cofiber sequence* is a sequence (finite or not)

$$\cdots \to X_i \to X_{i+1} \to X_{i+2} \to \cdots$$

where every pair of adjacent morphisms forms a cofiber sequence.

3.39. Proposition. *Let $f : X \to Y$ be an arbitrary map of pointed spaces, and let $g : Y \to Cf = Z$ be the canonical inclusion. Then $C(g) \simeq SX = S^1 \wedge X$. Moreover, there is a long cofiber sequence*

$$X \xrightarrow{f} Y \xrightarrow{g} Z \to SX \xrightarrow{Sf} SY \xrightarrow{Sg} \cdots \to S^n X \xrightarrow{S^n f} S^n Y \xrightarrow{S^n g} \cdots .$$

Proof. See Switzer [1], 2.36–2.37. □

Proposition 3.39 was originally proved by Puppe [1]. Because of this, the long cofiber sequence is often refered to as the Puppe sequence. Barratt [1] had obtained some preliminary results in this area.

3.40. Conventions about CW-complexes and CW-spaces. We use the definition of CW-complexes as in Switzer [1] and Fritsch–Piccinini [1]. A CW*-space* is a space which is homeomorphic to a CW-complex. Throughout this book, the word "cell" means "closed cell", i.e., the image of the closed disk under a characteristic map, see loc.cit.

We denote by $X^{(n)}$ the n*-skeleton* of a CW-complex X, i.e., $X^{(n)}$ is the union of all n-dimensional cells of X.

Recall that a map $f : X \to Y$ of CW-complexes is called *cellular* if $f(X^{(n)}) \subset Y^{(n)}$ for every n.

3.41. Theorem. (i) *Let $i : X^{(n)} \to X$ be the inclusion. Then $i_* : \pi_i(X^{(n)}, *) \to \pi_i(X, *)$ is an isomorphism for $i < n$ and an epimorphism for $i = n$.*

(ii) *Every map $f : X \to Y$ of CW-complexes is homotopic to a cellular map.*

Proof. See e.g. Switzer [1], 6.11 and 6.35 or Fritsch–Piccinini [1], 2.4.8 and 2.4.11. □

A finite (resp. finite dimensional) CW-space is a space which is homeomorphic to a finite (resp. finite dimensional) CW-complex. The category of CW-spaces and their maps is denoted by $\mathscr{C}$, and the category of pointed CW-spaces and pointed maps is denoted by $\mathscr{C}^\bullet$. Furthermore, $\mathscr{C}_{\mathrm{con}}$ (resp. $\mathscr{C}_{\mathrm{f}}$, resp. $\mathscr{C}_{\mathrm{fd}}$) denotes the full subcategory of $\mathscr{C}$ consisting of connected (resp. finite, resp. finite dimensional) CW-spaces. Similarly, $\mathscr{C}^\bullet_{\mathrm{con}}$, $\mathscr{C}^\bullet_{\mathrm{f}}$, $\mathscr{C}^\bullet_{\mathrm{fd}}$ are the corresponding subcategories of $\mathscr{C}^\bullet$. Finally, $\mathscr{H}\mathscr{C}$ (resp. $\mathscr{H}\mathscr{C}^\bullet$) denotes the category whose objects are the same as those of $\mathscr{C}$ (resp. $\mathscr{C}^\bullet$) but whose morphisms are the homotopy classes of maps (resp. pointed homotopy classes of pointed maps).

We recall that $i : A \to X$ is a cofibration for every CW-pair (X, A). In particular, every pointed CW-space is well-pointed, and so we can safely omit base points from notation.

Every CW-space is Hausdorff (and so weak Hausdorff) and compactly generated. Thus, when we talk about products (or smash products) of CW-complexes we follow 3.6. Then the direct product $X \times Y$ and the smash product $X \wedge Y$ of two CW-spaces X, Y are also CW-spaces.

Note that, for every cellular map $f : X \to Y$, the spaces Mf and Cf are CW-complexes in an obvious canonical way, see e.g. Fritsch–Piccinini [1]. In particular, the suspension SX of a CW-complex X is a CW-complex.

3.42. Definition. (a) Two maps $f, g : X \to Y$ of topological spaces are called *CW-homotopic* if $fi \simeq gi$ for every map $i : K \to X$ of a CW-space K. In this case we write $f \simeq^{CW} g$.

(b) A map $f : X \to Y$ is called a *Whitehead equivalence* if $f_* : \pi_n(X, x_0) \to \pi_n(Y, f(x_0))$ is a bijection for every $n \geq 0$ and every $x_0 \in X$.

(c) Let

$$X \overset{a}{\text{——}} Y$$

denote that either there is a map $a : X \to Y$ or $a : Y \to X$. We say that two spaces X, Y are *CW-equivalent* if there is a sequence

$$X = X_0 \overset{a_0}{\text{——}} X_1 \overset{a_1}{\text{——}} \cdots \overset{a_{i-1}}{\text{———}} X_i \overset{a_i}{\text{——}} \cdots \overset{a_{n-1}}{\text{———}} X_n = Y$$

where every a_i is a Whitehead equivalence.

3.43. Remark. Traditionally, CW-equivalences, as well as Whitehead equivalences, are called weak equivalences. We refrain from using this terminology in this book because these names are not quite compatible with the concept of weak homotopy (see Ch. II below).

Note that if X and Y are connected then $f : X \to Y$ is a Whitehead equivalence provided that $f_* : \pi_i(X, x_0) \to \pi_i(Y, f(x_0))$ is an isomorphism for some single point x_0, see Spanier [2], 7.3.4. Furthermore, every homotopy equivalence $X \to Y$ is a Whitehead equivalence, see e.g. Spanier [2], 7.3.15.

3.44. Proposition–Definition. *For every topological space X, there is a Whitehead equivalence $f : Y \to X$ where Y is a CW-space. Every such CW-space Y is called a CW-*substitute *for X.*

Proof. Without loss of generality, we can assume that X is connected. We construct a commutative diagram

$$\begin{array}{ccccccccc}
Y_0 & \xrightarrow{\subset} & Y_1 & \xrightarrow{\subset} & \cdots & \xrightarrow{\subset} & Y_n & \xrightarrow{\subset} & \cdots \\
{\scriptstyle f_0}\downarrow & & \downarrow{\scriptstyle f_1} & & & & \downarrow{\scriptstyle f_n} & & \\
X & = & X & = & \cdots & = & X & = & \cdots
\end{array}$$

such that $(f_n)_* : \pi_n(Y_n, y_0) \to \pi_n(X, x_0), x_0 = f_0(y_0)$, is an isomorphism and that the inclusion $Y_n \subset Y_{n+1}$ induces an isomorphism $\pi_i(Y_n, y_0) \to$

$\pi_i(Y_{n+1}, y_0)$ for every $i \leq n$. Then we define $Y := \cup Y_n$ and define $f : Y \to X$ by the condition $f|Y_n = f_n$.

We construct this diagram by induction on n. We set Y_0 to be a point. Suppose that $f_n : Y_n \to X$ is constructed. Let $\{x_\alpha\}$ be a family of generators of the group $\pi_{n+1}(X, x_0)$). Let S^{n+1}_α be a copy of S^{n+1}. Choose a map $g_\alpha : (S^{n+1}_\alpha, *) \to (X, x_0)$ which yields the element x_α. Now, we define $g : \vee_\alpha(S^{n+1}_\alpha, *) \to (X, x_0)$ by requiring $g|S^{n+1}_\alpha = g_\alpha$. Clearly,

$$g_* : \pi_{n+1}(\vee S^{n+1}_\alpha, *) \to \pi_{n+1}(X, x_0)$$

is epic. We set $(Z, *) := (Y_n, *) \vee (\vee S^{n+1}_\alpha, *)$. Define

$$h : Z = Y_n \vee (\vee S^{n+1}_\alpha) \xrightarrow{f_n \vee g} X \vee X \xrightarrow{\pi} X.$$

Note that $h_* : \pi_{n+1}(Z, *) \to \pi_{n+1}(X, x_0)$ is an epimorphism.

Let $\{z_\beta\}$ be a family of generators of the group

$$\operatorname{Ker}(h_* : \pi_{n+1}(Z, *) \to \pi_{n+1}(X, x_0)).$$

Let S^{n+1}_β be a copy of S^{n+1}. Choose a map $a_\beta : (S^{n+1}_\beta, *) \to (Z, *)$ which yields the element z_β. Now, we define $a : \vee_\beta S^{n+1}_\beta \to Z$ by requiring $a|S^{n+1}_\beta = a_\beta$, set $Y_{n+1} := C(a)$ and define $f_{n+1} : Y_{n+1} \to X$ to be any extension of h. (Note that h can be extended on Y_{n+1} because of the universal property of the cone.) By 3.41, the map f_{n+1} has the desired properties. The induction is confirmed. □

3.45. Lemma. *Let $h : Y \to Z$ be a Whitehead equivalence.*

(i) *Let (X, A) be a CW-pair, and let $f : A \to Y, u : X \to Z$ be two maps such that $hf = u|A$. Then there is a map $g : X \to Y$ such that $g|A = f$ and $hg \simeq u$.*

(ii) *Let (X, A) be a CW-pair, and let $f : A \to Y$ be a map such that hf can be extended to all of X. Then f can be extended to all of X.*

(iii) *Let K be a CW-space, and let $u, v : K \to Y$ be two maps such that $hu \simeq hv$. Then $u \simeq v$.*

Proof. (i) This is an exercise in elementary obstruction theory, see e.g. Switzer [1], 6.30.

(ii) This follows from (i).

(iii) This follows from (ii) if one puts $X = K \times I, A = K \times \{0, 1\}$ and defines $f : A \to Y$ by setting $f|K \times \{0\} = u, f|K \times \{1\} = v$. □

3.46. Corollary. *Let Z be a CW-space, and let $h : Y \to Z$ be a Whitehead equivalence. Then there exists a map $g : Z \to Y$ such that $hg \simeq 1_Z$ and $gh \simeq^{CW} 1_Y$.*

Proof. If we put $A = \emptyset, u = 1_Z$ in 3.45(i), we conclude that there is $g : Z \to Y$ such that $hg \simeq 1_Z$. We prove that $ghi \simeq i$ for every map

$i : K \to Y$ of a CW-space K. Indeed, $hi \simeq (hg)hi \simeq h(ghi)$, and thus, by 3.45(iii), $i \simeq ghi$. □

3.47. Corollary (the Whitehead Theorem). *If $h : Y \to Z$ is a Whitehead equivalence of CW-spaces then h is a homotopy equivalence.*

Proof. By 3.46, there is $g : Z \to Y$ such that $hg \simeq 1_Z$. We prove that $gh \simeq 1_Y$. Indeed, g is a Whitehead equivalence, and so, by 3.46, there is $f : Y \to Z$ such that $gf \simeq 1_Y$. Now, $f \simeq (hg)f \simeq h(gf) \simeq h$, i.e., $gh \simeq 1_Y$. □

3.48. Corollary. *If CW-spaces X, Y are CW-equivalent then they are homotopy equivalent. In particular, any two CW-substitutes for a given space are homotopy equivalent.*

Proof. It suffices to prove that if $A \overset{a}{\text{——}} B \overset{b}{\text{——}} C$ is a diagram where A is a CW-space and a, b are Whitehead equivalences then there is a Whitehead equivalence $j : A \to C$. Indeed, then one can construct a Whitehead equivalence $f : X \to Y$ by induction on such diagrams, and so, by 3.47, f is a homotopy equivalence.

So, we consider the diagram $A \overset{a}{\text{——}} B \overset{b}{\text{——}} C$. For the case $A \xrightarrow{a} B \xrightarrow{b} C$, the desired j is clear. In the case $A \xleftarrow{a} B \xrightarrow{b} C$ we can construct, by 3.46, a Whitehead equivalence $h : A \to B$, and we put $j = bh$. In the case $A \xleftarrow{a} B \xleftarrow{b} C$, the desired $j : A \to C$ exists by 3.46. Finally, in case $A \xrightarrow{a} B \xleftarrow{b} C$, the desired $j : A \to C$ exists by 3.45(i). □

Now we compare the homotopy types of kX and X. Of course, in these theorems X is assumed to be an arbitrary topological space, not necessary belonging to $\mathscr{W}$.

3.49. Theorem. *For every topological space X, the map $w : kX \to X$ is a Whitehead equivalence.*

Proof. We prove that $w_* : \pi_n(X, *) \to \pi_n(kX, *), n \geq 0$, is a bijection. Indeed, if $f : S^n \to X$ is a map then, by 3.5(vi), $w^{-1}f : S^n \to kX$ is, and hence w_* is surjective. Furthermore, if $f, g : S^n \to kX$ are two maps and $H : S^n \times I \to X$ is a pointed homotopy between kf and kg then $w^{-1}H$ is a pointed homotopy between f and g, and so w_* is injective. □

3.50. Theorem. *Let X be an arbitrary topological space having the homotopy type of a CW-space. Then $w : kX \to X$ is a homotopy equivalence. In particular, if $X \in \mathscr{W}$ and X has the homotopy type of a CW-space then ΩX has the homotopy type of a CW-space.*

Proof. Let Y be a CW-space, let $f : Y \to X$ be a homotopy equivalence, and let $g : Y \to X$ be homotopy inverse to f. By 3.5(vi), $w^{-1}f : Y \to kX$ is a map since $Y \in \mathscr{W}$. We prove that $w^{-1}f$ is homotopy inverse to gw. Firstly, $gww^{-1}f = gf \simeq 1_Y$. Furthermore, there is a homotopy $H : X \times^c I \to X$ such that $H|X \times^c \{0\} = f, H|X \times^c \{1\} = g$. Now,

$$kX \times^c I \xrightarrow{(w_X \times^c 1)^{-1}} X \times^c I \xrightarrow{w_{X \times^c I}} k(X \times^c I) \xrightarrow{kH} kX$$

is a homotopy between $k(fg) = w_X^{-1} fgw_X$ and 1_{kX}.

The last assertion follows from the above and the result of Milnor [3] that the space $C((S^1, *), (X, *))$ has the homotopy type of a CW-space. □

3.51. Exercise. Let T be the telescope of a filtration $\{X_n\}$ of a space X. We define a map $f : T \to X$ by setting $f(x,t) = x$. Prove that f is a Whitehead equivalence. Furthermore, prove that f is a homotopy equivalence provided that every inclusion $X_n \to X_{n+1}$ is a cofibration.

Chapter II. Spectra and (Co)homology Theories

In this chapter we discuss some preliminaries from stable homotopy theory. Sections 1–3 are concerned with basic properties of spectra and (co)homology theories. Here we mainly follow Adams [8] and Switzer [1]. Sections 4–7 contain an exposition of standard material, at a level suitable for students.

§1. Preliminaries on Spectra

In this section all spaces and maps are assumed to be pointed. Let SX denote the reduced suspension of a pointed space X, i.e., $SX = S^1 \wedge X$.

1.1. Definition. (a) A *spectrum* E is a sequence $\{E_n, s_n\}$, $n \in \mathbb{Z}$, of CW-complexes E_n and CW-embeddings $s_n : SE_n \to E_{n+1}$ (i.e., $s_n(SE_n)$ is a subcomplex of E_{n+1}).

(b) A *subspectrum* of a spectrum E is a spectrum $\{F_n, t_n\}$ such that F_n is a pointed CW-subcomplex of E_n and $t_n : SF_n \to F_{n+1}$ is the obvious restriction of s_n. In this case we also write $F \subset E$.

(c) Given a family $\{E(\alpha)\}$ of subspectra of E, we can form a subspectrum $\cup_\alpha E(\alpha)$ of E by setting $(\cup_\alpha E(\alpha))_n := \cup_\alpha E_n(\alpha)$, etc. A *filtration* of a spectrum E is a family

$$\{\cdots \subset E(i) \subset E(i+1) \subset \cdots \subset E\}$$

such that each $E(i)$ is a subspectum of E and, moreover, $\cup E(i) = E$.

(d) Given a spectrum E and an integer k, we define a spectrum $\Sigma^k E$ by setting $(\Sigma^k E)_n = E_{n+k}$ where the map $S(\Sigma^k E)_n \to (\Sigma^k E)_{n+1}$ is s_{n+k}.

(e) For every CW-complex X the spectrum $\Sigma^\infty X$ is defined as follows:

$$(\Sigma^\infty X)_n = \begin{cases} \text{pt}, & \text{if } n < 0, \\ S^n X, & \text{if } n \geq 0, \end{cases}$$

and $s_n : S(S^n X) \to S^{n+1} X$, for $n \geq 0$, are the identity maps.

For example, the spectrum $\Sigma^\infty S^0$ is the sphere spectrum $\{S^n, i_n\}$, where $i_n : SS^n \xrightarrow{=} S^{n+1}$.

It is convenient to regard SE_n as a subspace of E_{n+1}, i.e., to identify SE_n with its image $s_n(SE_n)$. Under this convention, if e is a cell of E_n then Se is a cell of E_{n+1}. We also write just $E = \{E_n\}$ rather than $E = \{E_n, s_n\}$ when the maps s_n are clear, and we denote the spectrum $\Sigma^\infty S^0$ just by S.

1.2. Definition. (a) A *cell* of a spectrum E is a sequence $\{e, Se, \dots, S^k e, \dots\}$ where e is a cell of any E_n such that e is not the suspension of any cell of E_{n-1}. If e is a cell of E_n of dimension d then the dimension of the cell $\{e, Se, \dots, S^k e, \dots\}$ of E is $d-n$. Furthermore, the base points of E_n's yield the cell of dimension $-\infty$.

(b) A subspectrum F of a spectrum E is *cofinal* (in E) if every cell of E is eventually in F, i.e., for every cell e of E_n there exists m such that $S^m e$ belongs to F_{n+m}.

(c) The *n-skeleton* of a spectrum E is the subspectrum $E^{(n)}$ of E consisting of all cells of dimensions $\leq n$.

(d) A spectrum E is *finite* if E has finitely many cells.

(e) A spectrum E has *finite type* if each skeleton of E is a finite spectrum.

(f) A spectrum E is *finite dimensional* if $E = E^{(n)}$ for some n.

(g) A *suspension spectrum* is a spectrum of the form $\Sigma^k \Sigma^\infty X$ where X is a pointed space and $k \in \mathbb{Z}$.

If $F = \{F_n\}, F' = \{F'_n\}$ are two subspectra of a spectrum E, we set $F \cap F' = \{F_n \cap F'_n\}$. It is obvious that $F \cap F'$ is a subspectrum. Furthermore, if F and F' are cofinal in E then so is $F \cap F'$.

1.3. Definition. (a) Let $E = \{E_n, s_n\}$ and $F = \{F_n, t_n\}$ be two spectra. A *map f from E to F* (i.e., a map $f : E \to F$) is a family of pointed cellular maps $f_n : E_n \to F_n$ such that $f_{n+1} s_n = t_n \circ S f_n$ for all n.

(b) Let $f : E \to F$ be a map of spectra. Given a subspectrum $G = \{G_n\}$ of E, the *restriction* of f to G is the map $f|G : G \to F$ of the form $\{f_n | G_n : G_n \to F_n\}$.

(c) Let E, F be two spectra. Consider the set $\mathscr{A}$ of pairs (f', E') where E' is cofinal in E and $f' : E' \to F$ is a map. Consider the equivalence relation $\sim$ on $\mathscr{A}$ such that $(f', E') \sim (f'', E'')$ iff $f'|B = f''|B$ for some $B \subset E' \cap E''$ with B cofinal in E. Every such equivalence class is called a *morphism from E to F*, and we use the notation $E \to F$ for morphisms as well as for maps.

(d) Given two maps $f : E \to F$ and $g : F \to G$, define the composition $gf : E \to G$ by setting $(gf)_n = g_n f_n$. It is straightforward to show that the composition $E \to G$ of morphisms $E \to F$ and $F \to G$ is also well defined. We can thus form a category $\mathscr{S}$ of spectra and their morphisms.

For every spectrum E and every integer n there is the embedding

$$i_n : \Sigma^{-n} \Sigma^\infty E_n \to E \qquad (1.4)$$

where $(i_n)_{n+k} : (\Sigma^{-n} \Sigma^\infty E_n)_{n+k} \to E_{n+k}$, for $k \geq 0$, is the composition

$$(i_n)_{n+k} : (\Sigma^{-n}\Sigma^\infty E_n)_{n+k} = S^k E_n \xrightarrow{s_{n+k-1}\circ\dots\circ S^{k-2}s_{n+1}S^{k-1}s_n} E_{n+k};$$

note that $(\Sigma^{-n}\Sigma^\infty E_n)_{n+k} = *$ for $k < 0$. We can thus regard $\Sigma^{-n}\Sigma^\infty E_n$ as a subspectrum of E.

1.5. Proposition. (i) *For every spectrum E and every $k \in \mathbb{Z}$, the spectrum $\{(E_n)^{(n+k)}\}$ is cofinal in the spectrum$(E^{(k)})$.*

(ii) *If E is a finite spectrum, then all the spaces E_k are finite spaces.*

(iii) *If E is a finite spectrum, then there is N such that $\Sigma^{-N}\Sigma^\infty E_N$ is cofinal in E.*

Proof. (i) This follows from the definitions.

(ii) The number of cells of E_k is bounded above by the number of cells of E.

(iii) Let a_k be the number of cells of E_k, and let a be the number of cells of E. Choose N such that $a_N = a$; this is possible because $a = \max_k a_k$. Now, $\Sigma^{-N}\Sigma^\infty E_N$ is cofinal in E because these two spectra have the same number of cells. □

Let $\mathscr{A}$ be as in 1.3(c). We can regard $\mathscr{A}$ as a poset as follows: $(f', E') \leq (f'', E'')$ iff E' is a subspectrum of E'' and $f''|E' = f'$.

1.6. Proposition. (i) *Let $f, g : E \to F$ be two maps such that $f|B = g|B$ for some cofinal subspectrum B of E. Then $f = g$.*

(ii) *Let E', E'' be cofinal in E, and let $f' : E' \to F$ and $f'' : E'' \to F$ be two equivalent maps. Then $f'|E' \cap E'' = f''|E' \cap E''$.*

(iii) *Every morphism contains a greatest element with respect to the above partial ordering.*

Proof. (i) Let $\{e, Se, \dots, S^k e, \dots\}$ be a cell of E where e is the cell of E_n. Since $f_{n+k}|S^k e = g_{n+k}|S^k e$ for some k, we have $f_n|e = g_n|e$.

(ii) This follows from (i).

(iii) Fix a morphism φ. If $(f', E') \in \varphi$ and $(f', E') \leq (f'', E'')$, then $(f'', E'') \in \varphi$. Hence, by Zorn's Lemma, φ has a maximal element. We denote this element by $(\bar{f}, \bar{E})$ and prove that $(\bar{f}, \bar{E})$ is the greatest element of φ. Suppose not. Then there exists $(f', E') \in \varphi$ which does not satisfy the inequality $(f', E') \leq (\bar{f}, \bar{E})$. Then, by (ii), $f'|E' \cap \bar{E} = \bar{f}|E' \cap \bar{E}$ since $(\bar{f}, \bar{E}) \sim (f', E')$. Hence, the map $\bar{f} \cup f' : \bar{E} \cup E' \to F$ is well defined. On the other hand, $(\bar{f}, \bar{E}) \sim (\bar{f} \cup f', \bar{E} \cup E')$. But this contradicts the maximality of $(\bar{f}, \bar{E})$ in φ. □

1.7. Definition. (a) Let $f : E \to F$ be a map of spectra. Define the *cone* of f to be the spectrum $Cf := \{Cf_n, s_n\}$ where Cf_n is the cone of the map $f_n : E_n \to F_n$ and s_n has the form

$$s_n : S(Cf_n) = SF_n \cup C(SE_n) \xrightarrow{s_n^F \cup C s_n^E} F_{n+1} \cup C(E_{n+1}) = Cf_{n+1}.$$

(b) Let $\varphi : E \to F$ be a morphism of spectra. Define the *cone* $C\varphi$ of φ by setting $C\varphi = Cf'$, where $f' : E' \to F$ is the greatest element of φ. The cone $C\varphi$ is also called the *cofiber* of φ. Furthermore, define the *canonical inclusion* $\psi : F \to C\varphi$ by setting $\psi = \{\psi_n : F_n \to Cf'_n\}$, where ψ_n is as in I.(3.17).

1.8. Proposition. *If $\varphi : E \to F$ is a morphism of finite spectra (resp. of spectra of finite type) then $C\varphi$ is a finite spectrum (resp. a spectrum of finite type).*

Proof. Decode the definitions. □

Given a spectrum E and a CW-complex X, we define spectra $E \wedge X := \{E_n \wedge X\}$ and $X \wedge E := \{X \wedge E_n\}$. In particular, the suspension $S^1 \wedge E$ of a spectrum E is defined.

1.9. Definition. (a) Two maps $g_0, g_1 : E \to F$ of spectra are called *homotopic* if there exists a map $G : E \wedge I^+ \to F$ (called a *homotopy*) such that G coincides with g_i on the subspectrum $E \wedge \{i, *\}, i = 0, 1$, of E. In this case we write $g_0 \simeq g_1$ or $G : g_0 \simeq g_1$.

(b) Two morphisms $\varphi_0, \varphi_1 : E \to F$ of spectra are called *homotopic*, if there exists a cofinal subspectrum E' of E and two maps $g_i : E' \to F, g_i \in \varphi_i, i = 0, 1$, such that $g_0|E' \simeq g_1|E'$. It is straightforward to show that homotopic morphisms form equivalence classes, and in particular we can define the homotopy class $[\varphi]$ of a morphism φ to be the set of all morphisms homotopic to φ. The set of all homotopy classes of morphisms $E \to F$ is denoted by $[E, F]$

(c) We say that a morphism of spectra is *trivial*, or *inessential*, if it is homotopic to the trivial morphism $\varepsilon = \{\varepsilon_n : E_n \to F_n\}$, $\varepsilon_n(E_n) = *$. Otherwise we say that it is *essential.*

One can prove that the homotopy class $[\varphi\psi]$ of the composition $\varphi\psi$ depends only on the homotopy classes of the morphisms φ, ψ, see e.g. Switzer [1]. So, we can define the composition of homotopy classes of morphisms by setting $[\varphi][\psi] = [\varphi\psi]$. Thus, we can define a category $\mathscr{HS}$ with spectra as objects and sets $[E, F]$ as sets of morphisms. Isomorphisms of $\mathscr{HS}$ are called *equivalences* (of spectra), and we use the notation $E \simeq F$ when E is equivalent to F.

It is straightforward to show that the cones of homotopic morphisms are equivalent.

Let $S = \Sigma^\infty S^0$ be the spectrum of spheres. The group $[\Sigma^k S, E]$ is called the *k-th homotopy group* of E and denoted by $\pi_k(E)$. It is easy to see that $\pi_k(E) = \varinjlim_{N \to \infty} \pi_{k+N}(E_N)$ where the direct limit is that of the direct system

$$\cdots \to \pi_{k+N}(E_N) \to \pi_{k+N+1}(SE_N) \xrightarrow{(s_N)_*} \pi_{k+N+1}(E_{N+1}) \to \cdots,$$

see Switzer [1], 8.21. In particular, if $E = \Sigma^\infty X$ then $\pi_k(E)$ is just the stable homotopy group $\Pi_k(X)$ (denoted also by $\pi_k^{\mathrm{st}}(X)$).

Given a morphism $\varphi : E \to F$, define $\varphi_* = \pi_k(\varphi) : \pi_k(E) \to \pi_k(F)$ by setting $\varphi_*(a) = [\varphi\psi]$ where $\psi : \Sigma^k S \to E$ is a morphism with $a = [\psi]$. Hence, π_k is a functor $\mathscr{HS} \to \mathscr{AG}$. Note that $\pi_k(E)$ can be non-zero even for $k < 0$, a simple example being the spectrum $E = \Sigma^{-N} S$.

An analog of the Whitehead Theorem is valid for spectra.

1.10. Theorem. *A morphism $\varphi : E \to F$ is an equivalence iff the induced homomorphism $\varphi_* : \pi_k(E) \to \pi_k(F)$ is an isomorphism for every integer k.*

Proof. See Switzer [1], 8.25. □

One of the important advantages of the category $\mathscr{HS}$ is that the suspension operator is invertible there.

1.11. Proposition. *The spectra $S^1 \wedge E$ and ΣE are equivalent.*

Proof. See Switzer [1], 8.26. □

1.12. Definition. (a) A *strict cofiber sequence* of spectra is a diagram

$$E \xrightarrow{\varphi} F \xrightarrow{\psi} G$$

where $\varphi : E \to F$ is a morphism of spectra (resp. map of spaces) and ψ is a canonical inclusion as in 1.7(b).

(b) A sequence

$$X \xrightarrow{f} Y \xrightarrow{g} Z$$

in $\mathscr{S}$ is called a *cofiber sequence* of spectra if there exists a homotopy commutative diagram in $\mathscr{S}$

$$\begin{array}{ccccc} X & \xrightarrow{f} & Y & \xrightarrow{g} & Z \\ {\scriptstyle a}\downarrow & & {\scriptstyle b}\downarrow & & {\scriptstyle c}\downarrow \\ E & \xrightarrow{\varphi} & F & \xrightarrow{\psi} & C\varphi \end{array}$$

such that all the vertical arrows are equivalences and the bottom row is a strict cofiber sequence of spectra.

(c) A *long cofiber sequence* of spectra is a sequence (finite or not)

$$\cdots \to X_i \to X_{i+1} \to X_{i+2} \to \cdots$$

where every pair of adjacent morphisms forms a cofiber sequence of spectra.

1.13. Lemma. (i) *If* $X \xrightarrow{f} Y \xrightarrow{g} Z$ *is a cofiber sequence of spectra, then there exists a map* $h : Z \to \Sigma X$ *such that* $X \xrightarrow{f} Y \xrightarrow{g} Z \xrightarrow{h} \Sigma X$ *is a long cofiber sequence.*

(ii) *Let* $X \xrightarrow{f} Y \xrightarrow{g} Z$ *and* $X' \xrightarrow{f'} Y' \xrightarrow{g'} Z'$ *be two cofiber sequences of spectra. For all morphisms* $\alpha : X \to X'$ *and* $\beta : Y \to Y'$ *with* $f'\alpha \simeq \beta f$, *there exist morphisms* γ, h *and* h' *with* h *and* h' *as above, such that the following diagram commutes up to homotopy:*

$$\begin{array}{ccccccc} X & \xrightarrow{f} & Y & \xrightarrow{g} & Z & \xrightarrow{h} & \Sigma X \\ {\scriptstyle\alpha}\downarrow & & {\scriptstyle\beta}\downarrow & & {\scriptstyle\gamma}\downarrow & & \downarrow{\scriptstyle\Sigma\alpha} \\ X' & \xrightarrow{f'} & Y' & \xrightarrow{g'} & Z' & \xrightarrow{h'} & \Sigma X'. \end{array}$$

(iii) *If* $X \xrightarrow{f} Y \xrightarrow{g} Z$ *is a cofiber sequence of CW-complexes, then* $\Sigma^\infty X \xrightarrow{\Sigma^\infty f} \Sigma^\infty Y \xrightarrow{\Sigma^\infty g} \Sigma^\infty Z$ *is a cofiber sequence of spectra.*

Proof. (i) Consider a diagram as in 1.12(b). We have $C\psi \simeq S^1 \wedge E$ (the proof is similar to the one for spaces). Therefore, in view of 1.11, we have an equivalence $u : C\psi \to \Sigma E$. So, we have the homotopy commutative diagram

$$\begin{array}{ccccccccc} X & \xrightarrow{f} & Y & \xrightarrow{g} & Z & & & & \Sigma X \\ {\scriptstyle a}\downarrow & & {\scriptstyle b}\downarrow & & {\scriptstyle c}\downarrow & & & & \downarrow{\scriptstyle\Sigma a} \\ E & \xrightarrow{\varphi} & F & \xrightarrow{\psi} & C\varphi & \xrightarrow{\xi} & C\psi & \xrightarrow{u} & \Sigma E \end{array}$$

where Σa is an equivalence. Thus, we can *define* the desired h by setting $h := (\Sigma a)^{-1} u \xi c$.

(ii) Proved in Switzer [1], 8.31, but the proof is implicitly based on (i).

(iii) This holds since $C(\Sigma^\infty h) \simeq \Sigma^\infty C(h)$ for every map $h : A \to B$ of CW-spaces. □

Given a family $E(\lambda), \lambda \in \Lambda$, of spectra, we define the *wedge* $\bigvee_\lambda E(\lambda)$ by setting $(\vee_\lambda E(\lambda))_n := \vee_\lambda (E_n(\lambda))$. Since $S(\vee_\lambda E_n(\lambda)) = \vee_\lambda S E_n(\lambda) \subset \vee_\lambda E_{n+1}(\lambda)$, we conclude that $\vee_\lambda E(\lambda)$ is a spectrum. Let $i_\lambda : E(\lambda) \to \vee_\lambda E(\lambda)$ be the obvious inclusion.

1.14. Proposition. *For every spectrum* F *the function*

$$\{i_\lambda^*\} : [\vee_\lambda E(\lambda), F] \to \prod_\lambda [E(\lambda), F], \text{ where } \{i_\lambda^*\}(f) = \{f i_\lambda\},$$

is a biection.

Proof. See Switzer [1], 8.18. □

Since $[E,F] = [\Sigma^2 E, \Sigma^2 F] = [S^2 \wedge E, \Sigma^2 F]$ (the last equality follows from 1.11), $[E,F]$ admits a natural structure of an abelian group. Indeed, let $\nu : S^2 \to S^2 \vee S^2$ be the usual comultiplication on S^2, the pinch map. Since $(S^2 \vee S^2) \wedge E \simeq (S^2 \wedge E) \vee (S^2 \wedge E)$, we obtain the function

$$\begin{aligned}[E,F] \oplus [E,F] &= [S^2 \wedge E, \Sigma^2 F] \oplus [S^2 \wedge E, \Sigma^2 F] \\ &= [(S^2 \wedge E) \vee (S^2 \wedge E), \Sigma^2 F] \\ &= [(S^2 \vee S^2) \wedge E, \Sigma^2 F] \xrightarrow{(\nu\wedge 1)^*} [S^2 \wedge E, \Sigma^2 F] = [E,F]\end{aligned}$$

which turns $[E,F]$ into an abelian group. Moreover, the composition

$$[E,F] \times [F,G] \to [E,G]$$

is biadditive, see Switzer [1], 8.27. Thus, $\mathscr{HS}$ is an additive category.

In view of 1.11 and 1.13(i), every cofiber sequence $X \xrightarrow{f} Y \xrightarrow{g} Z$ yields a long cofiber sequence

$$\ldots \to \Sigma^{-1} Y \xrightarrow{\Sigma^{-1} g} \Sigma^{-1} Z \to X \xrightarrow{f} Y \xrightarrow{g} Z \to \Sigma X \xrightarrow{\Sigma f} \Sigma Y \to \cdots .$$

1.15. Theorem. *For every spectrum E, the long cofiber sequence*

$$\ldots \to \Sigma^{-1} Y \xrightarrow{\Sigma^{-1} g} \Sigma^{-1} Z \to X \xrightarrow{f} Y \xrightarrow{g} Z \to \Sigma X \xrightarrow{\Sigma f} \Sigma Y \to \cdots$$

yields the exact sequences

$$\cdots \leftarrow [\Sigma^{-1} Z, E] \leftarrow [X,E] \xleftarrow{f^*} [Y,E] \xleftarrow{g^*} [Z,E] \leftarrow [\Sigma X, E] \leftarrow \cdots$$
$$\cdots \to [E, \Sigma^{-1} Z] \to [E,X] \xrightarrow{f_*} [E,Y] \xrightarrow{g_*} [E,Z] \to [E, \Sigma X] \to \cdots$$

of abelian groups and homomorphisms.

Proof. See Switzer [1], Proposition 8.32. □

The first of the above sequences is similar to a sequence which holds for a cofibration $X \to Y$ with cofiber Z, while the second one is similar to a sequence which holds for a fibration $X \to Y$ with fiber $\Sigma^{-1} Z$. Thus, the difference between fibrations and cofibrations disappears in the category $\mathscr{HS}$. For this reason we call $\Sigma^{-1} C\varphi$ the *fiber* of a morphism φ.

1.16. Proposition. (i) *For every spectrum F the function*

$$\{i_\lambda^*\} : [\vee_\lambda E(\lambda), F] \to \prod_\lambda [E(\lambda), F], \quad \{i_\lambda^*\}(f) = \{f i_\lambda\},$$

is an isomorphism of abelian groups.

(ii) *For every spectrum F the homomorphism*

$$\langle (i_k)_* \rangle : \bigoplus_{k=1}^{m} [F, E(k)] \to [F, \vee_{k=1}^{m} E(k)]$$

as in I.2.2 *is an isomorphism.*

(iii) *For every finite spectrum F the homomorphism*

$$\langle (i_\lambda)_* \rangle : \oplus_\lambda [F, E(\lambda)] \to [F, \vee_\lambda E(\lambda)]$$

is an isomorphism. In particular, $\pi_*(\vee_\lambda E(\lambda)) \cong \oplus_\lambda \pi_*(E(\lambda))$.

Proof. (i) It is easy to see that $\{i_\lambda^*\}$ is a homomorphism of abelian groups, and the result follows from 1.14.

(ii) It suffices to consider $m = 2$. Let $p : E_1 \vee E_2 \to E_1$ be the projection, $pi_1 = 1_{E_1}$. Then the cofiber sequence $E_2 \xrightarrow{i_2} E_1 \vee E_2 \xrightarrow{p} E_1$ induces an exact sequence $[F, E_1] \to [F, E_1 \vee E_2] \to [F, E_2]$ which splits by i_2 and/or p. Thus,

$$\langle (i_k)_* \rangle : [F, E_1] \oplus [F, E_2] \to [F, E_1 \vee E_2]$$

is an isomorphism.

(iii) Let $\mathscr{K} = \{K\}$ be the family of all finite subsets of the index set Λ. For every $K \in \mathscr{K}$ we have the monomorphism $l_K : [F, \vee_{k \in K} E(k)] \to [F, \vee_\lambda E(\lambda)]$. We set $K \le K'$ iff $K \subset K'$, consider the homomorphism

$$l = \langle l_K | \varinjlim \rangle : \varinjlim [F, \vee_{k \in K} E(k)] \to [F, \vee_\lambda E(\lambda)]$$

as in I.2.5 and prove that it is an isomorphism. Firstly, it is monic since l_K is monic for every K. Furthermore, F is finite, and so, for every $f : F \to \vee_\lambda E(\lambda)$, there exists K such that $f(F) \subset \vee_{k \in K} E(k)$. Thus, l is an isomorphism. Now, $\langle (i_\lambda)_* \rangle$ can be written as

$$\oplus_\lambda [F, E(\lambda)] = \varinjlim_{\mathscr{K}} \oplus_{k \in K} [F, E(k)] \cong \varinjlim_{\mathscr{K}} [F, \vee_{k \in K} E(k)] \xrightarrow{l} [F, \vee_\lambda E(\lambda)].$$

(The isomorphism holds by (ii).) □

By 1.16(ii), $[X, X \vee X] \cong [X, X] \oplus [X, X]$. Hence, the element $1_X \oplus 1_X$ of the right hand side yields a (unique up to homotopy) morphism $\nabla : X \to X \vee X$. We leave it to the reader to show that addition in $[X, E]$ is given by the composition

$$[X, E] \oplus [X, E] \xrightarrow{\cong} [X \vee X, E] \xrightarrow{\nabla^*} [X, E].$$

Because of this, we call ∇ *coaddition.*

1.17. Proposition. *Let $X \xrightarrow{f} Y \xrightarrow{g} Z$ be a cofiber sequence of spectra. The following two conditions are equivalent:*

(i) *The morphism g is inessential;*

(ii) *There is a morphism $s : Y \to X$ such that $fs \simeq 1_Y$.*

Furthermore, if these conditions hold then $X \simeq \Sigma^{-1}Z \vee Y$.

Proof. We prove that (i) $\Rightarrow$ (ii). Consider the exact sequence

$$[Y,X] \xrightarrow{f_*} [Y,Y] \xrightarrow{g_*} [Y,Z].$$

Now, f_* is epic since $g_* = 0$. Hence, there is $s : Y \to X$ such that $f_*[s] = [1_Y]$, i.e., $fs \simeq 1_Y$.

We prove that (ii) $\Rightarrow$ (i). Indeed, $g \simeq g(fs) \simeq (gf)s$, but gf is inessential.

Now we prove that $Y \simeq \Sigma^{-1}Z \vee X$. Indeed, the cofiber sequence $\Sigma^{-1}Z \to X \to Y$ induces the exact sequence

$$0 \to [E, \Sigma^{-1}Z] \to [E,X] \to [E,Y] \to 0,$$

and s gives us a natural splitting of this sequence. So, we have a natural in E isomorphism $[E,X] \cong [E,\Sigma^{-1}Z] \oplus [E,Y]$. On the other hand, by 1.16(ii), there is a natural isomorphism

$$[E,\Sigma^{-1}Z] \oplus [E,Y] \cong [E,\Sigma^{-1}Z \vee Y].$$

Hence, we have a natural isomorphism

$$[E,X] \cong [E,\Sigma^{-1}Z \vee Y], \ E \in \mathscr{S},$$

and thus, by the Yoneda Lemma I.1.5, $X \simeq \Sigma^{-1}Z \vee Y$. □

1.18. Definition. A *prespectrum* is a family $\{X_n, t_n\}, n \in \mathbb{Z}$, of pointed spaces X_n and pointed maps $t_n : SX_n \to X_{n+1}$.

A CW-prespectrum is a prespectrum $\{X_n, t_n\}$ such that every X_n is a CW-complex and every t_n is a cellular map.

1.19. Lemma–Definition. *For every prespectrum* $\{X_n, t_n\}$, *there exist a spectrum* $E = \{E_n, s_n\}$ *and pointed homotopy equivalences* $f_n : E_n \to X_n$ *such that the diagram*

$$\begin{array}{ccc} SE_n & \xrightarrow{Sf_n} & SX_n \\ \downarrow{s_n} & & \downarrow{t_n} \\ E_{n+1} & \xrightarrow{f_{n+1}} & X_{n+1} \end{array}$$

commutes. Every such spectrum E is called a spectral substitute *of the prespectrum X. Furthermore, if X_n are CW-complexes such that $(X_n)^{(n+k)} = *$ for all n and some fixed k, then E can be chosen so that $E^{(k)} = *$.*

Proof (cf. Switzer [1], Proposition 8.3.). Firstly, because of I.3.44 and I.3.45, we can replace X_n by its CW-substitute X'_n and construct a homotopy commutative diagram

$$\begin{array}{ccc} SX'_n & \xrightarrow{Sg_n} & SX_n \\ {\scriptstyle t'_n}\downarrow & & \downarrow{\scriptstyle t_n} \\ X'_{n+1} & \xrightarrow{g_{n+1}} & X_{n+1} \end{array}$$

where g_n, g_{n+1} are homotopy equivalences and t'_n is a cellular map. So, we can assume that $\{X_n, t_n\}$ is a CW-prespectrum. Now we use iterated mapping cylinders (i.e., telescopes) to convert the t_n into inclusions. Set

$$E_n = \left(\bigcup_{m<n} S^{n-m}X_m \wedge [m, m+1]^+\right) \cup X_n$$

with the following identifications: $(x, m+1) \in S^{n-m}X_m \wedge [m, m+1]^+$ is identified with $(S^{n-m-1}t_m(x), m+1) \in S^{n-m-1}X_{m+1} \wedge [m+1, m+2]^+$; and $(x, n) \in SX_{n-1} \wedge [n-1, n]^+$ is identified with $t_{n-1}(x) \in X_n$.

Now, the maps $t_m, m \leq n$, yield the obvious inclusion $s_n : SE_n \to E_{n+1}$, and the map $f_n : E_n \to X_n$, $f_n(x, s) = t_{n-1} \circ St_{n-2} \circ \ldots \circ S^{n-m-1}t_m(x)$ for $(x, s) \in S^{n-m}X_m \wedge [m, m+1]^+$, is a deformation retraction. Clearly, the diagram of the lemma commutes up to homotopy. But, since s_n is a cofibration, we can deform f such that the diagram turns out to be commutative, step by step.

Furthermore, if $A^{(n)} = *$ for a CW-complex A then $(SA)^{(n+1)} = *$ and $(A \wedge I^+)^{(n)} = *$. Thus, if $(X_n)^{(n+k)} = *$ for all n then $(E_n)^{(n+k)} = *$ for all n, and hence $E^{(k)} = *$. □

Let ΩX denote the loop space of a pointed space X.

1.20. Definition. A prespectrum $X = \{X_n, t_n\}$ is called an *Ω-prespectrum* if for every n the map $\tau_n : X_n \to \Omega X_{n+1}$ adjoint to t_n is a homotopy equivalence. A spectrum is called an *Ω-spectrum* if it is an Ω-prespectrum.

1.21. Proposition (cf. Adams [5]). *Every spectrum $E = \{E_n, s_n\}$ is equivalent to some Ω-spectrum.*

Proof. Let $\varepsilon_n : E_n \to \Omega E_{n+1}$ be the adjoint map to s_n, and let $\overline{\Omega^k E_m}$ be a CW-complex homotopy equivalent to $\Omega^k E_m$ (such a CW-complex exists by I.3.50). Fix some mutually inverse homotopy equivalences

$$\alpha = \alpha_{k,m} : \overline{\Omega^k E_m} \to \Omega^k E_m, \ \beta = \beta_{k,m} : \Omega^k E_m \to \overline{\Omega^k E_m}.$$

Fix some n and consider

$$\varphi_k : \overline{\Omega^k E_{n+k}} \xrightarrow{\alpha} \Omega^k E_{n+k} \xrightarrow{\Omega^k \varepsilon_{n+k}} \Omega^{k+1} E_{n+k+1} \xrightarrow{\beta} \overline{\Omega^{k+1} E_{n+k+1}}.$$

Let T_n be the reduced telescope (see I.3.23) of the sequence

$$E_n = \overline{E}_n \xrightarrow{\varphi_0} \overline{\Omega E_{n+1}} \to \ldots \to \overline{\Omega^k E_{n+k}} \xrightarrow{\varphi_k} \ldots .$$

Define ψ_k to be the composition

$$\begin{aligned}\psi_k : \overline{\Omega^k E_{n+k}} &\xrightarrow{\alpha} \Omega^k E_{n+k} \xrightarrow{\Omega^k \varepsilon_{n+k}} \Omega^{k+1} E_{n+k+1} = \Omega(\Omega^k E_{n+k+1}) \\ &\xrightarrow{\Omega\beta} \Omega(\overline{\Omega^k E_{n+k+1}}).\end{aligned}$$

Then we have a homotopy commutative diagram

$$\begin{array}{ccccccccc} E_n & \xrightarrow{\varphi_0} & \overline{\Omega E_{n+1}} & \longrightarrow & \dots & \longrightarrow & \overline{\Omega^k E_{n+k}} & \xrightarrow{\varphi_k} & \dots \\ \downarrow{\scriptstyle \varepsilon_n} & & \downarrow{\scriptstyle \psi_1} & & & & \downarrow{\scriptstyle \psi_k} & & \\ \Omega E_{n+1} & \xrightarrow{\Omega\varphi_0} & \Omega(\overline{\Omega E_{n+2}}) & \longrightarrow & \dots & \longrightarrow & \Omega(\overline{\Omega^k E_{n+k+1}}) & \xrightarrow{\Omega\varphi_k} & \dots \end{array}$$

where all the vertical arrows are homotopy equivalences. Passing to telescopes, we get an obvious map $\omega : T_n \to \Omega T_{n+1}$ induced by the ψ_k's. Since every compact set in T_n is contained in some finite union

$$\bigcup_{k=0}^{m} \overline{\Omega^k E_{n+k}} \times [k, k+1],$$

we conclude that $\pi_i(T_n) = \varinjlim \pi_i(\Omega^k E_{n+k}) = \pi_{i-n}(E)$. Hence, ω is a homotopy equivalence by the Whitehead theorem, and so T is an Ω-prespectrum. Let F be a spectral substitute of T. The inclusions $i_n : E_n \to T_n$ yield maps $h_n : E_n \to F_n$ (such that the composition $E_n \to F_n \to T_n$ is homotopic to i_n). The diagram

$$\begin{array}{ccc} SE_n & \xrightarrow{Sh_n} & SF_n \\ {\scriptstyle s_n}\downarrow & & \downarrow \\ E_{n+1} & \xrightarrow{h_{n+1}} & F_{n+1} \end{array}$$

commutes up to homotopy, and hence, since s_n is a cofibration, we can replace h_{n+1} by a homotopic map making the diagram commutative. Without loss of generality, we can assume that $E_n = \mathrm{pt}$ for $n < 0$, and so we can change the h_n's map by map so that each diagram as above will become commutative. Thus, we have constructed a morphism $E \to F$ which induces isomorphisms

$$\pi_i(E) = \varinjlim \pi_{i+n}(E_n) \to \varinjlim \pi_{i+n}(F_n) = \pi_i(F). \qquad \square$$

1.22. Proposition (cf. I.3.37). *Let A, B be two subspectra of a spectrum X such that $X = A \cup B$. Consider the map $h : A \vee B \to X$ such that $h(a) = a$ for every $a \in A$ and $h(b) = b$ for every $b \in B$. Then $Ch \simeq \Sigma(A \cap B)$.*

Proof. We set $C = A \cap B$. We have

$$(Ch)_n = C(h_n) \simeq (CA_n \vee CB_n) \cup_{h_n} X_n \simeq (CA_n \vee CB_n)/C_n \simeq S^1 \wedge C_n$$

because $CA_n \vee CB_n$ is contractible. Clearly, these homotopy equivalences yield the equivalence $Ch \simeq S^1 \wedge C \simeq \Sigma C$. □

1.23. Construction. We give an analog of the telescope construction for spectra. Let $E = \{E_n, s_n\}$ be a spectrum with $E_n = \mathrm{pt}$ for $n < 0$. We define a spectrum $\tau E = \{\tau_n E, t_n\}$ as follows: $\tau_n E$ is the reduced telescope of the sequence

$$\{S^n E_0 \to \cdots \to S^{n-i} E_i \xrightarrow{S^{n-i-1} s_i} \cdots \to E_n\};$$

in other words,

$$\begin{aligned}\tau_n E = (S^n E_0 \wedge [0,1]^+) \cup \cdots \cup (S^{n-k} E_k \wedge [k, k+1]^+) \cup \cdots \\ \cup (S^1 E_{n-1} \wedge [n-1, n]^+) \cup E_n,\end{aligned}$$

with the following identifications: $(x, k) \in S^{n-k+1} E_{k-1} \wedge [k-1, k]^+$ is identified with $(S^{n-k} s_{k-1}(x), k) \in S^{n-k} E_k \wedge [k, k+1]^+$ and $(x, n) \in S^1 E_{n-1} \wedge [n-1, n]^+$ is identified with $s_{n-1}(x) \in E_n$. Furthermore, the homeomorphism $i_k : S S^{n-k} E_k \to S^{n-k+1} E_k$ induces a homeomorphism

$$\varphi_k := i_k \wedge 1 : S S^{n-k} E_k \wedge [k, k+1]^+ = S^{n-k+1} E_k \wedge [k, k+1]^+, \; k = 0, 1, \ldots n-1,$$

and the inclusion $j : \{n\} \subset [n, n+1]^+$ induces an inclusion

$$\varphi_n := 1 \wedge j : SE_n = SE_n \times \{n\} = SE_n \wedge \{n\}^+ \subset SE_n \wedge [n, n+1]^+.$$

We define $t_n := \cup_{k=0}^{n} \varphi_k : S\tau_n E \to \tau_{n+1} E$. Thus, the spectrum $\tau E = \{\tau_n E, t_n\}$ is constructed.

The standard deformation retractions $\tau_n E \to E_n$ (which shrink each segment $[k, k+1]$) form a morphism $\tau E \to E$. Clearly, this morphism is an equivalence.

Define subspectra $\tau_{\mathrm{ev}} E, \tau_{\mathrm{od}} E$ of τE by setting

$$(\tau_{\mathrm{ev}})_n(E) := \bigcup_{i=0}^{[\frac{n}{2}]} (S^{n-2i+1} E_{2i-1} \wedge [2i-1, 2i]^+ \cup S^{n-2i} E_{2i}),$$

$$(\tau_{\mathrm{od}})_n(E) := \bigcup_{i=0}^{[\frac{n-1}{2}]} (S^{n-2i} E_{2i} \wedge [2i, 2i+1]^+ \cup S^{n-2i-1} E_{2i+1}).$$

It is clear that

$$\begin{aligned}&\tau_{\mathrm{ev}}(E) \cup \tau_{\mathrm{od}}(E) = \tau(E), \quad \tau_{\mathrm{ev}}(E) \cap \tau_{\mathrm{od}}(E) = \bigvee_{n=0}^{\infty} \Sigma^{-n} \Sigma^{\infty} E_n, \\ &\tau_{\mathrm{ev}}(E) \simeq \bigvee_{n=0}^{\infty} \Sigma^{-2n} \Sigma^{\infty} E_{2n}, \quad \tau_{\mathrm{od}}(E) \simeq \bigvee_{n=0}^{\infty} \Sigma^{-2n-1} \Sigma^{\infty} E_{2n+1}.\end{aligned} \tag{1.24}$$

1.25. Remark. The concept of a spectrum was in fact introduced by Lima [1], [2]. Later, different categories of spectra were constructed. We use the category suggested by Adams [8]. Some authors have developed a finer theory by indexing terms of a spectrum not by integers but by finite dimensional subspaces of $\mathbb{R}^\infty$. This approach was suggested by Puppe [2] and May [3]. Such spectra are very useful for working with some fine geometry. However, the foundations of this theory are quite complicated. For our purposes, the mass of preliminaries outweighs the gain; thus we do not use these spectra here and so do not dwell on them. However, they seem to be very useful for advanced homotopy theory. The reader who is interested in this theory is referred to the book Elmendorf–Kriz–Mandell–May [1].

§2. The Smash Product of Spectra, Duality, Ring and Module Spectra

One can introduce a smash product $E \wedge F$ of spectra E, F as a generalization of the smash product $E \wedge X$ of a spectrum and a space. The definition (construction) of the smash product of spectra can be found in Adams [8] or Switzer [1]. However, we do need to know the consrtuction; throughout the book we use only the properties listed in 2.1 below.

2.1. Theorem. *There is a construction which assigns to spectra E, F a certain spectrum denoted by $E \wedge F$. This construction is called the* smash product $E \wedge F$ *(of spectra) and has the following properties:*

(i) *It is a covariant functor of each of its arguments.*

(ii) *There are natural equivalences:*

$$\begin{aligned} a = a(E,F,G) &: (E \wedge F) \wedge G \to E \wedge (F \wedge G) \\ \tau = \tau(E,F) &: E \wedge F \to F \wedge E \\ l = l(E) &: S \wedge E \to E \\ r = r(E) &: E \wedge S \to E \\ \Sigma = \Sigma(E,F) &: \Sigma E \wedge F \to \Sigma(E \wedge F). \end{aligned}$$

(iii) *For every spectrum E and CW-complex X, there is a natural equivalence $e = e(E,X) : E \wedge X \to E \wedge \Sigma^\infty X$. In particular, $\Sigma^\infty(X \wedge Y) \simeq \Sigma^\infty X \wedge \Sigma^\infty Y$ for every pair of CW-complexes X, Y.*

(iv) *If $f : E \to F$ is an equivalence then $f \wedge 1_G : E \wedge G \to F \wedge G$ is.*

(v) *Let $\{E_\lambda\}$ be a family of spectra, and let $i_\lambda : E_\lambda \to \vee_\lambda E_\lambda$ be the inclusions. Then the morphism (see* 1.16(i)*)*

$$\{i_\lambda \wedge 1\} : \vee_\lambda(E_\lambda \wedge F) \to (\vee_\lambda(E_\lambda)) \wedge F$$

is an equivalence.

(vi) *If* $A \xrightarrow{f} B \xrightarrow{g} C$ *is a cofiber sequence of spectra, then so is the sequence* $A \wedge E \xrightarrow{f \wedge 1} B \wedge E \xrightarrow{g \wedge 1} C \wedge E$ *for every spectrum* E.

Proof. The proof can be found in Adams [8] or Switzer [1], but I want to say the following. The terms of the spectrum $E \wedge F$ are aggregated from the spaces $E_m \wedge F_n$, but, in order to get them as CW-complexes, we must follow definition I.3.6. By I.3.49, this modification keeps the theorem valid. □

2.2. Theorem. *The following diagrams commute up to homotopy:*

(i)

$$\begin{array}{ccccc} ((E \wedge F) \wedge G) \wedge H & \xrightarrow{a} & (E \wedge F) \wedge (G \wedge H) & \xrightarrow{a} & E \wedge (F \wedge (G \wedge H)) \\ \| & & & & \uparrow{\scriptstyle 1 \wedge a} \\ ((E \wedge F) \wedge G) \wedge H & \xrightarrow{a \wedge 1} & (E \wedge (F \wedge G)) \wedge H & \xrightarrow{a} & E \wedge ((F \wedge G) \wedge H)) \end{array}$$

(ii)

$$\begin{array}{ccc} E \wedge F & = & E \wedge F \\ {\scriptstyle \tau}\downarrow & & {\scriptstyle \tau}\uparrow \\ F \wedge E & = & F \wedge E \end{array}$$

(iii)

$$\begin{array}{ccccc} (E \wedge F) \wedge G & \xrightarrow{\tau \wedge 1} & (F \wedge E) \wedge G & \xrightarrow{a} & F \wedge (E \wedge G) \\ {\scriptstyle a}\downarrow & & & & \downarrow{\scriptstyle 1 \wedge \tau} \\ E \wedge (F \wedge G) & \xrightarrow{\tau} & (F \wedge G) \wedge E & \xrightarrow{a} & F \wedge (G \wedge E) \end{array}$$

(iv)

$$\begin{array}{ccc} (S \wedge E) \wedge F & \xrightarrow{a} & S \wedge (E \wedge F) \\ {\scriptstyle l \wedge 1}\downarrow & & \downarrow{\scriptstyle l} \\ E \wedge F & = & E \wedge F \end{array}$$

(v)

$$\begin{array}{ccc} (E \wedge F) \wedge S & \xrightarrow{a} & E \wedge (F \wedge S) \\ {\scriptstyle r}\downarrow & & \downarrow{\scriptstyle 1 \wedge r} \\ E \wedge F & = & E \wedge F \end{array}$$

(vi)

$$\begin{array}{ccc} (E \wedge S) \wedge F & \xrightarrow{a} & E \wedge (S \wedge F) \\ {\scriptstyle r \wedge 1}\downarrow & & \downarrow{\scriptstyle 1 \wedge l} \\ E \wedge F & = & E \wedge F \end{array}$$

(vii)
$$\begin{array}{ccc} S\wedge E & \xrightarrow{\tau} & E\wedge S \\ {\scriptstyle l}\downarrow & & \downarrow{\scriptstyle r} \\ E & = & E \end{array}$$

(viii)
$$\begin{array}{ccc} S\wedge S & \xrightarrow{1} & S\wedge S \\ \| & & \| \\ S\wedge S & \xrightarrow{\tau} & S\wedge S. \end{array}$$

Proof. See Adams [8] or Switzer [1]. □

2.3. Definition. (a) A morphism $u : S \to A \wedge A^{\perp}$ is called a *duality morphism*, or simply a *duality*, between spectra A and $A^{\perp}$ if for every spectrum E the homomorphisms

$$u_E : [A, E] \to [S, E \wedge A^{\perp}], \quad u_E(\varphi) = (\varphi \wedge 1_{A^{\perp}})u$$

and

$$u^E : [A^{\perp}, E] \to [S, A \wedge E], \quad u^E(\varphi) = (1_A \wedge \varphi)u$$

are isomorphisms.

(b) A spectrum $A^{\perp}$ is called *dual* to a spectrum A if there exists a duality $S \to A \wedge A^{\perp}$. By 2.1(ii), in this case A is dual to $A^{\perp}$. So, "to be dual" is a symmetric relation.

(c) Let $u : S \to A \wedge A^{\perp}$ and $v : S \to B \wedge B^{\perp}$ be two dualities, and let $f : A \to B$ be a morphism. Consider the isomorphism

$$D : [A, B] \xrightarrow{u_B} [S, B \wedge A^{\perp}] \xrightarrow{(v^{A^{\perp}})^{-1}} [B^{\perp}, A^{\perp}]$$

and define a dual morphism $f^{\perp} : B^{\perp} \to A^{\perp}$ by requiring $D[f] = [f^{\perp}]$. Thus, $f^{\perp}$ is defined uniquely up to homotopy.

2.4. Lemma. (i) *Let $u : S \to A \wedge A^{\perp}$ and $v : S \to B \wedge B^{\perp}$ be two dualities. Then, for every spectrum E and morphism $f : A \to B$, the following diagram is commutative:*

$$\begin{array}{ccc} [B, E] & \xrightarrow{v_E} & [S, E \wedge B^{\perp}] \\ {\scriptstyle f^*}\downarrow & & \downarrow{\scriptstyle (1_E \wedge f^{\perp})_*} \\ [A, E] & \xrightarrow{u_E} & [S, E \wedge A^{\perp}]. \end{array}$$

(ii) *Suppose that a spectrum A admits a dual spectrum $A^{\perp}$. Then $A^{\perp}$ is unique up to equivalence. In particular, $(A^{\perp})^{\perp} \simeq A$.*

Proof. (i) Decode the definitions.

(ii) Let $u_1 : S \to A \wedge A^{\perp}$, $u_2 : S \to A \wedge \bar{A}$ be two dualities. If we put $v = u_2, B = A, B^{\perp} = \bar{A}$ in (i), we get the homomorphism $D : [A, A] \to [\bar{A}, A^{\perp}]$. Let $\varphi : \bar{A} \to A^{\perp}$ be a morphism such that $[\varphi] = D(1_A) \in [\bar{A}, A^{\perp}]$. Then φ is an equivalence since $\varphi_* : [E, \bar{A}] \to [E, A^{\perp}]$ is an isomorphism for all E. □

2.5. Lemma. *Let A and $A^{\perp}$ be two finite spectra.*

(i) *Let $u : S \to A \wedge A^{\perp}$ be a morphism such that u_E and u^E are isomorphisms for $E = \Sigma^k S$, $k \in \mathbb{Z}$. Then u is a duality morphism.*

(ii) *If $u : S \to A \wedge A^{\perp}$ is a duality then for every pair of spectra E, F the homomorphisms*

$$_F u_E : [F \wedge A, E] \to [F, E \wedge A^{\perp}],\ _F u_E(\varphi) = (\varphi \wedge 1_A)(1_F \wedge u)$$

and

$$^F u^E : [A^{\perp} \wedge F, E] \to [F, A \wedge E],\ ^F u^E(\varphi) = (1_A \wedge \varphi)(u \wedge 1_F)$$

are isomorphisms.

(iii) *Let $u : S \to A \wedge A^{\perp}$ and $v : S \to B \wedge B^{\perp}$ be two dualities. Then*

$$w : S \xrightarrow{u} A \wedge A^{\perp} = A \wedge S \wedge A^{\perp} \xrightarrow{1 \wedge v \wedge 1} A \wedge B \wedge B^{\perp} \wedge A^{\perp}$$

is a duality between $A \wedge B$ and $B^{\perp} \wedge A^{\perp}$.

Proof. (i) Firstly, two remarks.

Remark 1. Let $\{E(\alpha)\}$ be a family of spectra. Set $E = \vee E(\alpha)$. If $u_{E(\alpha)}$ (resp. $u^{E(\alpha)}$) is an isomorphism for every α, then u_E (resp. u^E) is an isomorphism. This follows from 1.16.

Remark 2. If $F_1 \to F_2 \to F_3$ is a cofiber sequence of spectra and u^{F_1}, u^{F_3}, (resp. u_{F_1}, u_{F_3}) are isomorphisms, then u^{F_2} (resp. u_{F_2}) is an isomorphism. Indeed, by 2.1(vi), $A \wedge F_1 \to A \wedge F_2 \to A \wedge F_3$ is a cofiber sequence. Now consider the following commutative diagram:

$$\begin{array}{ccccccccc}
\cdots & \longrightarrow & [A^{\perp}, F_1] & \longrightarrow & [A^{\perp}, F_2] & \longrightarrow & [A^{\perp}, F_3] & \longrightarrow & \cdots \\
 & & \downarrow{\scriptstyle u^{F_1}} & & \downarrow{\scriptstyle u^{F_2}} & & \downarrow{\scriptstyle u^{F_3}} & & \\
\cdots & \longrightarrow & [S, A \wedge F_1] & \longrightarrow & [S, A \wedge F_2] & \longrightarrow & [S, A \wedge F_3] & \longrightarrow & \cdots
\end{array}$$

By 1.15, its rows are exact sequences. Now apply the Five Lemma.

Now we prove that u^E is an isomorphism.

Step 1. Let $E = \vee \Sigma^n S_\lambda$ where S_λ is a copy of S and n is a fixed integer number. Then, by Remark 1, u^E is an isomorphism.

Step 2. Let $E = \Sigma^\infty X$ where X is a finite dimensional CW-complex. Then $E = E^{(m)}$ for some m. We prove that $u^{E^{(n)}}$ is an isomorphism by induction. By Step 1, $u^{E^{(0)}}$ is an isomorphism. Suppose that $u^{E^{(n-1)}}$ is an isomorphism. Then, by Remark 2 and Step 1, $u^{E^{(n)}}$ is an isomorphism since there is a cofiber sequence $E^{(n-1)} \to E^{(n)} \to \vee \Sigma^n S_\lambda$. The induction is confirmed.

Step 3. Let $E = \vee_\lambda \Sigma^\infty X_\lambda$ where every X_λ is a finite dimensional space. Then, by Step 2 and Remark 1, u^E is an isomorphism.

Step 4. Let E be an arbitrary spectrum. Consider the spectrum $\tau = \tau E$ as in 1.23, $\tau \simeq E$. By 1.22, we have a cofiber sequence

$$\tau_{\mathrm{ev}} \vee \tau_{\mathrm{od}} \to \tau \to \Sigma(\tau_{\mathrm{ev}} \cap \tau_{\mathrm{od}}).$$

By Step 3 and Remark 1, u^F is an isomorphism for $F = \tau_{\mathrm{ev}} \vee \tau_{\mathrm{od}}$ and $F = \tau_{\mathrm{ev}} \cap \tau_{\mathrm{od}}$ Thus, by Remark 2, u^τ is an isomorphism.

(ii) This can be proved similarly to (i). We leave this to the reader.

(iii) The isomorphism

$$[A \wedge B, E] \xrightarrow{{}_A v_E} [A, E \wedge B^\perp] \xrightarrow{u_{E\wedge B^\perp}} [S, E \wedge B^\perp \wedge A^\perp]$$

coincides with w_E (prove it). Similarly, one can prove that the homomorphism w^E is an isomorphism for every E. □

2.6. Remarks. (a) Some authors define duality to be a morphism

$$v : A \wedge A^\perp \to S$$

such that ${}_F v_E : [E, A^\perp \wedge F] \to [A \wedge E, F]$ and ${}^F v^E : [E, F \wedge A] \to [E \wedge A^\perp, F]$ are isomorphisms, see Switzer [1], Husemoller [1]. This definition is equivalent to ours, at least for finite spectra. Namely, considering $u : S \to A \wedge A^\perp$ as in 2.3, we have a morphism

$$v : A \wedge A^\perp \simeq A^\perp \wedge A = (A \wedge A^\perp)^\perp \xrightarrow{u^\perp} S^\perp = S$$

such that ${}_F v_E$ and ${}^F v^E$ are isomorphisms, cf. Dold–Puppe [1]. Conversely (and similarly), any morphism $v : A \wedge A^\perp \to S$ as above yields a duality $u : S \to A \wedge A^\perp$. But 2.3 is preferable for our goals.

(b) Originally Spanier–Whitehead [1] considered a certain special case of duality, as in 2.8(a) below. Then Spanier [1] defined duality in terms of pairings between homology and cohomology. Later some authors proposed defining duality via the pairings $A \wedge A^\perp \to S$, cf. (a). A nice categorical approach to duality can be found in Dold–Puppe [1].

2.7. Definition. Spectra A, B are called *n-dual* if the spectra $A, \Sigma^{-n} B$ are dual. Spaces X, Y are called *stably n-dual* if the spectra $\Sigma^\infty X, \Sigma^\infty Y$ are n-dual.

2.8. Examples. (a) Let X be a finite cellular subspace of $\mathbb{R}^n$, and let U be a regular neighborhood of X. Then X is stably $(n-1)$-dual to $\mathbb{R}^n \setminus U$, Spanier–Whitehead [1]. Indeed, let $\mathfrak{o}$ denote the origin of $\mathbb{R}^n$. Without loss of generality we can assume that $\mathfrak{o} \in X$, and we agree that $\mathfrak{o}$ is the base point of X. Let ε be a positive number which is less than the distance between X and $\mathbb{R}^n \setminus U$, and let O_ε be the ε-neighborhood of $\mathfrak{o}$. Consider the map

$$f : X \times (\mathbb{R}^n \setminus O_\varepsilon) \to \mathbb{R}^n,\ f(a,b) = a - b.$$

Clearly, $f((X \times (\mathbb{R}^n \setminus U)) \cup \{o\} \times (\mathbb{R}^n \setminus O_\varepsilon)) \subset \mathbb{R}^n \setminus O_\varepsilon$, and so we get the quotient map

$$f' : \frac{X \times (\mathbb{R}^n \setminus O_\varepsilon)}{(X \times (\mathbb{R}^n \setminus U)) \cup \{o\} \times (\mathbb{R}^n \setminus O_\varepsilon)} \to \frac{\mathbb{R}^n}{\mathbb{R}^n \setminus O_\varepsilon}.$$

Now, there are canonical homeomorphisms

$$\frac{X \times (\mathbb{R}^n \setminus O_\varepsilon)}{(X \times (\mathbb{R}^n \setminus U)) \cup \{o\} \times (\mathbb{R}^n \setminus O_\varepsilon)} \cong X \wedge \frac{\mathbb{R}^n \setminus O_\varepsilon}{\mathbb{R}^n \setminus U} = X \wedge \frac{S^n \setminus O_\varepsilon}{S^n \setminus U}$$

where S^n is considered as the one-point compactification of $\mathbb{R}^n$. So, f' turns into the map

$$f'' : X \wedge ((S^n \setminus O_\varepsilon)/(S^n \setminus U)) \to \mathbb{R}^n/(\mathbb{R}^n \setminus O_\varepsilon) \cong S^n.$$

Finally, since $S^n \setminus O_\varepsilon$ is contractible, we conclude that $(S^n \setminus O_\varepsilon)/(S^n \setminus U) \simeq S(S^n \setminus U)$, and so f'' turns into the map

$$f''' : X \wedge S(S^n \setminus U) \to S^n.$$

We define the morphism

$$\begin{aligned} v :& \Sigma^{1-n}\Sigma^\infty X \wedge \Sigma^\infty(S^n \setminus U) = \Sigma^{-n}\Sigma^\infty X \wedge \Sigma^\infty S(S^n \setminus U) \\ \simeq& \Sigma^{-n}\Sigma^\infty(X \wedge S(S^n \setminus U)) \xrightarrow{\Sigma^{-n}\Sigma^\infty f'''} \Sigma^{-n}\Sigma^\infty S^n = S. \end{aligned}$$

Now, one can prove that v has the properties as in 2.6(a), see Dold–Puppe [1]. Thus, X and $S^n \setminus U$ are $(n-1)$-dual. In fact, Dold and Puppe proved that u_E and u^E are isomorphisms for every finite spectrum E, but this is sufficient because of 2.5(i).

(b) Let a finite CW-complex X be cellularly embedded in a sphere S^n, let U be a regular neighborhood of X in S^n, let $\overline{U}$ be the closure of U, and let ∂U be the boundary of U. It follows easily from (a) that X^+ is n-dual to $\overline{U}/\partial U$, but we want to construct here the duality morphism explicitly. Let $p : \overline{U} \to X$ be the standard projection (which is a deformation retraction). Define $\Delta : \overline{U} \to \overline{U} \times X, \Delta(a) = (a, p(a))$. Since $\Delta(\partial U) \subset \partial U \times X$, the map

$$\Delta' : \overline{U}/\partial U \to (U \times X)/(\partial U \times X) = (\overline{U}/\partial U) \wedge X^+$$

is defined. Let $c : S^n \to S^n/(S^n \setminus U) \cong (\overline{U}/\partial U)$ be a map which collapses $S^n \setminus U$. We have a map $f : S^n \xrightarrow{c} \overline{U}/\partial U \xrightarrow{\Delta'} (\overline{U}/\partial U) \wedge X^+$, and the morphism

$$u = \Sigma^{-n}\Sigma^\infty f : S \to \Sigma^{-n}(\Sigma^\infty(\overline{U}/\partial U) \wedge \Sigma^\infty X^+) \simeq \Sigma^{-n}\Sigma^\infty(\overline{U}/\partial U) \wedge \Sigma^\infty X^+$$

is a duality. A proof can be found in Dold–Puppe [1].

2.9. Corollary. (i) *Every finite CW-space X admits an n-dual finite CW-space X' for n large enough.*

(ii) *Every finite spectrum A admits a dual finite spectrum $A^\perp$.*

Proof. (i) This follows from 2.8 because every finite CW-space X can be embedded in a sphere S^n for some $n = n(X)$.

(ii) By 1.5(iii), $A \simeq \Sigma^{-m}\Sigma^{-\infty}A_m$ for m large enough. Since A_m is a finite CW-space, it admits an n-dual finite CW-space Y for n large enough. Now set $A^\perp = \Sigma^{-m-n}\Sigma^\infty Y$. □

2.10. Proposition. *If $A \xrightarrow{f} B \xrightarrow{g} C$ is a cofiber sequence of finite spectra, then $C^\perp \xrightarrow{g^\perp} B^\perp \xrightarrow{f^\perp} A^\perp$ is a cofiber sequence.*

Proof. This can be proved as in Switzer [1], 14.33. We leave it to the reader. □

If X, Y are finite CW-complexes such that $\Sigma^\infty X \simeq \Sigma^\infty Y$, then $S^N X \simeq S^N Y$ for N large enough. This follows from 1.5(iii). Furthermore, if $f, g : X \to Y$ are two maps of finite CW-complexes and $\Sigma^\infty f \simeq \Sigma^\infty g : \Sigma^\infty X \to \Sigma^\infty Y$, then there exists N such that $S^N f \simeq S^N g : S^N X \to S^N Y$. This follows from the Freudenthal Suspension Theorem. Thus, passing from spectra to spaces, we have the following fact.

2.11. Theorem. (i) *Let X be a finite CW-space. Choose a natural number n such that there exists a finite CW-space X' which is n-dual to X. Then the homotopy type of $S^N X'$ is uniquely determined by the homotopy type of X for N large enough.*

(ii) *Let $f : X \to Y$ be a map of finite CW-spaces. Choose a natural number n such that there exist finite CW-spaces X', Y' which are n-dual to X, Y respectively. Then there exist a natural number N and a map $f' : S^N Y' \to S^N X'$ such that $\Sigma^{-N}\Sigma^\infty f$ and $\Sigma^{-N}\Sigma^\infty f'$ are dual morphisms. Furthermore, f' is unique up to homotopy for N large enough.*

(iii) *Let $X \xrightarrow{f} Y \xrightarrow{g} Z$ be a cofiber sequence of finite CW-spaces. Choose N, $f' : S^N Y' \to S^N X'$ and $g' : S^N Z' \to S^N Y'$ as in* (ii). *Then the sequence $S^N Z' \xrightarrow{g'} S^N Y' \xrightarrow{f'} S^N X'$ is a cofiber sequence for N large enough.* □

2.12. Definition. (a) A *ring spectrum* is a triple (E, μ, ι) where E is a spectrum and $\mu : E \wedge E \to E$ (the *multiplication*) and $\iota : S \to E$ (the *unit morphism*, or the *unit*) are certain morphisms with the following properties:

(1) Associativity. The following diagram commutes up to homotopy:

$$\begin{array}{ccc} (E \wedge E) \wedge E & \xrightarrow{\mu \wedge 1} & E \wedge E \\ \downarrow{\scriptstyle a} & & \downarrow{\scriptstyle \mu} \\ E \wedge (E \wedge E) & \xrightarrow{1 \wedge \mu} & E \wedge E \xrightarrow{\mu} E. \end{array}$$

(2) Unitarity. The following diagram commutes up to homotopy:

$$\begin{array}{ccccc} S\wedge E & \xrightarrow{\iota\wedge 1} & E\wedge E & \xleftarrow{1\wedge \iota} & E\wedge S \\ {\scriptstyle l}\downarrow & & \downarrow{\scriptstyle \mu} & & \downarrow{\scriptstyle r} \\ E & = & E & = & E\ . \end{array}$$

A pair (μ, ι) is called a *ring structure* (on E).

The ring spectrum (E, μ, ι) is *commutative* if μ is commutative, i.e., if the following diagram commutes up to homotopy:

$$\begin{array}{ccc} E\wedge E & \xrightarrow{\tau} & E\wedge E \\ {\scriptstyle \mu}\downarrow & & \downarrow{\scriptstyle \mu} \\ E & = & E\ . \end{array}$$

(b) A *ring morphism* $\varphi : (E, \mu, \iota) \to (E', \mu', \iota')$ of ring spectra is a morphism $\varphi : E \to E'$ such that the following diagrams commute up to homotopy:

$$\begin{array}{ccc} E\wedge E & \xrightarrow{\varphi\wedge\varphi} & E'\wedge E' \\ {\scriptstyle \mu}\downarrow & & \downarrow{\scriptstyle \mu'} \\ E & \xrightarrow{\varphi} & E' \end{array} \qquad \begin{array}{ccc} S & \xrightarrow{\iota} & E \\ \| & & \downarrow{\scriptstyle \varphi} \\ S & \xrightarrow{\iota'} & E'. \end{array}$$

2.13. Definition. (a) A *module spectrum* over a ring spectrum (E, μ, ι), or an E-module spectrum, is a pair (F, m) where F is a spectrum and $m : E \wedge F \to F$ is a morphism such that the following diagrams commute up to homotopy:

$$\begin{array}{ccccc} (E\wedge E)\wedge F & & \xrightarrow{\mu\wedge 1} & & E\wedge F \\ {\scriptstyle a}\downarrow & & & & \downarrow{\scriptstyle m} \\ E\wedge(E\wedge F) & \xrightarrow{1\wedge m} & E\wedge F & \xrightarrow{m} & F \end{array} \qquad \begin{array}{ccc} S\wedge F & \xrightarrow{\iota\wedge 1} & E\wedge F \\ \downarrow{\scriptstyle l} & & \downarrow{\scriptstyle m} \\ F & = & F\ . \end{array}$$

(b) An *E-module morphism* $\varphi : (F, m) \to (F', m')$ of E-module spectra is a morphism $\varphi : F \to F'$ such that the following diagram commutes up to homotopy:

$$\begin{array}{ccc} E\wedge F & \xrightarrow{1\wedge\varphi} & E\wedge F' \\ {\scriptstyle m}\downarrow & & \downarrow{\scriptstyle m'} \\ F & \xrightarrow{\varphi} & F'\ . \end{array}$$

As usual, we shall simply say "a ring spectrum E", omitting μ and ι. Note that every ring spectrum E is an E-module spectrum with $m = \mu$.

Furthermore, if $\varphi: E \to E'$ is a ring morphism then the pairing $E \wedge E' \xrightarrow{\varphi\wedge 1} E' \wedge E' \xrightarrow{\mu'} E'$ turns E' into an E-module spectrum.

2.14. Construction-Definition. Let (F, m) be a module spectrum over the ring spectrum (E, μ, ι). Given a morphism $a: S^d \to E$, consider the morphism

$$a_\# : S^d \wedge F \xrightarrow{a\wedge 1} E \wedge F \xrightarrow{\mu} F.$$

This morphism $a_\#$ is called *multiplication by* a.

2.15. Proposition. *Let $\varphi: E \to E'$ be a ring morphism of ring spectra. If the morphism $S^d \xrightarrow{a} E \xrightarrow{\varphi} E'$ is inessential then so is the morphism $a_\#: S^d E' \to E'$.*

Proof. The morphism $a_\#$ has the form

$$S^d \wedge E' \xrightarrow{a\wedge 1} E \wedge E' \xrightarrow{\varphi\wedge 1} E' \wedge E' \xrightarrow{\mu'} E',$$

i.e., $a_\# = \mu'(\varphi a \wedge 1)$. But φa is inessential. □

§3. (Co)homology Theories and Their Connection with Spectra

In the early of the 1950's Eilenberg–Steenrod [1] discovered that the homology theory $H_*(-; G)$ as a functor on the category of finite CW-spaces is determined by certain axioms, called thereafter the Eilenberg–Steenrod axioms. Later (end of the 50's, beginning of the 60's) it was noticed that many useful constructions of algebraic topology (K-functor, (co)bordism, etc.) are formally similar to (co)homology theories. Afterwards the reason for this phenomenon was clarified: namely, most of these constructions satisfy all the Eilenberg–Steenrod axioms except the so-called dimension axiom. So, it seemed reasonable to consider the objects satisfying these axioms. These objects were called extraordinary (co)homology theories. However, later mathematicians came to call these objects just (co)homology theories, while $H(-; G)$ got the name ordinary (co)homology theory. [5] Now this terminology is commonly accepted, and we use it in this book.

Recall that $\mathscr{C}$ denotes the category of all CW-spaces and maps and that $\mathscr{C}_{\mathrm{f}}$ (resp. $\mathscr{C}_{\mathrm{fd}}$) denotes the full subcategory of $\mathscr{C}$ consisting of all finite (resp. finite

[5] In fact, the classical (co)homology should be called extraordinary, because it has a certain extraordinary property: it satisfies the dimension axiom. (This is note of Peter Hilton.)

dimensional) CW-spaces. Let $\mathscr{C}^2$ be the category of all CW-pairs (X, A) and maps $(X, A) \to (Y, B)$. The category $\mathscr{C}_{\mathrm{f}}^2, \mathscr{C}_{\mathrm{fd}}^2$ are defined similarly. Let $\mathscr{K}$ (resp. $\mathscr{K}^2$) denote one of the categories $\mathscr{C}$, $\mathscr{C}_{\mathrm{fd}}$, $\mathscr{C}_{\mathrm{f}}$ (resp $\mathscr{C}^2$, $\mathscr{C}_{\mathrm{fd}}^2$, $\mathscr{C}_{\mathrm{f}}^2$). We define a functor $R : \mathscr{K}^2 \to \mathscr{K}^2$ by setting $R(X, A) = (A, \emptyset)$.

3.1. Definition. (a). An *unreduced homology theory* on $\mathscr{K}^2$ is a family $\{h_n, \partial_n\}$, $n \in \mathbb{Z}$, of covariant functors $h_n : \mathscr{K}^2 \to \mathscr{AG}$ and natural transformations $\partial_n : h_n \to h_{n-1} \circ R$ satisfying the following axioms:

(1) The homotopy axiom. If $(X, A), (Y, B) \in \mathscr{K}^2$ and the maps $f, g : (X, A) \to (Y, B)$ are homotopic, then the induced homomorphisms

$$h_n(f), h_n(g) : h_n(X, A) \to h_n(Y, B)$$

coincide for every n.

(2) The exactness axiom. For every pair $(X, A) \in \mathscr{K}^2$, the sequence

$$\cdots \longrightarrow h_{n+1}(X, A) \xrightarrow{\partial_{n+1}} h_n(A, \emptyset) \xrightarrow{h_n(i)} h_n(X, \emptyset) \xrightarrow{h_n(j)} h_n(X, A) \longrightarrow \cdots$$

is exact. Here $i : (A, \emptyset) \to (X, \emptyset)$, $j : (X, \emptyset) \to (X, A)$ are the inclusions.

(3) The collapse axiom. For every pair $(X, A) \in \mathscr{K}^2$, the collapse $c : (X, A) \to (X/A, \{*\})$ induces an isomorphism $h_n(c) : h_n(X, A) \to h_n(X/A, \{*\})$.

(b) A *morphism*

$$T : \{h_n, \partial_n\} \to \{h'_n, \partial'_n\}$$

of homology theories is a family of natural transformations $\{T_n : h_n \to h'_n\}$ such that the following diagram commutes:

$$\begin{array}{ccc} h_n & \xrightarrow{\partial_n} & h_{n-1}R \\ {\scriptstyle T_n}\downarrow & & \downarrow{\scriptstyle T_{n-1}} \\ h'_n & \xrightarrow{\partial'_n} & h'_{n-1}R. \end{array}$$

Given a pair $(X, A) \in \mathscr{K}^2$, we use the notation $T_n^{(X,A)}$ for the corresponding homomorphism $h_n(X, A) \to h'_n(X, A)$.

It is easy to see that we have a category of homology theories and their morphisms. In particular, the equivalence (isomorphism) of homology theories is defined in the usual way: it is a morphism of homology theories which is also a natural equivalence of functors.

It is well known that the classical homology $\{H_n, \partial_n\}$ satisfyies 3.1, see e.g. Dold [5], Vick [1]. Moreover, Eilenberg–Steenrod [1] proved that a homology theory $\{h_n, \partial_n\}$ in $\mathscr{C}_{\mathrm{f}}^2$ with the additional property $h_n(\mathrm{pt}, \emptyset) = 0$ for $n \neq 0$ (the dimension axiom) is just a classical homology theory $\{H_n, \partial_n\}$.

3.2. Proposition. *Let $\{h_n, \partial_n\}$ be a homology theory on $\mathscr{K}^2$.*

(i) *If $f : (X,A) \to (Y,B)$ is a homotopy equivalence, $(X,A),(Y,B) \in \mathscr{K}^2$, then $h_n(f) : h_n(X,A) \to h_n(Y,B)$ is an isomorphism. In particular, the inclusion $t : (X \cup CA, \{*\}) \to (X \cup CA, CA)$ induces an isomorphism $h_n(t)$.*

(ii) *If the inclusion $i : A \subset X$, $(X,A) \in \mathscr{K}^2$, is a homotopy equivalence then $h_n(X,A) = 0$ for every n and $h_n(i) : h_n(A,\emptyset) \to h_n(X,\emptyset)$ is an isomorphism for every n. In particular, $h_n(X,\{x_0\}) = 0$ for every contractible $X \in \mathscr{K}$ and every $x_0 \in X$.*

(iii) *For every $(X,A) \in \mathscr{K}^2$ the inclusion $k : (X,A) \to (X \cup CA, CA)$ induces an isomorphism $h_n(k) : h_n(X,A) \to h_n(X \cup CA, CA)$.*

(iv) *For every CW-triple $A \subset X \subset Y$, $Y \in \mathscr{K}$ the sequence*

$$\cdots \to h_{n+1}(Y,X) \xrightarrow{d} h_n(X,A) \xrightarrow{h_n(I)} h_n(Y,A) \xrightarrow{h_n(J)} h_n(Y,X) \to \cdots$$

is exact. Here $I : (X,A) \to (Y,A)$, $J : (Y,A) \to (Y,X)$ are the inclusions and d is the composition $d : h_{n+1}(Y,X) \xrightarrow{\partial_{n+1}} h_n(X,\emptyset) \xrightarrow{h_n(i)} h_n(X,A)$.

(v) *Let $(X;A,B)$ be a CW-triad, $X \in \mathscr{K}$. Set $C = A \cap B$. Let $i_1 : A \to X, i_2 : B \to X, i_3 : C \to A, i_4 : C \to B$ be the inclusions. Define $\Delta : h_n(X,\emptyset) \to h_{n-1}(C,\emptyset)$ to be the composition*

$$\begin{aligned} h_n(X,\emptyset) \to h_n(X,A) &\xrightarrow{\cong} h_n(X/A,*) \\ &= h_n(B/C,*) \xleftarrow{\cong} h_n(B,C) \xrightarrow{\partial} h_{n-1}(C,\emptyset). \end{aligned}$$

Consider $\alpha : h_n(C,\emptyset) \xrightarrow{h_n(i_3)\oplus h_n(i_4)} h_n(A,\emptyset) \oplus h_n(B,\emptyset)$ and

$$\beta : h_n(A,\emptyset) \oplus h_n(B,\emptyset) \to h_n(X),\ \beta(x,y) = h_n(i_1)(x) - h_n(i_2)(y).$$

Then the sequence

$$\cdots \to h_n(C,\emptyset) \xrightarrow{\alpha} h_n(A,\emptyset) \oplus h_n(B,\emptyset) \xrightarrow{\beta} h_n(X,\emptyset) \xrightarrow{\Delta} h_{n-1}(C,\emptyset) \to \cdots$$

is exact.

Proof. (i) This follows from the homotopy axiom.

(ii) The homotopy axiom implies that $h_n(X,A) = 0$, and hence $h_n(i)$ is an isomorphism in view of the exactness axiom.

(iii) This follows from the collapse axiom.

(iv) For every $U \in \mathscr{K}$ we have a natural splitting $h_n(U,\emptyset) = h_n(U,*) \oplus h_n(\mathrm{pt},\emptyset)$ (consider the exact sequence of the pair $(U,*)$). Hence, for every $(U,V) \in \mathscr{K}^2$ the exact sequence 3.1(2) yields an exact sequence

$$\cdots \to h_{n+1}(U,V) \to h_n(V,*) \xrightarrow{\widetilde{h}_n(i)} h_n(U,*) \to h_n(U,V) \to \cdots,$$

where $\widetilde{h}_n(i)$ is the restriction of $h_n(i)$ to the direct summands.

Furthermore, the collapse axiom yields the isomorphisms $h_n(Y,X) \xrightarrow{\cong} h_n(Y/A, X/A)$ and $h_n(Z,A) \xrightarrow{\cong} h_n(Z/A, \{*\})$ for $Z = X, Y$. Because of this, the desired exact sequence turns out to be the above exact sequence of the pair $(Y/A, X/A)$

(v) Do this as an exercise, or see Switzer [1]. (Alternatively, this follows from 3.11 and 3.18 below.) □

The exact sequences as in 3.1(2), 3.2(iv), and 3.2(v) are known as the *exact sequence of a pair,* the *exact sequence of a triple,* and the *Mayer–Vietoris exact sequence .*

Given a pointed CW-space (X, x_0), let $(CX, \{*\})$, resp.$(SX, \{*\})$, denote the reduced cone, resp. suspension, over (X, x_0). Because of 3.2(iv), the triple $\{x_0\} \subset X \subset CX$ yields the exact sequence

$$h_n(CX, \{*\}) \to h_n(CX, X) \xrightarrow{d} h_{n-1}(X, \{*\}) \to h_{n-1}(CX, \{*\}).$$

By 3.2(ii), $h_i(CX, \{x_0\}) = 0$ for every i, and so we get the isomorphism

$$(3.3) \quad h_n(SX, \{x_0\}) \cong h_n(CX/X, \{x_0\}) \cong h_n(CX, X) \xrightarrow{d} h_{n-1}(X, \{x_0\}).$$

Recall that $\mathscr{C}^\bullet$ denotes the category of pointed CW-spaces and pointed maps and that $\mathscr{C}_{\mathrm{f}}^\bullet$, resp. $\mathscr{C}_{\mathrm{fd}}^\bullet$, denotes its full subcategories of finite, resp. finite dimensional, CW-spaces. Let $\mathscr{K}^\bullet$ denote one of the categories $\mathscr{C}^\bullet$, $\mathscr{C}_{\mathrm{fd}}^\bullet$, $\mathscr{C}_{\mathrm{f}}^\bullet$, and let $S : \mathscr{K}^\bullet \to \mathscr{K}^\bullet$ be the (reduced) suspension functor.

3.4. Construction-Definition. Let $\{h_n, \partial_n\}$ be an unreduced homology theory. Given $(X, x_0) \in \mathscr{K}^\bullet$, set $\widetilde{h}_n(X, x_0) = h_n(X, \{x_0\})$ and define the *suspension isomorphism*

$$\mathfrak{s}_n : \widetilde{h}_{n+1}(SX, *) = h_{n+1}(SX, \{*\}) \to h_n(X, \{x_0\}) = \widetilde{h}_n(X, x_0)$$

to be the composition (3.3). Thus, we get a family $\{\widetilde{h}_n, \mathfrak{s}_n\}$, $n \in \mathbb{Z}$, of covariant functors $\widetilde{h}_n : \mathscr{K}^\bullet \to \mathscr{AG}$ and natural equivalences $\mathfrak{s}_n : \widetilde{h}_{n+1}S \to \widetilde{h}_n$. The family $\{\widetilde{h}_n, \mathfrak{s}_n\}$ is called a *reduced homology theory* (on $\mathscr{K}^\bullet$) corresponding to $\{h_n, \partial_n\}$.

3.5. Proposition. *Let $\{h_n, \partial_n\}$ be an unreduced homology theory on $\mathscr{K}^2$.*

(i) *Homotopic maps $f, g : (X, x_0) \to (Y, y_0)$ in $\mathscr{K}^\bullet$ induce the same homomorphism $\widetilde{h}_n(f) = \widetilde{h}_n(g) : \widetilde{h}_n(X, x_0) \to \widetilde{h}_n(Y, y_0)$.*

(ii) *For every pointed CW-pair (X, A, x_0), $(X, A) \in \mathscr{K}^2$, the sequence*

$$\widetilde{h}_n(A, x_0) \to \widetilde{h}_n(X, x_0) \to \widetilde{h}_n(X/A, *)$$

is exact. Here the homomorphisms are induced by the inclusion $A \to X$ and the projection (collapsing map) $X \to X/A$.

Proof. (i) follows from the homotopy axiom, (ii) follows from 3.2(iv). □

3.6. Proposition. *Let* $\{F_n, t_n\}$, $n \in \mathbb{Z}$, *be a family of covariant functors* $F_n : \mathscr{K}^\bullet \to \mathscr{AG}$ *and natural equivalences* $t_n : F_{n+1}S \to F_n$ *satisfying the following properties:*

(i) *Homotopic maps* $f, g : (X, x_0) \to (Y, y_0)$ *in* $\mathscr{K}$ *induce the same homomorphism* $F_n(f) = F_n(g) : F_n(X, x_0) \to F_n(Y, y_0)$;

(ii) *For every pointed CW-pair* (X, A, x_0), $(X, A) \in \mathscr{K}^2$, *the sequence*

$$F_n(A, x_0) \to F_n(X, x_0) \to F_n(X/A, *)$$

is exact.

Then there exists a homology theory $\{h_n, \partial_n\}$ *such that its reduced homology theory* $\{\widetilde{h}_n, \mathfrak{s}_n\}$ *is equivalent to the family* $\{F_n, t_n\}$, *i.e., there are natural equivalences* $\varphi_n : F_n \to \widetilde{h}_n$ *such that the diagram*

$$\begin{array}{ccc} F_{n+1}(SX, *) & \xrightarrow{\varphi_{n+1}^{SX}} & \widetilde{h}_{n+1}(SX, *) \\ {\scriptstyle t_n^X}\downarrow & & \downarrow{\scriptstyle \mathfrak{s}_n^X} \\ F_n(X, x_0) & \xrightarrow{\varphi_n^X} & \widetilde{h}_n(X, x_0) \end{array}$$

commutes for every $(X, x_0) \in \mathscr{K}$. *Furthermore, this homology theory* $\{h_n, \partial_n\}$ *is unique up to equivalence.*

Proof (cf. Switzer [1], 7.33–7.42). Every pair (X, A) yields the long cofiber sequence

$$A^+ \to X^+ \to X^+ \cup C(A^+) \xrightarrow{l} S(A^+).$$

as in I.3.39. Recall that every space Y^+ is assumed to be pointed so that its base point is the added point. Set

$$h_n(X, A) = F_n(X^+ \cup C(A^+), *), \qquad h_n(X, \emptyset) = F_n(X^+, *)$$

and define $\partial_n : h_n(X, A) \to h_{n-1}(A, \emptyset)$ to be the composition

$$\begin{aligned} \partial_n : h_n(X, A) = F_n(X^+ \cup C(A^+), *) &\xrightarrow{F_n(l)} F_n(S(A^+), *) \\ &\xrightarrow{t_{n-1}} F_{n-1}(A^+, *) = h_{n-1}(A, \emptyset) \xrightarrow{-1} h_{n-1}(A, \emptyset), \end{aligned}$$

where -1 is multiplication by -1. Clearly, $\{h_n, \partial_n\}$ is an unreduced homology theory.

Because of (i), every pointed homotopy equivalence $f : (X, x_0) \to (Y, y_0)$ induces an isomorphism $F_n(f) : F_n(X, x_0) \to F_n(Y, y_0)$. By I.3.26 and I.3.29, the projection

$$p : (X^+ \cup C(\{x_0\}^+), *) \to (X, x_0), \quad p(x) = x, p(x_0, t) = t, x \in X, t \in I,$$

is a pointed homotopy equivalence. Hence, we have a canonical isomorphism

$$F_n(X, x_0) \cong F_n(X^+ \cup C(\{x_0\}^+), *).$$

We define $\varphi_n^X : F_n(X, x_0) \to \widetilde{h}_n(X, x_0)$, $(X, x_0) \in \mathscr{K}^\bullet$ to be the composition

$$F_n(X, x_0) \cong F_n(X^+ \cup C(\{x_0\}^+), *) = h_n(X, \{x_0\}) = \widetilde{h}_n(X, x_0).$$

We prove that $\mathfrak{s}_n^X \varphi_{n+1}^{SX} = \varphi_n^X t_n^X$. Consider the homomorphisms

$$u : F_{n+1}(SX, x_0) \xrightarrow{t_n} F_n(X, x_0) \cong F_n(X^+ \cup C(\{x_0\}^+), *)$$

and

$$\begin{aligned} v : F_{n+1}(SX, x_0) &\cong F_{n+1}((SX)^+ \cup C(\{x_0\}^+), *) \\ &\xrightarrow{(F_{n+1}(a))^{-1}} F_{n+1}((CX)^+ \cup C(X^+) \cup C(\{x_0\}^+), *) \\ &= F_{n+1}((CX)^+ \cup C(X^+), *) \xrightarrow{F_{n+1}(l)} F_{n+1}(S(X^+), *) \\ &\xrightarrow{t_n} F_n(X^+, *) \xrightarrow{F_n(\varepsilon)} F_n(X, x_0) \cong F_n(X^+ \cup C(\{x_0\}^+), *), \end{aligned}$$

where $a : (CX)^+ \cup C(X^+) \to (SX)^+$ collapses $C(X^+)$, $l : (CX)^+ \cup C(X^+) \to S(X^+)$ collapses $(CX)^+$, and $\varepsilon : (X^+, *) \to (X, x_0)$ collapses $\{x_0\}^+$, i.e., ε maps the added point to x_0 and $\varepsilon | X = 1_X$. Then $u = -v$ (prove it!). Now, the homomorphism $\varphi_n^X t_n^X$ has the form

$$\begin{aligned} F_{n+1}(SX, x_0) \xrightarrow{t_n} F_n(X, x_0) &\cong F_n(X^+ \cup C(\{x_0\}^+), *) \\ &= h_n(X, \{x_0\}) = \widetilde{h}_n(X, x_0), \end{aligned}$$

i.e., $\varphi_n^X t_n^X = u$. Furthermore, the homomorphism $\mathfrak{s}_n^X \varphi_{n+1}^{SX}$ has the form

$$\begin{aligned} F_{n+1}(SX, x_0) \cong F_{n+1}((SX)^+ \cup C(\{x_0\}^+), *) &= h_{n+1}(SX, \{x_0\}) \\ &= \widetilde{h}_{n+1}(SX, x_0) \xrightarrow{\mathfrak{s}_n} \widetilde{h}_n(X, x_0), \end{aligned}$$

i.e., the form

$$\begin{aligned} F_{n+1}(SX, *) \xrightarrow{v} F_n(X^+ \cup C(\{x_0\}^+), *) &= h_n(X, \{x_0\}) \\ &= \widetilde{h}_n(X, x_0) \xrightarrow{-1} \widetilde{h}_n(X, x_0). \end{aligned}$$

Thus, $\varphi_n^X t_n^X = \mathfrak{s}_n^X \varphi_{n+1}^{SX}$.

Suppose that there is another homology theory $\{h'_n, \partial'_n\}$ such that its reduced homology theory $\{\widetilde{h}'_n, \mathfrak{s}'_n\}$ is equivalent to $\{F_n, t_n\}$. So, we have equivalences $\psi_n : \widetilde{h}'_n(-) \to F_n(-)$. Given $(X, A) \in \mathscr{K}^2$, consider the isomorphism

$$\begin{aligned} h'_n(X, A) = \widetilde{h}'_n(X/A, *) \xrightarrow{\psi} F_n(X/A, *) &\xrightarrow{\varphi} \widetilde{h}_n(X/A, *) \\ &= h_n(X/A, \{*\}) \xleftarrow{h_n(c)} h_n(X, A). \end{aligned}$$

These isomorphisms constitute an equivalence $h'_* \to h_*$. □

This proposition shows that there is a bijective correspondence between unreduced and reduced homology theories. In other words, every unreduced homology theory is completely determined by its reduced form. Moreover, based on 3.6, sometimes one *defines* a reduced homology theory to be a family (F_n, t_n) as in 3.6. So, when we say "Let $\{\widetilde{h}_n, \mathfrak{s}_n\}$ be a reduced homology theory..." this means that $\{\widetilde{h}_n, \mathfrak{s}_n\}$ is a reduced form of an unreduced homology theory, but, on the other hand, one can think of $\{\widetilde{h}_n, \mathfrak{s}_n\}$ as a family satisfying 3.6; there is no contradiction. Moreover, sometimes it is more convenient to construct reduced homology theories and then corresponding unreduced ones rather than unreduced ones immediately.

The groups $h_i(\mathrm{pt}, \emptyset) = \widetilde{h}_i(S^0, *)$ are called the *coefficient groups* of the homology theory $\{h_n, \partial_n\}$. To justify this term, note that $H_*(\mathrm{pt}, \emptyset; A) = A$ for every abelian group A.

3.7. Proposition. *Let $\{\widetilde{h}_n, \mathfrak{s}_n\}$ be a reduced homology theory on $\mathscr{K}^\bullet$.*

(i) *Let* $(\mathrm{pt}, *)$ *denote the one-point pointed space. Then* $\widetilde{h}_n(\mathrm{pt}, *) = 0$ *for every* n.

(ii) *If* $f : (X, x_0) \to (Y, y_0)$ *is a pointed homotopy equivalence in* $\mathscr{K}^\bullet$, *then the homomorphism* $\widetilde{h}_n(f) : \widetilde{h}_n(X, x_0) \to \widetilde{h}_n(Y, y_0)$ *is an isomorphism for all* n. *In particular, the quotient map* $p : X \cup CA \to (X \cup CA)/CA = X/A$ *induces an isomorphism* $\widetilde{h}_n(p) : \widetilde{h}_n(X \cup CA, *) \to \widetilde{h}_n(X/A, *)$.

(iii) *Every cofiber sequence* $(X, x_0) \xrightarrow{f} (Y, y_0) \xrightarrow{g} (Z, z_0)$ *in* $\mathscr{K}^\bullet$ *induces an exact sequence*

$$\widetilde{h}_n(X, x_0) \xrightarrow{\widetilde{h}_n(f)} \widetilde{h}_n(Y, y_0) \xrightarrow{\widetilde{h}_n(g)} \widetilde{h}_n(Z, z_0).$$

(iv) *Let* (X, A) *be a pointed CW-pair,* $X \in \mathscr{K}$. *Define*

$$\widetilde{\partial}_n : \widetilde{h}_n(X/A, *) \xrightarrow{(\widetilde{h}_n(p))^{-1}} \widetilde{h}_n(X \cup CA, *) \xrightarrow{\widetilde{h}_n(k)} \widetilde{h}_n(SA, *) \xrightarrow{\mathfrak{s}_{n-1}} \widetilde{h}_{n-1}(A, *),$$

where $p : X \cup CA \to X/A$ *collapses* CA *(see* (ii)) *and* $k : X \cup CA \to SA$ *collapses* X. *Then the sequence*

$$\cdots \to \widetilde{h}_n(A, *) \to \widetilde{h}_n(X, *) \to \widetilde{h}_n(X/A, *) \xrightarrow{\widetilde{\partial}_n} \widetilde{h}_{n-1}(A, *) \to \cdots$$

is exact.

(v) *Let* $(X; A, B; x_0)$ *be a pointed CW-triad,* $X \in \mathscr{K}$. *Set* $C = A \cap B$. *Let* $i_1 : A \to X, i_2 : B \to X, i_3 : C \to A, i_4 : C \to B$ *be the inclusions. Define* $\Delta : \widetilde{h}_n(X, x_0) \to \widetilde{h}_{n-1}(C, x_0)$ *to be the composition*

$$\widetilde{h}_n(X, x_0) \to \widetilde{h}_n(X/A, x_0) = \widetilde{h}_n(B/C, x_0) \xrightarrow{\widetilde{\partial}_n} \widetilde{h}_{n-1}(C, x_0),$$

where the first homomorphism is induced by the collapsing map and $\widetilde{\partial}$ *is defined in* (iv). *Consider* $\alpha : \widetilde{h}_n(C, x_0) \xrightarrow{\widetilde{h}_n(i_3)\oplus\widetilde{h}_n(i_4)} \widetilde{h}_n(A, x_0) \oplus \widetilde{h}_n(B, x_0)$, *and* $\beta : \widetilde{h}_n(A, x_0) \oplus \widetilde{h}_n(B, x_0) \to \widetilde{h}_n(X)$, $\beta(x, y) = \widetilde{h}_n(i_1)(x) - \widetilde{h}_n(i_2)(y)$. *Then the Mayer–Vietoris sequence*

$$\cdots \to \widetilde{h}_n(C, x_0) \xrightarrow{\alpha} \widetilde{h}_n(A, x_0) \oplus \widetilde{h}_n(B, x_0) \xrightarrow{\beta} \widetilde{h}_n(X, x_0) \xrightarrow{\Delta} \widetilde{h}_{n-1}(C, x_0) \to \cdots$$

is exact.

(vi) $\widetilde{h}_n(X \vee Y, *) = \widetilde{h}_n(X, *) \oplus \widetilde{h}_n(Y, *)$ *for every* $X, Y \in \mathscr{K}$ *and every* n.

Proof. See Switzer [1], Ch. 7. □

3.8. Definition. (a) An *unreduced cohomology theory* on $\mathscr{K}^2$ is a family $\{h^n, \delta^n\}$, $n \in \mathbb{Z}$, of contravariant functors $h_n : \mathscr{K}^2 \to \mathscr{AG}$ and natural transformations $\delta^n : h^n \circ R \to h^{n+1}$ satisfying the following axioms:

(1) The homotopy axiom. If $(X, A), (Y, B) \in \mathscr{K}^2$ and the maps $f, g : (X, A) \to (Y, B)$ are homotopic, then the induced homomorphisms

$$h^n(f), h^n(g) : h^n(Y, B) \to h^n(X, A)$$

coincide for every n.

(2) The exactness axiom. For every pair $(X, A) \in \mathscr{K}^2$, the sequence

$$\cdots \longrightarrow h^{n-1}(A, \emptyset) \xrightarrow{\delta^{n-1}} h^n(X, A) \xrightarrow{h^n(j)} h^n(X, \emptyset) \xrightarrow{h^n(i)} h^n(A, \emptyset) \longrightarrow \cdots$$

is exact.

(3) The collapse axiom. For every pair $(X, A) \in \mathscr{K}^2$, the collapse $c : (X, A) \to (X/A, *)$ induces an isomorphism $h^n(c) : h^n(X/A, *) \to h^n(X, A)$.

(b) A *morphism* of cohomology theories

$$T : \{h^n, \delta^n\} \to \{(h^n)', (\delta^n)'\}$$

is a family of natural transformations $\{T^n : h^n \to (h^n)'\}$ such that the following diagram commutes:

$$\begin{array}{ccc} h^n R & \xrightarrow{\delta^n} & h^{n+1} \\ {\scriptstyle T^n}\downarrow & & \downarrow{\scriptstyle T^{n+1}} \\ (h^n)' R & \xrightarrow{(\delta^n)'} & (h^{n+1})' \end{array}$$

Given a pair (X, A), we use the notation $T^n_{(X,A)}$ for the corresponding homomorphism $h^n(X, A) \to (h^n)'(X, A)$.

For future reference, we formulate a cohomological analog of 3.11. The proof is similar.

3.12. Proposition. *Let $\mathscr{L}$ be as in 3.10, and let $\{h^n, \hat{\mathfrak{s}}^n\}$ be a cohomology theory on $\mathscr{L}$. Then:*

(i) $h^n(X \vee Y) \cong h^n(X) \oplus h^n(Y)$.

(ii) *For every cofiber sequence $X \xrightarrow{f} Y \xrightarrow{g} Z$ of spectra there is a natural exact sequence*

$$\cdots \to h^{n-1}(X) \to h^n(Z) \xrightarrow{g^*} h^n(Y) \xrightarrow{f^*} h^n(X) \to h^{n+1}(Z) \to \cdots .$$

(iii) *Let A, B be two subspectra of a spectrum X. Set $C = A \cap B$. Then there is a natural (Mayer-Vietoris) exact sequence*

$$\cdots \to h^n(X) \to h^n(A) \oplus h^n(B) \to h^n(C) \to h^{n-1}(X) \to \cdots . \qquad \square$$

3.13. Construction. (a) Given a homology theory $(h_n, \widehat{\mathfrak{s}}_n)$ on $\mathscr{S}$, set $\widetilde{h}_n := h_n \circ \Sigma^\infty : \mathscr{K}^\bullet \to \mathscr{AG}$ and define $\mathfrak{s}_n : h_n \to h_{n+1} \circ S$ to be the composition

$$\widetilde{h}_n = h_n \circ \Sigma^\infty \xrightarrow{\widehat{\mathfrak{s}}_n} h_{n+1} \circ \Sigma \circ \Sigma^\infty \cong h_{n+1} \circ \Sigma^\infty \circ S = \widetilde{h}_{n+1} \circ S.$$

In other words, $\widetilde{h}_n(X) = h_n(\Sigma^\infty X)$, etc. We leave it to the reader to check that $(\widetilde{h}_n, \mathfrak{s}_n)$ is a reduced homology theory on $\mathscr{K}^\bullet$.

(b) Similarly, given a cohomology theory $(h^n, \widehat{\mathfrak{s}}^n)$ on $\mathscr{S}$, we get a reduced cohomology theory $(\widetilde{h}^n, \mathfrak{s}^n)$, where $\widetilde{h}^n = h^n \circ \Sigma^\infty$ and $\mathfrak{s}^n$ is the composition

$$\widetilde{h}^{n+1} \circ S = h^{n+1} \circ \Sigma^\infty \circ S \cong h^{n+1} \circ \Sigma \circ \Sigma^\infty \xrightarrow{\widehat{\mathfrak{s}}^n} h^n \circ \Sigma^\infty = \widetilde{h}^n.$$

Thus, every (co)homology theory on $\mathscr{S}$ (resp. $\mathscr{S}_{\mathrm{s}}, \mathscr{S}_{\mathrm{fd}}, \mathscr{S}_{\mathrm{sfd}}, \mathscr{S}_{\mathrm{f}}$) yields a reduced (co)homology theory on $\mathscr{C}^\bullet$ (resp. $\mathscr{C}^\bullet, \mathscr{C}^\bullet_{\mathrm{fd}}, \mathscr{C}^\bullet_{\mathrm{fd}}, \mathscr{C}^\bullet_{\mathrm{f}}$).

3.14. Lemma. *Let $X(1) \xrightarrow{f_1} X(2) \to \cdots \xrightarrow{f_{n-1}} X(n)$ be a sequence of maps of spectra. Then there exists a set Ω and a family of sequences*

$$X(1)_\omega \xrightarrow{(f_1)_\omega} X(2)_\omega \to \cdots \xrightarrow{(f_{n-1})_\omega} X(n)_\omega,\ \omega \in \Omega$$

with the following properties:

(i) *$X(i)_\omega$ is a finite subspectrum of $X(i)$;*

(ii) *$(f_i)_\omega$ is the restriction of f_i, i.e., the following diagram commutes:*

$$\begin{array}{ccc} X(i)_\omega & \xrightarrow{(f_i)_\omega} & X(i+1)_\omega \\ {\scriptstyle\cap}\big\downarrow & & \big\downarrow{\scriptstyle\cap} \\ X(i) & \xrightarrow{f_i} & X(i+1); \end{array}$$

(iii) *For every i, every finite subspectrum of $X(i)$ is contained in some $X(i)_\omega$.*

We can turn Ω into a quasi-ordered set, by setting $\omega \leq \omega'$ iff $X(i)_\omega \subset X(i))_{\omega'}$ for every i. Then the family

$$X(1)_\omega \xrightarrow{(f_1)_\omega} X(2)_\omega \to \cdots \xrightarrow{(f_{n-1})_\omega} X(n)_\omega,\ \omega \in \Omega$$

can be considered as a direct Ω-system of sequences.

Proof. We prove this by induction on n. For $n = 1$ we can set $\{X_\omega\}$ to be the family of all finite subspectra of X. Suppose that the lemma holds for some $n > 1$ and consider the sequence

$$X(1) \xrightarrow{f_1} X(2) \to \cdots \xrightarrow{f_{n-1}} X(n) \xrightarrow{f_n} X(n+1).$$

Applying the inductive assumption to the sequence

$$X(2) \xrightarrow{f_2} \cdots \xrightarrow{f_{n-1}} X(n) \xrightarrow{f_{n+1}} X(n+1),$$

we find a quasi-ordered set A and sequences

$$X(2)_\alpha \xrightarrow{(f_2)_\alpha} \cdots \xrightarrow{(f_n)_\alpha} X(n+1)_\alpha,\ \alpha \in A$$

with the desired properties. Let $\{X(1)_\lambda\}$ be the family of all finite subspectra of $X(1)$, and let $X(1)_{(\alpha,\lambda)}$ be a maximal subspectrum of

$$X(1)_\lambda \cap (f_1^{-1}(X(2)_\alpha)).$$

We set $\Omega := \{(\alpha, \lambda)\}$ and $X(i)_{(\alpha,\lambda)} := X(i)_\alpha, (f_i)_{(\alpha,\lambda)} := (f_i)_\alpha$ for every (α, λ) and every $i \geq 2$. Clearly, $f_1(X(1))_\omega \subset X(2)_\omega$ for every $\omega \in \Omega$, and so we can form $(f_1)_\omega : X(1)_\omega \to X(2)_\omega$. Now, the family

$$X(1)_\omega \xrightarrow{(f_1)_\omega} X(2)_\omega \to \cdots \xrightarrow{(f_n)_\omega} X(n+1)_\omega,\ \omega \in \Omega$$

satisfies conditions (i)–(iii). □

3.15. Lemma. *Let $X \xrightarrow{f} Y \xrightarrow{g} Z$ be a strict cofiber sequence of maps of spectra. Then a family $\{X_\omega \xrightarrow{f_\omega} Y_\omega \xrightarrow{g_\omega} Z_\omega\}$ as in 3.14 can be chosen such that $Z_\omega = C(f_\omega)$.*

Proof. Consider a family $\{X_\omega \xrightarrow{f_\omega} Y_\omega \xrightarrow{g_\omega} Z_\omega\}$ as in 3.14 and set $(Z_\omega)_{\text{new}} := C(f_\omega)$. Let $h_\omega : Y_\omega \xrightarrow{h_\omega} (Z_\omega)_{\text{new}}$ be the canonical inclusion. Then $\{X_\omega \xrightarrow{f_\omega} Y_\omega \xrightarrow{h_\omega} (Z_\omega)_{\text{new}}\}$ is the desired family. □

Given a family $\{X_\lambda\}$ of spaces or spectra, let $i_\lambda : X_\lambda \to \vee X_\lambda$ denote the inclusion.

3.16. Definition. (a) Let $\mathscr{L}$ be as in 3.10. A homology theory h_* on $\mathscr{L}$ is called *additive* if

$$\langle (i_\lambda)_* \rangle : \oplus_\lambda h_*(X_\lambda) \to h_*(\vee X_\lambda)$$

is an isomorphism for every family $\{X_\lambda, \lambda \in \Lambda\}$ in $\mathscr{L}$ with $\vee X_\lambda \in \mathscr{L}$.

Similarly, a cohomology theory h^* on $\mathscr{L}$ is called *additive* if

$$\{i_\lambda^*\} : h^*(\vee X_\lambda) \to \prod_\lambda h^*(X_\lambda)$$

is an isomorphism for every family $\{X_\lambda, \lambda \in \Lambda\}$ in $\mathscr{L}$ with $\vee X_\lambda \in \mathscr{L}$.

(b) Let $\mathscr{K}^\bullet$ be as in 3.4. A reduced homology theory $\widetilde{h}_*$ on $\mathscr{K}^\bullet$ is called *additive* if

$$\langle (i_\lambda)_* \rangle : \oplus_\lambda \widetilde{h}_*(X_\lambda) \to \widetilde{h}_*(\vee X_\lambda).$$

is an isomorphism for every family $\{X_\lambda, \lambda \in \Lambda\}$ in $\mathscr{K}^\bullet$ with $\vee X_\lambda \in \mathscr{K}^\bullet$.

Similarly, a reduced cohomology theory $\widetilde{h}^*$ on $\mathscr{K}^\bullet$ is called *additive* if

$$\{i_\lambda^*\} : \widetilde{h}^*(\vee X_\lambda) \to \prod_\lambda \widetilde{h}^*(X_\lambda)$$

is an isomorphism for every family $\{X_\lambda, \lambda \in \Lambda\}$ in $\mathscr{K}^\bullet$ with $\vee X_\lambda \in \mathscr{K}^\bullet$.

(c) An unreduced (co)homology theory h on $\mathscr{K}^2$ is called *additive* if the corresponding reduced theory on $\mathscr{K}^\bullet$ is additive. In this case $h(\sqcup X_\lambda) = \prod h(X_\lambda)$. Indeed,

$$h(\sqcup X_\lambda) = \widetilde{h}\left((\sqcup X_\lambda)^+\right) = \widetilde{h}(\vee(X_\lambda^+)) = \prod \widetilde{h}(X_\lambda^+) = \prod h(X_\lambda).$$

Because of 3.7(vi), the homomorphisms $\{(i_\lambda)_*\}$, $\{i_\lambda^*\}$ are isomorphisms for every (co)homology theory if Λ is a *finite* set, and so the additivity condition gives no restrictions on the (co)homology theories on $\mathscr{C}_\mathrm{f}^\bullet$, as well as on $\mathscr{S}_\mathrm{f}$.

3.17. Example (James–Whitehead [1]). There are non-additive cohomology theories on $\mathscr{C}^\bullet$. For example, set

$$\widetilde{h}_k(X) := \frac{\prod_{n=0}^\infty \widetilde{H}_n(X)}{\sum_{n=0}^\infty \widetilde{H}_n(X)}$$

for every k. Then $\widetilde{h}_*(S^n) = 0$ for all n, but $\widetilde{h}_*\left(\bigvee_{n=1}^\infty S^n\right) \neq 0$.

3.18. Proposition. *Every (co)homology theory $\{\widetilde{h}, \mathfrak{s}\}$ on $\mathscr{C}^\bullet$ (resp. $\mathscr{C}_\mathrm{fd}^\bullet, \mathscr{C}_\mathrm{f}^\bullet$) can be obtained from a (co)homology $\{h, \widehat{\mathfrak{s}}\}$ on $\mathscr{S}_\mathrm{s}$ (resp. $\mathscr{S}_\mathrm{sfd}, \mathscr{S}_\mathrm{f}$) by Construction* 3.13, *and this (co)homology theory on $\mathscr{S}_\mathrm{s}$ (resp. $\mathscr{S}_\mathrm{sfd}, \mathscr{S}_\mathrm{f}$) is unique up to equivalence. Furthermore, $\{h, \widehat{\mathfrak{s}}\}$ is additive iff $\{\widetilde{h}, \mathfrak{s}\}$ is.*

Proof. If $X \in \mathscr{C}^\bullet$ and $Y = \Sigma^n \Sigma^\infty X$, we set $h_k(Y) := \widetilde{h}_{k-n}(X)$, etc. □

3.19. Proposition. (i) *Let* $\varphi : \widetilde{h}(X) \to \widetilde{k}(X)$ *be a morphism of reduced (co)homology theories on* $\mathscr{C}_{\mathrm{f}}^{\bullet}$. *If* φ *is an isomorphism for* $X = S^0$ *then* φ *is an isomorphism for every* $X \in \mathscr{C}_{\mathrm{f}}^{\bullet}$.

(ii) *Let* $\mathscr{K}^{\bullet}$ *be as in* 3.4, *and let* $\varphi : \widetilde{h}(X) \to \widetilde{k}(X)$ *be a morphism of reduced additive (co)homology theories on* $\mathscr{K}^{\bullet}$. *If* φ *is an isomorphism for* $X = S^0$ *then* φ *is an isomorphism for every* $X \in \mathscr{K}^{\bullet}$.

(iii) *Let* $\mathscr{L}$ *be as in* 3.10, *and let* $\varphi : h(X) \to k(X)$ *be a morphism of additive (co)homology theories on* $\mathscr{L}$. *If* φ *is an isomorphism for* $X = S$ *then* φ *is an isomorphism for every* $X \in \mathscr{L}$.

Proof. We prove this for the homology case only, because the cohomology case can be proved similarly.

(i) Note that φ is an isomorphism for every sphere S^n, and hence it is an isomorphism for every finite wedge $\bigvee S^n$. Given $X \in \mathscr{C}_{\mathrm{f}}^{\bullet}$, let X^n be the n-skeleton of X. Since $X^n/X^{n-1} \simeq \bigvee S^n$ (a finite wedge), we have the following commutative diagram:

$$\begin{array}{ccccccccc}
\widetilde{h}_*(\bigvee S^n) & \longrightarrow & \widetilde{h}_*(X^{n-1}) & \longrightarrow & \widetilde{h}_*(X^n) & \longrightarrow & \widetilde{h}_*(\bigvee S^n) & \longrightarrow & \cdots \\
\cong\downarrow\overline{\varphi} & & \downarrow\varphi_{n-1} & & \downarrow\varphi_n & & \cong\downarrow\overline{\varphi} & & \varphi_{n-1}\downarrow \\
\widetilde{k}_*(\bigvee S^n) & \longrightarrow & \widetilde{k}_*(X^{n-1}) & \longrightarrow & \widetilde{k}_*(X^n) & \longrightarrow & \widetilde{k}_*(\bigvee S^n) & \longrightarrow & \cdots,
\end{array}$$

whose rows are the exact sequence from 3.7(iv). We prove by induction that φ_n is an isomorphism. Note that φ_0 is an isomorphism since $X^0 = \bigvee S^0$. Now, if φ_{n-1} is an isomorphism then, by the Five Lemma, φ_n is an isomorphism. The induction is confirmed. It remains to note that $X = X^k$ for some k.

(ii) The case $\mathscr{K}^{\bullet} = \mathscr{C}_{\mathrm{f}}$ is proved in (i). Let $\mathscr{K}^{\bullet} = \mathscr{C}_{\mathrm{fd}}^{\bullet}$. Because of additivity, φ is an isomorphism for every wedge $\bigvee S^n$. Now, similarly to (i), we can prove that φ is an isomorphism for every finite dimensional CW-space Y. Thus, (ii) holds for $\mathscr{K}^{\bullet} = \mathscr{C}_{\mathrm{fd}}^{\bullet}$. Let $\mathscr{K}^{\bullet} = \mathscr{C}^{\bullet}$. Then, by additivity and what was proved above, φ is an isomorphism for every wedge $\bigvee Y_\lambda$ with finite dimensional Y_λ. Let T be the reduced telescope of the skeletal filtration of a CW-complex X, see I.3.23. We have $T_{\mathrm{ev}} \simeq \vee_n X^{2n}$, $T_{\mathrm{od}} \simeq \vee_n X^{2n+1}$, and $T_{\mathrm{ev}} \cap T_{\mathrm{od}} \simeq \vee_n X^n$. Since $T \simeq X$, it suffices to prove that φ is an isomorphism for T. Consider the commutative diagram

$$\begin{array}{ccccccccc}
\cdots & \longrightarrow & \widetilde{h}_*(T_{\mathrm{ev}}) \oplus \widetilde{h}_*(T_{\mathrm{od}}) & \longrightarrow & \widetilde{h}_*(T) & \longrightarrow & \widetilde{h}_*(T_{\mathrm{ev}} \cap T_{\mathrm{od}}) & \longrightarrow & \cdots \\
 & & \cong\downarrow\varphi'' & & \downarrow\varphi & & \cong\downarrow\varphi' & & \\
\cdots & \longrightarrow & \widetilde{k}_*(T_{\mathrm{ev}}) \oplus \widetilde{k}_*(T_{\mathrm{od}}) & \longrightarrow & \widetilde{k}_*(T) & \longrightarrow & \widetilde{k}_*(T_{\mathrm{ev}} \cap T_{\mathrm{od}}) & \longrightarrow & \cdots
\end{array}$$

of the Mayer–Vietoris sequences of the triad $(T; T_{\mathrm{ev}}, T_{\mathrm{od}})$. Now, using the Five Lemma, we conclude that φ is an isomorphism.

(iii) Because of 3.18, the assertion holds for $\mathscr{L} = \mathscr{S}_{\mathrm{f}}, \mathscr{S}_{\mathrm{s}}, \mathscr{S}_{\mathrm{sfd}}$. Now we give a proof for $\mathscr{L} = \mathscr{S}$ only, because the proof for $\mathscr{L} = \mathscr{S}_{\mathrm{fd}}$ is similar.

By (ii), φ is an isomorphism for every $Y \in \mathscr{C}$. Hence, by additivity, φ is an isomorphism for every spectrum of the form $\vee_\lambda \Sigma^n \Sigma^\infty Y_\lambda$, where each Y_λ is a CW-complex. Consider a spectrum $X = \{X_n\}$and the spectrum $\tau = \tau X$ as in 1.23. By (1.24) and what was proved above, φ is an isomorphism for $\tau_{\mathrm{ev}}, \tau_{\mathrm{od}}$, and $\tau_{\mathrm{ev}} \cap \tau_{\mathrm{od}}$. Since $\tau \simeq X$, it suffices to prove that φ is an isomorphism for τ. Consider the commutative diagram

$$\begin{array}{ccccccccc}
\cdots \longrightarrow & h_*(\tau_{\mathrm{ev}}) \oplus h_*(\tau_{\mathrm{od}}) & \longrightarrow & h_*(\tau) & \longrightarrow & h_*(\tau_{\mathrm{ev}} \cap \tau_{\mathrm{od}}) & \longrightarrow \cdots \\
& \cong \big\downarrow \varphi'' & & \big\downarrow \varphi & & \cong \big\downarrow \varphi' & \\
\cdots \longrightarrow & k_*(\tau_{\mathrm{ev}}) \oplus k_*(\tau_{\mathrm{od}}) & \longrightarrow & k_*(\tau) & \longrightarrow & k_*(\tau_{\mathrm{ev}} \cap \tau_{\mathrm{od}}) & \longrightarrow \cdots
\end{array}$$

of the Mayer–Vietoris sequences of the triad $(\tau; \tau_{\mathrm{ev}}, \tau_{\mathrm{od}})$, see 3.11. Now, using the Five Lemma, we conclude that φ is an isomorphism. □

3.20. Proposition. *Let $\{X_\lambda, \lambda \in \Lambda\}$ be the set of all finite subspectra of a spectrum X.*

(i) *Let h_* be a homology theory on $\mathscr{S}_{\mathrm{f}}$. Set $k_*(X) := \varinjlim\{h_*(X_\lambda)\}$ for any $X \in \mathscr{S}$. Then k_* is an additive homology theory on $\mathscr{S}$.*

(ii) *For every additive homology theory h_* on $\mathscr{S}$, the inclusions $\{i_\lambda : X_\lambda \to X\}$ induce an isomorphism $\langle (i_\lambda)_* | \varinjlim \rangle : \varinjlim\{h_*(X_\lambda)\} \cong h_*(X)$.*

(iii) *Every homology theory on $\mathscr{S}_{\mathrm{f}}$ can be extended to an additive homology theory on $\mathscr{S}$, and this extension is unique (up to equivalence).*

Proof. (i) Firstly, we show that the extension k_* is a functor on $\mathscr{S}$. Given a morphism $f : X \to Y$ of spectra, consider a family $\{f_\omega : X_\omega \to Y_\omega\}$ as in 3.14. Then $k_*(A) = \varinjlim\{h_*(A_\omega)\}$ for $A = X, Y$. Now, f induces a morphism of direct systems $\{\widetilde{h}_*(X_\omega)\} \to \{\widetilde{h}_*(Y_\omega)\}$ and, hence, a homomorphism

$$f_* : k_*(X) = \varinjlim\{h_*(X_\omega)\} \to \varinjlim\{h_*(Y_\omega)\} = k_*(Y).$$

Furthermore, the isomorphism $\mathfrak{s}$ extends from $\mathscr{S}_{\mathrm{f}}$ to $\mathscr{S}$ in an obvious manner.

The homotopy axiom holds obviously. To verify the exactness axiom, first consider a strict cofiber sequence $X \xrightarrow{f} Y \xrightarrow{g} Z = Cf$ of maps of spectra and take $X_\omega \xrightarrow{f_\omega} Y_\omega \xrightarrow{g_\omega} Z_\omega$ as in 3.15. Now, one has an exact sequence of direct systems

$$\{h_*(X_\omega)\} \to \{h_*(Y_\omega)\} \to \{h_*(Z_\omega)\},$$

and hence, by I.2.7, the sequence $k_*(X) \to k_*(Y) \to k_*(Z)$ is exact. Finally, given an arbitrary cofiber sequence $X \xrightarrow{f} Y \xrightarrow{g} Z$, we have a commutative diagram

$$\begin{array}{ccccc}
X & \xrightarrow{f} & Y & \longrightarrow & Cf \\
\| & & \| & & \big\downarrow h \\
X & \xrightarrow{f} & Y & \xrightarrow{g} & Z
\end{array}$$

where h is a homotopy equivalence. Hence, the sequence

$$k_*(X) \xrightarrow{f} k_*(Y) \xrightarrow{g} k_*(Z)$$

is exact.

The additivity property holds because $\varinjlim$ and $\oplus$ commute.

(ii) Given $X \in \mathscr{S}$, set $k_*(X) := \varinjlim\{h_*(X_\lambda)\}$. Consider the homomorphism $\varphi^X := \langle (i_\lambda)_* | \varinjlim \rangle : k_*(X) \to h_*(X)$ as I.2.5. In view of (i), the family $\{\varphi^X\}$ is a morphism $k_* \to h_*$ of additive homology theories on $\mathscr{S}$. It is an isomorphism for every finite X, and so, by 3.19(iii), for every $X \in \mathscr{S}$.

(iii) This follows from (i) and (ii). □

This proposition shows that, in fact, there is no difference between homology theories on $\mathscr{S}_\mathrm{f}$ and additive homology theories on $\mathscr{S}$. For cohomology theories the situation is similar, but more complicated: we discuss this in detail in Ch. III.

3.21. Construction-Definition. Given any cohomology theory h^* on $\mathscr{S}_\mathrm{f}$, one can construct a homology theory h_* on $\mathscr{S}_\mathrm{f}$ by setting $h_i(X) = h^{-i}(X^\perp)$. Since $\Sigma(X^\perp) = (\Sigma^{-1}X)^\perp$, the suspension isomorphism $h^{1-i}(\Sigma(X^\perp)) \to h^{-i}(X^\perp)$ induces a suspension isomorphism $h_{i-1}(\Sigma^{-1}X) \to h_i(X)$. By 2.4(ii), h_* is a well-defined homology theory on $\mathscr{S}_\mathrm{f}$. Conversely, in this manner one can construct a homology theory on $\mathscr{S}_\mathrm{f}$ starting from a cohomology theory on $\mathscr{S}_\mathrm{f}$. Moreover, the correspondences {homology} → {cohomology} and {cohomology} → {homology} are mutually inverse (if we assume $(X^\perp)^\perp = X$, etc). Cohomology and homology theories related in this manner are called *dual* (to each other, i.e., h^* is dual to h_*, and vice versa). In other words, k^* is dual to h_* if k_* is isomorphic to h_*. In this case we have the duality isomorphism $D : k^i(X) \to h_{-i}(X^\perp)$.

We leave it to the reader to transfer this construction to the category $\mathscr{C}_\mathrm{f}$.

3.22. Construction. Let E be an arbitrary spectrum.

(a) Define covariant functors $E_n : \mathscr{S} \to \mathscr{AG}$ where $E_n(X) := \pi_n(E \wedge X)$ for every $X \in \mathscr{S}$ and $E_n(f) := \pi_n(1_E \wedge f)$ for every morphism $f : X \to Y$ of spectra. Furthermore, define $\widehat{\mathfrak{s}}_n : E_n \to E_{n+1}\Sigma$ to be the composition

$$E_n(X) = \pi_n(E \wedge X) = \pi_{n+1}(\Sigma(E \wedge X)) \simeq \pi_{n+1}(E \wedge \Sigma X) = E_{n+1}(\Sigma X)$$

for every $X \in \mathscr{S}$. By 1.15 and 2.1(vi), $(E_n, \widehat{\mathfrak{s}}_n)$ is a homology theory on $\mathscr{S}$, and, by 1.16(iii) and 2.1(v), it is additive.

(b) Define contravariant functors $E^n : \mathscr{S} \to \mathscr{AG}$ by setting $E^n(X) := [X, \Sigma^n E]$ for every $X \in \mathscr{S}$ and

$$E^n(f) : [Y, \Sigma^n E] \to [X, \Sigma^n E], \quad E^n(f)[g] := [gf]$$

for every $f : X \to Y$ and $g : Y \to \Sigma^n E$. Furthermore, define $\widehat{\mathfrak{s}}^n : E^{n+1}\Sigma \to E^n$ to be the composition

$$E^{n+1}(\Sigma X) = [\Sigma X, \Sigma^{n+1} E] = [X, \Sigma^n E] = E^n(X).$$

By 1.15 and 1.16(i), $\{E^n, \widehat{\mathfrak{s}}^n\}$ is an additive cohomology theory on $\mathscr{S}$.

Thus, every spectrum yields a (co)homology theory on $\mathscr{S}$. Hence, by 3.6 and 3.13, every spectrum yields a reduced (co)homology theory on $\mathscr{K}^\bullet$ and an unreduced one on $\mathscr{K}^2$. For example, for every $X \in \mathscr{K}$ we have

$$E_n(X) = \widetilde{E}_n(X^+) = \pi_n(E \wedge X^+) \text{ and } E^n(X) = \widetilde{E}^n(X^+) = [\Sigma^\infty X^+, E].$$

Here the coefficient groups $E_i(S) = \widetilde{E}_i(S^0) = E_i(\text{pt}) = E^{-i}(\text{pt}) = \widetilde{E}^{-i}(S^0) = E^{-i}(S)$ are just the homotopy groups $\pi_i(E)$.

Notice that $E_i(X) \cong X_i(E)$ for any two spectra X, E.

Every morphism $\varphi : E \to F$ of spectra induces a morphism $\varphi : E_*(-) \to F_*(-)$ of homology theories and a morphism $\varphi : E^*(-) \to F^*(-)$ of cohomology theories on $\mathscr{S}$ (and, hence, on $\mathscr{K}^\bullet$ and $\mathscr{K}^2$). Here

$$\varphi = \{\varphi_i^X : E_i(X) \to F_i(X)\},\ \varphi[f] = [(\varphi \wedge 1_X) \circ (f)] \text{ for every } f : \Sigma^i S \to E \wedge X$$

for homology and

$$\varphi = \{\varphi_X^i : E^i(X) \to F^i(X)\},\ \varphi[f] = [(\Sigma^i \varphi) \circ f] \text{ for every } f : X \to \Sigma^i E$$

for cohomology. So, we have a functor from spectra to (co)homology theories. In particular, equivalent spectra yield isomorphic (co)homology theories.

According to 3.22, one can assign a (co)homology theory to a spectrum. This situation turns out to be invertible, see Ch. III, §3 below.

3.23. Proposition. *For every spectrum E, the cohomology theory E^* is dual to the homology theory E_*.*

Proof. We have a natural isomorphism

$$E_i(X^\perp) = [\Sigma^i S, E \wedge X^\perp] = [S, \Sigma^{-i} E \wedge X^\perp] = [X, \Sigma^{-i} E] = E^{-i}(X). \quad \square$$

3.24. Example. Let π be an abelian group, and let $H(\pi)$ (or simply $H\pi$) be a spectrum such that

$$\pi_i(H(\pi)) = \begin{cases} \pi & \text{if } i = 0 \\ 0 & \text{if } i \neq 0. \end{cases}$$

Firstly, we construct such a spectrum. Let $K(\pi, n)$ be an Eilenberg–Mac Lane space, $\pi_n(K(\pi, n)) = \pi$ and $\pi_i(K(\pi, n)) = 0$ for $i \neq n$. Obvious homotopy equivalences $\omega_n : K(\pi, n) \to \Omega K(\pi, n+1)$ yield an Ω-prespectrum, and, thus, an Ω-spectrum E with $E_n \simeq K(\pi, n)$. Of course, E satisfies the above conditions.

We prove that the conditions above determine $H(\pi)$ uniquely up to equivalence. Indeed, let F be another spectrum with $\pi_0(F) = \pi$ and $\pi_i(F) = 0$ for $i \neq 0$. By 1.21, we can assume that F is an Ω-spectrum. Then F_n must be an Eilenberg-Mac Lane space $K(\pi, n)$. The identity map 1_π induces a homotopy equivalence $K(\pi, n) = F_n \to E_n = K(\pi, n)$, and, clearly, these homotopy equivalences constitute an equivalence $F \to E$ of spectra.

Consider the (co)homology theory associated with $H(\pi)$. We have

$$H(\pi)_i(\mathrm{pt}) = \widetilde{H}(\pi)_i(S^0) = \pi_i(H(\pi)) = \begin{cases} 0 & \text{if } i \neq 0 \\ \pi & \text{if } i = 0. \end{cases}$$

Similarly, $H(\pi)^0(\mathrm{pt}) = \pi$, $H(\pi)^i(\mathrm{pt}) = 0$ for $i \neq 0$.

Thus, according to the Eilenberg–Steenrod Theorem, the spectrum $H(\pi)$ produces the ordinary (co)homology theory on $\mathscr{C}_{\mathrm{f}}$,

$$H(\pi)_i(X) \cong H_i(X;\pi),\ H(\pi)^i(X) = H^i(X;\pi) \text{ for every } X \in \mathscr{C}_{\mathrm{f}}.$$

Hence, for every $X \in \mathscr{C}^\bullet$,

$$H_i(X;\pi) \cong H(\pi)_i(X) = \pi_i(H(\pi) \wedge X^+) \cong \lim_{N\to\infty} \pi_{i+N}(K(\pi, N) \wedge X^+)$$

(the last isomorphism holds for every $X \in \mathscr{C}$ and can be proved directly or deduced from 3.19(ii), since the homomorphisms

$$\lim_{N\to\infty} \pi_{i+N}(K(\pi, N) \wedge X^+) \to \pi_i(H(\pi) \wedge X^+)$$

yield a morphism of homology theories on $\mathscr{C}^\bullet$).

In view of the above, we write $H^*(X;\pi)$ instead of $(H\pi)^*(X)$ and $H_*(X;\pi)$ instead of $(H\pi)_*(X)$ for every $X \in \mathscr{C}$ or $X \in \mathscr{S}$. Furthermore, we write $H(X)$ instead of $H(X;\mathbb{Z}) = H\mathbb{Z}(X)$.

3.25. Proposition. *If $E = \{E_n\}$ is an Ω-spectrum, then for every space X there is a natural equivalence $\widetilde{E}^i(X) \cong [X, E_i]^\bullet$.*

Proof. We have

$$\begin{aligned}\widetilde{E}^i(X) &= [\Sigma^\infty X, \Sigma^i E] \cong \lim_{N\to\infty} [S^N X, E_{i+N}]^\bullet \cong \lim_{N\to\infty} [X, \Omega^N E_{i+N}]^\bullet \\ &\cong \lim_{N\to\infty} [X, E_i]^\bullet = [X, E_i]^\bullet.\end{aligned}$$

□

3.26. Corollary. *For every spectrum E and every i, the functor*

$$\widetilde{E}^i : \mathscr{HC}^\bullet \to \mathscr{AG}, \quad X \mapsto \widetilde{E}^i(X),$$

is representable.

Proof. Let F be an Ω-spectrum equivalent to E. Then $\widetilde{E}^i(X) = \widetilde{F}^i(X) = [X, F_i]^\bullet$, i.e., the space F_i represents the functor $\widetilde{E}^i$. □

We are especially interested in the case $i = 0$.

3.27. Proposition-Definition. *Given a spectrum E, let $\Omega^\infty E$ denote a representing space for $\widetilde{E}^0 : \mathscr{HC}^\bullet \to \mathscr{AG}$ (i.e., $\Omega^\infty E = F_0$ for some Ω-spectrum F equivalent to E). This space $\Omega^\infty E$ is called the* infinite delooping *of E and has the following properties:*

(i) *It is uniquely defined up to homotopy equivalence.*

(ii) *Consider a pair of spectra E, F and fix certain spaces $\Omega^\infty E, \Omega^\infty F$ and equivalences $[-, \Omega^\infty E]^\bullet \cong \widetilde{E}^0(-)$, $[-, \Omega^\infty F]^\bullet \cong \widetilde{F}^0(-)$ of functors. For every morphism $\varphi : E \to F$ there exists a map $f : \Omega^\infty E \to \Omega^\infty F$ such that for every space X the diagram*

$$\begin{array}{ccc} \widetilde{E}^0(X) & \xrightarrow{\varphi_*} & \widetilde{F}^0(X) \\ \cong\downarrow & & \downarrow\cong \\ [X, \Omega^\infty E]^\bullet & \xrightarrow{f_*} & [X, \Omega^\infty F]^\bullet \end{array}$$

commutes, and such f is unique up to homotopy. This map f is called the infinite delooping *of φ and is denoted by $\Omega^\infty \varphi$.*

(iii) *For every $E \xrightarrow{\varphi} F \xrightarrow{\psi} G$ we have $\Omega^\infty(\psi\varphi) \simeq \Omega^\infty\psi\,\Omega^\infty\varphi$.*

Proof. (i) and (ii) follow from the Yoneda Lemma I.1.5, and (iii) follows from (ii). □

3.28. Remarks. (a) A space X is called an *infinite loop space* if it has the form $\Omega^\infty E$ for some E. If $F = \{F_n\}$ is an Ω-spectrum equivalent to E, then $X \simeq \Omega^n F_n$, i.e., an infinite loop space is an n-loop space for all n. This justifies the term "infinite loop space" (and the notation Ω^∞).

(b) Of course, the notation $\Omega^\infty E$, $\Omega^\infty f$ is not pedantically rigorous because, say, $\Omega^\infty E$ is defined only up to homotopy equivalence. In particular, Ω^∞ is not a functor $\mathscr{S} \to \mathscr{HC}^\bullet$. Nevertheless, we shall use this notation because it is very convenient, and there is no danger of confusion. However, if one wants to consider Ω^∞ as a functor, one can *choose* a space $\Omega^\infty E$ for every spectrum E, etc.

3.29. Corollary. (i) *Let F be an Ω-spectrum equivalent to E (see* 1.21*). Then $\Omega^\infty E \simeq F_0$.*

(ii) (an adjointness relation) *For every spectrum E and every pointed CW-complex X we have a natural equivalence $[\Sigma^\infty X, E] \cong [X, \Omega^\infty E]^\bullet$.* □

If we put $X = \Omega^\infty E$ in 3.29(ii), we obtain a map $j : \Sigma^\infty \Omega^\infty E \to E$ which is adjoint to $1_{\Omega^\infty E}$. If we put $E = \Sigma^\infty X$, we obtain a morphism $i : X \to \Omega^\infty \Sigma^\infty X$ which is adjoint to $1_{\Sigma^\infty X}$.

Consider the actions of Ω^∞ and Σ^∞ on homotopy groups. Firstly, $\pi_k(E) = \pi_k(\Omega^\infty E)$ (where on the left hand side we have the homotopy group of the spectrum, while on the right hand side we have that of the space). Furthermore, the group $\pi_k(\Sigma^\infty X) = \lim_{N\to\infty} \pi_{k+N}(S^N X)$ is just the stable homotopy group $\Pi_k(X)$ of X, and

$$i_*: \pi_k(X) \to \pi_k(\Omega^\infty\Sigma^\infty X) = \pi_k(\Sigma^\infty X) = \Pi_k(X)$$

is just the stabilization homomorphism.

Finally, if X is an infinite loop space, $X = \Omega^\infty E$, then the map $i: X \to \Omega^\infty\Sigma^\infty X$ has a homotopy left inverse $\tau: \Omega^\infty\Sigma^\infty X \to X$. Namely, τ is the composition

$$\Omega^\infty\Sigma^\infty X = \Omega^\infty\Sigma^\infty\Omega^\infty E \xrightarrow{\Omega^\infty j} \Omega^\infty E = X.$$

We leave it to the reader to prove (by purely categorical arguments) that $\tau i \simeq 1_X$. In particular,

$$\pi_k(X) \xrightarrow{i_*} \pi_k(\Omega^\infty\Sigma^\infty X) \xrightarrow{\tau_*} \pi_k(X)$$

coincides with $1_{\pi_k(X)}$. Since $\pi_k(\Omega^\infty\Sigma^\infty X) = \Pi_k(X)$, we have

3.30. Proposition. *If X is an infinite loop space, then $\pi_k(X)$ is a direct summand of $\Pi_k(X)$.* □

3.31. Remark. Given a spectrum E, it is possible and useful to extend the functors $\widetilde{E}^*$ and $\widetilde{E}_*$ to the whole category $\mathscr{W}^\bullet$. We define

$$\widetilde{E}^n(X) := [X, \Omega^\infty\Sigma^n E]^\bullet, \qquad \widetilde{E}_n(X) := \lim_{n\to\infty} \pi_{i+n}(E_i \wedge X)$$

for every $X \in \mathscr{W}^\bullet$. Clearly, the functors E^n and E_n are homotopy invariant, and there are exact sequences of pairs for every cofibered pair (X, A). Moreover, given a pointed triad $(X; A, B)$, there is the Mayer–Vietoris exact sequence provided that A, B is a numerable covering of X and A, B are well-pointed, cf. I.3.37.

3.32. Examples. Here we give some examples of spectra and (co)homology theories which will be discussed and used later.

(a) The sphere spectrum S. Clearly, $S_k(E) = \pi_k(E)$ for every spectrum E and $S_k(X) = \Pi_k(X^+)$ for every space X. In particular, S represents the stably (co)homotopy functor on $\mathscr{C}^\bullet$.

(b) The Moore spectrum $M(A)$ of an abelian group A, which is characterized by the conditions $\pi_i(M(A)) = 0$ for $i < 0$, $H_i(M(A)) = 0$ for $i \neq 0$, $H_0(M(A)) = A$, see 4.32 below.

(c) The Eilenberg–Mac Lane spectrum $H(\pi)$ of an abelian group π, see 3.24. This spectrum yields the ordinary (co)homology with coefficients in π.

(d) Let $G = \oplus G_i$ be a graded abelian group with homogeneous components G_i. The graded Eilenberg–Mac Lane spectrum of G is the spectrum

$H(G) := \vee\Sigma^i H(G_i)$. In particular, $\pi_*(H(G)) = G$. These spectra will be discussed in §7.

(e) Complex K-theory, see Atiyah [4], Karoubi [1], etc. It is represented by a spectrum K such that $\Sigma^2 K \simeq K$ and $\pi_*(K) = \mathbb{Z}[t, t^{-1}]$, $\dim t = 2$.

(f) There exist a spectrum k and a morphism $p : k \to K$ such that $\pi_*(k) = \mathbb{Z}[t]$, $\dim t = 2$ and $p_* : \pi_i(k) \to \pi_i(K)$ is an isomorphism for $i \geq 0$. This morphism $p : k \to K$ is constructed as a connective covering over K, see §4 below. The (co)homology theory given by k is called the *connected complex k-theory*. Sometimes one uses the notation **bu** instead of k.

(g) There are real analogs of examples (e), (f). Namely, there is a real KO-theory KO which is 8-periodic, $\Sigma^8 KO \simeq KO$, and it has a connective covering kO (also denoted by **bo**).

(h) Certain (co)bordism theories, which are represented by the so-called Thom spectra, see Ch. IV.

(i) The Brown–Peterson spectrum BP, see Ch. VII.

(j) Certain (co)bordism theories with singularities, see Ch. VIII, IX. For example, Morava K-theories $K(n)$ and $k(n)$, Morava–Johnson–Wilson spectra $P(n)$, Baas–Johnson–Wilson spectra $BP\langle n\rangle$, see Ch. IX.

Consider now a spectrum E and a morphism $\alpha : S \to E$. Let X be an arbitrary spectrum. The homomorphism

$$\alpha_* : \pi_*(X) \to E_*(X) \tag{3.33}$$

is called the *Hurewicz homomorphism* with respect to α and denoted by h_α. It has the following alternative description: $h_\alpha[f] = E_*(f)(\widehat{\mathfrak{s}}^d[\alpha])$, where $E_*(f) : E_*(\Sigma^d S) \to E_*(X)$ is induced by $f : \Sigma^d S \to X$ and $\widehat{\mathfrak{s}}^d$ is the iterated suspension. In order to see the equivalence of these two descriptions, note that $\alpha_*(f)$ is given by the composition $\Sigma^d S \wedge S \xrightarrow{f\wedge 1} X \wedge S \xrightarrow{1\wedge\alpha} X \wedge E$, while $h_\alpha[f]$ is given by $\Sigma^d S \wedge S \xrightarrow{1\wedge\alpha} \Sigma^d S \wedge E \xrightarrow{f\wedge 1} X \wedge E$. But each of these compositions is just $\alpha \wedge f$.

Now, let $E = (E, \mu, \iota)$ be a ring spectrum. The Hurewicz homomorphism h_ι we denote just by h. For every pair of spectra X, Y and every pair of integers m, n, we have pairings $\mu^{X,Y}$ and $\mu_{X,Y}$ induced by μ. Here

$$\mu^{X,Y} : E^m(X) \otimes E^n(Y) \to E^{m+n}(X \wedge Y) \tag{3.34}$$

maps $[f] \otimes [g]$ to (the homotopy class of)

$$\Sigma^{-m} X \wedge \Sigma^{-n} Y \xrightarrow{f\wedge g} E \wedge E \xrightarrow{\mu} E$$

where $[f] \in E^m(X)$, $[g] \in E^n(Y)$; and

$$\mu_{X,Y} : E_m(X) \otimes E_n(Y) \to E_{m+n}(X \wedge Y) \tag{3.35}$$

maps $[f] \otimes [g]$ to

$$\Sigma^m S \wedge \Sigma^n S \xrightarrow{f \wedge g} (E \wedge X) \wedge (E \wedge Y) \xrightarrow{\varphi} (E \wedge E) \wedge (X \wedge Y) \xrightarrow{\mu \wedge 1} E \wedge X \wedge Y$$

where $[f] \in E_m(X)$, $[g] \in E_n(Y)$ and φ is the composition

$$\begin{aligned}
&(E \wedge X) \wedge (E \wedge Y) \xrightarrow{a(E,X,E\wedge Y)} E \wedge (X \wedge (E \wedge Y)) \xrightarrow{1 \wedge a^{-1}(X,E,Y)} \\
&E \wedge ((X \wedge E) \wedge Y) \xrightarrow{1 \wedge \tau(X,E) \wedge 1} E \wedge ((E \wedge X) \wedge Y) \xrightarrow{1 \wedge a(E,X,Y)} \\
&E \wedge (E \wedge (X \wedge Y)) \xrightarrow{a^{-1}(E,E,X\wedge Y)} (E \wedge E) \wedge (X \wedge Y).
\end{aligned}$$

The pairings $\mu^{X,Y}$ and $\mu_{X,Y}$ are associative, and they commute with suspensions (i.e., with the shift of dimension). Here, for instance, the associativity of $\mu_{X,Y}$ means the commutativity of the diagram

$$(3.36) \qquad \begin{array}{ccc}
E_m(X) \otimes E_n(Y) \otimes E_p(Z) & \longrightarrow & E_m(X) \otimes E_{n+p}(Y \wedge Z) \\
\downarrow & & \downarrow \\
E_{m+n}(X \wedge Y) \otimes E_p(Z) & \longrightarrow & E_{m+n+p}(X \wedge Y \wedge Z)
\end{array} ,$$

while commuting with suspensions means the commutativity of the diagrams

$$(3.37) \qquad \begin{array}{ccc}
E_m(X) \otimes E_n(Y) & \longrightarrow & E_{m+n}(X \wedge Y) \\
\downarrow {\scriptstyle \sigma \otimes 1} & & \downarrow {\scriptstyle \sigma} \\
E_{m+1}(\Sigma X) \otimes E_n(Y) \to E_{m+n+1}(\Sigma X \wedge Y) & \cong & E_{m+n+1}(\Sigma(X \wedge Y)),
\end{array}$$

$$\begin{array}{ccc}
E_m(X) \otimes E_n(Y) & \longrightarrow & E_{m+n}(X \wedge Y) \\
\downarrow {\scriptstyle 1 \otimes \sigma} & & \downarrow {\scriptstyle \sigma} \\
E_m(X) \otimes E_{n+1}(\Sigma Y) & & E_{m+n+1}(\Sigma(X \wedge Y)) \\
\downarrow & & \downarrow {\scriptstyle (-1)^m} \\
E_{m+n+1}(X \wedge \Sigma Y) & \xrightarrow{\cong} & E_{m+n+1}(\Sigma(X \wedge Y)).
\end{array}$$

The commutativity of the diagrams (3.36) and (3.37) follows from the definition of ring spectra and the properties of the smash product, see §2. Furthermore, if E is a commutative ring spectrum then

$$(3.38) \qquad \tau_* \mu^{X,Y}(a \otimes b) = (-1)^{|a||b|} \mu^{X,Y}(b \otimes a),$$

where $\tau = \tau(X, Y) : X \wedge Y \to Y \wedge X$. The same is true for $\mu_{X,Y}$, cf. Adams [6], Switzer [1].

Moreover, there is a pairing

(3.39) $$\mu^X_{\bullet,Y} : E^m(X) \otimes E_n(X \wedge Y) \to E_{n-m}(Y)$$

where $\mu^X_{\bullet,Y}([f] \otimes [g])$ is represented by the morphism

$$\Sigma^n S \xrightarrow{g} E \wedge X \wedge Y \xrightarrow{1_E \wedge f \wedge 1_Y} E \wedge \Sigma^m E \wedge Y \xrightarrow{\Sigma^m \mu \wedge 1_Y} \Sigma^m E \wedge Y.$$

Similarly, one can construct a pairing

(3.40) $$\mu^{\bullet,Y}_X : E^m(X \wedge Y) \otimes E_n(X) \to E^{m-n}(Y).$$

If we put $Y = S$ in (3.39) and/or (3.40), we get the *Kronecker pairing*

(3.41) $$\langle -, - \rangle : E^m(X) \otimes E_n(X) \to E_{n-m}(S) = \pi_{n-m}(E).$$

If we put $X = Y = S$ in (3.35), we see that $E_*(S) = \pi_*(E)$ is a ring; its unit is given by $\iota : S \to E$. If we put $X = S$ in (3.35), we see that $E_*(Y)$ is a graded left $\pi_*(E)$-module. Of course, one can consider $E_*(Y)$ as a right $\pi_*(E)$-module: the equivalence $l : S \wedge Y \simeq Y$ yields the left $\pi_*(E)$-module structure, while the equivalence $r : Y \wedge S \simeq Y$ yields the right one. Similarly, $E^*(Y)$ is a (left) graded $E^*(S)$-module. Finally, if X is a ring spectrum with a multiplication $\nu : X \wedge X \to X$ then there is a pairing

$$E_*(X) \otimes E_*(X) \xrightarrow{\mu_{X,X}} E_*(X \wedge X) \xrightarrow{\nu_*} E_*(X),$$

turning $E_*(X)$ into a ring. In particular, $E_*(E)$ is a ring.

More generally, let (F, m) be a module spectrum over a ring spectrum (E, μ, ι). As above, m induces pairings

(3.42)
$$\begin{aligned}
m^{X,Y} &: E^m(X) \otimes F^n(Y) \to F^{m+n}(X \wedge Y),\\
m_{X,Y} &: E_m(X) \otimes F_n(Y) \to F_{m+n}(X \wedge Y),\\
m^X_{\bullet,Y} &: E^m(X) \otimes F_n(X \wedge Y) \to F_{n-m}(Y),\\
m^{\bullet,Y}_X &: E^m(X \wedge Y) \otimes F_n(X) \to F^{m-n}(Y),\\
{}'m^{\bullet,Y}_X &: F^m(X \wedge Y) \otimes E_n(X) \to F^{m-n}(Y),
\end{aligned}$$

and the Kronecker pairings

$$\begin{aligned}
\langle -, - \rangle &: E^m(X) \otimes F_n(X) \to F_{n-m}(S) = \pi_{n-m}(F),\\
\langle -, - \rangle &: F^m(X) \otimes E_n(X) \to F^{m-n}(S) = \pi_{n-m}(F).
\end{aligned}$$

Notice the following fact. Given a morphism $a : S^d \to E$, the morphism $a_\# : S^d F \to F$ defined in 2.14 induces the homomorphism $(a_\#)_* : F_n(X) \to F_{n+d}(X)$. This homomorphism coincides with multiplication by $[a] \in \pi_*(E)$ on the $\pi_*(E)$-module $F_*(X)$ (prove it). This justifies the term "multiplication" in 2.14.

3.43. Conventions. (a) We shall write ab instead of $\mu^{X,Y}(a \otimes b)$ as well as of $\mu_{X,Y}(a \otimes b)$. Similarly for $m^{X,Y}(a \otimes b)$, etc.

(b) The spectrum $\Sigma^n S$ will be denoted simply by S^n, when there is no danger of confusion.

(c) For any morphism $f : X \to \Sigma^i E$ we shall write just $f \in E^i(X)$ rather than $[f] \in E^i(X)$.

Let E and F be as above, and let Y be a module spectrum over a ring spectrum X. Then we have the homomorphism

$$E_*(X) \otimes F_*(Y) \xrightarrow{m_{X,Y}} F_*(X \wedge Y) \to F_*(Y)$$

(the right map is induced by the pairing $X \wedge Y \to Y$) turning $F_*(Y)$ into a (left) $E_*(X)$-module. In particular, $F_*(F)$ is an $E_*(E)$-module (put $X = E, Y = F$). Similarly, $F_*(E)$ is an $E_*(E)$-module.

Every morphism $a : S^d \to X$ yields an element $a \in \pi_d(X)$. The composition

$$a_\# : S^d \wedge Y \xrightarrow{a \wedge 1} X \wedge Y \to Y$$

induces a homomorphism $F_i(S^d \wedge Y) \to F_i(Y)$, i.e., a homomorphism $a_* : F_{i-d}(Y) \cong F_i(S^d \wedge Y) \to F_i(Y)$.

On the other hand, given $x \in E_*(X)$, the multiplication by x yields an additive homomorphism $x : F_*(Y) \to F_*(Y), a \mapsto xa, a \in F_*(Y)$.

3.44. Lemma. *The homomorphism a_* is the multiplication by $h(a) \in E_*(X)$.*

Proof. Consider the commutative diagram

$$\begin{array}{ccc} E_*(S) \otimes F_*(Y) & \xrightarrow{E_*(a) \otimes 1} & E_*(X) \otimes F_*(Y) \\ \downarrow & & \downarrow \\ F_*(S \wedge Y) = F_*(Y) & \xrightarrow{a_*} & F_*(Y). \end{array}$$

Since $E_*(a)(\iota) = h(a)$, the lemma follows. □

In the above discussion (from (3.34) to (3.42)) one can replace spectra X, Y by spaces. In this case all the formulae for reduced (co)homology remain valid (we only need to replace Σ by S). For unreduced (co)homology we must slightly modify the formulae. Recall that

$$X/A \wedge Y/B = (X \times Y)/(X \times B \cup A \times Y),$$

and so (3.34) must be replaced by a pairing

$$\mu = \mu^{(X,A),(Y,B)} : E^m(X, A) \otimes E^n(Y, B) \to E^{m+n}(X \times Y, X \times B \cup A \times Y),$$

etc.

Let $d : (X, A \cup B) \to (X \times X, X \times B \cup A \times X)$ be the diagonal. Define a pairing (called the *cup-multiplication*)

$$\cup : E^m(X,A) \otimes E^n(X,B) \xrightarrow{\mu} E^{m+n}(X \times X, X \times B \cup A \times X) \xrightarrow{d^*} E^{m+n}(X, A \cup B).$$

For $A = B$ this pairing yields a multiplication on $E^*(X, A)$ converting it into a graded ring. As with 3.43, we write ab instead of $a \cup b$. By (3.38), $ab = (-1)^{|a||b|}ba$ for every commutative ring spectrum E.

Furthermore, the pairing

$$\mu = \mu^{(X,A)}_{\bullet,(Y,B)} : E^m(X,A) \otimes E_n(X \times Y, X \times B \cup A \times Y) \to E_{n-m}(Y,B)$$

induces an inner operation (called the *cap-multiplication*)

$$\cap : E^m(X,A) \otimes E_n(X, A \cup B) \to E_{n-m}(X,B)$$

of the form

$$E^m(X,A) \otimes E_n(X, A\cup B) \xrightarrow{1 \otimes d_*} E^m(X,A) \otimes E_n(X \times X, X \times B \cup A \times X) \xrightarrow{\mu} E_{n-m}(X,B).$$

Below we shall meet many ring spectra. Note that sometimes it is difficult to prove the existence of a ring structure on a given spectrum, see Ch. VIII.

Now we formulate two useful technical theorems which enable us to construct ring morphisms of ring spectra. Let X be any spectrum, let E be a ring spectrum, and let F be an E-module spectrum. The Kronecker pairing $\langle -, - \rangle : F^*(X) \otimes E_*(X) \to \pi_*(F)$ yields the evaluation homomorphism

$$\text{ev} : F^n(X) \to \text{Hom}^n_{\pi_*(E)}(E_*(X), \pi_*(F)),$$
$$(\text{ev}(a))(b) = \langle a, b \rangle,\ a \in F^n(X), b \in E_*(X).$$

3.45. Theorem. *Suppose that there exists N such that $\pi_i(X) = 0$ for $i < N$. If the Atiyah–Hirzebruch spectral sequence* [6]

$$E^r_{**}(X) \Rightarrow E_*(X),\ E^2_{p,q}(X) = H_p(X; \pi_q(E))$$

collapses (i.e., all its differentials are trivial), and if the $\pi_*(E)$-module $E^2_{*,*}(X)$ is free, then

$$\text{ev} : F^n(X) \to \text{Hom}^n_{\pi_*(E)}(E_*(X), \pi_*(F)),\ (\text{ev}(a))(b) = \langle a, b \rangle$$

is an isomorphism.

---

[6] The necessary information about the Atiyah–Hirzebruch spectral sequence (hereafter denoted AHSS) can be found e.g. in Adams [8], Ch. III.

Proof. See Adams [5], p.20, Prop. 17 or Adams [8], p.48, Lemma 4.2. □

3.46. Theorem. *Let X be a ring spectrum, and let E be a commutative ring spectrum. Consider the evaluation*

$$\mathrm{ev}: E^n(X) \to \mathrm{Hom}^n_{\pi_*(E)}(E_*(X), \pi_*(E)), \quad (\mathrm{ev}(a))(b) = \langle a, b\rangle$$

and suppose that all the conditions of 3.45 *hold for X and $X \wedge X$. Then a morphism $f: X \to E$ is a ring morphism iff the homomorphism*

$$\mathrm{ev}(f) \in \mathrm{Hom}^0_{\pi_*(E)}(E_*(X), \pi_*(E))$$

is a homomorphism of $\pi_*(E)$-algebras.

Proof. This is a direct consequence of 3.45, but we want to demonstrate why E should be commutative. Let μ_X (resp. μ_E) be the multiplication on X (resp. E). Given a morphism $f: X \to E$, set $\mathrm{ev}(f) = e: E_*(X) \to \pi_*(E)$. We want to prove that the left square below commutes up to homotopy iff the right square commutes.

$$\begin{array}{ccc} X \wedge X & \xrightarrow{f \wedge f} & E \wedge E \\ {\scriptstyle \mu_X}\downarrow & & \downarrow{\scriptstyle \mu_E} \\ X & \xrightarrow{f} & E \end{array} \qquad \begin{array}{ccc} E_*(X) \otimes E_*(X) & \xrightarrow{e \otimes e} & \pi_*(E) \otimes \pi_*(E) \\ {\scriptstyle \mu''}\downarrow & & \downarrow{\scriptstyle \mu'} \\ E_*(X) & \xrightarrow{e} & \pi_*(E) \end{array} .$$

Assume that the left square commutes. Given $a: S^k \to E \wedge X$, $b: S^l \to E \wedge X$, consider the following commutative diagram where τ switches the factors:

$$\begin{array}{ccccccc} E \wedge X \wedge E \wedge X & \xrightarrow{1 \wedge \tau \wedge 1} & E \wedge E \wedge X \wedge X & \xrightarrow{\mu_E \wedge \mu_X} & E \wedge X & & \\ {\scriptstyle a \wedge b}\uparrow & & {\scriptstyle 1 \wedge 1 \wedge f \wedge f}\downarrow & & \downarrow{\scriptstyle 1 \wedge f} & & \\ S^k \wedge S^l & & E \wedge E \wedge E \wedge E & \xrightarrow{\mu_E \wedge \mu_E} & E \wedge E & \xrightarrow{\mu_E} & E. \end{array}$$

We have $e\mu''(a \otimes b) = \mu_E \circ (1 \wedge f) \circ (\mu_E \wedge \mu_X) \circ (1 \wedge \tau \wedge 1) \circ (a \wedge b)$, and this morphism is homotopic to $\mu_E \circ (\mu_E \wedge \mu_E) \circ (1 \wedge 1 \wedge f \wedge f) \circ (1 \wedge \tau \wedge 1) \circ (a \wedge b)$. On the other hand, $\mu'(e \otimes e)(a \otimes b)$ is represented by the composition

$$S^k \wedge S^l \xrightarrow{a \wedge b} E \wedge X \wedge E \wedge X \xrightarrow{1 \wedge f \wedge 1 \wedge f} E \wedge E \wedge E \wedge E \xrightarrow{\mu_E \wedge \mu_E} E \wedge E \xrightarrow{\mu_E} E.$$

Now, $\mu'(e \otimes e)(a \otimes b) = e\mu''(a \otimes b)$ *since E is commutative,* i.e., the right square above commutes. Similarly, one can prove that the left square above commutes if the right one commutes. □

3.47. Construction-Definition. Let E be an arbitrary spectrum. The graded group $E^*(E)$ admits a ring structure where the multiplication is given by the composition of morphisms $E \to \Sigma^? E$. In greater detail, if $a \in E^d(E)$

and $b \in E^n(E)$ then ab is given by the morphism $E \xrightarrow{b} \Sigma^d E \xrightarrow{\Sigma^n a} \Sigma^{n+d} E$. Furthermore, given $a \in E^d(E)$ and $x \in E^k(X)$, $X \in \mathscr{S}$, we define $a(x) \in E^{d+k}(X)$ to be the element which is represented by the morphism $X \xrightarrow{x} \Sigma^k E \xrightarrow{\Sigma^k a} \Sigma^{k+d} E$. So, we have a homomorphism

$$E^*(E) \otimes E^*(X) \to E^*(X),\ a \otimes x \mapsto a(x),$$

which turns $E^*(X)$ into an $E^*(E)$-module, and this module structure is natural in X.

Look at this from another point of view. Given a cohomology theory $\{h^n, \widehat{\mathfrak{s}}^n\}$ on $\mathscr{S}$, we define an *h-operation* (of degree d) to be a family $a = \{a^n : h^n(-) \to h^{n+d}(-)\}_{n=-\infty}^{\infty}$ of natural transformations such that $a^n \widehat{\mathfrak{s}}^n = \widehat{\mathfrak{s}}^{n+d} a^{n+1}$. Clearly, every operation a is completely determined by a^0.

Now, by the above, for every spectrum E each element $a \in E^d(E)$ yields an operation (of degree d). Moreover, by the Yoneda Lemma I.1.5, the set $E^*(E)$ is in a canonical bijective correspondence with the set of all E-operations. For this reason, $E^*(E)$ is called the *ring of E-operations.*

Finally, it makes sense to remark that $E^*(E)$ acts also on $E_*(X)$. Namely, given $a \in E^d(X)$ and $x \in E_k(X)$, we define $a(x) \in E_{k-d}(X)$ to be the element which is represented by the morphism $S^k \xrightarrow{x} X \wedge E \xrightarrow{1 \wedge \Sigma^d a} X \wedge \Sigma^d E$.

§4. Homotopy Properties of Spectra

In this section we develop the homotopy theory of spectra. Namely, we discuss Postnikov towers, Cartan killing constructions, Serre theory of classes of abelian group, etc., for spectra. (We assume that the reader knows these notions in the case of spaces; otherwise he can find them e.g. in Mosher–Tangora [1].) Closely related material is exposed in Dold [3] and Margolis [1].

4.1. Lemma. (i) *Let X, E be two spectra. Suppose that $\widetilde{E}^k(X_k) = 0 = \widetilde{E}^{k-1}(X_k)$ for every k. Then $E^0(X) = 0$.*

(ii) *Let E be a spectrum with $\pi_j(E) = 0$ for $j \le n+1$. Let X be a spectrum with $X^{(n)} = X$. Then $E^0(X) = 0$.*

(iii) *Let E be a spectrum and Y be a pointed CW–space. If $\widetilde{E}^0(Y^{(r)}) = 0 = \widetilde{E}^{-1}(Y^{(r)})$ for every r then $E^0(Y) = 0$.* [7]

(iv) *Let E be a spectrum with $\pi_j(E) = 0$ for $j > n$. Let X be a spectrum with $\pi_i(X) = 0$ for $i \le n$. Then $E^0(X) = 0$.*

Proof. (i) By (1.24) and 1.16(i),

$$E^k(\tau_{\mathrm{ev}}(X)) = E^k(\tau_{\mathrm{od}}(X)) = E^k(\tau_{\mathrm{ev}}(X) \cap \tau_{\mathrm{od}}(X)) = 0 \quad \text{for } k = 0, -1.$$

[7] This holds for a spectrum Y also, see III.4.18 below.

Hence, by 1.22 (or 3.12(iii)), $E^0(\tau(X)) = 0$, and thus $E^0(X) = 0$ because $\tau(X) \simeq X$.

(ii) We can assume that E is an Ω-spectrum, and so $\pi_i(E_k) = 0$ for $i \leq k+n+1$. By obstruction theory, $[X_k, E_k] = 0 = [X_k, E_{k-1}]$ since $X_k = X_k^{(k+n)}$. So, $\widetilde{E}^k(X_k) = 0 = \widetilde{E}^{k-1}(X_k)$ for every k. Thus, by (i), $E^0(X) = 0$.

(iii) Consider the reduced telescope T of the sequence

$$\cdots \subset Y^{(r)} \subset Y^{(r+1)} \subset \cdots .$$

We have

$$\widetilde{E}^k(T_{\rm ev}) = \widetilde{E}^k(T_{\rm od}) = \widetilde{E}^k(T_{\rm ev} \cap T_{\rm od}) = 0 \quad \text{for } k = 0, -1,$$

and $T = T_{\rm ev} \vee T_{\rm od}$. So, by 3.12(iii), $\widetilde{E}^0(T) = 0$. Thus, $\widetilde{E}^0(Y) = 0$ because $T \simeq Y$.

(iv) Assume that E and X are Ω-spectra. Fix any k. Then $X_k^{(k+n)} \simeq *$ and $\pi_i(E_k) = 0$ for $i > k+n$, and so, by obstruction theory, $[X_k, E_k] = 0 = [X_k, E_{k-1}]$. Now, we can finish the proof just as in case (ii). □

4.2. Lemma. *A spectrum E is equivalent to a spectrum F with $F^{(n)} = *$ if and only if $\pi_i(E) = 0$ for $i \leq n$.*

Proof. The "only if" part is trivial. So, let $\pi_i(E) = 0$ for $i \leq n$. Assuming E to be an Ω-spectrum, we have $\pi_{i+k}(E_k) = 0$ for $i \leq n$. Replacing E_k by a homotopy equivalent CW-complex E'_k with $(E'_k)^{(n+k)} = *$, we obtain a CW-prespectrum E'. Now, by 1.19, we can construct a spectral substitute F of E' with $F^{(n)} = *$. □

4.3. Proposition. *Let h_* be an additive homology theory on $\mathscr{S}$ such that $h_i(S) = 0$ for $i \leq m$. Let X be a spectrum such that $\pi_j(X) = 0$ for $j \leq n$. Then $h_i(X) = 0$ for $i \leq m+n+1$.*

Proof. Firstly, we prove by induction that $h_i(X^{(k)}) = 0$ for $i \leq m+n+1$ and every k. By 4.2, we can assume that $X^{(n)} = *$, and so $h_i(X^{(n)}) = 0$ for every i. Fix $k \geq n$ and suppose by induction that $h_i(X^{(k)}) = 0$ for $i \leq m+n+1$. Then the exactness of the sequence

$$h_i(X^{(k)}) \to h_i(X^{(k+1)}) \to \oplus h_i(S^{k+1})$$

(induced by the cofiber sequence $X^{(k)} \subset X^{(k+1)} \to \vee S^{k+1}$) implies that $h_i(X^{(k+1)}) = 0$ for $i \leq m+n+1$.

Now, every finite subspectrum of X is contained in some $X^{(k)}$, and so, by 3.20(ii), $h_i(X) = \varinjlim_k \{h_i(X^{(k)})\}$. Thus, $h_i(X) = 0$ for $i \leq m+n+1$. □

4.4. Definition. (a) A spectrum (space) E is called *n-connected* if $\pi_i(E) = 0$ for $i \leq n$. A spectrum is called *connected* if it is (-1)-connected. A spectrum is called *bounded below* if it is n-connected for some $n \in \mathbb{Z}$.

(b) A morphism $\varphi : E \to F$ of spectra (resp. map of spaces) is called *n-connected*, or an *n-equivalence*, if its cone $C\varphi$ is $(n+1)$-connected. In other words, $\varphi_* : \pi_i(E) \to \pi_i(F)$ is an isomorphism for $i \leq n$ and an epimorphism for $i = n+1$.

4.5. Proposition. (i) *If E is m-connected and F is n-connected, then $E \wedge F$ is $(m+n+1)$-connected.*

(ii) *If E is m-connected and $\varphi : F \to G$ is an n-equivalence, then $1_E \wedge \varphi : E \wedge F \to E \wedge G$ is an $(m+n+1)$-equivalence.*

(iii) *Given integers N, k, let $f : E \to F$ be a map of spectra such that $f_n : E_n \to F_n$ is an $(n+k)$-equivalence for every $n > N$. Then f is a k-equivalence.*

(iv) *For every spectrum E and every k the inclusion $E^{(k)} \subset E$ is a k-equivalence.*

Proof. (i) By 4.3, $E_i(F) = 0$ for $i \leq m+n+1$, i.e., $\pi_i(E \wedge F) = 0$ for $i \leq m+n+1$.

(ii) By (i), $C(1_E \wedge \varphi) = E \wedge C\varphi$ is $(m+n+2)$-connected since $C\varphi$ is $(n+1)$-connected.

(iii) The homomorphism $f_* : \pi_i(E) \to \pi_i(F)$ has the form

$$\pi_i(E) = \lim_{n\to\infty} \pi_{i+n}(E_n) \to \lim_{n\to\infty} \pi_{i+n}(F_n) = \pi_i(F).$$

(iv) By I.3.41, for every space X the inclusion $X^{(k)} \subset X$ is a k-equivalence. Now the result follows from (iii) and 1.5(i). □

4.6. Corollary. *Let $\alpha : S \to E$ be a 0-equivalence, and let X be a spectrum. If $\pi_i(X) = 0$ for $i < n$, then $E_i(X) = 0$ for $i < n$ and $\alpha_* : \pi_k(X) \to E_k(X)$ is an isomorphism for $k = n$ and is an epimorphism for $k = n+1$.*

Proof. The morphism $\alpha \wedge 1_X : S \wedge X \to E \wedge X$ is an n-equivalence. □

For every spectrum X we have the Hurewicz homomorphism $h = \iota_* : \pi_*(X) \to H_*(X)$, where $\iota : S \to H\mathbb{Z}$ yields the unit $1 \in \mathbb{Z} = \pi_0(H\mathbb{Z})$.

4.7. Corollary. (i) *Let X be a spectrum with $\pi_i(X) = 0$ for $i < n$. Then $H_i(X) = 0$ for $i < n$, and the Hurewicz homomorphism $h : \pi_k(X) \to H_k(X)$ is an isomorphism for $k = n$ and an epimorphism for $k = n+1$.*

(ii) *If X is a spectrum bounded below and such that $H_i(X) = 0$ for $i < n$, then $\pi_i(X) = 0$ for $i < n$.*

(iii) *Let E, F be two spectra bounded below. If $\varphi : E \to F$ is a morphism such that $\varphi_* : H_*(E) \to H_*(F)$ is an isomorphism, then φ is an equivalence.*

Proof. (i) This follows from 4.6, because the morphism $\iota : S \to H\mathbb{Z}$ is a 0-equivalence.

(ii) This follows from (i).

(iii) The cone $C\varphi$ is bounded below, and $H_*(C\varphi) = 0$. So, by (i), $\pi_*(C\varphi) = 0$. Thus, $\varphi_* : \pi_*(E) \to \pi_*(X)$ is an isomorphism. □

4.8. Remark. The boundedness below in 4.7(ii) is essential. Indeed, given a prime p and a natural number n, consider the spectrum $K(n)$ of the corresponding Morava K-theory, see Ch. IX, §7. By IX.7.27, we have $H_*(K(n)) = 0$, while $\pi_*(K(n)) \neq 0$.

4.9. Theorem (the Universal Coefficient Theorem). *For every spectrum E and every abelian group G, there are exact sequences*

$$0 \to \operatorname{Ext}(H_{n-1}(E), G) \to H^n(E; G) \to \operatorname{Hom}(H_n(E), G) \to 0$$

and

$$0 \to H_n(E) \otimes G \to H_n(E; G) \to \operatorname{Tor}(H_{n-1}(E), G) \to 0.$$

In particular, $H_0(H(A); B) \cong A \otimes B$, $H^0(H(A); B) \cong \operatorname{Hom}(A, B)$.

Proof. We prove the first formula in detail and indicate a proof of the second one. Given a morphism $\varphi : E \to \Sigma^n HG$, consider the homomorphism

$$\varphi_* : H_n(E) \to H_n(\Sigma^n HG) = H_0(HG) = G$$

(the last equality holds by 4.7(i)). In this way we get a homomorphism

$$\operatorname{ev} : H^n(E; G) \to \operatorname{Hom}(H_n(E), G), \quad \operatorname{ev}(\varphi) = \varphi_*.$$

Firstly, if G is an injective group I, then $\operatorname{Hom}(H_*(-), I)$ is an exact functor, and so $\operatorname{Hom}(H_*(X), I)$ is an additive cohomology theory on $\mathscr{S}$. Thus, by 3.19(iii), ev is an isomorphism for every $X \in \mathscr{S}$ because it is for $X = S$.

Given an arbitrary G, there is an exact sequence $0 \to G \to I \xrightarrow{f} J \to 0$ with injective I, J. It yields an exact sequence

$$\cdots \xrightarrow{f_{n-1}} H^{n-1}(E; J) \to H^n(E; G) \to H^n(E; I) \xrightarrow{f_n} H^n(E; J) \to \cdots,$$

i.e., the exact sequence

$$0 \to \operatorname{Coker} f_{n-1} \to H^n(E; G) \to \operatorname{Ker} f_n \to 0.$$

On the other hand, for every m, there is an exact sequence

$$0 \to \operatorname{Hom}(H_m(E), G) \to \operatorname{Hom}(H_m(E), I) \xrightarrow{f_\#} \operatorname{Hom}(H_m(E), J) \\ \to \operatorname{Ext}(H_m(E), G) \to 0,$$

and we have the commutative diagram

$$\begin{array}{ccc} H^m(E;I) & \xrightarrow{f_m} & H^m(E;J) \\ {\scriptstyle \mathrm{ev}}\big\downarrow & & \big\downarrow{\scriptstyle \mathrm{ev}} \\ \mathrm{Hom}(H_m(E),I) & \xrightarrow{f_\#} & \mathrm{Hom}(H_m(E),J) \end{array}$$

where every homomorphism ev is an isomorphism. Thus,

$$\mathrm{Ker}\, f_n = \mathrm{Hom}(H_n(E),G),\ \ \mathrm{Coker}\, f_{n-1} = \mathrm{Ext}(H_{n-1}(E),G),$$

and we get the desired formula.

We prove the second formula. By the above, we have

$$\begin{aligned} H^0(\mathbb{Z}\wedge HG;G) &= \mathrm{Hom}(H_0(H\mathbb{Z}\wedge HG);G) = \mathrm{Hom}(\pi_0(H\mathbb{Z}\wedge HG);G) \\ &= \mathrm{Hom}(H_0(HG),G) = \mathrm{Hom}(\pi_0(HG),G) = \mathrm{Hom}(G,G). \end{aligned}$$

In particular, the unit $1_G \in \mathrm{Hom}(G,G)$ corresponds to an element $m \in H^0(H\mathbb{Z}\wedge HG;G)$, i.e., to a morphism $m : H\mathbb{Z}\wedge HG \to HG$. Given two morphisms $f: S^n \to E\wedge H\mathbb{Z}$ and $g : S^0 \to HG$, consider the morphism

$$m(f,g) : S^n \xrightarrow{f\wedge g} E\wedge H\mathbb{Z}\wedge HG \xrightarrow{m} E\wedge HG.$$

We define a natural homomorphism

$$\varphi : H_*(E)\otimes G \to H_*(E;G), \varphi([f]\otimes[g]) = [m(f,g)], \quad E\in\mathscr{S}.$$

If G is a flat (e.g., free) abelian group then there is an additive homology theory in the domain, and thus, by 3.19(iii), φ is an isomorphism. Given an arbitrary G, there is an exact sequence $0 \to R \to F \to G \to 0$ with free abelian R, F, and the proof can be completed as in the previous case. □

4.10. Proposition. *For every ring R the spectrum HR admits a ring structure $\mu : HR\wedge HR \to HR$, $\iota : S \to R$ such that $\mu^{\mathrm{pt,pt}} : H^0(\mathrm{pt})\otimes H^0(\mathrm{pt}) \to H^0(\mathrm{pt})$ coincides with the multiplication $R\otimes R\to R$.*

Proof. Since HR and $HR\wedge HR$ are connected,

$$\begin{aligned} H_0(HR\wedge HR) &= \pi_0(HR\wedge HR) = H_0(HR;R) = H_0(HR)\otimes R \\ &= \pi_0(HR)\otimes R = R\otimes R. \end{aligned}$$

Hence, by 4.9,

$$H^0(HR\wedge HR;R) = \mathrm{Hom}(H_0(HR\wedge HR),R) = \mathrm{Hom}(R\otimes R,R).$$

Thus, the multiplication $R\otimes R\to R$ yields a pairing $\mu : HR\wedge HR \to HR$. Furthermore, the unit $1\in R = \pi_0(R)$ yields a morphism $\iota : S\to R$, and it is easy to see that (HR,μ,ι) is a ring spectrum. □

In particular, for every pair of spectra E, F we have the homomorphisms

$$\mu_{E,F} : H_*(E;R) \otimes H_*(F;R) \to H_*(E \wedge F;R)$$

and

$$\mu^{E,F} : H^*(E;R) \otimes H^*(F;R) \to H^*(E \wedge F;R).$$

4.11. Theorem (the Künneth Theorem). *Let k be a field, and let E, F be a pair of spectra.*

(i) *The homomorphism*

$$\mu_{E,F} : H_*(E;k) \otimes H_*(F;k) \to H_*(E \wedge F;k)$$

is an isomorphism. In particular,

$$H_n(E \wedge F;k) \cong \sum_{i+j=n} H_i(E;k) \otimes_k H_j(F;k).$$

(ii) *Assume that E is bounded below and F has finite type. Then the homomorphism*

$$\mu^{E,F} : H^*(E;k) \otimes H^*(F;k) \to H^*(E \wedge F;k)$$

is an isomorphism. In particular,

$$H^n(E \wedge F;k) \cong \sum_{i+j=n} H^i(E;k) \otimes_k H^j(F;k).$$

Proof. (i) Fixing E and considering F as indeterminate, we see that $\mu_{E,F}$ is a morphism of additive homology theories on $\mathscr{S}$, and so, by 3.19(iii), it is an isomorphism.

(ii) Below $H^*(-)$ denotes $H^*(-;k)$. Let $i_m : F^{(m)} \to F$ be the inclusion of the skeleton. Clearly, for every j there exists N such that $i_N^* : H^j(F) \to H^j(F^{(N)})$ is an isomorphism. Furthermore, since E is bounded below, for every n there existts N such that the morphism $1_E \wedge i_N : E \wedge F^{(N)} \to E \wedge F$ is an n-equivalence, see 4.5, and so $(1_E \wedge i_N)^* : H^n(E \wedge F) \to H^n(E \wedge F^{(N)})$ is an isomorphism. Now, since every skeleton of F is finite, it suffices to prove the theorem for finite spectra F. Now, fixing E and considering F as indeterminate, we see that $\mu^{E,F}$ is a morphism of homology theories on $\mathscr{S}_f$, and so, by 3.19(i), it is an isomorphism for every finite F. □

We need 4.11 only, but, of course, there is a Künneth exact sequence

$$0 \to \sum_{i+j=n} H_i(E;R) \otimes_R H_j(F;R) \to H_n(E \wedge F;R)$$
$$\to \sum_{i+j=n-1} \operatorname{Tor}^R(H_i(E;R), H_j(F;R)) \to 0$$

for every pair of spectra E, F and every ring R of homological dimension 1 (e.g., for a principal ideal domain R), see e.g. Margolis [1]. Moreover, under suitable conditions there is a spectral sequence

$$\mathrm{Tor}^R_{*,*}(H_*(E;R), H_*(F;R)) \Rightarrow H_*(E \wedge F;R)$$

for every ring R, see e.g. Adams [5].

4.12. Definition. A *Postnikov tower* of a spectrum E is a homotopy commutative diagram of spectra

$$\begin{array}{ccccccccc} \cdots & = & E & = & E & = & E & = & \cdots \\ & & \downarrow{\scriptstyle \tau_{n+1}} & & \downarrow{\scriptstyle \tau_n} & & \downarrow{\scriptstyle \tau_{n-1}} & & \\ \cdots & \longrightarrow & E_{(n+1)} & \xrightarrow{p_{n+1}} & E_{(n)} & \xrightarrow{p_n} & E_{(n-1)} & \longrightarrow & \cdots, \end{array}$$

where for every n we have:

(i) $\pi_i(E_{(n)}) = 0$ for $i > n$,

(ii) $(\tau_n)_* : \pi_i(E) \to \pi_i(E_{(n)})$ is an isomorphism for $i \le n$.

The spectrum $E_{(n)}$ is called the *n-coskeleton*, or the *Postnikov n-stage*, of E. We prove below that $E_{(n)}$ is uniquely determined by E up to equivalence.

4.13. Theorem. *Every spectrum E has a Postnikov tower.*

Proof. Step 1. Fix an integer n. We construct a sequence

$$E(n,0) \subset \cdots \subset E(n,i) \subset E(n,i+1) \subset \cdots$$

of spectra with the following properties:

(1) $E(n,0) = E$;

(2) $\pi_k(E(n,i)) = 0$ for $n < k \le n+i$;

(3) The homomorphism $\pi_k(E(n,i)) \to \pi_k(E(n,i+1))$ induced by the inclusion

$$E(n,i) \subset E(n,i+1)$$

is an isomorphism for $k \le n$.

We do this by induction on i. The case $i = 0$ is clear. Suppose that there is a finite sequence

$$E(n,0) \subset \cdots \subset E(n,i);$$

we construct the required inclusion $E(n,i) \subset E(n,i+1)$. Let $\{x_\lambda\}$ be a family of generators of $\pi_{n+i+1}(E(n,i))$, and let x_λ be represented by a map $f_\lambda : S^{n+i+1}_\lambda \to E(n,i)$. Consider the morphism

$$f : \vee S^{n+i+1}_\lambda \to E, f|S^{n+i+1}_\lambda = f_\lambda,$$

and let $E(n,i) \subset E(n,i+1)$ be the canonical inclusion in the cofiber sequence

$$\vee S_\lambda^{n+i+1} \xrightarrow{f} E(n,i) \subset E(n,i+1),$$

i.e., $E(n,i+1) = Cf$. Since $f_* : \pi_{n+i+1}(\vee S_\lambda^{n+i+1}) \to \pi_{n+i+1}(E)$ is epic, condition (2) holds, while (1) and (3) hold obviously. The induction is confirmed.

Step 2. We set $E_{(n)} := \cup_{i=0}^\infty E(n,i)$ and define $\tau_n : E = E(n,0) \to E_{(n)}$ to be the inclusion. It is clear that 4.12(i) and 4.12(ii) hold.

Step 3. We construct $p_n : E_{(n)} \to E_{(n-1)}$ such that $p_n \tau_n \simeq \tau_{n-1}$. Consider the commutative diagram

$$\begin{array}{ccccc} F & \xrightarrow{q} & E & \xrightarrow{\tau_n} & E_{(n)} \\ \| & & \| & & \\ F & \xrightarrow{q} & E & \xrightarrow{\tau_{n-1}} & E_{(n-1)} \end{array}$$

where the top line is a cofiber sequence. The exactness of the sequence

$$\pi_*(F) \xrightarrow{q_*} \pi_*(E) \xrightarrow{(\tau_n)_*} \pi_*(E_{(n)})$$

implies that $\pi_i(F) = 0$ for $i \le n$. Hence, by 4.1(iv), $[F, E_{(n-1)}] = 0$. In particular, $\tau_{n-1}q$ is inessential, and so there is $p_n : E_{(n)} \to E_{(n-1)}$ with $p_n \tau_n \simeq \tau_{n-1}$. □

4.14. Definition. A morphism $q = q_n^E : F \to E$ is called an $(n-1)$-*connective covering* of a spectrum E if $\pi_i(F) = 0$ for $i < n$ and $q_* : \pi_i(F) \to \pi_i(E)$ is an isomorphism for $i \ge n$. A *connective covering* is a (-1)-connective covering.

Every spectrum F as in 4.14 is called an $(n-1)$-*killing spectrum of* E and denoted by $E|n$. We prove below that $E|n$ is uniquely determined by E up to equivalence.

4.15. Theorem. *For every spectrum E and every n there exists an n-connective covering.*

Proof. (In fact, it was already proved in 4.13.) Consider the sequence

$$\pi_*(F) \xrightarrow{q_*} \pi_*(E) \xrightarrow{(\tau_n)_*} \pi_*(E_{(n)})$$

induced by a cofiber sequence $F \xrightarrow{q} E \xrightarrow{\tau_n} E_{(n)}$. It is clear that $\pi_i(F) = 0$ for $i \le n$ and that $q_* : \pi_k(F) \to \pi_k(E)$ is an isomorphism for every $i > n$, i.e., $q : F \to E$ is an n-connective covering. □

So, we have a cofiber sequence $E|(n+1) \to E \to E_{(n)}$.

4.16. Theorem. *Let $q = q_n^F : F|n \to F$ be an $(n-1)$-connective covering of a spectrum F, and let $\theta : D \to F$ be a morphism from an $(n-1)$-connected spectrum D. Then there exists $\hat{\theta} : D \to F|n$ with $q\hat{\theta} = \theta$, and $\hat{\theta}$ is unique up* *to homotopy. Moreover, if* $\theta \simeq \theta'$ *then* $\hat{\theta} \simeq \hat{\theta}'$. *Finally, for every morphism* $\varphi : E \to F$ *there exists a morphism* $\varphi|n : E|n \to F|n$ *such that the diagram*

$$\begin{array}{ccc} E|n & \xrightarrow{\varphi|n} & F|n \\ {\scriptstyle q_n^E}\downarrow & & \downarrow{\scriptstyle q_n^F} \\ E & \xrightarrow{\varphi} & F \end{array}$$

commutes up to homotopy, and such a morphism $\varphi|n$ *is unique up to homotopy. In particular, every two* $(n-1)$*-killing spectra of* E *are equivalent.*

Proof. By 4.1(iv), $[D, \Sigma^{-1}F_{(n-1)}] = 0 = [D, F_{(n-1)}]$. Now, the exactness of the sequence

$$[D, \Sigma^{-1}F_{(n-1)}] \to [D, F|n] \to [D, F] \to [D, F_{(n-1)}]$$

implies the existence and the uniqueness of $\hat{\theta}$ and the assertion about homotopy. To prove the second assertion, put $D = E|n, \theta = \varphi q_n^E$, and set $\varphi|n = \hat{\theta}$. To prove the uniqueness of $E|n$, put $\varphi = 1_E$. □

4.17. Theorem. *Let* $\theta : D \to F$ *be a morphism of spectra, where* $\pi_i(F) = 0$ *for* $i > n$. *Then there exists* $\bar{\theta} : D_{(n)} \to F$ *with* $\bar{\theta}\tau_n^D = \theta$, *and* $\bar{\theta}$ *is unique up to homotopy. Moreover, if* $\theta \simeq \theta'$ *then* $\bar{\theta} \simeq \bar{\theta}'$. *Finally, for every morphism* $\varphi : E \to F$ *there exists a morphism* $\varphi_{(n)} : E_{(n)} \to F_{(n)}$ *such that the diagram*

$$\begin{array}{ccc} E & \xrightarrow{\varphi} & F \\ {\scriptstyle \tau_n^E}\downarrow & & {\scriptstyle \tau_n^F}\downarrow \\ E_{(n)} & \xrightarrow{\varphi_{(n)}} & F_{(n)} \end{array}$$

commutes up to homotopy, and such a morphism $\varphi_{(n)}$ *is unique up to homotopy. In particular, every two* n*-coskeletons of a spectrum are equivalent.*

Proof. By 4.1(iv), $[\Sigma(D|n+1), F] = 0 = [D|n+1, F]$, so, the exactness of the sequence

$$[\Sigma(D|n+1), F] \to [D_{(n)}, F] \xrightarrow{(\tau_n^D)^*} [D, F] \to [D|n+1, F]$$

implies the existence and the uniqueness of $\bar{\theta}$. Now the proof can be completed as in 4.16. □

4.18. Corollary (naturality and uniqueness of Postnikov towers). *Let* $\varphi : E \to F$ *be a morphism of spectra, and let*

$$\begin{array}{ccccccccc} \cdots & = & E & = & E & = & E & = & \cdots \\ & & \downarrow & & \downarrow{\scriptstyle \tau_n} & & \downarrow & & \\ \cdots & \longrightarrow & E_{(n+1)} & \longrightarrow & E_{(n)} & \xrightarrow{p_n} & E_{(n-1)} & \longrightarrow & \cdots \end{array}$$

and

$$\begin{array}{ccccccccc} \cdots & = & F & = & F & = & F & = & \cdots \\ & & \downarrow & & \downarrow{\scriptstyle \sigma_n} & & \downarrow & & \\ \cdots & \longrightarrow & F_{(n+1)} & \longrightarrow & F_{(n)} & \xrightarrow{q_n} & F_{(n-1)} & \longrightarrow & \cdots \end{array}$$

be Postnikov towers of E and F. Then there exist morphisms $\varphi_{(n)} : E_{(n)} \to F_{(n)}$ such that the diagrams

$$\begin{array}{ccc} E & \xrightarrow{\varphi} & F \\ {\scriptstyle \tau_n}\downarrow & & \downarrow{\scriptstyle \sigma_n} \\ E_{(n)} & \xrightarrow{\varphi_{(n)}} & F_{(n)} \end{array} \qquad \begin{array}{ccc} E_{(n)} & \xrightarrow{\varphi_{(n)}} & F_{(n)} \\ {\scriptstyle p_n}\downarrow & & \downarrow{\scriptstyle q_n} \\ E_{(n-1)} & \xrightarrow{\varphi_{(n-1)}} & F_{(n-1)} \end{array}$$

commute up to homotopy.

Proof. By 4.17, there exists $\varphi_{(n)}$ such that the left diagram commutes, and this $\varphi_{(n)}$ is unique up to homotopy. Since the Postnikov tower of $E_{(n)}$ is a segment of the Postnikov tower of E, one can find $\psi_{(n-1)} : E_{(n-1)} \to F_{(n-1)}$ with $\psi_{(n-1)}p_n = q_n\varphi_{(n)}$. But then $\psi_{(n-1)}\tau_{n-1} = \sigma_{n-1}\varphi$, and hence, again by 4.17, $\psi_{(n-1)} \simeq \varphi_{(n-1)}$. □

Consider a Postnikov tower of a spectrum E. It is easy to see that the cone of the morphism $p_n : E_{(n)} \to E_{(n-1)}$ is the graded Eilenberg–Mac Lane spectrum $\Sigma^{n+1}H(\pi_n(E))$. So, we have a cofiber sequence

$$p_n : E_{(n)} \xrightarrow{p_n} E_{(n-1)} \xrightarrow{\kappa_n} \Sigma^{n+1}H(\pi_n(E)).$$

4.19. Definition. The element $\kappa_n \in H^{n+1}(E_{(n-1)}; \pi_n(E))$ is called the *n-th Postnikov invariant* of E.

To be precise, the morphism κ_n is defined up to self-equivalence of the spectra $\Sigma^{n+1}H(\pi_n(E))$ and $E_{(n-1)}$, i.e., the real invariant is the corresponding orbit in $H^{n+1}(E_{(n-1)}; \pi_n(E))$. However, the above terminology is commonly accepted and does not lead to confusion.

4.20. Proposition. *The Postnikov invariant κ_n is trivial iff p_n admits a homotopy right inverse morphism*

$$s : E_{(n-1)} \to E_{(n)}, \quad p_n s \simeq 1_{E_{(n-1)}}.$$

Furthermore, in this case $E_{(n)} \simeq E_{(n-1)} \vee \Sigma^n H(\pi_n(E))$.

Proof. This follows from 1.17. □

Now we apply the Serre class theory (see Serre [1], Mosher–Tangora [1]) to spectra. Cf. also Margolis [1].

4.21. Definition. (a) A *Serre class* is a family of abelian groups $\mathfrak{C}$ satisfying the following axiom: If

$$0 \to A' \to A \to A'' \to 0$$

is a short exact sequence, then A is in $\mathfrak{C}$ iff both A' and A'' are in $\mathfrak{C}$.

(b) Let $H(A)$ denote the Eilenberg–Mac Lane spectrum of an abelian group A. A Serre class is called *stable* if it satisfies the following axiom: If $A \in \mathfrak{C}$, then $H_i(H(A)) \in \mathfrak{C}$ for every i.

(c) A homomorphism $f : A \to B$ of abelian groups is called a $\mathfrak{C}$-*monomorphism* if $\operatorname{Ker} f \in \mathfrak{C}$, and f is called a $\mathfrak{C}$-*epimorphism* if $\operatorname{Coker} f \in \mathfrak{C}$. Furthermore, f is called a $\mathfrak{C}$-*isomorphism* if $\operatorname{Ker} f \in \mathfrak{C}$ and $\operatorname{Coker} f \in \mathfrak{C}$.

Notice that if $\mathfrak{C}$ is a Serre class and $A \to B \to C$ is an exact sequence of abelian groups then $B \in \mathfrak{C}$ provided $A, C \in \mathfrak{C}$.

Recall that the Five Lemma mod $\mathfrak{C}$ holds. This means that the Five Lemma remains valid if we replace the words "monomorphism, epimorphism, isomorphism" by the words "$\mathfrak{C}$-monomorphism, $\mathfrak{C}$-epimorphism, $\mathfrak{C}$-isomorphism".

4.22. Proposition. *Let* $\mathfrak{C}$ *be a Serre class with the following properties:*

(i) *If* $A, B \in \mathfrak{C}$ *then* $A \otimes B \in \mathfrak{C}$ *and* $\operatorname{Tor}(A, B) \in \mathfrak{C}$;

(ii) *If* $A \in \mathfrak{C}$ *then* $H_i(K(A, 1)) \in \mathfrak{C}$ *for every* $i > 0$.

Then $\mathfrak{C}$ *is a stable Serre class.*

Here $K(A, 1)$ is the Eilenberg–Mac Lane space, $\pi_1(K(A, 1)) = A$ and $\pi_i(K(A, 1)) = 0$ for $i > 1$.

Proof. Let $\mathfrak{C}$ be a class in question. Serre [1] proved the following theorem (the so-called Hurewicz Theorem mod $\mathfrak{C}$): Given a simply connected space X, suppose that $\pi_i(X) \in \mathfrak{C}$ for every i. Then $H_i(X) \in \mathfrak{C}$ for every $i > 0$.

Now, let $A \in \mathfrak{C}$. Then $\pi_i(K(A, n)) \in \mathfrak{C}$ for every i and every $n > 1$, and so $H_i(K(A, n)) \in \mathfrak{C}$ for every $i > 0$ and every $n > 0$. It remains to note that $H_i(H(A)) = H_{i+N}(K(A, N))$ for N large enough. □

4.23. Proposition. (i) *The class of all finite abelian groups is a stable Serre class.*

(ii) *The class of all finitely generated abelian groups is a stable Serre class.*

(iii) *Given a prime* p, *let* $\mathfrak{C}$ *be the class of all abelian groups having* p-*primary exponents (i.e., for every* $A \in \mathfrak{C}$ *there exists* k *such that* $p^k A = 0$). *Then* $\mathfrak{C}$ *is a stable Serre class.*

(iv) *Given a prime p, the class of all finite p-primary abelian groups is a stable Serre class.*

Proof. (i) It is well known that $H_i(K(\mathbb{Z}/m,1))$ is $\mathbb{Z}/m$ for i odd and 0 for i even, $i > 0$, see e.g. Mac Lane [2]. Hence, because of the Künneth Theorem, $H_i(K(A,1)), i > 0$, is a finite abelian group if A is. Now apply 4.22.

(ii) This can be proved similarly to (i), with the additional remark that $K(\mathbb{Z},1)$ is the circle S^1.

(iii) It suffices to prove that $H_i(H(A)) \in \mathfrak{C}$ whenever $A \in \mathfrak{C}$. By 4.9, $[H(A), H(A)] = \mathrm{Hom}(A, A)$ for every abelian group A. Given $m \in \mathbb{Z}$, let $m : H(A) \to H(A)$ be a morphism which corresponds to the element $m1_A \in \mathrm{Hom}(A, A)$. It is easy to see that, for every i,

$$m_* : H_i(H(A)) \to H_i(H(A))$$

is the multiplication by m. Now, let $A \in \mathfrak{C}$ have exponent p^k. Then the morphism $p^k : H(A) \to H(A)$ is inessential, and so $(p^k)_* : H_*(H(A)) \to H_*(H(A))$ is the zero homomorphism. Thus, for every i, $p^k H_i(H(A)) = 0$, i.e., $H_i(H(A)) \in \mathfrak{C}$.

(iv) This class is the intersection of the classes from (i) and (iii). □

4.24. Theorem. *Let $\mathfrak{C}$ be a stable Serre class. Let E be a spectrum bounded below such that $\pi_i(E) \in \mathfrak{C}$ for $i < n$. Then $H_i(E) \in \mathfrak{C}$ for $i < n$, and the Hurewicz homomorphism $h : \pi_n(E) \to H_n(E)$ is a $\mathfrak{C}$-isomorphism. In particular, $\pi_i(E) \in \mathfrak{C}$ for every i iff $H_i(E) \in \mathfrak{C}$ for every i provided E is bounded below.*

Proof. Let π_k denote $\pi_k(E)$, and let E_n denote $E_{(n)}$. Fix any m such that $\pi_i = 0$ for $i < m$. We can assume that $m < n$, and so $H_k(E_m) = H_k(H(\pi_m)) \in \mathfrak{C}$ for all k. For every s, there is a cofiber sequence $\Sigma^s H(\pi_s) \to E_s \to E_{s-1}$. Now, using the exactness of the sequence

$$\cdots \to H_k(\Sigma^s H(\pi_s)) \to H_k(E_s) \to H_k(E_{s-1}) \to \cdots,$$

one can prove by induction on s (starting with $s = m$) that $H_k(E_s) \in \mathfrak{C}$ for $s < n$ and every k. Furthermore, for $i < n$ we have an exact sequence

$$0 = H_i(\Sigma^n H(\pi_n)) \to H_i(E_n) \to H_i(E_{n-1}) \to H_{i-1}(\Sigma^n H(\pi_n)) = 0$$

(the first and the last groups are trivial by 4.7(i)). So, $H_i(E_n) \in \mathfrak{C}$ for $i < n$. Finally, for $i < n$ the exactness of the sequence

$$0 = H_i(E|(n+1)) \to H_i(E) \to H_i(E_n)$$

implies that $H_i(E) \in \mathfrak{C}$ for $i < n$.

We prove that $h : \pi_n(E) \to H_n(E)$ is a $\mathfrak{C}$-isomorphism. Notice that $\pi_i(E|(n+1)) = 0 = H_i(E|(n+1))$ for $i \leq n$, and so we have the commutative diagram

$$\begin{array}{ccc} \pi_n(E) & \xrightarrow{\cong} & \pi_n(E_n) \\ {\scriptstyle h}\downarrow & & \downarrow{\scriptstyle h'} \\ H_n(E) & \xrightarrow{\cong} & H_n(E_n) \end{array}$$

where the horizontal arrows are isomorphisms. Hence, it suffices to prove that $h' : \pi_n(E_n) \to H_n(E_n)$ is a $\mathfrak{C}$-isomorphism. The cofiber sequence $\Sigma^n H(\pi_n) \to E_n \to E_{n-1}$ induces the commutative diagram with exact rows (where the vertical arrows are the Hurewicz homomorphisms):

$$\begin{array}{ccccccccc} 0 & \longrightarrow & \pi_n & \to & \pi_n(E_n) & \to & 0 & \longrightarrow & 0 \\ {\scriptstyle \alpha_1}\downarrow & & {\scriptstyle \alpha_2}\downarrow & & {\scriptstyle \alpha_3}\downarrow & & {\scriptstyle \alpha_4}\downarrow & & {\scriptstyle \alpha_5}\downarrow \\ H_{n+1}(E_{n-1}) & \longrightarrow & H_n(\Sigma^n H(\pi_n)) & \to & H_n(E_n) & \to & H_n(E_{n-1}) & \longrightarrow & 0. \end{array}$$

Here α_2 and α_5 are isomorphisms. Furthermore, since $H_*(E_{n-1}) \in \mathfrak{C}$, we conclude that α_1 and α_4 are $\mathfrak{C}$-isomorphisms. Thus, α_3 is a $\mathfrak{C}$-isomorphism because of the Five Lemma mod $\mathfrak{C}$. □

4.25. Proposition. (i) *Let $\mathfrak{C}$ be a Serre class of abelian groups. If E is a spectrum such that $\pi_i(E) \in \mathfrak{C}$ for every i then $E_i(X) \in \mathfrak{C}$ and $E^i(X) \in \mathfrak{C}$ for every finite spectrum X and all i.*

(ii) *Let E be a spectrum such that all the groups $\pi_i(E)$ are finite. Then the groups $E^i(X)$ and $E_i(X)$ are finite for every finite spectrum X and all i.*

(iii) *Let R be a commutative Noetherian ring, and let E be a spectrum such that $E_i(X)$ and $E^i(X)$ are natural in X R-modules. Suppose that each group $\pi_i(E)$ is a finitely generated R-module. Then $E_i(X)$, as well as $E^i(X)$, is a finitely generated R-module for every finite spectrum X and all i.*

Proof. (i) By duality, it suffices to prove only that $E_i(X) \in \mathfrak{C}$ for every finite spectrum X. Now, because of 1.5(iii), it suffices to prove that $\widetilde{E}_i(X) \in \mathfrak{C}$ for every finite CW-complex X. Clearly, this holds for $X = \text{pt}$. Suppose by induction that $\widetilde{E}_i(X) \in \mathfrak{C}$ whenever X has $\leq n$ cells. Consider a CW-complex Y which has $n+1$ cells. Then $Y = X \cup e_k$ where X has n cells and e_k is an attached cell. We have the exact sequence

$$\widetilde{E}_i(X) \to \widetilde{E}_i(Y) \to \widetilde{E}_i(S^k)$$

where the outside groups are in $\mathfrak{C}$. Thus, $\widetilde{E}_i(X) \in \mathfrak{C}$. The induction is confirmed.

(ii) This follows from (i) and 4.23(i).

(iii) Let $A \to B \to C$ be an exact sequence of R-modules. Since R is Noetherian, B is finitely generated over R provided A and C are. Now the proof can be completed similarly to (i). □

4.26. Proposition. (i) *If a spectrum X has finite type then all the groups $\pi_i(X)$ are finitely generated.*

(ii) *If X is a spectrum bounded below and such that all the groups $\pi_i(X)$ are finitely generated then X is equivalent to a spectrum of finite type.*

Proof. (i) Since $\pi_i(X) = \pi_i(X^{(N)})$ for $N > i$, it suffices to prove the proposition for finite spectra X. But this follows from 4.25(i), since, by the Serre theorem, all the groups $\pi_i(S)$ are finitely generated (and even finite for $i > 0$, see Serre [1], Mosher–Tangora [1]).

(ii) We construct a sequence $\cdots \subset Y(0) \subset \cdots \subset Y(n) \subset Y(n+1) \subset \cdots$ of spectra and morphisms $f_n : Y(n) \to X$ with the following properties:

(1) Each spectrum $Y(n)$ is finite;

(2) Each morphism $f_n : Y(n) \to X$ is an n-equivalence;

(3) $f_{n+1}|Y(n) = f_n$.

Since X is bounded below, there is k such that $\pi_i(X) = 0$ for $i \leq k$. We put $Y(k-1) = *$. Now, suppose by induction that we have constructed a finite spectrum $Y(n)$ and an n-equivalence $f_n : Y(n) \to X$. Then $(f_n)_* : \pi_{n+1}(Y(n)) \to \pi_{n+1}(X)$ is an epimorphism. Note that, by (i), the group $K_n := \operatorname{Ker}(f_n)_*$ is finitely generated, and choose generators $a_1, \dots, a_m$ of K_n. Let $S_i^{n+1}, i = 1, \dots, m$, be a copy of the spectrum S^{n+1}. Consider a map $g : \vee_{i=1}^m S_i^{n+1} \to Y(n)$ such that $g|S_i^{n+1}$ represents a_i, and set $Z := C(g)$. Then f_n can be extended to a morphism $h : Z \to X$, and $h_* : \pi_i(Z) \to \pi_i(X)$ is an isomorphism for $i \leq n+1$. Now, let $b_1, \dots, b_l$ be generators of $\pi_{n+2}(X)$. We set $Y(n+1) := Z \vee (\vee_{i=1}^l S_i^{n+2})$ and define $f_{n+1} : Y(n+1) \to X$ by requiring $f_{n+1}|Y(n) = f_n$ and $f_{n+1}|S_i^{n+2}$ represents b_i. The induction is confirmed.

Now, we set $Y := \bigcup Y(n)$ and define $f : Y \to X$ by setting $f|Y(n) = f_n$. Clearly, f is an equivalence, and Y has finite type. □

Now we explain how to equip connective coverings and Postnikov towers of ring spectra with ring structures.

4.27. Lemma. *For every pair of spectra E, F and every pair of integers m, n, there exists a morphism $\alpha_{m,n} : E|m \wedge F|n \to (E \wedge F)|(m+n)$ such that the diagram*

$$\begin{array}{ccc} E|m \wedge F|n & \xrightarrow{\alpha_{m,n}} & (E \wedge F)|(m+n) \\ {\scriptstyle q_m \wedge q_n}\downarrow & & \downarrow{\scriptstyle q_{m+n}} \\ E \wedge F & = & E \wedge F \end{array}$$

commutes up to homotopy (here the q's are the connective coverings). Furthermore, such a morphism α is unique up to homotopy.

Proof. By 4.5(i), $E|m \wedge F|n$ is $(m+n-1)$-connected. Now apply 4.16. □

4.28. Theorem. *Fix any $n \leq 0$.*

(i) *If E is a ring spectrum, then $E|n$ admits a ring structure such that the $(n-1)$-connective covering $q: E|n \to E$ is a ring morphism.*

(ii) *Let $\varphi: D \to E$ be a ring morphism of ring spectra. Equip $D|n$ and $E|n$ with any ring structures as in* (i). *Then $\varphi|n: D|n \to E|n$ is a ring morphism. In particular, $E|n$ admits only one (up to ring equivalence) ring structure such that $q: E|n \to E$ is a ring morphism.*

(iii) *Let F be an E-module spectrum with the pairing (module structure) $m: E \wedge F \to F$. Then there exists a pairing $\widetilde{m}: E|n \wedge F|n \to F|n$ turning $F|n$ into a $E|n$-module spectrum such that the diagram*

$$\begin{array}{ccc} E|n \wedge F|n & \xrightarrow{\widetilde{m}} & F|n \\ {\scriptstyle q_n^E \wedge q_n^F}\downarrow & & \downarrow{\scriptstyle q_n^F} \\ E \wedge F & \xrightarrow{m} & F \end{array}$$

commutes up to homotopy, and this morphism $\widetilde{m}$ is unique up to homotopy.

(iv) *If E is a commutative ring spectrum, then so is $E|n$.*

Proof. (i) Firstly, the unit $\iota: S \to E$ admits a q-lifting $\widetilde{\iota} := i|n : S \to E|n$ because $n \leq 0$. Furthermore, let $\mu: E \wedge E \to E$ be the multiplication on E. Consider the following diagram where q_n is the $(n-1)$-connective covering and $\alpha_{n,n}$ as in 4.27:

$$\begin{array}{ccccc} E|n \wedge E|n & \xrightarrow{\beta} & (E \wedge E)|n & \xrightarrow{\mu|n} & E|n \\ \| & & \downarrow{\scriptstyle q_n} & & \downarrow \\ E|n \wedge E|n & \xrightarrow{\alpha_{n,n}} & (E \wedge E)|2n & \xrightarrow{\mu|2n} & E|2n \\ \downarrow & & \downarrow & & \downarrow \\ E \wedge E & = & E \wedge E & \xrightarrow{\mu} & E\ . \end{array}$$

By 4.16, there exists $\beta = \hat{\alpha}_{n,n}$ such that the diagram commutes (up to homotopy). We prove that the pairing $\widetilde{\mu} := (\mu|n) \circ \beta$ is associative. Indeed, the morphisms $\widetilde{\mu} \circ (1 \wedge \widetilde{\mu}) \circ a$ and $\widetilde{\mu} \circ (\widetilde{\mu} \wedge 1): E|n \wedge E|n \wedge E|n \to E|n$ (where $\widetilde{a}: (E|n \wedge E|n) \wedge E|n \to E|n \wedge (E|n \wedge E|n)$ is as in 2.1(ii)) are homotopic because they cover the homotopic morphisms $\mu \circ (1 \wedge \mu) \circ a$ and $\mu \circ (\mu \wedge 1)$ respectively. By the above, $(E|n, \widetilde{\mu}, \widetilde{\iota})$ is a ring spectrum and $q: E|n \to E$ is a ring morphism.

(ii) Let μ^D, μ^E be the multiplications on D, E respectively, let μ', μ'' be the multiplications (as in (i)) on $D|n, E|n$ respectively, and let $q^D: D|n \to D$, $q^E: E|n \to E$ be the $(n-1)$-connective coverings. We must prove that the diagram

$$\begin{array}{ccc} D|n \wedge D|n & \xrightarrow{\varphi|n \wedge \varphi|n} & E|n \wedge E|n \\ {\scriptstyle \mu'}\downarrow & & {\scriptstyle \mu''}\downarrow \\ D|n & \xrightarrow{\varphi|n} & E|n \end{array}$$

commutes up to homotopy. By 4.16, it suffices to prove that $q^E\mu''(\varphi|n\wedge\varphi|n) \simeq q^E(\varphi|n)\mu'$. But

$$q^E\mu''(\varphi|n \wedge \varphi|n) \simeq \mu^E(q^E \wedge q^E)(\varphi|n \wedge \varphi|n) \simeq \mu^E(\varphi \wedge \varphi)(q^D \wedge q^D)$$
$$\simeq \varphi\mu^D(q^D \wedge q^D) \simeq \varphi q^D \mu' \simeq q^E(\varphi|n)\mu'.$$

(iii) This can be proved as (i) was if one considers the diagram

$$\begin{array}{ccccc} E|n \wedge F|n & \xrightarrow{\beta} & (E \wedge F)|n & \xrightarrow{m|n} & F|n \\ \downarrow & & \downarrow & & \downarrow \\ E \wedge F & = & E \wedge F & \xrightarrow{m} & F\,. \end{array}$$

(iv) This holds because the morphism $E|n \wedge E|n \xrightarrow{\tau_{E|n}} E|n \wedge E|n \xrightarrow{\widetilde{\mu}} E|n$ covers the morphism $E\wedge E \xrightarrow{\tau_E} E\wedge E \xrightarrow{\mu} E$. But $\mu\tau_E \simeq \mu$, and so $\widetilde{\mu}\tau_{E|n} \simeq \widetilde{\mu}$. □

4.29. Lemma. *For every pair of connected spectra E, F the morphism*

$$(\tau_n^E \wedge \tau_n^F)_{(n)} : (E \wedge F)_{(n)} \to (E_{(n)} \wedge F_{(n)})_{(n)}$$

is an equivalence.

Proof. By 2.1(vi), we have a cofiber sequence

$$E \wedge (F|(n+1)) \to E \wedge F \xrightarrow{1\wedge\tau_n^F} E \wedge F_{(n)}.$$

By 4.5(i), $E \wedge (F|(n+1))$ is n-connected. So,

$$(1 \wedge \tau_n^F)_* : \pi_i(E \wedge F) \to \pi_i(E \wedge F_{(n)})$$

is an isomorphism for $i \le n$. Similarly,

$$(\tau_n^E \wedge 1)_* : \pi_i(E \wedge F_{(n)}) \to \pi_i(E_{(n)} \wedge F_{(n)})$$

is an isomorphism for $i \le n$. But $(\tau_n^E \wedge \tau_n^F)_{(n)} = ((\tau_n^E \wedge 1)\circ(1 \wedge \tau_n^F))_{(n)}$. □

4.30. Theorem. *Let $E = (E, \mu, \iota)$ be a connected ring spectrum. Fix any $n \ge 0$.*

(i) *$E_{(n)}$ admits a ring structure such that $\tau_n : E \to E_{(n)}$ is a ring morphism.*

(ii) *Let $\varphi : D \to E$ be a ring morphism of ring spectra. Equip $D_{(n)}$ and $E_{(n)}$ with any ring structures as in* (i). *Then $\varphi_{(n)} : D_{(n)} \to E_{(n)}$ is a ring morphism. In particular, $E_{(n)}$ admits only one (up to ring equivalence) ring structure such that $\tau_n : E \to: E_{(n)}$ is a ring morphism.*

(iii) *Let F be an E-module spectrum with the pairing (module structure) $m : E \wedge F \to F$. Then there exists a pairing $\widetilde{m} : E_{(n)} \wedge F_{(n)} \to F_{(n)}$ turning $F_{(n)}$ into a $E_{(n)}$-module spectrum and such that the diagram*

$$\begin{array}{ccc} E_{(n)} \wedge F_{(n)} & \xrightarrow{\widetilde{m}} & F_{(n)} \\ {\scriptstyle \tau_n^E \wedge \tau_n^F}\downarrow & & \downarrow{\scriptstyle \tau_n^F} \\ E \wedge F & \xrightarrow{m} & F \end{array}$$

commutes up to homotopy, and this morphism $\widetilde{m}$ is unique up to homotopy.

(iv) *If E is a commutative ring spectrum then so is $E_{(n)}$.*

Proof. Define $\widetilde{\mu} : E_{(n)} \wedge E_{(n)} \to E_{(n)}$ to be the composition

$$E_{(n)} \wedge E_{(n)} \xrightarrow{\tau_n} (E_{(n)} \wedge E_{(n)})_{(n)} \xrightarrow{h} (E \wedge E)_{(n)} \xrightarrow{\mu} E_{(n)},$$

where h is an equivalence inverse to that of 4.29. Furthermore, define $\widetilde{\iota} := \tau_n \iota : S \to E_{(n)}$. Following 4.28(i), one can prove that $(E_{(n)}, \widetilde{\mu}, \widetilde{\iota})$ is a ring spectrum and that τ_n is a ring morphism. All the other assertions can be proved similarly to those of 4.28. □

4.31. Remark. The connectedness of E in 4.30 cannot be omitted or replaced by the boundedness below of E. Indeed, given a prime p and a natural number n, consider the spectrum $K(n)$ of the corresponding Morava K-theory, see Ch. IX, §7. Let $H(-)$ denote $H(-;\mathbb{Z}/p)$. We prove in IX.7.27(ii) that $H_0(K(n)_{(0)}) = 0$ while $H_*(K(n)_{(0)}) \neq 0$. But, for every ring spectrum E with $H_0(E) = 0$ we have $H_n(E) = 0$ for all n. Indeed, the homomorphism

$$\pi_0(S) \otimes H_n(E) \xrightarrow{\iota_* \otimes 1} \pi_0(E) \otimes H_n(E) \xrightarrow{h \otimes 1} H_0(E) \otimes H_n(E) \xrightarrow{\mu_{E,E}} H_n(E)$$

is an isomorphism, and hence $H_0(E) = 0$ implies $H_n(E) = 0$. Thus, $K(n)_{(0)}$ is not a ring spectrum. Similarly, considering $K(n)|N$ instead of $K(n)$, $N << 0$, one obtains a counterexample for a spectrum E bounded below.

4.32. Theorem-Definition. *For every abelian group A, there exists a spectrum $M(A)$ with the following properties:*

(i) $\pi_i(M(A)) = 0$ *for* $i < 0$*;*

(ii) $\pi_0(M(A)) = A = H_0(M(A))$*;*

(iii) $H_i(M(A)) = 0$ *for* $i \neq 0$.

Moreover, these properties determine $M(A)$ uniquely up to equivalence. This spectrum $M(A)$ is called the Moore spectrum *of the abelian group A.*

Proof. Consider an exact sequence $0 \to R \xrightarrow{i} F \xrightarrow{q} A \to 0$ with free abelian groups F, R. Let $\{r_\beta\}_{\beta \in B}$ and $\{f_\gamma\}_{\gamma \in \Gamma}$ be certain systems of free abelian generators of R and F, respectively. Let S_β and S_γ be copies of the sphere spectrum S. Consider any morphism $\varphi : \vee_{\beta\in B} S_\beta \to \vee_{\gamma\in\Gamma} S_\gamma$ such that $\varphi|S_\beta$ yields the element $i(r_\beta) \in F = \pi_0(\vee_{\gamma\in\Gamma} S_\gamma)$. Such a morphism φ exists, because, by 1.16(i),

$$[\vee_{\beta\in B} S_\beta, E] = \prod_{\beta\in B} [S_\beta, E] \text{ and } \pi_i(\vee_{\gamma\in\Gamma} S_\gamma) = F.$$

It is clear that $M(R) = \vee_{\beta\in B} S_\beta$ and $M(F) = \vee_{\gamma\in\Gamma} S_\gamma$. Now the exactness of the sequences

$$\cdots \to H_{n+1}(C\varphi) \to H_n(\vee_{\beta\in B} S_\beta) \to H_n(\vee_{\gamma\in\Gamma} S_\gamma) \to H_n(C\varphi) \to \cdots$$
$$\cdots \to \pi_{n+1}(C\varphi) \to \pi_n(\vee_{\beta\in B} S_\beta) \to \pi_n(\vee_{\gamma\in\Gamma} S_\gamma) \to \pi_n(C\varphi) \to \cdots$$

implies that $C\varphi$ satisfies (i)–(iii). Thus, we can set $M(A) := C\varphi$.

Now, let N be a spectrum satisfying (i)–(iii). Consider a morphism $g : \vee_{\gamma\in\Gamma} S_\gamma \to N$ such that $g|S_\gamma$ yields $q(f_\gamma)$. Then $g\varphi|S_\beta$ is inessential for every $\beta \in B$, and so, by 1.16(i), $g\varphi$ is inessential. Hence, there exists $f : M(A) = C\varphi \to N$ with $g \simeq f\psi$, where $\psi : \vee_{\gamma\in\Gamma} S_\gamma \to C\varphi$ is the canonical morphism. Clearly, $f_* : H_*(M(A)) \to H_*(N)$ is an isomorphism, and thus, by 4.7(iii), f is an equivalence. □

4.33. Proposition. *Let E be a spectrum, and let $h : A \to \pi_n(E)$ be a homomorphism from an abelian group A. Then there exists a morphism $f : \Sigma^n M(A) \to E$ such that $h = f_* : \pi_n(\Sigma^n M(A)) \to \pi_n(E)$.*

Proof. We use the notation of 4.32. Consider a morphism $g : \vee_{\gamma\in\Gamma} S^n_\gamma \to E$ such that $g|S^n_\gamma$ yields $hq(f_\gamma)$. Since $g\varphi$ is inessential, there exists $f : \Sigma^n M(A) \to E$ with $g \simeq f\psi$, and, clearly, $h = f_*$. □

4.34. Remark. Given a spectrum F and a spectrum (or a space) X, we can consider the exact couple (given by the cofiber sequences $F_{(n)} \to F_{(n-1)} \to \Sigma^{n+1} H(\pi_n)$, $n \in \mathbb{Z}$)

$$\begin{array}{ccc} [X, F_{(n)}]_* & \xrightarrow{\ i\ } & [X, F_{(n-1)}]_* \\ {\scriptstyle k}\uparrow & & \downarrow{\scriptstyle j} \\ [X, \Sigma^{n+1} H(\pi_n)]_* & = & [X, \Sigma^{n+1} H(\pi_n)]_*, \end{array}$$

where $\pi_n = \pi_n(F)$ and $\deg k = -1$ (cf. Mosher–Tangora [1], Ch. 14); here $[A, B]_* := \oplus_n [A, \Sigma^n B]$. This exact couple yields a spectral sequence $E_r^{*,*}(X)$ with $E_2^{p,q}(X) = H^p(X; F^q(S)) = H^p(X; \pi_{-q})$, which converges (under certain conditions) to $F^*(X)$. (The diagram above gives us the term E_2.) One can prove (see Mosher–Tangora [1], Ch. 14) that this spectral sequence coincides with the Atiyah–Hirzebruch spectral sequence. Hence, the differential $d_r^{p,q} : H^p(X; \pi_{-q}) \to H^{p+r}(X; \pi_{r-q-1})$ in the Atiyah–Hirzebruch spectral sequence has the form

$$[X, \Sigma^p H(\pi_{-q})] \xrightarrow{k} [X, \Sigma^{p+q} F_{(-q)}] \xleftarrow{i} [X, \Sigma^{p+q} F_{(1-q)}] \leftarrow \cdots$$
$$\leftarrow [X, \Sigma^{p+q} F_{(r-q-2)}] \xrightarrow{j} [X, \Sigma^{p+q}\Sigma^{r-q} H(\pi_{r-q-1})] = [X, \Sigma^{p+r} H(\pi_{r-q-1})].$$

Thus, $d_r^{p,q}$ (considered as a higher cohomology operation) is the r-th Postnikov invariant of the spectrum $F|q$. In particular, if F is connected then $d_r^{p,0}$ is the r-th Postnikov invariant of F.

§5. Localization

Let $\mathbb{Q}$ be the field of rational numbers. Let p be a prime, and let $\mathbb{Z}[p]$ be the subring of $\mathbb{Q}$ consisting of all irreducible fractions with denominators relatively prime to p. The $\mathbb{Z}[p]$-localization of an abelian group A is the homomorphism $A \to A \otimes \mathbb{Z}[p]$, $a \mapsto a \otimes 1$. The group $A \otimes \mathbb{Z}[p]$ is simpler than A in a certain sense: for example, it has no q-torsion if $(p, q) = 1$. On the other hand, if we know the groups $A \otimes \mathbb{Z}[p]$ for all p then we can obtain a lot of information about A; for example, if A is finitely generated then it is completely determined by the groups $A \otimes \mathbb{Z}[p]$, where p runs through all primes. So, we can describe an abelian group A via descriptions of the simpler groups $A \otimes \mathbb{Z}[p]$, and this trick is very effective. For example, it is very convenient to describe the ring $H^*(H\mathbb{Z})$ of cohomology operations via the rings $H^*(H\mathbb{Z}[p]; \mathbb{Z}[p])$. Also, localization enables us to ignore the torsions which are irrelevant to a particular problem.

More generally, it makes sense to consider subrings Λ of $\mathbb{Q}$. In this case the localization $A \to A \otimes \Lambda$ deletes the q-torsion with $q \in S$, where S is the set of denominators of all irreducible fractions of Λ.

It is remarkable that the localization can be transferred from algebra to topology, and, in particular, one can consider the $\mathbb{Z}[p]$-homotopy type of a space and a spectrum. As usual, not one but several mathematicians (J.F. Adams, F.P. Peterson) proposed the idea of this transfer, while Serre [1] asked about developing a $\mathfrak{C}$-homotopy types theory (where $\mathfrak{C}$ is a Serre class of abelian groups) in 1953. Nevertheless, usually Sullivan is treated as the author of the theory of localization of topological spaces, because he amplified the language and theory with useful applications. Localization theory for spaces is discussed, e.g., in Sullivan [2], Postnikov [1], Hilton–Mislin–Roitberg [1]. Localization theory for spectra is similar (but simpler), and we expose it here, see also Margolis [1].

Let Λ be a subring of $\mathbb{Q}$; its additive group is also denoted by Λ. Let π be an abelian group.

5.1. Definition. The homomorphism $l = l_\Lambda^\pi : \pi \to \pi \otimes \Lambda$, $a \mapsto a \otimes 1$ is called the Λ-*localization* of π. The abelian group π is called Λ-*local* if l is an isomorphism. A homomorphism $u : \pi \to \tau$ *Λ-localizes* π if there exists an isomorphism $v : \pi \otimes \Lambda \to \tau$ with $u = vl$.

It is clear that $l_\Lambda^\Lambda : \Lambda \to \Lambda \otimes \Lambda$ is an isomorphism. So, $\pi \otimes \Lambda$ is a Λ-local group for every abelian group π.

Let $\iota : S \to M(\Lambda)$ be the morphism given by the unit $1 \in \Lambda = \pi_0(M(\Lambda))$, where $M(\Lambda)$ is the Moore spectrum of Λ. We set $E_\Lambda := E \wedge M(\Lambda)$ for every spectrum E.

5.2. Definition. The morphism $j = j_\Lambda^E : E = E \wedge S \xrightarrow{1 \wedge \iota} E \wedge M(\Lambda) = E_\Lambda$ is called the *Λ-localization* of a spectrum E. Furthermore, E is called *Λ-local* if j is an equivalence. A morphism $f : E \to F$ *Λ-localizes* E if there exists an equivalence $g : E_\Lambda \to F$ with $f = gj$.

Sometimes (for simplicity) one says that the Λ-localization of E is just the spectrum E_Λ, keeping in mind the morphism j implicitly.

Given a morphism $f : E \to F$, we define $f_\Lambda := f \wedge 1_{M(\Lambda)} : E_\Lambda \to F_\Lambda$, and it is clear that $(gf)_\Lambda = g_\Lambda f_\Lambda$. So, Λ-localization is a functor.

5.3. Proposition. *If $E \xrightarrow{f} F \xrightarrow{g} G$ is a cofiber sequence of spectra then $E_\Lambda \xrightarrow{f_\Lambda} F_\Lambda \xrightarrow{g_\Lambda} G_\Lambda$ is. In particular, $C(f_\Lambda) = (Cf)_\Lambda$ for every morphism $f : E \to F$ of spectra.*

Proof. This follows from 2.1(vi). □

5.4. Theorem. *For every pair of spectra X, E there is an isomorphism*

$$t : (E_\Lambda)_*(X) \cong E_*(X) \otimes \Lambda$$

which is natural with respect to X and E, and this isomorphism can be chosen such that the diagram

$$\begin{array}{ccc} E_*(X) & \xrightarrow{j_*} & (E_\Lambda)_*(X) \\ {\scriptstyle l}\downarrow & & \cong\downarrow{\scriptstyle t} \\ E_*(X) \otimes \Lambda & = & E_*(X) \otimes \Lambda \end{array}$$

commutes. In other words, j_* Λ-localizes $E_*(X)$.

Similarly, there is a natural isomorphism $t : (E_\Lambda)^*(X) \cong E^*(X) \otimes \Lambda$, and this isomorphism can be chosen such that $j_* : E^*(X) \to E_\Lambda^*(X)$ Λ-localizes $E^*(X)$.

Proof. We consider the case of homology only; cohomology can be considered similarly. Consider an exact sequence $0 \to R \to F \xrightarrow{\varepsilon} \Lambda \to 0$, where R, F are free abelian groups. The cofiber sequence $M(R) \to M(F) \to M(\Lambda)$ (see the proof of 4.32) induces a cofiber sequence

$$E \wedge M(R) \xrightarrow{\varkappa} E \wedge M(F) \xrightarrow{\sigma} E \wedge M(\Lambda).$$

Let $\{f_\alpha\}$ be a free basis of F, and let S_α, E_α be copies of S, E respectively. We have $M(F) \simeq \vee_\alpha S_\alpha$, and so, by 2.1(v), $E \wedge M(F) \simeq \vee_\alpha E_\alpha$. We define the isomorphism

$$b : (E \wedge M(F))_*(X) \cong (\vee_\alpha E_\alpha)_*(X) \cong \oplus_\alpha (E_\alpha)_*(X) \xrightarrow{B, \cong} E_*(X) \otimes F,$$

where $B(x) = x \otimes f_\alpha$ for $x \in (E_\alpha)_*(X)$. Consider the following commutative diagram (where c is defined just as b) with exact rows:

$$\begin{array}{ccccccc}
 & (E \wedge M(R))_*(X) & \xrightarrow{\varkappa_*} & (E \wedge M(F))_*(X) & \xrightarrow{\sigma_*} & (E \wedge M(\Lambda))_*(X) \\
 & \cong \downarrow c & & \cong \downarrow b & & \\
0 \to & E_*(X) \otimes R & \longrightarrow & E_*(X) \otimes F & \longrightarrow & E_*(X) \otimes \Lambda.
\end{array}$$

The bottom row is exact because $\mathrm{Tor}(A, \Lambda) = 0$ for every abelian group A, see e.g. Bourbaki [3]. So, $\varkappa_*$ is monic, and hence σ_* is epic. So, there exists an isomorphism $t : (E \wedge M(\Lambda))_*(X) \to E_*(X) \otimes \Lambda$ which preserves the commutativity of the diagram.

In order to construct t with $tj_* = l$ we shall assume that there is $f_0 \in \{f_\alpha\}$ with $\varepsilon(f_0) = 1 \in \Lambda$. Then there exists a commutative diagram

$$\begin{array}{ccc}
S & \xrightarrow{f_0} & M(F) \\
\iota \downarrow & & \downarrow M(\varepsilon) \\
M(\Lambda) & = & M(\Lambda),
\end{array}$$

and the following diagram commutes:

$$\begin{array}{ccccc}
(E \wedge S)_*(X) & \xrightarrow{(1 \wedge f_0)_*} & (E \wedge M(F))_*(X) & \xrightarrow{(1 \wedge M(\varepsilon))_*} & (E \wedge M(\Lambda))_*(X) \\
\downarrow & & b \downarrow & & t \downarrow \\
E_*(X) \otimes \mathbb{Z} & \xrightarrow{(f_0)_*} & E_*(X) \otimes F & \xrightarrow{1 \otimes \varepsilon} & E_*(X) \otimes \Lambda.
\end{array}$$

Now, $tj_* = l$ because $E_*(X) = (E \wedge S)_*(X)$. □

Below we fix such a natural isomorphism t with $tj_* = l$ and use it without any mention; e.g., the formula $(E_\Lambda)_*(X) \xrightarrow{\cong} E_*(X) \otimes \Lambda$ means $t : (E_\Lambda)_*(X) \xrightarrow{\cong} E_*(X) \otimes \Lambda$.

5.5. Corollary. *There are natural isomorphisms*

$$\pi_i(E_\Lambda) \cong \pi_i(E) \otimes \Lambda, \ H_i(E_\Lambda) \cong H_i(E) \otimes \Lambda$$

such that the homomorphisms $\pi_i(E) \xrightarrow{j_*} \pi_i(E_\Lambda) \cong \pi_i(E) \otimes \Lambda$, $H_i(E) \xrightarrow{j_*} H_i(E_\Lambda) \cong H_i(E) \otimes \Lambda$ *have the form* $x \mapsto x \otimes 1$. *So,* j Λ*-localizes homotopy and homology groups. In particular, every* Λ*-local spectrum has* Λ*-local homotopy and homology groups.*

Proof. We have

$$\pi_i(E_\Lambda) = (E_\Lambda)_i(S) \cong E_i(S) \otimes \Lambda = \pi_i(E) \otimes \Lambda,$$
$$H_i(E_\Lambda) = (E_\Lambda)_i(H\mathbb{Z}) \cong E_i(H\mathbb{Z}) \otimes \Lambda = H_i(E) \otimes \Lambda. \qquad \square$$

5.6. Lemma. *Let* τ *be any flat abelian group (e.g.,* τ *is an additive subgroup of* $\mathbb{Q}$*). Then:*

(i) $H\tau \wedge M(\pi) \simeq H(\tau \otimes \pi) \simeq H\pi \wedge M(\tau)$ *for every abelian group* π*. In particular,* $H\mathbb{Z} \wedge M(\pi) \simeq H\pi$.

(ii) $M(\tau) \wedge M(\pi) \simeq M(\tau \otimes \pi)$. *In particular,* $M(\Lambda) \wedge M(\Lambda) \simeq M(\Lambda)$.

(iii) *If* π *is a* Λ*-local group then the* Λ*-localizations* $j^H : H(\pi) \to H(\pi)_\Lambda$ *and* $j^M : M(\pi) \to M(\pi)_\Lambda$ *are equivalences. In other words,* $H(\pi)$ *and* $M(\pi)$ *are* Λ*-local spectra.*

Proof. (i) If π is a free abelian group F with a basis $\{f_\alpha\}$, then $\tau \otimes F = \oplus_\alpha \tau$, and hence

$$H(\tau \otimes F) = H(\oplus_\alpha \tau) \simeq \vee_\alpha H\tau \simeq (H\tau) \wedge (\vee_\alpha S_\alpha) = H\tau \wedge M(F),$$

where S_α is a copy of S. For arbitrary π consider an exact sequence

$$0 \to R \to F \to \pi \to 0$$

with free abelian groups R and F. Then the cofiber sequence $M(R) \to M(F) \to M(\pi)$ induces a cofiber sequence

$$H\tau \wedge M(R) \to H\tau \wedge M(F) \to H\tau \wedge M(\pi).$$

Since τ is a flat abelian group, the sequence $0 \to \tau \otimes R \to \tau \otimes F \to \tau \otimes \pi \to 0$ is exact, and hence we have a cofiber sequence

$$H(\tau \otimes R) \to H(\tau \otimes F) \to H(\tau \otimes \pi).$$

Thus, we have a homotopy commutative diagram

$$\begin{array}{ccccccc} H(\tau \otimes R) & \longrightarrow & H(\tau \otimes F) & \longrightarrow & H(\tau \otimes \pi) & \longrightarrow & \Sigma H(\tau \otimes R) \\ {\scriptstyle \alpha}\downarrow & & {\scriptstyle \beta}\downarrow & & & & \downarrow{\scriptstyle \Sigma\alpha} \\ H\tau \wedge M(R) & \longrightarrow & H\tau \wedge M(F) & \longrightarrow & H\tau \wedge M(\pi) & \longrightarrow & \Sigma(H\tau \wedge M(R)) \end{array}$$

where the rows are long cofiber sequences and α, β are equivalences. By 1.13(ii), there exists a morphism $H(\tau \otimes \pi) \to H(\tau) \wedge M(\pi)$ preserving the commutativity of the diagram, and this morphism is an equivalence since it induces an isomorphism of homotopy groups.

In particular, $H\mathbb{Z} \wedge M(\pi) \simeq H\pi$. So, $HF \wedge M(\tau) \simeq H(\tau \otimes F)$ for every free abelian group F. As above, we have the homotopy commutative diagram

$$\begin{array}{ccccccc} H(\tau \otimes R) & \longrightarrow & H(\tau \otimes F) & \longrightarrow & H(\tau \otimes \pi) & \longrightarrow & \Sigma H(\tau \otimes R) \\ \simeq\downarrow & & \simeq\downarrow & & \downarrow\simeq & & \\ HR \wedge M(\tau) & \longrightarrow & HF \wedge M(\tau) & \longrightarrow & H\pi \wedge M(\tau) & \longrightarrow & \Sigma(HR \wedge M(\tau)) \end{array}$$

where the rows are long cofiber sequences, etc. Thus, $H(\tau \otimes \pi) \simeq H(\pi) \wedge M(\tau)$.

(ii) We have

$$\begin{aligned} H_i(M(\tau) \wedge M(\pi)) &= \pi_i(H\mathbb{Z} \wedge (M(\tau) \wedge M(\pi))) = \pi_i((H\mathbb{Z} \wedge M(\pi)) \wedge M(\tau)) \\ &= \pi_i((H\pi) \wedge M(\tau)) = \pi_i(H(\tau \otimes \pi)) \ . \end{aligned}$$

This group is $\tau \otimes \pi$ for $i = 0$ and 0 for $i \neq 0$. Moreover, by 4.5(i), we have $\pi_i(M(\tau) \wedge M(\pi)) = 0$ for $i < 0$. Thus, $M(\pi) \wedge M(\tau) = M(\pi \otimes \tau)$.

(iii) By 5.5,

$$j_*^H : \pi = \pi_*(H(\pi)) = \pi_*(H(\pi)_\Lambda) \to \pi \otimes \Lambda$$

Λ-localizes π, and so j_*^H is an isomorphism since π is Λ-local. Thus, j^H is an equivalence. Similarly, $j_*^M : \pi = H_*(M(\pi)) \to H_*(M(\pi)_\Lambda) = \pi \otimes \Lambda$ is an isomorphism, and thus j^M is an equivalence. □

5.7. Lemma. *Let E be an arbitrary spectrum, and let Cj be the cone of the localization $j : E \to E_\Lambda$. Then $H^*(Cj; \pi) = 0$ for every Λ-local group π.*

Proof. Firstly, consider the localization $\iota : S \to M(\Lambda)$. Set $C = C\iota$. We prove that $\iota^* : H^d(M(\Lambda); \pi) \to H^d(S; \pi)$ is an isomorphism for all d. Clearly, both groups are trivial for $d \neq 0, 1$. In view of 4.9, the homomorphism ι^* for $d = 0$ has the form $k^* : \operatorname{Hom}(\Lambda, \pi) \to \operatorname{Hom}(\mathbb{Z}, \pi)$ where $k : \mathbb{Z} \to \Lambda$ is the inclusion. But $k^*$is an isomorphism since π is Λ-local. Finally, for $d = 1$ we have $H^1(M(\Lambda); \pi) = \operatorname{Ext}(\Lambda, \pi)$, but $\operatorname{Ext}(\Lambda, \pi) = 0$ for every Λ-local π. So, $H^*(C; \pi) = 0$.

For arbitrary E we have $Cj = E \wedge C$. By the above, $H^*(S^n \wedge C; \pi) = 0$ for every n. Choose any $k \in \mathbb{Z}$. By 4.1(ii), $H^k(E^{(m)} \wedge C; \pi) = 0$ for $m << k$. Considering the cofiber sequences $E^{(n-1)} \to E^{(n)} \to \vee S^n$, $n = m, m+1, \dots$, we obtain by induction that $H^k(E^{(n)} \wedge C; \pi) = 0$ for all n.

By 4.5(ii), $i_n \wedge 1 : E^{(n)} \wedge C \to E \wedge C$ is an $(n-1)$-equivalence because $i_n : E^{(n)} \to E$ is. Let X be the cone of $i_n \wedge 1$. We have $\pi_i(X) = 0$ for $i \leq n$, and so, by 4.7(i) and 4.9, $H^i(X; \pi) = 0$ for $i \leq n$. Hence, $H^k(E \wedge C; \pi) = H^k(E^{(n)} \wedge C; \pi)$ for $n \geq k$. Thus, $H^*(E \wedge C) = 0$. □

5.8. Theorem. (i) *For every spectrum E the morphism $(j_\Lambda^E)_\Lambda : E_\Lambda \to (E_\Lambda)_\Lambda$ is an equivalence. In particular, E_Λ is Λ-local.*

(ii) *If F is a Λ-local spectrum then for every morphism $f: E \to F$ there exists a morphism $g: E_\Lambda \to F$ with $gj_\Lambda^E = f$. So, $j^* = (j_\Lambda^E)^* : [E_\Lambda, F] \to [E, F]$ is epic.*

(iii) *If in* (ii) *at least one of the spectra E, F is bounded below then $j^*: [E_\Lambda, F] \to [E, F]$ is an isomorphism.*

Proof. (i) Indeed, $(j_\Lambda^E)_\Lambda$ has the form

$$1_E \wedge j_\Lambda^{M(\Lambda)} : E \wedge M(\Lambda) \to E \wedge M(\Lambda) \wedge M(\Lambda).$$

But, by 5.6(iii), $j_\Lambda^{M(\Lambda)}$ is an equivalence.

(ii) The morphism g can be given by the composition $E_\Lambda \xrightarrow{f_\Lambda} F_\Lambda \xrightarrow{h} F$, where h is an equivalence inverse to j_Λ^F.

(iii) Firstly, consider the case when F is bounded below. It suffices to prove that $F^*(Cj) = 0$, i.e., that $F^*(E \wedge C) = 0$ where $C = C\iota$ and $\iota : S \to M(\Lambda)$ is the localization.

Consider an exact sequence $0 \to R \xrightarrow{i} F \xrightarrow{q} \Lambda \to 0$ with free abelian groups F, R. Let $\{r_\beta\}_{\beta \in B}$ and $\{f_\gamma\}_{\gamma \in \Gamma}$ be certain systems of free abelian generators of R and F, respectively. The spectrum $M(\Lambda)$ is the cone of the morphism $\varphi : \vee_{\beta \in B} S_\beta \to \vee_{\gamma \in \Gamma} S_\gamma$ such that $\varphi|S_\beta$ yields $i(r_\beta)$, see the proof of 4.32. Consider a map

$$\psi : \vee_{\beta \in B} S_\beta^2 \to \vee_{\gamma \in \Gamma} S_\gamma^2$$

such that $\psi|S_\beta^2$ gives us $i(r_\beta) \in F = \pi_2(\vee S_\gamma^2)$ (here S^2 is the space, not a spectrum). Denoting $C(\psi)$ by M_Λ, we conclude that $M(\Lambda) = \Sigma^{-2}\Sigma^\infty M_\Lambda$. Let $u : S^2 \to M_\Lambda$ yield the unit 1 of $\Lambda = \pi_2(M_\Lambda)$. Set $Y := C(u)$. Then $C \simeq \Sigma^{-2}\Sigma^\infty Y$, where $\dim Y = 3$.

Step 1. Let $E = \Sigma^\infty X$ for some finite dimensional CW-complex X. By 5.5, every group $\pi_i(F)$ is Λ-local. Hence, by 5.7, $H^*(Cj; \pi_*(F)) = 0$. Now we prove by induction that $F_{(n)}^*(Cj) = 0$ for every n. Firstly, F is bounded below, and so $F_{(k)} = \Sigma^k H(\pi_k(F))$ for some $k \in \mathbb{Z}$. Hence, by 5.7, $F_{(k)}^*(E \wedge C) = 0$. Now, suppose that $F_{(n)}^*(E \wedge C) = 0$ for some n. Considering the cofiber sequence

$$\Sigma^{n+1} H(\pi_{n+1}(F)) \to F_{(n+1)} \to F_{(n)}$$

and using 5.7, we conclude that $F_{(n+1)}^*(E \wedge C) = 0$. The induction is confirmed.

In other words, $F_{(n)}^i(E \wedge C) = 0$ for every i, n. But the spectrum $E \wedge C = \Sigma^{-2}\Sigma^\infty(X \wedge Y)$ is finite dimensional, i.e., $(E \wedge C)^{(r)} = E \wedge C$ for some r. Hence, by 4.1(ii), for every i we can find n such that $(\tau_n)_* : F^i(E \wedge C) \to F_{(n)}^i(E \wedge C)$ is an isomorphism. Thus, $F^*(E \wedge C) = 0$.

Step 2. Let $E = \Sigma^\infty X$ for some CW-complex X. Because of Step 1, $F^*(E^{(r)} \wedge C) = 0$ for every r. Now,

$$\begin{aligned} E^{(r)} \wedge C = & E^{(r)} \wedge \Sigma^{-2} Y = \Sigma^{-2}(E^{(r)} \wedge Y) = \Sigma^{-2}\Sigma^{\infty}(X^{(r)} \wedge Y) \\ = & \Sigma^{-2}\Sigma^{\infty}(X \wedge Y)^{(r+3)}. \end{aligned}$$

Hence, $\widetilde{F}^*((X \wedge Y)^{(r)}) = 0$ for every r, and so, by 4.1(iii), $\widetilde{F}^*(X \wedge Y) = 0$. Thus, $F^*(E \wedge C) = 0$ because $E \wedge C \simeq \Sigma^{-2}\Sigma^{\infty}(X \wedge Y)$.

Step 3. Any spectrum E. (Roughly speaking, $\widetilde{F}^*((E \wedge C)_n) = 0$ for every n, and so, by 4.1(i), $F^*(E \wedge C) = 0$.) By Step 2, $F^*(\Sigma^{-n}\Sigma^{\infty}E_n \wedge C) = 0$ for every n. Now,

$$\begin{aligned} \Sigma^{-n}\Sigma^{\infty}(E_n \wedge C) & \simeq \Sigma^{-n}\Sigma^{\infty}E_n \wedge \Sigma^{-2}\Sigma^{\infty}Y \simeq \Sigma^{-n-2}\Sigma^{\infty}(E_n \wedge Y) \\ & \simeq \Sigma^{-n-2}\Sigma^{\infty}((E \wedge Y)_n). \end{aligned}$$

Hence, $\widetilde{F}^*((E \wedge Y)_n) = 0$ for every n. By 4.1(i), $F^*(E \wedge Y) = 0$, and thus $F^*(E \wedge C) = 0$.

Finally, if E is bounded below, $E = E|k$, then, by 4.16, $[E, F] = [E, F|k]$, $[E_\Lambda, F] = [E_\Lambda, F|k]$. Thus, the situation is reduced to the case when F is bounded below. □

5.9. Corollary. *Let E be an arbitrary spectrum, and let F be a Λ-local spectrum.*

(i) *The morphism $f : E \to F$ of spectra Λ-localizes E iff $f_* : \pi_*(E) \to \pi_*(F)$ Λ-localizes homotopy groups.*

(ii) *Suppose that both E, F are bounded below. Then the morphism $f : E \to F$ Λ-localizes E iff $f_* : H_*(E) \to H_*(F)$ Λ-localizes homology groups.*

Proof. The "only if" part was proved in 5.5, so we shall prove the "if" part. By 5.8(ii), there exists $h : E_\Lambda \to F$ with $f = hj_\Lambda^E$. In case (i), $h_* : \pi_*(E_\Lambda) \to \pi_*(F)$ is an isomorphism. In case (ii), $h_* : H_*(E_\Lambda) \to H_*(F)$ is an isomorphism, and hence, by 4.7(iii), h is an equivalence. □

5.10. Corollary. *Let E be a spectrum such that either $\pi_i(E)$ is Λ-local for every i, or E is bounded below and $H_i(E)$ is Λ-local for every i. Then E is a Λ-local spectrum. In particular, if a spectrum F is Λ-local then so are $F_{(n)}$ and $F|n$ for every n.*

Proof. By 5.8(i), E_Λ is Λ-local. By 5.5, $j : E \to E_\Lambda$ Λ-localizes homotopy and homology groups. So, j is an equivalence provided $\pi_i(E)$ (resp. $H_i(E)$ for E bounded below) are Λ-local. □

5.11. Proposition. *Let*

$$\begin{array}{ccccccccc} \cdots & = & E & = & E & = & E & = & \cdots \\ & & \downarrow{\scriptstyle r_{n+1}} & & \downarrow{\scriptstyle r_n} & & \downarrow{\scriptstyle r_{n-1}} & & \\ \cdots & \longrightarrow & E_{(n+1)} & \xrightarrow{p_{n+1}} & E_{(n)} & \xrightarrow{p_n} & E_{(n-1)} & \longrightarrow & \cdots, \end{array}$$

be the Postnikov tower of E. Then

$$\begin{array}{ccccccccc}
\cdots & = & E_\Lambda & = & E_\Lambda & = & E_\Lambda & = & \cdots \\
 & & \downarrow{\scriptstyle (\tau_{n+1})_\Lambda} & & \downarrow{\scriptstyle (\tau_n)_\Lambda} & & \downarrow{\scriptstyle (\tau_{n-1})_\Lambda} & & \\
\cdots & \longrightarrow & (E_{(n+1)})_\Lambda & \xrightarrow{(p_{n+1})_\Lambda} & (E_{(n)})_\Lambda & \xrightarrow{(p_n)_\Lambda} & (E_{(n-1)})_\Lambda & \longrightarrow & \cdots,
\end{array}$$

is the Postnikov tower of E_Λ.

Proof. Clearly, $\pi_i((E_{(n)})_\Lambda) = 0$ for $i > n$. Furthermore, by 5.5,

$$((\tau_n)_\Lambda)_* : \pi_i(E_\Lambda) = \pi_i(E) \otimes \Lambda \xrightarrow{(\tau_n)_* \otimes 1} \pi_i(E_{(n)}) \otimes \Lambda = \pi_i((E_{(n)})_\Lambda)$$

is an isomorphism for $i \leq n$. □

This proposition enables us to construct localizations via Postnikov towers. Firstly, we have $H(\pi)_\Lambda \simeq H(\pi \otimes \Lambda)$, see 5.6(i). Furthermore, consider a Postnikov tower of E:

$$\begin{array}{ccccccc}
\cdots & \longrightarrow & E_{(n)} & \longrightarrow & E_{(n-1)} & \longrightarrow & \cdots \\
 & & & & \downarrow{\scriptstyle \kappa} & & \\
 & & & & \Sigma^{n+1} H(\pi_n(E)). & &
\end{array}$$

If a Λ-localization of $E_{(n-1)}$ is already constructed, we define $(E_{(n)})_\Lambda$ to be the cone of $(\Sigma^{-1}\kappa)_\Lambda : (\Sigma^{-1}E_{(n-1)})_\Lambda \to \Sigma^n H(\pi_n(E) \otimes \Lambda)$, and the localization morphism j can be constructed in an obvious manner based on 5.11. Here we have described the inductive step, but we must suppose that E is bounded below in order to organize a base of the induction: namely, to set $(E_{(m)})_\Lambda = H(\pi_m(E) \otimes \Lambda)$, where $\pi_i(E) = 0$ for $i < m$. Finally, one can prove that there is a spectrum E_Λ such that $(E_\Lambda)_{(m)}$ is $(E_{(m)})_\Lambda$ for every m, cf. III.6.3(ii) below.

This approach enables us to construct the localization of *spaces* also. The main results of this theory are Theorems 5.12 and 5.13 below. The proofs can be found in Hilton–Mislin–Roitberg [1], Postnikov [1], Sullivan [2]. (These theorems hold for so-called nilpotent spaces, but we formulate them for a more special case: simple spaces.) Recall that a connected space X is called *simple* if the action of $\pi_1(X)$ on $\pi_n(X)$ is trivial for every n. In particular, $\pi_1(X)$ must be an abelian group.

5.12. Theorem–Definition. *For every simple space X there exist a simple space X_Λ and a map $j = j_\Lambda^X : X \to X_\Lambda$ such that the homomorphisms $\pi_i(X) \xrightarrow{j_*} \pi_i(X_\Lambda) \cong \pi_i(X) \otimes \Lambda$ and $H_i(X) \xrightarrow{j_*} H_i(X_\Lambda) \cong H_i(X) \otimes \Lambda$ have the form $x \mapsto x \otimes 1$. So, j localizes homotopy and homology groups. Every such a map j is called* localization *of X.* □

As in 5.2, a simple space X is called Λ-*local* if $j : X \to X_\Lambda$ is a homotopy equivalence.

5.13. Theorem–Definition. *For every two simple spaces X, Y the following conditions are equivalent:*

(i) *The map $f : X \to Y$ Λ-localizes homotopy groups;*

(ii) *The map $f : X \to Y$ Λ-localizes homology groups;*

(iii) *For every Λ-local space Z the map $f^* : [Y, Z] \to [X, Z]$ is a bijection.*

If some (and hence all) of these conditions hold then there exists a homotopy equivalence $h : X_\Lambda \to Y$ with $f = hj_\Lambda^X$. Furthermore, in this case we say that f localizes *X.* □

So, by 5.9, if $f : X \to Y$ localizes a space X then $\Sigma^\infty j : \Sigma^\infty X \to \Sigma^\infty Y$ localizes the spectrum $\Sigma^\infty X$. In particular, $j^* : E^*(X_\Lambda) \to E^*(X)$ is an isomorphism for every Λ-local spectrum E, cf. 5.8(iii). Similarly, if $\varphi : E \to F$ localizes a spectrum E then $\Omega^\infty \varphi : \Omega^\infty E \to \Omega^\infty F$ localizes the space $\Omega^\infty E$.

Now we show that localization respects multiplicative structures.

5.14. Lemma. *For every two spectra E, F there exists an equivalence*

$$\varphi : (E \wedge F)_\Lambda \to E_\Lambda \wedge F_\Lambda$$

such that the diagram

$$\begin{array}{ccc} E \wedge F & \xrightarrow{j^{E\wedge F}} & (E \wedge F)_\Lambda \\ {\scriptstyle j^E \wedge j^F}\downarrow & & \downarrow{\scriptstyle \varphi} \\ E_\Lambda \wedge F_\Lambda & = & E_\Lambda \wedge F_\Lambda \end{array}$$

is homotopy commutative.

Proof. We have

$$(E_\Lambda \wedge F_\Lambda)_\Lambda = ((E \wedge M(\Lambda)) \wedge (F \wedge M(\Lambda))) \wedge M(\Lambda) \simeq (E \wedge M(\Lambda)) \wedge (F \wedge M(\Lambda)).$$

Hence, $E_\Lambda \wedge F_\Lambda$ is a local spectrum, and so, by 5.8(ii), there exists a morphism φ such that the diagram commutes. Now, the morphism $j^E \wedge j^F$ induces a homomorphism

$$\begin{aligned} h : F_*(E) = \pi_*(E \wedge F) &\xrightarrow{(j^E \wedge j^F)_*} \pi_*(E_\Lambda \wedge F_\Lambda) = (F_\Lambda)_*(E_\Lambda) \simeq F_*(E_\Lambda) \otimes \Lambda \\ &\simeq (E_\Lambda)_*(F) \otimes \Lambda \simeq E_*(F) \otimes \Lambda \otimes \Lambda = E_*(F) \otimes \Lambda \simeq F_*(E) \otimes \Lambda, \end{aligned}$$

where $h(a) = a \otimes 1$. Thus, φ induces an isomorphism of homotopy groups. □

5.15. Theorem. (i) *If* (E, μ, ι) *is a ring spectrum bounded below then* E_Λ *admits a ring structure such that* $j : E \to E_\Lambda$ *is a ring morphism, and this ring structure is unique up to ring equivalence.*

(ii) *Let* (E, μ, ι) *be a ring spectrum, let* (F, m) *be an* E*-module spectrum, and let* E *and* F *be bounded below. Then there exists a pairing* $\overline{m} : E_\Lambda \wedge F_\Lambda \to F_\Lambda$ *turning* F_Λ *into* E_Λ*-module spectrum such that the diagram*

$$\begin{array}{ccc} E \wedge F & \xrightarrow{j^E \wedge j^F} & E_\Lambda \wedge F_\Lambda \\ {\scriptstyle m}\downarrow & & \downarrow{\scriptstyle \overline{m}} \\ F & \xrightarrow{j} & F_\Lambda \end{array}$$

commutes up to homotopy. Furthermore, this $\overline{m}$ *is unique up to homotopy.*

Proof. We prove only (i) because (ii) can be proved similarly. By 5.8(iii) and 5.14, there exists a morphism $\bar{\mu} : E_\Lambda \wedge E_\Lambda \to E_\Lambda$ (unique up to homotopy) such that the diagram

$$\begin{array}{ccccc} E \wedge E & \xrightarrow{\mu} & E & \xrightarrow{j} & E_\Lambda \\ \downarrow & & & & \| \\ (E \wedge E)_\Lambda & \underset{\varphi}{\xrightarrow{\simeq}} & E_\Lambda \wedge E_\Lambda & \xrightarrow{\bar{\mu}} & E_\Lambda \end{array}$$

is homotopy commutative. Likewise, there exist $g_i, i = 1, 2$, unique up to homotopy, such that the diagrams

$$\begin{array}{ccc} E \wedge E \wedge E & \xrightarrow{c_i} & E \\ \downarrow & & \downarrow{\scriptstyle j} \\ E_\Lambda \wedge E_\Lambda \wedge E_\Lambda & \xrightarrow{g_i} & E_\Lambda, \end{array}$$

$i = 1, 2$, are homotopy commutative. Here $c_1 = \mu \circ (\mu \wedge 1)$, $c_2 = \mu \circ (1 \wedge \mu)$. So, $g_1 \simeq \bar{\mu} \circ (\bar{\mu} \wedge 1)$, $g_2 \simeq \bar{\mu} \circ (1 \wedge \bar{\mu}) \circ a$. Now, $g_1 \simeq g_2$ because $c_1 \simeq c_2$. Hence, $\bar{\mu}$ is associative. □

Most frequently one uses the cases $\Lambda = \mathbb{Q}, \Lambda = \mathbb{Z}[p]$, $\Lambda = \mathbb{Z}[1/p]$, where p is a prime (and $\mathbb{Z}[1/p]$ is the subgroup of $\mathbb{Q}$ consisting of the fractions m/p^k). In these cases, X_Λ (where X is a space or a spectrum) is denoted by $X[0]$, $X[p]$, $X[1/p]$ respectively.

5.16. Definition. A spectrum E has *finite* Λ*-type* if it is bounded below and every group $\pi_i(E)$ is a finitely generated Λ-module.

5.17. Remarks. (a) Any spectrum of finite type has finite $\mathbb{Z}$-type. Every spectrum of finite $\mathbb{Z}$-type is equivalent to a spectrum of finite type, see 4.26. A spectrum of finite Λ-type is a Λ-local spectrum.

(b) It is easy to see that every submodule of a finitely generated Λ-module is finitely generated. So, a Λ-module is finitely generated iff it is finitely presented. Furthermore, Λ is a principal ideal domain, and hence every finitely presented Λ-module splits into a direct sum of cyclic Λ-modules. In particular, every finitely generated $\mathbb{Z}[p]$-module splits into a direct sum of $\mathbb{Z}[p]$-modules $\mathbb{Z}[p]$ and $\mathbb{Z}/p^k$.

5.18. Lemma. *Let p be a prime.*

(i) *If E is a spectrum of finite $\mathbb{Z}[p]$-type with $H^i(E;\mathbb{Z}/p) = 0$ for $i \leq n$, then $\pi_i(E) = 0$ for $i \leq n$. In particular, a spectrum F of finite $\mathbb{Z}[p]$-type is contractible iff $H^i(F;\mathbb{Z}/p) = 0$ for all i.*

(ii) *Let E, F be two spectra of finite $\mathbb{Z}[p]$-type. Then a morphism $f : E \to F$ is an equivalence iff the homomorphism $f^* : H^*(F;\mathbb{Z}/p) \to H^*(E;\mathbb{Z}/p)$ is an isomorphism.*

Proof. (i) Suppose that $\pi_i(E) = 0$ for $i < m < n$ and $\pi_m(E) \neq 0$. Then $\pi_m(E)$ contains a direct summand $A = \mathbb{Z}[p]$ or $A = \mathbb{Z}/p^k$. Thus,

$$H^m(E;\mathbb{Z}/p) = \mathrm{Hom}(H_m(E),\mathbb{Z}/p) = \mathrm{Hom}(\pi_m(E),\mathbb{Z}/p) \supset \mathrm{Hom}(A,\mathbb{Z}/p) \neq 0\,.$$

(ii) The "only if" part is clear, so we prove the "if" part. By 5.3 and 5.17(b), the cone Cf of f is a spectrum of finite $\mathbb{Z}[p]$-type. Now, by (i), Cf is contractible since $H^*(Cf;\mathbb{Z}/p) = 0$. □

5.19. Proposition. (i) *Let E be a spectrum such that every group $\pi_i(E)$ is finitely generated. If $E[p]$ is contractible for every prime p then E is contractible.*

(ii) *Let E, F be two spectra such that every group $\pi_i(E), \pi_i(F)$ is finitely generated. If $f : E \to F$ is such that $f[p]$ is an equivalence for every prime p, then f is an equivalence.*

Proof. (i) We have $0 = \pi_i(E[p]) = \pi_i(E) \otimes \mathbb{Z}[p]$ for every p and every i. Thus, $\pi_i(E) = 0$ for every i.

(ii) By (i), Cf is contractible. □

5.20. Proposition. *Let p be a prime, and let E be a ring spectrum such that $1 \in \pi_0(E)$ has order p. Then E is a $\mathbb{Z}[p]$-local spectrum.*

Proof. Since all $\pi_i(E)$ are $\pi_0(E)$-modules, they are $\mathbb{Z}/p$-vector spaces and thus $\mathbb{Z}[p]$-local groups. Now apply 5.10. □

Finally, we remark that one can localize spaces (and spectra bounded below) cell by cell, see Sullivan [2].

§6. Algebras, Coalgebras, and Hopf Algebras

Here we discuss the notions mentioned in the title in order to use them in the next section. We mainly follow Mac Lane [2] and Milnor–Moore [1]. Recall that every ring is assumed to be associative and unitary.

Let R be a commutative ring. In this section $\otimes$ denotes $\otimes_R$. We consider R as a graded ring, where $R_0 = R, R_i = 0$ for $i \neq 0$. The words "an R-module" mean "a left graded R-module". Given two R-modules M, N, the words "R-homomorphism $f : M \to N$" mean that f is a homomorphism of R-modules with $\deg f(m) = \deg m, m \in M$. We denote by T the switching homomorphism $T = T_{M,N} : M \otimes N \to N \otimes M$, $T(m \otimes n) = (-1)^{|m||n|}(n \otimes m)$.

6.1. Definition. (a) An *algebra over* R (or simply an R-*algebra*) is a triple (A, μ, η), where A is an R-module and $\mu : A \otimes A \to A$, $\eta : R \to A$ are R-homomorphisms such that the diagrams

$$\begin{array}{ccc} A \otimes A \otimes A & \xrightarrow{\mu\otimes 1} & A \otimes A \\ {\scriptstyle 1\otimes\mu}\downarrow & & \downarrow{\scriptstyle\mu} \\ A \otimes A & \xrightarrow{\mu} & A \end{array} \qquad \begin{array}{ccccc} R \otimes A & \xrightarrow{\eta\otimes 1} & A \otimes A & \xleftarrow{1\otimes\eta} & A \otimes R \\ {\scriptstyle\cong}\uparrow & & \downarrow{\scriptstyle\mu} & & \uparrow{\scriptstyle\cong} \\ A & = & A & = & A \end{array}$$

commute. Here $\cong$ denotes the canonical isomorphisms $R \otimes A \cong A \cong A \otimes R$, e.g., $A \cong A \otimes R$ has the form $a \mapsto a \otimes 1$. Furthermore, μ is called the *multiplication* and η is called the *unit homomorphism*. An algebra (A, μ, η) is *commutative* if $\mu T_{A,A} = \mu$.

(b) A (*left*) module over an R-algebra (A, μ, η) is a pair (M, φ), where M is an R-module and $\varphi : A \otimes M \to M$ is an R-homomorphism such that the following diagrams commute:

$$\begin{array}{ccc} A \otimes A \otimes M & \xrightarrow{\mu\otimes 1} & A \otimes M \\ {\scriptstyle 1\otimes\varphi}\downarrow & & \downarrow{\scriptstyle\varphi} \\ A \otimes M & \xrightarrow{\varphi} & M \end{array} \qquad \begin{array}{ccc} A \otimes M & \xrightarrow{\varphi} & M \\ {\scriptstyle\eta\otimes 1}\uparrow & & \uparrow{\scriptstyle\cong} \\ R \otimes M & = & R \otimes M. \end{array}$$

As usual, we shall simply say "algebra A" or "A-module M", omitting μ, η, φ, and we shall write ab instead of $\mu(a \otimes b)$ and am instead of $\varphi(a \otimes m)$.

It is clear that A is a ring with multiplication μ and unit $\eta(e_R)$, where e_R is the unit of R. Furthermore, R can be tautologically considered as the R-algebra $(R, \mu_R, 1_R)$ where $\mu_R(r \otimes r') = rr'$.

Note that every ring is a $\mathbb{Z}$-algebra.

6.2. Definition. A *homomorphism* $f : A \to B$ *of* R-*algebras* is an R-homomorphism such that the first two of the three diagrams below commute.

A *homomorphism* $h : M \to N$ *of A-modules* is an R-homomorphism such that the third diagram commutes.

$$\text{(6.3)}\qquad \begin{array}{ccc} A\otimes A & \xrightarrow{f\otimes f} & B\otimes B \\ {\scriptstyle \mu_A}\downarrow & & \downarrow{\scriptstyle \mu_B} \\ A & \xrightarrow{f} & B \end{array} \qquad \begin{array}{ccc} R & \xrightarrow{\eta_A} & A \\ \| & & \downarrow{\scriptstyle f} \\ R & \xrightarrow{\eta_B} & B \end{array} \qquad \begin{array}{ccc} A\otimes M & \xrightarrow{\varphi_M} & M \\ \downarrow{\scriptstyle 1\otimes h} & & \downarrow{\scriptstyle h} \\ A\otimes N & \xrightarrow{\varphi_N} & N \end{array}$$

6.4. Definition. An *augmented* R-algebra is a quadruple $(A, \mu, \eta, \varepsilon)$, where (A, μ, η) is an R-algebra and $\varepsilon : A \to R$ is a homomorphism of R-algebras (called the *augmentation*). An augmented algebra $(A, \mu, \eta, \varepsilon)$ is called *connected* if $A_i = 0$ for $i < 0$ and $\varepsilon|A_0 : A_0 \to R$ is an isomorphism.

6.5. Definition. Let A be a connected R-algebra, and let $\bar{A}$ be the R-submodule consisting of all elements of positive degrees. The ideal $\bar{A}\bar{A}$ is denoted by Dec A, and its elements are called *decomposable elements* of A. Given an A-module M, let GM denote the factor module $M/\bar{A}M$. Furthermore, $G\bar{A} := \bar{A}/\operatorname{Dec} A$ usually is denoted by QA and is called (not very aptly) the set of *indecomposable elements* of A.

Sometimes we write Dec rather than Dec A when A is clear.

The following lemma can be proved by an obvious induction on dimension.

6.6. Lemma. (i) *Let $h : M \to N$ be a homomorphism of A-modules bounded below over a connected R-algebra A. If the R-homomorphism $Gh : GM \to GN$ is epic then so is h.*

(ii) *Let $f : A \to B$ be a homomorphism of connected R-algebras. If the function $Qf : QA \to QB$ is onto then f is epic.* □

The concept of a coalgebra is dual to the concept of an algebra.

6.7. Definition. (a) A *coalgebra* over R is a triple (C, Δ, ε), where C is an R-module and $\Delta : C \to C \otimes C$, $\varepsilon : C \to R$ are R-homomorphisms such that the diagrams

$$\begin{array}{ccc} C & \xrightarrow{\Delta} & C\otimes C \\ {\scriptstyle \Delta}\downarrow & & \downarrow{\scriptstyle \Delta\otimes 1} \\ C\otimes C & \xrightarrow{1\otimes\Delta} & C\otimes C\otimes C \end{array} \qquad \begin{array}{ccccc} C & = & C & = & C \\ {\scriptstyle \cong}\downarrow & & \downarrow{\scriptstyle \Delta} & & \downarrow{\scriptstyle \cong} \\ R\otimes C & \xleftarrow{\varepsilon\otimes 1} & C\otimes C & \xrightarrow{1\otimes\varepsilon} & C\otimes R \end{array}$$

commute. Here Δ is called the *comultiplication*, or the *diagonal map*, or just the *diagonal*, and ε is called the *augmentation*, or the *counit homomorphism*. A coalgebra (C, Δ, ε) is *cocommutative* if $T_{C,C}\Delta = \Delta$.

(b) A *comodule* over a coalgebra (C, Δ, ε) (or, briefly, a *C-comodule*) is a pair (V, ψ), where V is a graded R-module and $\psi = \psi_V$ is an R-homomorphism $V \to C \otimes V$ such that the following diagrams commute:

$$\begin{array}{ccc} V & \xrightarrow{\psi} & C \otimes V \\ {\scriptstyle\psi}\big\downarrow & & \big\downarrow{\scriptstyle 1\otimes\psi} \\ C \otimes V & \xrightarrow{\Delta\otimes 1} & C \otimes C \otimes V \end{array} \qquad \begin{array}{ccc} V & \xrightarrow{\psi} & C \otimes V \\ {\scriptstyle\cong}\big\downarrow & & \big\downarrow{\scriptstyle \varepsilon\otimes 1} \\ R \otimes V & = & R \otimes V. \end{array}$$

(c) A *homomorphism $h : C \to D$ of R-coalgebras* is an R-homomorphism such that $\Delta_D \circ h = (h \otimes h) \circ \Delta_C$ and $\varepsilon_D \circ h = \varepsilon_C$. A *homomorphism $f : V \to W$ of comodules* over a coalgebra (C, Δ, ε) is an R-homomorphism such that $(1_C \otimes f) \circ \psi_V = \psi_W \circ f$.

The set of all homomorphisms of C-comodules $V \to W$ will be denoted by $\mathrm{Hom}_C(V, W)$.

The duality between algebras and coalgebras is exhibited not only in the defining diagrams. For instance, let R be a field k. Given a k-vector space C, consider the dual vector space $C^* = \mathrm{Hom}_k(C, k)$. If C is a k-algebra (C, μ, η) then C^* obtains a natural k-coalgebra structure $(C^*, \Delta, \varepsilon)$ provided that every component C_n of C is a finite dimensional k-vector space. Namely, $\Delta(f)(a \otimes b) = f(ab)$, $\varepsilon = \eta^* : C^* \to k^* = k$. Conversely, if (C, Δ, ε) is a coalgebra over k then C^* obtains a k-algebra structure with $(fg)(a) = \sum f(a'_i) g(a''_i)$, where $\Delta(a) = \sum a'_i \otimes a''_i$. Moreover, a homomorphism $h : C \to D$ of algebras induces a homomorphism $h^* : D^* \to C^*$ of the coalgebras, and vice versa. There is also a similar duality between modules and comodules.

Let (C, Δ, ε) be a coalgebra over R, and let M be an R-module. We turn $C \otimes M$ into a C-comodule $(C \otimes M, \psi_M)$ by setting $\psi_M(c \otimes m) = \Delta(c) \otimes m$.

6.8. Lemma. *For every C-comodule (V, ψ), the function*

$$t : \mathrm{Hom}_C(V, C \otimes M) \to \mathrm{Hom}_R(V, M), \quad t(f) = (\varepsilon \otimes 1) f,$$

is bijective.

Proof. Define $s : \mathrm{Hom}_R(V, M) \to \mathrm{Hom}_C(V, C \otimes M)$ by setting $s(g) = (1 \otimes g) \circ \psi$. Then $ts(g) = (\varepsilon \otimes 1)(1 \otimes g)\psi = g$. On the other hand, the diagram

$$\begin{array}{ccccc} V & \xrightarrow{f} & C \otimes M & = & C \otimes M \\ {\scriptstyle\psi}\big\downarrow & & \big\downarrow{\scriptstyle\Delta} & & \big\| \\ C \otimes V & \xrightarrow{1\otimes f} & C \otimes C \otimes M & \xrightarrow{1\otimes\varepsilon\otimes 1} & C \otimes M \end{array}$$

commutes, and thus

$$st(f) = s((\varepsilon \otimes 1) f) = (1 \otimes (\varepsilon \otimes 1) f)\psi = (1 \otimes \varepsilon \otimes 1)(1 \otimes f)\psi = f. \qquad \square$$

6.9. Definition. A coalgebra (C, Δ, ε) is called *connected* if $C_i = 0$ for $i < 0$ and $\varepsilon | C_0 : C_0 \to R$ is an isomorphism. In this case the element $v = \varepsilon^{-1}(1)$ is called the *counit* of C.

Given two coalgebras C, D, we can turn $C \otimes D$ into a coalgebra by defining $\Delta_{C\otimes D}$ and $\varepsilon_{C\otimes D}$ to be the compositions

$$\Delta_{C\otimes D} : C \otimes D \xrightarrow{\Delta_C \otimes \Delta_D} C \otimes C \otimes D \otimes D \xrightarrow{1\otimes T\otimes 1} C \otimes D \otimes C \otimes D,$$

$$\varepsilon_{C\otimes D} : C \otimes D \xrightarrow{\varepsilon_C \otimes \varepsilon_D} R \otimes R \cong R.$$

It is easy to see that $\Delta : C \to C \otimes C$ is a homomorphism of coalgebras if the coalgebra (C, Δ, ε) is cocommutative.

Similarly, given two algebras A, B, we can turn $A \otimes B$ into an algebra by setting $\mu_{A\otimes B} = (\mu_A \otimes \mu_B) \circ (1 \otimes T \otimes 1)$ and $\eta_{A\otimes B} = \eta_A \otimes \eta_B$. Again, $\mu : A \otimes A \to A$ is a homomorphism of algebras if (A, μ, η) is commutative.

6.10. Lemma. *Let (C, Δ, ε) be a connected coalgebra with counit v. Then:*

(i) $\Delta(v) = v \otimes v$.

(ii) *For every $c \in C$, $|c| > 0$, we have $\Delta(c) = v \otimes c + c \otimes v + \sum c_i' \otimes c_i''$ with $|c_i'| < |c|, |c_i''| < |c|, |c'| + |c''| = |c|$.*

(iii) *Let (V, ψ) be a comodule over C. Then, for every $x \in V$, we have $\psi(x) = v \otimes x + \sum c' \otimes x'$ with $|x'| < |x|, |c'| + |x'| = |x|$.*

Proof. (i) Since C is connected, $\Delta(v) = v \otimes \lambda v$, $\lambda \in R$. Furthermore, $\lambda = 1$ because $v = (\varepsilon \otimes 1)(\Delta(v)) = (\varepsilon \otimes 1)(v \otimes \lambda v) = \varepsilon(v) \otimes \lambda v = \lambda v$.

(ii) Because of the commutativity of the right diagram of 6.7(a), and since C is connected, we conclude that $\Delta(c) = \lambda v \otimes c + c \otimes \mu v + \sum c_i' \otimes c_i''$ with $|c_i'| < |c|, |c_i''| < |c|$. Now, the equalities $\lambda = 1 = \mu$ can be proved as in (i).

(iii) This can be proved as (ii), using the commutativity of the right diagram of 6.7(b). □

6.11. Examples. (a) Let p be a prime, and let (E, μ, ι) be a ring spectrum of finite $\mathbb{Z}[p]$-type. Then $H^*(E; \mathbb{Z}/p)$ has the natural structure of a $\mathbb{Z}/p$-coalgebra. Indeed, consider the homomorphism

$$\Delta : H^*(E; \mathbb{Z}/p) \xrightarrow{\mu^*} H^*(E \wedge E; \mathbb{Z}/p) \xleftarrow[\mu^{E,E}]{\cong} H^*(E; \mathbb{Z}/p) \otimes H^*(E; \mathbb{Z}/p)$$

(by 4.11(ii), $\mu^{E,E}$ is an isomorphism) and set $\varepsilon := \iota^* : H^*(E; \mathbb{Z}/p) \to H^*(S; \mathbb{Z}/p)$. It is easy to see that $(H^*(E; \mathbb{Z}/p), \Delta, \varepsilon)$ is a coalgebra. The naturality is clear.

(b) If the ring spectrum (E, μ, ι) in (a) is commutative then $H^*(E; \mathbb{Z}/p)$ is a cocommutative coalgebra.

(c) If E is a spectrum as in (a) and F is any E-module spectrum of finite $\mathbb{Z}[p]$-type, then $H^*(F; \mathbb{Z}/p)$ admits a structure of a comodule over $H^*(E; \mathbb{Z}/p)$.

Case (c) can be considered in the same manner as (a).

(d) Dually, given a ring spectrum (E, μ, ι), consider the homomorphism

$$\mu_{\text{alg}} : H_*(E; \mathbb{Z}/p) \otimes H_*(E; \mathbb{Z}/p) \xrightarrow{\mu_{E,E}} H_*(E \wedge E) \xrightarrow{\mu_*} H_*(E)$$

and set $\eta := \iota_* : H_*(S; \mathbb{Z}/p) \to H_*(E; \mathbb{Z}/p)$. Then $(H_*(E; \mathbb{Z}/p), \mu_{\text{alg}}, \eta)$ is a $\mathbb{Z}/p$-algebra. Similarly, $H_*(F; \mathbb{Z}/p)$ is an $H_*(E; \mathbb{Z}/p)$-module for every E-module spectrum F.

(e) For every CW-space X and every field k we have a coalgebra $H_*(X; k)$, where the diagonal $d : X \to X \times X$ yields a comultiplication

$$\Delta : H_*(X; k) \xrightarrow{d_*} H_*(X \times X; k) \xleftarrow{\cong} H_*(X; k) \otimes H_*(X; k)$$

and the map $X \to \text{pt}$ yields an augmentation $H_*(X; k) \to k$. This coalgebra is connected iff X is connected.

(f) Dually to (e), $H^*(X; k)$ is a k-algebra for every field k and CW-space X of finite type.

6.12. Definition. Let (C, Δ, ε) be a connected coalgebra with counit v.

(a) An element $m \in C$ is called *primitive* if $\Delta(m) = m \otimes v + v \otimes m$. The set (in fact, R-submodule) of all primitive elements of C is denoted by $\operatorname{Pr} C$.

(b) Let (V, ψ) be a C-comodule. An element $m \in V$ is called *simple* if $\psi(m) = v \otimes m$. The R-submodule of simple elements of V is denoted by $\operatorname{Si} V$.

6.13. Remarks. (a) Under the duality between algebras and coalgebras over a field, $\operatorname{Pr} C$ is dual to QC^*.

(b) Sometimes simple elements are also called primitive, but we do not like this because of the danger of confusion: $\operatorname{Pr} C \neq \operatorname{Si} C$ where C is regarded as a coalgebra on the left and as comodule on the right.

6.14. Lemma. *Let $h : C \to D$ be a homomorphism of connected coalgebras over a field k. If the map $h|\operatorname{Pr} C$ is injective in dimensions $\leq d$ then is h is.*

Proof. Let v be the counit of C. Since $h(v)$ is the counit of D, we conclude that $h(v) \neq 0$, i.e., $h|C_0$ is monic. If h is monic on the subgroup of all elements of dimension $\leq d$, then $h \otimes h : C \otimes C \to D \otimes D$ is monic on the subgroup A_d generated by elements $m \otimes m'$ with $|m| < d, |m'| < d$. This is true because C and D are k-vector spaces. Now, let $x \in \operatorname{Ker} h$ be a non-zero element of minimal dimension. If $\dim x < d$ then x is not primitive, and so $\Delta(x) = x \otimes v + v \otimes x + \sum x' \otimes x''$ where $|x'| < |x|, |x''| < |x|$ and $\sum x' \otimes x'' \neq 0$. Now

$$\begin{aligned} 0 = \Delta(h(x)) &= (h \otimes h)\Delta(x) = (h \otimes h)(v \otimes x + x \otimes v + \sum x' \otimes x'') \\ &= h(v) \otimes h(x) + h(x) \otimes h(v) + \sum h(x') \otimes h(x'') = \sum h(x') \otimes h(x''). \end{aligned}$$

However, $\sum h(x') \otimes h(x'') \neq 0$, since $\sum x' \otimes x'' \neq 0$ and $h \otimes h$ is monic on A_d. This is a contradiction. □

6.15. Construction. (Boardman [1]). Given a connected coalgebra C, we define a filtration $F_m C$ by setting

$$(F_m C)_n = \begin{cases} C_n & \text{if } n \leq m, \\ 0 & \text{otherwise.} \end{cases}$$

Given a C-comodule (V, ψ), we set

$$F_{-1}V = 0,\ F_m V = \{x \in V | \psi(x) \in F_m C \otimes V\}.$$

6.16. Proposition. (i) $\bigcup_m F_m V = V$ *for every* V.
(ii) $F_0 V = \operatorname{Si} V$.
(iii) $f(F_m V) \subset F_m W$ *for every* C*-comodule homomorphism* $f : V \to W$.

Proof. This is obvious. □

6.17. Lemma. *Let* (V, ψ) *be a comodule over a connected coalgebra* (C, Δ, ε) *over a field. Then* $\psi(F_m V) \subset \sum_{j=0}^{m} C_j \otimes F_{m-j} V$.

Proof. Choose $x \in F_m V$. We have $\psi(x) = v \otimes x + \sum c_i \otimes x_i$, where $c_i \in C$ are assumed to be linearly independent. Since $(\Delta \otimes 1) \circ \psi = (1 \otimes \psi) \circ \psi$, we conclude that

$$\sum c_i \otimes \psi(x_i) = \sum (\Delta(c_i) - v \otimes c_i) \otimes x_i$$

in $C \otimes C \otimes V$. Since $x \in F_m V$, we conclude that $c_i \in F_m C$, and so

$$\Delta(c_i) - v \otimes c_i \in \sum_{r+s \leq m} C_r \otimes C_s$$

for every i. Hence, $c_i \otimes \psi(x_i) \in \sum_{r+s \leq m} C_r \otimes C_s \otimes V$. Thus, if $c_i \in C_j$ then $\psi(x_i) \in \sum_{s \leq m-j} C_s \otimes V = F_{m-j} C \otimes V$, i.e., $x_i \in F_{m-j} V$. □

6.18. Definition. Let M be a free R-module. Given a coalgebra (C, Δ, ε), define its *cofree* M*-extension* to be the C-comodule (V, ψ) where $V = C \otimes M$ and $\psi(c \otimes x) = \Delta(c) \otimes x, c \in C, x \in M$. A C-comodule V is called *cofree* if there is M such that V is isomorphic to the cofree M-extension of C.

Let C and V be as in 6.17. Define $\psi' : F_m V \to C_m \otimes \operatorname{Si} V$ to be the composition

$$F_m V \xrightarrow{\psi} \sum_{j=0}^{m} C_j \otimes F_{m-j} V \xrightarrow{q} C_m \otimes F_0 V = C_m \otimes \operatorname{Si} V$$

where q is the quotient map. Since $\psi'(F_{m-1}V) = 0$, ψ' passes through the homomorphism $\overline{\psi} = \overline{\psi}_m : F_mV/F_{m-1}V \to C_m \otimes \operatorname{Si} V$.

6.19. Lemma. *For every m, the homomorphism $\overline{\psi} : F_mV/F_{m-1}V \to C_m \otimes \operatorname{Si} V$ is a monomorphism, and it is an isomorphism if V is a cofree C-comodule. Furthermore, $\overline{\psi}$ is natural in V, i.e., for every C-homomorphism $f : V \to W$ and every m, the following diagram commutes:*

(6.20)
$$\begin{array}{ccc} F_mV/F_{m-1}V & \xrightarrow{f_*} & F_mW/F_{m-1}W \\ \Big\downarrow{\scriptstyle \overline{\psi^V}} & & \Big\downarrow{\scriptstyle \overline{\psi^W}} \\ C_m \otimes \operatorname{Si} V & \xrightarrow{1\otimes \operatorname{Si} f} & C_m \otimes \operatorname{Si} W \end{array} .$$

Proof. If $\overline{\psi}(x) = 0$ then $\psi(x) \in \sum_{j=0}^{m-1} C_j \otimes F_{m-j}V \subset F_{m-1}C \otimes V$, i.e., $x \in F_{m-1}V$. Thus, $\overline{\psi}$ is monic. Furthermore, if V is cofree, $V \cong C \otimes M$, then $F_mV \cong F_mC \otimes M$. So, $\overline{\psi}$ is an isomorphism. The naturality is clear. □

6.21. Corollary. *Let C be as in* 6.17, *and let $f : V \to W$ be a homomorphism of C-comodules.*

(i) *If* $\operatorname{Si} f : \operatorname{Si} V \to \operatorname{Si} W$ *is a monomorphism then so is f.*

(ii) *If V is cofree and* $\operatorname{Si} f : \operatorname{Si} V \to \operatorname{Si} W$ *is an isomorphism, then so is f.*

Proof. (i) By 6.16(iii), the homomorphism f induces homomorphisms $F_mf : F_mV \to F_mW$. By 6.16(i), it suffices to prove that F_mf is monic. We prove this by induction. By 6.16(ii), F_0f is monic. Suppose that $F_{m-1}f$ is monic. But $1\otimes \operatorname{Si} f$ is monic, and so, by 6.19, $F_mV/F_{m-1}V \to F_mW/F_{m-1}W$ is a monomorphism. Thus, by the Five Lemma, F_mf is a monomorphism. The induction is confirmed.

(ii) By 6.16(i), it suffices to prove that F_mf is an isomorphism. Since V is cofree, $\overline{\psi^V}$ is an isomorphism for every m, and so $\overline{\psi^W}$ is epic, and so it is an isomorphism. Now the proof can be completed similarly to (i). □

6.22. Definition. A *Hopf algebra* over R is a quintuple $(A, \mu, \eta, \Delta, \varepsilon)$ such that $(A, \mu, \eta, \varepsilon)$ is an augmented R-algebra, (A, Δ, ε) is an R-coalgebra, $\Delta : A \to A \otimes A$ and $\varepsilon : A \to R$ are homomorphisms of algebras, and $\eta : R \to A$ is a homomorphism of coalgebras.

It is easy to see that $\mu : A \otimes A \to A$ is a homomorphism of coalgebras. Also, if A is connected then its unit 1 coincides with its counit v, i.e., $v = 1$.

Note that if A is a Hopf algebra over a field k and if $\dim_k A_n < \infty$ for every n then $A^* = \operatorname{Hom}(A, k)$ is a Hopf algebra also (see the text after 6.7), but the multiplication and comultiplication interchange roles (e.g. if the multiplication on A is commutative then the comultiplication on A^* is cocommutative).

The homomorphism $A \otimes R \xrightarrow{\varepsilon \otimes 1} R \otimes R = R$ turns R into an A-module; the homomorphism $R = R \otimes R \xrightarrow{\eta \otimes 1} A \otimes R$ turns R into an A-comodule.

6.23. Constructions. Let $(A, \mu, \eta, \Delta, \varepsilon)$ be a Hopf algebra.

(a) Given two A-modules (M, φ), (N, ψ), we form an A-module $(M \otimes N, \theta)$ by setting θ to be the composition

$$A \otimes M \otimes N \xrightarrow{\Delta \otimes 1} A \otimes A \otimes M \otimes N \xrightarrow{1 \otimes T \otimes 1} A \otimes M \otimes A \otimes N \xrightarrow{\varphi \otimes \psi} M \otimes N.$$

In other words, $a(m \otimes n) = \sum a_i' m \otimes a_i'' n$ where $\Delta(a) = \sum a_i' \otimes a_i''$.

(b) Given two A-comodules (V, ψ) and (W, φ), we form an A-comodule $(V \otimes W, \theta)$ by setting θ to be the composition

$$V \otimes W \xrightarrow{\psi \otimes \varphi} A \otimes V \otimes A \otimes W \xrightarrow{1 \otimes T \otimes 1} A \otimes A \otimes V \otimes W \xrightarrow{\mu \otimes 1} A \otimes V \otimes W.$$

6.24. Definition. Let A be a Hopf algebra over R.

(a) An *A-module algebra* is a quadruple (M, μ, η, φ), where (M, μ, η) is an R-algebra and (M, φ) is an A-module such that $\mu : M \otimes M \to M$ is a homomorphism of A-modules and $\varphi : A \otimes M \to M$ is a homomorphism of R-algebras. A *homomorphism of A-module algebras* is a homomorphism of A-modules which is at the same time a homomorphism of R-algebras.

(b) An *A-comodule algebra* is a quadruple (M, μ, η, ψ), where (M, μ, η) is an R-algebra and (M, ψ) is an A-comodule such that $\mu : M \otimes M \to M$ is a homomorphism of A-comodules and $\psi : M \to A \otimes M$ is a homomorphism of R-algebras.

(c) An *A-module coalgebra* is a quadruple $(V, \Delta, \varepsilon, \varphi)$, where (V, Δ, ε) is an R-coalgebra and (V, φ) is an A-module such that $\Delta : V \to V \otimes V$ is a homomorphism of A-modules and $\varphi : A \otimes V \to V$ is a homomorphism of R-coalgebras.

(d) We leave it to the reader to define the A-comodule coalgebras (and homomorphisms in cases (b), (c), (d)).

6.25. Recollection. A very important example of a Hopf algebra over the field $\mathbb{Z}/p$, p prime, is the Steenrod algebra

$$\mathscr{A}_p = H^*(H\mathbb{Z}/p; \mathbb{Z}/p) = \bigoplus_d [H\mathbb{Z}/p, \Sigma^d H\mathbb{Z}/p].$$

The multiplication $\mu : H\mathbb{Z}/p \wedge H\mathbb{Z}/p \to H\mathbb{Z}/p$ induces the diagonal map

$$\begin{aligned} H^*(H\mathbb{Z}/p; \mathbb{Z}/p) &\to H^*(H\mathbb{Z}/p \wedge H\mathbb{Z}/p; \mathbb{Z}/p) \\ &\cong H^*(H\mathbb{Z}/p; \mathbb{Z}/p) \otimes H^*(H\mathbb{Z}/p; \mathbb{Z}/p); \end{aligned}$$

the algebra structure is given by the composition of cohomology operations $H\mathbb{Z}/p \to \Sigma^n H\mathbb{Z}/p$.

All that we need to know about $\mathscr{A}_p$ can be found in Steenrod–Epstein [1], Margolis [1]. The Steenrod algebra $\mathscr{A}_2$ is generated by elements Sq^i of dimension i, where $i = 1, 2, \dots$, and all relations among Sq^i follow from the Adem relations

$$Sq^a Sq^b = \sum_{c=0}^{[a/2]} \binom{b-c-1}{a-2c} Sq^{a+b-c} Sq^c \text{ for } a < 2b, \tag{6.26}$$

where $Sq^0 := 1$. The comultiplication has the form $\Delta(Sq^k) = \sum_{i=0}^{k} Sq^i \otimes Sq^{k-i}$.

The Steenrod algebra $\mathscr{A}_p, p > 2$, is generated by elements β and P^i, $\dim \beta = 1, \dim P^i = 2i(p-1), i = 1, 2, \dots$. Again, all relations in $\mathscr{A}_p$ follow from the Adem relations (for explicit formulae see Steenrod–Epstein [1]). The comultiplication has the form $\Delta(\beta) = \beta \otimes 1 + 1 \otimes \beta$, $\Delta(P^k) = \sum_{i=0}^{k} P^i \otimes P^{k-i}$, where $P^0 := 1$.

Now we describe the primitive elements of $\mathscr{A}_p$. Milnor [2] described the Hopf algebra $\mathscr{A}_p^* = \operatorname{Hom}^*(\mathscr{A}_p, \mathbb{Z}/p)$. For $p > 2$ we have

$$\mathscr{A}_p^* = \mathbb{Z}/p\,[\xi_1, \xi_2, \dots, \xi_n, \dots] \otimes \Lambda(\tau_0, \tau_1, \dots, \tau_n, \dots),$$

where $\dim \xi_i = 2(p^i - 1), \dim \tau_i = 2p^i - 1$. The comultiplication ∇ on $\mathscr{A}_p^*$ has the form

$$\nabla(\xi_k) = \sum_{i=0}^{k} \xi_{k-i}^{p^i} \otimes \xi_i, \qquad \nabla(\tau_k) = \sum_{i=0}^{k} \xi_{k-i}^{p^i} \otimes \tau_i,$$

where $\xi_0 := 1$. Let $\mathscr{R}$ be the set of all sequences $R = \{r_1, \dots, r_n, \dots\}$ of integers such that $r_i \geq 0$ and $r_i = 0$ for all but a finite number of i's. The algebra $\mathscr{A}_p^*$ has an additive basis $\{\xi^R \tau_0^{\varepsilon_0} \cdots \tau_k^{\varepsilon_k} \cdots\}$, where $\varepsilon_i = 0$ or 1 and $\xi^R = \xi_1^{r_1} \cdots \xi_n^{r_n}, R = \{r_1, \dots, r_n, \dots\} \in \mathscr{R}$. The element of $\mathscr{A}_p$ which is dual to τ_i is denoted by $Q_i, \dim Q_i = 2p^i - 1$; the dual to ξ^R is denoted by $\mathscr{P}^R, \dim \mathscr{P}^R = \sum 2r_i(p^i - 1)$. A $\mathbb{Z}/p$-basis of $\mathscr{A}_p$ is just $\{\mathscr{P}^R Q_0^{\varepsilon_0} \cdots Q_k^{\varepsilon_k} \cdots\}$, where $\varepsilon_i = 0$ or 1. There are the relations $\mathscr{P}^R Q_k - Q_k \mathscr{P}^R = \sum Q_{k+1} \mathscr{P}^{R - p^k \Delta_i}$, where Δ_i is the sequence with 1 in the i-th place and zeros elsewhere. Furthermore, $Q_i^2 = 0$ and $Q_i Q_j + Q_j Q_i = 0$. The expansion of $\mathscr{P}^R \mathscr{P}^S$ with respect to this basis can be found in Milnor [2].

Under this notation the comultiplication Δ on $\mathscr{A}_p$ has the form

$$\Delta(Q_i) = Q_i \otimes 1 + 1 \otimes Q_i, \ \Delta(\mathscr{P}^R) = \sum_{R' + R'' = R} \mathscr{P}^{R'} \otimes \mathscr{P}^{R''}.$$

So, Q_i and $\mathscr{P}^{\Delta_i}$ are primitive elements. Furthermore, there are no other primitive elements (up to multiplication by constants), i.e.,

$$\Pr \mathscr{A}_p = \mathbb{Z}/p\{Q_0, Q_1, \dots, Q_k, \dots, \mathscr{P}^{\Delta_1}, \mathscr{P}^{\Delta_2}, \dots, \mathscr{P}^{\Delta_n}, \dots\},$$

where $\mathbb{Z}/p\{\dots\}$ denotes "the $\mathbb{Z}/p$-vector space spanned by ...". This is true because $\dim \Pr \mathscr{A}_p = \dim Q\mathscr{A}_p^*$.

Mind the relations

$$Q_0 = \beta,\ Q_i = [P^{p^{i-1}}, Q_{i-1}],\ \mathscr{P}^{\Delta_1} = P^1,\ \mathscr{P}^{\Delta_i} = [P^{p^{i-1}}, \mathscr{P}^{\Delta_{i-1}}],$$

where $[x,y] = xy - (-1)^{|x||y|}yx$. Furthermore, the two-sided ideal (Q_0) of $\mathscr{A}_p$ coincides with the left ideal $\mathscr{A}_p(Q_0, \dots, Q_n, \dots)$.

For $p = 2$ we have $\mathscr{A}_2^* = \mathbb{Z}/2[\zeta_1, \zeta_2, \dots, \zeta_n, \dots]$, $\dim \zeta_n = 2^n - 1$, and $\nabla(\zeta_k) = \sum_{i=0}^k \zeta_{k-i}^{2^i} \otimes \zeta_i$. As in the case $p > 2$, there is a $\mathbb{Z}/2$-basis $\{\zeta^R\}$ of $\mathscr{A}_2^*$, and so $\mathscr{A}_2$ has a $\mathbb{Z}/2$-basis $\{Sq^R\}$, $\dim Sq^R = \sum r_n(2^n - 1)$, and

$$\Pr \mathscr{A}_2 = \mathbb{Z}/2\{Sq^{\Delta_1}, \dots, Sq^{\Delta_n}, \dots\}.$$

Moreover, $Sq^{\Delta_1} = Sq^1$, $Sq^{\Delta_i} = [Sq^{2^{i-1}}, Sq^{\Delta_{i-1}}]$.

It is possible and useful to introduce the notation $Q_i = Sq^{\Delta_{i+1}}$, $i = 0, 1, \dots$, similar to the case $p > 2$. The relevance of this notation lies in the fact that in this case some formulae for $p = 2$ look like those for $p > 2$. For example, $Q_i^2 = 0$ and $Q_iQ_j + Q_jQ_i = 0$, and the left ideal $\mathscr{A}_2(Q_0, \dots, Q_n, \dots)$ coincides with the two-sided ideal (Q_0) (prove this).

Note that the Hopf algebra $\mathscr{A}_p^*$ can be described as $H_*(H)$, where $H = H\mathbb{Z}/p$. The multiplication has the form

$$H_*(H) \otimes H_*(H) \xrightarrow{\cong} H_*(H \wedge H) \xrightarrow{\mu_*} H_*(H)$$

where the isomorphism is $\mu_{H,H}$, and the comultiplication has the form

$$\begin{aligned} H_*(H) = \pi_*(H \wedge H) \to \pi_*(H \wedge S \wedge H) &\xrightarrow{1 \wedge \iota \wedge 1} \pi_*(H \wedge H \wedge H) \\ \cong H_*(H \wedge H) \cong H_*(H) \otimes H_*(H). \end{aligned}$$

6.27. Examples. (a) For every H-space X and every field k, we have a Hopf algebra $H_*(X;k)$: the multiplication is induced by the H-structure $X \times X \to X$ while the comultiplication is induced by the diagonal $d : X \to X \times X$, cf. 6.11(e). Similarly, $H^*(X;k)$ is a Hopf algebra for every H-space X of finite type, and $H_*(X;k)$ is the Hopf algebra dual to $H^*(X;k)$.

(b) Let $H(-)$ denote $H(-;\mathbb{Z}/p)$. For every spectrum E, the group $H^*(E)$ admits a natural $\mathscr{A}_p$-module structure. The action

$$\varphi : \mathscr{A}_p \otimes H^*(E) \to H^*(E),\ \varphi(a \otimes x) = a(x),$$

is given by the evaluation of a cohomology operation a on an element x, see 3.47. The detailed description of this action can be found in Steenrod–Epstein [1], Margolis [1], Mosher–Tangora [1]. Here we recall that $Sq^i(x)=0$ if $|x|<i$, $Sq^i(x)=x^2$ if $|x|=i$, $P^i(x)=0$ if $|x|<2i(p-1)$, $P^i(x)=x^p$ if $|x|=2i(p-1)$. Finally, if $f: S^n \to S^n$ is a map of degree p and x is a generator of the group $H^n(Cf;\mathbb{Z}/p)=\mathbb{Z}/p$ then $\beta(x)\neq 0$.

Furthermore, if E is a ring spectrum of finite $\mathbb{Z}$- or $\mathbb{Z}[p]$-type then $H^*(E)$ is a coalgebra over the Hopf algebra $\mathscr{A}_p$, see 7.19(ii) below. Moreover, one can easily prove that a ring morphism $E \to F$ of such spectra induces a homomorphism $H^*(F)\to H^*(E)$ of $\mathscr{A}_p$-module coalgebras.

(c) Dually, let $\iota: S \to H$ be the unit. Given a spectrum E, the morphism $b^E: E \xleftarrow{l,\cong} S\wedge E \xrightarrow{\iota\wedge 1} H\wedge E$ induces a homomorphism

$$\psi: H_*(E)=H_*(S\wedge E)\xrightarrow{(\iota\wedge 1)_*} H_*(H\wedge E)=H_*(H)\otimes H_*(E)=\mathscr{A}_p^*\otimes H_*(E),$$

and this homomorphism turns $H_*(E)$ into a comodule over the Hopf algebra $\mathscr{A}_p^*$. Furthermore, every ring morphism $E\to F$ induces a homomorphism $H_*(E)\to H_*(F)$ of $\mathscr{A}_p^*$-comodule algebras.

Let C be a connected module coalgebra over a Hopf algebra A, and let v be the counit of C.

6.28. Lemma. *The map $\nu: A\to C$, $\nu(a)=av$, is a homomorphism of coalgebras.*

Proof. Let $\Delta: C\to C\otimes C$ be the diagonal. We have

$$(\nu\otimes\nu)(\Delta a)=(\nu\otimes\nu)\left(\sum a'\otimes a''\right)=\sum \nu(a')\otimes\nu(a'')=\sum a'v\otimes a''v\,.$$

On the other hand, by 6.10(i), $\Delta(v)=v\otimes v$, and so

$$\Delta\nu(a)=\Delta(av)=a\Delta(v)=a(v\otimes v)=\sum a'v\otimes a''v\,.$$

Thus, $(\nu\otimes\nu)(\Delta a)=\Delta\nu(a)$. □

6.29. Theorem. *Let A be a connected Hopf algebra over a field k, let C be a connected module coalgebra over A, and let v be the counit of C. Let $p: C\to GC$ be the canonical epimorphism (see 6.5), and let $\lambda: GC\to C$ be a k-homomorphism with $p\lambda=1_{GC}$. Let B_m be the subspace of $A\otimes GC$ generated by all elements $a\otimes x$ with $|a|\leq m$. If the map $\nu: A\to C$, $\nu(a)=av$, is monic for $|a|\leq m$ then the A-homomorphism $\eta: A\otimes GC\to C$, $\eta(a\otimes x)=a(\lambda x)$ is monic on B_m. Furthermore, η is epic and thus is an isomorphism in dimensions $\leq m$.*

Proof. This theorem is a version of the Milnor–Moore Theorem, and the proof below follows its proof contained in Stong [3]. Consider the composition of A-homomorphisms

$$A \otimes GC \xrightarrow{\eta} C \xrightarrow{\Delta} C \otimes C \xrightarrow{1\otimes p} C \otimes GC.$$

For every $y, x \in C$ and for every $a \in A$ (with $\Delta(a) = \sum a' \otimes a''$), we have

$$(1 \otimes p)(a \cdot (y \otimes x)) = (1 \otimes p)\left(\sum a'y \otimes a''x\right) = (1 \otimes p)(ay \otimes x)$$
$$= ay \otimes p(x)$$

(here $\cdot$ denotes the A-action on $C \otimes C$) because $p(a''x) = 0$ for $|a''| > 0$. Now,

$$(1 \otimes p)\Delta\eta(a \otimes x) = (1 \otimes p)a \cdot (\Delta(\lambda x))$$
$$=(1 \otimes p)a \cdot (v \otimes \lambda x + \lambda x \otimes v + \sum (\lambda x)' \otimes (\lambda x)'') = av \otimes x + b,$$

where $b \in \bigcup_{k<|x|} C \otimes (GC)_k$. Since $av \neq 0$ for $|a| \leq m$, we conclude that $av \otimes x + b \neq 0$ (for dimensional reasons). Hence, $(1 \otimes p)\Delta\eta$ is monic on B_m, and thus so is η.

To prove that η is epic, consider any k-basis $\{e_i\}$ of GC and set $c_i = \lambda e_i$. So, $c_i \in \operatorname{Im} \eta$. Let $c \in (C \setminus \operatorname{Im} \eta)$ be a homogeneous element of minimal dimension. We have $pc = \sum n_i e_i$, $n_i \in k$. So, $p(c - \sum n_i c_i) = 0$, i.e., $c - \sum n_i c_i = \sum a_k x_k$, $a_k \in \bar{A}, x_k \in C$. So, $\dim a_k > 0$. Hence, $\dim x_k < \dim c$, and therefore $x_k \in \operatorname{Im} \eta$. Thus, $c \in \operatorname{Im} \eta$. □

6.30. Corollary (The Milnor–Moore Theorem). *Let A and C be as in 6.29. If $\nu : A \to C$, $\nu(a) = av$, is monic then there is an isomorphism of A-modules $C \cong A \otimes GC$. In particular, C is a free A-module.* □

6.31. Corollary. *Let A and C be as in 6.29. If $\nu(x) \neq 0$ for every primitive $x \in A$, $x \neq 0$, then C is a free A-module.*

Proof. This follows from 6.28, 6.14, and 6.30. □

6.32. Remark. In fact, the proof of 6.29 shows that 6.29 and 6.30, 6.31 are valid also for "non-coassociative coalgebras", i.e., for triples (C, Δ, ε) not satisfying the commutativity of the left diagram from 6.7. This remark is useful for applications.

6.33. Lemma. *If (V, ψ) is a comodule algebra over a connected Hopf algebra A then $\operatorname{Si} V$ is a comodule subalgebra of V.*

Proof. Only that $\operatorname{Si} V$ is a subalgebra needs proof. But $\psi : V \to C \otimes V$ is a homomorphism of algebras, and thus

$$\psi(xy) = \psi(x)\psi(y) = (v \otimes x)(v \otimes y) = v \otimes xy. \qquad \square$$

6.34. Theorem (cf. Boardman [1], Milnor–Moore [1]). *Let V be a commutative comodule algebra over a connected Hopf algebra A over a field k. Let $b : A \to V$ be a homomorphism of A-comodule algebras. Then the composition*

$$f : A \otimes \operatorname{Si} V \xrightarrow{b\otimes 1} V \otimes \operatorname{Si} V \to V \otimes V \xrightarrow{\mu_V} V$$

is an isomorphism of A-comodule algebras.

Proof. Since V is commutative, μ is a homomorphism of algebras, and therefore so is f. Clearly, f is a homomorphism of A-comodules. Furthermore, $\operatorname{Si} A = k = k\{v\}$, and so $\operatorname{Si} f : \operatorname{Si}(A \otimes \operatorname{Si} V) = k \otimes \operatorname{Si} V \to \operatorname{Si} V$ is an isomorphism. But $A \otimes \operatorname{Si} V$ is a cofree A-comodule, and thus, by 6.21(ii), f is an isomorphism. □

6.35. Definition. Given a connected Hopf algebra $(A, \mu, \eta, \Delta, \varepsilon)$, we define a *canonical antiautomorphism* $c : A \to A$ (called also an *antipode*) as follows. Firstly, $c(1) := 1$. Now, if

$$\Delta(x) = x \otimes 1 + 1 \otimes x + \sum x' \otimes x'', \quad |x'| < |x|, |x''| < |x|$$

then $c(x) := -x - \sum c(x')x''$.

One can prove that $c^2 = 1_A$ and $c(ab) = c(b)c(a)$. Moreover, c can be characterized by commutativity of the diagrams

$$\begin{array}{ccc} A & \xrightarrow{\eta\varepsilon} & A \\ {\scriptstyle\Delta}\downarrow & & \uparrow{\scriptstyle\mu} \\ A \otimes A & \xrightarrow{1\otimes c} & A \otimes A \end{array} \qquad \begin{array}{ccc} A & \xrightarrow{\eta\varepsilon} & A \\ {\scriptstyle\Delta}\downarrow & & \uparrow{\scriptstyle\mu} \\ A \otimes A & \xrightarrow{c\otimes 1} & A \otimes A, \end{array}$$

see Milnor–Moore [1].

The first example of such an antiautomorphism was found by Thom [1]; this was the canonical antiautomorphism $\chi : \mathscr{A}_p \to \mathscr{A}_p$ of the Steenrod algebra $\mathscr{A}_p$. For the dual Hopf algebra $\mathscr{A}_p^*$, this antiautomorphism has the form

$$\mathscr{A}_p^* = H_*(H) = \pi_*(H \wedge H) \xrightarrow{\tau_*} \pi_*(H \wedge H) = H_*(H) = \mathscr{A}_p^*$$

where $H = H\mathbb{Z}/p$ and τ switches the factors, see Switzer [1], Th. 17.8.

Every morphism $\theta : E \to F$ of spectra induces a morphism $\theta_* : E_*(X) \to F_*(X)$ of the corresponding homology theories. In particular, every operation $\varphi \in \mathscr{A}_p$ gives us a morphism $\varphi_* : H_*(X; \mathbb{Z}/p) \to H_*(X; \mathbb{Z}/p)$, cf. 3.47. On the other hand, we can define another morphism $\varphi_\bullet : H_*(X; \mathbb{Z}/p) \to H_*(X; \mathbb{Z}/p)$ of homology theories by setting $\langle \varphi_\bullet(x), y \rangle = \langle x, \varphi(y) \rangle$ for every $x \in H_*(X, \mathbb{Z}/p), y \in H^*(X; \mathbb{Z}/p)$. In this case we have the *right* $\mathscr{A}_p^*$-action on $H_*(X; \mathbb{Z}/p)$.

6.36. Proposition. $\varphi_* = (\chi(\varphi))_\bullet$ *for every* $\varphi \in \mathscr{A}_p$.

Proof. Given $\varphi \in \mathscr{A}_p$, define an operation

$$\lambda(\varphi) : H^*(X; \mathbb{Z}/p) \to H^*(X; \mathbb{Z}/p)$$

via the formula

$$\langle \varphi_*(x), y \rangle = \langle x, \lambda(\varphi)(y) \rangle, \quad x \in H_*(X; \mathbb{Z}/p), y \in H^*(X; \mathbb{Z}/p).$$

Now check that $\lambda : \mathscr{A}_p \to \mathscr{A}_p$ preserves the commutativity of the above diagrams, i.e., $\lambda = \chi$, cf. Thom [1], Th. III.23. □

6.37. Remarks. Hopf [1] found that ordinary (co)homology rings of Lie groups (in fact, H-spaces) had certain specific algebraic properties. (For example, the rational cohomology ring of a Lie group is a free commutative algebra.) Afterwards Borel [1] clarified the situation: every algebra A (over a field) admitting a diagonal $\Delta : A \to A \otimes A, \Delta(ab) = \Delta(a)\Delta(b)$, has such properties. Borel did not use the term "Hopf algebra"; he referred to the "algebraic theorem of Hopf". A systematic treatment of Hopf algebras was given by Milnor–Moore [1].

Milnor discovered that the Steenrod algebra is a Holf algebra. In this way he got a new description of $\mathscr{A}_p$, and this enabled him (and some others) to compute initial terms of certain Adams spectral sequences.

Milnor–Moore [1] proved 6.30. Furthermore, they proved 6.34 for V bounded below. Boardman [1] got rid of this restriction. To do this, he introduced the filtration $F_m V$ as in 6.15 and proved its properties 6.16–6.21.

§7. Graded Eilenberg–Mac Lane Spectra

A graded Eilenberg–Mac Lane spectrum $H(G)$ of a graded abelian group G was defined in 3.32(d). For future reference we mention the following fact.

7.1. Proposition. *For every two graded abelian groups G, G' the homomorphism*

$$[H(G), H(G')] \to \mathrm{Hom}^0(G, G'), \ [f] \mapsto \pi_*(f)$$

is epic.

Proof. Let G_i be the component of degree i of G. We have $H(G) = \vee \Sigma^i H(G_i)$. Similarly for G'. Consider the inclusion $j_i : \Sigma^i H(G_i) \to H(G)$ and the projection $p_i : H(G') \to \Sigma^i H(G'_i)$. We have the homomorphism

$$\begin{aligned} h_i : [H(G), H(G')] \xrightarrow{j_i^*} [\Sigma^i H(G_i), H(G')] \xrightarrow{(p_i)_*} [\Sigma^i H(G_i), \Sigma^i H(G'_i)] \\ = [H(G_i), H(G'_i)] = \mathrm{Hom}(G_i, G'_i); \end{aligned}$$

the last equality follows from 4.9. Note that h_i is an epimorphism since both j_i^* and $(p_i)_*$ are. These epimorphisms h_i yield the epimorphism

$$h := \{h_i\} : [H(G), H(G')] \to \prod_i \mathrm{Hom}(G_i, G'_i) = \mathrm{Hom}^0(G, G').$$

We leave it to the reader to prove that this homomorphism h coincides with the homomorphism in question. □

Every graded abelian group G can be realized as the total homotopy group $\pi_*(H(G))$ of the graded Eilenberg–Mac Lane spectrum $H(G)$, but not every spectrum E is (equivalent to) the graded Eilenberg–Mac Lane spectrum $H(\pi_*(E))$. (For example, the sphere spectrum S is not, because otherwise $H_*(H\mathbb{Z})$ would be a direct summand of $H_*(S)$.) It is clear that it is useful to know whether a spectrum is a graded Eilenberg–Mac Lane spectrum. For example, Thom [2] proved that the spectrum TBO of the non-oriented (co)bordism is a graded Eilenberg–Mac Lane spectrum, and this enabled him to compute the group $\pi_*(TBO)$ (i.e., non-oriented cobordism group) and to prove the realizability of all $\mathbb{Z}/2$-homology classes by singular manifolds, see Ch. IV. In this section we give some sufficient conditions for a spectrum E to be a graded Eilenberg–Mac Lane spectrum, i.e., $E \simeq H(\pi_*(E))$; these conditions will be used in next chapters.

7.2. Lemma. *Let Y be a graded Eilenberg–Mac Lane spectrum, and let $f : Y \to Z$ be a morphism of spectra such that $f_* : \pi_*(Y) \to \pi_*(Z)$ is a split epimorphism. Then Z is a graded Eilenberg–Mac Lane spectrum, and f has a homotopy right inverse $s : Z \to Y$, $fs \simeq 1_Z$.*

Proof. Since f_* splits, there is a subgroup G of $\pi_*(Y)$ such that $f_*|G : G \to \pi_*(Z)$ is an isomorphism. By 7.1, the inclusion $G \subset \pi_*(Y)$ is induced by a morphism $j : H(G) \to Y$, and fj is an equivalence. Now set $s = jg$, where $g : Z \to H(G)$ is a homotopy inverse to fj. □

7.3. Proposition. (i) *If E is a graded Eilenberg–Mac Lane spectrum then so is each of its coskeletons $E_{(n)}$. In particular, every Postnikov invariant of E is trivial.*

(ii) *If all Postnikov invariants of a spectrum E are trivial then E is a graded Eilenberg–Mac Lane spectrum.*

(iii) *If every coskeleton of a spectrum E is a graded Eilenberg–Mac Lane spectrum then E is a graded Eilenberg–Mac Lane spectrum.*

Proof. (i) This follows from 7.2.

(ii) Set $\pi_k = \pi_k(E)$. By 1.17, $E_{(n)} = E_{(n-1)} \vee \Sigma^n H(\pi_n)$ for every n. The morphism $E \xrightarrow{\tau} E_{(n)} \xrightarrow{\mathrm{proj}} \Sigma^n H(\pi_n)$ induces a homomorphism $\varphi_n : [X, E] \to H^n(X; \pi_n)$, $X \in \mathscr{S}$. So, we get a homomorphism $\varphi := \{\varphi_n\} : [X, E] \to \prod_n H^n(X; \pi_n)$. Furthermore, φ yields a morphism

$$E^i(X) = E^0(\Sigma^{-i} X) \xrightarrow{\varphi} \prod_n H^n(\Sigma^{-i} X; \pi_n) = \prod_n H^{n+i}(X; \pi_n)$$

of additive cohomology theories on $\mathscr{S}$. Now, by 3.19(iii), this is an isomorphism of cohomology theories on $\mathscr{S}$.

Similarly, by setting $F := \bigvee_{n=-\infty}^{\infty} \Sigma^n H(\pi_n)$, we get a natural isomorphism $[X, F] \cong \prod_n H^n(X; \pi_n), X \in \mathscr{S}$. So, there is a natural isomorphism $[X, E] \cong [X, F]$, and thus, by general categorical reasons, $E \simeq F$.

(iii) This follows from (ii). □

7.4. Proposition. *Let E be a spectrum of finite $\mathbb{Z}$-type such that its localization $E[p]$ is a graded Eilenberg–Mac Lane spectrum for every prime p. Then E is a graded Eilenberg–Mac Lane spectrum.*

Proof. By 7.3(i), every Postnikov invariant of $E[p]$ is trivial. So, by 5.11, the $\mathbb{Z}[p]$-localization of every Postnikov invariant of E is trivial for every p. Since E has finite $\mathbb{Z}$-type, each of its Postnikov invariants belongs to a finitely generated group, and so, by the above, every Postnikov invariant of E is trivial. Thus, by 7.3(ii), E is a graded Eilenberg–Mac Lane spectrum. □

Recall that $H\mathbb{Z}$ is a ring spectrum by 4.10. Let $\iota : S \to H\mathbb{Z}$ be its unit.

7.5. Theorem. *Let E be a spectrum. Suppose that there exists a morphism $m : H\mathbb{Z} \wedge E \to E$ such that the diagram*

$$\begin{array}{ccc} H\mathbb{Z} \wedge E & \xrightarrow{m} & E \\ {\scriptstyle \iota\wedge 1}\uparrow & & \| \\ S \wedge E & \underset{\simeq}{\xrightarrow{l(E)}} & E \end{array}$$

commutes up to homotopy. Then E is a graded Eilenberg–Mac Lane spectrum.

Proof. By 4.33, for every k there exists a morphism $f_k : \Sigma^k M(\pi_k(E)) \to E$ such that $(f_k)_* : \pi_k(\Sigma^k M(\pi_k(E))) \to \pi_k(E)$ is an isomorphism. Consider the morphism

$$g_k : H\mathbb{Z} \wedge \Sigma^k M(\pi_k(E)) \xrightarrow{1\wedge f_k} H\mathbb{Z} \wedge E \xrightarrow{m} E.$$

By 5.6(i), $H\mathbb{Z} \wedge M(\pi) \simeq H(\pi)$, and so g_k has the form $g_k : \Sigma^k H(\pi_k(E)) \to E$. Furthermore, $(g_k)_* : \pi_k(\Sigma^k H(\pi_k(E))) \to \pi_k(E)$ is an isomorphism, since the diagram

$$\begin{array}{ccccc} S \wedge \Sigma^k M(\pi_k(E)) & \xrightarrow{1\wedge f_k} & S \wedge E & \xrightarrow{\simeq} & E \\ {\scriptstyle \iota\wedge 1}\downarrow & & \downarrow{\scriptstyle \iota\wedge 1} & & \| \\ H\mathbb{Z} \wedge \Sigma^k M(\pi_k(E)) & \xrightarrow{1\wedge f_k} & H\mathbb{Z} \wedge E & \xrightarrow{m} & E \end{array}$$

commutes up to homotopy. Let $i_k : \Sigma^k H(\pi_k(E)) \to \vee_k \Sigma^k H(\pi_k(E))$ be the inclusion. By 1.16(i), there is a morphism $g : \vee_k \Sigma^k H(\pi_k(E)) \to E$ such that

$gi_k \simeq g_k$ for every k. Then g induces an isomorphism of homotopy groups, and thus it is an equivalence. □

7.6. Corollary. *Every $H\mathbb{Z}$-module spectrum is a graded Eilenberg–Mac Lane spectrum.* □

7.7. Corollary. *Let E be a ring spectrum with the unit $\iota_E : S \to E$. If there exists a morphism $f : H\mathbb{Z} \to E$ such that the composition $S \xrightarrow{\iota} H\mathbb{Z} \xrightarrow{f} E$ is homotopic to ι_E, then E is a graded Eilenberg–Mac Lane spectrum.*

Proof. The composition $H\mathbb{Z} \wedge E \xrightarrow{f \wedge 1} E \wedge E \xrightarrow{\mu} E$ satisfies the conditions of 7.5. □

7.8. Corollary. *If a ring spectrum E admits a ring morphism $H\mathbb{Z} \to E$ then E is a graded Eilenberg–Mac Lane spectrum.* □

7.9. Corollary. *For every spectrum E, the spectrum $H\mathbb{Z} \wedge E$ is a graded Eilenberg–Mac Lane spectrum.*

Proof. Let μ be the multiplication on $H\mathbb{Z}$. The associativity of μ implies that the morphism

$$m : H\mathbb{Z} \wedge (H\mathbb{Z} \wedge E) \xrightarrow{\simeq} (H\mathbb{Z} \wedge H\mathbb{Z}) \wedge E \xrightarrow{\mu} H\mathbb{Z} \wedge E$$

yields a $H\mathbb{Z}$-module structure on $H\mathbb{Z} \wedge E$. □

7.10. Corollary. (i) *For every abelian group π and every spectrum E, the spectrum $H(\pi) \wedge E$ is a graded Eilenberg–Mac Lane spectrum.*

(ii) *If F is a graded Eilenberg–Mac Lane spectrum then $E \wedge F$ is a graded Eilenberg–Mac Lane spectrum for every spectrum E.*

Proof. (i) $H(\pi) \wedge E \simeq H\mathbb{Z} \wedge M(\pi) \wedge E$.

(ii) This follows from (i) and 2.1(v). □

7.11. Theorem. (i) *The $\mathbb{Q}$-localization of the sphere spectrum S is $H\mathbb{Q}$. In particular, the Hurewicz homomorphism $h : \pi_*(E) \otimes \mathbb{Q} \to H_*(E) \otimes \mathbb{Q}$ is an isomorphism for every spectrum E.*

(ii) *The $\mathbb{Q}$-localization $E[0]$ of every spectrum E is the graded Eilenberg–Mac Lane spectrum $H(\pi_*(E) \otimes \mathbb{Q})$.*

(iii) *Let G, G' be a pair of graded vector spaces over $\mathbb{Q}$. Then the homomorphism $[H(G), H(G')] \to \mathrm{Hom}^0(G, G')$ in 7.1 is an isomorphism.*

(iv) *For every two $\mathbb{Q}$-local spectra E, F, the homomorphism $\mu_{E,F} : \pi_*(E) \otimes \pi_*(F) \to \pi_*(E \wedge F)$ is an isomorphism.*

(v) *Given two ring $\mathbb{Q}$-local spectra E, F, a morphism $f : E \to F$ is a ring morphism iff $f_* : \pi_*(E) \to \pi_*(F)$ is a ring homomorphism.*

Proof. (i) The groups $\pi_i(S)$ are finite for $i \neq 0$ by the Serre Theorem, see Serre [3] or Mosher–Tangora [1]. So, $\pi_i(S[0]) = \pi_i(S) \otimes \mathbb{Q} = 0$ for $i \neq 0$, and $\pi_0(S[0]) = \mathbb{Q}$.

(ii) Because of (i), $\pi_*(M\mathbb{Q}) \cong \pi_*(S) \otimes \mathbb{Q} \cong \pi_*(H\mathbb{Q})$, and so $M\mathbb{Q} \simeq H\mathbb{Q}$. Hence, $H\mathbb{Q} \wedge H\mathbb{Z} \simeq H\mathbb{Q}$. Thus,

$$E[0] = E \wedge M\mathbb{Q} \simeq E \wedge H\mathbb{Q} \simeq E \wedge H\mathbb{Q} \wedge H\mathbb{Z} \simeq E[0] \wedge H\mathbb{Z}.$$

Now the result follows from 7.9.

(iii) By 1.16(i), we have $[H(G), H(G')] = \prod_i [\Sigma^i H(G_i), H(G')]$. By 5.8(iii),

$$[\Sigma^i H\mathbb{Q}, HG'] = [S^i, HG'] = G'_i = [S^i, HG'_i] = [\Sigma^i H\mathbb{Q}, HG'_i].$$

Hence,

$$[\Sigma^i H(G_i), H(G')] = [\Sigma^i H(G_i), \Sigma^i H(G'_i)] = \mathrm{Hom}(G_i, G'_i),$$

and thus $[HG, HG'] = \mathrm{Hom}^0(G, G')$.

(iv) By 5.14, $E \wedge F$ is a $\mathbb{Q}$-local spectrum, and hence $H_*(X) = H_*(X; \mathbb{Q})$ for $X = E, F, E \wedge F$. Consider the commutative diagram

$$\begin{array}{ccc}
\pi_*(E) \otimes \pi_*(F) & \xrightarrow{\mu^\pi} & \pi_*(E \wedge F) \\
{\scriptstyle h\otimes h}\downarrow & & \downarrow{\scriptstyle h} \\
H_*(E) \otimes H_*(F) & \xrightarrow{\mu^H} & H_*(E \wedge F).
\end{array}$$

Now, by (i), h is an isomorphism, and, by 4.11(i), μ^H is an isomorphism. Thus, μ^π is an isomorphism.

(v) We must prove that the left hand diagram below

$$\begin{array}{ccc}
E \wedge E & \xrightarrow{f \wedge f} & F \wedge F \\
\downarrow & & \downarrow \\
E & \xrightarrow{f} & F
\end{array}
\qquad
\begin{array}{ccc}
\pi_*(E) \otimes \pi_*(E) & \xrightarrow{f_* \otimes f_*} & \pi_*(F) \otimes \pi_*(F) \\
\downarrow & & \downarrow \\
\pi_*(E) & \xrightarrow{f_*} & \pi_*(F)
\end{array}$$

commutes up to homotopy iff the right hand diagram commutes. But this follows from (iii) and (iv). □

7.12. Corollary. *Let F be a spectrum of finite Λ-type with Λ as in §5. Then:*

(i) *Each Postnikov invariant of F has finite order.*

(ii) *Let X be a spectrum bounded below. Consider the Atiyah–Hirzebruch spectral sequence $E_r^{**}(X) \Rightarrow F^*(X)$, $E_2^{p,q}(X) = H^p(X; F^q(\mathrm{pt}))$. Then every differential in this spectral sequence for $E^*(X)$ has finite order (i.e., it becomes trivial after tensoring by $\mathbb{Q}$). Simiarly, every differential in the homology spectral sequence $E^r_{**}(X) \Rightarrow F_*(X)$ has finite order.*

Proof. (i) It follows from 4.25(iii) that each group $(F_{(k)})_i(H\Lambda)$ is a finitely generated Λ-module. Hence, each group $H_i(F_{(k)})$ is a finitely generated Λ–module. So by 4.9, each group $H^i(F_{(k)};\pi_j(F))$ is a finitely generated Λ–module.

Consider a Postnikov invariant $\kappa \in H^{n+1}(F_{(n-1)};\pi_n(F))$ of F. By 5.11,

$$\kappa \otimes 1 \in H^{n+1}(F_{(n-1)};\pi_n(F)) \otimes \mathbb{Q} = H^{n+1}((F_{(n-1)})[0];\pi_n(F) \otimes \mathbb{Q})$$

is the Postnikov invariant of $F[0]$. So, by 7.11(ii) and 7.3(i), $\kappa\otimes 1 = 0$. Thus, κ has finite order since $H^{n+1}(F_{(n-1)};\pi_n(F))$ is a finitely generated Λ-module.

(ii) This follows from (i) and 4.34. □

7.13. Theorem-Definition (cf. Dold [1]). *For every ring spectrum E there exists a ring equivalence $E[0] \to H(\pi_*(E)\otimes\mathbb{Q})$. This equivalence is called the* Chern–Dold character *with respect to E and is denoted by ch_E.*

Proof. There is a ring isomorphism $h : \pi_*(E[0]) \to \pi_*(H(\pi_*(E)\otimes\mathbb{Q}))$. By 7.11(ii,iii), h is induced by a morphism $f : E[0] \to H(\pi_*(E) \otimes \mathbb{Q})$, and, by 7.11(v), f is a ring equivalence. □

7.14. Theorem. *Let p be an odd prime, and let E be a $\mathbb{Z}[p]$-local spectrum of finite $\mathbb{Z}[p]$-type. If $E \wedge M(\mathbb{Z}/p)$ is a graded Eilenberg–Mac Lane spectrum then so is E.*

Proof. For simplicity, denote $M(\mathbb{Z}/p)$ by $M(p)$. The spectrum $M(p)$ admits a ring structure for $p > 3$, while for $p = 3$ it admits a pairing (non-associative) $M(3) \wedge M(3) \to M(3)$. This can be proved directly, just by considering the group $[M(p) \wedge M(p), M(p)]$, see Araki–Toda [1], or, alternatively, this can be deduced from certain general results, see Ch. VIII of this book. Since $\pi_0(M(p)) = \mathbb{Z}/p$, every group $\pi_i(M(p))$ has exponent p. Furthermore, if $p > 3$ then the group $M(p)_*(X)$ is a $\pi_*(M(p))$-module for every spectrum (space) X. Moreover, for $p = 3$ we have a pairing $(M(3))_*(X) \otimes (M(3))_*(X) \to M(3)_*(X)$. So, the group $M(p)_*(X)$ is a $\mathbb{Z}/p$-vector space. [8]

Let

$$\begin{array}{ccccccc}
\cdots & = & E & = & E & = & \cdots \\
 & & \downarrow{\scriptstyle \tau_n} & & \downarrow{\scriptstyle \tau_{n-1}} & & \\
\cdots & \longrightarrow & E_n & \xrightarrow{p_n} & E_{n-1} & \longrightarrow & \cdots
\end{array}$$

be a Postnikov tower of E (we are writing simply E_n instead of $E_{(n)}$).

7.15. Lemma. *The homomorphism $(\tau_n\wedge 1)_* : \pi_*(E\wedge M(p)) \to \pi_*(E_n\wedge M(p))$ is epic for every n.*

[8] This is not true for $p = 2$: the element $1_{M(2)} \in M(2)_*(M(2))$ has order 4, see Araki–Toda [1].

Proof. The cofiber sequence $S \xrightarrow{p} S \xrightarrow{j} M(p)$ yields the commutative diagram

$$\begin{array}{ccccccccc}
\pi_i(E) & \xrightarrow{p} & \pi_i(E) & \longrightarrow & \pi_i(E \wedge M(p)) & \longrightarrow & \pi_{i-1}(E) & \xrightarrow{p} & \pi_{i-1}(E) \\
\downarrow{\scriptstyle a_1} & & \downarrow{\scriptstyle a_2} & & \downarrow{\scriptstyle (\tau_n \wedge 1)_*} & & \downarrow{\scriptstyle a_3} & & \downarrow{\scriptstyle a_4} \\
\pi_i(E_n) & \xrightarrow{p} & \pi_i(E_n) & \longrightarrow & \pi_i(E_n \wedge M(p)) & \longrightarrow & \pi_{i-1}(E_n) & \xrightarrow{p} & \pi_{i-1}(E_n)
\end{array}$$

with exact rows. If $i \leq n$ then a_k, $k = 1, 2, 3, 4$, is an isomorphism, and hence $(\tau_n \wedge 1)_*$ is an isomorphism. If $i > n + 1$ then $\pi_i(E_n) = 0 = \pi_{i-1}(E_n)$, and hence $\pi_i(E_n \wedge M(p)) = 0$. Finally, if $i = n+1$ then $(\tau_n \wedge 1)_*$ is an epimorphism because $\pi_i(E_n) = 0$ and a_3, a_4 are isomorphisms. ⊞

We continue the proof of the theorem. By 7.15, the homomorphism

$$(\tau_n \wedge 1)_* : \pi_*(E \wedge M(p)) \to \pi_*(E_n \wedge M(p))$$

is epic, and it splits since $\pi_*(E \wedge M(p))$ is a $\mathbb{Z}/p$-vector space. Hence, by 7.2, for every n, $E_n \wedge M(p)$ is a graded Eilenberg–Mac Lane spectrum and $p_n \wedge 1 : E_n \wedge M(p) \to E_{n-1} \wedge M(p)$ admits a homotopy right inverse morphism.

Suppose that E is not a graded Eilenberg–Mac Lane spectrum. Then, by 7.3(iii), there exists a minimal n such that E_n is not a graded Eilenberg–Mac Lane spectrum. Hence, the first non-trivial Postnikov invariant of E is $k \in H^{n+1}(E_{n-1}; \pi_n)$, where $\pi_n = \pi_n(E)$. We have

$$\begin{aligned}
&H^0(H(\pi_n) \wedge M(p); \pi_n \otimes \mathbb{Z}/p) = \operatorname{Hom}(H_0(H(\pi_n) \wedge M(p); \mathbb{Z}), \pi_n \otimes \mathbb{Z}/p) \\
&= \operatorname{Hom}(\pi_0(H(\pi_n \wedge M(p))), \pi_n \otimes \mathbb{Z}/p) = \operatorname{Hom}(H_0(M(p); \pi_n), \pi_n \otimes \mathbb{Z}/p) \\
&= \operatorname{Hom}(\pi_n \otimes \mathbb{Z}/p, \pi_n \otimes \mathbb{Z}/p).
\end{aligned}$$

Let $u \in H^0(H(\pi_n) \wedge M(p); \pi_n \otimes \mathbb{Z}/p)$ correspond to

$$1_{\pi_n \otimes \mathbb{Z}/p} \in \operatorname{Hom}(\pi_n \otimes \mathbb{Z}/p, \pi_n \otimes \mathbb{Z}/p).$$

Consider the following commutative diagram:

$$\begin{array}{ccccc}
E_n & \xrightarrow{p_n} & E_{n-1} & \xrightarrow{k} & \Sigma^{n+1} H(\pi_n) \\
\downarrow & & \downarrow & & \downarrow{\scriptstyle r} \\
E_n \wedge M(p) & \xrightarrow{p_n \wedge 1} & E_{n-1} \wedge M(p) & \xrightarrow{k \wedge 1} & \Sigma^{n+1} H(\pi_n) \wedge M(p) \\
& & & & \downarrow{\scriptstyle \Sigma^{n+1} u} \\
& & & & \Sigma^{n+1} H(\pi_n \otimes \mathbb{Z}/p) .
\end{array}$$

Here the rows are cofiber sequences and the morphism r has the form

$$\Sigma^{n+1} H(\pi_n) = \Sigma^{n+1} H(\pi_n) \wedge S \xrightarrow{1 \wedge j} \Sigma^{n+1} H(\pi_n) \wedge M(p).$$

Let π, τ be two cyclic $\mathbb{Z}[p]$-modules (i.e., π, as well as τ, is isomorphic to $\mathbb{Z}/p^m$ or $\mathbb{Z}[p]$). It is well-known that the group $H^i(H\pi;\tau), i > 0$ has exponent p, see e.g. Cartan [1]. Hence, the group $H^*(E_{n-1};\pi_n)$ has the exponent p, since E_{n-1} is a graded Eilenberg–Mac Lane spectrum and π_n is a direct sum of the groups $\mathbb{Z}[p]$ and $\mathbb{Z}/p^m$. So, the reduction $\bar{k} \in H^{n+1}(E_{n-1};\pi_n \otimes \mathbb{Z}/p)$ of k is non-zero. However, $\bar{k} = \Sigma^{n+1}u \circ r \circ k$, and so $k \wedge 1$ is non-zero. Hence, by 4.20, $p_n \wedge 1$ does not admit a homotopy right inverse morphism. This is a contradiction. □

Fix a prime p. *Until the end of the section, H denotes $H\mathbb{Z}/p$ and $H(-)$ denotes $H(-;\mathbb{Z}/p)$.* Let $\mathscr{A}_p$ be the mod p Steenrod algebra, $\mathscr{A}_p = H^*(H)$, and let $\mathscr{A}_p^*$ be the Hopf algebra dual to $\mathscr{A}_p$, $\mathscr{A}_p^* = H_*(H)$. Finally, let $\mu : H \wedge H \to H$ be the multiplication on H.

7.16. Theorem. *Let E be a spectrum of finite $\mathbb{Z}[p]$-type. If $H^*(E)$ is a free $\mathscr{A}_p$-module then E is a graded Eilenberg–Mac Lane spectrum.*

Proof. Note that every group $H^k(E)$ is finite. Let $a_1, \dots, a_s, \dots, |a_s| \le |a_{s+1}|$, be a family of free $\mathscr{A}_p$-generators of $H^*(E)$. Every element a_i yields a morphism $a_i : E \to \Sigma^{|a_i|}H$. Set

$$F(s) = \vee_{i=1}^{s} \Sigma^{|a_i|}H, \; F = \vee_{i=1}^{\infty} \Sigma^{|a_i|}H.$$

We have obvious inclusions $i_s^k : \Sigma^{|a_k|}H\mathbb{Z}/p \to F(s), k \le s$, of direct summands, and we have projections $p_s : F(s+1) \to F(s), q_s : F \to F_s$ onto direct summands. By 1.16(ii), the morphisms a_i form a morphism $h_s : E \to F(s)$ with $p_s h_{s+1} \simeq h_s$ and $i_s^k a_k = h_s$. Clearly, for every N there exists s such that $h_s^* : H^i(F(s)) \to H^i(E)$ is an isomorphism for $i \le N+1$. Hence, $H^i(Ch_s) = 0$ for $i \le N$, and so, by 5.18(i), $\pi_i(Ch_s) = 0$ for $i \le N$. Hence, h_s is an N-equivalence, and so $(h_s)_{(N)} : E_{(N)} \to (F(s))_{(N)}$ is an equivalence.

By 7.3(i), $(F(s))_{(N)}$ is a graded Eilenberg–Mac Lane spectrum. Hence, $E_{(n)}$ is a graded Eilenberg–Mac Lane spectrum for every n, and thus, by 7.3(iii), E is a graded Eilenberg–Mac Lane spectrum. □

One can give a stronger version of the previous theorem.

7.17. Corollary (of the proof). *Let E be a spectrum of finite $\mathbb{Z}[p]$-type. Suppose that there exist a free finitely generated $\mathscr{A}_p$-module V and an $\mathscr{A}_p$-homomorphism $f : V \to H^*(E)$ such that f is an isomorphism in dimensions $\le m+1$. Then the coskeleton $E_{(m)}$ is a graded Eilenberg–Mac Lane spectrum.*

Proof. Let $\{a_i\}$ be a family of free generators of V. As in 7.16, one can construct a morphism $h : E \to F := \vee\Sigma^{|a_i|}H$ such that $h^* : H^*(F) \to H^*(E)$ is an isomorphism in dimensions $\le m+1$. Thus, $E_{(m)} \simeq F_{(m)}$. But, by 7.3(i), $F_{(m)}$ is a graded Eilenberg–Mac Lane spectrum. □

7.18. Lemma. *Let E, F be a pair of spectra. Then the following hold:*

(i) *The homomorphism* $\mu_{E,F} : H_*(E) \otimes H_*(F) \to H_*(E \wedge F)$ *is an isomorphism of comodules over the Hopf algebra* $\mathscr{A}_p^*$;

(ii) *Assume that* E *is bounded below and* F *has finite* $\mathbb{Z}$- *or* $\mathbb{Z}[p]$-*type. Then* $\mu^{E,F} : H^*(E) \otimes H^*(F) \to H^*(E \wedge F)$ *is an isomorphism of modules over the Hopf algebra* $\mathscr{A}_p$.

Proof. (i) Because of 4.11(i), it suffices to prove that $\mu_{E,F}$ is a homomorphism of comodules over $\mathscr{A}_p^*$. Consider the morphisms $b^E : E \to H \wedge E$ and $b^F : F \to H \wedge F$ as in 6.27(c). Because of the naturality of $\mu_{X,Y}$ in X, Y, we have the following commutative diagram:

$$
\begin{array}{ccc}
H_*(E) \otimes H_*(F) & \xrightarrow{\mu_{E,F}} & H_*(E \wedge F) \\
{\scriptstyle (b^E)_* \otimes (b^F)_*} \Big\downarrow & & \Big\downarrow {\scriptstyle (b^E \wedge b^F)_*} \\
H_*(H \wedge E) \otimes H_*(H \wedge F) & \xrightarrow{\mu_{H \wedge E, H \wedge F}} & H_*(H \wedge E \wedge H \wedge F) \\
{\scriptstyle \mu_{H,E} \otimes \mu_{H,F}} \Big\uparrow {\scriptstyle \cong} & & \Big\| \\
H_*(H) \otimes H_*(H) \otimes H_*(E) \otimes H_*(F) & & H_*(H \wedge E \wedge H \wedge F) \\
{\scriptstyle 1 \otimes T \otimes 1} \Big\downarrow & & \Big\downarrow {\scriptstyle (1 \wedge \tau \wedge 1)_*} \\
H_*(H) \otimes H_*(E) \otimes H_*(H) \otimes H_*(F) & \xrightarrow{\nu} & H_*(H \wedge H \wedge E \wedge F) \\
{\scriptstyle \mu_{H,H} \otimes 1 \otimes 1} \Big\downarrow & & \Big\downarrow {\scriptstyle (\mu \wedge 1 \wedge 1)_*} \\
H_*(H) \otimes H_*(E) \otimes H_*(F) & & H_*(H \wedge E \wedge F) \\
\Big\| & & {\scriptstyle \cong} \Big\uparrow {\scriptstyle \mu_{H, E \wedge F}} \\
H_*(H) \otimes H_*(E) \otimes H_*(F) & \xrightarrow{1 \otimes \mu_{E,F}} & H_*(H) \otimes H_*(E \wedge F)
\end{array}
$$

where

$$\nu := \mu_{H \wedge H, E \wedge F} \circ (\mu_{H,H} \otimes \mu_{E,F}).$$

Now, the aggregated left vertical homomorphism

$$H_*(E) \otimes H_*(F) \to H_*(H) \otimes H_*(E) \otimes H_*(F)$$

is the A_p^*-comodule structure map for $H_*(E) \otimes H_*(F)$, and it easy to see that the aggregated right vertical homomorphism $H_*(E \wedge F) \to H_*(H) \otimes H_*(E \wedge F)$ is the A_p^*-comodule structure map for $H_*(E \wedge F)$.

(ii) Similarly to the above, one can prove that $\mu^{E,F}$ is a homomorphism of modules over the Hopf algebra $\mathscr{A}_p$. So, we must check that $\mu^{E,F}$ is an isomorphism (of groups). If F has finite $\mathbb{Z}$-type then this follows from 4.11(ii); we leave it to the reader. Now, let F have finite $\mathbb{Z}[p]$-type. Firstly, let $F = H(\pi)$ where π is a finitely generated $\mathbb{Z}[p]$-module. Since π is a finite direct sum of the groups isomorphic to $\mathbb{Z}[p]$ and $\mathbb{Z}/p^k$, we conclude that $\pi \cong \tau \otimes \mathbb{Z}[p]$

where τ is a finitely generated abelian group. Since $H(\tau)$ has finite $\mathbb{Z}$-type, $\mu^{E,H(\tau)}$ is an isomorphism. But $H_*(H(\tau)) = H_*(H(\pi))$, and so $\mu^{E,H(\pi)}$ is an isomorphism.

Now, given F as above, we prove that $\mu^{E,F_{(n)}}$ is an isomorphism for every n. We prove this by induction on n. For brevity, we write F_n instead of $F_{(n)}$ and π_k instead of $\pi_k(F)$. Since F is bounded below, there is k such that $F_k = \Sigma^k H(\pi_k)$ and so, by the above, our claim is true for F_k. Now, suppose that $\mu^{E,F_{n-1}}$ is an isomorphism, and consider the cofiber sequence

$$\Sigma^n H(\pi_n) \to F_n \to F_{n-1}.$$

It yields the commutative diagram

$$\begin{array}{ccc}
\vdots & & \vdots \\
\downarrow & & \downarrow \\
H^*(E) \otimes H^*(F_{n-1}) & \xrightarrow{\cong} & H^*(E \wedge F_{n-1}) \\
\downarrow & & \downarrow \\
H^*(E) \otimes H^*(F_n) & \xrightarrow{\mu^{E,F_n}} & H^*(E \wedge F_n) \\
\downarrow & & \downarrow \\
H^*(E) \otimes H^*(\Sigma^n H(\pi_n)) & \xrightarrow{\cong} & H^*(E \wedge \Sigma^n H(\pi_n)) \\
\downarrow & & \downarrow \\
\vdots & & \vdots
\end{array}$$

with exact columns. Now, the Five Lemma implies that μ^{E,F_n} is an isomorphism. The induction is confirmed.

Finally, for every k there is N such that $(1 \wedge \tau_N)_* : H^k(E \wedge F_{(N)}) \to H^k(E \wedge F)$ is an isomorphism. Thus, $\mu^{E,F}$ is an isomorphism for every spectrum F of finite $\mathbb{Z}[p]$-type. □

7.19. Corollary. *Let (E, μ, ι) be a ring spectrum.*

(i) *Define the homomorphisms*

$$\mu_{alg} : H_*(E) \otimes H_*(E) \xrightarrow{\mu_{E,E}} H_*(E \wedge E) \xrightarrow{\mu_*} H_*(E)$$

and $\iota_* : \mathbb{Z}/p = H_*(S) \to H_*(E)$. Then $(H_*(E), \mu_{\text{alg}}, \iota_*)$ is a comodule algebra over the Hopf algebra $\mathscr{A}_p^*$.

(ii) *In addition, suppose that E has finite $\mathbb{Z}$- or $\mathbb{Z}[p]$-type. Define the homomorphisms*

$$\mu^{alg} : H^*(E) \xrightarrow{\mu^*} H^*(E \wedge E) \xleftarrow{\mu^{E,E}, \cong} H^*(E) \otimes H^*(E)$$

and $\iota^* : H^*(E) \to H^*(S) = \mathbb{Z}/p$. *Then* $(H^*(E), \mu^{\text{alg}}, \iota^*)$ *is a module coalgebra over the Hopf algebra* $\mathscr{A}_p$. □

7.20. Lemma. *If* E *is a connected spectrum and* $\pi_0(E) = \mathbb{Z}/p$ *then* $H^0(E) = \mathbb{Z}/p$. *Furthermore,* $Q_0(u) \neq 0$ *for every* $u \in H^0(E), u \neq 0$.

Proof. By 4.7, the Hurewicz homomorphism $h : \pi_0(E) \to H_0(E;\mathbb{Z})$ is an isomorphism, i.e., $H_0(E;\mathbb{Z}) = \mathbb{Z}/p$. Furthermore, by 4.9, the homomorphism

$$H^0(E) \to \operatorname{Hom}(H_0(E;\mathbb{Z}), \pi_0(H)) \xrightarrow{h^*} \operatorname{Hom}(\pi_0(E), \pi_0(H)) = \mathbb{Z}/p$$

is an isomorphism. In particular, $H^0(E) = \mathbb{Z}/p$.

Let $u : E \to H$ represent a non-zero element $u \in H^0(E)$. Then, by the above, the induced homomorphism $u_* : \pi_0(E) \to H_0(E;\mathbb{Z})$ is an isomorphism, and so u is a 0-equivalence. Hence, $u^* : H^0(H) \to H^0(E)$ is an isomorphism and $u^* : H^1(H) \to H^1(E)$ is a monomorphism. Thus, $Q_0(u) \neq 0$. □

7.21. Lemma. *Let* E *be a connected ring spectrum* (E, μ, ι) *of finite* $\mathbb{Z}[p]$*-type with* $\pi_0(E) = \mathbb{Z}/p$. *Then* $H^*(E)$ *is a connected* $\mathscr{A}_p$*-coalgebra, and its counit* $v \in H^0(E)$ *yields a ring morphism* $v : E \to H$.

Proof. By 7.19(ii), $H^*(E)$ is an $\mathscr{A}_p$-coalgebra. Furthermore, the arguments in the proof of 7.20 show that the augmentation $\iota^* : \mathbb{Z}/p = H^0(E) \to H^0(S) = \mathbb{Z}/p$ is an isomorphism. So, $H^*(E)$ is a connected coalgebra. Its counit v is defined by the condition $\iota^*(v) = 1 \in \mathbb{Z}/p = H^0(S)$. We must prove that $v : E \to H$ is a ring morphism, i.e., that the diagram

$$\begin{array}{ccc} E \wedge E & \xrightarrow{\mu} & E \\ {\scriptstyle v\wedge v}\downarrow & & \downarrow{\scriptstyle v} \\ H \wedge H & \xrightarrow{\mu_H} & H. \end{array}$$

commutes (up to homotopy). The morphism

$$S = S \wedge S \xrightarrow{\iota\wedge\iota} E \wedge E \xrightarrow{\mu} E \xrightarrow{v} H$$

coincides (up to homotopy) with $v\iota$, while the morphism

$$S = S \wedge S \xrightarrow{\iota\wedge\iota} E \wedge E \xrightarrow{v\wedge v} H \wedge H \xrightarrow{\mu_H} H$$

coincides with the unit ι_H of H. Also, $\iota_H \simeq v\iota$ since $\iota^*(v) = 1 \in H^0(S)$. Hence,

$$v\circ\mu\circ(\iota \wedge \iota) \simeq v\iota \simeq \iota_H \simeq \mu_H\circ(v \wedge v)\circ(\iota \wedge \iota).$$

Since $(\iota\wedge\iota)^* : H^0(E\wedge E) \to H^0(S\wedge S)$ is an isomorphism, $v\circ\mu \simeq \mu_H\circ(v\wedge v)$. □

Let E be a spectrum as in 7.21. We define $\nu : \mathscr{A}_p \to H^*(E)$ by setting $\nu(a) = a(v)$.

7.22. Theorem. *If ν is monic in dimension $\leq m+1$ then the coskeleton $E_{(m)}$ of E is a graded Eilenberg–Mac Lane spectrum. Furthermore, if ν is monic then E is a graded Eilenberg–Mac Lane spectrum.*

Proof. By 6.29, the homomorphism $\eta : \mathscr{A}_p \otimes G(H^*(E)) \to H^*(E)$ is an isomorphism in dimensions $\leq m+1$. Now, by 7.17, $E_{(m)}$ is a graded Eilenberg–Mac Lane spectrum. Furthermore, the last claim follows from the above and 7.3(iii). □

7.23. Corollary. *Fix a natural number m. Suppose either*

(i) $p=2$ *and* $\nu(Q_i) \neq 0$ *for* $2^{i+1}-1 \leq m+1$, *or*

(ii) $p>2$ *and* $\nu(Q_i) \neq 0$ *for* $2p^i - 1 \leq m+1$, $\nu(\mathscr{P}^{\Delta_j}) \neq 0$ *for* $2(p^j-1) \leq m+1$.

Then $E_{(m)}$ is a graded Eilenberg–Mac Lane spectrum.

Proof. This follows from 6.14 and 7.22, because the elements Q_i for $p=2$ and the elements $Q_i, \mathscr{P}^{\Delta_j}$ for $p>2$ form a $\mathbb{Z}/p$-basis of the vector space $\operatorname{Pr}\mathscr{A}_p$ of primitives. □

7.24. Corollary. *Suppose either*

(i) $p=2$ *and* $\nu(Q_i) \neq 0$ *for* $i=0,1,\ldots$, *or*

(ii) $p>2$ *and* $\nu(Q_i) \neq 0$ *for* $i=0,1,\ldots$, $\nu(\mathscr{P}^{\Delta_j}) \neq 0$ *for* $j=1,2,\ldots$.

Then E is a graded Eilenberg–Mac Lane spectrum, $E \simeq H(\pi_*(E))$. In other words, E is a wedge of iterated suspensions over $H\mathbb{Z}/p$. □

7.25. Remark. It follows from 6.32 that 7.22–7.24 are valid for "non-associative ring spectra" also, i.e., for spectra which satisfy Definition 2.12 with condition (1) omitted.

Now we consider ring structures on graded Eilenberg–Mac Lane spectra (following Boardman [1]). We work here with homology rather than with cohomology because the homology Künneth formula holds without any restrictions, unlike the cohomology one.

7.26. Lemma. *If a commutative ring spectrum E is a graded Eilenberg–Mac Lane spectrum with $p\pi_*(E) = 0$, then there exists a homomorphism $b : \mathscr{A}_p^* \to H_*(E)$ of $\mathscr{A}_p^*$-comodule algebras.*

Proof. We consider the case $p=2$ only; the case $p>2$ can be proved similarly. By 7.1, the inclusion $\mathbb{Z}/p = \pi_0(E) \to \pi_*(E)$ is induced by a morphism $\varphi : H \to E$, and $f = \varphi_* : \mathscr{A}_p^* = H_*(H) \to H_*(E)$ is a homomorphism of $\mathscr{A}_p^*$-comodules.

Recall that $\mathscr{A}_2^* = \mathbb{Z}/2[\zeta_1, \dots, \zeta_n, \dots]$ and define a homomorphism of $\mathbb{Z}/2$-algebras $b : \mathscr{A}_2^* \to H_*(E; \mathbb{Z}/2)$ by setting $b(\zeta_i) = f(\zeta_i)$. Such a homomorphism of algebras exists (and is unique) because E is commutative. We check that b is a homomorphism of $\mathscr{A}_2^*$-comodules, i.e., that the diagram

$$\begin{array}{ccc} \mathscr{A}_2^* & \xrightarrow{b} & H_*(E;\mathbb{Z}/2) \\ \nabla\downarrow & & \downarrow\psi \\ \mathscr{A}_2^* \otimes \mathscr{A}_2^* & \xrightarrow{1\otimes b} & \mathscr{A}_2^* \otimes H_*(E;\mathbb{Z}/2) \end{array}$$

commutes. In this diagram all arrows are homomorphisms of $\mathbb{Z}/2$-algebras (b and $1 \otimes b$ by construction, ∇ by definition and ψ by general reasons, cf. 6.25). So, it suffices to prove that $(1 \otimes b)\nabla(\zeta_i) = \psi b(\zeta_i)$ for every i. We have

$$\begin{aligned}(1 \otimes b)\nabla(\zeta_k) &= (1 \otimes b)\left(\sum \zeta_{k-i}^{2^i} \otimes \zeta_i\right) = \sum \zeta_{k-i}^{2^i} \otimes b(\zeta_i) = \sum \zeta_{k-i}^{2^i} \otimes f(\zeta_i) \\ &= (1 \otimes f)\nabla(\zeta_k) = \psi f(\zeta_k) = \psi b(\zeta_k).\end{aligned}$$

The fifth equality holds because f is a homomorphism of comodules. □

Let E be a ring spectrum with $\pi_0(E) = \mathbb{Z}/p$. We turn $\pi_*(E)$ into an $\mathscr{A}_p^*$-comodule by requiring $\pi_*(E) = \mathrm{Si}(\pi_*(E))$. In this way $\pi_*(E)$ turns into a $\mathscr{A}_p^*$-comodule algebra, and the Hurewicz homomorphism $h : \pi_*(E) \to H_*(E)$ is a homomorphism of $\mathscr{A}_p^*$-comodule algebras.

7.27. Lemma. *If a commutative ring spectrum E is a graded Eilenberg–Mac Lane spectrum with $p\pi_*(E) = 0$, then the Hurewicz homomorphism $h : \pi_*(E) \to H_*(E)$ is a monomorphism, and $h(\pi_*(E)) = \mathrm{Si}(H_*(E))$.*

Proof. Since $E \simeq \vee \Sigma^d H$, h is monic. Furthermore, because of this equivalence, $H_*(E)$ is just a cofree extension $H_*(H) \otimes h(\pi_*(E))$. Finally, $\mathrm{Si}(H_*(H)) = \mathbb{Z}/p = H_0(H)$, and thus $\mathrm{Si}(H_*(E)) = h(\pi_*(E))$. □

7.28. Corollary. *If a commutative ring spectrum E is a graded Eilenberg–Mac Lane spectrum with $p\pi_*(E) = 0$, then there is an isomorphism $H_*(E) \cong \mathscr{A}_p^* \otimes \pi_*(E)$ of $\mathscr{A}_p$-comodule algebras.*

Proof. The homomorphism $b : \mathscr{A}_p^* \to H_*(E)$ in 7.26 yields, by 6.34, an isomorphism $H_*(E) \cong \mathscr{A}_p^* \otimes \mathrm{Si}(H_*(E))$. But $\mathrm{Si}(H_*(E)) \cong \pi_*(E)$. □

7.29. Proposition. *Let E, F be two graded Eilenberg–Mac Lane spectra with $p\pi_*(E) = 0 = p\pi_*(F)$.*

(i) *The homomorphism $\varphi : [E, F] \to \mathrm{Hom}_{\mathscr{A}_p^*}(H_*(E), H_*(F))$, $\varphi(f) = f_*$, is an isomorphism.*

(ii) *In addition, suppose that E, F are ring spectra. Then $f : E \to F$ is a ring morphism iff $f_* : H_*(E) \to H_*(F)$ is a homomorphism of $\mathscr{A}_p^*$-comodule algebras.*

Proof. (i) The spectrum E is a wedge of spectra of the form $\Sigma^d H$. Since each of the groups $[E, F]$ and $\operatorname{Hom}_{\mathscr{A}_p^*}(H_*(E), H_*(F))$ is additive with respect to E, it suffices to prove (i) for $E = H$.

Since H is a ring spectrum and $F = \vee \Sigma^d H$, we conclude that F is an H-module spectrum. By 7.28 and 6.8, we have an isomorphism

$$\begin{aligned} h : \operatorname{Hom}_{\mathscr{A}_p^*}(H_*(H), H_*(F)) &\cong \operatorname{Hom}_{\mathscr{A}_p^*}(H_*(H), \mathscr{A}_p^* \otimes \pi_*(F)) \\ &\cong \operatorname{Hom}_{\mathbb{Z}/p}(H_*(H), \pi_*(F)). \end{aligned}$$

We leave it to the reader to check that $h\varphi : F^*(H) \to \operatorname{Hom}_{\mathbb{Z}/p}(H_*(H), \pi_*(F))$ coincides with the homomorphism ev as in 3.45. Hence, by 3.45, $h\varphi$ is an isomorphism, and thus φ is an isomorphism.

(ii) If f_* is a homomorphism of comodule algebras then the diagram

$$\begin{array}{ccc} H_*(E) \otimes H_*(E) & & H_*(F) \otimes H_*(F) \\ \cong\downarrow & & \cong\downarrow \\ H_*(E \wedge E) & \xrightarrow{(f\wedge f)_*} & H_*(F \wedge F) \\ \downarrow \mu_*^E & & \mu_*^F \downarrow \\ H_*(E) & \xrightarrow{f_*} & H_*(F) \end{array}$$

commutes. By 7.10(ii), $E \wedge E$ and $F \wedge F$ are graded Eilenberg–Mac Lane spectra, and so, by (i), the morphisms $E \wedge E \xrightarrow{\mu_E} E \xrightarrow{f} F$ and $E \wedge E \xrightarrow{f\wedge f} F \wedge F \xrightarrow{\mu_F} F$ are homotopic. Furthermore, the Hurewicz homomorphism $h : \pi_*(X) \to H_*(X)$, $X = E, F$, is a ring monomorphism, and so f preserves the units. Thus, f is a ring morphism. The converse is obvious. □

7.30. Theorem (Boardman [1]). *Let E, F be two commutative ring spectra. Suppose that E, F are graded Eilenberg–Mac Lane spectra with $p\pi_*(E) = 0 = p\pi_*(F)$. Then every ring homomorphism $r : \pi_*(E) \to \pi_*(F)$ is induced by a ring morphism $f : E \to F$. So, if there exists a ring isomorphism $\pi_*(E) \cong \pi_*(F)$ then there exists a ring equivalence $E \simeq F$.*

In particular, there is a ring equivalence $E \simeq H(\pi_*(E))$.

Proof. The composition

$$H_*(E) \xrightarrow{\cong} \mathscr{A}_p^* \otimes \pi_*(E) \xrightarrow{1\otimes r} \mathscr{A}_p^* \otimes \pi_*(F) \xrightarrow{\cong} H_*(F)$$

(where the first and the last isomorphisms come from 7.28) is a homomorphism of comodule algebras. By 7.29(i), it is induced by a morphism $E \to F$, which is a ring morphism by 7.29(ii). The last assertion follows if we put $F = H(\pi_*(E))$. □

Chapter III. Phantoms

A *phantom*, or a *phantom map*, is an essential map $f : X \to Y$ of a CW-complex X such that $f|X^{(n)}$ is inessential for every n. Adams–Walker [1] found an example of a phantom, and many other authors found phantoms later. The existence of phantoms was very exotic at that time and adorned (and adorns now, by the way) any results. However, as usual, the other tendency occured afterwards: phantoms began to frustrate mathematicians because they appeared (or could appear) in very unexpected situations. Keeping in mind the two above tendencies, we give examples of phantoms and some sufficient conditions for the absence of phantoms. In fact, this chapter can be treated as an exposition of some effects arising when we pass from finite dimensional spaces (spectra) to infinite dimensional ones. In this context it is also natural to consider spaces (spectra) which have the same n-type for all n.

Many other things about phantoms one can find in McGibbon [1].

§1. Phantoms and the Inverse Limit Functor

Let $\mathscr{X} = \{X_\alpha\}$ be a family of subspaces of a space X (or subspectra of a spectrum X). Given a space (spectrum) Y, we say that maps $f, g : X \to Y$ are $\mathscr{X}$-homotopic if $f|X_\alpha \simeq g|X_\alpha$ for every α. Similarly, elements $a, b \in E^*(X)$ are called $\mathscr{X}$-equivalent if $a|X_\alpha = b|X_\alpha$ for every α. The classes of $\mathscr{X}$-homotopic maps (or $\mathscr{X}$-equivalent elements) form a set $[X, Y]_{\mathscr{X}}$ (or a group $E^*_{\mathscr{X}}(X)$) with the distinguished element $*$ given by a constant map . There are the obvious quotient functions $\sigma : [X, Y] \to [X, Y]_{\mathscr{X}}$ and $\sigma : E^*(X) \to E^*_{\mathscr{X}}(X)$.

1.1. Definition. A map $f : X \to Y$ of spaces (or a morphism of spectra) is called an $\mathscr{X}$*-phantom* if $\sigma[f] = *$ while $[f] \neq *$. Similarly, an element $a \in E^*(X)$ is an $\mathscr{X}$-phantom if $\sigma(a) = 0$ while $a \neq 0$.

This definition is given for an arbitrary family $\mathscr{X}$, but really interesting are families with $\cup X_\alpha = X$. Moreover, to justify the term "phantom" the subspaces X_α must be sufficiently "massive", otherwise phantoms do not live up to their name. For example, if $\{X_\alpha\}$ is a family of charts of a manifold X, then every non-trivial element of $\widetilde{E}^*(X)$ is a phantom. Usually one considers the families $\{X_\lambda\}$ of all finite CW-subcomplexes (subspectra) and $\{X^{(n)}\}$ of skeletons of a CW-complex (spectrum) X. We fix these cases in the following definition.

1.2. Definition. (a) Given a CW-complex (spectrum) X, let $\mathscr{X}$ be the family $\{X^{(n)}\}$ of all skeleta of X. Then an $\mathscr{X}$-phantom is called just a *phantom.*

(b) Given a CW-complex (spectrum) X, let $\mathscr{X}$ denote the family $\{X_\lambda\}$ of all finite subcomplexes (subspectra) of X. Then an $\mathscr{X}$-phantom is called a *weak phantom.*

Propositions 1.3 and 1.4 below are formulated for a spectrum X and phantoms in $E^*(X)$. We leave it to the reader to consider the case of spaces X and sets $[X, Y]$.

1.3. Proposition. *Let $h : X \to Y$ be an equivalence of spectra, and let $h^* : E^*(Y) \to E^*(X)$ be the induced isomorphism. Then h^* maps phantoms to phantoms and weak phantoms to weak phantoms.*

Proof. Exercise. Use II.3.14 (cf. 1.14 and 1.15 below). □

1.4. Proposition. *Let X, E be a pair of spectra. Then every phantom in $E^*(X)$ is a weak phantom. Furthermore, if X has finite $\mathbb{Z}$-type then every weak phantom in $E^*(X)$ is a phantom.*

Proof. The first assertion is trivial. The second assertion is clear if X has finite type, since in this case each skeleton $X^{(n)}$ is finite. Finally, by II.4.26(ii), every spectrum of finite $\mathbb{Z}$-type is equivalent to a spectrum of finite type, and the result follows from 1.3. □

1.5. Example of a weak phantom. Let $X = S^n[1/3]$ be a $\mathbb{Z}[1/3]$-localized sphere S^n, $n > 1$, i.e., the telescope of a sequence

$$S^n \xrightarrow{f} S^n \xrightarrow{f} \cdots \xrightarrow{f} S^n \xrightarrow{f} \cdots,$$

where $f : S^n \to S^n$ is a map of degree 3. If we regard S^n as a CW-complex with two cells, we obtain a cellular decomposition of X with 0-, n- and $(n+1)$-dimensional cells. This gives us a chain complex $\{C_*(X), \partial_*\}$, where $C_n(X)$ has $\mathbb{Z}$-basis $\{a_1, \dots, a_i, \dots\}$, and $C_{n+1}(X)$ has $\mathbb{Z}$-basis $\{b_1, \dots, b_n, \dots\}$, and $\partial_{n+1} b_i = a_i - 3a_{i+1}$. Let $\mathscr{X} = \{X_\lambda\}$ be the family of all finite CW-subcomplexes of X. It is clear that $H^{n+1}_{\mathscr{X}}(X) = H^{n+1}(S^n) = 0$. On the other hand, $H^{n+1}(X) \neq 0$ because the cocycle $\varphi : C_{n+1}(X) \to \mathbb{Z}$, $\varphi(b_i) = 1$ for every i, is not a coboundary. Indeed, if $\varphi = \delta\psi$ for some $\psi : C_n(X) \to \mathbb{Z}$, then $\psi(a_i) - 3\psi(a_{i+1}) = 1$. In particular,

$$\psi(a_1) = \frac{3^k - 1}{2} + 3^k \psi(a_{k+1})$$

for every k. Hence, 3^k divides $2\psi(a_1) + 1$ for every k, and so $2\psi(a_1) = -1$. This is a contradiction. Thus, the subgroup of weak phantoms of $H^{n+1}(X)$ is nontrivial (and even uncountable, see 5.1 below).

1.6. Example of a phantom (Adams–Walker [1]). Let $X = S^1 \wedge CP^\infty$. Consider the space $T = S^3[0]$, the telescope of the sequence

$$S_1^3 \xrightarrow{\varphi_1} S_2^3 \xrightarrow{\varphi_2} \cdots \to S_n^3 \xrightarrow{\varphi_n} S_{n+1}^3 \to \cdots,$$

where S_n^3 is a copy of S^3 and $\deg \varphi_n = n$. As in 1.5, we have $C_3(T) = \{a_1, \dots, a_n, \dots\}$, $C_4(T) = \{b_1, \dots, b_n, \dots\}$ and $\partial b_i = a_i - i a_{i+1}$. Let $f_0 : X \to T$ be any essential map: they exist because $T \simeq K(\mathbb{Q}, 3)$. (This holds, in turn, since $\pi_i(S^3)$ is finite for $i > 3$, cf. II.7.11(i).) We regard every sphere S_i^3 as a subspase of T: namely, $S_i^3 = S_i^3 \times \{i+1\} \subset T$. Let Y be the space obtained by attaching a cone $C_i = C(S_i^3)$ across each sphere S_i^3, and let $i : T \to Y$ be the inclusion. Set $f = i f_0$; we now prove that f is a phantom.

Firstly, $f|X^{(m)}$ is inessential for every m. Indeed, let T_k be the telescope of the finite sequence

$$S_1^3 \xrightarrow{\varphi_1} S_2^3 \xrightarrow{\varphi_2} \cdots \xrightarrow{\varphi_{k-1}} S_k^3;$$

it is clear that T_k contracts to S_k^3. The space $X^{(m)}$ is compact, and hence $f(X^{(m)})$ is contained in some T_N. But the sphere S_N^3 is coned off in Y, and hence $f|X^{(m)}$ is inessential.

On the other hand, f is essential. Indeed, otherwise it can be extended to a map $F : (CX, X) \to (Y, T)$. Let z be a generator of the group $H_4(CX, X) = \mathbb{Z}$, and let c_i be a generator of the group $H_4(C_i, C_i \cap T) = \mathbb{Z}$. Since the elements c_i generate $H_4(Y, T)$, there exists N such that $F_*(z) = \sum_{i=1}^{N} n_i c_i$, where $n_i \in \mathbb{Z}$. Let $\partial : H_4(CX, X) \to H_3(X)$ be the connecting homomorphism. It is clear that ∂z generates $H_3(X)$, and hence $F_*(z) = \pm (f_0)_*(\partial z)$. But $(f_0)_*(\partial z) \neq 0$ by construction, and so there exists k such that $n_k \neq 0$. Let p be any prime which does not divide n_k, and let $\bar{c}_i$ be the mod p reduction of c_i. Consider the class $c_k^* \in H^4(Y, T; \mathbb{Z}/p)$ dual to $\bar{c}_k$ with respect to the basis $\{\bar{c}_i\}$. We have $F^* c_k^* \neq 0$. Since $P^1 : H^2(CP^\infty; \mathbb{Z}/p) \to H^{2p}(CP^\infty; \mathbb{Z}/p)$ is an isomorphism, so is $P^1 : H^4(CX, X; \mathbb{Z}/p) \to H^{2p+2}(CX, X; \mathbb{Z}/p)$. Hence, $P^1 F^* c_k^* \neq 0$. But $P^1 c_k^* \in H^{2p+2}(Y, T; \mathbb{Z}/p) = 0$. This contradiction proves that f is essential.

1.7. Remarks. (a) There is no commonly accepted terminology concerning phantoms. For example, some people use the term "phantoms" for what we call weak phantoms; Margolis [1] uses the term "**f**-map" for weak phantoms, etc.

(b) Let $\mathscr{X}$ be the family of all suspension subspectra of a spectrum X. Margolis [1] introduced the term "hyperphantom" for $\mathscr{X}$-phantoms. I think it is still unknown whether hyperphantoms exist.

(c) Probably, "theoretically" it is preferably to define, say, weak phantoms as essential maps $f : X \to Y$ such that $f\varphi$ is inessential for every map $\varphi : A \to X$ of a finite CW-complex. In this way $\mathscr{X}$ in 1.1 should be a (small?) category with the terminal object X, etc. However, we do need this flavor, preferring the style "for working mathematicians".

Now we want to describe the groups $E^*_{\mathscr{X}}(X)$. Sometimes the inverse limit concept helps to do it.

1.8. Definition. (a) Let $\Lambda = (\Lambda, \leq)$ be a quasi-ordered directed set. Let $\mathscr{K}$ be an arbitrary category. An *inverse system* over Λ, or, briefly, an inverse Λ-system, in $\mathscr{K}$ is a contravariant functor $\mathscr{M} : \Lambda \to \mathscr{K}$. In other words, $\mathscr{M}$ is a family $\mathscr{M} = \{M_\lambda, j^\mu_\lambda\}_{\lambda,\mu\in\Lambda}$ where $M_\lambda \in \mathscr{K}$ and where $j^\mu_\lambda : M_\mu \to M_\lambda$, $\lambda \leq \mu$, are morphisms such that $j^\mu_\lambda j^\nu_\mu = j^\nu_\lambda$ for $\lambda \leq \mu \leq \nu$ and $j^\lambda_\lambda = 1_{M_\lambda}$.

(b) A *morphism* $f : \{M_\lambda, j^\mu_\lambda\} \to \{N_\lambda, h^\mu_\lambda\}$ of inverse Λ-systems is a natural transformation of the functors, i.e., a family $\{f_\lambda : M_\lambda \to N_\lambda\}$ with $h^\mu_\lambda f_\mu = f_\lambda j^\mu_\lambda$.

It is clear that there arises a category $\mathscr{K}_\Lambda$ of inverse Λ-systems in $\mathscr{K}$.

1.9. Definition. Let $\mathscr{K}$ be one of the following categories: $\mathscr{E}ns$, $\mathscr{W}$, groups and homomorphisms, R-modules over some ring R and R-homomorphisms, topological groups (in $\mathscr{W}$) and continuous homomorphisms. Given a quasi-ordered set Λ, let $\mathscr{M} = \{M_\lambda\}$ be an inverse Λ-system in $\mathscr{K}$. An element $\{a_\lambda\} \in \prod_\lambda M_\lambda$ is called a *string* if $j^\mu_\lambda a_\mu = a_\lambda$ for every $\lambda, \mu \in \Lambda$ with $\lambda \leq \mu$. The set of all strings is called the *inverse limit* or *projective limit* of the inverse system $\mathscr{M}$ and is denoted by $\varprojlim \mathscr{M}$ or $\varprojlim\{M_\lambda\}$ or $\varprojlim_\Lambda \{M_\lambda\}$.

Clearly, $\varprojlim\{M_\lambda\} = \prod M_\lambda$ if Λ is a discrete quasi-ordered set.

It is obvious that $\varprojlim$ is a functor $\mathscr{K}_\Lambda \to \mathscr{K}$. Furthermore, $\varprojlim$ commutes with the forgetful functor $\mathscr{K} \to \mathscr{E}ns$.

Note that the projections $p_\lambda : \prod\limits_\lambda M_\lambda \to M_\lambda$ yield functions

$$q_\lambda : \varprojlim \mathscr{M} \to M_\lambda,\ q_\lambda = p_\lambda | \varprojlim \mathscr{M}. \tag{1.10}$$

1.11. Proposition. *Let $\{f_\lambda : N \to M_\lambda\}$ be a family of morphisms such that $j^\mu_\lambda f_\mu = f_\lambda$ for every $\lambda \leq \mu$. Then there exists a morphism $f : N \to \varprojlim \mathscr{M}$ with $q_\lambda f = f_\lambda$, and this morphism is unique.*

Similarly to I.2.5, we denote this f by $\{f_\lambda | \varprojlim\}$.

Proof. Set $f(n) = \{f_\lambda(n)\} \in \prod_\lambda M_\lambda$. Then $\{f_\lambda(n)\}$ is a string, and $p_\lambda f = f_\lambda$. The uniqueness of f is obvious. □

As an illustration, consider the following important example.

1.12. Example. Let E be a spectrum, let X be a CW-complex (resp. a spectrum), and let $\mathscr{X} = \{X_\lambda\}$ be a family of CW-subcomplexes (resp. subspectra) of X ordered with respect to inclusions. We also assume that $X_\lambda \cup X_\mu \in \mathscr{X}$, $X_\lambda \cap X_\mu \in \mathscr{X}$ for every $\lambda, \mu \in \Lambda$. Let $i_\lambda : X_\lambda \to X$ and $i_\lambda^\mu : X_\lambda \to X_\mu$ be the inclusions. Then we have an inverse system $\{E^*(X_\lambda), (i_\lambda^\mu)^*\}$. Considering homomorphisms $q_\lambda : \varprojlim\{E^*(X_\lambda)\} \to E^*(X_\lambda)$ as in (1.10), we conclude that the family $\{f_\lambda := (i_\lambda)^* : E^*(X) \to E^*(X_\lambda)\}$ satisfies the conditions of 1.11. Thus, there exists a unique homomorphism $\rho = \{i_\lambda^* | \varprojlim\} : E^*(X) \to \varprojlim\{E^*(X_\lambda)\}$ with $q_\lambda \rho = i_\lambda^*$.

1.13. Proposition. *The morphism $\rho : E^*(X) \to \varprojlim\{E^*(X_\lambda)\}$ can be decomposed as*

$$E^*(X) \xrightarrow{\sigma} E^*_{\mathscr{X}}(X) \xrightarrow{\varkappa} \varprojlim\{E^*(X_\lambda)\},$$

where σ is the epimorphism defined at the beginning of this section and $\varkappa$ is a monomorphism.

Proof. It is easy to see that $i_\lambda^* : E^*(X) \to E^*(X_\lambda)$ can be decomposed as $E^*(X) \xrightarrow{\sigma} E^*_{\mathscr{X}}(X) \xrightarrow{\varkappa_\lambda} E^*(X_\lambda)$ with some $\varkappa_\lambda$. Furthermore, the homomorphisms $\varkappa_\lambda$ satisfy 1.11, and so there exists $\varkappa : E^*_{\mathscr{X}}(X) \to \varprojlim\{E^*(X_\lambda)\}$ with $q_\lambda \varkappa = \varkappa_\lambda$. The equality $\rho = \varkappa\sigma$ follows from 1.11.

We prove that $\varkappa$ is monic. Suppose that $\varkappa(a) = 0$ and choose $b \in E^*(X)$ with $\sigma(b) = a$. Then $i_\lambda^*(b) = 0$ for every λ. Thus, b is $\mathscr{X}$-equivalent to 0, i.e., $a = 0$. □

In particular, $\operatorname{Ker}\sigma = \operatorname{Ker}\rho$. Hence, $\mathscr{X}$-phantoms are just (nontrivial) elements of the group $\operatorname{Ker}\{\rho : E^*(X) \to \varprojlim\{E^*(X_\lambda)\}\}$.

1.14. Construction. Given two spectra X, Y, let $\{X_\lambda\}$, resp. $\{Y_\mu\}$ be the family of all finite subspectra of X, resp. Y, and $f : X \to Y$ be a map of spectra. By II.3.14, there are families $\{X_\omega\} \subset \{X_\lambda\}$, $\{Y_\omega\} \subset \{Y_\mu\}$ and maps $f_\omega : X_\omega \to Y_\omega, \omega \in \Omega$, such that $\{X_\omega\}$ is cofinal in $\{X_\lambda\}$, $\{Y_\omega\}$ is cofinal in $\{Y_\mu\}$, and the composition

$$X_\omega \xrightarrow{f_\omega} Y_\omega \subset Y$$

coincides with $f|X_\omega$. Now, for every spectrum E we have the homomorphisms $f_\omega^* : E^*(Y_\omega) \to E^*(X_\omega)$. They yield the homomorphism

$$f^* := \varprojlim f_\omega^* : \varprojlim\{E^*(Y_\mu)\} = \varprojlim\{E^*(Y_\omega)\} \to \varprojlim\{E^*(X_\omega)\} = \varprojlim\{E^*(X_\lambda)\}.$$

Furthermore, if $f : X \to Y$ is not a map but a morphism of spectra, we can consider a map $f' : X \to Y$ where X' is cofinal in X, and get the similar homomorphism $f^* : \varprojlim\{E^*(Y_\mu)\} \to \varprojlim\{E^*(X_\lambda)\}$.

1.15. Proposition. *Let X, Y, E be three spectra, and let $\{X_\lambda\}$, resp. $\{Y_\mu\}$ be the family of all finite subspectra of X, resp. Y.*

(i) *Given a morphism $f : X \to Y$, the homomorphism*

$$f^* : \varprojlim\{E^*(Y_\mu)\} \to \varprojlim\{E^*(X_\lambda)\}$$

in 1.14 does not depend on the choice of $\{X_\omega\}$ and $\{Y_\omega\}$. In particular, the group $\varprojlim\{E^*(X_\lambda)\}$ is natural in X.

(ii) *If $f \simeq g : X \to Y$ then $f^* = g^* : \varprojlim\{E^*(Y_\mu)\} \to \varprojlim\{E^*(X_\lambda)\}$. In particular, if $h : X \to Y$ is a homotopy equivalence then $h^* : \varprojlim\{E^*(Y_\mu)\} \to \varprojlim\{E^*(X_\lambda)\}$ is an isomorphism.*

Proof. We leave it to the reader. □

1.16. Proposition. *Let $\cdots \subset X_n \subset X_{n+1} \subset \cdots$ be a CW-filtration of a CW-complex X. Then for every space Y the function $\rho : [X, Y] \to \varprojlim\{[X_n, Y]\}$ is surjective. Similarly, if, in addition, we equip X and Y with base points and assume that $\{X_n\}$ is a pointed filtration, then the function $\rho : [X, Y]^\bullet \to \varprojlim\{[X_n, Y]^\bullet\}$ is surjective.*

Proof. We consider only the case of non-pointed spaces. Let an element $a \in \varprojlim\{[X_n, Y]\}$ be represented by a family of maps $\{f_n : X_n \to Y\}$ with $f_{n+1}|X_n \simeq f_n$. Using the homotopy extension property for CW-pairs, one can construct (by induction) a family of maps $\{g_n : X_n \to Y\}$ with $g_{n+1}|X_n = g_n$ and $g_n \simeq f_n$. If we define $g : X \to Y$ by setting $g|X_n = g_n$, we conclude that $\rho[g] = \{[g_n]\} = \{[f_n]\} = a$. □

1.17. Lemma. *Let $\cdots \to K_n \xrightarrow{f_n} K_{n-1} \to \cdots$ be an inverse sequence of non-empty finite sets. Then $\varprojlim\{K_n\} \neq \emptyset$.*

Proof. We set $P_n := \bigcap_{m=n}^{\infty} \operatorname{Im}\{K_m \to K_n\}$. Clearly, $P_n \neq \emptyset$ for every n, and every function $g_n : P_n \to P_{n-1}$ (the restriction of f_n) is surjective. So, we can find elements $x_n \in P_n, n \in \mathbb{Z}$, such that $g_n(x_n) = x_{n-1}$. Now, $\{x_n\}$ is a string in $\{K_n\}$. □

1.18. Theorem. *Let $Y = (Y, y_0)$ be a pointed space which is connected and simple.*

(i) *Let (Z, A) be a CW-pair such that the group $H^k(Z, A; \pi_k(Y))$ is finite for every $k > 0$. Given a map $u : A \to Y$, suppose that u can be extended to $Z^{(n)} \cup A$ for every n. Then u can be extended to the whole space Z.*

(ii) *Let $X = (X, x_0)$ be a pointed CW-complex such that the group $H^{k-1}(X; \pi_k(Y))$ is finite for every $k > 0$. Then both functions $\rho : [X, Y]^\bullet \to \varprojlim\{[X^{(n)}, Y]^\bullet\}$ and $\rho : [X, Y] \to \varprojlim\{[X^{(n)}, Y]\}$ are bijections.*

Proof. (i) It suffices to construct a family $\{v_n : Z^{(n)} \cup A \to Y\}$ such that $v|A = u$ and $v_{n+1}|Z^{(n)} \simeq v_n$. Indeed, then, deforming v_n map by map, we can construct a family $\{v'_n\}$ such that $v'_{n+1}|Z^{(n)} = v'_n$ and $v'_n|A = u$, cf. the proof of 1.16. Then we define $v(x) := v'_n(x)$ if $x \in Z^{(n)}$.

We set $k_n := \{v : Z^{(n)} \cup A \to Y \mid v|A = u\}$ and $K_n := \{[v] \mid v \in k_n\}$ (where $[v]$ denotes, as usual, the homotopy class of v). Then we have the inverse sequence

$$\cdots \to K_n \xrightarrow{f_n} K_{n-1} \to \cdots \to K_0$$

where $f_n[v] = [v|Z^{n-1} \cup A]$. Since every group $H^k(Z, A; \pi_k(Y))$ is finite, we conclude, using the elementary obstruction theory, that every set K_n is finite. So, by 1.17, there is a string $[v_n] \in \varprojlim\{K_n\}$, i.e., the desired family $\{v_n\}$.

(ii) The surjectivity of ρ is proved in 1.16. We prove the injectivity of ρ.

Firstly, we consider pointed maps. Let $f, g : (X, x_0) \to (Y, y_0)$ be two pointed maps such that $\rho[f] = \rho[g]$. We must prove that there a pointed homotopy between f and g.

We set $Z := X \times I, A = X \times \{0\} \cup X \times \{1\} \cup \{x_0\} \times I$ and define $u : A \to Y$ by setting $u(x, 0) = f(x), u(x, 1) = g(x), u(x_0, t) = y_0$. Since $\rho[f] = \rho[g]$, we conclude that $f|X^{(n)}$ and $g|X^{(n)}$ are homotopic for every n, and so u can be extended to $(X \times I)^{(n)} \cup A$ for every n. Now,

$$\begin{aligned}\widetilde{H}^k(Z, A; \pi_k(Y)) &= \widetilde{H}^k(X \times I/(X \times \{0,1\} \cup \{x_0\} \times I); \pi_k(Y)) \\ &= \widetilde{H}^k(SX; \pi_k(Y)) \cong H^{k-1}(X; \pi_k(Y)).\end{aligned}$$

So, every group $H^k(Z, A; \pi_k(Y))$ is finite, and hence, by (i), u can be extended to X, i.e., f and g are homotopic as pointed maps.

Now we consider non-pointed maps, i.e., we prove the injectivity of $\rho : [X, Y] \to \varprojlim\{[X^{(n)}, Y]\}$. So, we consider two maps $f, g : X \to Y$ and prove that $f \simeq g$ whenever $\rho[f] = \rho[g]$. Without loss of generality we can assume that $f(x_0) = g(x_0) = y_0$. Since $\rho[f] = \rho[g]$, for every n there is a homotopy $H_n : X^{(n)} \times I \to Y$ between f and g. We fix such a family $\{H_n\}$. Then every loop $H_n(x_0, t), t \in I$ gives us an element $\alpha_n \in \pi_1(Y, y_0)$. Since $(X \times I, \{x_0\} \times I)$ is a cofibered pair, we can assume that $H_m(x_0, t) = H_n(x_0, t)$ whenever $\alpha_m = \alpha_n$. Note that $\pi_1(Y)$ is finite since the group $H^0(X; \pi_1(Y))$ is finite. So, there is an infinite subset $M = \{n_1, \dots, n_k, \dots\}$ of $\mathbb{N}$ such that $\alpha_{n_i} = \alpha_{n_1}$ for every $i \in M$. Now we define a new family $\{H'_n : X^{(n)} \times I \to Y\}$ of homotopies by setting $H'_m := H_n|X^{(m)} \times I$ where $n := \min\{k | k \in M \text{ and } m \leq k\}$. In particular, $H'_n(x_0, t)$ does not depend on n. We set $Z := X \times I, A = X \times \{0\} \cup X \times \{1\} \cup \{x_0\} \times I$ and define $u : A \to Y$ by setting $u(x, 0) = f(x), u(x, 1) = g(x), u(x_0, t) = H'_n(x_0, t)$. Now the proof can be completed similarly to the previous case. □

Let $\mathscr{M}$ be an inverse Λ-system, and let Λ' be a subset of Λ with the quasi-ordering inherited from Λ. Consider the morphisms

$$q_\lambda : \varprojlim_{\Lambda} \mathcal{M} \to M_\lambda, \lambda \in \Lambda \text{ and } q'_\lambda : \varprojlim_{\Lambda'} \mathcal{M} \to M_\lambda, \lambda \in \Lambda'$$

as in (1.10). Based on 1.11, we define

$$\mathcal{M}^{\Lambda}_{\Lambda'} : \varprojlim_{\Lambda}\{M_\lambda\} \to \varprojlim_{\Lambda'}\{M_\lambda\} \tag{1.19}$$

to be the unique morphism such that $q'_\lambda \mathcal{M}^{\Lambda}_{\Lambda'} = q_\lambda$ for every $\lambda \in \Lambda'$.

1.20. Lemma. *If Λ' is cofinal in Λ then $\mathcal{M}^{\Lambda}_{\Lambda'}$ is an isomorphism for every inverse Λ-system $\mathcal{M}$.*

Proof. This is obvious. □

We finish this section with some algebraic remarks. Let Λ, Λ' be two quasi-ordered directed sets. Consider the quasi-ordered set $\Lambda \times \Lambda'$, where $(\lambda, \lambda') \le (\mu, \mu')$ iff $\lambda \le \mu$ and $\lambda' \le \mu'$.

1.21. Lemma. $\varprojlim_{\Lambda\times\Lambda'}\{A_{\lambda,\lambda'}\} = \varprojlim_{\Lambda}\{\varprojlim_{\Lambda'}\{A_{\lambda,\lambda'}\}\} = \varprojlim_{\Lambda'}\{\varprojlim_{\Lambda}\{A_{\lambda,\lambda'}\}\}$. *In particular, if Λ' is a discrete ordered set, then*

$$\varprojlim_{\Lambda}\left\{\prod_{\Lambda'} A_{\lambda,\lambda'}\right\} = \prod_{\Lambda'}\varprojlim_{\Lambda}\{A_{\lambda,\lambda'}\}.$$

Proof. Routine. □

1.22. Definition. Let R be a commutative ring. Let $\{M_\lambda\}$, resp. $\{N_{\lambda'}\}$, be inverse systems of R-modules. Set $M = \varprojlim\{M_\lambda\}, N = \varprojlim\{N_{\lambda'}\}$. Define the $\Lambda - \Lambda'$*-completed tensor product* $M \otimes^{\Lambda-\Lambda'} N := \varprojlim\{M_\lambda \otimes N_{\lambda'}\}$.

Given $a \in M, b \in N$, it is useful to denote the string $\{a_\lambda \otimes b_{\lambda'}\}$ by $a \otimes^{\Lambda-\Lambda'} b$.

1.23. Examples. (a) Let A, B be two graded abelian groups, and let $A_{[k]}$ be the subgroups of elements of degree $\le k$, $A_{[k]} = \oplus_{i\le k} A_i$. We have the inverse $\mathbb{Z}$-system $\{A_{[k]}, j^l_k\}$, where $j^l_k | A_i = 1_{A_i}$ for $i \le k$ and $j^l_k | A_r = 0$ for $k < r \le l$. In this way we have the *completed graded tensor product*

$$A \otimes^{\text{grad}} B := \varprojlim\{A_{[k]} \otimes B_{[l]}\}.$$

(b) Let E, F, X, Y be any spectra. Suppose that $E^*(X) = \varprojlim\{E^*(X_\lambda)\}$ and $F^*(Y) = \varprojlim\{F^*(Y_{\lambda'})\}$, where $\{X_\lambda\}, \{Y_{\lambda'}\}$ are the families of all finite subspectra of X, Y respectively. (See §4 about conditions when this holds.) Then we have the *profinitely completed tensor product*

$$E^*(X)\widehat{\otimes}F^*(Y) := \varprojlim\{E^*(X_\lambda) \otimes F^*(Y_{\lambda'})\}.$$

Furthermore, if E is a ring spectrum and F is an E-module spectrum, we define

$$E^*(X)\widehat{\otimes}_{E^*(S)}F^*(Y) := \varprojlim\{E^*(X_\lambda) \otimes_{E^*(S)} F^*(Y_{\lambda'})\}.$$

§2. Derived Functors of the Inverse Limit Functor

From here to the end of this chapter the words "inverse system" mean "inverse system in $\mathscr{AG}$" unless something else is said explicitly. Moreover, if we use a script letter (say, $\mathscr{A}$) in order to denote an inverse Λ-system then we use the same capital letter (A_λ in this case) in order to denote terms of this system.

2.1. Definition. A sequence $\cdots \to \mathscr{A} \to \mathscr{B} \to \mathscr{C} \to \cdots$ of inverse Λ-systems is called *exact* if for every $\lambda \in \Lambda$ the sequence

$$\cdots \to A_\lambda \to B_\lambda \to C_\lambda \to \cdots$$

is exact.

2.2. Theorem. *If the sequence*

$$0 \to \mathscr{A} \to \mathscr{B} \to \mathscr{C} \to 0$$

is exact, then the induced sequence

$$0 \to \varprojlim \mathscr{A} \to \varprojlim \mathscr{B} \to \varprojlim \mathscr{C}$$

is exact. In other words, the functor $\varprojlim$ *is a left exact functor.*

Proof. This follows immediately from the definitions, see e.g. Switzer [1], 7.63, or Eilenberg–Steenrod [1], VIII.5.3. □

However, the functor $\varprojlim$ is *not* a right exact functor, i.e., the homomorphism $\varprojlim \mathscr{B} \to \varprojlim \mathscr{C}$ in 2.2 is not epic in general. There is the following well-known example.

2.3. Example. Consider the following short exact sequence of inverse systems (the latter are vertically situated)

$$
\begin{array}{ccccccccc}
 & & \vdots & & \vdots & & \vdots & & \\
 & & \downarrow{\scriptstyle 3} & & \downarrow{\scriptstyle 3} & & \| & & \\
0 & \longrightarrow & \mathbb{Z} & \xrightarrow{2} & \mathbb{Z} & \longrightarrow & \mathbb{Z}/2 & \longrightarrow & 0 \\
 & & \downarrow{\scriptstyle 3} & & \downarrow{\scriptstyle 3} & & \| & & \\
0 & \longrightarrow & \mathbb{Z} & \xrightarrow{2} & \mathbb{Z} & \longrightarrow & \mathbb{Z}/2 & \longrightarrow & 0
\end{array}
$$

where the number at the arrow means the multiplication by this number. Applying the functor $\varprojlim$ to this sequence, we get a non-exact sequence

$$0 \to 0 \to 0 \to \mathbb{Z}/2 \to 0.$$

This suggests the existence of right derived functors $\varprojlim^k$ of $\varprojlim$, converting the exact sequence $0 \to \mathscr{A} \to \mathscr{B} \to \mathscr{C} \to 0$ into an exact sequence

$$
\begin{aligned}
&0 \to \varprojlim \mathscr{A} \to \varprojlim \mathscr{B} \to \varprojlim \mathscr{C} \to \varprojlim{}^1 \mathscr{A} \to \cdots \to \varprojlim{}^{k-1} \mathscr{C} \\
&\to \varprojlim{}^k \mathscr{A} \to \varprojlim{}^k \mathscr{B} \to \varprojlim{}^k \mathscr{C} \to \varprojlim{}^{k+1} \mathscr{A} \to \cdots .
\end{aligned} \tag{2.4}
$$

These functors really exist, and now we describe them.

2.5. Definition. An inverse Λ-system $\mathscr{A}$ is called *flabby* if the homomorphism $\mathscr{A}^{\Lambda}_{\Lambda'} : \varprojlim_{\lambda\in\Lambda} A_\lambda \to \varprojlim_{\lambda\in\Lambda'} A_\lambda$ as in (1.19) is epic for every $\Lambda' \subset \Lambda$.

Note that $\{A_\lambda | \lambda \in \Lambda'\}$ is flabby for every $\Lambda' \subset \Lambda$ if $\{A_\lambda | \lambda \in \Lambda\}$ is flabby. Furthermore, if the system $\{A_\lambda\}$ is flabby, then every homomorphism $j^\mu_\lambda : A_\mu \to A_\lambda, \lambda \le \mu$, is epic.

2.6. Theorem (cf. Godement [1], Th. 3.1.2). *Let*

$$0 \to \mathscr{A} \to \mathscr{B} \to \mathscr{C} \to 0$$

be an exact sequence of inverse systems. If $\mathscr{A}$ is flabby, then the sequence

$$0 \to \varprojlim \mathscr{A} \to \varprojlim \mathscr{B} \to \varprojlim \mathscr{C} \to 0$$

is exact.

Proof. Given a string $\{c_\lambda\} \in \varprojlim \mathscr{C}$, we prove that it is the image of a string $\{b_\lambda\} \in \varprojlim \mathscr{B}$. Consider the set E of all "substrings" $\{b_\lambda \in B_\lambda | \lambda \in \Lambda' \subset \Lambda\}$ such that $b_\lambda \mapsto c_\lambda$ for every $\lambda \in \Lambda'$ and Λ' runs over all subsets of Λ. We say that $\{b'_\lambda | \lambda \in \Lambda' \subset \Lambda\} \le \{b''_\lambda | \lambda \in \Lambda'' \subset \Lambda\}$ iff $\Lambda' \subset \Lambda''$ and $b'_\lambda = b''_\lambda$ for every $\lambda \in \Lambda'$. It is clear that E is an inductive set. Therefore, by Zorn's Lemma, it has a maximal element $\{b_\lambda | \lambda \in \Omega\}$. We prove that $\Omega = \Lambda$. Indeed, let $\nu \in \Lambda$, $\nu \notin \Omega$. Set $[\nu] := \{\alpha | \alpha \le \nu\} \subset \Lambda$. Choose an element $\bar{b}_\nu$ with

$\overline{b}_\nu \mapsto c_\nu$ and set $\overline{b}_\alpha = j_\alpha^\nu \overline{b}_\nu$ for every $\alpha \leq \nu$. It is clear that the difference of the substrings $\{b_\beta | \beta \in [\nu] \cap \Omega\}$ and $\{\overline{b}_\beta | \beta \in [\nu] \cap \Omega\}$ is the image of some substring $\{a_\beta \in A_\beta | \beta \in [\nu] \cap \Omega\}$. Since $\mathscr{A}$ is flabby, we can find $a_\nu \in A_\nu$ with $a_\beta = j_\beta^\nu a_\nu$ for every $\beta \leq \nu$. Set $b'_\nu = \overline{b}_\nu + a_\nu$ and $b'_\beta = j_\beta^\nu b'_\nu$ for $\beta \leq \nu$. Then $\{b_\lambda | \lambda \in \Omega\}$ and $\{b'_\beta | \beta \in [\nu]\}$ agree on $[\nu] \cap \Omega$. But in this case they produce a substring indexed by $[\nu] \cup \Omega$. Hence, $\{b_\lambda | \lambda \in \Omega\}$ is not a maximal element. This is a contradiction. □

2.7. Corollary. *Let* $0 \to \mathscr{A} \xrightarrow{f} \mathscr{B} \xrightarrow{g} \mathscr{C} \to 0$ *be an exact sequence of inverse systems. If* $\mathscr{A}$ *and* $\mathscr{B}$ *are flabby then so is* $\mathscr{C}$.

Proof. In the commutative diagram

$$\begin{array}{ccc}
\varprojlim_{\lambda\in\Lambda} B_\lambda & \longrightarrow & \varprojlim_{\lambda\in\Lambda} C_\lambda \\
{\scriptstyle \mathscr{B}^\Lambda_{\Lambda'}}\downarrow & & \downarrow{\scriptstyle \mathscr{C}^\Lambda_{\Lambda'}} \\
\varprojlim_{\lambda\in\Lambda'} B_\lambda & \xrightarrow{\varprojlim\{g_\lambda\}} & \varprojlim_{\lambda\in\Lambda'} C_\lambda
\end{array}$$

the homomorphisms $\varprojlim\{g_\lambda\}$ and $\mathscr{B}^\Lambda_{\Lambda'}$ are epic. Thus, $\mathscr{C}^\Lambda_{\Lambda'}$ is epic. □

2.8. Corollary. *If* $0 \to \mathscr{A}^0 \xrightarrow{f^0} \mathscr{A}^1 \xrightarrow{f^1} \cdots \xrightarrow{f^n} \mathscr{A}^{n+1} \to \cdots$ *is an exact sequence of flabby inverse systems, then the induced sequence*

$$0 \to \varprojlim \mathscr{A}^0 \to \varprojlim \mathscr{A}^1 \to \cdots \to \varprojlim \mathscr{A}^n \to \cdots$$

is exact.

Proof. It follows from 2.7 (inductively) that

$$0 \to 0 \to \mathscr{A}^0 \to \operatorname{Im} f^0 \to 0, \quad 0 \to \operatorname{Im} f^n \to \mathscr{A}^{n+1} \to \operatorname{Im} f^{n+1} \to 0$$

are exact sequences of flabby inverse systems. Now use 2.6. □

Let $\mathscr{A} = \{A_\lambda, j_\lambda^\mu\}$ be an inverse system. Following Roos [1], consider an inverse system $\mathscr{R}(\mathscr{A}) = \{R_\lambda(\mathscr{A}), \pi_\lambda^\mu\}$, where $R_\lambda = \prod_{\alpha \leq \lambda} A_\alpha$ and π_λ^μ is defined as follows: $\pi_\lambda^\mu | A_\alpha = 1_{A_\alpha}$ for $\alpha \leq \lambda$ and $\pi_\lambda^\mu | A_\alpha = 0$ otherwise. There is a canonical embedding

$$r : \mathscr{A} \to \mathscr{R}(\mathscr{A}), \quad r_\lambda = \{j_\alpha^\lambda\} : A_\lambda \to \prod_{\alpha \leq \lambda} A_\alpha = R_\lambda(\mathscr{A}).$$

It is easy to see that $\mathscr{R}$ is an autofunctor on the category of inverse Λ-systems and that $r : 1 \to \mathscr{R}$ is a morphism of functors.

2.9. Proposition. *The inverse system $\mathscr{R}(\mathscr{A})$ is flabby for every inverse system $\mathscr{A}$. Thus, every inverse system can be embedded in a flabby one.*

Proof. This follows immediately from the definition of $\mathscr{R}$. □

Given an inverse system $\mathscr{A}$, we set $\mathscr{R}^0(\mathscr{A}) = \mathscr{R}(\mathscr{A})$ and $\mathscr{R}^1(\mathscr{A}) = \mathscr{R}(\mathscr{R}^0(A)/r(\mathscr{A}))$. We define the morphism

$$e^0 : \mathscr{R}^0(\mathscr{A}) \xrightarrow{\text{quotient}} \mathscr{R}^0(\mathscr{A})/r(\mathscr{A}) \xrightarrow{r} \mathscr{R}(\mathscr{R}^0(\mathscr{A})/r(\mathscr{A})) = \mathscr{R}^1(\mathscr{A}).$$

Now, by induction, for $n = 0, 1, 2, \dots$ we define

$$\mathscr{R}^{n+1}(\mathscr{A}) = \mathscr{R}(\mathscr{R}^n(\mathrm{A})/e^{n-1}(\mathscr{R}^{n-1}(\mathrm{A})))$$

and

$$\begin{aligned} e^n : \mathscr{R}^n(\mathscr{A}) &\to \mathscr{R}^n(\mathscr{A})/e^{n-1}(\mathscr{R}^{n-1}(\mathscr{A})) \\ &\xrightarrow{r} \mathscr{R}(\mathscr{R}^n(\mathscr{A})/e^{n-1}(\mathscr{R}^{n-1}(\mathscr{A}))) = \mathscr{R}^{n+1}(\mathscr{A}). \end{aligned}$$

2.10. Definition. The *Roos resolution* of an inverse system $\mathscr{A}$ is the exact sequence

$$0 \to \mathscr{A} \xrightarrow{r} \mathscr{R}^0(\mathscr{A}) \xrightarrow{e^0} \mathscr{R}^1(\mathscr{A}) \to \cdots \to \mathscr{R}^n(\mathscr{A}) \xrightarrow{e^n} \cdots .$$

It is clear that the Roos resolution is natural with respect to $\mathscr{A}$.

2.11. Proposition. *If a sequence*

$$0 \to \mathscr{A} \to \mathscr{B} \to \mathscr{C} \to 0$$

of inverse systems is exact, then for every n the sequence

$$0 \to \mathscr{R}^n(\mathscr{A}) \to \mathscr{R}^n(\mathscr{B}) \to \mathscr{R}^n(\mathscr{C}) \to 0$$

is exact. More generally, if a sequence

$$0 \to \mathscr{A}^0 \to \mathscr{A}^1 \to \cdots \to \mathscr{A}^k \to \cdots$$

is exact, then the sequence

$$0 \to \mathscr{R}^n(\mathscr{A}^0) \to \mathscr{R}^n(\mathscr{A}^1) \to \cdots \to \mathscr{R}^n(\mathscr{A}^k) \to \cdots$$

is exact.

Proof. The exactness of the sequence $0 \to \mathscr{R}(\mathscr{A}) \to \mathscr{R}(\mathscr{B}) \to \mathscr{R}(\mathscr{C}) \to 0$ follows from the definition of $\mathscr{R}$. Furthermore, the sequence

$$0 \to \mathscr{R}(\mathscr{A})/\mathscr{A} \to \mathscr{R}(\mathscr{B})/\mathscr{B} \to \mathscr{R}(\mathscr{C})/\mathscr{C} \to 0$$

is exact, and so the short sequence of the proposition is exact for $n = 1$. By iteration of these arguments, we can prove the exactness of this short sequence for every n. This implies the last assertion of the proposition: it can be deduced from the previous one just as we deduced 2.8 from 2.7. □

2.12. Definition. Given an arbitrary inverse system $\mathscr{A}$, we set

$$\delta^n := \varprojlim\{e^n\} : \varprojlim \mathscr{R}^n(\mathscr{A}) \to \varprojlim \mathscr{R}^{n+1}(\mathscr{A}).$$

Thus, we get a cochain complex

$$\varprojlim \mathscr{R}^0(\mathscr{A}) \xrightarrow{\delta^0} \varprojlim \mathscr{R}^1(\mathscr{A}) \xrightarrow{\delta^1} \cdots \to \varprojlim \mathscr{R}^n(\mathscr{A}) \xrightarrow{\delta^n} \varprojlim \mathscr{R}^{n+1}(\mathscr{A}) \to \cdots .$$

We define

$$\varprojlim{}^0 \mathscr{A} := \operatorname{Ker} \delta^0, \ \varprojlim{}^n \mathscr{A} := \operatorname{Ker} \delta^n / \operatorname{Im} \delta^{n-1}.$$

It is clear that $\varprojlim^n$ is a functor on the category of inverse systems.

2.13. Proposition. (i) $\varprojlim^0 \mathscr{A} = \varprojlim \mathscr{A}$.

(ii) *For every short exact sequence* $0 \to \mathscr{A} \xrightarrow{f} \mathscr{B} \xrightarrow{g} \mathscr{C} \to 0$ *of inverse systems, there is a natural long exact sequence*

$$0 \to \varprojlim \mathscr{A} \xrightarrow{\varprojlim f} \varprojlim \mathscr{B} \xrightarrow{\varprojlim g} \varprojlim \mathscr{C} \to \varprojlim{}^1 \mathscr{A} \to \cdots \to \varprojlim{}^{n-1} \mathscr{C} \to$$
$$\varprojlim{}^n \mathscr{A} \xrightarrow{\varprojlim^n f} \varprojlim{}^n \mathscr{B} \xrightarrow{\varprojlim^n g} \varprojlim{}^n \mathscr{C} \to \varprojlim{}^{n+1} \mathscr{A} \to \cdots .$$

(iii) *If an inverse system* $\mathscr{A}$ *is flabby then* $\varprojlim^n \mathscr{A} = 0$ *for every* $n > 0$.

Proof. (i) We have the exact sequences

$$0 \to \varprojlim \mathscr{A} \to \varprojlim \mathscr{R}(\mathscr{A}) \xrightarrow{h} \varprojlim(\mathscr{R}(\mathscr{A})/r(\mathscr{A}))$$

and

$$0 \to \varprojlim(\mathscr{R}(\mathscr{A})/r(\mathscr{A})) \xrightarrow{\varprojlim(r)} \varprojlim \mathscr{R}^1(\mathscr{A}),$$

where h is induced by the quotient map and $\varprojlim(r)h = \delta^0$. Thus, $\varprojlim^0 \mathscr{A} = \varprojlim \mathscr{A}$.

(ii) Consider the following commutative diagram with exact rows and columns:

$$\begin{array}{ccccccccc}
 & & 0 & & 0 & & 0 & & \\
 & & \downarrow & & \downarrow & & \downarrow & & \\
0 & \longrightarrow & \mathscr{A} & \longrightarrow & \mathscr{B} & \longrightarrow & \mathscr{C} & \longrightarrow & 0 \\
 & & \downarrow & & \downarrow & & \downarrow & & \\
0 & \longrightarrow & \mathscr{R}(\mathscr{A}) & \longrightarrow & \mathscr{R}(\mathscr{B}) & \longrightarrow & \mathscr{R}(\mathscr{C}) & \longrightarrow & 0 \\
 & & \downarrow & & \downarrow & & \downarrow & & \\
 & & \vdots & & \vdots & & \vdots & & \\
 & & \downarrow & & \downarrow & & \downarrow & & \\
0 & \longrightarrow & \mathscr{R}^n(\mathscr{A}) & \longrightarrow & \mathscr{R}^n(\mathscr{B}) & \longrightarrow & \mathscr{R}^n(\mathscr{C}) & \longrightarrow & 0 \\
 & & \downarrow & & \downarrow & & \downarrow & & \\
 & & \vdots & & \vdots & & \vdots & &
\end{array}$$

It induces a commutative diagram

$$\begin{array}{ccccccccc}
0 & \longrightarrow & \varprojlim \mathscr{R}^0(\mathscr{A}) & \longrightarrow & \varprojlim \mathscr{R}^0(\mathscr{B}) & \longrightarrow & \varprojlim \mathscr{R}^0(\mathscr{C}) & \longrightarrow & 0 \\
 & & \downarrow & & \downarrow & & \downarrow & & \\
0 & \longrightarrow & \varprojlim \mathscr{R}^1(\mathscr{A}) & \longrightarrow & \varprojlim \mathscr{R}^1(\mathscr{B}) & \longrightarrow & \varprojlim \mathscr{R}^1(\mathscr{C}) & \longrightarrow & 0 \\
 & & \downarrow & & \downarrow & & \downarrow & & \\
 & & \vdots & & \vdots & & \vdots & & \\
 & & \downarrow & & \downarrow & & \downarrow & & \\
0 & \longrightarrow & \varprojlim \mathscr{R}^n(\mathscr{A}) & \longrightarrow & \varprojlim \mathscr{R}^n(\mathscr{B}) & \longrightarrow & \varprojlim \mathscr{R}^n(\mathscr{C}) & \longrightarrow & 0 \\
 & & \downarrow & & \downarrow & & \downarrow & & \\
 & & \vdots & & \vdots & & \vdots & &
\end{array}$$

By 2.6 and 2.9, its rows are exact. Furthermore, each of its columns is a complex, i.e., this diagram is a short exact sequence of complexes. The homology exact sequence of this short exact sequence is the desired sequence.

(iii) This follows from 2.8. □

One says that an inverse system $\mathscr{A}$ is $\varprojlim$-*acyclic*, or simply *acyclic*, if $\varprojlim^i \mathscr{A} = 0$ for all $i > 0$. For example, every flabby system is acyclic. One expects that one can compute $\varprojlim^i$ via any acyclic resolution, and in fact this is true.

2.14. Theorem. *Let*

$$0 \to \mathscr{A} \to \mathscr{A}^0 \xrightarrow{f^0} \mathscr{A}^1 \xrightarrow{f^1} \dots \to \mathscr{A}^n \xrightarrow{f^n} \dots$$

be an exact sequence of inverse systems with acyclic $\mathscr{A}^i$ *for every* $i \geq 0$. *Then* $\varprojlim^0 \mathscr{A} = \operatorname{Ker} \delta^0$, $\varprojlim^n \mathscr{A} = \operatorname{Ker} \delta^n / \operatorname{Im} \delta^{n-1}$, *where*

$$\delta^n = \varprojlim f^n : \varprojlim \mathscr{A}^n \to \varprojlim \mathscr{A}^{n+1}.$$

Proof. Consider the following commutative diagram:

$$\begin{array}{ccccccccc}
 & & 0 & & 0 & & & & 0 & & \\
 & & \downarrow & & \downarrow & & & & \downarrow & & \\
0 & \longrightarrow & \mathscr{A} & \longrightarrow & \mathscr{A}^0 & \longrightarrow & \cdots & \longrightarrow & \mathscr{A}^j & \longrightarrow & \cdots \\
 & & \downarrow & & \downarrow & & & & \downarrow & & \\
0 & \longrightarrow & \mathscr{R}^0(\mathscr{A}) & \longrightarrow & \mathscr{R}^0(\mathscr{A}^0) & \longrightarrow & \cdots & \longrightarrow & \mathscr{R}^0(\mathscr{A}^j) & \longrightarrow & \cdots \\
 & & \downarrow & & \downarrow & & & & \downarrow & & \\
 & & \vdots & & \vdots & & & & \vdots & & \\
 & & \downarrow & & \downarrow & & & & \downarrow & & \\
0 & \longrightarrow & \mathscr{R}^i(\mathscr{A}) & \longrightarrow & \mathscr{R}^i(\mathscr{A}^0) & \longrightarrow & \cdots & \longrightarrow & \mathscr{R}^i(\mathscr{A}^j) & \longrightarrow & \cdots \\
 & & \downarrow & & \downarrow & & & & \downarrow & & \\
 & & \vdots & & \vdots & & & & \vdots & &
\end{array}$$

We apply $\varprojlim$ to this diagram and obtain the bicomplex $\varprojlim \mathscr{R}^i(\mathscr{A}^j)$. As usual, there are two spectral sequences ${}'E_r^{p,q}$ and ${}''E_r^{p,q}$, both converging to the same limit. Here for ${}'E_r^{p,q}$ (given by the horizontal level filtration) we have ${}'E_2^{p,q} = 0$ for $q \neq 0$ and ${}'E_2^{p,0} = \varprojlim^p \mathscr{A}$, i.e., ${}'E_\infty^{p,0} = \varprojlim^p \mathscr{A}$. On the other hand, by 2.8 and 2.9, ${}''E_2^{p,q} = 0$ for $p \neq 0$, while ${}''E_2^{0,q}$ coincides with the group $\operatorname{Ker} \delta^q / \operatorname{Im} \delta^{q-1}$ of the theorem. Thus, $\varprojlim^n \mathscr{A} = \operatorname{Ker} \delta^n / \operatorname{Im} \delta^{n-1}$. □

Now we consider the important special case: the index set Λ is the set $\mathbb{Z}$ of integers. So, let $\mathscr{A} = \{A_n, j_n^m\}$ be an arbitrary inverse system over $\mathbb{Z}$. Define the endomorphism

$$\delta : \prod_{n=-\infty}^{\infty} A_n \to \prod_{n=-\infty}^{\infty} A_n$$

by setting

$$\delta(\dots, a_1, \dots, a_n, \dots) = (\dots, a_1 - j_1^2 a_2, \dots, a_n - j_n^{n+1} a_{n+1}, \dots), \quad a_k \in A_k.$$

2.15. Theorem. $\varprojlim \mathscr{A} = \operatorname{Ker} \delta$, $\varprojlim^1 \mathscr{A} = \operatorname{Coker} \delta$, $\varprojlim^i \mathscr{A} = 0$ *for* $i > 1$.

Proof. Consider the inverse system $\mathscr{S}(\mathscr{A}) = \{S_n(\mathscr{A}), p_n^m\}$ where

$$S_n(\mathscr{A}) = \prod_{k \le n-1} A_k$$

and the homomorphisms $p_n^m : S_m(\mathscr{A}) \to S_n(\mathscr{A}), m \ge n$, have the form

$$p_n^m | A_k = \begin{cases} 1_{A_k} & \text{if } k \le n-1, \\ 0 & \text{if } n-1 < k \le m. \end{cases}$$

Define $\varphi = \{\varphi_n\} : \mathscr{R}(\mathscr{A}) \to \mathscr{S}(\mathscr{A})$ by setting $\varphi_n(\dots, a_1, \dots, a_n) = (\dots, a_1 - j_1^2 a_2, \dots, a_{n-1} - j_{n-1}^n a_n)$. Obviously, the sequence of the inverse systems

$$0 \to \mathscr{A} \xrightarrow{r} \mathscr{R}(\mathscr{A}) \xrightarrow{\varphi} \mathscr{S}(\mathscr{A}) \to 0$$

is exact. Also, it is easy to see that $\mathscr{S}(\mathscr{A})$ is flabby, and hence this sequence is an acyclic resolution of $\mathscr{A}$, i.e., it satisfies the condition of 2.14. It remains to note that $\varprojlim\{\varphi : \mathscr{R}(\mathscr{A}) \to \mathscr{S}(\mathscr{A})\} = \delta : \prod A_n \to \prod A_n$. □

Now we describe a useful class of acyclic resolutions.

2.16. Definition. (a) An inverse system $\{A_\lambda, j_\lambda^\mu\}$ of topological abelian groups and continuous homomorphisms is called *compact* if all the groups A_λ are compact.

(b) An inverse system $\mathscr{A}$ of abelian groups is called *algebraically compact* or, briefly, *a-compact* if it can be obtained from a compact one by ignoring the topology. Similarly, a group is called *a-compact* if it can be obtained from a compact topological group by ignoring the topology.

(c) Given an inverse system $\mathscr{A}$, an exact sequence

$$0 \to \mathscr{A} \to \mathscr{A}^0 \xrightarrow{\varphi^0} \dots \to \mathscr{A}^n \xrightarrow{\varphi^n} \dots$$

of inverse systems is called an *a-compact resolution* of $\mathscr{A}$ if every inverse system $\mathscr{A}^i, i \ge 0$, is a-compact. An a-compact resolution of a group is defined similarly.

Clearly, every finite group is a-compact. The group $\mathbb{Z}$ is not a-compact.

2.17. Theorem. *Let* $\{A_\lambda\} \xrightarrow{\{f_\lambda\}} \{B_\lambda\} \xrightarrow{\{g_\lambda\}} \{C_\lambda\}$ *be an exact sequence of compact inverse systems such that all the homomorphisms* f_λ, g_λ *are continuous. Then the induced sequence* $\varprojlim\{A_\lambda\} \to \varprojlim\{B_\lambda\} \to \varprojlim\{C_\lambda\}$ *is exact.*

Proof. See Eilenberg–Steenrod [1], Theorem VIII.5.6. □

2.18. Corollary. (i) *Every compact inverse system $\mathscr{A}$ is acyclic.*

(ii) *Every a-compact inverse system is acyclic. In particular, every system of finite abelian groups is acyclic.*

(iii) *Let*

$$0 \to \mathscr{A} \to \mathscr{A}^0 \xrightarrow{\varphi^0} \cdots \to \mathscr{A}^n \xrightarrow{\varphi^n} \cdots$$

be an a-compact resolution of $\mathscr{A}$. Then $\varprojlim^n \mathscr{A} = \ker \delta^n / \operatorname{Im} \delta^{n-1}$, $\varprojlim^0 \mathscr{A} =$ $\operatorname{Ker} \delta^0$, where $\delta^n = \varprojlim \varphi^n : \varprojlim \mathscr{A}^n \to \varprojlim \mathscr{A}^{n+1}$.

Proof. (i) Since the product of compact topological spaces is compact, all the groups $R_\lambda(\mathscr{A})$ are compact. [9] It is easy to see that all the projections $\pi^\mu_\lambda : R_\mu(\mathscr{A}) \to R_\lambda(\mathscr{A})$ are continuous, as well as the maps $r_\lambda : A_\lambda \to R_\lambda(\mathscr{A})$. Hence, the Roos resolution $0 \to \mathscr{A} \to \mathscr{R}^0(\mathscr{A}) \to \cdots \xrightarrow{e^n} \mathscr{R}^{n+1}(\mathscr{A}) \to \cdots$ of $\mathscr{A}$ consists of compact inverse systems, and all the homomorphisms e^n_λ are continuous. Hence, by 2.17, the sequence

$$\varprojlim \mathscr{R}^0(\mathscr{A}) \to \cdots \to \varprojlim \mathscr{R}^n(\mathscr{A}) \xrightarrow{\delta^n} \varprojlim \mathscr{R}^{n+1}(\mathscr{A}) \to \cdots$$

is exact. Thus, $\varprojlim^n(\mathscr{A}) = 0$ for every $n > 0$.

(ii) This follows from (i).

(iii) This follows from 2.14 and (ii). □

2.19. Lemma. *Every abelian group can be naturally embedded in an a-compact one. In other words, there exist a functor $\Phi : \mathscr{AG} \to \mathscr{AG}$ and a natural transformation $c : 1_{\mathscr{AG}} \to \Phi$ such that, for every abelian group A, $\Phi(A)$ is algebraically compact and $c_A : A \to \Phi(A)$ is a monomorphism.*

Proof. Given a topoogical group G, let $\chi(G)$ be the character group of G, i.e., the topological group of all continuous homomorphisms $G \to SO(2)$. Recall that the canonical map $\omega : G \to \chi(\chi(G))$, $\omega(g)(\varphi) = \varphi(g)$, is an isomorphism and that $\chi(A)$ is a compact group for every discrete group A, see Pontrjagin [1], Th. 39 and 36. Clearly, the correspondence $A \mapsto \chi(A)$ yields a contravariant functor $\mathscr{AG} \to \mathscr{AG}$. Let $\chi^\delta(G)$ be the discrete group which has the same underlying set as $\chi(G)$. Then $\varepsilon : \chi^\delta(G) \to \chi(G)$ is a continuous epimorphism, and hence $\chi(\varepsilon) : \chi(\chi(G)) \to \chi(\chi^\delta(G))$ is a continuous monomorphism. Now, given an abelian group A, we set $\Phi(A) = \chi^\delta\chi^\delta(A)$, and we define c by setting c_A to be the composition

$$c_A : A \xrightarrow{\omega} \chi(\chi(A)) \xrightarrow{\chi(\varepsilon)} \chi(\chi^\delta(A)) \xrightarrow{\cong} \chi^\delta\chi^\delta(A) = \Phi(A) \tag{2.20}$$

[9] Recall that we use definition I.3.6 of the product of topological spaces. It is well-known that $\prod^c X_i$ is compact provided every X_i is compact, see e.g. Bourbaki [2]. However, kY is compact if Y is compact.

where the last arrow is the additive isomorphism $\varepsilon^{-1}_{\chi^\delta(A)}$. □

2.21. Definition. In future we write $\widehat{A}$ instead of $\Phi(A)$ and call the homomorphism $c = c_A : A \to \widehat{A}$ the *canonical a-compactification* of A.

2.22. Definition. Let A be an abelian group, and let $c : A \to A^0$ be the canonical a-compactification of A. We consider the canonical a-compactification $c_0 : \operatorname{Coker} c \to A^1$ and define $c^0 : A^0 \xrightarrow{h_0} \operatorname{Coker} c \xrightarrow{c_0} A^1$, where h_0 is the canonical epimorphism. Inductively, let $c_n : \operatorname{Coker} c^{n-1} \to A^{n+1}$ be the canonical a-compactification, and let $h_n : A^n \to \operatorname{Coker} c^{n-1}$ be the canonical epimorphism. Set $c^n = c_n h_n : A^n \to A^{n+1}$. Then we have an exact sequence

$$0 \to A \xrightarrow{c} A^0 \xrightarrow{c^0} A^1 \to \cdots \xrightarrow{c^{n-1}} A^n \xrightarrow{c^n} \cdots$$

where all the groups A^i are a-compact. This exact sequence is called the *canonical a-compact resolution of* A.

2.23. Theorem. *Every inverse system $\mathscr{A}$ admits an a-compact resolution. Furthermore, this resolution can be constructed naturally with respect to $\mathscr{A}$.*

Proof. Given an inverse Λ-system $\mathscr{A}$, consider the canonical a-compact resolution of A_λ

$$0 \to A_\lambda \to A^0_\lambda \to \cdots \to A^n_\lambda \xrightarrow{c^n_\lambda} A^{n+1}_\lambda \to \cdots$$

for every $\lambda \in \Lambda$. By naturality, these sequences form an exact sequence of inverse systems

$$0 \to \mathscr{A} \to \mathscr{A}^0 \to \cdots \to \mathscr{A}^n \xrightarrow{\{c^n_\lambda\}} \mathscr{A}^{n+1} \to \cdots .$$

Clearly, this is the desired a-compact resolution of $\mathscr{A}$. □

§3. Representability Theorems

According to II.3.22, one can assign a (co)homology theory to a spectrum. This situation turns out to be invertible.

3.1. Notation. As in Ch. II, §3, let $\mathscr{K}^\bullet$ denote one of the categories $\mathscr{C}^\bullet, \mathscr{C}^\bullet_{\text{fd}}, \mathscr{C}^\bullet_{\text{f}}$, and let $\mathscr{L}$ denote one of the categories $\mathscr{S}, \mathscr{S}_{\text{s}}, \mathscr{S}_{\text{fd}}, \mathscr{S}_{\text{sfd}}, \mathscr{S}_{\text{f}}$.

3.2. Definition. (a) One says that a reduced cohomology theory $\widetilde{h}^*$ (resp. homology theory $\widetilde{h}_*$) on $\mathscr{K}^\bullet$ is represented by a spectrum E if there is an isomorphism $\widetilde{h}^* \cong \widetilde{E}^*$ of cohomology theories (resp. an isomorphism $\widetilde{h}_* \cong \widetilde{E}_*$ of homology theories) on $\mathscr{K}^\bullet$.

(b) One says that a cohomology theory h^* (resp. homology theory h_*) on $\mathscr{L}$ is represented by a spectrum E if there is an isomorphism $h^* \cong E^*$ of cohomology theories (resp. an isomorphism $h_* \cong E_*$ of homology theories) on $\mathscr{L}$.

3.3. Definition. (a) Let $\mathscr{K}^\bullet, \mathscr{L}$ be as in 3.1. Two morphisms $f, g : E \to F$ of spectra (resp. maps of spaces) are called *$\mathscr{L}$-homotopic* (resp. *$\mathscr{K}^\bullet$-homotopic*) iff $fi \simeq gi : A \to F$ for every morphism (resp. map) $i : A \to E$ of every spectrum A of $\mathscr{L}$ (resp. every space A of $\mathscr{K}^\bullet$).

(b) Two $\mathscr{S}_\mathrm{f}$-homotopic morphisms (resp. $\mathscr{C}_\mathrm{f}^\bullet$-homotopic maps) are called *weakly homotopic*. In this case we write $f \simeq^w g$.

Clearly, $f \simeq^w g : E \to F$ iff $f|E_\lambda \simeq g|E_\lambda$ for every finite subspectrum (subspace) E_λ of E.

3.4. Theorem–Definition. *Let h^* be an additive cohomology theory on $\mathscr{S}$, and let $\cdots \to X(n) \xrightarrow{f_n} X(n+1) \to \cdots$ be a sequence of morphisms of spectra. Then there exist a spectrum X and morphisms $i_n : X(n) \to X$ with the following properties:*

(i) $i_{n+1} f_n \simeq i_n$;

(ii) *The homomorphism*

$$\langle (i_n)_* | \varinjlim \rangle : \varinjlim \{\pi_*(X(n))\} \to \pi_*(X)$$

as in I.2.5 *is an isomorphism;*

(iii) *There is an exact sequence*

$$(3.5) \qquad 0 \to \varprojlim{}^1 \{h^{k-1}(X(n))\} \to h^k(X) \xrightarrow{\rho} \varprojlim \{h^k(X(n))\} \to 0$$

where $\rho = \{i_n^* | \varprojlim\}$ as in 1.11. *Such a spectrum X is called a* weak homotopy direct limit *of the sequence $\{X(n)\}$.*

Proof (cf. Milnor [5], Margolis [1], Ch. 3). In fact, X is a suitable telescope of the sequence $\{X(n)\}$. Consider the morphism

$$g_n : X(n) \xrightarrow{\nabla} X(n) \vee X(n) \xrightarrow{1 \vee (-f_n)} X(n) \vee X(n+1) \subset \bigvee_n X(n),$$

where ∇ is the coaddition. Let $g : \bigvee_n X(n) \to \bigvee_n X(n)$ be the morphism such that $g|X(n) = g_n$. Set $X := C(g)$ and define i_n to be the morphism $X(n) \subset \bigvee_n X(n) \to X$. We prove properties (i)–(iii).

The property (i) follows from the definition of g_n.

To prove (ii), consider the homomorphism

$$g_*: \oplus_n \pi_*(X(n)) = \pi_*\left(\bigvee_n X(n)\right) \to \pi_*\left(\bigvee_n X(n)\right) = \oplus_n \pi_*(X(n)).$$

Then there is the following commutative diagram with the exact rows:

$$\begin{array}{ccccccc} 0 \longrightarrow \oplus_n \pi_*(X(n)) & \xrightarrow{g_*} & \oplus_n \pi_*(X(n)) & \longrightarrow & \varinjlim\{\pi_*(X(n))\} & \longrightarrow & 0 \\ \| & & \| & & \downarrow{\scriptstyle \langle (i_n)_* | \varinjlim \rangle} & & \\ 0 \longrightarrow \oplus_n \pi_*(X(n)) & \xrightarrow{g_*} & \oplus_n \pi_*(X(n)) & \longrightarrow & \pi_*(X) & \longrightarrow & 0, \end{array}$$

and so $\langle (i_n)_* | \varinjlim \rangle$ is an isomorphism, and (ii) is proved. (Here the top row is exact because of the definition of $\varinjlim$, see I.2.4.)

The cofiber sequence $\bigvee_n X(n) \xrightarrow{g} \bigvee_n X(n) \to X$ induces the exact sequence

$$\prod h^{k-1}(X(n)) \xrightarrow{\delta} \prod h^{k-1}(X(n)) \to h^k(X) \\ \to \prod h^k(X(n)) \xrightarrow{\delta} \prod h^k(X(n))$$

where δ is as in 2.15. Now, by 2.15, the exact sequence

$$0 \to \operatorname{Coker} \delta \to h^k(X) \to \operatorname{Ker} \delta \to 0$$

yields the desired exact sequence (3.5). □

3.6. Theorem. (i) *Every additive cohomology theory h^* on $\mathscr{S}$ can be represented by a spectrum.*

(ii) *Let E, F be two spectra. Every morphism $E^*(-) \to F^*(-)$ of cohomology theories on $\mathscr{S}$ can be induced by a morphism $E \to F$ of spectra, and this morphism of spectra is unique up to homotopy. In particular, a representing spectrum for a cohomology theory on $\mathscr{S}$ is unique up to equivalence.*

Proof. (i) Given a spectrum F, every element $f \in h^0(F)$ yields a morphism $\widehat{f}: F^*(X) \to h^*(X)$ of cohomology theories on $\mathscr{S}$, $\widehat{f}(a) = a^*(f)$ for every $a: X \to \Sigma^n F$. We construct a spectrum E and an element $e \in h^0(E)$ such that $\widehat{e}: E^0(S^n) \to h^0(S^n)$ is an isomorphism for every $n \in \mathbb{Z}$. Then, by II.3.19(iii), E represents h^*.

Let $A_r = \{a_i(r)\}$ be a family of generators of the group $h^0(S^r), r \in \mathbb{Z}$. Consider a spectrum $E(0) := \bigvee_{r \in \mathbb{Z}} \left(\bigvee_{A_r} S^r_{a_i(r)}\right)$, where S^r_a is a copy of the spectrum S^r. By additivity, there exists $e_0 \in h^0(E(0))$ such that $e_0|S^r_{a_i(r)} = a_i(r)$ for every r, i. It is clear that $\widehat{e}_0 : E(0)^0(S^r) \to h^0(S^r)$ is epic. By induction, suppose that we have constructed a sequence

$$E(0) \to \cdots \xrightarrow{\varphi_{n-1}} E(n)$$

and elements $e_i \in h^0(E(i)), i \leq n$, with the following properties:

(1) $\varphi_k^*(e_{k+1}) = e_k$ for every $k < n$. In particular, $\widehat{e}_n : E(n)^0(S^r) \to h^0(S^r)$ is epic for every r.
(2) For every $r \in \mathbb{Z}$ and every $k \leq n$, $\operatorname{Ker} \widehat{e}_k \subset \operatorname{Ker}(\varphi_k)_*$ in the diagram

$$h^0(S^r) \xleftarrow{\widehat{e}_k} E(k)^0(S^r) \xrightarrow{(\varphi_k)_*} E(k+1)^0(S^r).$$

In order to construct φ_n, E_{n+1} and e_{n+1}, let $B_r = \{b_i(r)\}$ be a family of generators of $\operatorname{Ker}(\widehat{e}_n : E(n)^0(S^r) \to h^0(S^r))$. Set $Y := \bigvee_r \left(\bigvee_{B_r} S^r_{b_i(r)}\right)$, and let $x \in h^0(Y)$ be such that $x|S^r_{b_i(r)} = b_i(r)$ for every r, i. By (1), there exists $y \in E(n)^0(Y)$ such that $\widehat{e}_n(y) = x$. Constructing a cofiber sequence

$$Y \xrightarrow{y} E(n) \xrightarrow{\varphi_n} E(n+1),$$

we get an exact sequence

$$h^0(E(n+1)) \xrightarrow{\varphi_n^*} h^0(E(n)) \xrightarrow{y^*} h^0(Y).$$

Now, $y^*(e_n) = 0$, and hence there exists $e_{n+1} \in h^0(E(n+1))$ such that $\varphi_n^*(e_{n+1}) = e_n$. The induction is confirmed.

Consider the sequence

$$E(0) \to \cdots \xrightarrow{\varphi_{n-1}} E(n) \xrightarrow{\varphi_n} \cdots$$

and its weak homotopy direct limit E. Then, by 3.4(iii), $\rho : h^0(E) \to \varprojlim\{h^0(E(n))\}$ is an epimorphism, and so there exists $e \in h^0(E)$ such that $e|E(n) = e_n$. Hence, $\widehat{e} : E^0(S^r) \to h^0(S^r)$ is an epimorphism for every r.

Now we prove that $\widehat{e} : E^0(S^r) \to h^0(S^r)$ is monic. Let $f : S^r \to E$ be such that $\widehat{e}(f) = 0 \in h^0(S^r)$. Since $\pi_*(E) = \varinjlim\{\pi_*(E(n))\}$, there is n such that f can be decomposed as $S^r \xrightarrow{g} E(n) \to E$ with $\widehat{e}_n(g) = 0$. But then, in view of (2), $(\varphi_n)_*(g) = 0$, and so $f : S^r \xrightarrow{g} E(n) \to E$ is inessential.

(ii) This follows from the Yoneda Lemma I.1.5. □

Now we prove the Representability Theorem for cohomology theories on $\mathscr{S}_{\mathrm{f}}$. We need some preliminaries.

3.7. Definition. We say that an inverse Λ-system $\mathscr{I} = \{I_\lambda, j_\lambda^\mu\}$ of sets is *totally surjective* if every function j_λ^μ is surjective and every set I_λ is nonempty.

3.8. Definition. A *quasi-ordered sequence* is a quasi-ordered set $\{b_i\}_{i=1}^\infty$ such that $b_i \leq b_{i+1}$ for every i. (Of course, it may happen that $b_i \geq b_{i+1}$ for some i, or that $b_i = b_j$ for $i \neq j$.)

3.9. Lemma. *If $\mathscr{I}$ is a totally surjective inverse system over a quasi-ordered sequence then $\varprojlim \mathscr{I} \neq \emptyset$.* □

3.10. Lemma. *Every countable directed quasi-ordered set $\mathscr{A}$ contains a cofinal quasi-ordered sequence.*

Proof. Let $\mathscr{A} = \{a_i\}_{i=1}^{\infty}$. By induction, define a sequence $\{b_i\}$ by setting $b_1 = a_1$ and choosing b_n so that $b_n \geq a_n$ and $b_n \geq b_{n-1}$. It is clear that $\{b_i\}$ is a desired quasi-ordered sequence. □

3.11. Lemma. *Let $\mathscr{I} = \{I_\lambda, j_\lambda^\mu\}$ be a totally surjective inverse Λ-system of sets. If Λ contains a countable cofinal subset then $\varprojlim \mathscr{I} \neq \emptyset$.*

Proof. Indeed, by 3.10, Λ contains a cofinal quasi-ordered sequence, and so, by 1.20 and 3.9, $\varprojlim \mathscr{I} \neq \emptyset$. □

3.12. Construction. Let $\mathscr{I}$ be a totally surjective inverse Λ-system of sets. Given $\lambda, \mu \in \Lambda$, we say that $\lambda \preceq \mu$ iff there exists a commutative diagram

$$\begin{array}{ccc} I_\nu & = & I_\nu \\ {\scriptstyle j_\mu^\nu}\downarrow & & \downarrow{\scriptstyle j_\lambda^\nu} \\ I_\mu & \xrightarrow{k_\lambda^\mu} & I_\lambda \end{array}$$

with some k_λ^μ. It is clear that $\lambda \preceq \mu$ if $\lambda \leq \mu$, and for every $\lambda \preceq \mu$ there is just one function $k_\lambda^\mu : I_\mu \to I_\lambda$ (and $k_\lambda^\mu = j_\lambda^\mu$ for $\lambda \leq \mu$). Now we define a quasi-ordered set $\bar{\Lambda}$, which has the same objects as Λ and where a morphism $\lambda \to \mu$ exists iff $\lambda \preceq \mu$. Let $\overline{\mathscr{I}}$ be the inverse system $\{I_\lambda, k_\lambda^\mu\}$ over $\bar{\Lambda}$.

3.13. Lemma. *Let $\mathscr{I} = \{I_\lambda, j_\lambda^\mu\}$ be a totally surjective inverse Λ-system of sets. If $\bar{\Lambda}$ contains a countable cofinal subset then $\varprojlim \mathscr{I} \neq \emptyset$.*

Proof. It is clear that $\overline{\mathscr{I}}$ is totally surjective. Hence, by 3.11, $\varprojlim \overline{\mathscr{I}} \neq \emptyset$. But every string $\{x_\lambda\} \in \varprojlim \overline{\mathscr{I}}$ is at the same time an element of $\varprojlim \mathscr{I}$. □

3.14. Lemma. *Let $\mathscr{A} \xrightarrow{f} \mathscr{B} \xrightarrow{g} \mathscr{C} \xrightarrow{h} \mathscr{D}$ be an exact sequence of inverse Λ-systems of abelian groups. Suppose that $\mathscr{A}$ is a system such that $A_\lambda = A$ and $j_\lambda^\mu = 1_A$ for every $\lambda \leq \mu$. Set $P_\lambda := \operatorname{Ker}(A = A_\lambda \xrightarrow{f_\lambda} B_\lambda)$. Suppose that there exists a countable set $\{\lambda_i\}_{i=1}^{\infty}$ such that every P_λ contains some P_{λ_i}. Then the sequence*

$$\varprojlim \mathscr{B} \to \varprojlim \mathscr{C} \to \varprojlim \mathscr{D}$$

is exact.

Proof. Choose a string $\{y_\lambda\} \in \varprojlim \mathscr{C}$ and set $I_\lambda := g_\lambda^{-1}(y_\lambda)$. Let $\mathscr{B} = \{B_\lambda, \varphi_\lambda^\mu\}$. Clearly, $\varphi_\lambda^\mu(I_\mu) \subset I_\lambda$; we define $j_\lambda^\mu : I_\mu \to I_\lambda$ to be the restriction of φ_λ^μ and set $\mathscr{I} := \{I_\lambda, j_\lambda^\mu\}$. We must prove that $\varprojlim\{I_\lambda\} \neq \emptyset$. So, it suffices to prove that $\mathscr{I}$ satisfies 3.13. Firstly, we prove that j_λ^μ is surjective for every $\lambda \le \mu$. Consider the commutative diagram

$$\begin{array}{ccccc} A & \xrightarrow{f_\mu} & B_\mu & \longrightarrow & C_\mu \\ \| & & \downarrow{\scriptstyle \varphi_\lambda^\mu} & & \downarrow \\ A & \xrightarrow{f_\lambda} & B_\lambda & \longrightarrow & C_\lambda . \end{array}$$

Take any $x_\lambda \in I_\lambda$. Given $x_\mu \in I_\mu$, we have $\varphi_\lambda^\mu x_\mu - x_\lambda = f_\lambda(a)$, and so $\varphi_\lambda^\mu(x_\mu - f_\mu(a)) = x_\lambda$, i.e., $x_\lambda \in \operatorname{Im} j_\lambda^\mu$.

3.15. Sublemma. *If $P_\mu \subset P_\lambda$ then there is a function $k_\lambda^\mu : I_\mu \to I_\lambda$ as in* 3.12.

Proof. Choose ν such that $\nu \ge \mu, \lambda$ and consider the commutative diagram

$$\begin{array}{ccccc} A & \xrightarrow{f_\mu} & B_\mu & \longrightarrow & C_\mu \\ \| & & \uparrow{\scriptstyle \varphi_\mu^\nu} & & \uparrow \\ A & \xrightarrow{f_\nu} & B_\nu & \longrightarrow & C_\nu \\ \| & & \downarrow{\scriptstyle \varphi_\lambda^\nu} & & \downarrow \\ A & \xrightarrow{f_\lambda} & B_\lambda & \longrightarrow & C_\lambda . \end{array}$$

Given $x', x'' \in I_\nu$, we must prove that $\varphi_\lambda^\nu(x') = \varphi_\lambda^\nu(x'')$ if $\varphi_\mu^\nu(x') = \varphi_\mu^\nu(x'')$. We have $x' - x'' = f_\nu(a)$ for some $a \in A$. Hence, $\varphi_\lambda^\nu(x' - x'') = f_\lambda(a)$ and $\varphi_\mu^\nu(x' - x'') = f_\mu(a)$. But $f_\lambda(a) = 0$ if $f_\mu(a) = 0$, since $P_\mu \subset P_\lambda$. ⊞

Now, given $\lambda \in \Lambda$, choose λ_i such that $P_{\lambda_i} \subset P_\lambda$. Then, by 3.15, $\lambda \preceq \lambda_i$. Thus, $\{\lambda_i\}$ is a countable cofinal subset of $\bar{\Lambda}$. □

3.16. Lemma. (i) *There is a countable family $\mathscr{F}$ of finite spectra such that every finite spectrum is equivalent to some spectrum of $\mathscr{F}$.*

(ii) *For every pair of finite spectra F, A, the set $[A, F]$ is countable.*

Proof. (i) Every finite spectrum is equivalent to one of the form $\Sigma^n \Sigma^\infty X$ for some $n \in \mathbb{Z}$ and some $X \in \mathscr{C}_{\mathrm{f}}^\bullet$ (cf. II.1.5(iii)). Furthermore, every finite CW-space is homotopy equivalent to a finite polyhedron. But every polyhedron is completely determined (up to homeomorphism) by its scheme of vertices. Thus, finite polyhedra form a countable set.

(ii) Following (i), we can assume that $F = \Sigma^\infty X, A = \Sigma^\infty Y$ where X and Y are finite polyhedra. Then $[F, A] \cong [X, Y]^\bullet$. But every map $X \to Y$ of finite polyhedra is homotopic to a simplicial map $X^r \to Y$ where X^r is the r-th barycentric subdivision of X, see e.g. Hilton–Wiley [1]. But a simplicial map is determined by its values on vertices, and so the set M_r of simplicial maps $X^r \to Y$ is countable, and hence $\bigcup M_r$ is. □

3.17. Corollary. *Let $\mathscr{F}$ be as in* 3.16(i). *Then the set $\bigcup_{F\in\mathscr{F}}[F,A]$ is countable for every finite spectrum A.* □

Given a cohomology theory h^* on $\mathscr{S}$, set

$$\widehat{h}^*(X) := \varprojlim\{h^*(X_\lambda)\}$$

where $\{X_\lambda\}$ is the family of all finite subspectra of a spectrum X. Because of 1.15, $\widehat{h}^i$ is a functor $\mathscr{S} \to \mathscr{AG}$.

3.18. Proposition. *Let $\{X_\alpha\}, \bigcup X_\alpha = X$, be a family of subspectra of a spectrum X. Then $\widehat{h}^*(X) = \varprojlim\{\widehat{h}^*(X_\alpha)\}$. In particular, $\widehat{h}^*(\vee X_\beta) = \prod \widehat{h}^*(X_\beta)$ for every family $\{X_\beta\}$ of spectra.*

Proof. This follows from 1.21. □

3.19. Theorem. *Let $X \xrightarrow{f} Y \to Cf$ be a strict cofiber sequence of maps of spectra such that $X = \vee X_\alpha$ with finite X_α. Then the induced sequence*

$$\widehat{h}^*(Cf) \to \widehat{h}^*(Y) \to \widehat{h}^*(X)$$

is exact.

Proof. Firstly, let X be a finite spectrum. Put $Z = Cf$ and consider a cofiber sequence

$$X \xrightarrow{f} Y \to Z \xrightarrow{k} \Sigma X.$$

By II.3.15, we can form an exact sequence of the inverse systems

$$\{h^*(X_\omega)\} \leftarrow \{h^*(Y_\omega)\} \leftarrow \{h^*(Z_\omega)\} \leftarrow \{h^*(\Sigma(X_\omega))\},\ \omega \in \Omega,$$

with finite A_ω for every ω and such that every finite CW-subspectrum of A is contained in some A_ω, where A denotes X, Y, Z or ΣX. Passing to a certain cofinal system Λ of Ω, we can assume that $X_\lambda = X$ for every $\lambda \in \Lambda$ and get an exact sequence of inverse systems

$$\{h^*(X)\} \leftarrow \{h^*(Y_\lambda)\} \leftarrow \{h^*(Z_\lambda)\} \xleftarrow{k_\lambda^*} \{h^*(\Sigma X)\}$$

with $\widehat{h}^*(A) = \varprojlim\{\widehat{h}^*(A_\lambda)\}$ for $A = X, Y, Z$. By 3.14, it suffices to prove that the set of subgroups $\operatorname{Ker}\{k_\lambda^* : h^*(\Sigma X) \to h^*(Z_\lambda)\}$ of $h^*(\Sigma X)$ is countable. But this holds because, by 3.17, the set $\bigcup_\lambda [Z_\lambda, X]$ is countable.

Now, let $X = \vee X_\alpha$ with finite X_α. Note that Y is a subspectrum of Cf. Let $y \in \widehat{h}^*(Y)$ be such that $f^*(y) = 0$. We can and shall assume that $f : \vee X_\alpha \to Y$ is an inclusion. Given a subset $\Gamma \subset \{\alpha\}$, set $Z_\Gamma := C(f|\vee_{\alpha\in\Gamma} X_\alpha)$. Consider the set $\mathscr{Z}$ of all pairs (Γ, z) where Γ runs over all subsets of $\{\alpha\}$ and $z \in \widehat{h}^*(Z_\Gamma)$ is such that $z|Y = y$. We say that $(\Gamma, z) \le (\Gamma', z')$ if $\Gamma \subset \Gamma'$

and $z'|Z_\Gamma = z$. In this way $\mathscr{Z}$ becomes a partially ordered set. By 3.18, $\mathscr{Z}$ is inductive, and so it contains a maximal element (Γ_0, z_0). We prove that $Z_{\Gamma_0} = Cf$, and this will complete the proof.

We have $Cf = Y \cup C(\vee X_\alpha)$. So, if $Z_{\Gamma_0} \neq Cf$ then there exists α with $C(X_\alpha) \not\subset Z_{\Gamma_0}$. We let $\Omega = \Gamma \cup \{\alpha\}$ and consider the inclusion $g : Z_{\Gamma_0} \to Z_\Omega$. Consider the composition $i : X_\alpha \to \vee_\alpha X_\alpha \xrightarrow{f} Y \subset Z_{\Gamma_0}$. Since $C(\vee_\beta E_\beta) \simeq \vee_\beta C E_\beta$ for every family $\{E_\beta\}$ of spectra, we conclude that

$$X_\alpha \xrightarrow{i} Z_{\Gamma_0} \xrightarrow{g} Z_\Omega$$

is a cofiber sequence. By the above, the sequence

$$\widehat{h}^0(Z_\Omega) \xrightarrow{g^*} \widehat{h}^0(Z_{\Gamma_0}) \xrightarrow{i^*} h^0(X_\alpha)$$

is exact. Since $f^*(y) = 0$, we conclude that $i^*(z_0) = 0$, and so there is $z \in \widehat{h}^0(Z_\Omega)$ with $g^* z = z_0$, i.e., $z|Z_{\Gamma_0} = z_0$, i.e., (Γ_0, z_0) is not a maximal element. This is a contradiction. □

3.20. Theorem. (i) *Every cohomology theory h^* on $\mathscr{S}_{\mathrm{f}}$ can be represented by a spectrum.*

(ii) *For every cohomology theory h^* on $\mathscr{S}$ and every spectrum X the homomorphism $\rho : h^*(X) \to \widehat{h}^*(X)$ is an epimorphism.*

(iii) *Given two spectra E, F, every morphism $\varphi : E^*(-) \to F^*(-)$ of cohomology theories on $\mathscr{S}_{\mathrm{f}}$ can be induced by a morphism $E \to F$ of spectra, and this morphism of spectra is unique up to weak homotopy. Furthermore, a representing spectrum for a cohomology theory on $\mathscr{S}_{\mathrm{f}}$ is unique up to equivalence.*

Proof. (i) Because of 3.18 and 3.19, we can follow the proof of 3.6. So, we get a spectrum E and an element $x \in \widehat{h}^0(E)$ such that the homomorphism $\widehat{x} : E^0(S^r) \to \widehat{h}^0(S^r)$ is an isomorphism for every $r \in \mathbb{Z}$. Thus, by II.3.19(iii), $\widehat{x} : E^*(X) \to \widehat{h}^*(X) = h^*(X)$ is an isomorphism for every $X \in \mathscr{S}_{\mathrm{f}}$, i.e., E represents h^*.

(ii) (Brown's trick.) Given $Y \in \mathscr{S}$ and $y \in \widehat{h}^0(Y)$, we prove that $y \in \operatorname{Im}\{\rho : h^0(Y) \to \widehat{h}^0(Y)\}$. Let E and $e \in h^0(E)$ be as in 3.6. Set $E'(0) := Y \vee E$, and let $e_0' \in \widehat{h}^0(E'(0))$ be such that $e_0'|E = \rho(e)$ and $e_0'|Y = y$. Such e_0' exists by 3.18. Note that $\widehat{e}_0' : E'(0)^0(S^r) \to h^0(S^r)$ is an epimorphism for every r. Now, we can follow the proof of 3.6 and construct a sequence

$$E'(0) \to \cdots \to E'(n) \to \cdots,$$

its weak homotopy direct limit E' and an element $e' \in \widehat{h}^0(E')$ such that $\widehat{e'} : (E')^0(X) \to \widehat{h}^0(X) = h^0(X)$ is an isomorphism for every $X \in \mathscr{S}_{\mathrm{f}}$.

Let $a : Y \to E'(0), b : E \to E'(0), c : E'(0) \to E'$ be the obvious inclusions. Consider the commutative diagram

$$\begin{array}{ccc} E^0(S^r) & \xrightarrow{\widehat{e}} & h^0(S^r) \\ {\scriptstyle b_*}\downarrow & & \| \\ (E'(0))^0(S^r) & \xrightarrow{\widehat{e'_0}} & h^0(S^r) \\ {\scriptstyle c_*}\downarrow & & \| \\ (E')^0(S^r) & \xrightarrow{\widehat{e'}} & h^0(S^r). \end{array}$$

Since $\widehat{e}$ and $\widehat{e}'$ are isomorphisms, we conclude that $cb : E \to E'$ is an equivalence. Let $i : E' \to E$ be homotopy inverse to cb. We set $f := ica : Y \to E$ and prove that $\rho(f) = y$.

We have $b^*c^*(e') = \rho(e)$. Furthermore, $cbf = cbica = ca$. Now,

$$y = a^*e'_0 = a^*c^*e' = f^*b^*c^*e' = f^*\rho(e) = \rho(f).$$

(iii) Let $\{E_\lambda\}$ be the family of all finite subspectra of E. The inclusions $i_\lambda : E_\lambda \subset E$ form a string $\{i_\lambda\} \in \widehat{E}^0(E)$. Since $\rho : F^0(E) \to \widehat{F}^0(E)$ is an epimorphism, there exists $\theta \in F^0(E)$ such that $\rho(\theta) = \{\varphi(i_\lambda)\}$. It is clear that θ induces φ.

If there is another morphism θ' which induces φ then $\theta i_\lambda \simeq \theta' i_\lambda$ for every λ, and so θ and θ' are weak homotopic.

Finally, if both spectra E, F represent h^*, then every morphism $\theta : E \to F$ inducing 1_h is an equivalence. □

3.21. Theorem. *Let $\mathscr{L}$ be as in 3.1. Every additive cohomology theory h^* on $\mathscr{L}$ can be represented by a spectrum, and this representing spectrum is unique up to equivalence. Furthermore, every morphism of cohomology theories on $\mathscr{L}$ can be induced by a morphism of the representing spectra, and this morphism of spectra is unique up to $\mathscr{L}$-homotopy.*

Proof. The case $\mathscr{L} = \mathscr{S}_{\mathrm{f}}$ is proved in 3.20. We consider the case $\mathscr{L} = \mathscr{S}_{\mathrm{sfd}}$ only, all the other cases can be considered similarly. Given a spectrum $X \in \mathscr{S}_{\mathrm{sfd}}$, set $\widehat{h}^*(X) = \varprojlim_{m,n} \{h^*(\Sigma^{-n}\Sigma^\infty(X_n^{(m)}))\}$. It is clear that the analog of 3.18 holds for $\widehat{h}^*$. Furthermore, if $\vee_\alpha X_\alpha \to Y \to Z$ is a cofiber sequence with $X_\alpha \in \mathscr{S}_{\mathrm{sfd}}$ then the sequence $\widehat{h}^*(\vee X_\alpha) \leftarrow \widehat{h}^*(Y) \leftarrow \widehat{h}^*(Z)$ is exact. This can be proved just as 3.19, but we do not need 3.17 because the family $\{\Sigma^{-n}\Sigma^\infty(X_n^{(m)})\}$ is countable for every spectrum X. Now we can complete the proof just as the one of 3.20. □

3.22. Corollary. *Let $\mathscr{K}^\bullet$ be as in 3.1. Then every additive cohomology theory (reduced) on $\mathscr{K}^\bullet$ can be represented by a spectrum, and this representing spectrum is unique up to equivalence. Furthermore, every morphism of cohomology theories on $\mathscr{K}^\bullet$ can be induced by a morphism of the representing spectra.*

Proof. This follows from 3.21 in view of II.3.18. □

Now we turn to homology theories. First, if a homology theory on $\mathscr{S}_{\mathrm{f}}^{\bullet}$ is represented by some spectrum, then its unique additive extension to $\mathscr{S}$ (see II.3.20(iii)) is represented by the same spectrum. This is true because each spectrum produces an additive homology theory on $\mathscr{S}$. Second, given any homology theory on $\mathscr{S}_{\mathrm{f}}$, one can construct the dual cohomology theory on $\mathscr{S}_{\mathrm{f}}^{\bullet}$ and use 3.20 and II.3.23 in order to represent the homology theory. In other words, we have the following theorem.

3.23. Theorem. (i) *Every additive homology theory on $\mathscr{K}^{\bullet}$, as well as on $\mathscr{L}$, can be represented by a certain spectrum. Furthermore, its representing spectrum is unique up to equivalence.*

(ii) *Every morphism of additive homology theories on $\mathscr{K}^{\bullet}$ or on $\mathscr{L}$ can be induced by a morphism of corresponding spectra, and this morphism of spectra is unique up to weak homotopy.* □

Let $\mathscr{H}\mathscr{C}_{\mathrm{con}}^{\bullet}$ be the homotopy category for the category $\mathscr{C}_{\mathrm{con}}^{\bullet}$ of all connected pointed CW-spaces and maps. Let $\mathscr{H}$ be the category $\mathscr{H}\mathscr{C}_{\mathrm{con}}^{\bullet}$ or its full subcategory consisting of all finite dimensional CW-spaces.

3.24. Definition. A contravariant functor $F : \mathscr{H} \to \mathscr{E}ns^{\bullet}$ is called *half-exact* if it satisfies the following axioms:

(i) Let $(X; A, B)$ be a (pointed, CW-) triad such that $X, A, B, A \cap B \in \mathscr{H}$. Then for every $a \in F(A), b \in F(B)$ with $a|A \cap B = b|A \cap B$ there exists an element $x \in F(X)$ such that $x|A = a, x|B = b$ (the Mayer–Vietoris axiom, below simply the MV-axiom).

(ii) Let X_α be a family of objects of $\mathscr{H}$ such that $X := \vee_\alpha X_\alpha \in \mathscr{H}$, and let $i_\alpha : X_\alpha \to X$ be the inclusions. Then $\{F(i_\alpha)\} : F(X) \to \prod_\alpha F(X_\alpha)$ is a bijection (the wedge axiom, or the additivity axiom).

It is clear that every representable functor $F : \mathscr{H} \to \mathscr{E}ns^{\bullet}$ is half-exact.

3.25. Theorem (the Brown Representability Theorem)**.** *For every half-exact functor $F : \mathscr{H} \to \mathscr{E}ns^{\bullet}$ there exists a connected CW-space B such that there is a natural equivalence $F(-) \to [-, B]^{\bullet}$ of functors on $\mathscr{H}$. In other words, every half-exact functor $F : \mathscr{C}_{\mathrm{con}}^{\bullet} \to \mathscr{E}ns^{\bullet}$ is representable.*

Proof (some words about). In Switzer [1], Ch. 9 this theorem is stated for functors on the category $\mathscr{H}\mathscr{C}^{\bullet}$, not only on $\mathscr{H}\mathscr{C}_{\mathrm{con}}^{\bullet}$ (Theorem 9.12 there). In this general form the theorem is wrong, see Heller [1], Matveev [1], but the proof in Switzer [1] can be used and appears correct for the formulation above.

The case when $\mathscr{H}$ is the category of finite dimensional connected spaces can be considered similarly; or see Dold [3]. □

3.26. Example. Let E_α be a family of spectra. Then $F^*(X) := \prod_\alpha E^*_\alpha(X)$ is an additive cohomology theory on $\mathscr{S}$, and so, by 3.6(i), it can be represented by a spectrum F. This spectrum is called the *product* of spectra E_α and is denoted by $\prod_\alpha E_\alpha$.

By 3.6(ii), for every α the projections $p^X_\alpha : F^*(X) \to E^*_\alpha(X), X \in \mathscr{S}$ yield a morphism $p_\alpha : F \to E_\alpha$.

By 3.6(ii), the inclusion $E^*_\alpha(X) \subset (\prod_\alpha E_\alpha)^*(X)$ can be induced by a morphism $j_\alpha : E_\alpha \to \prod_\alpha E_\alpha$ of spectra. So, by II.1.16(i), we get a morphism $j : \bigvee_\alpha E_\alpha \to \prod_\alpha E_\alpha$ with $j|E_\alpha = j_\alpha$. It induces a homomorphism

$$j_* : \bigoplus_\alpha \pi_k(E_\alpha) = \pi_k\left(\bigvee_\alpha E_\alpha\right) \to \pi_k\left(\prod_\alpha E_\alpha\right) = \prod_\alpha \pi_k(E_\alpha)$$

of homotopy groups, which coincides with the standard inclusion. Hence, j is not an equivalence in general. For example, j is not an equivalence if $E_\alpha = H\mathbb{Z}$ and $\{\alpha\}$ is a countable set. On the other hand, the morphism $j : \bigvee_n \Sigma^n H(\pi_n) \to \prod_n \Sigma^n H(\pi_n)$ is an equivalence.

3.27. Proposition. *Given a family $\{f_\alpha : X \to E_\alpha\}$ of morphisms of spectra, there is a morphism $f : X \to \prod_\alpha E_\alpha$ such that $p_\alpha f \simeq f_\alpha$ for every α, and this f is unique up to homotopy.*

Proof. Left as an exercise, based on what was said above. □

3.28. Example. Given two spectra X, E, we have an additive cohomology theory $[X \wedge Y, E]$ on $\mathscr{S}$; here Y is a variable. Hence, by 3.21, there is a unique spectrum $F(X, E)$ such that $[X \wedge Y, E] = [Y, F(X, E)]$. This spectrum $F(X, E)$ is called the *functional spectrum.* (Note that the equality looks like the exponential law.)

3.29. Remarks. (a) The Representability Theorems enable us to reduce any research of (co)homology theories and interconnections between them to an investigation of universal objects – namely, spectra. Spectra, in turn, can be studied by the powerful machinery of stable homotopy theory. This approach was originally demonstrated by Serre [2] and Thom [2], and the further development of algebraic topology affirms the fruitfulness of this methodology.

(b) It was Brown [1] who discovered that half-exactness implies representability. He proved 3.22 for $\mathscr{K}^\bullet = \mathscr{C}^\bullet$. Furthermore, Adams [7] proved 3.23 for $\mathscr{K}^\bullet = \mathscr{C}^\bullet_\mathrm{f}$ (without uniqueness of a representing spectrum).

(c) In the proofs of the Representability Theorems we followed Margolis [1], which, in turn, followed the original papers of Brown [1] (in case 3.6) and Adams [7] (in case 3.20).

§4. A Spectral Sequence

Throughout this section $\{X_\lambda\}$ denotes the family of all finite subspectra of a spectrum X. The goal of this section is to express $E^*(X)$ in terms of $E^*(X_\lambda)$, see 4.11–4.22.

By 3.20(ii), the homomorphism $\rho = \rho_X^E : E^*(X) \to \varprojlim\{E^*(X_\lambda)\}$ is epic for every $X, E \in \mathscr{S}$, and, by 1.15, ρ_X^E is natural with respect to X and E.

4.1. Definition. A cohomology theory E on $\mathscr{S}_\mathrm{f}$ is called *compact* if all groups $E^n(X)$, $X \in \mathscr{S}_\mathrm{f}$, are compact topological groups and all induced homomorphisms $f^* : E^n(X) \to E^n(Y)$ and suspension isomorphisms $E^n(X) \cong E^{n+1}(\Sigma X)$ are continuous. A cohomology theory E on $\mathscr{S}_\mathrm{f}$ is called *algebraically compact*, or *a-compact*, if it can be obtained from some compact cohomology theory by ignoring the topology. The spectrum F is called *a-compact* if it represents an a-compact cohomology theory on $\mathscr{S}_\mathrm{f}$.

For example, by II.4.25(ii), F is a-compact if every group $\pi_i(F)$ is finite.

4.2. Theorem. *If a spectrum F is a-compact then the homomorphism $\rho = \rho_X : F^*(X) \to \varprojlim\{F^*(X_\lambda)\}$ as in 1.12 is an isomorphism for every $X \in \mathscr{S}$. In other words, $F^*(X)$ does not contain weak phantoms.*

Proof. Considering X as a variable, we prove that $G^*(X) := \varprojlim\{F^*(X_\lambda)\}$ is an additive cohomology theory on $\mathscr{S}$. Firstly, we prove that G^* is a functor. Indeed, given $f : X \to Y$, consider a family $\{f_\omega : X_\omega \to Y_\omega\}$ as in II.3.14. Then $G^*(A) = \varprojlim\{F^*(A_\omega)\}$ for $A = X, Y$, and hence f induces a homomorphism

$$f^* : G^*(Y) = \varprojlim\{F^*(Y_\omega)\} \to \varprojlim\{F^*(X_\omega)\} = G^*(X).$$

Now we prove the exactness of G^*. Firstly, consider a strict cofiber sequence $X \xrightarrow{f} Y \xrightarrow{g} Z = Cf$ of maps of spectra. Now, the cofiber sequences $X_\omega \to Y_\omega \to Z_\omega$ as in II.3.15 yield an exact sequence of the inverse systems

$$\{F^*(Z_\omega)\} \to \{F^*(Y_\omega)\} \to \{F^*(X_\omega)\}.$$

Furthermore, 2.17 and the a-compactness of F imply the exactness of the sequence

$$\varprojlim\{F^*(Z_\omega)\} \to \varprojlim\{F^*(Y_\omega)\} \to \varprojlim\{F^*(X_\omega)\},$$

i.e., the sequence

$$G^*(Z) \xrightarrow{g^*} G^*(Y) \xrightarrow{f^*} G^*(X)$$

is exact.

Finally, given an arbitrary cofiber sequence $X \xrightarrow{f} Y \xrightarrow{g} Z$, we have a commutative diagram

$$\begin{array}{ccccc} X & \xrightarrow{f} & Y & \longrightarrow & Cf \\ \| & & \| & & \downarrow h \\ X & \xrightarrow{f} & Y & \xrightarrow{g} & Z \end{array}$$

where h is a homotopy equivalence. Hence, the sequence

$$G^*(Z) \xrightarrow{g^*} G^*(Y) \xrightarrow{f^*} G^*(X)$$

is exact.

Clearly, the homotopy axiom is valid.

The additivity holds because $\varprojlim$ and $\prod$ commute, see 1.21.

Thus, G^* is an additive cohomology theory on $\mathscr{S}$. Therefore, the family

$$\{\rho_X : F^*(X) \to \varprojlim\{F^*(X_\lambda)\} = G^*(X)\}$$

is a morphism $F^*(-) \to G^*(-)$ of additive cohomology theories. Because of II.3.19(iii), this morphism is an equivalence. □

4.3. Corollary. *Let F, X be as in 4.2. Let $\{X_\mu\}$ be a family of subspectra of X such that every finite subspectrum of X is contained in some X_μ. Then $F^*(X) = \varprojlim\{F^*(X_\mu)\}$. In particular, $F^*(X) = \varprojlim\{F^*(X^{(n)})\}$.*

Proof. This follows from 4.2 and 1.20. □

Let c and $\widehat{A}$ be as in 2.21.

4.4. Lemma. *If $A \xrightarrow{f} B \xrightarrow{g} C$ is an exact sequence of abelian groups, then the sequence*

$$\widehat{A} \xrightarrow{\widehat{f}} \widehat{B} \xrightarrow{\widehat{g}} \widehat{C}$$

is exact. In other words, the functor c is exact.

Proof. The functor $\operatorname{Hom}(-, SO(2))$ is exact because of the infinite divisibility of the group $SO(2)$. Thus, c is exact. □

4.5. Theorem. *For every spectrum F, there exists a morphism $c : F \to \widehat{F}$ such that the spectrum $\widehat{F}$ is a-compact and c induces the canonical a-compactification $F^*(X) \to \widehat{F}^*(X)$ for every $X \in \mathscr{S}_{\mathrm{f}}$. In other words, for every $n \in \mathbb{Z}$ there is a natural isomorphism $a_X : \widehat{F}^n(X) \to \widehat{F^n(X)}$ such that the following diagram commutes for every $X \in \mathscr{S}_{\mathrm{f}}$:*

$$\begin{array}{ccc} F^n(X) & \xrightarrow{c_X} & \widehat{F}^n(X) \\ \| & & a_X \downarrow \cong \\ F^n(X) & \xrightarrow{c_{F^n(X)}} & \widehat{F^n(X)}. \end{array} \tag{4.6}$$

Proof. For every $X \in \mathscr{S}_{\mathrm{f}}$ we consider the canonical a-compactification $c_{F^n(X)} : F^n(X) \to \widehat{F^n(X)}$. By 2.20 and 4.4, $\widehat{F^*(-)}$ is a cohomology theory on $\mathscr{S}_{\mathrm{f}}$. Hence, it can be represented by a certain spectrum $\widehat{F}$, and this spectrum is a-compact. Now, given $X \in \mathscr{S}$, consider the homomorphism

$$c_X : F^*(X) \xrightarrow{\rho} \varprojlim\{F^*(X_\lambda)\} \xrightarrow{\varprojlim\{c^*_{X_\lambda}\}} \varprojlim\{\widehat{F}^*(X_\lambda)\} = \widehat{F}^*(X),$$

the last equality holding by 4.2. Arguments like 4.2 show that c_X is natural with respect to X, and in fact the family $\{c_X\}$ is a morphism of cohomology theories on $\mathscr{S}$. Hence, by 3.20(iii), it is induced by a morphism $c : F \to \widehat{F}$. It is clear that c is the desired morphism. □

4.7. Definition. A morphism $c : F \to \widehat{F}$ as in 4.5 is called an *a-compactification* of the spectrum F.

4.8. Remark. One can ask why we do not use the a-compactification $F^*(X) \to \widehat{F^*(X)}$ with $X \in \mathscr{S}$. The answer is that the cohomology theory $\widehat{F^*(X)}$ is not additive in general, and so it cannot be represented.

4.9. Lemma. *Let G be a spectrum such that the canonical epimorphism $\rho : G^*(X) \to \varprojlim\{G^*(X_\lambda)\}$ splits naturally with respect to X for every $X \in \mathscr{S}$. Then ρ is an isomorphism for every $X \in \mathscr{S}$.*

Proof. It suffices to prove that $\varprojlim\{G^*(X_\lambda)\}$ is an additive cohomology theory on $\mathscr{S}$, cf. 4.2. Only the exactness axiom needs to be verified. Firstly, consider a strict cofiber sequence $X \xrightarrow{f} Y \xrightarrow{g} Z = Cf$ of maps of spectra. Choose $\{X_\omega \to Y_\omega \to Z_\omega\}$ as in II.3.15 and consider the commutative diagram

$$\begin{array}{ccccc}
\varprojlim\{G^*(Z_\omega)\} & \xrightarrow{\bar g} & \varprojlim\{G^*(Y_\omega)\} & \xrightarrow{\bar f} & \varprojlim\{G^*(X_\omega)\} \\
\downarrow{\scriptstyle \tau'} & & \downarrow{\scriptstyle \tau} & & \downarrow \\
G^*(Z) & \xrightarrow{g^*} & G^*(Y) & \xrightarrow{f^*} & G^*(X) \\
\downarrow{\scriptstyle \rho'} & & \downarrow{\scriptstyle \rho} & & \downarrow \\
\varprojlim\{G^*(Z_\omega)\} & \xrightarrow{\bar g} & \varprojlim\{G^*(Y_\omega)\} & \xrightarrow{\bar f} & \varprojlim\{G^*(X_\omega)\}
\end{array}$$

with the exact middle row, where τ' and τ are natural splittings (here $\bar g = \varprojlim\{g_\omega\}$, $\bar f = \varprojlim\{f_\omega\}$). Let $a \in \varprojlim\{G^*(Y_\omega)\}$, $\bar f(a) = 0$. Then $f^*(\tau(a)) = 0$, i.e., $\tau(a) = g^*(b)$. Now, $a = \rho\tau a = \rho g^* b = \bar g \rho' b$, i.e., $a \in \operatorname{Im} \bar g$. Thus, the bottom (as well as top) row is exact. The exactness for a general cofiber sequence $X \to Y \to Z$ can be proved as in 4.2. □

4.10. Lemma. *Let F be a spectrum such that $\varprojlim^1\{F^*(X_\lambda)\} = 0$ for every $X \in \mathscr{S}$. Then the epimorphism $\rho : F^*(X) \to \varprojlim\{F^*(X_\lambda)\}$ is an isomorphism for every $X \in \mathscr{S}$.*

Proof. Let $c : F \to \widehat{F}$ be an a-compactification of F. The cofiber sequence

$$F \xrightarrow{c} \widehat{F} \xrightarrow{\sigma} G := C(c)$$

induces the following exact sequence of inverse systems:

$$0 \to \{F^*(X_\lambda)\} \xrightarrow{c} \{\widehat{F}^*(X_\lambda)\} \xrightarrow{\sigma} \{G^*(X_\lambda)\} \to 0.$$

Note that

$$\bar{\sigma} := \varprojlim \sigma : \varprojlim\{\widehat{F}^*(X_\lambda)\} \to \varprojlim\{G^*(X_\lambda)\}$$

is an epimorphism since $\varprojlim^1\{F^*(X_\lambda)\} = 0$. Consider the following commutative diagram with exact rows:

$$\begin{array}{ccccccccc}
0 \longrightarrow \varprojlim\{F^*(X_\lambda)\} & \xrightarrow{\bar{c}} & \varprojlim\{\widehat{F}^*(X_\lambda)\} & \xrightarrow{\bar{\sigma}} & \varprojlim\{G^*(X_\lambda)\} \longrightarrow 0 \\
\uparrow \rho & & \cong \uparrow \rho_1 & & \uparrow \rho_2 \\
\cdots \longrightarrow F^*(X) & \xrightarrow{c_*} & \widehat{F}^*(X) & \xrightarrow{\sigma_*} & G^*(X) \longrightarrow \cdots.
\end{array}$$

Since ρ is epic, $\rho_1(\operatorname{Im} c_*) = \operatorname{Im} \bar{c}$. By 4.2, ρ_1 is an isomorphism. Hence, the monomorphism $\widehat{F}^*(X)/\operatorname{Im} c_* \to G^*(X)$ yields a monomorphism

$$\begin{aligned}
\varprojlim\{G^*(X_\lambda)\} &= \varprojlim\{F^*(X_\lambda)\}/\operatorname{Im}\bar{c} \\
&= \rho_1(\widehat{F}^*(X)/\operatorname{Im} c_*) \xrightarrow[\cong]{\rho_1^{-1}} \widehat{F}^*(X)/\operatorname{Im} c_* \to G^*(X).
\end{aligned}$$

This is a natural splitting of ρ_2. Hence, by 4.9, ρ_2 is an isomorphism. Thus, by the Five Lemma, $\rho : F^*(X) \to \varprojlim\{F^*(X_\lambda)\}$ is an isomorphism. □

4.11. Theorem. *Given a spectrum F, suppose that there exists a number N such that $\varprojlim^q\{F^*(X_\lambda)\} = 0$ for every $q > N$ and every $X \in \mathscr{S}$ (resp. $X \in \mathscr{C}$). Then for every $X \in \mathscr{S}$ (resp. $X \in \mathscr{C}$) there is a spectral sequence $E_r^{*,*}(X)$ converging to $F^*(X)$ and such that $E_2^{p,q}(X) = \varprojlim^p\{F^q(X_\lambda)\}$.*

Proof. We consider the case of a spectrum X only. Let $c_0 : F \to F_0$ be an a-compactification of F. Let G_1 be the fiber of c_0 (i.e., ΣG_1 is the cone of c_0). Consider an a-compactification $c_1 : G_1 \to F_1$. Let G_2 be the fiber of c_1, consider an a-compactification $c_2 : G_2 \to F_2$, and so on. For every n we get a long cofiber sequence

$$(4.12) \qquad \cdots \Sigma^{-1} F_n \xrightarrow{\delta_n} G_{n+1} \xrightarrow{p_n} G_n \xrightarrow{c_n} F_n \xrightarrow{\Sigma\delta_n} \Sigma G_{n+1} \to \cdots,$$

where c_n is an a-compactification of G_n and ΣG_{n+1} is the cone of c_n. Consider the following diagram:

$$\begin{array}{ccccccccc}
\Sigma^{-1}F_n & & \Sigma^{-1}F_{n-1} & & & & \Sigma^{-1}F_0 & & \\
{\scriptstyle\delta_n}\big\downarrow & & \big\downarrow{\scriptstyle\delta_{n-1}} & & & & \big\downarrow{\scriptstyle\delta_0} & & \\
\cdots\to G_{n+1} & \longrightarrow & G_n & \longrightarrow & G_{n-1}\to\cdots & \to & G_1 & \longrightarrow & G_0 = F \\
 & & \big\downarrow{\scriptstyle c_n} & & \big\downarrow{\scriptstyle c_{n-1}} & & \big\downarrow{\scriptstyle c_1} & & \big\downarrow{\scriptstyle c_0} \\
 & & F_n & & F_{n-1} & & F_1 & & F_0\ .
\end{array}$$

For every $Y \in \mathscr{S}_{\mathrm{f}}$ the sequence (4.12) induces the short exact sequence

$$0 \to G_n^i(Y) \to F_n^i(Y) \to G_{n+1}^{i+1}(Y) \to 0.$$

Hence, there arises an exact sequence

$$0 \to F^n(Y) \to F_0^n(Y) \to F_1^{n+1}(Y) \to \cdots \xrightarrow{c_k\delta_{k-1}} F_k^{n+k}(Y) \to \cdots$$

which is an a-compact resolution of the group $F^n(Y)$.

We can compute the groups $F^*(X)$, $X \in \mathscr{S}$, using the spectral sequence

$$\{E_r^{p,q}, d_r^{p,q} : E_r^{p,q} \to E_r^{p+r,q-r+1}\}$$

based on the tower

$$\{\cdots \to G_{n+1} \xrightarrow{p_n} G_n \to \cdots \to G_1 \to F\}.$$

Here $E_1^{p,q} = F_p^{p+q}(X)$, and the differential $d_1 : E_1^{p,q} \to E_1^{p+1,q}$ is $(c_{p+1}\delta_p)_* : F_p^{p+q}(X) \to F_{p+1}^{p+q+1}(X)$. Hence, $E_2^{p,q}$ is the p-th cohomology group of the cochain complex

$$\cdots \to F_{p-1}^{p+q-1}(X) \to F_p^{p+q}(X) \to F_{p+1}^{p+q+1}(X) \to \cdots.$$

Now, by 4.2, $F_p^*(X) = \varprojlim\{F_p^*(X_\lambda)\}$, and so this complex has the form

$$\text{(4.14)} \qquad \cdots \to \varprojlim\{F_{p-1}^{p+q-1}(X_\lambda)\} \to \varprojlim\{F_p^{p+q}(X_\lambda)\} \to \cdots,$$

where $\{F^q(X_\lambda)\} \to \{F_0^q(X_\lambda)\} \to \cdots \to \{F_p^{p+q}(X_\lambda)\} \to \cdots$ is an a-compact resolution of the inverse system $\{F^q(X_\lambda)\}$. Hence, $E_2^{p,q}$ is the p-th cohomology of the cochain complex (4.14), i.e., $E_2^{p,q} = \varprojlim^p\{F^q(X_\lambda)\}$.

We prove that this spectral sequence converges to $F^*(X)$. It suffices to prove that for every m there exists $M = M(m)$ such that the homomorphism $G_M^*(X) \to G_m^*(X)$ is trivial. We prove more: namely, the homomorphism $(p_n)_* : G_{n+1}^*(X) \to G_n^*(X)$ is trivial for $n > N$. The exactness of the sequence

$$0 \to \{G_k^*(X_\lambda)\} \to \{F_k^*(X_\lambda)\} \to \{G_{k+1}^*(X_\lambda)\} \to 0$$

and the equality $\varprojlim^i\{F_k^*(X_\lambda)\} = 0, i > 0$, imply (using 2.13(ii)) that $\varprojlim^i\{G_{k+1}^*(X_\lambda)\} = \varprojlim^{i+1}\{G_k^*(X_\lambda)\}$, $i > 0$. Hence,

$$\varprojlim^i\{G_q^*(X_\lambda)\} = \varprojlim^{i+q}\{F^*(X_\lambda)\} = 0$$

for $i+q > N$. Therefore, if $q > N$ then $\varprojlim^i\{G_q^*(X_\lambda)\} = 0$ for every $i > 0$ and every X. Thus, by 4.10, $G_n^*(X) = \varprojlim\{G_n^*(X_\lambda)\}$ for $n > N$, and it remains to note that the homomorphism $(p_k)_* : G_{k+1}^*(X_\lambda) \to G_k^*(X_\lambda)$ is trivial for every k, since $X_\lambda \in \mathscr{S}_{\mathrm{f}}$. □

The following proposition gives us a sufficient condition for the existence of the number N in 4.11. (Observe that the existence of such N is the convergence condition of the spectral sequence.)

4.15. Proposition (see Roos [2], Jensen [1]). *Let R be a commutative Noetherian ring of homological dimension d, and let $\mathscr{M}$ be any inverse system of finitely generated R-modules. Then $\varprojlim^i\mathscr{M} = 0$ for $i > d$.* □

4.16. Corollary. *Let R be a commutative Noetherian ring, and let F be a spectrum such that $F^i(X), i \in \mathbb{Z}$ are natural in X R-modules. Suppose that $\pi_k(F)$ is a finitely generated R-module for every k.*

(i) *If R has homological dimension ≤ 1 (for example, a subring of $\mathbb{Q}$), then for every $X \in \mathscr{S}$ there is an exact sequence*

$$0 \to \varprojlim^1\{F^{k-1}(X_\lambda)\} \to F^k(X) \xrightarrow{\rho} \varprojlim\{F^k(X_\lambda)\} \to 0.$$

(ii) *If R has homological dimension* 0 *(e.g., R is a field), then the homomorphism $\rho : F^k(X) \to \varprojlim\{F^k(X_\lambda)\}$ is an isomorphism for every $X \in \mathscr{S}$. In particular, $\varprojlim\{F^k(X_\lambda)\}$ is a cohomology theory on $\mathscr{S}$.*

Proof. (i) By II.4.25(iii), $F^k(Y)$ is a finitely generated R module for every $Y \in \mathscr{S}_{\mathrm{f}}$. So, by 4.15, $\varprojlim^i\{F^*(X_\lambda)\} = 0$ for every $i > 1$ and every $X \in \mathscr{S}$. Now apply 4.11.

(ii) This follows from (i), since $\varprojlim^1\{F^{k-1}(X_\lambda)\} = 0$. □

4.17. Corollary. *Let F be a spectrum such that $\pi_i(F)$ is a finite abelian group for every i. Then the homomorphism $\rho : F^k(X) \to \varprojlim\{F^k(X_\lambda)\}$ is an isomorphism for every $X \in \mathscr{S}$. In particular, $\varprojlim\{F^k(X_\lambda)\}$ is a cohomology theory on $\mathscr{S}$.*

Proof. If Y is finite then, by II.4.25(ii), $F^k(Y)$ is a finite abelian group for every k. Hence, by 2.18(iii), $\varprojlim^i\{F^*(X_\lambda)\} = 0$ for every $i > 0$ and every $X \in \mathscr{C}$. Now the result follows from 4.11. □

Now, let $\cdots \subset X(0) \subset \cdots \subset X(r) \subset \cdots \subset X$, $X = \bigcup_{r=-\infty}^{\infty} X(r)$ be a filtration of a spectrum X by spectra $X(r)$. Then for every a-compact spectrum F the homomorphism $\rho : F^*(X) \to \varprojlim\{F^*(X(r))\}$ is an isomorphism by 4.3. Furthermore, by 2.15, $\varprojlim^i\{F^*(X(r))\} = 0$ for $i > 1$. Thus, we can replace $\{X_\lambda\}$ by $\{X(r)\}$ in the proof of 4.11 and obtain the following fact.

4.18. Corollary. *For every spectrum F and every $X \in \mathscr{S}$ with a filtration as above there is an exact sequence*

$$0 \to \varprojlim{}^1\{F^{k-1}(X(r))\} \to F^k(X) \xrightarrow{\rho} \varprojlim\{F^k(X(r))\} \to 0.$$

In particular, there is an exact sequence

$$0 \to \varprojlim{}^1\{F^{k-1}(X^{(n)})\} \to F^k(X) \xrightarrow{\rho} \varprojlim\{F^k(X^{(n)})\} \to 0. \tag{4.19}$$ □

Thus, phantoms are just elements ($\neq 0$) of the group $\varprojlim^1\{F^*(X^{(n)})\}$.

4.20. Remarks. (a) Milnor [5] first proved 4.18. Moreover, 4.18 holds for an arbitrary filtration $X(0) \subset \cdots \subset X(n) \subset \cdots X = \cup X(n)$ (as defined in I.3.1) of an arbitrary space X, if we define $F^k(X)$ as in II.3.31. The proof can be done just as in Milnor [5].

(b) The spectral sequence in 4.11 was constructed (for a space X) by Bousfield–Kan [1] and Araki–Yosimura [1] (in different manners).

4.21. Theorem. *For every two spectra E, F, there is an exact sequence*

$$0 \to \varprojlim{}^1\{F^{k+n-1}(E_n)\} \to F^k(E) \to \varprojlim\{F^{k+n}(E_n)\} \to 0.$$

Proof. Consider the filtration $\{\Sigma^{-n}\Sigma^\infty E_n\}$ of E and apply 4.18. □

4.22. Theorem. *For every two spectra E, F, the homomorphism*

$$r : \varinjlim F_{k+n}(E_n) \to F_k(E)$$

is an isomorphism.

Proof. This holds if E is a suspension spectrum. Hence, this holds for a wedge of suspension spectra. In particular, r is an isomorphism for the spectra $\tau_{\text{ev}}E$ and $\tau_{\text{od}}E$ as in II.1.23. The Mayer–Vietoris sequence of the triad $(\tau E; \tau_{\text{ev}}E, \tau_{\text{od}}E)$, see II.3.12(iii), yields a commutative diagram

$$\begin{array}{ccccccc}
\cdots \to & L_{k+1}(\tau_{\text{ev}}E \cap \tau_{\text{od}}E) & \to L_k(\tau E) \to & L_k(\tau_{\text{ev}}E) \oplus L_k(\tau_{\text{od}}E) & \to \cdots \\
\downarrow{\scriptstyle r''} & \downarrow{\scriptstyle r'} & \downarrow{\scriptstyle r} & \downarrow{\scriptstyle r''} & \downarrow{\scriptstyle r'} \\
\cdots \to & F_{k+1}(\tau_{\text{ev}}E \cap \tau_{\text{od}}E) & \to F_k(\tau E) \to & F_k(\tau_{\text{ev}}E) \oplus F_k(\tau_{\text{od}}E) & \to \cdots
\end{array}$$

where $L_i(A) := \varinjlim\{F_{n+i}(A_n)\}$. Here the top row is exact because $\varinjlim$ preserves exactness, see I.2.7. Since r' and r'' are isomorphisms, r is an isomorphism by the Five Lemma. $\square$

4.23. Proposition. *Let $F_{(n)}$ be the Postnikov n-stage of a spectrum F. Then for every $X \in \mathscr{S}$ there is an exact sequence*

$$0 \to \varprojlim_n{}^1\{F^{i-1}_{(n)}(X)\} \to F^i(X) \to \varprojlim_n\{F^i_{(n)}(X)\} \to 0,$$

where

$$\cdots \to F^*_{(n)}(X) \xrightarrow{(p_n)_*} F^*_{(n-1)}(X) \to \cdots$$

is the obvious inverse system.

Proof. Given $Y \in \mathscr{S}$, consider the homomorphism

$$\prod_n F^*_{(n)}(Y) \to \prod_n F^*_{(n)}(Y),$$
$$\{\dots, a_n, a_{n-1}, \dots\} \longmapsto \{\dots, a_n - (p_{n+1})_* a_{n+1}, a_{n-1} - (p_n)_* a_n, \dots\}$$

where $a_n \in F^*_{(n)}(Y)$. Considering Y as an indeterminate, we get a morphism of additive cohomology theories on $\mathscr{S}$. By 3.6(ii), this morphism is represented by a morphism $\delta : \prod_n F_{(n)} \to \prod_n F_{(n)}$ of spectra, cf. 3.26. Let us form a cofiber sequence $G \xrightarrow{k} \prod_n F_{(n)} \xrightarrow{\delta} \prod_n F_{(n)}$. Then we have the exact sequence

$$\prod_n F^{i-1}_{(n)}(X) \xrightarrow{\delta_*} \prod_n F^{i-1}_{(n)}(X) \to G^i(X) \xrightarrow{k_*} \prod_n F^i_{(n)}(X) \xrightarrow{\delta_*} \prod_n F^i_{(n)}(X).$$

By 2.15, it yields the exact sequence

$$0 \to \varprojlim_n{}^1\{F^{i-1}_{(n)}(X)\} \to G^i(X) \xrightarrow{\bar{k}_*} \varprojlim_n\{F^i_{(n)}(X)\} \to 0.$$

We must prove that $G \simeq F$. Given $Y \in \mathscr{S}$, we define the homomorphism $F^*(Y) \to \prod_n F^*_{(n)}(Y), a \mapsto \{(\tau_n)_*(a)\}$. In this way we obtain a morphism of cohomology theories. By 3.6(ii), this morphism is represented by a morphism $\mu : F \to \prod F_{(n)}$ of spectra, and, by 3.6(ii) again, the morphism $F \xrightarrow{\mu} \prod_n F_{(n)} \xrightarrow{\delta} \prod_n F_{(n)}$ is inessential. Hence, there is $f : F \to G$ such that $kf = \mu$. So, we have the commutative diagram

$$\begin{array}{ccc} F^i(X) & \xrightarrow{\bar{\mu}_*} & \varprojlim_n \{F^i_{(n)}(X)\} \\ {\scriptstyle f_*}\downarrow & & \| \\ G^i(X) & \xrightarrow{\bar{\mu}_*} & \varprojlim_n \{F^i_{(n)}(X)\}. \end{array}$$

Clearly, both horizontal arrows are isomorphisms for $X = S^n, n \in \mathbb{Z}$. Thus, f is an equivalence. □

4.24. Remarks. (a) Let $\mathscr{F}$ be a sheaf of abelian groups over a space X. Let $\mathscr{U} = \{U\}$ be the family of all open subsets of X ordered with respect to inclusion. By definition, $\mathscr{F}$ is a functor $\mathscr{U} \to \mathscr{AG}$, and so it is just an inverse $\mathscr{U}$-system. Namely, $F_U = \mathscr{F}(U)$ for every $U \in \mathscr{U}$. Now, by the definitions, we have $\varprojlim^i \mathscr{F} = H^i(X; \mathscr{F})$, see Godement [1].

(b) I did not prove yet that $\varprojlim^i$ can be non-zero for every i. But now this is clear because of (a). Indeed, if $\mathscr{A}$ is a constant sheaf over $\vee_{i=1}^{\infty} S^i$ then $\varprojlim^i \mathscr{A} \neq 0$ for every i.

(c) Let $X, \mathscr{U}$ be as in (a), and let $f: Y \to X$ be a map. Given a spectrum F, define a sheaf $\mathscr{F}^k$ by setting $\mathscr{F}^k(U) = F^k((f^{-1}U)^+)$. (In fact, $f^{-1}U$ might not be a CW-space, but here I do not care about it.) Since the set $\{f^{-1}(U)\}$ is cofinal in the quasi-ordered set $\{Y_\lambda\}$, we get (under suitable conditions) a spectral sequence

$$(4.25) \qquad E_r^{p,q}(Y) \Rightarrow F^*(Y),\ E_2^{p,q} = \varprojlim^p \{F^q((f^{-1}U)^+)\} = H^p(X; \mathscr{F}^q).$$

In particular, if $f = 1_X$ then there is a spectral sequence

$$(4.26) \qquad E_r^{p,q}(X) \Rightarrow F^*(X),\ E_2^{p,q} = \varprojlim^p \{F^q(U^+)\} = H^p(X; \mathscr{F}^q).$$

For a good space X (defined below), we have $H^p(X; \mathscr{F}^q) = \check{H}^p(X; F^q(S^0))$, where $\check{H}$ means the Čech cohomology, and the spectral sequence (4.26) is just the Atiyah–Hirzebruch spectral sequence. We recommend also comparing the spectral sequence (4.25) with that in 15.27 from Switzer [1]. Probably, there is some folklore about the spectral sequences (4.25) and (4.26), but, as far as I know, nobody has written this down accurately. (Of course, the case $F = H$ in (4.25) is well known, see e.g. Godement [1].)

I want to explain what is meant to be a good space. We say that a covering $\mathscr{U} = \{U\}$ is strongly contractible if it is locally finite and every finite intersection $U_1 \cap \cdots \cap U_k$, $U_i \in \mathscr{U}$ is contractible. If $\mathscr{U}$ is strongly contractible then $H^i(X; \mathscr{F}) = \check{H}^i(N_{\mathscr{U}}; \mathscr{F}_{\mathscr{U}})$, where $N_{\mathscr{U}}$ is the nerve of $\mathscr{U}$ and $\mathscr{F}_{\mathscr{U}}$ is the local system given by $\mathscr{F}$, see e.g. Godement [1]. We say that X is good if every covering of it admits a strongly contractible refinement. Since $F^i(U^+) = F^i(S^0)$ for every contractible U, we conclude that $H^p(X; \mathscr{F}^q) = \check{H}^p(X; F^q(S^0))$ for every good space X.

§5. A Sufficient Condition for the Absence of Phantoms

The results of this section are due to Anderson [1], cf. also Atiyah [2].

As well as in §4, in this section $\{X_\lambda\}$ denotes the family of all finite subspectra of a spectrum X.

5.1. Lemma. *Let $\{\cdots \to A_{n+1} \to A_n \to \cdots \to A_0\}$ be an inverse system of monomorphisms of countable abelian groups. If $\varprojlim^1\{A_r\} = 0$, then there exists m such that $A_n \to A_m$ is an isomorphism for every $n \geq m$.*

Proof. Set $C_r = \operatorname{Coker}\{A_r \to A_0\}$. Since $\varprojlim^1\{A_r\} = 0$, the sequence $0 \to \varprojlim\{A_r\} \to A_0 \to \varprojlim\{C_r\} \to 0$ is exact by 2.13(ii). Hence, $\varprojlim\{C_r\}$ is countable. It is clear that the inverse system

$$\{\cdots \to C_n \xrightarrow{q_n} C_{n-1} \to \cdots\}$$

consists of epimorphisms. If it does not stabilize then there exists an arbitrarily large n with $q_n \neq 0$. Thus, $\varprojlim\{C_r\}$ is not countable. This is a contradiction. □

5.2. Definition. We say that an inverse system $\{\cdots \to A_{n+1} \to A_n \to \cdots \to A_0\}$ satisfies the *Mittag-Leffler condition* if for every r there exists $m = m(r)$ such that $\operatorname{Im}\{A_n \to A_r\} = \operatorname{Im}\{A_m \to A_r\}$ for all $n \geq m$.

5.3. Lemma. (i) *If an inverse system $\mathscr{A}$ satisfies the Mittag-Leffler condition then $\varprojlim^1 \mathscr{A} = 0$.*

(ii) *If all the groups A_r are countable and $\varprojlim^1 \mathscr{A} = 0$, then $\mathscr{A}$ satisfies the Mittag-Leffler condition.*

(iii) *Let $\mathscr{A}$ satisfy the Mittag-Leffler condition. Suppose that a sequence $\mathscr{A} \xrightarrow{\varkappa} \mathscr{B} \xrightarrow{\sigma} \mathscr{C}$ of inverse systems be exact (at $\mathscr{B}$). Then the sequence $\varprojlim \mathscr{A} \to \varprojlim \mathscr{B} \to \varprojlim \mathscr{C}$ is exact.*

Proof. (i) See Switzer [1], Th. 7.75.

(ii) If $\mathscr{A}$ is an inverse system of monomorphisms then the assertion follows from 5.1. Given an arbitrary system $\mathscr{A}$, we set $B_n^m = A_n$ for $m \leq n$ and $B_n^m = \operatorname{Im}\{A_m \to A_n\}$ for $m > n$. For every r we have the inverse system $\cdots \subset B_r^{m+1} \subset B_r^m \subset \cdots \subset B_r^r = A_r$ of monomorphisms, and, by 5.1, it stabilizes as $m \to \infty$. Hence, there exists m such that $B_r^n \to B_r^m$ is an isomorphism for $n \geq m$. Thus, $\mathscr{A}$ satisfies the Mittag-Leffler condition.

(iii) We prove that for every string $\{b_n\} \in \mathscr{B}$ with $\sigma\{b_n\} = 0$ there exists $\{a_n\}$ with $\varkappa_n(a_n) = b_n$. Set $P_n = \varkappa_n^{-1}(b_n)$, $Q_n = \cap_{m=n}^{\infty} \operatorname{Im}\{P_m \to P_n\}$. The homomorphisms p_n of the inverse system $\mathscr{A}$ induce epimorphisms $q_n : Q_n \to$

Q_{n-1}. The Mittag-Leffler condition for $\mathscr{A}$ implies that $Q_r \neq \emptyset$ for every r. Thus, there exist elements $a_n \in Q_n$ with $q_n a_n = a_{n-1}$ for every n, and it is clear that $\{a_n\}$ is a string with $\varkappa_n(a_n) = b_n$. □

Now let $\mathscr{A} = \{\cdots \to A_n \xrightarrow{p_n} A_{n-1} \to \cdots \to A_0\}$ be an inverse system of countable abelian groups. Suppose that for every n there is a filtration

$$A_n = A_n(-1) \supset A_n(0) \supset \cdots \supset A_n(n) \supset A_n(n+1) = 0$$

with $p_n(A_n(i)) \subset A_{n-1}(i)$, i.e., that a certain decreasing filtration of $\mathscr{A}$ is given. Consider the inverse system $\mathscr{B}(i) = \{\cdots \to B_n(i) \to B_{n-1}(i) \to \cdots\}$ where $B_n(i) = A_n(i)/A_n(i+1)$.

5.4. Lemma. *If* $\varprojlim \mathscr{B}(i) = 0$ *for every* i *then* $\varprojlim^1 \mathscr{A} = 0$.

Proof. We prove that $\mathscr{A}$ satisfies the Mittag-Leffler condition. Since $A_n(n+1) = 0$, it suffices to prove that $\{A_n/A_n(i)\}$ satisfies the Mittag-Leffler condition for every i. Since $A_n/A_n(i)$ is a finite extension of the groups $B_n(j)$ and $\varprojlim^1 B_n(j) = 0$, we conclude that $\varprojlim^1\{A_n/A_n(i)\} = 0$. Thus, by 5.3(ii), $\{A_n/A_n(i)\}$ satisfies the Mittag-Leffler condition. □

5.5. Theorem. *Let* X *be a spectrum of finite* $\mathbb{Z}$*-type. Fix a natural number* m*. Let* F *be a spectrum such that* $F^m(X^{(n)})$ *is a finitely generated abelian group for every* n*. Consider the Atiyah–Hirzebruch spectral sequence* $E_r^{p,q}(X,F)$, $E_2^{p,q}(X,F) = H^p(X;\pi_{-q}(F))$*. Suppose that for every* (p,q) *with* $p+q = m$ *the differentials*

$$d_r^{p,q} : E_r^{p,q}(X,F) \to E_r^{p+r,q-r+1}(X,F)$$

are trivial for all but a finite number of r*. Then*

$$F^m(X) = \varprojlim_n \{F^m(X^{(n)})\} = \varprojlim_\lambda \{F^m(X_\lambda)\}.$$

In particular, $F^m(X)$ *does not contain phantoms.*

Proof. Because of II.4.26(ii) and 1.15(ii), we can assume that X has finite type. So, the family $\{X_\lambda\}$ is cofinal in $\{X^{(n)}\}$, and it suffices to prove that $F^m(X) = \varprojlim_n \{F^m(X^{(n)})\}$. Now, because of 4.18, it suffices to prove that $\varprojlim^1 F^m(X^{(n)}) = 0$. Set $A_n = F^m(X^{(n)})$ and consider the filtration

$$A_n(i) := \operatorname{Ker}\{F^m(X^{(n)}) \to F^m(X^{(i)})\}$$

of A_n. Then $A_n(i)/A_n(i-1) = E_\infty^{i,m-i}(X^{(n)},F)$. Since the groups A_n are finitely generated, it suffices to prove (by 5.4) that $\varprojlim^1 E_\infty^{i,m-i}(X^{(n)},F) = 0$ for every i.

Recall that $E_1^{p,q}(X,F) = F^{p+q}(X^{(p)}, X^{(p-1)})$. Hence, $E_1^{p,q}(X^{(n)}, F) = 0$ for $p > n$, and the restriction $E_1^{p,q}(X,F) \to E_1^{p,q}(X^{(n)}, F)$ is an isomorphism for $p \le n$. This implies that $E_r^{p,q}(X,F) \to E_r^{p,q}(X^{(n)}, F)$ is an isomorphism for $p \le n-r+1$, i.e., for $n \ge p+r-1$. If $p+q = m$, then $E_r^{p,q}(X,F)$ does not depend on r for r large enough. Hence, $E_\infty^{p,q}(X^{(n)}, F)$ stabilizes as $n \to \infty$. Thus, $\varprojlim^1 E_\infty^{p,q}(X^{(n)}, F) = 0$. □

5.6. Corollary. *If the conditions of* 5.5 *hold for all* m, *then* $E_r^{p,q}(X,F)$ *converges to* $F^*(X)$.

Proof. By 5.5, $\varprojlim\{F^*(X^{(n)})\} = F^*(X)$. But $E_\infty^{*,*}(X,F)$ is associated to the graded group $\varprojlim\{F^*(X^{(n)})\}$. □

5.7. Corollary. *Let* X, F *be two spectra of finite* $\mathbb{Z}$*-type. Then*

$$F^m(X) = \varprojlim_n\{F^m(X^{(n)})\} = \varprojlim_\lambda\{F^m(X_\lambda)\}$$

provided at least one of the following conditions holds:

(i) *All the groups* $\pi_*(F)$ *are finite;*

(ii) *All the groups* $H_*(X)$ *are finite;*

(iii) *All the groups* $H^*(X)$ *and* $\pi_*(F)$ *are torsion free.*

Proof. (i), (ii) This follows from 5.5.

(iii) All the groups $E_2^{p,q}(X,F)$ are torsion free, and so, by II.7.12, all the differentials are trivial. □

5.8. Proposition. *Let* X, F *be two spectra, and suppose that there exists* N *such that* $\pi_i(F) = 0$ *for* $i \ge N$. *Then:*

(i) $\rho : F^k(X) \to \varprojlim\{F^k(X^{(n)})\}$ *is an isomorphism for every* k.

(ii) *Two morphisms* $\varphi, \psi : X \to F$ *are homotopic iff the homomorphisms* $\varphi_*, \psi_* : X^*(Y) \to F^*(Y)$ *coincide for every* $Y \in \mathscr{S}_{\mathrm{fd}}$.

Proof. Fix any k. The cofiber sequence $X^{(n)} \to X^{(n+1)} \to \vee S^{n+1}$ yields the exact sequence

$$F^{k-1}(\vee S^{n+1}) \to F^{k-1}(X^{(n+1)}) \to F^{k-1}(X^{(n)}) \to F^k(\vee S^{n+1}).$$

Since $\pi_i(F) = 0$ for $i \ge N$, we conclude that $F^i(S^m) = 0$ for $m - i \ge N$, and so $F^{k-1}(\vee S^{n+1}) = 0 = F^k(\vee S^{n+1})$ for n large enough. Hence, the sequence $\{F^{k-1}(X^{(n)})\}$ stabilizes as $n \to \infty$, and so it satisfies the Mittag-Leffler condition. Hence, by 5.3, $\varprojlim^1\{F^{k-1}(X^{(n)})\} = 0$, and thus, by 4.19, ρ is an isomorphism.

(ii) Only "if" needs proving. Let $f : X \to F$ be a morphism with $[f] = [\varphi] - [\psi]$. If $\varphi_*, \psi_* : X^*(Y) \to F^*(Y)$ coincide for every $Y \in \mathscr{S}_{\mathrm{fd}}$ then the morphism $X^{(n)} \subset X \xrightarrow{f} F$ is inessential. But, by (i), $\varprojlim^1\{F^0(X^{(n)})\} = 0$, and thus $\varphi \simeq \psi$. □

§6. Almost Equivalent Spectra (Spaces)

6.1. Definition. We say that two spectra (spaces) E, F are *almost equivalent* if the coskeletons $E_{(n)}$ and $F_{(n)}$ are equivalent for all n, i.e., if E and F have the same n-type for all n.

Let $[F]$ denotes the equivalence class of a spectrum F, and let

$$ALEQ(E) := \{[F] \mid F \text{ is almost equivalent to } E\}.$$

The term "almost equivalent" and notation $ALEQ$ are innovations. Traditionally one uses the notation $SNT(X)$ (same n-type) instead of $ALEQ$.

Here we describe the set $ALEQ(E)$ for a spectrum (space) E, but first we give an example of almost equivalent but inequivalent spaces.

6.2. Example (Adams [1]). Let A_m, $m = 1, 2, \dots$, be a countable family of pointed CW-complexes with finite skeletons. Consider the subset A of $\prod_{m=1}^{\infty} A_m$ consisting of the elements $(x_1, \dots, x_n, \dots)$, $x_n \in A_n$, such that all but a finite number of x_k coincide with the base points. Products of cells of A_m give the cells in A, and we introduce the weak topology with respect to these cells. We call A the direct sum of A_m and denote it by $\oplus A_m$.

Consider the sphere S^d, $d \geq 2$. Set $X = \oplus_{m=d}^{\infty}(S^d_{(m)})$, $Y = S^d \times X$. Then $X_{(n)} \simeq Y_{(n)}$ for every n. Indeed, given n, let Z be the direct sum of a countable set of copies of $S^d_{(n)}$. One has $S^d_{(n)} \times Z = Z$, and therefore $X_{(n)} \simeq \prod_{d \leq m < n} S^d_{(m)} \times Z \simeq Y_{(n)}$. Hence, X and Y are almost equivalent.

We prove that they are not equivalent. Suppose that there are $f: X \to Y, g: Y \to X$ with $fg \simeq 1_Y, gf \simeq 1_X$. Consider the direct summand S^d of $Y = S^d \times X$. The subspace $g(S^d)$ of X is contained in a finite cellular subspace of X, and so $g(S^d) \subset \prod_{d \leq m < N} S^d_{(m)}$ for some N. Then the map

$$S^d \xrightarrow{g} \prod_{d \leq m < N} S^d_{(m)} \subset X \xrightarrow{f} Y \xrightarrow{\text{proj}} S^d$$

is homotopic to 1_{S^d}. Since $\pi_i\left(\prod_{d \leq m < N} S^d_{(m)}\right) = 0$ for $i \geq N$, we conclude that $\pi_i(S^d) = 0$ for $i \geq N$. On the other hand, by a well-known theorem of Serre [3], the groups $\pi_i(S^d)$ are non-trivial for arbitrarily large i (namely, the dimension of the first nontrivial p-component increases with increasing p). This is a contradiction.

Now we pass to a description of $ALEQ(E)$.

6.3. Lemma. (i) *Let E, F be two spectra (resp. spaces), and let*

$$\begin{array}{ccccccccc}
\cdots & = & E & = & E & = & E & = & \cdots \\
 & & \downarrow & & \downarrow{\scriptstyle \tau_n} & & \downarrow & & \\
\cdots & \longrightarrow & E_{(n+1)} & \longrightarrow & E_{(n)} & \xrightarrow{p_n} & E_{(n-1)} & \longrightarrow & \cdots
\end{array}$$

and

$$\begin{array}{ccccccccc}
\cdots & = & F & = & F & = & F & = & \cdots \\
 & & \downarrow & & \downarrow{\scriptstyle \sigma_n} & & \downarrow & & \\
\cdots & \longrightarrow & F_{(n+1)} & \longrightarrow & F_{(n)} & \xrightarrow{q_n} & F_{(n-1)} & \longrightarrow & \cdots
\end{array}$$

be Postnikov towers of E and F. Let $\varphi_n : E_{(n)} \to F_{(n)}$ be morphisms (resp. maps) such that every diagram

$$\begin{array}{ccc}
E_{(n)} & \xrightarrow{\varphi_n} & F_{(n)} \\
{\scriptstyle p_n}\downarrow & & \downarrow{\scriptstyle q_n} \\
E_{(n-1)} & \xrightarrow{\varphi_{n-1}} & F_{(n-1)}
\end{array}$$

commutes up to homotopy. Then there exists $\varphi : E \to F$ such that every diagram

$$\begin{array}{ccc}
E & \xrightarrow{\varphi} & F \\
{\scriptstyle \tau_n}\downarrow & & \downarrow{\scriptstyle \sigma_n} \\
E_{(n)} & \xrightarrow{\varphi_n} & F_{(n)}
\end{array}$$

commutes up to homotopy.

(ii) *Given a sequence of morphisms of spectra*

$$\cdots \leftarrow E(0) \leftarrow E(1) \leftarrow \cdots \leftarrow E(n-1) \xleftarrow{p_n} E(n) \leftarrow \cdots$$

such that $\pi_i(E(0)) = 0$ for $i > 0$ and that the cone of p_n is an Eilenberg–Mac Lane spectrum $\Sigma^{n+1}H(\pi_n)$ for some π_n, there exist a spectrum E and morphisms $\tau_n : E \to E(n)$ such that

$$\begin{array}{ccccccccc}
\cdots & = & E & = & E & = & E & = & \cdots \\
 & & \downarrow & & \downarrow{\scriptstyle \tau_n} & & \downarrow & & \\
\cdots & \longrightarrow & E(n+1) & \longrightarrow & E(n) & \xrightarrow{p_n} & E(n-1) & \longrightarrow & \cdots
\end{array}$$

is a Postnikov tower of E, and this spectrum E is unique up to equivalence.

(iii) *Given a sequence of maps of spaces*

$$\mathrm{pt} \leftarrow X(1) \leftarrow \cdots \leftarrow X(n-1) \xleftarrow{p_n} X(n) \leftarrow \cdots,$$

suppose that for every n the following holds: $\pi_i(X(n)) = 0$ for $i > n$ and $(p_n)_* : \pi_i(X(n)) \to \pi_i(X(n-1))$ is an isomorphism for $i < n$. Then there exist a space X and maps $\tau_n : X \to X(n)$ such that

$$\begin{array}{ccccccccc}
\cdots & = & X & = & X & = & X & = & \cdots \\
 & & \downarrow & & \downarrow{\scriptstyle \tau_n} & & \downarrow & & \\
\cdots & \longrightarrow & X(n+1) & \longrightarrow & X(n) & \xrightarrow{p_n} & X(n-1) & \longrightarrow & \cdots
\end{array}$$

is a Postnikov tower of X, and this X is unique up to homotopy equivalence.

Proof. (i) Since the inclusion $i_k : E^{(k)} \to E$ is a $(k-1)$-equivalence, for fixed n the function $i_k^* : [E, F_{(n)}] \to [E^{(k)}, F_{(n)}]$ is a bijection for k large enough (for spectra this follows from II.4.1(iv), for spaces from the obstruction theory). Similarly, since $\sigma_n : F \to F_{(n)}$ is an n-equivalence, for fixed k the function $(\sigma_n)_* : [E^{(k)}, F] \to [E^{(k)}, F_{(n)}]$ is a bijection for n large enough. So, we have the commutative diagram

$$\begin{array}{ccc}
[E, F] & \xrightarrow{h} & \varprojlim_n [E, F_{(n)}] \\
{\scriptstyle \rho}\downarrow & & \downarrow{\scriptstyle \cong} \\
\varprojlim_k [E^{(k)}, F] & \xrightarrow{\cong} & \varprojlim_{k,n} [E^{(k)}, F_{(n)}]
\end{array}$$

where $h = \{(\sigma_n)_* | \varprojlim\}$. By 3.20(ii) (for spectra) and 1.16 (for spaces), ρ is a surjection. Hence, $h : [E, F] \to \varprojlim_n \{[E, F_{(n)}]\}$ is a surjection. Thus, there is $\varphi : E \to F$ with $h(\varphi) = \{\varphi_n \tau_n\}$

(ii) Set $E^*(X) = \varprojlim\{E(n)^*(X)\}$ for every finite dimensional spectrum X. Fixing X, by II.4.1(ii) we have $E^i(X) = E(N)^i(X)$ for N large enough. So, E^* is a cohomology theory on $\mathscr{S}_{\mathrm{fd}}$. By 3.21, it can be represented by a spectrum E. Furthermore, also by 3.21, the morphisms $E^*(-) \to E(n)^*(-)$ of cohomology theories are induced by certain morphisms $\tau_n : E \to E(n)$. By 5.8(ii), $p_n \tau_n \simeq \tau_{n-1}$. Hence, the diagram of the lemma is a Postnikov tower of E. Finally, if there is a spectrum F with the same Postnikov tower, then, by (i), $E \simeq F$.

(iii) We can assume spaces and maps to be pointed. Given a tower as in the lemma, define $F(Y) := \varprojlim[Y, X(n)]^\bullet$ for every finite dimensional connected CW-space Y. Then F is a half-exact functor on the category $\mathscr{H}$ of pointed connected finite dimensional pointed CW-spaces, and so, by 3.25, $F(Y) = [Y, X]^\bullet$ for a certain pointed connected CW-space X and every $Y \in \mathscr{H}$. Because of the universality of $\varprojlim$, we have certain natural maps $a_n : [Y, X]^\bullet \to [Y, X(n)]^\bullet$.

Given a CW-space Z, consider the diagram

$$\begin{array}{ccc}
[Z,X]^{\bullet} & & [Z,X(n)]^{\bullet} \\
\Big\downarrow{\scriptstyle\rho} & & \Big\downarrow{\scriptstyle\cong} \\
\varprojlim_m [Z^{(m)},X]^{\bullet} & \xrightarrow{\varprojlim_m a_n} & \varprojlim_m [Z^{(m)},X(n)]^{\bullet}.
\end{array}$$

Here the right arrow is a bijection because $\pi_i(X(n)) = 0$ for $i > n$. Hence, we get a natural map $(\varprojlim_m a_n)\rho : [Z,X]^{\bullet} \to [Z,X(n)]^{\bullet}$. It yields a certain map $\tau_n : X \to X(n)$, and $p_n\tau_n \simeq \tau_{n-1}$. Thus, the tower above is a Postnikov tower of X. The uniqueness of X follows from (i). □

Consider an inverse system of groups (not necessary abelian)

$$\cdots \leftarrow G_0 \xleftarrow{j_1} G_1 \leftarrow \cdots \leftarrow G_{n-1} \xleftarrow{j_n} G_n \leftarrow \cdots .$$

The group $G = \prod_{n=-\infty}^{\infty} G_n$ acts on the set $\prod_{n=-\infty}^{\infty} G_n$ as follows:

$$\{g_n\}\{\alpha_n\} = \{g_n\alpha_n j_{n+1}(g_{n+1}^{-1})\}.$$

Define $\varprojlim^1\{G_n\}$ to be the set of all orbits of this G-action. This construction coincides with that given above (e.g. in 2.15) for abelian groups G_n.

6.4. Proposition (cf. 2.15). *If $\{G_n, j_n\}$ is an inverse system of compact topological groups and continuous homomorphisms, then $\varprojlim^1\{G_n\}$ is trivial (i.e., it is just a one-point set). In particular, $\varprojlim^1\{G_n\}$ is trivial for every system of finite groups G_n.*

Proof. We leave it to the reader, but give a hint. (Also, see Wilkerson [1].) Firstly, one must prove that $\{G_n\}$ satisfies the Mittag-Leffler condition (using the criterion for compactness in terms of the nested systems of closed sets). This implies that $\varprojlim^1\{G_n\}$ is trivial, cf. Switzer [1], Th. 7.75. □

Let $\operatorname{aut} E$ be the group (under the composition law) of the homotopy classes of all self-equivalences $E \to E$ of a spectrum (space) E. Because of II.4.18, we obtain the inverse system of groups (non-abelian)

$$\cdots \leftarrow \operatorname{aut} E_{(n)} \leftarrow \operatorname{aut} E_{(n+1)} \leftarrow \cdots .$$

6.5. Theorem (cf. Wilkerson [1]). *There is a bijective correspondence between $ALEQ(E)$ and $\varprojlim^1\{\operatorname{aut} E_{(n)}\}$.*

Proof. We write E_k instead of $E_{(k)}$. Firstly, we construct a map $\varphi : \prod_{n=-\infty}^{\infty} \operatorname{aut} E_n \to ALEQ(E)$. Consider a Postnikov tower of E

$$\cdots \leftarrow E_0 \xleftarrow{p_1} E_1 \leftarrow \cdots \leftarrow E_{n-1} \xleftarrow{p_n} E_n \leftarrow \cdots .$$

Given $\{\alpha_n\} \in \prod_{n=-\infty}^{\infty} \operatorname{aut} E_n$, consider the tower

$$\cdots \leftarrow E_0 \xleftarrow{\alpha_0 p_1} E_1 \leftarrow \cdots \leftarrow E_{n-1} \xleftarrow{\alpha_{n-1} p_n} E_n \leftarrow \cdots .$$

By 6.3(ii), it is a Postnikov tower of a certain spectrum (space) F, and it is clear that the spectra (spaces) F and E are almost equivalent. Define $\varphi(\{\alpha_n\})$ to be the equivalence class $[F]$ of F.

We prove that φ is surjective. Let F be almost equivalent to E, and let $f_n : F_n \to E_n$ be the corresponding equivalences. By II.4.18, the morphism f_{n+1} induces a morphism $(f_{n+1})_n : F_n = (F_{n+1})_n \to (E_{n+1})_n = E_n$. Set $\alpha_n = f_n \circ ((f_{n+1})_n)^{-1}$. It is easy to see that $\varphi(\{\alpha_n\}) = [F]$.

We prove that φ induces a well-defined map $\varprojlim^1\{\operatorname{aut} E_n\} \to ALEQ(E)$. Let $\{\beta_n\} = \{g_n\}\{\alpha_n\}$. For every n the diagram

$$\begin{array}{ccc}
E_{n+1} & \xrightarrow{g_{n+1}} & E_{n+1} \\
{\scriptstyle p_{n+1}}\downarrow & & \downarrow{\scriptstyle p_{n+1}} \\
E_n & & E_n \\
{\scriptstyle \alpha_n}\downarrow & & \downarrow{\scriptstyle \beta_n} \\
E_n & \xrightarrow{g_n} & E_n
\end{array}$$

commutes up to homotopy. Now one can construct an equivalence $\varphi(\{\alpha_n\}) \to \varphi(\{\beta_n\})$ in the same manner as in 6.3.

Finally, given spectra F, G with $[F] = \varphi(\{\alpha_n\}), [G] = \varphi(\{\beta_n\})$, there are equivalences $a_n : E_n \to F_n$ and $b_n : G_n \to E_n$ such that in the diagram below the left and the right squares are commutative. Suppose that there exists an equivalence $h : F \to G$. By II.4.18, it induces maps $h_n : F_n \to G_n$ such that the middle square of the diagram below commutes. Hence, there is the homotopy commutative diagram

$$\begin{array}{ccccccc}
E_{n+1} & \xrightarrow{a_{n+1}} & F_{n+1} & \xrightarrow{h_{n+1}} & G_{n+1} & \xrightarrow{b_{n+1}} & E_{n+1} \\
{\scriptstyle \alpha_n p^E_{n+1}}\downarrow & & {\scriptstyle p^F_{n+1}}\downarrow & & \downarrow{\scriptstyle p^G_{n+1}} & & \downarrow{\scriptstyle \beta_n p^E_{n+1}} \\
E_n & \xrightarrow{a_n} & F_n & \xrightarrow{h_n} & G_n & \xrightarrow{b_n} & E_n .
\end{array}$$

So, $\{g_n\}\{\alpha_n\} = \{\beta_n\}$, where $g_n = b_n h_n a_n$. Hence, $\varphi : \varprojlim^1\{\operatorname{aut} E_{(n)}\} \to ALEQ(E)$ is injective, and thus it is bijective. □

6.6. Corollary. *If E is bonuded below and every group $\pi_i(E)$ is finite, then $ALEQ(E) = \{[E]\}$.*

Proof. Every group $\operatorname{aut} E_{(n)}$ is finite, being a subset of the finite set $E^0_{(n)}(E_{(n)})$. Thus, by 6.4, $\varprojlim^1\{\operatorname{aut} E_{(n)}\} = 0$. □

6.7. Theorem. *Let E be a spectrum of finite $\mathbb{Z}$-type. Given a sequence*

$$E \xrightarrow{\varphi} F \xrightarrow{\alpha} G \xrightarrow{\beta} \Sigma E \xrightarrow{\Sigma\varphi} \Sigma F \xrightarrow{\Sigma\alpha} \cdots$$

of spectra such that the sequence

$$\cdots \to E_i(X) \xrightarrow{\varphi_*} F_i(X) \xrightarrow{\alpha_*} G_i(X) \xrightarrow{\beta_*} E_{i-1}(X) \to \cdots$$

is exact for every CW-space X, suppose that $\varphi_* : \pi_*(E) \to \pi_*(F)$ is monic. Then the following hold:

(i) *The spectra G and $C\varphi$ are almost equivalent;*

(ii) *If $G^0(E)$ does not contain phantoms, then $G \simeq C\varphi$, and $E \xrightarrow{\varphi} F \xrightarrow{\alpha} G$ is a cofiber sequence.*

Proof. (i) By duality, for every finite spectrum Y we have the exact sequence $\cdots \to E^i(Y) \to F^i(Y) \to \cdots$. By II.4.26(ii), we can assume that $E^{(n)}$ is a finite spectrum for every n. Putting $Y = E^{(n)}, i = 0$ in this exact sequence, we conclude that the morphism $E^{(n)} \xrightarrow{\varphi^{(n)}} F^{(n)} \xrightarrow{\alpha^{(n)}} G^{(n+1)}$ is trivial. Hence, there exists $h : C(\varphi^{(n)}) \to G^{(n+1)}$ such that the following diagram commutes:

$$\begin{array}{ccccc}
E^{(n)} & \xrightarrow{\varphi^{(n)}} & F^{(n)} & \longrightarrow & C(\varphi^{(n)}) \\
\| & & \| & & \downarrow h \\
E^{(n)} & \xrightarrow{\varphi^{(n)}} & F^{(n)} & \xrightarrow{\alpha^{(n)}} & G^{(n+1)} .
\end{array}$$

For $i < n-1$ this diagram induces the following diagram with the exact rows:

$$\begin{array}{ccccccccc}
0 & \longrightarrow & \pi_i(E^{(n)}) & \longrightarrow & \pi_i(F^{(n)}) & \longrightarrow & \pi_i(C(\varphi^{(n)})) & \longrightarrow & 0 \\
\| & & \| & & \| & & \downarrow h_* & & \| \\
0 & \longrightarrow & \pi_i(E^{(n)}) & \longrightarrow & \pi_i(F^{(n)}) & \longrightarrow & \pi_i(G^{(n+1)}) & \longrightarrow & 0.
\end{array}$$

Hence, $C(\varphi^{(n)})$ and $G^{(n+1)}$ are $(n-2)$-equivalent for every n.

This implies that $C\varphi$ and G are almost equivalent. Indeed, the inclusion $G^{(n+1)} \to G$ is an $(n+1)$-equivalence, and so the induced morphism $G^{(n+1)}_{(n)} \to G_{(n)}$ is an equivalence. Similarly, there is an $(n-2)$-equivalence $C(\varphi^{(n)}) \to C(\varphi)$, and so we obtain an equivalence $C(\varphi^{(n)})_{(n-2)} \to C(\varphi)_{(n-2)}$. So, since $C(\varphi^{(n)})$ and $G^{(n+1)}$ are $(n-2)$-equivalent, $C(\varphi)_{(n-2)}$ and $G_{(n-2)}$ are equivalent. Since this holds for every n, we conclude that $C(\varphi)$ and G are almost equivalent.

(ii) The morphism $\alpha\varphi f$ is inessential for every morphism $f : Y \to E$ of a finite spectrum Y. In particular, for every n the morphism $\alpha\varphi|E^{(n)}$ is inessential because $E^{(n)}$ is a finite spectrum. Hence, $\alpha\varphi$ is inessential because

$G^0(E)$ does not contain phantoms. Therefore, there exists $f : C\varphi \to G$ such that the diagram (where $E \xrightarrow{\varphi} F \xrightarrow{\psi} C\varphi$ is the strict cofiber sequence)

$$\begin{array}{ccccc} E & \xrightarrow{\varphi} & F & \xrightarrow{\psi} & C\varphi \\ \| & & \| & & \downarrow f \\ E & \xrightarrow{\varphi} & F & \xrightarrow{\alpha} & G \end{array}$$

commutes. Hence, in the diagram

$$\begin{array}{ccccccccc} 0 & \longrightarrow & \pi_*(E) & \longrightarrow & \pi_*(F) & \longrightarrow & \pi_*(C\varphi) & \longrightarrow & 0 \\ \| & & \| & & \| & & \downarrow f_* & & \| \\ 0 & \longrightarrow & \pi_*(E) & \longrightarrow & \pi_*(F) & \longrightarrow & \pi_*(G) & \longrightarrow & 0 \end{array}$$

f_* is an isomorphism, and thus f is an equivalence. □

§7. Multiplications and Quasi-multiplications

Every ring spectrum (E, μ, ι) yields a family

$$\{\mu_{(X,A),(Y,B)} : E_i(X, A) \otimes E_j(Y, B) \to E_{i+j}(X \times Y, X \times B \cup A \times Y)\},$$
$$(X, A), (Y, B) \in \mathscr{C}^2.$$

However, for certain homology theories $E_*(-)$, one can easily construct a family $\{\mu_{X,Y}\}$ even when knowing neither the multiplication μ nor the spectrum E. Typical examples are geometrically defined homology theories, like bordism and bordism with singularities, see Ch. VIII, IX. Moreover, sometimes one can construct a family $\{\mu_{X,Y}\}$ with suitable properties even if we do not know whether the multiplication μ exists. We call such a family $\{\mu_{X,Y}\}$ a quasi-multiplication. So, every multiplication yields a quasi-multiplication, and we are interested in the converse of this situation.

Recall that every spectrum E is an S-module spectrum. Hence, for all CW-pairs (X, A), (Y, B) there is a natural pairing

$$\varphi = \varphi_{(X,A),(Y,B)} : \Pi_i(X, A) \otimes E_j(Y, B) \to E_{i+j}(X \times Y, X \times B \cup A \times Y)$$

where $\varphi(\alpha \otimes \beta)$ is given by the morphism

$$\begin{aligned} S^i \wedge S^j \xrightarrow{\alpha\wedge\beta} & \Sigma^\infty(X/A) \wedge \Sigma^\infty(Y/B) \wedge E \simeq \Sigma^\infty(X/A \wedge Y/B) \wedge E \\ & = \Sigma^\infty((X \times Y)/(X \times B \cup A \times Y)) \wedge E \end{aligned}$$

for $\alpha : S^i \to \Sigma^\infty(X/A)$, $\beta : S^j \to \Sigma^\infty(Y/B) \wedge E$.

Similarly, there is a natural pairing

$$\psi = \psi_{(X,A),(Y,B)} : E_i(X,A) \otimes \Pi_j(Y,B) \to E_{i+j}(X \times Y, X \times B \cup A \times Y).$$

7.1. Definition. (a) A quasi-ring spectrum is a triple $(E, \{\mu_{(X,A),(Y,B)}\}, \iota)$, where:

E is a spectrum;

$$\{\mu_{(X,A),(Y,B)} : E_i(X,A) \otimes E_j(Y,B) \to E_{i+j}(X \times Y, X \times B \cup A \times Y), i,j \in \mathbb{Z}\}$$

is a certain family (called the *quasi-multiplication*) of natural pairings, defined for all CW-pairs (X,A), (Y,B);

$\iota : S \to E$ is a certain morphism (called the *unit*).

Furthermore, we require that the following four diagrams are commutative:

$$\begin{array}{ccc} \Pi_i(X,A) \otimes E_j(Y,B) & \xrightarrow{\varphi} & E_{i+j}(X \times Y, X \times B \cup A \times Y) \\ \downarrow{\scriptstyle \iota^X \otimes 1} & & \| \\ E_i(X,A) \otimes E_j(Y,B) & \xrightarrow{\mu} & E_{i+j}(X \times Y, X \times B \cup A \times Y) \end{array}$$

and the similar diagram for ψ;

$$\begin{array}{ccc} E_i(X,A) \otimes E_j(Y,B) & \xrightarrow{\mu_{(X,A),(Y,B)}} & E_{i+j}(X \times Y, X \times B \cup A \times Y) \\ \downarrow{\scriptstyle \partial \otimes 1} & & \downarrow{\scriptstyle d} \\ E_{i-1}(A) \otimes E_j(Y,B) & & E_i(X \times B \cup A \times Y, X \times B) \\ \| & & {\scriptstyle c}\downarrow{\scriptstyle \cong} \\ E_{i-1}(A) \otimes E_j(Y,B) & \xrightarrow{\mu_{A,(Y,B)}} & E_{i+j-1}(A \times Y, A \times B) \end{array}$$

and the similar for $\partial : E_j(Y,B) \to E_{j-1}(B)$. Here d is the (boundary) homomorphism in the exact sequence II.3.2(iv) of the triple $(X \times Y, X \times B \cup A \times Y, X \times B)$ and

$$c : E_*(X \times B \cup A \times Y, X \times B) \xrightarrow{\cong} E_*(A \times Y/A \times B) \xleftarrow{\cong} E_*(A \times Y, A \times B)$$

is the composition of the collapsing isomorphisms.

A quasi-multiplication is *commutative* if the diagram

$$\begin{array}{ccc} E_i(X,A) \otimes E_j(Y,B) & \xrightarrow{\mu_{(X,A),(Y,B)}} & E_{i+j}(X \times Y, X \times B \cup A \times Y) \\ \downarrow{\scriptstyle \chi} & & \downarrow{\scriptstyle \tau_*} \\ E_j(Y,B) \otimes E_i(X,A) & \xrightarrow{\mu_{Y,X}} & E_{i+j}(Y \times X, Y \times A \cup B \times X). \end{array}$$

commutes. Here $\chi(a \otimes b) = b \otimes a$ and $\tau = \tau(X,Y) : X \times Y \to Y \times X$. We leave it to the reader to define the *associativity* condition for a quasi-multiplication.

(b) Let E, F be two quasi-ring spectra. A morphism $\varphi = \{\varphi^{(X,A)} : E_*(X,A) \to F_*(X,A)\}$ is a *quasi-ring morphism* if the following diagrams commute for all pairs $(X,A), (Y,B)$:

$$\begin{array}{ccc} \Pi_*(X,A) & \xrightarrow{\iota_*^E} & E_*(X.A) \\ \| & & \downarrow{\scriptstyle \varphi^{(X,A)}} \\ \Pi_*(X,A) & \xrightarrow{\iota_*^F} & F_*(X,A) \end{array}$$

$$\begin{array}{ccc} E_*(X,A) \otimes E_*(Y,B) & \xrightarrow{\mu^E} & E_*((X,A) \times (Y,B)) \\ {\scriptstyle \varphi^{(X,A)} \otimes \varphi^{(Y,B)}}\downarrow & & \downarrow{\scriptstyle \varphi^{(X,A)\times(Y,B)}} \\ F_*(X,A) \otimes F_*(Y,B) & \xrightarrow{\mu^F} & F_*((X,A) \times (Y,B)) \end{array}$$

where, as usual, $(X,A) \times (Y,B) := (X \times Y, X \times B \cup A \times Y)$.

7.2. Construction. Let $(E, \{\mu_{(X,A),(Y,B)}\}, \iota)$ be a quasi-ring spectrum.

(a) Given two pointed CW-spaces X, Y, we define a pairing

$$\begin{aligned} \widetilde{E}_i(X) \otimes \widetilde{E}_j(Y) = E_i(X,*) \times E_j(Y,*) &\xrightarrow{\mu_{(X,*),(Y,*)}} E_{i+j}(X \times Y, X \vee Y) \\ &= \widetilde{E}_{i+j}(X \wedge Y). \end{aligned}$$

We leave it to the reader to prove that these pairings commute with the suspension isomorphisms, i.e., that the diagrams like II.(3.37) (with Σ replaced by S) commutes.

(b) Because of (a), we have a pairing

$$\mu_{A,B} : E_i(A) \otimes E_j(A) \to E_{i+j}(A \wedge B)$$

for every pair of finite spectra A, B. Now, given two arbitrary spectra X, Y, let X_λ, resp.$Y_{\lambda'}$ be the family of all finite subspectra of X, resp. Y. By II.3.20(ii), there is a canonical isomorphism $E_i(X) \cong \varinjlim\{E_i(X_\lambda)\}$, and similarly for $E_*(Y)$. Now, we define the pairing

$$\begin{aligned} \mu_{X,Y} : &E_i(X) \otimes E_j(Y) = \varinjlim\{E_i(X_\lambda)\} \otimes \varinjlim\{E_i(Y_{\lambda'})\} \\ &= \varinjlim\{E_i(X_\lambda) \otimes E_j(Y_{\lambda'})\} \to \varinjlim\{E_{i+j}(X_\lambda \wedge Y_{\lambda'})\} \to E_{i+j}(X \wedge Y) \end{aligned}$$

(recall that $\varinjlim$ and $\otimes$ commute, see e.g. Bourbaki [1], §6, n^{0}7).

As we have already remarked, every multiplication induces a quasi-multiplication, and we are interested in the converse of this situation.

7.3. Theorem (cf. Switzer [1], 13.80 ff.). *Let $(E, \{\mu_{(X,A),(Y,B)}\}, \iota)$ be a quasi-ring spectrum of finite $\mathbb{Z}$-type. Let E^k denote the k-skeleton of E. Then :*

(i) *There exists a pairing $\mu : E \wedge E \to E$ inducing the pairings $\mu_{(X,A),(Y,B)}$.*

(ii) *If* $\varprojlim^1\{E^{-1}(E^n \wedge E^n)\} = 0$ *then the pairing* μ *as in* (i) *is unique up to homotopy.*

(iii) *If* $\varprojlim^1\{E^{-1}(E^n)\} = 0$ *then the morphism* $S \wedge E \xrightarrow{\iota\wedge 1} E \wedge E \xrightarrow{\mu} E$ (*with* μ *as in* (i)) *is homotopic to* $l(E) : S \wedge E \to E$. *Thus, the diagrams as in* 7.1 *commute for every* μ *as in* (i).

(iv) *If* $\varprojlim^1\{E^{-1}(E^n \wedge E^n \wedge E^n)\} = 0$ *and the quasi-multiplication on* E *is associative then every pairing* μ *as in* (i) *is associative.*

(v) *If* $\varprojlim^1\{E^{-1}(E^n \wedge E^n)\} = 0$ *and the quasi-multiplication on* E *is commutative then every pairing* μ *as in* (i) *is commutative.*

In particular, if E *is a spectrum of finite* $\mathbb{Z}$*-type with finite groups* $\pi_i(E)$ *for every* i *then every associative quasi-multiplication on* E *is induced by a multiplication* $\mu : E \wedge E \to E$. *This multiplication is unique up to homotopy, and it is commutative if the quasi-multiplication is.*

Proof. (i) By 7.2(a) and duality, we have pairings $\mu^{A,B} : E^i(A) \otimes E^j(B) \to E^{i+j}(A \wedge B)$ for all *finite spectra* A, B. By II.4.26(ii), we can assume that all skeletons of the spectra $E, E\wedge E, E\wedge E\wedge E$ are finite. Hence, there are certain pairings

$$\mu^{E^n,E^n} : E^0(E^n) \otimes E^0(E^n) \to E^0(E^n \wedge E^n).$$

Let $a_n : E^n \to E$ be the inclusion, and let a morphism $v_n : E^n \wedge E^n \to E$ give the element $\mu^{E^n,E^n}(a_n \otimes a_n)$. Since the inclusions $a_n \wedge a_n : E^n \wedge E^n \to E \wedge E$ and $b_{2n} : (E \wedge E)^{2n} \to E \wedge E$ are n-equivalences, there exists a unique n-equivalence $h_n : (E \wedge E)^{2n} \to E^n \wedge E^n$ such that $(a_n \wedge a_n)h \simeq b_{2n}$. Set $u_{2n} := v_n h_n : (E \wedge E)^{2n} \to E$. Then $u_{2n+2}|(E \wedge E)^{2n} \simeq u_{2n}$. Hence, the family $\{u_{2n}\}$ gives an element $u \in \varprojlim\{E^0((E \wedge E)^n)\}$. Now, the exactness of the sequence

$$(7.4)\quad 0 \to \varprojlim^1\{E^{-1}((E \wedge E)^n)\} \to E^0(E \wedge E) \to \varprojlim\{E^0((E \wedge E)^n)\} \to 0$$

(see (4.19)) implies the existence of the pairing $\mu : E \wedge E \to E$.

(ii) This follows from the exactness of (7.4).

We only prove (iv) because the assertions (iii)–(v) can be proved in a similar way. Let $f : E \wedge E \wedge E \to E$ be a morphism with $[f] = [\mu\circ(\mu \wedge 1)] - [\mu\circ(1 \wedge \mu)]$. In the diagram

$$\begin{array}{ccccc} E^n \wedge E^n \wedge E^n & \xrightarrow{a_n\wedge a_n\wedge a_n} & E \wedge E \wedge E & \xrightarrow{1\wedge\mu} & E \wedge E \\ & & \downarrow{\scriptstyle \mu\wedge 1} & & \downarrow{\scriptstyle \mu} \\ & & E \wedge E & \xrightarrow{\mu} & E \end{array}$$

the morphisms $\mu\circ(\mu \wedge 1)\circ(a_n \wedge a_n \wedge a_n)$ and $\mu\circ(1 \wedge \mu)\circ(a_n \wedge a_n \wedge a_n)$ are homotopic (by associativity of the quasi-multiplication) for every n. Hence, if f is essential then it must be a phantom. But $E^0(E \wedge E \wedge E)$ does not contain phantoms because $\varprojlim^1\{E^{-1}(E^n \wedge E^n \wedge E^n)\} = 0$.

Finally, the validity of (ii)–(v) for spectra with finite homotopy groups follows from 5.7(i). □

The following proposition can be proved similarly to 7.3(ii).

7.5. Proposition. *Let E, F be two ring spectra. Let $\theta : E \to F$ be a morphism such that $\{\theta^X : E_*(X) \to F_*(X)\}$ is a quasi-ring morphism. If $F^0(E \wedge E)$ does not contain phantoms, i.e., $\varprojlim^1\{F^{-1}(E^{(n)} \wedge E^{(n)})\} = 0$ (for example, E has finite $\mathbb{Z}$-type and every group $\pi_i(F)$ is finite), then the diagram* (7.6) *below commutes up to homotopy.*

$$
\begin{array}{ccc}
E \wedge E & \xrightarrow{\theta\wedge\theta} & F \wedge F \\
{\scriptstyle \mu_E}\downarrow & & \downarrow{\scriptstyle \mu_F} \\
E & \xrightarrow{\theta} & F
\end{array} \tag{7.6}
$$

In other words, θ is a ring morphism. □

Similarly to quasi-ring spectra, we can consider quasi-module spectra.

7.7. Definition. Let $(E, \{\mu_{(X,A),(Y,B)}\}, \iota)$ be an associative quasi-ring spectrum. A *quasi-module spectrum* over E is a pair $(F, \{m_{(X,A),(Y,B)}\})$ where F is a spectrum and

$$\{m_{(X,A),(Y,B)} : E_i(X,A) \otimes F_j(Y,B) \to F_{i+j}(X \times Y, X \times B \cup A \times Y), i, j \in \mathbb{Z}\}$$

is a family of natural pairings, defined for all CW-pairs $(X,A), (Y,B)$. Furthermore, we require that the following diagram commutes:

$$
\begin{array}{ccc}
E_m(X,A) \otimes E_n(Y,B) \otimes F_p(Z,C) & \longrightarrow & E_m(X,A) \otimes F_{n+p}((Y,B) \times (Z,C)) \\
\downarrow & & \downarrow \\
E_{m+n}((X,A) \times (Y,B)) \otimes F_p(Z,C) & \longrightarrow & F_{m+n+p}((X,A) \times (Y,B) \times (Z,C)).
\end{array}
$$

Clearly, every module spectrum over a ring spectrum E is a quasi-module spectrum over E. We suggest that the reader carries out an analog of 7.3 for quasi-module spectra. We formulate its special case which will be used below.

7.8. Theorem. *Let E be a ring spectrum of finite $\mathbb{Z}$-type, and let F be a quasi-module spectrum over E. If every group $\pi_i(F)$ is finite, then the quasi-module structure on F extends to a unique E-module structure on F.* □

Chapter IV. Thom Spectra

In the introduction we discussed the importance and usefulness of Thom spaces (spectra). In this chapter we develop a general theory of Thom spectra, investigate some special Thom spectra and apply this to certain geometrical problems. Some aspects of a general theory of Thom spectra are also considered in Lewis–May–Steinberger [1]. Now it is clear that a proper theory of Thom spaces occurs in the context of sectioned spherical fibrations, and so we pay a lot of attention to sectioned fibrations; they are discussed at the beginning of the chapter.

§1. Fibrations and Their Classifying Spaces

Following Husemoller [1], we treat a bundle as "just a map viewed as an object of a particular category".

1.1. Definition. A *bundle* ξ over a space B is a map $p: E \to B$. The spaces E, B are called the *total space* and the *base* (or *base space*) of ξ, respectively, and the map p is called the *projection*. The subspace $F_b := p^{-1}(b)$ of E is called the *fiber* of ξ over $b \in B$.

We use the notation $\operatorname{ts}\xi$ for E, $\operatorname{bs}\xi$ for B and proj_ξ (or simply p_ξ) for p.

A *subbundle* of a given bundle ξ is just a map $q: Y \to X$, where $Y \subset \operatorname{ts}(\xi)$, $X \subset \operatorname{bs}(\xi)$ and $q(y) = p_\xi(y)$ for every $y \in Y$.

1.2. Definition. (a) Given a bundle $\xi = \{p: E \to B\}$ and a map $f: X \to B$, a *p-lifting* of f to E, or a *lifting of f with respect to p*, is an arbitrary map $g: X \to E$ with $pg = f$. If such a lifting exists, we say that *the map f can be lifted to a map g*. Two p-liftings g, g' of f are called *vertically homotopic* if there exists a map $H: X \times I \to E$ (called a *vertical homotopy*) with $H(x, 0) = g(x)$, $H(x, 1) = g'(x)$ and $pH(x, t) = f(x)$ for every $x \in X, t \in I$. The set of all p-liftings of f we denote by $\operatorname{Lift}_p f$, and the set of the vertical homotopy classes of all p-liftings of f we denote by $[\operatorname{Lift}_p f]$.

(b) A p-lifting of the map 1_B is called a *section* of ξ. In other words, a section is a map $s: \operatorname{bs}\xi \to \operatorname{ts}\xi$ such that $ps = 1_B$. We use the notation $\operatorname{Sec}\xi := \operatorname{Lift}_p 1_B$, $[\operatorname{Sec}\xi] := [\operatorname{Lift}_p 1_B]$.

(c) A *sectioned bundle* is a pair (ξ, s_ξ) where ξ is a bundle and s_ξ is a section of ξ.

1.3. Definition. Let $\xi = \{p : E \to B\}, \eta = \{p' : E' \to B'\}$ be two bundles.

(a) A *fiberwise map* is a map $g : E \to E'$ such that $p'gx = p'gy$ whenever $px = py$.

(b) A *bundle morphism* $\varphi : \xi \to \eta$ is a pair $\varphi = (g, f)$ of maps such that the diagram

$$\begin{array}{ccc} E & \xrightarrow{g} & E' \\ {\scriptstyle p}\downarrow & & \downarrow{\scriptstyle p'} \\ B & \xrightarrow{f} & B' \end{array}$$

commutes.

We use the notation $\operatorname{ts}\varphi$ for g and $\operatorname{bs}\varphi$ for f. So, $\operatorname{ts}\varphi$ is always a fiberwise map.

In particular, there is the *identity bundle morphism* $1_\xi := (1_E, 1_B)$. Furthermore, a *bundle isomorphism* is a morphism $\varphi : \xi \to \xi'$ such that there exists a morphism $\psi : \xi' \to \xi$ with $\psi\varphi = 1_\xi$ and $\varphi\psi = 1_{\xi'}$.

(c) A bundle morphism $\varphi : \xi \to \eta$ of the form $(g, 1_B)$ is called a *morphism over* B. In this case we say also that g is a *map over* B.

(d) Given two sectioned bundles $(\xi, s_\xi), (\eta, s_\eta)$, a *sectioned bundle morphism* is a bundle morphism $\varphi : \xi \to \eta$ which respects the sections, i.e., $(\operatorname{ts}\varphi)s_\xi = s_\eta \operatorname{bs}\varphi$. A sectioned bundle morphism of the form $(g, 1_B)$ is called a *sectioned morphism over* B.

Let ξ, η be two bundles, and let ζ be a subbundle of ξ. Given a bundle morphism $\varphi : \zeta \to \eta$, define a bundle $p : \operatorname{ts}(\xi) \cup_{\operatorname{ts}(\varphi)} \operatorname{ts}(\eta) \to \operatorname{bs}(\xi) \cup_{\operatorname{bs}(\varphi)} \operatorname{bs}(\eta)$ by setting $p(x) = p_\xi(x)$ for $x \in \operatorname{ts}(\xi)$, $p(x) = p_\eta(x)$ for $x \in \operatorname{ts}(\eta)$. This bundle is denoted by $\xi \cup_\varphi \eta$ and called a *gluing of* ξ *and* η *via* φ.

1.4. Constructions–Definitions. (a) The *product* of two bundles ξ, η is the bundle $\xi \times \eta := \{p \times p' : E \times E' \to B \times B'\}$. Given two morphisms $\varphi : \xi \to \xi'$ and $\psi : \eta \to \eta'$ of bundles, define a morphism $\varphi \times \psi : \xi \times \eta \to \xi' \times \eta'$ by setting $\operatorname{ts}(\varphi \times \psi) := \operatorname{ts}\varphi \times \operatorname{ts}\psi$.

(b) In particular, we can consider a space P as a bundle P over a point and get the bundle $\xi \times P$ over $\operatorname{bs}\xi$. On the other hand, we can consider the bundle 1_P and construct the bundle $\xi \times 1_P$ over $(\operatorname{bs}\xi) \times P$.

(c) Given a morphism $\psi : \xi \to \eta$ over B, we define the *mapping cylinder* $\operatorname{Cyl}(\psi)$ *over* B to be the bundle $\xi \times I \cup_\psi \eta$, where ψ is considered as a morphism $\psi : \xi \times \{0\} \to \eta$.

(d) Given a diagram $\eta_1 \xleftarrow{\varphi_1} \xi \xrightarrow{\varphi_2} \eta_2$ of morphisms over $B = \operatorname{bs}\xi = \operatorname{bs}\eta_i, i = 1, 2$, we define its *double mapping cylinder over* B to be the bundle

$$\operatorname{DCyl}(\varphi_1, \varphi_2) := \xi \times [0, 2] \cup_\psi (\eta_1 \sqcup \eta_2),$$

where ψ is the morphism $\psi : (\xi \times \{0\}) \sqcup (\xi \times \{2\}) = \xi \sqcup \xi \xrightarrow{\varphi_1 \sqcup \varphi_2} \eta_1 \sqcup \eta_2$. As in I.3.18, there are morphisms $i_{\text{left}} : \eta_1 \subset \text{DCyl}(\varphi_1, \varphi_2)$, $i_{\text{right}} : \eta_2 \subset \text{DCyl}(\varphi_1, \varphi_2)$ and $i_{\text{mid}} : \xi = \xi \times \{1\} \subset \text{DCyl}(\varphi_1, \varphi_2)$.

(e) Note that p_ξ yields a morphism $\widehat{p_\xi} : \xi \to 1_{\text{bs}\,\xi}$ of bundles. Define the *bundle join* $\xi * \eta$ of bundles ξ, η to be the double mapping cylinder of the diagram

$$1_{\text{bs}\,\xi} \times \eta \xleftarrow{\widehat{p_\xi} \times 1} \xi \times \eta \xrightarrow{1 \times \widehat{p_\eta}} \xi \times 1_{\text{bs}\,\eta}$$

of bundles over $\text{bs}\,\xi \times \text{bs}\,\eta$. It is easy to see that the fiber of $\xi * \eta$ over $(b_1, b_2) \in \text{bs}\,\xi \times \text{bs}\,\eta = \text{bs}(\xi * \eta)$ is $F_{b_1} * F_{b_2}$ where $*$ is the usual join of the spaces.

(f) Note that s_ξ yields a morphism $\widehat{s_\xi} : 1_{\text{bs}\,\xi} \to \xi$ of bundles for every sectioned bundle (ξ, s_ξ). Given two sectioned bundles ξ, η over B, we define the *bundle h-wedge*

$$\xi \vee^h \eta := \text{DCyl}(\xi \xleftarrow{\widehat{s_\xi}} 1_B \xrightarrow{\widehat{s_\eta}} \eta).$$

For every $b \in B$ the fiber of $\xi \vee^h \eta$ over b is the h-wedge of the (pointed) fibers of ξ, η. We equip $\xi \vee^h \eta$ with the section $\text{ts}(i_{\text{mid}}) : B \to \text{ts}(\xi \vee^h \eta)$.

There is the following fiberwise analog of I.3.33. Let $\varphi : \xi \to \zeta, \psi : \eta \to \zeta$ be sectioned morphisms over B. Then there exists a unique sectioned morphism $\varphi \top \psi : \xi \vee^h \eta \to \zeta$ over B such that $(\varphi \top \psi) i_{\text{left}} = \varphi, (\varphi \top \psi) i_{\text{right}} = \psi$ and $\text{ts}(\varphi \top \psi)(b, t) = s_\zeta(b)$ for every $b \in B, t \in [0, 2]$.

(g) Given two sectioned bundles $(\xi, s_\xi), (\eta, s_\eta)$, we define the *bundle h-smash product* $\xi \wedge^h \eta$ as follows. Set $\bar{\xi} = \xi \times 1_{\text{bs}\,\eta}$, $\bar{\eta} = 1_{\text{bs}\,\xi} \times \eta$. Define

$$\varphi : \bar{\xi} = \xi \times 1_{\text{bs}\,\eta} \xrightarrow{1 \times \widehat{s_\eta}} \xi \times \eta, \ \psi : \bar{\eta} = 1_{\text{bs}\,\xi} \times \eta \xrightarrow{\widehat{s_\xi} \times 1} \xi \times \eta,$$

and let $\varphi \top \psi : \bar{\xi} \vee^h \bar{\eta} \to \xi \times \eta$ be as in (f). We set

$$\xi \wedge^h \eta := \text{DCyl}(1_{\text{bs}\,\xi \times \text{bs}\,\eta} \xleftarrow{\widehat{p_{\bar{\xi} \vee^h \bar{\eta}}}} \bar{\xi} \vee^h \bar{\eta} \xrightarrow{\varphi \top \psi} \xi \times \eta).$$

We equip $\xi \wedge^h \eta$ with the section $\text{ts}(i_{\text{left}}) : \text{bs}\,\xi \times \text{bs}\,\eta \to \text{ts}(\xi \wedge^h \eta)$. Clearly, the fiber of $\xi \wedge^h \eta$ over (b_1, b_2) is $F_{b_1} \wedge^h F_{b_2}$.

Given morphisms $\varphi : \xi \to \xi'$ and $\psi : \eta \to \eta'$, the diagram

$$\begin{array}{ccccc}
1_{\text{bs}\,\xi} \times \eta & \xleftarrow{\widehat{p_\xi} \times 1} & \xi \times \eta & \xrightarrow{1 \times \widehat{p_\eta}} & \xi \times 1_{\text{bs}\,\eta} \\
\downarrow{\scriptstyle \text{bs}\,\varphi \times \psi} & & \downarrow{\scriptstyle \varphi \times \psi} & & \downarrow{\scriptstyle \varphi \times \text{bs}\,\psi} \\
1_{\text{bs}\,\xi'} \times \eta' & \xleftarrow{\widehat{p_{\xi'}} \times 1} & \xi' \times \eta' & \xrightarrow{1 \times \widehat{p_{\eta'}}} & \xi' \times 1_{\text{bs}\,\eta'}
\end{array}$$

induces a morphism of the double mapping cylinders. We denote this morphism by $\varphi * \psi : \xi * \eta \to \xi' * \eta'$. Moreover, if φ and ψ preserve the given sections, then $\varphi \vee^h \psi$ and $\varphi \wedge^h \psi$ can be defined in an obvious way (do it). In other words, the constructions $\times, *, \vee^h, \wedge^h$ are natural with respect to bundle morphisms.

1.5. Definition. (a) A *bundle homotopy* between two morphisms $\varphi, \varphi' : \xi \to \eta$ is a morphism $H : \xi \times 1_I \to \eta$ such that $H|\xi \times 1_{\{0\}} = \varphi$ and $H|\xi \times 1_{\{1\}} = \varphi'$. If such a bundle homotopy H exists, we say that φ and φ' are *bundle homotopic* and write $\varphi \simeq^{\text{bun}} \varphi'$.

A bundle morphism $\varphi : \xi \to \eta$ is a *bundle homotopy equivalence* if there exists $\psi : \eta \to \xi$ such that $\varphi\psi$ and $\psi\varphi$ are bundle homotopic to the corresponding identity maps.

(b) A *homotopy over B* between two morphisms $\varphi, \varphi' : \xi \to \eta$ over B is a morphism $H : \xi \times I \to \eta$ over B such that $H|\xi \times \{0\} = \varphi, H|\xi \times \{1\} = \varphi'$. If such a homotopy exists, we say that φ and φ' are *homotopic over B* and write $\varphi \simeq_B \varphi'$ or $H : \varphi \simeq_B \varphi'$. A morphism $\varphi : \xi \to \eta$ over B is a *homotopy equivalence over B* if there exists $\psi : \eta \to \xi$ over B such that $\varphi\psi \simeq_B 1_\eta$, $\psi\varphi \simeq_B 1_\xi$.

(c) If s_ξ is a section of ξ, then $s_\xi \times 1_I : \operatorname{bs}\xi \times I \to \operatorname{ts}\xi \times I$ is a section of $\xi \times 1_I$. A *sectioned bundle homotopy* between two sectioned morphisms $\varphi, \varphi' : (\xi, s_\xi) \to (\eta, s_\eta)$ is a bundle homotopy $H : \xi \times 1_I \to \eta$ between them which respects the sections, i.e., $(\operatorname{ts} H)(s_\xi \times 1_I) = s_\eta \operatorname{bs} H$; furthermore, one can define a *sectioned bundle homotopy equivalence.* Similarly, one can define a *sectioned homotopy over B* and a *sectioned homotopy equivalence over B* (do it). If there is a sectioned homotopy H over B between two morphisms φ, φ' over B, we use the notation $\varphi \simeq^\bullet_B \varphi'$ or $H : \varphi \simeq^\bullet_B \varphi'$.

(d) Let ξ be a subbundle of a bundle $\eta, \operatorname{bs}\xi = \operatorname{bs}\eta = B$. We say that the inclusion $i : \xi \subset \eta$ is a *cofibration over B* if every morphism $\xi \times I \cup \eta \to \zeta$ over B can be extended to a morphism $\eta \times I \to \zeta$ over B.

1.6. Remark. If B is a point then a bundle over B is just a space, and a sectioned bundle over B is just a pointed space; furthermore, a homotopy (resp. a cofibration) over B is just an ordinary homotopy (resp. cofibration). So, notions "over B" can be regarded as fiberwise versions of ordinary notions. Note that categorists call "a space over B" what we call "a bundle over B", etc., but this categorical flavor is irrelevant in our context.

1.7. Lemma. (i) *Let (ξ, s_ξ) be a sectioned bundle over B. We set $\eta := \operatorname{Cyl}\widehat{s}_\xi$ and define $s_\eta : B \to \operatorname{ts}\eta, s_\eta(b) := (s_\xi(b), 1)$. Then the inclusion $(\xi, s_\xi) \to (\eta, s_\eta)$ is a sectioned homotopy equivalence over B, and $\widehat{s}_\eta : 1_B \to \eta$ is a cofibration over B.*

(ii) *If $i : \xi \subset \eta$ is a cofibration over B, then $\operatorname{ts} i : \operatorname{ts}\xi \subset \operatorname{ts}\eta$ is a cofibration.*

Proof. (i) This is a fiberwise version of I.3.26(i) (more precisely, of the special case of I.3.26(i) with $X = \text{pt}$). The proof is left to the reader.

(ii) We set $A = \operatorname{ts}\xi, X = \operatorname{ts}\eta$. We prove that every map $f : A \times I \cup X \to Y$ can be extended to a map $g : X \times I \to Y$. We set $\zeta = \{p_2 : Y \times B \to B\}$

and define $\varphi : \xi \times I \cup \eta \to \zeta$ by setting $\operatorname{ts}\varphi(x) = (f(x), p_\eta(x))$. Now, there is $\psi : \eta \times I \to \zeta$ which extends φ, and we define $g : X \times I \to Y$ to be the composition

$$X \times I \xrightarrow{\operatorname{ts}\psi} Y \times B \xrightarrow{p_1} Y. \qquad \square$$

1.8. Definition. For every bundle $\xi = \{p : E \to B\}$ and every map $f : X \to B$ the *induced bundle* $f^*(\xi) = \{p' : E' \to X\}$ is defined via the pull-back diagram

$$\begin{array}{ccc} E' & \longrightarrow & E \\ {\scriptstyle p'}\downarrow & & \downarrow{\scriptstyle p} \\ X & \xrightarrow{f} & B. \end{array}$$

In other words, $E' = \{(x, e) \in X \times E | f(x) = p(e)\}$, and $p'(x, e) = x$. We define the *canonical morphism* $\mathfrak{I} = \mathfrak{I}_{f,\xi} : f^*\xi \to \xi$, $\operatorname{ts}\mathfrak{I}(x, e) = e$.

1.9. Proposition. (i) *There are canonical bijections*

$$\operatorname{Lift}_p f = \operatorname{Sec} f^*(\xi), \ [\operatorname{Lift}_p f] = [\operatorname{Sec} f^*(\xi)].$$

(ii) *Given a morphism $\varphi : \xi \to \eta$, there exists a unique morphism $\mathfrak{F}_\varphi : \xi \to (\operatorname{bs}\varphi)^*\eta$ over $\operatorname{bs}\xi$ such that $\varphi = \mathfrak{I}_{\operatorname{bs}\varphi,\eta}\mathfrak{F}_\varphi$.*

(iii) *Let $\varphi : \xi \to \eta$ be a morphism over B, and let $f : X \to B$ be a map. Then there exists a unique morphism $f^*\varphi : f^*\xi \to f^*\eta$ over X such that the following diagram commutes:*

$$\begin{array}{ccc} f^*\xi & \xrightarrow{f^*\varphi} & f^*\eta \\ {\scriptstyle \mathfrak{I}_{f,\xi}}\downarrow & & \downarrow{\scriptstyle \mathfrak{I}_{f,\eta}} \\ \xi & \xrightarrow{\varphi} & \eta\ . \end{array}$$

(iv) *If (ξ, s_ξ) is a sectioned bundle, then $s : X \to E', s(x) = (x, s_\xi(f(x)))$ is a section of $f^*\xi$, and $\mathfrak{I}_{f,\xi} : f^*\xi \to \xi$ maps s to s_ξ, i.e., $\mathfrak{I}_{f,\xi}$ is a sectioned morphism.*

(v) *If ξ, η are homotopy equivalent bundles (resp. sectioned bundles) over B, then $f^*\xi$, $f^*\eta$ are homotopy equivalent bundles (resp. sectioned bundles) over X for every map $f : X \to B$.*

Proof. Decode the definitions. $\square$

1.10. Notation and Convention. (a) If $i : A \subset B$ is an inclusion, we write $\xi|A$ rather than $i^*\xi$ for a bundle ξ and $\varphi|A$ rather than $i^*\varphi$ for a morphism φ in 1.9(iii).

(b) Sometimes E' is denoted by $X \times_B E$.

(c) We shall say just "equivalence over B" rather than "homotopy equivalence over B".

1.11. Definition. Given a space F, the *product F-bundle*, or the *standard trivial F-bundle* $\theta_B = \theta_B^F$ over B is just the projection $p_1 : B \times F \to B$. A *trivial* F-bundle over B is a bundle which is isomorphic to the product F-bundle. An isomorphism of a trivial F-bundle ξ with the product F-bundle is called a *trivialization* of ξ. A *trivial bundle* is a bundle which is a trivial F-bundle for some F. A *locally trivial* bundle over B is a bundle ξ such that, for some covering $\{U_i\}$ of B, the bundle $\xi|U_i$ is trivial for every i. A *fiberwise homotopy trivial bundle* is a bundle which is equivalent over the base to a trivial bundle.

1.12. Definition. (a) A *fibration* (or a *Hurewicz fibration*) is a bundle $\xi = \{p : E \to B\}$ which satisfies the following covering homotopy property: For every map $F : X \times I \to B$ and every $g : X \to E$ with $pg(x) = F(x, 0)$, there exists $G : X \times I \to E$ with $G(x, 0) = g(x)$ and $pG = F$.

(b) A *Dold fibration* (a weak fibration in the terminology of Dold [2]) is a bundle $\xi = \{p : E \to B\}$ which satisfies the following weak covering homotopy property: For every map $F : X \times [0, 1] \to B$ and every $g : X \to E$ with $pg(x) = F(x, 0)$ there exists $G : X \times [-1, 1] \to E$ such that $G(x, -1) = g(x)$, $pG(x, t) = F(x, 0)$ for $t \in [-1, 0]$, $pG(x, t) = F(x, t)$ for $t \in [0, 1]$.

(c) A *quasi-fibration* is a bundle $\xi = \{p : E \to B\}$ such that for every $x \in B$ and for every $a \in F_x$ the map $p_* : \pi_k(E, F_x, a) \to \pi_k(B, x)$ is bijective for every $k \geq 1$. In particular, one has the *homotopy exact sequence* (induced by the homotopy exact sequence of the pointed pair (E, F_x, a_0))

$$\cdots \to \pi_i(F_x, a_0) \to \pi_i(E, a_0) \to \pi_i(B, x) \to \pi_{i-1}(F_x, a_0) \to \cdots$$

for every $x \in B$.

(d) A *sectioned fibration* is a sectioned bundle (ξ, s_ξ) such that ξ is a fibration and $\widehat{s}_\xi : 1_{\operatorname{bs}\xi} \subset \xi$ is a cofibration over $\operatorname{bs}\xi$.

1.13. Proposition. *Every fibration is a Dold fibration. Every Dold fibration is a quasi-fibration.*

Proof. Only the second claim needs proof. This is well known for fibrations, see e.g. Switzer [1], Hu [1], and the proof can be immediately generalized for Dold fibrations. □

1.14. Proposition. *Let $f : X \to B$ be a map. If $\xi = \{p : E \to B\}$ is a fibration (resp. Dold fibration, resp. sectioned fibration) then so is $f^*\xi$.*

Proof. Exercise. □

The advantage of Dold fibrations is that they are invariant under bundle homotopy equivalences, unlike fibrations.

1.15. Proposition. *Let $\xi \to \eta$ be an equivalence over a space B. Then ξ is a Dold fibration if η is.*

Proof. See Dold [2], 5.2 or tom Dieck–Kamps–Puppe [1], 6.7. □

1.16. Example. Consider the bundle $p : [-1, 1] \to [0, 1]$, $p(t) = 0$ for $t \leq 0, p(t) = t$ for $t \geq 0$. This bundle is equivalent over $[0, 1]$ to the fibration $1_{[0,1]}$, but it is not a fibration (prove this). On the other hand, by 1.15, it is a Dold fibration.

1.17. Theorem (Strøm [1]). *Let $p : E \to B$ be a fibration.*

(i) *Let (Y, Z) be a pair such that Z is a strong deformation retract of Y. Suppose that there exists a map $h : Y \to I$ with $h^{-1}(0) = Z$. Then for all maps $u'' : Z \to E$ and $u' : Y \to B$ with $pu'' = u'$ there exists $u : Y \to E$ with $pu = u'$ and $u|Z = u''$.*

(ii) *Let (X, A) be a cofibered pair (e.g., a CW-pair). Then for every map $F : X \times I \to B$ and every map $g : X \cup A \times I \to E$ with $pg = F|(X \cup A \times I)$ there exists $G : X \times I \to E$ with $G|(X \cup A \times I) = g$ and $pG = F$.*

Proof. (i) Let $D : Y \times I \to Y$ be a map such that $D(y, 1) = y$ for every y, $D(y, 0) \in Z$ and $D(z, t) = z$ for every $z \in Z, t \in I$. Such a map D exists because Z is a strong deformation retract of Y. Define

$$\overline{D}(y, t) := \begin{cases} D(y, t/h(y)) & \text{if } t < h(y), \\ y & \text{otherwise.} \end{cases}$$

It is easy to see that $\overline{D}$ is continuous.

Since $p : E \to B$ is a fibration, the map $u'\overline{D} : Y \times I \to B$ can be lifted to $F : Y \times I \to E$ such that $F(y, 0) = u''D(y, 0)$ for every $y \in Y$. Now set $u(y) = F(y, h(y))$.

(ii) This follows from (i), if we put $Y = X \times I, Z = X \cup A \times I, u'' = g, u' = F$ and prove that Z is a strong deformation retract of Y. Let $p_1 : X \times I \to X$, $p_2 : X \times I \to I$ be the projections. Firstly, Z is a retract of Y by I.3.25(ii). We choose a retraction $r : X \times I \to X \cup A \times I$, and set $h(x, t) := \sup_{t \in I} |t - p_2 r(x, t)|$. Considering the homotopy $D : ir \simeq 1_{X \times I}$ rel $X \cup A \times I$, $D(x, t, s) := (p_1 r(x, (1 - s)t), (1 - s)p_2 r(x, t) + st), s \in [0, 1]$, we conclude that Z is a strong deformation retract of Y. □

1.18. Lemma. *Let $\xi = \{p : E \to B, s : B \to E\}$ be a sectioned bundle such that $\widehat{s} : 1_B \to \xi$ is a cofibration over B. The following conditions are equivalent:*

(i) *ξ is a fibration;*

(ii) *Let $\zeta = \{q : D \to A, r : A \to D\}$ be an arbitrary sectioned bundle such that $\widehat{r} : 1_A \to \zeta$ is a cofibration over A, and let $\varphi = (g, f) : \zeta \to \xi$ be a sectioned morphism. Then for every homotopy $h : A \times I \to B, h(a, 0) = f(a)$ there is a homotopy $H : D \times I \to E$ such that $H|D \times \{0\} = g, pH = h(q \times 1_I)$ and H preserves the sections, i.e., $H(r(a), t) = s(h(a, t))$.*

Proof. The implication (i) ⇒ (ii) follows from 1.17(ii) and 1.7(i). To prove that (ii) ⇒ (i), consider certain maps $F : X \times I \to B$ and $u : X \to E, pu(x) =$

$F(x, 0)$. We must find $G : X \times I \to E$ such that $pG = F$ and $G|X \times \{0\} = u$. Let $X_i, i = 1, 2$, be a copy of X. We put

$$D := X_1 \sqcup X_2 = X \sqcup X,\ A := X, h := F,$$
$$q : X_1 \sqcup X_2 \to X,\ q|X_i := 1_X,$$
$$r : X \to X_1 \sqcup X_2, r(x) := x_1, \text{ i.e., } r \text{ is the embedding on } X_1$$
$$g : X_1 \sqcup X_2 \to E, g(x) := s(F(x, 0)) \text{ for } x \in X_1,\ g(x) := u(x) \text{ for } x \in X_2.$$

If (ii) holds, then there is a homotopy $H : D \times I \to E$, and we set $G = H|X_2 \times I$. □

1.19. Remark. Why do I need this lemma? Here I want to quote some results of May [2], but he defined sectioned fibrations as in 1.18(ii), see May [2], 2.1 and 5.2. So, 1.18 just shows the equivalence of the definitions.

1.20. Lemma. (i) *If $\xi = \{p : E \to X \times I\}$ is a Dold fibration, then $\xi|(X \times \{0\})$ and $\xi|(X \times \{1\})$ are Dold fibrations and equivalent over X.*

(ii) *If $\xi = \{p : E \to X \times I\}$ is a sectioned fibration, then $\xi|(X \times \{0\})$ and $\xi|(X \times \{1\})$ are sectioned fibrations and are equivalent over X.*

Proof. See Dold [2], 6.6, May [2], 2.4. □

1.21. Corollary. (i) *Let $f \simeq g : X \to B$, and let η be a Dold fibration (resp. a sectioned fibration) over B. Then $f^*\eta$ and $g^*\eta$ are equivalent Dold fibrations (resp. equivalent sectioned fibrations).*

(ii) *Let ξ be a Dold fibration over a connected base. Then all its fibers are homotopy equivalent.*

Proof. (i) Let $H : X \times I \to B$ be a homotopy between f and g, $H(x, 0) = f(x), H(x, 1) = g(x)$. Set $\xi = H^*\eta$. Then $f^*\eta = \xi|(X \times \{0\})$ and $g^*\eta = \xi|(X \times \{1\})$. Now apply 1.20.

(ii) For every two points $b_1, b_2 \in B$, the inclusions $\{b_1\} \subset B, \{b_2\} \subset B$ are homotopic maps. Now the result follows from (i). □

1.22. Lemma. *Let ξ be a Dold fibration over a space B.*

(i) *If $f \simeq 1_B : B \to B$ then $\mathfrak{I}_{f,\xi} : f^*\xi \to \xi$ is a bundle homotopy equivalence.*

(ii) *If $f : X \to B$ is a homotopy equivalence then $\mathfrak{I}_{f,\xi} : f^*\xi \to \xi$ is a bundle homotopy equivalence.*

Proof. (i) We let $\xi = \{p : E \to B\}$ and $f^*\xi = \{p' : E' \to B\}$ where

$$E' = \{(b, e) | b \in B, e \in E,\ f(b) = p(e)\}.$$

Consider a homotopy

$$F : B \times [-1, 1] \to B, \quad F(b, t) = b \text{ for } t \in [-1, 0], F(b, 1) = f(b).$$

It can be covered by a homotopy

$$G : E \times [-1, 1] \to E, \; G|E \times \{-1\} = 1_E.$$

We set $g = G|E \times \{1\}$ and $\varphi = (g, f) : \xi \to \xi$. By 1.9(ii), φ can be decomposed as

$$\xi \xrightarrow{\mathfrak{F}_\varphi} f^*\xi \xrightarrow{\mathfrak{I}_{f,\xi}} \xi.$$

It is clear that φ is bundle homotopy equivalent to 1_ξ, i.e., $\mathfrak{I}_{f,\xi}\mathfrak{F}_\varphi \simeq 1_\xi$. On the other hand, the map $\operatorname{ts}(\mathfrak{F}_\varphi \mathfrak{I}_{f,\xi})$ has the form $(b, e) \mapsto (f(b), g(e))$. Now, the homotopy

$$E' \times [-1, 1] \to E', \quad ((b, e), t) \longmapsto (F(b, t), G(e, t))$$

yields a bundle homotopy $1_{f^*\xi} \simeq \mathfrak{F}_\varphi \mathfrak{I}_{f,\xi}$.

(ii) Let $g : B \to X$ be homotopy inverse to f. We set $\mathfrak{I} = \mathfrak{I}_{f,\xi}, \mathfrak{I}' = \mathfrak{I}_{g,f^*\xi}$. By (i),

$$g^* f^* \xi \xrightarrow{\mathfrak{I}'} f^*\xi \xrightarrow{\mathfrak{I}} \xi$$

is a bundle homotopy equivalence; let α be bundle homotopy inverse to $\mathfrak{I}\mathfrak{I}'$. Note that $\mathfrak{I}'\alpha$ is bundle homotopy right inverse to $\mathfrak{I}$. Furthermore, similarly, $\mathfrak{I}'$ has a bundle homotopy right inverse, say, β. Now, $1 \simeq \alpha\mathfrak{I}\mathfrak{I}'$, and so

$$\mathfrak{I}'\alpha\mathfrak{I} \simeq \mathfrak{I}'\alpha\mathfrak{I}\mathfrak{I}'\beta \simeq \mathfrak{I}'\beta \simeq 1.$$

Thus, $\mathfrak{I}$ and $\mathfrak{I}'\alpha$ are bundle homotopy inverse. □

1.23. Definition. A covering (not necessarily open) $\{C_i\}$ of a space X is *numerable* if there exists a family $\{f_i\}$ of maps $f_i : X \to [0, 1]$ such that

1. For every $x \in X$, $f_i(x) = 0$ for all but finitely many indices.
2. $\sum_i f_i(x) \equiv 1$.
3. $\overline{f_i^{-1}(0, 1]} \subset C_i$ for every i.

In other words, f_i is a partition of unity such that $\overline{f_i^{-1}(0, 1]}$ refines $\{C_i\}$.

1.24. Recollection. Recall that every locally finite covering of a paracompact space is numerable, see e.g. Munkres [2]. Every *CW*-space is paracompact, Miyazaki [1], see also Fritsch-Piccinini [1].

1.25. Theorem (Dold [2]). *Let $\{C_i\}$ be a numerable covering of a space B, and let ξ be a bundle over B. If $\xi|C_i$ is a fibration for every i, then so is ξ. If $\xi|C_i$ is a Dold fibration for every i, then so is ξ.*

Proof. See Dold [2], 4.8 or tom Dieck–Kamps–Puppe [1], 9.4 and 9.5. □

1.26. Corollary. *Every locally trivial bundle over a paracompact space (e.g., over a CW-space) is a fibration.*

Proof. Note that every trivial bundle is a fibration. Now, let ξ be a locally trivial bundle, and let $\{U_i\}$ be a covering of $\operatorname{bs}\xi$ such that $\xi|U_i$ is trivial for every i. Then $\{U_i\}$ admits a locally finite refinement $\{V_j\}$, which is numerable. Thus, by 1.25, ξ is a fibration since $\xi|V_j$ is a fibration for every j. □

1.27. Theorem (Dold [2]). *Let ξ, η be two Dold fibrations over a space B, and let $\varphi : \xi \to \eta$ be a morphism over B such that $\operatorname{ts}\varphi : \operatorname{ts}\xi \to \operatorname{ts}\eta$ is a homotopy equivalence. Then φ is an equivalence over B.*

Proof. See Dold [2], 6.1 or tom Dieck–Kamps–Puppe [1], 6.21. □

1.28. Corollary. *Let $\varphi : \xi_1 \to \xi_2$ be a morphism of Dold fibrations such that $\operatorname{ts}\varphi : \operatorname{ts}\xi_1 \to \operatorname{ts}\xi_2$ and $\operatorname{bs}\varphi : \operatorname{bs}\xi_1 \to \operatorname{bs}\xi_2$ are homotopy equivalences. Then φ is a bundle homotopy equivalence.*

Proof. Set $f = \operatorname{bs}\varphi$. By 1.9(ii), φ can be decomposed as

$$\xi_1 \xrightarrow{\mathfrak{F}_\varphi} f^*\xi_2 \xrightarrow{\mathfrak{I}_{f,\xi_2}} \xi_2.$$

By 1.22(ii), $\mathfrak{I}_{f,\xi_2}$ is a bundle homotopy equivalence, and so $\operatorname{ts}\mathfrak{F}_\varphi$ is a fiberwise homotopy equivalence. Hence, by 1.27, $\mathfrak{F}_\varphi$ is a bundle homotopy equivalence, and thus $\varphi = \mathfrak{I}_{f,\xi_2}\mathfrak{F}_\varphi$ is a bundle homotopy equivalence. □

1.29. Theorem. *Let $\{U_i\}$ be a numerable covering of a space B such that every inclusion $U_i \subset B$ is inessential. Then the following hold.*

(i) *Let $\varphi : \xi \to \eta$ be a morphism of Dold fibrations over B. If $\varphi_b : p_\xi^{-1}(b) \to p_\eta^{-1}(b)$ is a homotopy equivalence for every $b \in B$ then φ is a homotopy equivalence over B.*

(ii) *Let $\varphi : (\xi, s_\xi) \to (\eta, s_\eta)$ be a sectioned morphism of sectioned fibrations over B. If $\varphi_b : (p_\xi^{-1}(b), s_\xi(b)) \to (p_\eta^{-1}(b), s_\eta(b))$ is a pointed homotopy equivalence for every $b \in B$ then φ is a sectioned equivalence over B.*

Proof. The proof of (i) can be found in Dold [2] or tom Dieck–Kamps–Puppe [1], the proof of (ii) can be found in May [2], §§2, 5. □

1.30. Corollary. (i) *Let B be a CW-space, and let $\varphi : \xi \to \eta$ be a morphism of Dold fibrations over B. If $\varphi_b : p_\xi^{-1}(b) \to p_\eta^{-1}(b)$ is a homotopy equivalence for every $b \in B$ then φ is a homotopy equivalence over B.*

(ii) *Let B be a CW-space, and let $\varphi : (\xi, s_\xi) \to (\eta, s_\eta)$ be a sectioned morphism of sectioned fibrations over B. If $\varphi_b : (p_\xi^{-1}(b), s_\xi(b)) \to (p_\eta^{-1}(b), s_\eta(b))$ is a pointed homotopy equivalence for every $b \in B$ then φ is a sectioned homotopy equivalence over B.*

(iii) *Let $\varphi : \xi \to \eta$ be a morphism of Dold fibrations. Suppose that $\operatorname{bs}\varphi : \operatorname{bs}\xi \to \operatorname{bs}\eta$ is a homotopy equivalence, that $\operatorname{bs}\xi$ and $\operatorname{bs}\eta$ are CW-spaces, and that $\varphi_b : p_\xi^{-1}(b) \to p_\eta^{-1}(\operatorname{bs}\varphi(b))$ is a homotopy equivalence for every $b \in B$. Then φ is a bundle homotopy equivalence.*

Proof. (i), (ii) These follows from 1.29 because B admits a covering of the type required by 1.29. Indeed, every CW-space is locally contractible, i.e., it admits a covering $\{U_i\}$ such that every U_i is contractible. Since B is paracompact, $\{U_i\}$ admits a locally finite refinement $\{V_j\}$. Now, $\{V_j\}$ is numerable, and the inclusion $V_j \subset B$ is inessential.

(iii) For simplicity, we put $f = \operatorname{bs}\varphi$. By 1.9(ii), φ can be decomposed as

$$\xi \xrightarrow{\mathfrak{F}_\varphi} f^*\eta \xrightarrow{\mathfrak{I}_{f,\eta}} \eta.$$

By 1.22(ii), $\mathfrak{I}_{f,\eta}$ is a bundle homotopy equivalence, while, by (i), $\mathfrak{F}_\varphi$ is an equivalence over $\operatorname{bs}\xi$. □

1.31. Remarks. (a) For every CW-space B, a simple direct construction of its covering required by 1.29 is given in Dold [2] and is credited to Puppe.

(b) Using 1.17 or 1.18, one can prove a sectioned analog of 1.22 and, based on this, deduce a sectioned analog of 1.30(iii). We do not need this, but the reader can do it as an exercise.

1.32. Lemma. *Let (ξ, s_ξ) and (ξ, s'_ξ) be two sectioned fibrations over a CW-space B. If $s_\xi \simeq_B s'_\xi : \operatorname{bs}\xi \to \operatorname{ts}\xi$ then (ξ, s_ξ) and (ξ, s'_ξ) are equivalent sectioned fibrations over B.*

Proof. Let $\xi = \{p : E \to B\}$, and let $H : B \times I \to E$ be a vertical homotopy $H : s_\xi \simeq_B s'_\xi$. We consider the maps

$$F : E \times I \to B, \quad F(e,t) = p(e)$$

and

$$g : E \cup s_\xi(B) \times I \to E, \quad g(e) = e, g(s_\xi(b), t) = H(b,t).$$

By 1.17(ii) and 1.7(i), there is a map $G : E \times I \to E$ which extends g and covers F. Now we have the map

$$G|E \times \{1\} : (E, s_\xi(B)) \to (E, s'_\xi(B))$$

which yields a bundle morphism $(\xi, s_\xi) \to (\xi, s'_\xi)$ over B. By 1.30(ii), this map is a sectioned equivalence over B. □

1.33. Definition. The *homotopy fiber* of a Dold fibration ξ over a connected base is the homotopy type of its fibers. By 1.21(ii), this is well–defined.

1.34. Proposition. *Let ξ, η be two Dold fibrations over connected bases. If ξ and η are bundle homotopy equivalent then they have the same homotopy fiber.* □

1.35. Proposition–Definition–Construction. *For every bundle $\xi = \{p : E \to B\}$, there exists a morphism $\varphi : \xi \to \overline{\xi}$ over B such that $\overline{\xi}$ is a fibration* *and* ts φ *is a homotopy equivalence. In other words, there exists a commutative diagram*

$$\begin{array}{ccc} E & \xrightarrow{h} & \overline{E} \\ {\scriptstyle p}\downarrow & & \downarrow{\scriptstyle \overline{p}} \\ B & = & B \end{array}$$

where $\overline{\xi} = \{\overline{p} : \overline{E} \to B\}$ *is a fibration and* $h :=$ ts φ *is a homotopy equivalence. Every such fibration* $\overline{\xi}$ *is called a* fibrational substitute *of the bundle* ξ *(or the map* $p : E \to B$*).*

Proof. Following Serre [1], we set

$$\overline{E} = \{(e, \omega) | e \in E, \omega : [0, 1] \to B, \omega(0) = p(e)\}$$

and define $\overline{p} : \overline{E} \to B, \overline{p}(e, \omega) = \omega(1)$. We define $h : E \to \overline{E}$ by setting $h(e) := (e, \omega_e)$, where $\omega_e(t) = p(e)$ for every $t \in I$. It is easy to see that $\overline{p} : \overline{E} \to B$ is a fibration and h is a homotopy equivalence, see e.g. Fuks–Rokhlin [1]. □

1.36. Proposition. (i) *If two maps* $p_i : E_i \to B_i, i = 1, 2$, *are homotopy equivalent then their fibrational substitutes are bundle homotopy equivalent. Moreover, every two fibrational substitutes of a map* $p : E \to B$ *are equivalent over* B.

(ii) *Let* $\xi = \{p : E \to B\}$ *be a bundle, let* $\overline{\xi}$ *be a fibrational substitute of* ξ, *and let* $u : A \to B$ *be a map. Then* $u^*\overline{\xi}$ *is a fibrational substitute of* $u^*\xi$.

Proof. (i) Let $\overline{\xi}_i = \{\overline{p}_i : \overline{E}_i \to B_i\}, i = 1, 2$, be a fibrational substitute of ξ_i. Then there is a diagram

$$\begin{array}{ccc} \overline{E}_1 & \xrightarrow{v} & \overline{E}_2 \\ {\scriptstyle \overline{p}_1}\downarrow & & \downarrow{\scriptstyle \overline{p}_2} \\ B_1 & \xrightarrow{u} & B_2 \end{array}$$

which commutes up to homotopy and where u, v are homotopy equivalences. Since $\overline{\xi}_2$ is a fibration, we can replace v by a homotopic map $\widehat{v} : \overline{E}_1 \to \overline{E}_2$ such that the diagram will commute strictly. Now, by 1.28, $\widehat{v}$ is a bundle homotopy equivalence.

Now, if we have two fibrational substitutes of a map $f : E \to B$ then there is a diagram as above with $u = 1_B$. Thus, by 1.27, v is an equivalence over B.

(ii) By (i) and 1.9(v), it suffices to prove the assertion for some particular fibrational substitute $\overline{\xi}$. We choose $\overline{\xi}$ as in the proof of 1.35 and use the same notation. Let $u^*\xi = \{g : Y \to A\}$, where $Y = \{(a, e) | u(a) = p(e)\}$ and $g(a, e) = a$. Furthermore, $u^*\overline{\xi} = \{q : V \to A\}$ where

$$V = \{(a, e, \omega) \mid u(a) = \omega(1), p(e) = \omega(0)\},\ a \in A, e \in E, \omega \in B^I$$

and $q(a, e, \omega) = a$. We consider the map $j : Y \to V, j(a, e) = (a, e, \omega_e)$ and prove that j is a homotopy equivalence. Indeed, we define $k : V \to Y, k(a, e, \omega) = (a, e)$ and

$$G : V \times I \to V,\ G((a, e, \omega), s) = (a, e, \omega_s) \text{ where } \omega_s(t) = \omega(st), s \in I.$$

Clearly, $kj = 1_Y$. Furthermore, $G|V \times \{1\} = 1_V$ and $G|V \times \{0\} = jk$, i.e., $jk \simeq 1_V$. □

1.37. Definition. The *homotopy fiber of a bundle* (or, if you prefer, of a map) $p : E \to B$ over a connected base B is the homotopy fiber of its fibrational substitute. By 1.36(i) and 1.34, this is well defined.

Given a space F, we say "F is the homotopy fiber of p" meaning that the homotopy type of F is the homotopy fiber of p.

1.38. Proposition. *Let B be a connected CW-space, and let $p : E \to B$ be a fibration such that every (or equivalently, some single) fiber of p has the homotopy type of a CW-space. Then E has the homotopy type of a CW-space.*

Proof. See e.g. Fritsch–Piccinini [1], Appendix. □

1.39. Examples (Serre [1]). (a) Let X be a connected space, and let $i : \{x_0\} \to X$ be the inclusion of a point. What is the homotopy fiber of i? Using the Serre construction as in 1.35, we get a fibration $PX \to X$ with contractible PX, and its fiber over x_0 is just $\Omega(X, x_0)$. Thus, the homotopy fiber of i is ΩX.

(b) Similarly to II.4.14, we define an $(m-1)$-*connective covering* of a space X to be a map $q = q_m : Y \to X$ such that $\pi_i(Y) = 0$ for $i < m$ and $q_* : \pi_i(Y) \to \pi_i(X)$ is an isomorphism for $i \geq m$. As in II.4.14, we denote Y by $X|m$, and call every such Y a *killing space*. For example, the universal covering $\widetilde{X} \to X$ of a connected space X is its 1-connective covering.

How to construct an m-connective covering for an arbitrary m? Consider a space X such that $\pi_i(X) = 0$ for $i < n$, where $n > 0$, and set $\pi = \pi_n(X)$. Then there is a map $f : X \to K(\pi, n)$ such that $f_* : \pi_n(X) \to \pi_n(K(\pi, n))$ is an isomorphism. (In fact, f is given by the element $1_\pi \in \operatorname{Hom}(\pi, \pi) = H^n(X, \pi)$.) Let $p : E \to K(\pi, n)$ be a fibrational substitute of f, and let $i : F \to E$ be the inclusion of a fiber. Then i is an n-connective covering of E, and so $q_n : F \xrightarrow{i} E \simeq X$ is an n-connective covering of X, i.e., $F = X|(n+1)$. Similarly, we can consider an $(n+1)$-connective covering $F_1 \to X|(n+1)$, and the composition

$$F_1 \to X|(n+1) \xrightarrow{q_{n+1}} X$$

is an $(n+1)$-connective covering of X, i.e., $F_1 = X|(n+2)$. And so on.

Summing up, in this way we get a tower of fibrations

$$\cdots \to X|(n+k) \xrightarrow{q_{n+k}} X|(n+k-1) \to \cdots \to X|(n+1) \xrightarrow{q_{n+1}} X$$

where the (homotopy) fiber of q_{n+k} is $K(\pi_{n+k}(X), n+k)$. Furthermore, the composition $q_{n+1} \circ \cdots \circ q_{n+k} : X|(n+k) \to X$ is an $(n+k-1)$-connective covering of X. This tower is natural with respect to X, and the obvious analog of II.4.16 holds, see e.g. Hu [1], Whitehead [1].

Serre [1] suggested the following program to compute homotopy groups. Suppose that we are able to compute homology of killing spaces (e.g., using the Leray–Serre spectral sequence). Then, because of the Hurewicz Theorem,

$$\begin{aligned} \pi_n(X) &= H_n(X), \\ \pi_{n+1}(X) &= \pi_{n+1}(X|(n+1)) = H_{n+1}(X|(n+1)), \\ \pi_{n+2}(X) &= \pi_{n+2}(X|(n+2)) = H_{n+2}(X|(n+2)), \\ &\cdots \end{aligned}$$

Unfortunately, really we can't proceed to the very end, since the computation of $H_*(X|k)$ becomes more and more complicated as k increases, but, for example, in this way Serre proved the finiteness of $\pi_i(S^{2n+1}), i \neq 2n+1$.

1.40. Proposition. *Every Dold fibration ξ over B is equivalent over B to its fibrational substitute $\overline{\xi}$. In particular, every Dold fibration over B is equivalent over B to a fibration.*

Proof. Let $\varphi : \xi \to \overline{\xi}$ be as in 1.35. Then, because of 1.27, φ is an equivalence over B. □

1.41. Proposition. (i) *If ξ is a quasi-fibration over a connected base B, then each of its fibers is CW-equivalent to the homotopy fiber of ξ. In particular, every two fibers of ξ are CW-equivalent.*

(ii) (Dold–Thom [1]) *Consider the commutative diagram*

$$\begin{array}{ccc} E' & \xrightarrow{\subset} & E \\ {\scriptstyle p'}\downarrow & & \downarrow{\scriptstyle p} \\ B' & \xrightarrow{\subset} & B \end{array}$$

where p' is a quasi-fibration and $p' = p|E'$. Assume that there exist deformations $R_t : E \to E, r_t : B \to B, t \in I$, with the following properties: $pR_t = r_t p, R_0 = 1_E, R_1(E) \subset E', r_0 = 1_B, r_1(B) \subset B'$. Furthermore, suppose that $(R_1)_* : \pi_i(F_x) \to \pi_i(F_{r_1(x)})$ is an isomorphism for every $x \in B$ and every i. Then p is a quasi-fibration.

(iii) *Consider the commutative diagram*

$$\begin{array}{ccccccc} \cdots & \longrightarrow & E_n & \xrightarrow{g_n} & E_{n+1} & \longrightarrow & \cdots \\ & & {\scriptstyle p_n}\downarrow & & \downarrow{\scriptstyle p_{n+1}} & & \\ \cdots & \longrightarrow & B_n & \xrightarrow{f_n} & B_{n+1} & \longrightarrow & \cdots. \end{array}$$

Assume that all maps g_n, f_n are inclusions. Set $E = \cup E_n, B = \cup B_n$ and define $p : E \to B$, $p|E_n = p_n$. Assume that every compact subset of B is contained in some B_n. Then p is a quasi-fibration provided that every p_n is a quasi-fibration.

(iv) *Given the diagram in* (iii), *assume that g_n, f_n are arbitrary maps (not necessary inclusions). Suppose that every space B_k is connected. Given a CW-space F, suppose that every vertical map in the diagram is a quasi-fibration such that each of its fibers is homotopy equivalent to F. Moreover, suppose that $g_k|F_x : F_x \to F_{g_k(x)}$ is a homotopy equivalence for every k and every $x \in B_k$. Let E, resp. B, be the telescope of the top, resp. bottom, sequence. Define $p : E \to B$ to be the telescope of the maps p_n. Then p is a quasi-fibration. Furthermore, every fiber of p is CW-equivalent to F.*

Proof. (i) Consider $\varphi : \xi \to \overline{\xi}$ as in 1.35, and set $\overline{F}_x = (\overline{p})^{-1}(x)$. Let $\varphi_x : F_x \to \overline{F}_x, \varphi_x(f) = (\text{ts}\,\varphi)(f)$, be the induced map of fibers. Choose $a_0 \in F_x$ and set $b_0 = \varphi_x(a_0)$. Consider the following commutative diagram of exact sequences:

$$\begin{array}{ccccccccc}
\cdots \to & \pi_{k+1}(B,x) & \to & \pi_k(F_x,a_0) & \to & \pi_k(E,a_0) & \to & \pi_k(B,x) & \to \cdots \\
 & \downarrow{\scriptstyle 1} & & \downarrow{\scriptstyle (\varphi_x)_*} & & \cong\downarrow{\scriptstyle \varphi_*} & & \downarrow{\scriptstyle 1} & \\
\cdots \to & \pi_{k+1}(B,x) & \to & \pi_k(\overline{F}_x,b_0) & \to & \pi_k(\overline{E},b_0) & \to & \pi_k(B,x) & \to \cdots .
\end{array}$$

By the Five Lemma, $(\varphi_x)_*$ is an isomorphism for every $k \geq 1$ and a bijection for $k = 0$.

(ii) Consider the deformation retractions $a := R_1 : E \to E'$ and $b := r_1 : B \to B'$. They induces certain isomorphisms $a_* : \pi_i(E,y) \to \pi_i(E', a(y))$, $b_* : \pi_i(B,x) \to \pi_i(B', b(x))$. Now, a yields a map of pointed pairs $(E, F_x, y) \to (E', F_{b(x)}, a(y))$, and so, by the Five Lemma, we get isomorphisms $a_* : \pi_i(E, F_x, y) \to \pi_i(E', F_{b(x)}, a(y))$. Hence, in the diagram

$$\begin{array}{ccc}
\pi_*(E,F_x,y) & \xrightarrow{a_*} & \pi_*(E',F_{b(x)},a(y)) \\
\downarrow{\scriptstyle p_*} & & \downarrow{\scriptstyle p'_*} \\
\pi_*(B,x) & \xrightarrow{b_*} & \pi_*(B',b(x))
\end{array}$$

the horizontal arrows are isomorphisms, and the right arrow is an isomorphism because p' is a fibration. Thus, the left arrow is an isomorphism.

(iii) Given $f : (S^n, *) \to (B, x)$, we conclude that $f(S^n, *) \subset (B_m, x)$ for some m, and so $f \simeq pg$ for some $g : (D^n, S^{n-1}) \to (E_m, F_x)$ (because p_m is a quasi-fibration). Thus, p_* is epic. Furthermore, let $g : (D^n, S^{n-1}) \to (E, F_x)$ be such that $p_*[g] = 0 \in \pi_n(B, x)$. Then the map pg extends to $h : (D^{n+1}, 0) \to (B, x)$. Now, $[g] = 0 \in \pi_n(E, F_x)$ since $h(D^{n+1}, 0) \subset (B_m, x)$ for some m.

(iv) For every k there is a commutative diagram

$$\begin{array}{ccc} E_k & \xrightarrow{\subset} & Mg_k \\ {\scriptstyle p_k}\downarrow & & \downarrow{\scriptstyle M(p_k)} \\ B_k & \xrightarrow{\subset} & Mf_k, \end{array}$$

where M denotes the ordinary mapping cylinder. By (ii), the right-hand map is a quasi-fibration. Thus, by (iii), p is a quasi-fibration. The last assertion follows from (i). □

1.42. Definition. (a) Let F be a topological space (in $\mathscr{W}$, as usual). An *F-fibration*, resp. a *Dold F-fibration*, is a fibration, resp. a Dold fibration, such that all its fibers are homotopy equivalent to F. A *morphism* $\varphi = (g, f) : \xi \to \eta$ of (Dold) F-fibrations, or simply an *F-morphism*, is a bundle morphism such that

$$g|F_x : F_x \to F_{f(x)}$$

is a homotopy equivalence for every $x \in \operatorname{bs}\xi$. An *equivalence* of (Dold) F-fibrations over B is just an equivalence over B of them.

(b) Let $(F, *)$ be a well-pointed space. Define an *$(F, *)$-fibration* to be a sectioned F-fibration (ξ, s_ξ) such that $(F_x, s(x))$ is pointed homotopy equivalent to $(F, *)$ for every $x \in \operatorname{bs}\xi$. A *morphism* $\varphi = (g, f) : (\xi, s_\xi) \to (\eta, s_\eta)$ of $(F, *)$-fibrations, or simply an *$(F, *)$-morphism*, is a sectioned morphism such that φ is an F-morphism and

$$g|F_x : (F_x, s_\xi(x)) \to (F_{f(x)}, s_\eta(f(x)))$$

is a pointed homotopy equivalence for every $x \in \operatorname{bs}\xi$. An equivalence of $(F, *)$-fibrations over B is just a sectioned homotopy equivalence over B.

Sometimes we shall say "a fibration $F \to E \to B$" instead of "an F-fibration $E \to B$". Recall that a (Dold) fibration over a connected base is a (Dold) F-fibration for some F.

Given a space X, we define $\pi = \pi_X : X^I \to X$ by setting $\pi(\omega) = \omega(0)$.

1.43. Proposition. (i) *Given a bundle $p : E \to B$, consider the pull-back diagram*

$$\begin{array}{ccc} Y & \xrightarrow{\tau} & E \\ {\scriptstyle q}\downarrow & & \downarrow{\scriptstyle p} \\ B^I & \xrightarrow{\pi_B} & B. \end{array}$$

The bundle $p : E \to B$ is a fibration iff there exists a map $h : Y \to E^I$ such that $\pi_E h = \tau$ and $p^I h = q$, where $p^I : E^I \to B^I$, $p^I(\omega) = p\omega$.

(ii) *The product of two (Dold) fibrations is a (Dold) fibration.*

(iii) *Let $\varphi : \xi_1 \to \xi_2$ be a morphism of fibrations over B. Then* $\operatorname{Cyl}(\varphi)$ *is a fibration over B.*

(iv) *The double mapping cylinder of two fibrations over B is a fibration over B.*

(v) *If ξ is an F-fibration and η is a G-fibration then $\xi * \eta$ is an $F * G$-fibration.*

(vi) *If ξ is an $(F,*)$-fibration and η is a $(G,*)$-fibration then $\xi \wedge^h \eta$ is an $(F \wedge^h G, *)$-fibration.*

Proof. (i) Suppose that there exists h as required. Consider maps $F : X \times I \to B$ and $g : X \to E$ with $F(x,0) = pg(x)$. We must construct a p-lifting G of F with $G(x,0) = g(x)$. Define $\overline{F} : X \to B^I$ by setting $\overline{F}(x)(t) = F(x,t)$. Since $\pi_B \overline{F} = pg$, there is a map $k : X \to Y$ such that $qk = \overline{F}$ and $\tau k = g$. Now, the map $hk : X \to E^I$ yields the desired map $G : X \times I \to E$, $G(x,t) = hk(x)(t)$.

Conversely, if p is a fibration, we define the map $F : Y \times I \to B, F(y,t) = q(y)(t)$. Then $p\tau(y) = F(y,0)$. Since p is a fibration, there exists a p-lifting $G : Y \times I \to E$ of F. Now we define the required $h : Y \to E^I$, $h(y)(t) = G(y,t)$.

(ii) This is obvious.

(iii) (cf. Clapp–Puppe [1].) Let $\xi_i = \{p_i : E_i \to B\}, i = 1,2$, and let $\operatorname{Cyl}\varphi = \{p : E \to B\}$. Let $h_i : Y_i \to E_i^I, i = 1,2$, be the maps as in (i). We construct $h : Y := E \times_B B^I \to E^I$ as follows. Firstly, let $(e,t) \in E$, $e \in E_1, t \in I$, and let $\omega \in B^I$. We define

$$h(e,t,\omega)(s) = \begin{cases} (h_1(e,\omega)(s), t - s + st) & \text{if } t \geq 1/2, \\ (h_1(e,\omega)(s), t - s/2) & \text{if } t \leq 1/2 \text{ and } s \leq 2t, \\ h_2(\operatorname{ts}\varphi(h_1(e,\omega)(2t)), \omega_{2t})(s - 2t) & \text{if } t \leq 1/2 \text{ and } s \geq 2t, \end{cases}$$

where $\omega_{2t}(r) = \omega(\min\{2t + r, 1\})$. Finally, we define $h(e,\omega) = h_2(e,\omega)$ for $(e,\omega) \in Y_2$. Thus, by (i), $\operatorname{Cyl}\varphi$ is a fibration.

(iv) This follows from (iii). Let $\xi = \operatorname{DCyl}(\xi_2 \xleftarrow{\varphi_1} \xi_1 \xrightarrow{\varphi_2} \xi_3)$. We set $\xi_4 := \operatorname{Cyl}(\varphi_2)$, and let $\xi_i = \{p_i : E_i \to B\}$. Let $h_i : Y_i \to E_i^I, i = 1,2,3$, be as (i), and let $h_4 : Y_4 \to E_4^I$ be as in (iii). Let $\xi = \{p : E \to B\}$. We construct $h : Y \to E^I$ as follows. Firstly, let $e \in E_1, t \in [0,2], \omega \in B^I$. Set

$$h(e,t,\omega) = \begin{cases} h_4(e,t,\omega) & \text{if } 0 \leq t \leq 1, \\ h_4(e,2-t,\omega) & \text{if } 1 \leq t \leq 2. \end{cases}$$

Finally, we set $h(e,\omega) = h_i(e,\omega)$ if $(e,\omega) \in Y_i, i = 2,3$.

(v) This follows from (iv) because the join is a double mapping cylinder.

(vi) By (iv), $(\xi \times 1_{\operatorname{bs}\eta}) \vee^h (1_{\operatorname{bs}\xi} \times \eta)$ is a fibration, and thus, again by (iv), $\xi \wedge^h \eta$ is. □

Let $t(X) = t_F(X)$ (resp. $w(X) = w_F(X)$) be the class of all classes of equivalent over X F-fibrations (resp. Dold F-fibrations) over X. We regard t and w as functors on $\mathscr{HC}$ (namely, $t(f) := \mathfrak{I}_{\xi,f} : f^*(\xi) \mapsto \xi$ for $f : Y \to X$, etc.), and we want to prove the representability of the functors t and w.

We want to apply the Brown Representability Theorem III.3.25, but we can't do it directly. Recall that III.3.25 deals with functors $\mathscr{H}\mathscr{C}^{\bullet}_{\text{con}} \to \mathscr{E}ns^{\bullet}$. So, preliminarily, we should treat t and w as functors on $\mathscr{H}\mathscr{C}^{\bullet}_{\text{con}}$ and prove that t, w are set-valued functors.

We need some brief preliminaries about set theory. Here we follow Kelley [1], Addendum. We use the notion of a class, which is primitive and a wider notion than a set. Furthermore, there are two primitive constants (besides logical constants): $\in$ (belongs to) and $\{ \dots \mid \dots \}$ (the class of all ...| such that ...). The operations $\cup$ and $\cap$ and the relation $\subset$ are defined in the usual way. A *set* is defined to be a class which belongs to some other class, i.e., A is a set iff, for some B, $A \in B$. (An example of a class which is not a set is the class of all sets. One can prove this, using the Hilbert–Bernays–von Neumann–Gödel axioms following Kelley [1], Addendum. Informally, if it were a set, then one would have the well-known Russell Paradox, and in fact the classes were introduced in order to avoid paradoxes like this one.) The *singleton* $\{X\}$ of a set X is defined to be a one-element class containing as an element only the set X. An *ordered pair* (X, Y) of sets is a class $\{\{X\}, \{X\} \cup \{Y\}\}$. The *Cartesian product* $X \times Y$ of classes X, Y is defined to be a class of ordered pairs $\{ (x, y) \mid x \in X, y \in Y \}$. Given two classes X, Y, a *function* $f : X \to Y$ is a class f in $X \times Y$ with the following property: if $(x, y) \in f$ and $(x, z) \in f$ then $y = z$; this y is denoted by $f(x)$. The class of all functions $X \to Y$ is denoted by $\operatorname{Fun}(X, Y)$. A *relation* on a class X is a subclass of $X \times X$. The notions of equivalence relation is defined in the usual way.

1.44. Theorem. (i) *A class contained in a set is a set.*

(ii) *If X and Y are sets, then $X \times Y$ is a set.*

(iii) *If X and Y are sets, then* $\operatorname{Fun}(X, Y)$ *is a set.*

(iv) *If Λ is a set and $\{A_\lambda\}_{\lambda \in \Lambda}$ is a set of sets, then $\cup_\lambda A_\lambda$ is a set.*

(v) *If R is an equivalence relation on a set X then the class X/R is a set.*

Proof. (i)–(iv) See Kelley [1], Addendum, Theorems 33, 74, 77 and Axiom VI respectively.

(v) Given $x \in X$, we set $[x] := \{y \in X \mid (x, y) \in R\}$. Let $\{[x]\}$ be the singleton of $[x]$. By (iv), $\bigcup_{x \in X} \{[x]\}$ is a set. But this set is just X/R. □

Define a *rooted* Dold F-fibration ξ over a pointed space (X, x_0) to be a Dold F-fibration with a fixed homotopy equivalence (called a *root*) $i = i_\xi : F \to p_\xi^{-1}(x_0)$ (cf Milnor [7], §7). A *rooted equivalence* of rooted Dold F-fibrations ξ, η is an equivalence $\varphi : \xi \to \eta$ over X such that $\varphi i_\xi \simeq i_\eta$. Let $r(X)$ denote the class of all rooted equivalence classes of rooted Dold F-fibrations over the pointed space X.

1.45. Lemma. *Let (X, x_0) be a pointed space, and let $\{U, V\}$ be a numerable covering of X such that $x_0 \in U \cap V$. Let ξ be a rooted Dold fibration over U,* *let η be a rooted Dold fibration over V, and let $\varphi : \xi|U \cap V \to \eta|U \cap V$ be a rooted equivalence over $U \cap V$. Then there exist a rooted Dold fibration ζ over X and rooted equivalences $a : \zeta|U \to \xi$, $b : \zeta|V \to \eta$ such that the following diagram commutes up to homotopy over $U \cap V$:*

$$\begin{array}{ccc} \zeta|U \cap V & \xrightarrow{a|U\cap V} & \xi|U \cap V \\ \| & & \downarrow{\scriptstyle \varphi} \\ \zeta|U \cap V & \xrightarrow{b|U\cap V} & \eta|U \cap V. \end{array}$$

Furthermore, $a|U \setminus V : \zeta|U \setminus V \to \xi|U \setminus V$ and $b|U \setminus V : \zeta|U \setminus V \to \eta|U \setminus V$ are fiberwise homeomorphisms over the bases.

Proof. For simplicity, denote $U \cap V$ by W. Consider the map

$$\xi|W \sqcup \eta|W \xrightarrow{\psi} \eta|W, \ \psi|(\xi|W) = \varphi, \psi|(\eta|W) = 1_\eta.$$

Set $\omega = \operatorname{Cyl}\psi$, $\operatorname{bs}\omega = W$. We have the standard mapping cylinder inclusions

$$r : \xi|W \to \omega, \ s : \eta|W \to \omega$$

which both are homotopy equivalences over W (this is clear for s, and one can prove this for r following Fox [1], cf. also Kamps [1], 8.2). Moreover, r and s are cofibrations over W. Set $\xi' := \xi \cup_r \omega$, $\eta' := \eta \cup_s \omega$. Then the inclusions $i : \xi \to \xi'$, $j : \eta \to \eta'$ are equivalences over the bases because r and s are equivalences over W. (Note that $\operatorname{bs}(\xi') = U$, $\operatorname{bs}(\eta') = V$.) So, by 1.15, ξ' and η' are Dold fibrations. We set

(1.46) $$\zeta = \zeta_\varphi := \xi' \cup_{1_\omega} \eta', \operatorname{bs}\zeta = X.$$

By 1.25, ζ is a Dold fibration because ξ' and η' are.

Let $a : \zeta|U \to \xi$ (resp. $b : \zeta|V \to \eta$) be an equivalence over U (resp. over V) inverse to the one $\xi \to \zeta|U$ (resp. to $\eta \to \zeta|V$). Since all the diagrams

$$\begin{array}{ccc} \xi' & \xleftarrow{\simeq} & \omega \\ {\scriptstyle i}\uparrow & & \uparrow{\scriptstyle r} \\ \xi & \xleftarrow{\supset} & \xi|W \end{array} \qquad \begin{array}{ccc} \omega & = & \omega \\ {\scriptstyle r}\uparrow & & \uparrow{\scriptstyle s} \\ \xi|W & \xrightarrow{\varphi} & \eta|W \end{array} \qquad \begin{array}{ccc} \omega & \xrightarrow{\simeq} & \eta' \\ {\scriptstyle s}\uparrow & & \uparrow{\scriptstyle j} \\ \eta|W & \xrightarrow{\subset} & \eta \end{array}$$

are commutative up to homotopy over W, so is the diagram of the lemma. The last assertion follows from the construction of ζ. □

1.47. Lemma (the MV property for r). *Let $(X; A, B; x_0)$ be a pointed CW-triad. Let ξ be a rooted Dold fibration over A, let η be a rooted Dold fibration over B, and let $\varphi : \xi|A \cap B \to \eta|A \cap B$ be a rooted equivalence. Then there exist a rooted Dold fibration ζ over X and rooted equivalences $a : \zeta|A \to$*

ξ, $b : \zeta|B \to \eta$ such that the following diagram commutes up to homotopy over $A \cap B$:

$$\begin{array}{ccc} \zeta|A\cap B & \xrightarrow{a|A\cap B} & \xi|A\cap B \\ \Vert & & \downarrow{\scriptstyle\varphi} \\ \zeta|A\cap B & \xrightarrow{b|A\cap B} & \eta|A\cap B. \end{array} \tag{1.48}$$

Proof. Let $C = A \cap B$. Consider the double mapping cylinder

$$Y = A \times \{0\} \cup C \times [0,1] \cup B \times \{1\} \subset X \times [0,1]$$

with the base point $(x_0, 1/2)$. Put $U = A\times\{0\}\cup C\times[0,2/3)$; $V = C\times(1/3,1]\cup B \times \{1\}$. Consider the map $f : Y \to X$, $f(a,0) = a, f(b,1) = b, f(c,t) = c$, where $t \in [0,1]$. It is easy to see that f is a homotopy equivalence. Indeed, f has the form

$$A \times \{0\} \cup C \times [0,1] \cup B \times \{1\} \xrightarrow{\alpha} A \times [0,1/2] \cup B \times [1/2,1] \xrightarrow{\beta} A \cup B,$$

where α is the obvious inclusion and $\beta(a,t) = a$, $\beta(b,t) = b$. It is clear that α is the inclusion of a deformation retract (because the inclusions $C \subset A$ and $C \subset B$ are cofibrations) and that β is a deformation retraction.

We define $f_1 : U \to A, f_1(u) = f(u)$, and $f_2 : V \to B, f_2(v) = f(v)$. We set $\xi' := f_1^*\xi$, $\eta' := f_2^*\eta$, $\psi = \varphi \times 1_{[1/3,2/3]} : \xi'|(U \cap V) \to \eta'|(U \cap V)$. By 1.45, there exist a Dold F-fibration ζ' over Y and equivalences $a' : \zeta'|U \to \xi', b' : \zeta'|V \to \eta'$ such that $\psi a'|U \cap V = b'|U \cap V$. Now, set $\zeta = g^*\zeta'$ where $g : X \to Y$ is homotopy inverse to f. □

1.49. Lemma (the wedge property for r). *Let $\{(X_\lambda, x_\lambda)\}$ be a set of pointed CW-spaces. Suppose that $r(X_\lambda, x_\lambda)$ is a set for every λ. Then*

$$h : r(\vee_\lambda (X_\lambda, x_\lambda)) \to \prod_\lambda r(X_\lambda, x_\lambda),\ h(\xi) = \{\xi|X_\lambda\},$$

is a bijection. In particular, $r(\vee_\lambda(X_\lambda, x_\lambda))$ is a set.

Proof. Throughout the proof "equivalence" means "rooted equivalence" and "F-fibration" means "rooted Dold F-fibration". Firstly, some constructions. Let $(I_\lambda, 0)$ be a copy of the pointed space $(I, 0)$. Let $Y_\lambda := X_\lambda \vee I_\lambda$, let $y_\lambda \in Y_\lambda$ be the image of $1 \in I_\lambda$, and let $f_\lambda : (Y_\lambda, y_\lambda) \to (X_\lambda, x_\lambda)$ collapse I_λ; by I.3.26(iii) and I.3.29, f_λ is a pointed homotopy equivalence. Let g_λ be a homotopy equivalence which is pointed homotopy inverse to f_λ, and let $g := \vee_\lambda g_\lambda : \vee_\lambda(X_\lambda, x_\lambda) \xrightarrow{\simeq} (Y, y_0)$, where $(Y, y_0) := \vee_\lambda(Y_\lambda, y_\lambda)$.

Given a family of F-fibrations $\{\xi_\lambda\}$, $\mathrm{bs}(\xi_\lambda) = X_\lambda$, let F_λ be the fiber of ξ_λ over x_λ, and let $i_\lambda : F \to F_\lambda$ be the root. Set $\overline{\xi}_\lambda = f_\lambda^*\xi_\lambda$, $\mathrm{bs}(\overline{\xi}_\lambda) = Y_\lambda$. Let $\overline{F}_\lambda \simeq F$ be the fiber of $\overline{\xi}_\lambda$ over y_λ; then the root i_λ yields a root $h_\lambda : F \to \overline{F}_\lambda$

of $\bar{\xi}_\lambda$. Consider a bundle morphism $\varphi = \varphi_\lambda : F \to \xi_\lambda$ where $\operatorname{ts}\varphi$ has the form $F \xrightarrow{h_\lambda} \bar{F}_\lambda \subset \operatorname{ts}\xi_\lambda$, and set $\xi'_\lambda = \xi_\lambda \cup_\varphi F$. By 1.15, ξ'_λ is an F-fibration. The fiber of ξ'_λ over y_λ is the mapping cylinder $M(h_\lambda)$, and the inclusion $F = F \times \{0\} \to M(h_\lambda)$ is the root of ξ'_λ. Furthermore, we have the equivalence $\xi'_\lambda \simeq \bar{\xi}_\lambda \simeq f^*_\lambda \xi_\lambda$ over (Y_λ, y_λ).

Let Φ_λ denote the subspace $F \times \{0\}$ of $M(h_\lambda)$, and let $j_\lambda = 1_F : \Phi_\lambda \to F$. Let E be the push-out of the diagram

$$\begin{array}{ccc} F & & \\ \uparrow{\scriptstyle \langle j_\lambda \rangle} & & \\ \sqcup \Phi_\lambda & \xrightarrow{\sqcup k_\lambda} & \sqcup \operatorname{ts}(\xi_\lambda), \end{array}$$

where $k_\lambda : \Phi_\lambda \to \operatorname{ts}(\xi_\lambda)$ is the inclusion and $\langle j_\lambda \rangle | \Phi_\lambda = j_\lambda$. Then there exists a map $p : E \to Y$ such that the diagram (where p_λ is the projection in ξ_λ and $\langle p_\lambda \rangle | \xi_\lambda = p_\lambda$)

$$\begin{array}{ccc} \sqcup \operatorname{ts}(\xi_\lambda) & \longrightarrow & E \\ {\scriptstyle \langle p_\lambda \rangle}\downarrow & & \downarrow{\scriptstyle p} \\ Y & = & Y \end{array}$$

commutes. We set $\xi' = \{p : E \to Y\}$. Now, $\xi' | ([0, 2/3)_\lambda) = \xi'_\lambda | ([0, 2/3)_\lambda)$ and $\xi' | (X_\lambda \vee (1/3, 1]_\lambda) = \bar{\xi}_\lambda | (X_\lambda \vee (1/3, 1]_\lambda)$ are F-fibrations. So, by 1.25, ξ' is an F-fibration.

We prove that h is surjective. Consider a family $\{\xi_\lambda\}, \operatorname{bs}(\xi_\lambda) = X_\lambda$. Then h maps $g^*(\xi')$ to $\{\xi_\lambda\}$, because

$$(g^*\xi') | X_\lambda = g^*_\lambda(\xi' | X_\lambda) = g^*_\lambda(\xi'_\lambda) \simeq g^*_\lambda f^*_\lambda \xi_\lambda \simeq \xi_\lambda .$$

We prove that h is injective. Let ξ, η be two F-fibrations over $X := \vee X_\lambda$, and suppose that for every λ an equivalence $e_\lambda : \xi | X_\lambda \to \eta | X_\lambda$ over X_λ is given. We set $\xi_\lambda := \xi | X_\lambda, \eta_\lambda := \eta | X_\lambda$ and construct ξ' and η' as above. The equivalences $e_\lambda : \xi_\lambda \to \eta_\lambda$ yield equivalences

$$e'_\lambda : \xi'_\lambda \simeq f^*_\lambda \xi_\lambda \to f^*_\lambda \eta_\lambda \simeq \eta'_\lambda$$

over X_λ, and there is a morphism $e : \xi' \to \eta'$ over X such that $e | \eta_\lambda$ coincides with e'_λ. By 1.30(ii), e is an equivalence over X. Since $\xi'_\lambda \simeq f^*_\lambda \xi_\lambda$, we conclude that $\xi' \simeq f^*\xi$, i.e., $g^*\xi' \simeq \xi$. Now, $\xi \simeq g^*\xi' \simeq g^*\eta' \simeq \eta$. □

Now we prove that r is a set-valued (contravariant) functor. This means that for every $(X, *)$ there exists a set $\{\xi_\lambda\}$ such that every rooted Dold F-fibration over $(X, *)$ is equivalent to some ξ_λ and $\xi_\lambda \not\simeq \xi_\mu$ for $\lambda \neq \mu$. Every such set $\{\xi_\lambda\}$ is called a *representing set* for $r(X, *)$.

1.50. Lemma. (i) *Suppose that there is a set $\{\xi_\lambda\}$ of rooted Dold F-fibrations over $(X, *)$ such that every rooted Dold F-fibration over $(X, *)$ is equivalent over X_λ (as a rooted fibration) to some ξ_λ. Then $r(X, *)$ is a set.*

(ii) *If $f : X \to Y$ is a pointed homotopy equivalence and $r(Y,*)$ is a set, then $r(X,*)$ is a set.*

(iii) *$r(SX,*)$ is a set for every pointed CW-space X.*

(iv) *Let $f : X \to Y$ be a map of CW-spaces. If $r(Y,*)$ is a set, then $r(Cf,*) = r(Y \cup_f CX,*)$ is a set.*

Proof. Again, equivalence means "rooted equivalence over the base", and "F-fibration" means "rooted Dold F-fibration".

(i) This follows from 1.44(ii).

(ii) Let $\{\xi_\lambda\}$ be a representing set for $r(Y,*)$. Then $\{f^*(\xi_\lambda)\}$ is a representing set for $r(X,*)$. Indeed, let $g : Y \to X$ be homotopy inverse to f. If $f^*(\xi_\lambda) = f^*(\xi_\mu)$, then $g^*f^*(\xi_\lambda) = g^*f^*(\xi_\mu)$, and thus $\xi_\lambda = \xi_\mu$. Furthermore, given η over X, one has $\eta \simeq f^*g^*\eta$, but $g^*\eta \simeq \xi_\lambda$ for some λ.

(iii) Let X_-, resp. X_+ be the subspace of SX given by $X \times [0,2/3)$, resp. $X \times (1/3,1]$. Note that, for every F-fibration ξ over SX, the fibration $\xi|X_+$ is fiberwise homotopy trivial since the inclusion $X_+ \subset SX$ is inessential. Similarly for $\xi|X_-$. Let θ_-, resp. θ_+ be the product F-bundle over X_-, resp. X_+. Let $a_+ : \theta_+ \to \xi|X_+$, $a_- : \theta_- \to \xi|X_-$ be fixed equivalences over the bases. We set $X_0 := X_- \cap X_+$. Choose a morphism $\varphi : \theta_+|X_0 \to \theta_-|X_0$ over X_0 such that $a_-\varphi \simeq_{X_0} a_+|X_0$.

Consider the bundle $\theta_+|X_0 \sqcup_{X_0} \theta_-|X_0 := \{p : X_0 \times F \sqcup X_0 \times F \to X_0\}$ where $p|X_0 \times F$ is the projection on X_0. We define the morphism

$$\psi : \theta_+|X_0 \sqcup_{X_0} \theta_-|X_0 \to \theta_-|X_0,\ \psi \mid (\theta_+|X_0) = \varphi, \psi \mid (\theta_-|X_0) = 1$$

and set $\omega = \operatorname{Cyl}\psi$, $\operatorname{bs}\omega = X_0$. A homotopy $a_-\varphi \simeq a_+|X_0$ over X_0 yields a morphism $a_0 : \omega \to \xi|X_0$ over X_0, and, by 1.30(i), it is an equivalence. We consider the inclusions-equivalences

$$r_+ : \theta_+|X_0 = (\theta_+|X_0) \times \{1\} \to \omega,\ r_- : \theta_-|X_0 = (\theta_-|X_0) \times \{1\} \to \omega$$

and set $\zeta_+ = \theta_+ \cup_{r_+} \omega$, $\zeta_- = \theta_- \cup_{r_-} \omega$ and $\zeta = \zeta_+ \cup_{1_\omega} \zeta_-$. As in 1.45, one can prove that ζ is an F-fibration. Furthermore, by 1.30(i), the morphism $a_+ \cup a_0 \cup a_- : \zeta \to \xi$ is an equivalence over SX.

Note that ζ is completely determined by φ, i.e., $\zeta = \zeta_\varphi$. By 1.44(iii), all functions $X_0 \times F \to X_0 \times F$ form a set, and so, by 1.44(i), all maps $X_0 \times F \to X_0 \times F$ form a set. So, all F-fibrations ζ_φ form a set. We have proved that every F-fibration ξ over SX is equivalent to some F-fibration ζ_φ. So, by (i), $r(SX,*)$ is a set.

(iv) Let $j : Y \to Y \cup_f CX$ be the inclusion. Roughly speaking, we consider an "exact sequence of classes" $r(Y,*) \leftarrow r(Y \cup_f CX,*) \leftarrow r(SX,*)$, where the group $r(SX,*)$ acts on $r(Y \cup_f CX,*)$. So, orbits of the action are sets, and the class of orbits is contained in the set $r(Y,*)$, and thus it is a set, etc.

Consider the map $l : Y \cup_f CX \to (Y \cup_f CX) \vee SX$ which pinches $X \times \{1/2\}$, see the picture.

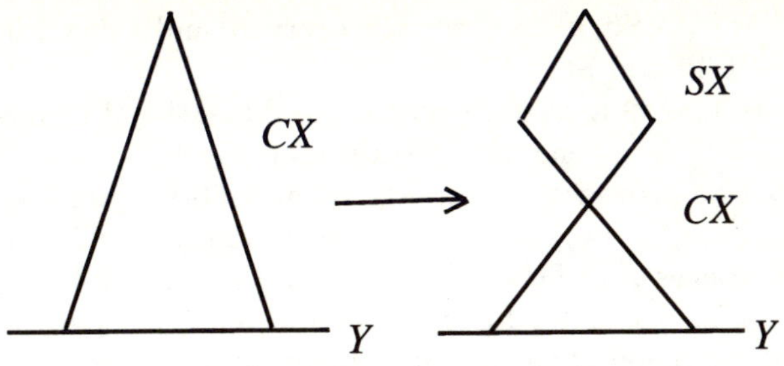

If you prefer formulae, we parametrize $(SX, *)$ as

$$X \times [1,2]/(X \times \{1,2\} \cup \{*\} \times I)$$

and define l by setting

$$l(y) = y,\ l(x,t) = \begin{cases} (x, 2t) \in CX & \text{if } 0 \leq t \leq 1/2, \\ (x, 2t) \in SX & \text{if } 1/2 \leq t \leq 1. \end{cases}$$

Given two F-fibrations μ over $Y \cup_f CX$ and λ over SX, there is (by 1.49) just one F-fibration (μ, λ) over $(Y \cup_f CX) \vee SX$ which restricts to μ and λ. Set $\lambda\mu := l^*(\mu, \lambda)$. Clearly, the equivalence class of $\lambda\mu$ depends only on the equivalence classes of λ, μ.

Consider two F-fibrations ξ, η over $Y \cup_f CX$ such that $\xi|Y \simeq \eta|Y$. Then there is an equivalence $c : \xi|Y \cup_f X \times (1/3, 1] \to \eta|Y \cup_f X \times (1/3, 1]$. Set $X_0 = X \times (1/3, 2/3)$ and define $h : X_0 \to X_0, h(x,t) = (x, 1-t)$. Let X_+ be the image of $X \times [0, 2/3)$ in $Y \cup_f CX$. Since X_+ is contractible in $Y \cup_f CX$, there are equivalences $a : \theta_{X_+} \to \xi|X_+$ and $b : \eta|X_+ \to \theta_{X_+}$. Let φ be the composition

$$\text{(1.51)} \qquad \theta_{X_0} \xrightarrow{a|X_0} \xi|X_0 \xrightarrow{c|X_0} \eta|X_0 \xrightarrow{b|X_0} \theta_{X_0} \xrightarrow{\psi} \theta_{X_0}$$

where $\psi = \mathfrak{I}_{h,\theta_{X_0}}$. One can check that $\xi \simeq \zeta_\varphi \eta$, where ζ_φ is as in (iii). In other words, $\xi \simeq \lambda\eta$ for some λ over SX.

Let Λ be a representing set for $r(SX, *)$, and let Γ be a representing set for $\operatorname{Im}\{j^* : r(Y \cup_f CX, *) \to r(Y, *)\}$. Given $\gamma \in \Gamma$, choose $\eta = \eta_\gamma$ over $Y \cup_f CX$ with $j^*\eta \simeq \gamma$. By 1.44(iv), $V_\gamma := \{\lambda\eta \mid \lambda \in \Lambda\} = \cup_\lambda \lambda\eta$ is a set, and so, by 1.44(iv), $\cup_\gamma V_\gamma$ is a set. Now, given ξ over $Y \cup_f CX$, one has $j^*\xi \simeq \gamma$ for some $\gamma \in \Gamma$, and so $\xi \simeq \lambda\eta_\gamma$ for some λ. Thus, ξ is equivalent to $\xi' \in \cup_\gamma V_\gamma$, and so, by (i), $r(Y \cup_f CX, *)$ is a set. □

1.52. Theorem. *$r(X, *)$ is a set for every connected pointed CW-space $(X, *)$. Furthermore, it can be turned into a pointed set naturally with respect to $(X, *)$. So, r is a functor $\mathscr{H}\mathscr{C}^\bullet \to \mathscr{E}ns^\bullet$.*

Proof. Firstly, we prove that $r(X,*)$ is a set.

Step 1. $r(S^n,*)$ is a set for every n. This follows from 1.50(iii).

Step 2. $r(\vee_{\lambda\in\Lambda} S^n_\lambda,*)$ is a set for every n and every index set Λ. This follows from 1.49 and Step 1.

Step 3. $r(X,*)$ is a set for every finite dimensional connected CW-space X. Indeed, $X^{(n)}$ is a cone of a certain map $f:\vee S^{n-1}\to X^{(n-1)}$. Now the assertion can be proved by induction, using 1.50(iv) and Step 2.

Step 4. $r(\vee_\lambda(X_\lambda,x_\lambda))$ is a set for every family of finite dimensional connected CW-spaces X_λ. This follows from Step 3 and 1.49.

Step 5. $r(X,*)$ is a set for every connected CW-space X. Let $(T,*)$ be the reduced telescope of the skeletal filtration of X, see I.3.23(d). Note that T_{ev}, as well as T_{od}, is the wedge of the finite dimensional summands, and hence by Step 4, $r(T_{\text{ev}},*)$ and $r(T_{\text{od}},*)$ are sets. Since $(X,*)$ is homotopy equivalent to $T_{\text{ev}}(X)\vee T_{\text{od}}(X)$, $r(X,*)$ is a set.

Now, we turn $r(X,*)$ into a pointed set, if we define the distinguished element of $r(X,*)$ to be the equivalence class of the trivial fibration. □

The restriction of r to $\mathscr{HC}^\bullet_{\text{con}}$ is also denoted by r.

Now III.3.25, 1.47, 1.49 and 1.52 imply

1.53. Corollary. *The functor* $r:\mathscr{HC}^\bullet_{\text{con}}\to\mathscr{E}ns^\bullet$ *is representable. In other words, there exists a pointed CW-space* $(B,*)$ *such that for every pointed connected space* $(X,*)$ *one has a natural equivalence*

$$r(X,*)=[(X,*),(B,*)]. \tag{1.54}$$ □

1.55. Theorem. $w_F(X)$ *is a set for every* X. *Furthermore, the space* B *from* 1.53 *represents the functor* $w=w_F:\mathscr{HC}\to\mathscr{E}ns$. *In other words, for every CW-space* X *we have a natural bijection* $w(X)=[X,B]$. *Finally, the forgetful transformation* $t_F\to w_F$ *is a natural equivalence, and so the functor* $t_F:\mathscr{HC}\to\mathscr{E}ns$ *is representable by the same space* B.

Proof. Because of 1.49, the standard map $S^1\to S^1\vee S^1$ (pinching S^0) turns $r(S^1,*)$ into a group. This group acts on $r(X,*)$, and now we describe this action.

Recall that there is the well-known $\pi_1(Y,*)$-action on $[(X,*),(Y,*)]$ for all spaces X,Y, see e.g. Hu [1], Spanier [2]. In particular, $\pi_1(B,*)$ acts on $[(X,*),(B,*)]$, and the orbit set of this action is just $[X,B]$. Consider the elements

$$\alpha\in\pi_1(S^1\vee X,*)=[(S^1,*),(S^1\vee X,*)] \text{ and } x\in[(X,*),(S^1\vee X,*)]$$

given by the inclusions of the direct summands $S^1\to S^1\vee X$ and $X\to S^1\vee X$. The $\pi_1(S^1\vee X,*)$-action on $[(X,*),(S^1\vee X,*)]$ gives us the element $\alpha x\in[(X,*),(S^1\vee X,*)]$. Consider

$$r(\alpha x):r(S^1,*)\times r(X,*)=r(S^1\vee X,*)\to r(X,*)$$

This is the action mentioned above. The orbit set of this action is $w(X)$. In particular, $w(X)$ is a set for every X. Furthermore, this action is compatible with the $\pi_1(B,*)$-action on $[(X,*),(B,*)]$ under the equivalence (1.54). Hence, we have a natural equivalence $w(X) = [X,B]$ for every connected X. Finally, if a CW-space X is a disjoint union of connected spaces, $X = \sqcup X_\alpha$, then $w(X) = \prod w(X_\alpha)$. So, $w(X) = [X,B]$ for every CW-space X.

Since every F-fibration is a Dold F-fibration, we have a natural forgetful transformation $t \to w$. Conversely, let $\bar{\xi}$ be a fibrational substitute of a Dold fibration ξ. By 1.36 and 1.40, the correspondence $\xi \mapsto \bar{\xi}$ is a well-defined natural transformation $w \to t$, which is inverse to the forgetful transformation $t \to w$. So, the forgetful transformation is a natural equivalence. □

1.56. Definition. (a) A *universal F-fibration* is an F-fibration

$$\gamma^F = \{p_F : E_F \to B_F\}$$

with the following properties:

(1) Every F-fibration over a CW-space X is equivalent to a fibration $f^*\gamma^F$ for some $f : X \to B_F$.

(2) Let $f, g : X \to B_F$ be two maps of a CW-space X. Then F-fibrations $f^*\gamma^F$ and $g^*\gamma^F$ are equivalent iff $f \simeq g$.

(b) The base B_F of a universal F-fibration is called a *classifying space* for F-fibrations. If an F-fibration ξ is equivalent to $f^*\gamma^F$ for some $f : \operatorname{bs}\xi \to B_F$, we say that f *classifies* ξ or that f is a *classifying map* for ξ.

(c) A *classifying morphism* for an F-fibration ξ is any F-morphism $\varphi : \xi \to \gamma^F$.

1.57. Theorem. *There exists a universal F-fibration γ^F. Furthermore, the base B_F of γ^F can be chosen to be a CW-space, and in this case B_F is uniquely defined up to homotopy equivalence. Moreover, γ^F can be chosen so that F_{b_0} is the space F for some point $b_0 \in B_F$.*

Proof. Considering B as in 1.55, we see that B is a CW-space. Furthermore, under the bijection $t_F(B) \cong [B,B]$ the element $1_B \in [B,B]$ corresponds to an equivalence class of a certain F-fibration over B. By 1.55, every fibration in this class is a universal F-fibration. So, we have proved the existence of a universal F-fibration over a CW-base. The homotopy uniqueness of B follows from the Yoneda Lemma I.1.5.

We prove the last assertion. Consider any universal F-fibration $\gamma = \{p : E \to B\}$ over a CW-base B. Choosing a point $b \in B$, consider $B' := (B,b) \vee (I,0)$, and let $b_0 \in B'$ be the image of $1 \in I$. Let $p : B' \xrightarrow{\simeq} B$ collapse I. Set $U := B \vee [0, 2/3)$, $V := (1/3, 1] \subset I \subset B'$, $W := U \cap V$, $\eta := (p|U)^*\gamma$, $\xi := \theta_V^F$. One has $\operatorname{ts}(\xi|W) = W \times F$, $\operatorname{ts}(\eta|W) = W \times F_b$. Choose $h : F \xrightarrow{\simeq} F_b$ and define $\varphi : \xi|W \to \eta|W, \varphi(w,f) = (w, h(f)), w \in W, f \in F$. Constructing ζ as in 1.47,

$\operatorname{bs}\zeta = B'$, we see that $\zeta \simeq p^*\gamma$, and the fiber of ζ over b_0 is F. Since p is a homotopy equivalence, ζ is a universal F-fibration. Now put $\gamma^F := \zeta$. $\square$

In future we always assume that γ^F is a fibration as in the last phrase of 1.57. In particular, B_F is a CW-space.

1.58. Proposition. *Let ξ be an F-fibration over a CW-base X.*

(i) *If $\varphi : \xi \to \gamma^F$ is a classifying morphism for ξ then $\mathfrak{F}_\varphi : \xi \to (\operatorname{bs}\varphi)^*\gamma^F$ is an equivalence over X.*

(ii) *If $\varphi : \xi \to \gamma^F$ is a classifying morphism for ξ then $\operatorname{bs}\varphi$ is a classifying map for ξ.*

(iii) *If $f : X \to B_F$ is a classifying map for ξ then there exists a classifying morphism $\varphi : \xi \to \gamma^F$ with $\operatorname{bs}\varphi = f$.*

Proof. (i) This follows from 1.30(i) since $\mathfrak{F}_\varphi$ induces a homotopy equivalence of fibers.

(ii) By (i), $(\operatorname{bs}\varphi)^*\gamma^F$ and ξ are equivalent over X.

(iii) Since $f^*\gamma^F$ and ξ are equivalent over X, we have the F-morphism

$$\varphi : \xi \to f^*\gamma^F \xrightarrow{\mathfrak{I}_{f,\gamma}} \gamma^F, \quad \operatorname{bs}\varphi = f. \qquad \square$$

We have proved that a classifying space for F-fibrations exists. However, sometimes one prefers to have a more or less explicit construction of B_F. To do this, it is useful to use classifying spaces for monoids.

1.59. Definition. A *topological monoid* is a triple (M, μ, e) where M is a topological space, $\mu : M \times M \to M$ is an associative multiplication and $e \in M$ is a two-sided unit of μ. A monoid is *well-pointed* if the inclusion $\{e\} \subset M$ of the unit e is a cofibration. A monoid is *grouplike* if μ induces a group structure on $\pi_0(M)$.

1.60. Definition. (a) Let M be a monoid with the unit e. A *principal M-bundle* is a pair (ξ, ν), where $\xi = \{p : E \to B\}$ is a bundle and $\nu : E \times M \to E$ is a map such that (below $y \in E, h, h' \in M$ and yh means $\nu(y, h)$, hh' means $\mu(h, h')$):

(1) $ye = y$, $(yh)h' = y(hh')$ for every y, h, h';

(2) $p(yh) = p(y)$ for every y;

(3) For every y the map $M \to F_{p(y)}, h \mapsto yh$ is a Whitehead equivalence.

(b) The map ν is called an *M-action* on ξ.

(c) A *principal M-(quasi-)fibration* is a principal M-bundle which is at the same time a (quasi-)fibration.

If M is grouplike, then condition (3) of (a) holds automatically. Moreover, if M is a topological group, then the maps $h \mapsto yh$ are homeomorphisms.

1.61. Proposition. *Let M be a topological monoid. For every principal M-quasi-fibration $\xi = \{p : E \to B\}$ there is a principal M-fibration $\bar{\xi} = \{\bar{p} : \bar{E} \to B\}$ which is a fibrational substitute of ξ.*

Proof. Consider the fibrational substitute $\overline{\xi} = \{\overline{p} : \overline{E} \to B\}$ of ξ as in 1.35, i.e., $\overline{E} := \{(e, \omega) | e \in E, \omega : [0,1] \to B, \omega(0) = p(e)\}$ and $\overline{p}(e, \omega) := \omega(1)$. We define an M-action $\overline{\nu}$ on $\overline{\xi}$ by setting $\overline{\nu}((e, \omega), h) := (\nu(e, h), \omega)$ where $h \in M$ and ν is the M-action on ξ. We leave it to the reader to check that $(\overline{\xi}, \overline{\nu})$ is a principal M-bundle (to prove 1.60(a,3), use the proof of 1.41(i)). □

1.62. Definition. A *classifying space* for a grouplike monoid M is any space B which is the base of a principal M-quasi-fibration $EM \to BM$ such that EM is an aspherical space. [10]

Let Δ^n be the standard n-simplex,

$$\Delta^n = \{(t_0, \dots, t_n) \in \mathbb{R}^{n+1} | 0 \le t_i \le 1, \sum t_i = 1\}.$$

For every $i = 0, 1, \dots, n$ we define

$$\delta_i : \Delta^{n-1} \to \Delta^n,\ \delta_i(t_0, \dots, t_{n-1}) = (t_0, \dots, t_{i-1}, 0, t_i, \dots, t_{n-1}),$$
$$\sigma_i : \Delta^{n+1} \to \Delta^n,\ \sigma_i(t_0, \dots, t_{n+1}) = (t_0, \dots, t_{i-1}, t_i + t_{i+1}, t_{i+2}, \dots, t_n).$$

Given a well-pointed topological monoid M, consider left and right M-spaces X and Y, respectively. Following May [2], [4], we set $B_n(Y, M, X) = Y \times M^n \times X$ and define the maps $\partial_i : B_n(Y, M, X) \to B_{n-1}(Y, M, X), i = 0, 1, \dots, n$ and $s_i : B_n(Y, M, X) \to B_{n+1}(Y, M, X), i = 0, 1, \dots, n$ as follows. Let $[y|m_1|\cdots|m_n|x]$ be the typical element of $B_n(Y, M, X)$. Put

$$\partial_i[y|m_1|\cdots|m_n|x] = \begin{cases} [ym_1|m_2|\cdots|m_n|x] & \text{if } i = 0, \\ [y|m_1|\cdots|m_i m_{i+1}|m_{i+2}|\cdots|m_n|x] & \text{if } 1 \le i \le n, \\ [y|m_1|\cdots|m_{n-1}|m_n x] & \text{if } i = n. \end{cases}$$

and $s_i[y|m_1|\cdots|m_n|x] = [y|m_1|\cdots|m_i|e|m_{i+1}|\cdots|m_n|x]$.

Consider the disjoint union $\overline{B} = \coprod_{n=0}^{\infty} B_n(Y, M, X) \times \Delta^n$ and define an equivalence relation $\sim$ on $\overline{B}$ to be that generated by:

$$(\partial_i u, v) \sim (u, \delta_i v) \text{ for } u \in B_n(Y, M, X), v \in \Delta^{n-1},$$
$$(s_i u, v) \sim (u, \sigma_i v) \text{ for } u \in B_n(Y, M, X), v \in \Delta^{n+1}.$$

We set

$$B_\bullet(Y, M, X) := \overline{B}/(\sim). \tag{1.63}$$

---

[10] Recall that a space Z is called aspherical if $\pi_i(Z, z_0) = 0$ for every $z_0 \in Z$ and every i. In particular, every contractible space is aspherical, and, for CW-spaces, the converse is also true.

The construction $B_\bullet(-,-,-)$ is natural: let $h: M \to M'$ be a monoid homomorphism, and let X (resp. Y) be a left (resp. right) M-space. Let $f: X \to X'$, $g: Y \to Y'$ be two maps such that the diagrams

$$\begin{array}{ccc} M \times X & \xrightarrow{h\times f} & M' \times X' \\ \downarrow & & \downarrow \\ X & \xrightarrow{f} & X' \end{array} \qquad \begin{array}{ccc} Y \times M & \xrightarrow{g\times h} & Y' \times M' \\ \downarrow & & \downarrow \\ Y & \xrightarrow{g} & Y' \end{array}$$

commute (here the vertical arrows are the actions). Then the maps

$$Y \times M^n \times X \xrightarrow{g\times h^n \times f} Y' \times (M')^n \times X'$$

induce a map $B_\bullet(g,h,f): B_\bullet(Y,M,X) \to B_\bullet(Y',M',X')$ with the usual functorial properties.

Given a map $\alpha: Y \times X \to Z$ such that $\alpha(ym,x) = \alpha(y,mx)$ for every $m \in M$, we define

$$b_n: B_n(Y,M,X) \times \Delta^n \xrightarrow{p_1} B_n(Y,M,X) \xrightarrow{a_n} Z$$

where $a_n[y|m_1|\cdots|m_n|x] = \alpha(ym_1\cdots m_n, x)$. The family $b_n, n = 0,1,\dots,$ yields a map $b: \sqcup B_n \to Z, b|B_n = b_n$ which, in turn, induces a well-defined quotient map

$$\varepsilon_\alpha : B_\bullet(Y,M,X) \to Z.$$

This construction is natural in the following sense. Let $f: X \to X', g: Y \to Y', h: M \to M'$ be as above. Given $\alpha': Y' \times X' \to Z'$ with $\alpha'(ym,x) = \alpha'(y,mx)$, let $k: Z \to Z'$ be a map such that the left hand diagram below commutes. Then the right hand diagram commutes.

$$\begin{array}{ccc} Y \times X & \xrightarrow{\alpha} & Z \\ {\scriptstyle g\times f}\downarrow & & \downarrow{\scriptstyle k} \\ Y' \times X' & \xrightarrow{\alpha'} & Z' \end{array} \qquad \begin{array}{ccc} B_\bullet(Y,M,X) & \xrightarrow{\varepsilon_\alpha} & Z \\ {\scriptstyle B_\bullet(g,h,f)}\downarrow & & \downarrow{\scriptstyle k} \\ B_\bullet(Y',M',X') & \xrightarrow{\varepsilon_{\alpha'}} & Z'. \end{array}$$

Finally, for every right M-space Y we define the right M-action $B_\bullet(Y,M,M)\times M \to B_\bullet(Y,M,M)$ of the form: $[y|m_1|\cdots|m_n|m]m' = [y|m_1|\cdots|m_n|mm']$.

We set $EM := B_\bullet(\mathrm{pt}, M, M), B_\bullet M := B_\bullet(\mathrm{pt}, M, \mathrm{pt})$, and

(1.64) $$p_M := B_\bullet(1_{\mathrm{pt}}, 1_M, c_M): EM \to B_\bullet M,$$

where $c_M: M \to \mathrm{pt}$.

1.65. Theorem. *For every well-pointed monoid M the following hold:*

(i) *Let* $\nu : Y \times M \to Y$ *be an action. Then the map* $\varepsilon_\nu : B_\bullet(Y, M, M) \to Y$ *is a map of right* M*-spaces and a homotopy equivalence. In particular,* EM *is contractible.*

(ii) *If* M *is grouplike, then* $B_\bullet(c_Y, 1_M, 1_X) : B_\bullet(Y, M, X) \to B_\bullet(\mathrm{pt}, M, X)$, *resp.* $B_\bullet(1_Y, 1_M, c_X) : B_\bullet(Y, M, X) \to B_\bullet(Y, M, \mathrm{pt})$, *is a quasi-fibration with fiber (homotopy equivalent to)* Y, *resp.* X. *Furthermore,* $p_M : EM \to B_\bullet M$ *is a principal* M*-quasi-fibration.*

(iii) *If* M *is a topological group (not necessary well-pointed) then* $p_M : EM \to B_\bullet M$ *is a locally trivial principal* M*-bundle.*

Proof. See May [2], [4]. □

1.66. Corollary. *Let* M *be either a well-pointed topological monoid or a topological group. Then the following hold:*

(i) *The space* $B_\bullet M$ *is a classifying space for* M.

(ii) *Let* $p : E \to B$ *be a principal* M*-quasi-fibration. Consider the map* $\alpha : E \times \mathrm{pt} = E \xrightarrow{p} B$. *Then* $\varepsilon_\alpha : B_\bullet(E, M, \mathrm{pt}) \to B$ *is a Whitehead equivalence.*

Proof. (i) This follows from 1.65.

(ii) Consider the commutative diagram

$$\begin{array}{ccc} E & \xleftarrow{\varepsilon_\nu} & B_\bullet(E, M, M) \\ {\scriptstyle p}\downarrow & & \downarrow{\scriptstyle \pi} \\ B & \xleftarrow{\varepsilon_\alpha} & B_\bullet(E, M, \mathrm{pt}) \end{array}$$

where $\pi = B_\bullet(1_E, 1_M, c_M)$. Both vertical maps are quasi-fibrations with fibers CW-equivalent to M, and $\varepsilon_\nu, \varepsilon_\alpha$ induce Whitehead equivalences of fibers. By 1.65(i), ε_ν is a homotopy equivalence, and so ε_α is a Whitehead equivalence. □

Because of 1.66(i), we call $B_\bullet M$ *May's model* of a classifying space for M.

1.67. Corollary. *Let* M *be either a well-pointed grouplike monoid, or a topological group. Then the classifying space for* M *is defined uniquely up to* CW*-equivalence. Furthermore, every* CW*-substitute of any classifying space for* M *is a classifying space for* M.

Proof. Let $p : E \to B$ be a principal M-quasi-fibration with aspherical E, i.e., B is a classifying space for M. Consider the diagram

$$B \xleftarrow{\varepsilon_\alpha} B_\bullet(E, M, \mathrm{pt}) \xrightarrow{B_\bullet(c_E, 1_M, 1_{\mathrm{pt}})} B_\bullet(\mathrm{pt}, M, \mathrm{pt}) = B_\bullet M$$

Now, by 1.66(ii) and 1.65(ii), ε_α and $B_\bullet(c_E, 1_M, 1_{\mathrm{pt}})$ are Whitehead equivalences.

Furthermore, let Y be a CW-space and $f : Y \to B$ be a Whitehead equivalence. By 1.61, there is a principal M-fibration ξ over B with aspherical $\operatorname{ts}\xi$. Now, it is easy to see that $f^*\xi$ is a principal M-fibration with aspherical total space. □

1.68. Corollary. *Let M be a grouplike monoid or a topological group. Then there is a Whitehead equivalence $M \to \Omega B_\bullet M$. In particular, if BM is a classifying space for M then $\pi_i(M) \cong \pi_{i+1}(BM)$.*

Proof. Let $p_M : EM \to B_\bullet M$ be the M-quasi-fibration (1.64) and let $PB_\bullet M \to B_\bullet M$ be as in 1.39(a). Since EM is contractible, there exists a commutative diagram

$$\begin{array}{ccc} EM & \xrightarrow{\widehat{p}} & PB_\bullet M \\ {\scriptstyle p_M}\downarrow & & \downarrow \\ B_\bullet M & = & B_\bullet M . \end{array}$$

This diagram is a morphism of quasi-fibrations with contractible total spaces. Thus, the induced map $M \to \Omega B_\bullet M$ of the fibers is a Whitehead equivalence of fibers (consider the ladder of the homotopy exact sequences).

The last assertion follows from 1.66(i) and 1.67. □

Let $\mathscr{H}(F)$ be the monoid of all homotopy equivalences $F \to F$ topologized as the subspace of F^F. Let $\xi = \{p : E \to B\}$ be an F-fibration. Following Dold–Lashof [1], we define a bundle

$$\operatorname{Prin}\xi = \{\operatorname{Prin}p : \operatorname{Prin}E \to B\}$$

as follows. $\operatorname{Prin}E$ is the subspace of E^F consisting of all maps $\varphi : F \to E$ such that $p\varphi(F)$ is a point $x = x(\varphi) \in B$ and $\varphi : F \to p^{-1}(x)$ is a homotopy equivalence; and $(\operatorname{Prin}p)(\varphi) = x(\varphi)$. We define the action $\nu : \operatorname{Prin}E \times \mathscr{H}(F) \to \operatorname{Prin}E, \nu(\varphi, h) = \varphi h$. By I.3.10(iii), ν is continuous.

1.69. Proposition (cf. Stasheff [1]). *$(\operatorname{Prin}\xi, \nu)$ is a principal $\mathscr{H}(F)$-fibration.*

Proof. We recall the exponential law $(A^F)^Y = A^{Y\times F}$, see I.3.10(ii). Given maps $f : X \to \operatorname{Prin}E$, $h : X \times I \to B$ with $(\operatorname{Prin}p)f(x) = h_0(x) := h(x,0)$, consider the map $\pi : X \times I \xrightarrow{p_1} X = X \times \{0\} \subset X \times I$ and set $a := \mathfrak{I}_{\pi,h^*\xi} : (\pi h)^*\xi \to h^*\xi$. Then a is an equivalence over $X \times I$. Define $g : X \times F \to \operatorname{ts}(h_0^*\xi)$, $g(x,y) = (x, f(x)y)$. Then the map

$$X \times F \times I \xrightarrow{g\times 1} \operatorname{ts}(h_0^*\xi) \times I = \operatorname{ts}((\pi h)^*\xi) \xrightarrow{\operatorname{ts} a} \operatorname{ts}(h^*\xi) \to \operatorname{ts}\xi$$

induces a homotopy equivalence of fibers. So, the adjoint map can be decomposed as $X \times I \xrightarrow{H} \operatorname{Prin}E \subset (\operatorname{ts}\xi)^F$, and H is a $(\operatorname{Prin}p)$-lifting of h with

$H|X \times \{0\} = f$. Thus, Prin ξ is a fibration. Furthermore, it is easy to check that ν turns Prin ξ into a principal $\mathscr{H}(F)$-fibration. □

1.70. Theorem (cf. Allaud [1], [2]). *An F-fibration $\xi = \{p : E \to B\}$ over a CW-base is a universal F-fibration iff* Prin(E) *is aspherical.*

Proof. Consider a universal F-fibration as in 1.57. The inclusion $i_0 : F = F_{b_0} \to E_F$ induces the inclusion $I_0 : \mathscr{H}(F) \to \text{Prin}(E_F)$. We prove that $\text{Prin}(E_F)$ is aspherical. We do it in two steps. Firstly, we prove that the map $(I_0)_* : [X, \mathscr{H}(F)] \to [X, \text{Prin}(E_F)]$ is onto for every $X \in \mathscr{C}$. Then we prove that $\text{Im}(I_0)_*$ consists of just one element for every X.

Step 1. Consider a map $\varphi : X \to \text{Prin}(E_F)$. It yields the adjoint map $\overline{\varphi} : X \times F \to E_F$, and we have the commutative diagram

$$\begin{array}{ccc} X \times F & \xrightarrow{\overline{\varphi}} & E_F \\ {\scriptstyle p_1}\big\downarrow & & \big\downarrow{\scriptstyle p_F} \\ X & \xrightarrow{f} & B_F. \end{array}$$

Since $\overline{\varphi}$ yields a homotopy equivalence of fibers, $f^*\gamma^F$ is fiberwise homotopy trivial. So, $f \simeq *$. Deforming f to the constant map $X \to b_0$, we can cover this deformation by a deformation $\overline{\varphi}_t$ of φ such that $\overline{\varphi}_0 = \overline{\varphi}$ and $\overline{\varphi}_1(X \times F) \subset F$. This implies that $(I_0)_*$ is onto.

Step 2. Given $\rho : X \to F^F$ with $\rho(x) \in \mathscr{H}(F)$, consider the equivalence

$$\psi : X \times (1/3, 2/3) \times F \to X \times (1/3, 2/3) \times F, \ \psi(x, t, f) = (x, t, \rho(x)(f)).$$

Let $U \subset SX$ (resp. $V \subset SX$) be the image of $X \times (0, 2/3)$ (resp. of $X \times (1/3, 1)$). If θ_- (resp. θ_+) is the product F-bundle over U (resp. over V) then ψ induces an equivalence $\varphi : \theta_-|U \cap V \to \theta_+|U \cap V$. Let $\zeta = \zeta_\varphi$, $\text{bs}\,\zeta = SX$, be the (Dold) F-fibration which was constructed in the proof of 1.45. This ζ is classified by the diagram

$$\begin{array}{ccc} \text{ts}\,\zeta & \xrightarrow{g} & E_F \\ \big\downarrow & & \big\downarrow \\ SX & \xrightarrow{f} & B_F, \end{array}$$

and we can assume that $f(X \times [1/2, 1]) = b_0 \in B$. Consider the inclusion $i_0 : F = F_{b_0} \subset E_F$. Clearly, the composition $X \times \{3/4\} \times F \to \text{ts}\,\zeta \xrightarrow{g} E_F$ is homotopic to $X \times F \xrightarrow{p_2} F \xrightarrow{i_0} E_F$, while the composition $X \times \{1/4\} \times F \to \text{ts}\,\zeta \xrightarrow{g} E_F$ is homotopic to $X \times F \xrightarrow{\psi} F \xrightarrow{i_0} E_F$. Thus, $\psi : X \to F^F \to \text{Prin}(E_F)$ is homotopic to the constant map.

Conversely, suppose that Prin E is aspherical. Let $f : B \to B_F$ classify ξ. Then f can be covered by a fiberwise map $\widehat{f} : E \to E_F$, which yields a fiberwise map $\mathrm{Prin}(E) \to \mathrm{Prin}(E_F)$. This map induces a homotopy equivalence of fibers, and thus, because of the asphericity of the total spaces, f is a homotopy equivalence of the bases. □

1.71. Corollary (Stasheff [1]). *If F is a CW-space then $B\mathscr{H}(F) \simeq^{CW} B_F$.*

Proof. Lewis [1] proved that $\mathscr{H}(X)$ is a well-pointed monoid for every CW-space X. So, because of 1.67, it suffices to prove that B_F is a classifying space for $\mathscr{H}(F)$. But this follows from 1.69 and 1.70. □

In fact, for every finite CW-space F, Stasheff [1] constructed a classifying space $B\mathscr{H}(F)$ and proved that $B\mathscr{H}(F)$ classifies F-fibrations.

Thus, any CW-substitute for $B\mathscr{H}(F)$ can play the role of B_F, i.e., $B_\bullet\mathscr{H}(F)$ gives us a more or less explicit construction of B_F.

1.72. Theorem (cf. Allaud [2]). *Let ξ be an F-fibration over a CW-space X, and let A be a CW-subspace of X. Then every F-morphism $\varphi : \xi|A \to \gamma^F$ can be extended to an F-morphism $\psi : \xi \to \gamma^F$.*

Proof. It suffices to consider the case $X = D^n, A = S^{n-1}$. (Then we can perform transfinite induction on cells.) Firstly, let ξ be the product F-bundle θ^F. The map $\mathrm{ts}\,\varphi : A \times F \to E_F$ yields the adjoint map

$$\varphi^{\mathrm{ad}} : A \to \mathrm{Prin}\, E_F \quad \varphi^{\mathrm{ad}}(a)(u) = (\mathrm{ts}\,\varphi)(a, u), a \in A, u \in F.$$

But, by 1.70, $\mathrm{Prin}\, E_F$ is aspherical, and so φ^{ad} can be extended to a map $b : X \to \mathrm{Prin}\, E_F$. Now, define $\psi : \xi \to \gamma^F$ by setting

$$(\mathrm{ts}\,\psi)(x, u) = b(x)(u), x \in X, u \in F.$$

Clearly, ψ is an extension of φ.

In the general case ξ is not trivial, but it is fiberwise homotopy trivial since $X = D^n$ is a contractible space. So, we have fiberwise homotopy equivalences

$$\theta^F_X \xrightarrow{\alpha} \xi \xrightarrow{\beta} \theta^F_X.$$

By the above, the morphism $\theta^F_A \xrightarrow{\alpha|A} \xi \xrightarrow{\varphi} \gamma^F$ can be extended to a morphism $\psi : \theta^F_X \to \gamma^F$, and there is the commutative diagram

$$\begin{array}{ccccc} \xi & \xrightarrow{\beta} & \theta^F_X & \xrightarrow{\psi} & \gamma^F \\ \cup\uparrow & & \uparrow\cup & & \| \\ \xi|A & \xrightarrow{\beta|A} & \theta^F_A & \xrightarrow{\psi|A} & \gamma^F. \end{array}$$

Notice that $(\psi|A)(\beta|A) = \varphi(\alpha|A)(\beta|A) \simeq^{\mathrm{bun}} \varphi$. Now, we can deform the morphism $(\psi|A)(\beta|A)$ to φ, and, by 1.17(ii), this deformation can be extended to a deformation of $\psi\beta$. Thus, in the end of this extended deformation we get the desired extension of φ. □

In fact, the property of γ^F formulated in 1.72 can be treated as a criterion of the universality, cf. Steenrod [1], §19.

Let θ be the product S^0-bundle over pt, and let $\sigma : B_F \to B_{SF}$ classify the SF-fibration $\gamma^F * \theta$. Consider the map

$$i : \mathscr{H}(F) \to \mathscr{H}(SF),\ i(h)[x,t] = [h(x),t], h \in \mathscr{H}(F)\,, x \in F, t \in I.$$

1.73. Proposition. (i) *There are Whitehead equivalences g, f such that the diagram*

$$\begin{array}{ccc} E_F & \xrightarrow{g} & B_\bullet(\mathrm{pt}, \mathscr{H}(F), F) \\ {\scriptstyle p_F}\downarrow & & \downarrow{\scriptstyle r} \\ B_F & \xrightarrow{f} & B_\bullet(\mathrm{pt}, \mathscr{H}(F), \mathrm{pt}), \end{array}$$

$r = B_\bullet(1_{\mathrm{pt}}, 1_{\mathscr{H}(F)}, c_F)$, *commutes up to homotopy.*

(ii) *There are Whitehead equivalences h_1 and h_2 such that the diagram*

$$\begin{array}{ccc} B_F & \xrightarrow{\sigma} & B_{SF} \\ {\scriptstyle h_1}\downarrow & & \downarrow{\scriptstyle h_2} \\ B_\bullet(\mathrm{pt}, \mathscr{H}(F), \mathrm{pt}) & \xrightarrow{Bi} & B_\bullet(\mathrm{pt}, \mathscr{H}(SF), \mathrm{pt}) \end{array}$$

commutes up to homotopy.

Proof. (i) We consider the commutative diagram

$$\begin{array}{ccccc} E_F & \xleftarrow{\varepsilon_\mu} & B_\bullet(\mathrm{Prin}\, E_F, \mathscr{H}(F), F) & \xrightarrow{B_\bullet(c,1_H,1_F)} & B_\bullet(\mathrm{pt}, \mathscr{H}(F), F) \\ {\scriptstyle p_F}\downarrow & & \downarrow{\scriptstyle q} & & \downarrow{\scriptstyle r} \\ B_F & \xleftarrow{\varepsilon_\alpha} & B_\bullet(\mathrm{Prin}\, E_F, \mathscr{H}(F), \mathrm{pt}) & \xrightarrow{B_\bullet(c,1_H,1_{\mathrm{pt}})} & B_\bullet(\mathrm{pt}, \mathscr{H}(F), \mathrm{pt}) \end{array}$$

where

$$1_H = 1_{\mathscr{H}(F)}, q = B_\bullet(1_{\mathrm{Prin}\, E_F}, 1_H, c_F),$$
$$\alpha = \mathrm{Prin}\, p_F : \mathrm{Prin}\, E_F \to B_F, c = c_{\mathrm{Prin}\, E_F}$$

and $\mu : \mathrm{Prin}\, E_F \times F \to E_F$ has the form $\mu(\varphi, f) = \varphi(f), \varphi \in \mathrm{Prin}\, E_F, f \in F$. All vertical maps are quasi-fibrations with fibers CW-equivalent to F, and, by 1.66(ii), ε_α is a Whitehead equivalence. Hence, ε_μ is a Whitehead equivalence. Now, by I.3.46, there are Whitehead equivalences

$$v : E_F \to B_\bullet(\mathrm{Prin}\, E_F, \mathscr{H}(F), F) \text{ and } u : B_F \to B_\bullet(\mathrm{Prin}\, E_F, \mathscr{H}(F), \mathrm{pt})$$

such that $v\varepsilon_\mu \simeq^{CW} 1, \varepsilon_\mu v \simeq 1$, $u\varepsilon_\alpha \simeq^{CW} 1, \varepsilon_\alpha u \simeq 1$. Hence, $qv \simeq^{CW} up_F$. Moreover, $qv \simeq up_F$ since E_F has the homotopy type of a CW-space. Finally, both maps $B_\bullet(c,1,1)$ are Whitehead equivalences, and we set

$$g := B_\bullet(c, 1_H, 1_{\text{pt}})v, \qquad f := B_\bullet(c, 1_H, 1_F)u.$$

(ii) Let $S^*E_F := \text{ts}(\gamma^F * \theta)$. Consider the commutative diagram

$$\begin{array}{ccc}
S^*E_F & \xrightarrow{\;p\;} & B_F \\
{\scriptstyle\varepsilon_\rho}\uparrow & & \uparrow{\scriptstyle\varepsilon_\alpha} \\
B_\bullet(\text{Prin}\,E_F, \mathscr{H}(F), SF) & \xrightarrow{\;q\;} & B_\bullet(\text{Prin}\,E_F, \mathscr{H}(F), \text{pt}) \\
{\scriptstyle B_\bullet(d,i,1_{SF})}\downarrow & & \downarrow{\scriptstyle B_\bullet(d,i,1_{\text{pt}})} \\
B_\bullet(\text{Prin}\,E_{SF}, \mathscr{H}(SF), SF) & \xrightarrow{\;r\;} & B_\bullet(\text{Prin}\,E_{SF}, \mathscr{H}(SF), \text{pt}) \\
{\scriptstyle\varepsilon_\nu}\downarrow & & \downarrow{\scriptstyle\varepsilon_\beta} \\
E_{SF} & \xrightarrow{\;p_{SF}\;} & B_{SF}
\end{array}$$

where α is as in (i), β is similar to α, ν is similar to μ as in (i),

$$q = B_\bullet(1_{\text{Prin}\,E_F}, 1_{\mathscr{H}(F)}, c_{SF}), \;\; r = B_\bullet(1_{\text{Prin}\,E_{SF}}, 1_{\mathscr{H}(SF)}, c_{SF}),$$

$d : \text{Prin}\,E_F \to \text{Prin}\,E_{SF}$ maps $\varphi : F \to E_F$ to the composition $SF \xrightarrow{S\varphi} S^*E_F \xrightarrow{\text{ts}\,\mathfrak{I}_{\sigma,\gamma^F}} E_{SF}$, and $\rho : \text{Prin}\,E_F \times SF \to S^*E_F$ has the form $\rho(\varphi, [f,t]) = [\varphi(f), t], \varphi \in \text{Prin}\,E_F, f \in F, t \in I$.

Similarly to (i), we conclude that all the ε's are Whitehead equivalences. Now, let $u : B_F \to B_\bullet(\text{Prin}\,E_F, \mathscr{H}(F), \text{pt})$ be as in (i), and let $v : S^*E_F \to B_\bullet(\text{Prin}\,E_F, \mathscr{H}(F), SF)$ be such that $\varepsilon_\rho v \simeq 1, v\varepsilon_\rho \simeq^{CW} 1$. Then, as in (i), $qv \simeq up$. This implies easily that

$$\varepsilon_\beta B_\bullet(d, i, 1_{\text{pt}})u \simeq \sigma.$$

(Indeed, the left-hand map classifies $\gamma^F * \theta$.) Furthermore, we construct

$$w : B_{SF} \to B_\bullet(\text{Prin}\,E_{SF}, \mathscr{H}(SF), \text{pt})$$

analogous to u, i.e., such that $\varepsilon_\beta w \simeq 1$, $w\varepsilon_\beta \simeq^{CW} 1$.

Now, we consider the Whitehead equivalences

$$B_\bullet(c_{\text{Prin}\,E_F}, 1_{\mathscr{H}(F)}, 1_{\text{pt}}) : B_\bullet(\text{Prin}\,E_F, \mathscr{H}(F), \text{pt}) \to B_\bullet(\text{pt}, \mathscr{H}(F), \text{pt})$$
$$B_\bullet(c_{\text{Prin}\,E_{SF}}, 1_{\mathscr{H}(SF)}, 1_{\text{pt}}) : B_\bullet(\text{Prin}\,E_{SF}, \mathscr{H}(SF), \text{pt}) \to B_\bullet(\text{pt}, \mathscr{H}(SF), \text{pt})$$

and notice that

$$(Bi)B_\bullet(c_{\text{Prin}\,E_F}, 1_{\mathscr{H}(F)}, 1_{\text{pt}}) = B_\bullet(c_{\text{Prin}\,E_{SF}}, 1_{\mathscr{H}(SF)}, 1_{\text{pt}})B_\bullet(d, i, 1_{\text{pt}}).$$

Now we complete the proof by setting $h_1 = B_\bullet(c_{\text{Prin}\,E_F}, 1_{\mathscr{H}(F)})u$, $h_2 = B_\bullet(c_{\text{Prin}\,E_{SF}}, 1_{\mathscr{H}(SF)}, 1_{\text{pt}})w$. □

Now we pass to $(F,*)$-fibrations. Let $\xi = \{Y \to X\}$ be an $(F,*)$-fibration with the section $s : X \to Y$. Let $t_{(F,*)}(X)$ be the class of the equivalence classes of all the $(F,*)$-fibrations over X. Similarly to (and based on) the above, one can prove that $t_{(F,*)}(X)$ is a set for every X. By 1.20(ii), $t_{(F,*)}$ is a functor on $\mathscr{HC}$.

1.74. Theorem. *The functor* $t_{(F,*)} : \mathscr{HC} \to \mathscr{E}ns$ *is representable.*

Proof. The proof is similar to that of 1.55, therefore we give only a sketch. Given a well-pointed space (F, f_0), we define a rooted $(F,*)$-fibration over (X, x_0) to be an $(F,*)$-fibration $\xi = \{p : Y \to X,\ s : X \to Y, ps = 1\}$ with a given pointed homotopy equivalence (root) $\iota_\xi : (F, f_0) \to (F_{x_0}, s(x_0))$. A rooted equivalence of rooted $(F,*)$-fibrations ξ, η over (X, x_0) is an equivalence $\varphi : \xi \to \eta$ of $(F,*)$-fibrations such that $\varphi i_\xi \simeq i_\eta$ rel $s(x_0)$. Consider an auxiliary functor $r^\bullet : \mathscr{HC}^\bullet_{\text{con}} \to \mathscr{E}ns^\bullet$ such that $r(X, x_0)$ is the set of the equivalence classes of rooted $(F,*)$-fibrations over (X, x_0), and $r^\bullet$ acts on maps as $t_{(F,*)}$ acts.

As in 1.55, the representability of $t_{(F,*)}$ follows from the representability of $r^\bullet$.

We prove that $r^\bullet$ satisfies the MV axiom. Given a pointed CW-triad $(X; A, B; x_0)$, let ξ (resp. η) be a rooted $(F,*)$-fibration over A (resp. B). Given an $(F,*)$-equivalence $\varphi : \xi|C \simeq \eta|C$, $C := A \cap B$, which preserves roots, we get a rooted F-fibration ζ over X such that the diagram (1.48) commutes. (Indeed, take the Dold fibration ζ from 1.47 and consider its fibrational substitute.) Now, the equivalence $\xi \to \zeta|A$ gives us the section $s_A : A \xrightarrow{s_\xi} \text{ts}\,\xi \to \text{ts}(\zeta|A)$. Similarly, we get a section $s_B : B \to \text{ts}(\zeta|B)$. Moreover, $s_A|C \simeq_C s_B|C$ because $\xi|C$ and $\eta|C$ are rooted equivalent $(F,*)$-fibrations. By 1.17(ii), one can construct a section $s = s_X : X \to \text{ts}\,\zeta$ such that $s|A = s_A$ and $s|B \simeq_B s_B$.

Let $\widehat{s} : 1_X \to \zeta$ be the bundle morphism given by s. It would be good if (ζ, s) were a cofibration over X, but we can't claim this. So, we consider the $(F,*)$-fibration $\zeta' := \text{Cyl}\,\widehat{s}$. Denoting by $i : \zeta = \zeta \times 1_{\{1\}} \subset \zeta'$ the inclusion, we get the sections

$$s'_Y : Y \xrightarrow{s_Y} \text{ts}(\zeta|Y) \xrightarrow{\text{ts}(i|Y)} \text{ts}(\zeta'|Y)$$

for $Y = A, B, X$. In particular, $s' = s'_X$ is a section of ζ'. Now, $s'_A \simeq_A s_A$, and so, by 1.7(i) and 1.32,

$$(\zeta'|A, s'|A) \simeq_A (\zeta'|A, s'_A) \simeq_A (\zeta|A, s_A) \simeq_A (\xi, s_\xi).$$

Similarly, $(\zeta'|B, s'|B) \simeq_B (\eta, s_\eta)$, i.e., ζ' is the desired rooted $(F,*)$-fibration.

We leave it to the reader to prove that $r^\bullet$ satisfies the wedge axiom. □

Similarly to 1.57, we can prove the existence of a universal $(F,*)$-fibration $\gamma^{(F,*)} = \{p_{(F,*)} : E_{(F,*)} \to B_{(F,*)}\}$. Here $B_{(F,*)}$ is a classifying space for $t_{(F,*)}$. Let $\mathscr{H}(F,*)$ be the monoid of self-equivalences $(F,*) \to (F,*)$ topologized as the subspace of $(F,*)^{(F,*)}$.

Given a map $f : (F,*) \to (F,*)$, consider the map

$$f \wedge 1 : (SF,*) = (F \wedge S^1, *) \to (F \wedge S^1, *) = (SF,*).$$

We define $i : \mathscr{H}(F,*) \to \mathscr{H}(SF,*)$, $i(f) = f \wedge 1$. Let θ be the product $(S^1,*)$-bundle over pt, and let $\sigma : B_{(F,*)} \to B_{(SF,*)}$ classify the $(SF,*)$-fibration $\gamma^{(F,*)} * \theta$.

Given an $(F,*)$-fibration $\xi = \{p : E \to B\}$ with the section s, we define

$$\operatorname{Prin}^\bullet \xi = \{\operatorname{Prin}^\bullet p : \operatorname{Prin}^\bullet E \to B\}$$

as follows. $\operatorname{Prin}^\bullet E$ is the subspace of E^F consisting of all maps $\varphi : (F,*) \to (E, s(B))$ such that $p\varphi(F)$ is a point $x = x(\varphi) \in B$ and $\varphi : (F,*) \to (p^{-1}(x), s(x))$ is a homotopy equivalence; and $\operatorname{Prin}^\bullet p(\varphi) = x(\varphi)$.

The following pointed analog of 1.69, 1.70, 1.71 and 1.73 holds; we leave the proof to the reader.

1.75. Theorem. (i) $\operatorname{Prin}^\bullet \xi$ *is a principal* $\mathscr{H}(F,*)$*-fibration for every* $(F,*)$*-fibration* ξ.

(ii) *An* $(F,*)$*-fibration* ξ *over a CW-base is universal iff* $\operatorname{ts}(\operatorname{Prin}^\bullet \xi)$ *is aspherical.*

(iii) $B_{(F,*)}$ *and* $B\mathscr{H}(F,*)$ *are CW-equivalent.*

(iv) *There are Whitehead equivalences* g, f *such that the diagram*

$$\begin{array}{ccc} E_{(F,*)} & \xrightarrow{g} & B_\bullet(\mathrm{pt}, \mathscr{H}(F,*), F) \\ {\scriptstyle p_{(F,*)}}\downarrow & & \downarrow{\scriptstyle r} \\ B_{(F,*)} & \xrightarrow{f} & B_\bullet(\mathrm{pt}, \mathscr{H}(F,*), \mathrm{pt}), \end{array}$$

$r = B_\bullet(1_{\mathrm{pt}}, 1_{\mathscr{H}(F,*)}, c_F)$, *commutes up to homotopy.*

(v) *There are Whitehead equivalences* h_1 *and* h_2 *such that the diagram*

$$\begin{array}{ccc} B_{(F,*)} & \xrightarrow{\sigma} & B_{(SF,*)} \\ {\scriptstyle h_1}\downarrow & & \downarrow{\scriptstyle h_2} \\ B_\bullet(\mathrm{pt}, \mathscr{H}(F,*), \mathrm{pt}) & \xrightarrow{Bi} & B_\bullet(\mathrm{pt}, \mathscr{H}(SF,*), \mathrm{pt}) \end{array}$$

commutes up to homotopy. □

Since every $(F,*)$-fibration is an F-fibration, we have a forgetful map $B_{(F,*)} \to B_F$. Because of 1.35, we can consider this map as a fibration $q : B_{(F,*)} \to B_F$.

1.76. Theorem (cf. Gottlieb [1]). *The fibration q is a universal F-fibration. In particular, the homotopy fiber of q is F.*

Proof. Consider a universal F-fibration $\gamma^F = \{p_F : E_F \to B_F\}$. Clearly, $p_F^*\gamma^F$ has a canonical section s (given by the p-lifting 1_{E_F} of p_F). Let λ be the $(F,*)$-fibration $(p_F^*\gamma^F, s)$. We set $D = \operatorname{ts}\lambda$ and $P = \operatorname{ts}(\operatorname{Prin}^\bullet \lambda)$. Firstly, we prove that P is aspherical. In fact, given a map $f : S^n \to P$, we prove that it can be extended to a map $CS^n \to P$. Indeed, let $g : S^n \times F \to D$ be the adjoint map to f. Regarding D as the subset of $E_F \times E_F$, see 1.8, define $h : S^n \times F \to E_F$ by setting $h = p_2 g$. The adjoint map to h has the form $S^n \to \operatorname{Prin}(E_F) \subset (E_F)^F$, and it is inessential because $\operatorname{Prin}(E_F)$ is aspherical. Thus, h can be extended to a fiberwise map $\overline{h} : CS^n \times F \to E_F$. Define $\overline{g} : CS^n \times F \to D, \overline{g}(x,y) = (\overline{h}(x,*), \overline{h}(x,y))$. Let $\overline{f} : CS^n \to P$ be the map adjoint to $\overline{g}$. Since $\overline{g}$ extends g, $\overline{f}$ extends f. So, P is aspherical.

By 1.75(ii), λ is the universal $(F,*)$-fibration over E_F. Let a homotopy equivalence $u : E_F \to B_{(F,*)}$ classify λ. Then, clearly, $qu : E_F \to B_F$ classifies $p_F^*\gamma^F$, i.e., $qu \simeq p_F$. Thus, $p_F : E_F \to B_F$ and $q : B_{(F,*)} \to B_F$ are homotopy equivalent. □

1.77. Remarks. (a) The Representability Theorem 1.53 was proved by Allaud [1] and Dold [3] (without proof that t is set-valued), in the proof of 1.53 we also followed Schön [1]; the Representability Theorem 1.55 was remarked by Dold [3].

(b) The construction of $B\Pi$ for a topological group Π was originally given by Milnor [1]. In fact, he constructed a locally trivial principal Π-bundle with aspherical total space. The classifying space for a monoid M was considered by Dold–Lashof [1] and Stasheff [1]. The construction (1.63) of the classifying space $B_\bullet M$ is taken from May [2]. Similar constructions are in Boardman–Vogt [1] and Milgram [1]. Each of these constructions can be treated as a geometric realization of a certain bar-construction. All these constructions are homotopy equivalent to one another.

(c) Every locally trivial bundle has a structure group. What is an analog of structure group for F-fibrations? Of course, the monoid $\mathscr{H}(F)$ can play this role, but this is not perfect because the fibers of the fibration are different. It seems better to consider the category $\mathscr{I}(F)$ with objects homotopy equivalent to F and the homotopy equivalences as morphisms. May [2] developed this approach, cf. also Segal [1].

§2. Structures on Fibrations

Now we discuss structures on fibrations. We will define not one but several notions of structures. While all are equivalent each has certain advantages in different situations.

Consider an F-fibration ξ over a CW-space X.

2.1. Definition (cf. Browder [2], [3]). (a) Let $\lambda = \{q : E \to B\}$ be an F-fibration. A *λ-prestructure on ξ* is a morphism of F-fibrations $a : \xi \to \lambda$ such that $\mathfrak{F}_a : \xi \to \mathrm{bs}(a)^*\lambda$ is an equivalence over X. Two λ-prestructures $a_0 : \xi \to \lambda$, $a_1 : \xi \to \lambda$ are *equivalent* if there exists a prestructure $b : \xi \times 1_I \to \lambda$ such that $b|(\xi \times \{i\}) = a_i$, $i = 0, 1$. An equivalence class of prestructures is called a *λ-structure* on ξ.

In particular, a γ^F-prestructure on ξ is just a classifying morphism for ξ.

(b) Given a map $\varphi : B \to B_F$, a *(B,φ)-structure on ξ* is a λ-structure on it where $\lambda := \varphi^*\gamma^F$.

2.2. Proposition. *If $\varphi : B \to B_F$ and $\psi : C \to B_F$ are homotopy equivalent maps then (B,φ)-structures on ξ are in a bijective correspondence with (C,ψ)-structures on it.*

Proof. This is obvious. □

Because of 2.2 and 1.35, it suffices to consider (B,φ)-structures such that $\varphi : B \to B_F$ is a fibration.

2.3. Theorem (cf. Browder [3]). *Let ξ be an F-fibration over a CW-space X, and let $\varphi : B \to B_F$ be an arbitrary fibration.*

(i) *Every classifying morphism $\omega : \xi \to \gamma^F$ induces a bijection*

$$\Phi_\omega : [\mathrm{Lift}_\varphi \, \mathrm{bs}\, \omega] \to \{(B,\varphi)\text{-}structures\ on\ \xi\}.$$

(ii) *Let $f : X \to B_F$ classify ξ. Then the set of all (B,φ)-structures on ξ is in a bijective correspondence with the set* $[\mathrm{Lift}_\varphi f]$.

Proof. (i) For simplicity, we denote γ^F by γ and set $\lambda := \varphi^*\gamma$. We say that a λ-prestructure $a : \xi \to \lambda$ is *special* if

$$\xi \xrightarrow{a} \lambda \xrightarrow{\mathfrak{I}_{\varphi,\gamma}} \gamma$$

coincides with ω. Furthermore, we say that two special λ-prestructures $a_0, a_1 : \xi \to \lambda$ are *specially equivalent* if there is an equivalence $b : \xi \times 1_I \to \lambda$ between them which is itself a special prestructure on $\xi \times 1_I$.

Let $g : X \to B$ be a φ-lifting of $f := \mathrm{bs}\, \omega$. Then

$$a_g : \xi \xrightarrow{\mathfrak{F}_\omega} f^*\gamma = g^*\lambda \xrightarrow{\mathfrak{I}_{g,\lambda}} \lambda$$

is a λ-prestructure on ξ, and it is clear that vertically homotopic liftings yield equivalent λ-prestructures. So, we get a correspondence

$$\Phi = \Phi_\omega : [\mathrm{Lift}_\varphi f] \to \{\lambda\text{-structures on } \xi\}, \quad g \mapsto a_g.$$

It suffices to prove that every λ-prestructure is equivalent to a special one (the surjectivity of Φ) and that equivalent special λ-prestructures are specially equivalent (the injectivity of Φ).

We prove the surjectivity of Φ. We are given a λ-prestructure $a : \xi \to \lambda$. By 1.72, there is an F-morphism $H : \xi \times 1_I \to \gamma$ such that $H|\xi \times 1_{\{0\}} = \mathfrak{I}_{\varphi,\gamma} a$ and $H|\xi \times 1_{\{1\}} = \omega$. Furthermore, by 1.9(ii), H can be decomposed as

$$\xi \times 1_I \xrightarrow{\mathfrak{F}_H} u^*\gamma \xrightarrow{\mathfrak{I}_{u,\gamma}} \gamma$$

where $u = \mathrm{bs}\, H$. Since φ is a fibration, there is a φ-lifting $v : X \times I \to B$ of u such that $v|X \times \{0\} = \mathrm{bs}\, a$. Now, consider the F-morphism

$$b : \xi \times 1_I \xrightarrow{\mathfrak{F}_H} u^*\gamma = v^*\lambda \xrightarrow{\mathfrak{I}_{v,\lambda}} \lambda.$$

Clearly, $b|\xi \times 1_{\{1\}}$ is a special λ-prestructure. Finally, $\mathfrak{I}_{\varphi,\gamma} \circ (b|\xi \times 1_{\{0\}}) = \mathfrak{I}_{\varphi,\gamma} a$, and so $b|\xi \times 1_{\{0\}} = a$ (we use the claim about uniqueness from 1.9(ii)).

We prove the injectivity of Φ. Let $a_0, a_1 : \xi \to \lambda$ be two special λ-prestructures on ξ, and let $b : \xi \times 1_I \to \lambda$ be an equivalence between a_0 and a_1. Consider the subspaces $Y := I \times \{0\} \cup I \times \{1\} \cup \{0\} \times I$, $Z := I \times \{0\} \cup I \times \{1\} \cup \{1\} \times I$ of $I \times I$. We define an F-morphism $c : \xi \times 1_{I^2}|(X \times \partial I^2) \to \gamma$ as follows: $c|\xi \times 1_Y$ is just the morphism $\xi \times 1_Y \xrightarrow{\mathrm{proj}} \xi \xrightarrow{\omega} \gamma$ and $c|\xi \times \{1\} \times 1_I = \mathfrak{I}_{\varphi \times 1,\gamma} b : \xi \times 1_I \to \gamma$. Then, by 1.72, there is an F-morphism $d : \xi \times 1_{I^2} \to \gamma$ which extends c. By 1.9(ii), d can be uniquely decomposed as

$$\xi \times 1_{I^2} \xrightarrow{\mathfrak{F}_d} k^*\gamma \xrightarrow{\mathfrak{I}_{k,\gamma}} \gamma$$

where $k = \mathrm{bs}\, d$. Since φ is a fibration, there is a φ-lifting $l : X \times I^2 \to B$ of k such that $l|Z = (\mathrm{bs}\, c)|Z$. Consider the morphism

$$R : \xi \times 1_{I^2} \xrightarrow{\mathfrak{F}_d} k^*\gamma = l^*\lambda \xrightarrow{\mathfrak{I}_{l,\lambda}} \lambda.$$

Now, $R|\xi \times 1_{\{0\} \times I} : \xi \times 1_I \to \lambda$ is a special equivalence between a_0 and a_1.

(ii) This follows from (i), because, by 1.58(iii), $f = \mathrm{bs}\, \omega$ for some classifying morphism $\omega : \xi \to \gamma$. □

Lashof [1] and Stong [3] considered a fibration $\varphi : B \to B_F$ and defined a (B, φ)-structure on (ξ, f) to be an element of $[\mathrm{Lift}_\varphi f]$ where f classifies ξ. Because of 2.3, their definition is equivalent (in some sense) to 2.1. (I said "in some sense" since their definition deals not just with ξ but with the pair (ξ, f).)

2.4. Definition (cf. May [2]). Given a space K, consider the space K^F of all maps $F \to K$. We define the right $\mathscr{H}(F)$-action $K^F \times \mathscr{H}(F) \to K^F, (\varphi, h) \mapsto \varphi h$. Consider any $\mathscr{H}(F)$-invariant subset $\mathscr{M}$ of K^F. An $(\mathscr{M}, K)$*-prestructure* on ξ is a map $l : \operatorname{ts}\xi \to K$ such that for every $x \in \operatorname{bs}\xi$ and every homotopy equivalence $u : F \to p^{-1}(x)$ the composition

$$F \xrightarrow{u} p^{-1}(x) \subset \operatorname{ts}\xi \xrightarrow{l} K$$

belongs to $\mathscr{M}$. Two $(\mathscr{M}, K)$-prestructures $l_0, l_1 : \operatorname{ts}\xi \to K$ on ξ are called *equivalent* if there exists a prestructure $L : \operatorname{ts}\xi \times I \to K$ on $\xi \times 1_I$ such that $L| \operatorname{ts}\xi \times \{i\} = l_i$, i =0,1. An $(\mathscr{M}, K)$*-structure* on ξ is an equivalence class of $(\mathscr{M}, K)$-prestructures on ξ.

Every F-morphism $\xi \to \eta$ induces a function

$$\{\text{structures on } \eta\} \to \{\text{structures on } \xi\}$$

in an obvious way. So, the correspondence $\xi \mapsto \{\text{structures on } \xi\}$ is natural in ξ.

One says that two structured fibrations over the same base X are equivalent if there exists an equivalence over X which carries one structure to another. More precisely, we have the following definition.

2.5. Definition. Consider two F-fibrations $\xi_i, i = 1, 2$, over X and two $(\mathscr{M}, K)$-prestructures $l_i : \operatorname{ts}\xi_i \to K$, i =1,2, one says that (ξ_1, l_1) and (ξ_2, l_2) are $(\mathscr{M}, K)$*-equivalent structured fibrations* if there exists an equivalence $\alpha : \xi_1 \to \xi_2$ over X such that $l_2 \operatorname{ts}(\alpha)$ and l_1 yield equivalent prestructures on ξ_1. Similarly, given two λ-prestructures $a_i : \xi_i \to \lambda, i = 1, 2$, one says that (ξ_1, a_1) and (ξ_2, a_2) are λ*-equivalent structured fibrations* if there exists an equivalence $\alpha : \xi_1 \to \xi_2$ over X such that $a_2 \operatorname{ts}(\alpha)$ and a_1 yield equivalent prestructures on ξ_1.

Given λ as in 2.1, set $K_\lambda := E$ and $\mathscr{M}_\lambda := \operatorname{ts}(\operatorname{Prin}\lambda)$.

2.6. Proposition. *For every F-fibration ξ over a CW-base, there is a natural bijective correspondence $\{\lambda$-structures on $\xi\} \to \{(\mathscr{M}_\lambda, K_\lambda)$-structures on $\xi\}$.*

Proof. Clearly, any λ-prestructure $\xi \to \lambda$ yields an $(\mathscr{M}_\lambda, K_\lambda)$-prestructure on ξ. Conversely, every $(\mathscr{M}_\lambda, K_\lambda)$-prestructure on ξ yields a fiberwise map $f : \operatorname{ts}\xi \to E$ which induces a homotopy equivalence of fibers. Let $a : \xi \to \lambda$ be a morphism with $\operatorname{bs} a = f$. Then $\mathfrak{F}_a : \xi \to (\operatorname{bs} a)^*\lambda$ is an equivalence over $\operatorname{bs}\xi$, i.e., we get a λ-prestructure on ξ. It is easy to see that in this way we obtain the desired correspondence. □

Now we prove that every $(\mathscr{M}, K)$-structure can be regarded as a certain λ-structure. Let $t_{(\mathscr{M},K)}(X)$ be the set of all equivalence classes of $(\mathscr{M}, K)$-structured F-fibrations over X. (It is a set because $(\mathscr{M}, K)$-structures on every fibration ξ form a set, say, by 1.52; and if ξ, η are equivalent then there is an obvious bijective correspondence between $(\mathscr{M}, K)$-structures on ξ and those on η.) Clearly, in this way we get a functor $t = t_{(\mathscr{M},K)} : \mathscr{H}\mathscr{C} \to \mathscr{E}ns$.

2.7. Theorem. *The functor* $t_{(\mathscr{M},K)} : \mathscr{H}\mathscr{C} \to \mathscr{E}ns$ *is representable.*

Proof. The proof is similar to that of 1.55, and so we give only a sketch. In the proof "F-fibration" means "rooted F-fibration".

1. Fix a map $f : F \to K$ in $\mathscr{M}$. Given an F-fibration with a root i, we define a rooted $(\mathscr{M}, K)$-prestructure on ξ to be an $(\mathscr{M}, K)$-prestructure $l : \operatorname{ts}\xi \to K$ such that the composition li is homotopic to f. Then one can define a rooted $(\mathscr{M}, K)$-structure on ξ and equivalence of rooted $(\mathscr{M}, K)$-structured F-fibrations. Let $s(X, x_0) = s_{(\mathscr{M},K)}(X, x_0)$ be the class of all equivalence classes of rooted $(\mathscr{M}, K)$-structured F-fibrations over (X, x_0). Since $r(X, x_0)$ as in 1.53 is a set for every (X, x_0), we conclude that $s(X, x_0)$ is a set. To prove the representability of t, it suffices to prove the representability of $s : \mathscr{H}\mathscr{C}^\bullet_{\text{con}} \to \mathscr{E}ns^\bullet$.

2. We prove that s satisfies the MV-axiom. Let $(X; A, B; x_0)$ be a CW-triad, set $C := A \cap B$. Let $(\xi, l), \operatorname{bs}\xi = A$ and $(\eta, m), \operatorname{bs}\eta = B$, be two structured F-fibrations, and let $\varphi : \xi|C \to \eta|C$ be an equivalence of structured F-fibrations. By 1.47, there exists ζ over X and the rooted equivalences $a : \zeta|A \to \xi$, $b : \zeta|B \to \eta$ such that the diagram (1.48) commutes up to homotopy over C. (More precisely, we take a fibrational substitute of a Dold fibration ζ from 1.47.) In particular, $l \circ \operatorname{ts}(a|C) \simeq m \circ \operatorname{ts}(b|C)$. So, using the homotopy extension property for CW-pairs, one can construct an $(\mathscr{M}, K)$-prestructure $\operatorname{ts}(\zeta) \to K$ which extends both l and m.

We leave it to the reader to check that s satisfies the wedge axiom. □

So, we have the universal $(\mathscr{M}, K)$-structured F-fibration

$$\lambda^{(\mathscr{M},K)} = \{p_{(\mathscr{M},K)} : E_{(\mathscr{M},K)} \to B_{(\mathscr{M},K)}\},$$

where $B_{(\mathscr{M},K)}$ is the classifying space for $t_{(\mathscr{M},K)}$. Of course, $\lambda^{(\mathscr{M},K)}$ is classified by the forgetful map $p : B_{(\mathscr{M},K)} \to B_F$. Below we assume that p is a fibration and that $\lambda^{(\mathscr{M},K)} = p^*\gamma^F$.

2.8. Theorem. (i) *For every F-fibration ξ over a CW-base, there is a natural bijective correspondence between $(\mathscr{M}, K)$-structures on ξ and $\lambda^{(\mathscr{M},K)}$-structures on ξ.*

(ii) *For every F-fibration ξ over a CW-base, each classifying morphism $\omega : \xi \to \gamma^F$ induces a bijective correspondence*

$$\Psi_\omega : (\mathscr{M}, K)\text{-structures on } \xi \to [\operatorname{Lift}_p \operatorname{bs}\omega].$$

(iii) *The homotopy fiber of $p : B_{(\mathscr{M},K)} \to B_F$ is $\mathscr{M}$, i.e., p is an $\mathscr{M}$-fibration.*

Proof. (i) Let an $(\mathscr{M},K)$-structure on ξ be classified by $h : \operatorname{bs}\xi \to B_{(\mathscr{M},K)}$. By 1.58(iii), there is a classifying morphism $\omega : \xi \to \gamma^F$ with $\operatorname{bs}\omega = ph$. We define $a_h = a_{h,\omega} : \xi \to \lambda := \lambda^{(\mathscr{M},K)}$ to be the composition

$$\xi \xrightarrow{\mathfrak{F}_\omega} h^*p^*\gamma^F = h^*\lambda \xrightarrow{\mathfrak{I}_{h,\lambda}} \lambda.$$

Clearly, a_h is just a λ-prestructure on ξ. Furthermore, if $h_1 \simeq h_2 : \operatorname{bs}\xi \to B_{(\mathscr{M},K)}$, then the λ-prestructures a_{h_1} and a_{h_2} are equivalent. Conversely, every λ-prestructure $a : \xi \to \lambda$ yields a map $\operatorname{ts}\xi \to \operatorname{ts}\lambda \xrightarrow{l} K$, where l is given by the universal $(\mathscr{M},K)$-structure. One can check that these correspondences are mutually inverse.

(ii) This follows from (i) and 2.3(i).

(iii) Let Φ be the homotopy fiber of p. Consider the product F-bundle θ_X over a CW-space X. We require that θ_X is classified by a morphism ω such that $\operatorname{ts}\omega : X \times F \to E_F$ has the form $X \times F \xrightarrow{p_2} F = F_{b_0} \subset E_F$ with b_0 as in 1.57. Then, by (ii), $(\mathscr{M},K)$-structures on θ_X are in a natural bijective correspondence with $[X,\Phi]$. On the other hand, the $(\mathscr{M},K)$-prestructures on θ_X are just the maps $l : X \times F \to K$ such that $l|(\{x\}\times F) : F \to K$ belongs to $\mathscr{M}$ for every $x \in X$. Furthermore, under the exponential law $K^{X\times F} = (K^F)^X$ the $(\mathscr{M},K)$-prestructures on θ_X are in a bijective correspondence with the maps $X \to \mathscr{M}$. Moreover, the set of all $(\mathscr{M},K)$-structures on θ is in a bijective correspondence with the set $[X,\mathscr{M}]$. In other words, we have the natural equivalence $[X,\Phi] = [X,\mathscr{M}]$, and hence $\Phi \simeq \mathscr{M}$. □

Thus, by 2.6 and 2.8, Definitions 2.1 and 2.4 are equivalent. We can refine and develop this equivalence as follows.

2.9. Lemma. *Every F-fibration λ over a CW-base is bundle equivalent to the F-fibration $\lambda^{(\mathscr{M}_\lambda,K_\lambda)}$.*

Proof. By 2.6 and 2.8(i), we have a natural in ξ bijection

$$\{\lambda\text{-structures on } \xi\} \longleftrightarrow \{\lambda^{(\mathscr{M}_\lambda,K_\lambda)}\text{-structures on } \xi\}.$$

Now, let a fibration $\varphi : B \to B_F$ (resp. $\psi : C \to B_F$) classify λ (resp $\lambda^{(\mathscr{M}_\lambda,K_\lambda)}$). By 2.3(ii), for every $f : X \to B_F$ we have a bijection $[\operatorname{Lift}_\varphi f] \cong [\operatorname{Lift}_\psi f]$, and this bijection is natural in the category of bundles over B_F. Now, standard category-theoretical arguments imply that φ and ψ are equivalent over B_F, and the result follows. □

Let λ be an F-fibration over a CW-space B, and let $\varphi : B \to B_F$ classify λ. Similarly to the above, we define a functor $t_\lambda : \mathscr{HC} \to \mathscr{E}ns$ by setting $t_\lambda(X)$ to be the set of equivalence classes of λ-structured F-fibrations.

2.10. Theorem. *The functor t_λ is representable. Moreover, $t_\lambda(X) = [X,B]$.*

Proof. By 2.9, $\lambda \simeq \lambda^{(\mathscr{M}_\lambda, K_\lambda)}$, and so $B \simeq B_{(\mathscr{M}_\lambda, K_\lambda)}$. Furthermore, similarly to 2.6, one can prove that $t_\lambda(X) = t_{(\mathscr{M}_\lambda, K_\lambda)}(X)$. Now,

$$t_\lambda(X) = t_{(\mathscr{M}_\lambda, K_\lambda)}(X) = [X, B_{(\mathscr{M}_\lambda, K_\lambda)}] = [X, B]. \qquad \square$$

2.11. Comments. (a) Let ξ be classified by a map $f : X \to B_F$, and let $\lambda := \varphi^* \gamma^F$ where $\varphi : B \to B_F$ is a fibration. Because of 2.9, we have a commutative square

$$\begin{array}{ccc} \{\lambda\text{-structures on } \xi\} & \longleftrightarrow & [\operatorname{Lift}_\varphi f] \\ \downarrow & & \downarrow \\ \{\text{equivalence classes of } \lambda\text{-structured } F\text{-fibrations over } X\} & \longleftrightarrow & [X, B]. \end{array}$$

Here the horizontal arrows are bijections and the vertical arrows are forgetful maps. Namely, the left-hand vertical arrow sends a structured fibration ξ to its equivalence class, the right-hand vertical arrow sends a vertical homotopy class of a map $X \to B$ to its homotopy class.

Similarly, let $p : B_{(\mathscr{M}, K)} \to B_F$ be the forgetful map. Then we have the commutative square

$$\begin{array}{ccc} \{(\mathscr{M}, K)\text{-structures on } \xi\} & \longleftrightarrow & [\operatorname{Lift}_p f] \\ \downarrow & & \downarrow \\ \{\text{classes of } (\mathscr{M}, K)\text{-structured } F\text{-fibrations over } X\} & \longleftrightarrow & [X, B_{(\mathscr{M}, K)}]. \end{array}$$

(b) Of course, the space $B_\bullet(\mathscr{M}, \mathscr{H}(F), \mathrm{pt})$ in (1.63) can play the role of the classifying space $B_{(\mathscr{M}, K)}$. In fact, May [2] proved that $B_\bullet(\mathscr{M}, \mathscr{H}(F), \mathrm{pt})$ represents $t_{(\mathscr{M}, K)}$, and this was the original proof of 2.7.

Structures on $(F, *)$-fibrations can be introduced similarly to structures on F-fibrations. (By the way, observe that the section can be treated as a structure, see 1.74.) Definition 2.1 can be reformulated for $(F, *)$-fibrations word for word, with the replacement of F by $(F, *)$. Definition 2.4 changes in the following way.

2.12. Definition. Consider a *pointed* space $(K, *)$ and a pointed $\mathscr{H}(F, *)$-invariant subspace $\mathscr{N}$ of $(K, *)^{(F, *)}$. An $(\mathscr{N}, (K, *))$-prestructure on an $(F, *)$-fibration (ξ, s) over X is a map $l : (\operatorname{ts} \xi, s(X)) \to (K, *)$ such that for every $x \in \operatorname{bs} \xi$ and every homotopy equivalence $u : (F, *) \to (p^{-1}(x), s(x))$ the composition

$$(F, *) \xrightarrow{u} (p^{-1}(x), s(x)) \subset (\operatorname{ts} \xi, s(X)) \xrightarrow{l} (K, *)$$

belongs to $\mathscr{N}$.

The equivalence of $(\mathscr{N}, (K, *))$-prestructures can be defined just as in 2.5, and an equivalence class of $(\mathscr{N}, (K, *))$-prestructures is an $(\mathscr{N}, (K, *))$-structure on ξ. As above, there is a functor $t = t_{(\mathscr{N}, (K, *))}$, where $t(X)$ is the set of all equivalence classes of $(\mathscr{N}, (K,*))$-structured $(F,*)$-fibrations over X. The following theorem holds and can be proved as 2.7 and 2.8 were.

2.13. Theorem. (i) *The functor* $t_{(\mathscr{N},(K,*))} : \mathscr{H}\mathscr{C} \to \mathscr{E}ns$ *is representable.*

(ii) *Let* $B_{(\mathscr{N},(K,*))}$ *be the representing space for* $t_{(\mathscr{N},(K,*))}$. *Then the homotopy fiber of the forgetful map* $p : B_{(\mathscr{N},(K,*))} \to B_{(F,*)}$ *is* $\mathscr{N}$.

(iii) *Let* ξ *be an* $(F,*)$*-fibration classified by* $f : X \to B_{(F,*)}$ *where* X *is a CW-space. Then the set of all* $(\mathscr{N}, (K,*))$*-structures on* ξ *is in a bijective correspondence with the set* $[\mathrm{Lift}_p\, f]$. □

§3. A Glance at Locally Trivial Bundles

3.1. Recollections. Let Π be a topological group. For a definition of a locally trivial principal Π-bundle and their morphisms, see e.g. Fuks–Rokhlin [1], Husemoller [1], Steenrod [1], Switzer [1]. The equivalence (=isomorphism) of locally trivial principal Π-bundles over the same base is defined. Given a left Π-space F, the term "(F,Π)-bundle" means a locally trivial bundle with fiber F and structure group Π. So, every (F,Π)-bundle ξ is associated with a unique locally trivial principal Π-bundle η, $\mathrm{ts}\,\xi = (\mathrm{ts}\,\eta) \times_\Pi F$, $\mathrm{bs}\,\xi = \mathrm{bs}\,\eta$. One says that two (F,Π)-bundles over the same base are equivalent if the corresponding locally trivial principal Π-bundles are equivalent.

Given two locally trivial principal Π-bundles η, η', a Π-bundle morphism $\eta \to \eta'$ is just a bundle morphism $\varphi : \eta \to \eta'$ such that $\mathrm{ts}\,\varphi : \mathrm{ts}\,\eta \to \mathrm{ts}\,\eta'$ is a Π-equivariant map.

Let ξ (resp. ξ') be the (F,Π)-bundle associated with the locally trivial principal Π-bundle η (resp. η'). Clearly, every Π-bundle morphism $\varphi : \eta \to \eta'$ induces a bundle morphism $\bar{\varphi} : \xi \to \xi'$ where

$$\mathrm{bs}\,\bar{\varphi} = \mathrm{bs}\,\varphi,\ \mathrm{ts}\,\bar{\varphi} = \mathrm{ts}\,\varphi \times_\Pi F : \mathrm{ts}\,\eta \times_\Pi F \to \mathrm{ts}\,\eta' \times_\Pi F.$$

We define an (F,Π)-bundle morphism to be a bundle morphism $\psi : \xi \to \xi'$ which has the form $\psi = \bar{\varphi}$ for some Π-bundle morphism $\varphi : \eta \to \eta'$. In particular, every (F,Π)-bundle morphism yields a Π-equivariant homeomorphism of fibers.

Given an (F,Π)-bundle ξ, an *admissible inclusion* $i : F \to \mathrm{ts}\,\xi$ is any map $F \to F_x \subset \mathrm{ts}\,\xi$ such that $F \to F_x$ is a morphism of (F,Π)-bundles, where F at the domain is the (F,Π)-bundle over pt. Admissible inclusions can also be described as follows. Let $\eta = \{p : E \to B\}$ be the principal Π-bundle associated with ξ. Then admissible inclusions are just maps of the form

$$F \xrightarrow{i_e} E \times F \xrightarrow{\text{quotient}} E \times_\Pi F = \mathrm{ts}\,\xi,\ i_e(f) = (e,f), e \in E, f \in F.$$

Finally, we notice that the space of all admissible inclusions is equivariantly homotopy equivalent to E (prove this).

As in 1.56, the universal locally trivial principal Π-bundle is defined to be a locally trivial principal Π-bundle $\gamma = \{p : E \to B\}$ such that every locally trivial principal Π-bundle ξ over a CW-space X is equivalent to a locally trivial principal Π-bundle of the form $f^*\gamma$ for some $f : X \to B$, and that two maps $f, g : X \to B$ are homotopic iff $f^*\gamma$ and $g^*\gamma$ are equivalent.

According to 1.66, there exists a classifying space $B\Pi$ for Π, and, by 1.67, $B\Pi$ is uniquely defined up to CW-equivalence.

Let $u_{F,\Pi}$, resp. u_Π, be the class of all equivalence classes of (F, Π)-bundles, resp. locally trivial principal Π-bundles. By definition, the functors $u_{F,\Pi}$ and u_Π are equivalent. Furthermore, one can check that u_Π is homotopy invariant.

3.2. Theorem. (i) *$u_\Pi(X)$ is a set for every CW-space X, and the functor $u_\Pi : \mathscr{H}\mathscr{C} \to \mathscr{E}ns$ is representable. In particular, there exists a universal locally trivial principal Π-bundle.*

(ii) *A locally trivial principal Π-bundle is universal iff its total space is aspherical.*

(iii) *Let γ be the universal locally trivial principal Π-bundle, let ξ be a locally trivial principal Π-bundle over a CW-space X, and let A be a CW-subspace of X. Then every morphism $\xi|A \to \gamma$ of Π-bundles can be extended to the whole of ξ.*

(iv) *Every CW-substitute for $B\Pi$ represents u_Π.*

Proof (Sketch). (i) Analogs of 1.47 and 1.45 can be proved in this case more easily than for fibrations. One can just glue bundles, without any homotopy tricks, see e.g. Switzer [1], Ch. 11. So, the representability holds.

(ii) See Steenrod [1], Switzer [1], Ch.11, Husemoller [1], Ch.4.

(iii) See Steenrod [2], §19.

(iv) This follows from (ii). □

3.3. Proposition. *If a locally trivial principal Π-bundle ξ admits a section then ξ is trivial.*

Proof. The map $f : \operatorname{bs}\xi \times \Pi \to \operatorname{ts}\xi$, $f(x, g) = s(x)g, x \in \operatorname{bs}\xi, g \in \Pi$ gives an equivalence of ξ with the product Π-bundle over $\operatorname{bs}\xi$. □

Let $i : \Sigma \to \Pi$ be an inclusion of a closed subgroup. Any locally trivial principal Σ-bundle $E \to B$ gives us a locally trivial principal Π-bundle $E \times_\Sigma \Pi \to B$. So, there is a natural transformation $i_* : u_\Sigma \to u_\Pi$, and thus we get a map $Bi : B\Sigma \to B\Pi$.

3.4. Theorem. *Let Σ be a closed subgroup of a Lie group Π. Then Bi is homotopy equivalent to a $(\Pi/\Sigma, \Pi)$-bundle $B\Sigma \to B\Pi$.*

Proof. Let $E\Pi \to B\Pi$ be the universal locally trivial principal Π-bundle. Since Σ is a closed Lie subgroup of Π, the quotient map

(3.5) $$p : E\Pi \to E\Pi/\Sigma$$

turns out to be a locally trivial bundle, and hence $E\Pi/\Sigma$ can be regarded as $B\Sigma$. Now, the further factorization yields a $(\Pi/\Sigma, \Pi)$-bundle $B\Sigma \to B\Pi$. Clearly, this map yields the transformation $i_* : u_\Sigma \to u_\Pi$. □

3.6. Definition. Let ξ be an (F, Π)-bundle.

(a) Given an (F, Π)-bundle λ, a *λ-prestructure on ξ* is an (F, Π)-bundle morphism $a : \xi \to \lambda$. Two λ-prestructures $a_0, a_1 : \xi \to \lambda$ are *equivalent* if there exists a λ-prestructure $b : \xi \times 1_I \to \lambda$ such that $b|(\xi \times \{i\}) = a_i$, i =0,1. An equivalence class of prestructures is called a *λ-structure* on ξ.

(b) Given a map $\varphi : B \to B_\Pi$, a *(B, φ)-structure on ξ* is a λ-structure on it where $\lambda := \varphi^*\bar{\gamma}$ and $\bar{\gamma}$ is the (F, Π)-bundle associated with the universal principal π-bundle γ.

(c) An analog of 2.4 can also be formulated for (F, Π)-bundles: here $\mathscr{M}$ is required to be a Π-invariant subset of K^F and the map $l : \operatorname{ts}\xi \to K$ is required to be such that the map $F \xrightarrow{i} \operatorname{ts}\xi \xrightarrow{l} K$ belongs to $\mathscr{M}$ for every admissible inclusion i.

3.7. Example. Let Σ be a subgroup of Π. We can regard every Π-space F as a Σ-space, and so every (F, Σ)-bundle can be regarded as an (F, Π)-bundle. Now, let λ be an (F, Σ)-bundle associated with the universal principal Σ-bundle. Then, by the above, we can consider λ-structures on any (F, Π)-bundle ξ. If a l-structure on ξ exists, one says that the structure group Π of ξ can be reduced to Σ, cf, Steenrod [1], Husemoller [1].

Another examples (namely, orientations) appear in §5 and Ch. V below.

3.8. Theorem. *Let $f : X \to B\Pi$ be a map of a CW-space, and let $\xi := f^*\bar{\gamma}$ where $\bar{\gamma}$ is the (F, Π)-bundle associated with the universal principal Π-bundle.*

(i) *For every map $\varphi : B \to B\Pi$ the set of all (B, φ)-structures on ξ is in a natural bijective correspondence with the set* $[\operatorname{Lift}_\varphi f]$.

(ii) *Given an (F, Π)-bundle λ, set $K_\lambda := \operatorname{ts}\lambda$ and let $\mathscr{M}_\lambda$ be the set of all admissible inclusions $F \to \operatorname{ts}\lambda$. Then the set of all λ-structures on ξ is in a natural bijective correspondence with the set of all $(\mathscr{M}_\lambda, K_\lambda)$-structures on ξ.*

(iii) *Given a pair $(\mathscr{M}, K)$ as in 3.6(c), we turn the right Π-space M into a left Π-space by setting $gx = xg^{-1}, x \in \mathscr{M}, g \in \Pi$. Consider the $(\mathscr{M}, \Pi)$-bundle $\lambda^{(\mathscr{M},K)} = \{q : B \to B\Pi\}$ associated with the universal Π-bundle. Then the set of all $\lambda^{(\mathscr{M},K)}$-structures on ξ is in a natural bijective correspondence with the set of all $(\mathscr{M}, K)$-structures on ξ.*

Proof. (i) This is similar to 2.3.

(ii) This is similar to 2.6.

(iii) By (i), it suffices to prove that $(\mathscr{M}, K)$-structures on $\xi = f^*\bar{\gamma}$ are in a natural (with respect to f) bijective correspondence with $[\operatorname{Lift}_q f]$. Consider a principal Π-bundle $\eta = \{E \to X\}$ associated with ξ. Then $f^*\lambda$ is an $(\mathcal{M}, \Pi)$-bundle associated with η. Furthermore,

$$\mathrm{Lift}_q f = \mathrm{Sec}\, f^*\lambda = \{\text{ all } \Pi\text{-equivariant maps } E \to \mathcal{M}\},$$

see Husemoller [1], Ch. 4. Under the exponential law $(K^F)^E = K^{F\times E}$, the last set transforms into

{all maps $\varphi : E \times F \to K$ such that $\varphi(eg, y) = \varphi(e, gy)$ for every $g \in \Pi, e \in E, y \in F$, and $\varphi|\{e\} \times F : \{e\} \times F \to K$ belongs to $\mathcal{M}$ for every $e \in E$}

={all maps $\psi : E \times_\Pi F \to K$ such that $F \xrightarrow{i} E \times_\Pi F \xrightarrow{\psi}$ belongs to $\mathcal{M}$ for every admissible inclusion i}

={all $\mathcal{M}$-prestructures on ξ},

because $\mathrm{ts}\, \xi = E \times_\Pi F$. So, $\mathrm{Lift}_q f =$ {all $\mathcal{M}$-prestructures on ξ}.

Clearly, under this correspondence vertical homotopy classes of q-liftings of f correspond to $\mathcal{M}$-structures on ξ. □

We leave it to the reader to define equivalent λ-structured, resp. $(\mathcal{M}, K)$-structured, (F, Π)-bundles following 2.5. Let $u_\lambda(X)$, resp. $u_{(\mathcal{M},K)}(X)$ be the set of all equivalence classes of λ-structured, resp. $(\mathcal{M}, K)$-structured, (F, Π)-bundles over X. The following analog of 2.10 holds.

3.9. Theorem. *There are natural equivalences*

$$u_\lambda(X) = [X, \mathrm{bs}\,\lambda],\ u_{(\mathcal{M},K)}(X) = [X, B_{(\mathcal{M},K)}]$$

where $B_{(\mathcal{M},K)} := \mathrm{bs}\,\lambda^{(\mathcal{M},K)}$ *with* $\lambda^{(\mathcal{M},K)}$ *as in* 3.8(iii).

Proof. This is similar to that of 2.10, so we give only a sketch. Firstly, we can prove that $u_\lambda : \mathcal{HC} \to \mathcal{E}ns$ is a representable functor, i.e., $u_\lambda(X) = [X, B]$ for some $B \in \mathcal{C}$. So, there exists a unversal λ-structured (F, Π)-bundle μ over B. Let μ be classified by a map $q : B \to B\Pi$; we assume that q is a fibration. Now, for every $f : X \to B\Pi$, we have natural bijections

$$\{\lambda\text{-structures on } f^*\bar\gamma\} \cong [\mathrm{Lift}_q f] \cong \{\mu\text{-structures on } f^*\bar\gamma\}$$

where $\bar\gamma$ is the (F, Π)-bundle associated with the universal principal Π-bundle. The existence of the first bijection can be proved as in 2.8(i), the existence of the second bijection follows from 3.7(i). Hence, λ and μ are equivalent (F, Π)-bundles, i.e., $B \simeq \mathrm{bs}\,\lambda$, cf. 2.9. Similarly for $u_{(\mathcal{M},K)}$. □

3.10. Remarks. (a) The obvious analog of 2.11(a) holds for (F, Π)-bundles.

(b) Clearly, 3.8(iii) is an analog of 2.8, but for bundles we are able to give, and have given, an explicit construction of $\lambda^{(\mathcal{M},K)}$, cf. 2.11(b).

§4. $\mathbb{R}^n$-Bundles and Spherical Fibrations

4.1. Recollection. We consider the following classes of objects arising in geometric topology.

(a) S^{n-1}-fibrations. They are classified by a space $B_{S^{n-1}}$. By 1.71, $B_{S^{n-1}} \simeq B\mathcal{G}_n$, where $\mathcal{G}_n := \mathscr{H}(S^{n-1})$.

(b) Locally trivial $\mathbb{R}^n$-bundles equipped with sections. These are just $((\mathbb{R}^n, 0), \mathcal{TOP}_n)$-bundles, where $\mathcal{TOP}_n$ is the topological group of homeomorphisms $(\mathbb{R}^n, 0) \to (\mathbb{R}^n, 0)$ topologized as in I.3.9(a). Thus, by 3.2(i), they can be classified by the space $B\mathcal{TOP}_n$. Note that two $((\mathbb{R}^n, 0), \mathcal{TOP}_n)$-bundles ξ, η over B are equivalent iff there exists a homeomorphism

$$(\operatorname{ts} \xi, s_\xi) \to (\operatorname{ts} \eta, s_\eta)$$

over B. Indeed, consider the space $E(\xi)$ of all maps $f : (\mathbb{R}^n, 0) \to (\operatorname{ts}(\xi), s_\xi)$ such that f is a homeomorphism onto a fiber. Then

$$P(\xi) := \{p : E(\xi) \to B, \, p(f) = p_\xi f(0)\}$$

is a locally trivial principal $\mathcal{TOP}_n$-bundle, and ξ is associated with $P(\xi)$. Now, the fiberwise homeomorphism $\operatorname{ts}\xi \to \operatorname{ts}\eta$ induces an equivalence $P(\xi) \to P(\eta)$ over B.

(c) Piecewise linear (in future PL) $\mathbb{R}^n$-bundles. (See Rourke–Sanderson [1] about PL notions.) These are $((\mathbb{R}^n, 0), \mathcal{TOP}_n)$-bundles $p : Y \to X$, where X and Y are simplicial complexes (not necessarily finite, see Hilton–Wiley [1], I.10) and the projection p and the section $X \to Y$ are PL maps, and, moreover, for every simplex $\Delta \subset B$ there exists a PL isomorphism φ such that the diagram

$$\begin{array}{ccc} p^{-1}(\Delta) & \xrightarrow{\varphi} & \Delta \times \mathbb{R}^n \\ {\scriptstyle p}\big\downarrow & & \big\downarrow{\scriptstyle p_1} \\ \Delta & = & \Delta \end{array}$$

commutes. Equivalence of such bundles is defined to be a fiberwise PL isomorphism which preserves the sections. One can formulate (and prove) a variant of the Brown Representability Theorem for the category of simplicial complexes and check that both MV and wedge properties hold for piecewise linear $\mathbb{R}^n$-bundles, cf. Kirby–Siebenmann [1], Essay 4, § 8. The corresponding classifying space is denoted by $B\mathcal{PL}_n$.

Justification of the last notation. Let Δ^k be be the standard k-dimensional simplex. Let pl_n be a simplicial group such that its k-simplices are PL isomorphisms $\Delta^k \times \mathbb{R}^n \to \Delta^k \times \mathbb{R}^n$ preserving the zero section and commuting with the projections on Δ^k. The faces and the degeneracies are induced by the corresponding maps of Δ^i. Let $\mathcal{PL}_n$ be the geometric realization of pl_n. Then the classifying space $B\mathcal{PL}_n$ for the group $\mathcal{PL}_n$ classifies piecewise linear $\mathbb{R}^n$-bundles, see Lashof–Rothenberg [1], Kuiper–Lashof [1].

(d) Vector bundles. These are well-known objects, see Fuks–Rokhlin [1], Atiyah [4], Karoubi [1], Switzer [1], Fuks–Rokhlin [1], Husemoller [1], etc. Let $\mathcal{O}_n$ be the group of all orthogonal transformations of $\mathbb{R}^n$. Every $((\mathbb{R}^n,0),\mathcal{O}_n)$-bundle can be regarded as an n-dimensional vector bundle, and every vector bundle over a CW-space admits a Riemannian metric and so can be turned into an $((\mathbb{R}^n,0),\mathcal{O}_n)$-bundle. Thus, vector bundles can be classified by the space $B\mathcal{O}_n$, see loc cit.

4.2. Conventions. (a) Sometimes, when it is possible, we say just $\mathcal{PL}$-bundle rather than $((\mathbb{R}^n,0),\mathcal{PL}_n)$-bundle, and so on. We also call objects of the four classes above $\mathcal{G}_n$-, $\mathcal{TOP}_n$-, $\mathcal{PL}_n$- and $\mathcal{O}_n$-objects, respectively. The reason is that frequently we shall consider these four classes simultaneously. Therefore we introduce the uniform symbol $\mathcal{V}$ in order to denote any of the four symbols $\mathcal{G},\mathcal{TOP},\mathcal{PL},\mathcal{O}$. For example, we can (and shall) speak about $\mathcal{V}_n$-objects, $\mathcal{V}$-equivalences of $\mathcal{V}$-objects, classifying space $B\mathcal{V}_n$, etc. The universal $\mathcal{V}$-object over $B\mathcal{V}_n$ will be denoted by $\gamma^n_{\mathcal{V}}$.

(b) We denote by θ^n_B the standard trivial $\mathcal{V}_n$-object over a space B; there is no necessity to specify $\mathcal{V}$. Moreover, sometimes we shall omit the subscript B if it is clear from the context.

(c) Because of 1.67, we can and shall assume that every space $B\mathcal{V}_n$ is a CW-space.

Traditionally there arise PL and topological *microbundles* in geometric topology (e.g., as tangent and normal microbundles of the corresponding manifolds). However, one can prove that they are equivalent (as microbundles) to bundles of the corresponding classes, cf. 7.7 below.

There is a hierarchy of the four classes above. Every n-dimensional vector bundle over a simplicial complex is a $\mathcal{PL}_n$-object, every $\mathcal{PL}_n$-object is a $\mathcal{TOP}_n$-object by definition, and every $((\mathbb{R}^n,0),\mathcal{TOP}_n)$-bundle can be turned into a spherical fibration by deleting the section. This hierarchy induces the sequence of forgetful maps

$$B\mathcal{O}_n \xrightarrow{a^{\mathcal{O}}_{\mathcal{PL}}} B\mathcal{PL}_n \xrightarrow{a^{\mathcal{PL}}_{\mathcal{TOP}}} B\mathcal{TOP}_n \xrightarrow{a^{\mathcal{TOP}}_{\mathcal{G}}} B\mathcal{G}_n, \tag{4.3}$$

where, say, $a^{\mathcal{O}}_{\mathcal{PL}} = a^{\mathcal{O}}_{\mathcal{PL}}(n)$ classifies the universal $((\mathbb{R}^n,0),\mathcal{O}_n)$-bundle regarded as an $((\mathbb{R}^n,0),\mathcal{PL}_n)$-bundle. We introduce the ordering

$$\mathcal{O}\le\mathcal{PL}\le\mathcal{TOP}\le\mathcal{G};$$

e.g., $\mathcal{V}\le\mathcal{TOP}$ means that $\mathcal{V}=\mathcal{O}$, $\mathcal{PL}$ or $\mathcal{TOP}$. Thus, for $\mathcal{V}'\le\mathcal{V}$ one has a forgetful map $a^{\mathcal{V}'}_{\mathcal{V}}(n): B\mathcal{V}'_n\to B\mathcal{V}_n$. The homotopy fiber of $a^{\mathcal{V}'}_{\mathcal{V}}(n)$ is denoted by $\mathcal{V}_n/\mathcal{V}'_n$.

4.4. Definition. Let ξ be a $\mathcal{V}_n$-object.

(a) Given a $\mathcal{V}_n$-object λ, we define a λ-prestructure on ξ to be a $\mathcal{V}_n$-morphism $a : \xi \to \lambda$. We say that two λ-prestructures are equivalent if there is a $\mathcal{V}_n$-morphism $b : \xi \times 1_I \to \lambda$ such that $b|(\xi \times \{i\}) = a_i$, $i = 0, 1$. An equivalence class of prestructures is called a λ-*structure* on ξ.

(b) Given a map $\varphi : B \to B\mathcal{V}_n$, we define a (B, φ)-structure on ξ to be a $(\varphi^* \gamma^n_{\mathcal{V}})$-structure on it.

4.5. Remarks. (a) If $\mathcal{V} \le \mathcal{V}'$ then λ can be canonically regarded as a $\mathcal{V}'$-object, cf. (4.3). So, we can talk about λ-structures on $\mathcal{V}'$-objects provided $\mathcal{V} \le \mathcal{V}'$. For example, we can consider $\gamma^n_{\mathcal{O}}$-structures on $\mathcal{G}_n$-objects, i.e., vector structures on spherical fibrations, cf. Browder [3].

(b) We leave it to the reader to prove the following analog of 2.3 for $\mathcal{V}$-objects: *Every classifying morphism* $\omega : \xi \to \gamma^n_{\mathcal{V}}$ *induces a bijection*

$$\Phi_\omega : [\mathrm{Lift}_\varphi \, \mathrm{bs}\, \omega] \to \{(B, \varphi)\text{-}structures\ on\ \xi\}.$$

(c) Every morphism $\sigma : \eta \to \xi$ of $\mathcal{V}_n$-objects induces a function

$$\begin{aligned} \sigma^\flat : \{\lambda\text{-structures on } \xi\} &\longrightarrow \{\lambda\text{-structures on } \eta\}, \\ \{a : \xi \to \lambda\} &\longmapsto \{\sigma a : \eta \to \lambda\}. \end{aligned}$$

Consider a $\mathcal{V}_m$-object ξ over X and a $\mathcal{V}_n$-object η over Y. If $\mathcal{V} \le \mathcal{TOP}$ then the product $\xi \times \eta$ is a $\mathcal{V}_{m+n}$-object. If $\mathcal{V} = \mathcal{G}$ then, by 1.43(v), the bundle join $\xi * \eta$ is a $\mathcal{G}_{m+n}$-object. Given two $\mathcal{V}$-objects ξ, η over the same base X, we define the *Whitney sum* $\xi \oplus \eta$ to be $d^*(\xi \times \eta)$ for $\mathcal{V} \le \mathcal{TOP}$ and $d^*(\xi * \eta)$ for $\mathcal{V} = \mathcal{G}$, where $d : X \to X \times X$ is the diagonal.

Let $\theta^n = \theta^n_{B\mathcal{V}}$ denote the standard trivial $\mathcal{V}_n$-object over $B\mathcal{V}$, and let a morphism $\rho_n = \rho_n^{\mathcal{V}} : \gamma^n_{\mathcal{V}} \oplus \theta^1 \to \gamma^{n+1}_{\mathcal{V}}$ classify $\gamma^n_{\mathcal{V}} \oplus \theta^1$. We set $r_n := \mathrm{bs}\, \rho_n : B\mathcal{V}_n \to B\mathcal{V}_{n+1}$. Furthermore, let $\mu^{\mathcal{V}}_{m,n} : B\mathcal{V}_m \times B\mathcal{V}_n \to B\mathcal{V}_{m+n}$ classify the $\mathcal{V}$-object $\gamma^n_{\mathcal{V}} \times \gamma^m_{\mathcal{V}}$ for $\mathcal{V} \le \mathcal{TOP}$ and $\gamma^n_{\mathcal{G}} * \gamma^m_{\mathcal{G}}$ for $\mathcal{V} = \mathcal{G}$.

4.6. Proposition. (i) *For every* m, n, p *the following diagram commutes up to homotopy:*

$$\begin{array}{ccc} B\mathcal{V}_m \times B\mathcal{V}_n \times B\mathcal{V}_p & \xrightarrow{\mu_{m,n} \times 1} & B\mathcal{V}_{m+n} \times B\mathcal{V}_p \\ {\scriptstyle 1 \times \mu_{n,p}} \downarrow & & \downarrow {\scriptstyle \mu_{m+n,p}} \\ B\mathcal{V}_m \times B\mathcal{V}_{n+p} & \xrightarrow{\mu_{m,n+p}} & B\mathcal{V}_{m+n+p} \end{array} .$$

(ii) *For every* m, n, p, q, *the following diagram commutes up to homotopy:*

$$\begin{array}{ccc} B\mathcal{V}_m \times B\mathcal{V}_n & \xrightarrow{R^m_p \times R^n_q} & B\mathcal{V}_p \times B\mathcal{V}_q \\ {\scriptstyle \mu} \downarrow & & \downarrow {\scriptstyle \mu} \\ B\mathcal{V}_{m+n} & \xrightarrow{R^{m+n}_{p+q}} & B\mathcal{V}_{p+q} \end{array} .$$

Here $R^a_b = r_{b-1} \circ r_{b-2} \circ \cdots \circ r_{a+1} \circ r_a : B\mathcal{V}_a \to B\mathcal{V}_b$, $a \le b$.

(iii) *For every* m, n *and every* $\mathcal{V}' \le \mathcal{V}$, *the following diagram commutes up to homotopy:*

$$\begin{array}{ccc} B\mathcal{V}'_m \times B\mathcal{V}'_n & \xrightarrow{\mu'_{m,n}} & B\mathcal{V}'_{m+n} \\ {\scriptstyle a\times a}\downarrow & & \downarrow{\scriptstyle a} \\ B\mathcal{V}_m \times B\mathcal{V}_n & \xrightarrow{\mu_{m,n}} & B\mathcal{V}_{m+n}. \end{array}$$

(iv) *For every* m, n, *the following diagram commutes up to homotopy:*

$$\begin{array}{ccc} B\mathcal{V}_m \times B\mathcal{V}_n & \xrightarrow{\mu_{m,n}} & B\mathcal{V}_{m+n} \\ {\scriptstyle T}\downarrow & & \| \\ B\mathcal{V}_n \times B\mathcal{V}_m & \xrightarrow{\mu_{n,m}} & B\mathcal{V}_{m+n}. \end{array}$$

Here T *switches the factors.*

Proof. (i) This is clear for $\mathcal{V} \le \mathcal{TOP}$, since $(\xi \times \eta) \times \zeta = \xi \times (\eta \times \zeta)$ for all $\mathcal{TOP}$-objects ξ, η, ζ. So, it remains to prove that $(\xi * \eta) * \zeta$ is equivalent to $\xi * (\eta * \zeta)$ for every spherical fibrations ξ, η, ζ. Recall that every point of $\operatorname{ts}(\xi * \eta)$ can be written as a suitable equivalence class $[x, t, y], x \in \operatorname{ts}\xi, y \in \operatorname{ts}\eta, t \in [0, 2]$. Given any three bundles ξ, η, ζ, we define a bundle morphism $\varphi : (\xi * \eta) * \zeta \to \xi * (\eta * \zeta)$ by setting

$$\operatorname{ts}\varphi((x, t, y), s, z) = (x, t, (y, s, z)), x \in \operatorname{ts}\xi, y \in \operatorname{ts}\eta, z \in \operatorname{ts}\zeta, s, t \in [0, 2].$$

Now, if ξ, η, ζ are spherical fibrations, then $\operatorname{ts}\varphi$ induces a homotopy equivalence of fibers (prove it!), and so, by 1.30(i), φ is a homotopy equivalence over the base.

(ii) This can be proved as (i), so we leave it to the reader.

(iii) This is clear for $\mathcal{V} \le \mathcal{TOP}$, and so it suffices to consider $\mathcal{V}' = \mathcal{TOP}, \mathcal{V} = \mathcal{G}$. Let ξ, η be two $\mathcal{TOP}$-objects over CW-bases, $\dim \xi = i$, $\dim \eta = j$. Given a $\mathcal{TOP}$-object ζ, let ζ' be the spherical fibration (in fact, a locally trivial bundle) obtained from ζ by deleting the zero section. We must prove that $\xi' * \eta' \simeq_B (\xi \times \eta)'$ where $B = \operatorname{bs}(\xi \times \eta)$.

The group $\mathcal{TOP}_m \times \mathcal{TOP}_n$ acts on $\mathbb{R}^m \times \mathbb{R}^n$, as well as on $(\mathbb{R}^m \times \mathbb{R}^n) \setminus 0$, in the obvious way. Furthermore, it acts on $(\mathbb{R}^m \setminus 0) * (\mathbb{R}^n \setminus 0)$ as follows:

$$(g, g')[x, t, y] = [gx, t, g'y],$$
$$g \in \mathcal{TOP}_m, g' \in \mathcal{TOP}_n, [x, t, y] \in (\mathbb{R}^m \setminus 0) * (\mathbb{R}^n \setminus 0).$$

Now, $(\xi \times \eta)'$ (resp. $\xi' \times \eta'$) is the $((\mathbb{R}^m \times \mathbb{R}^n) \setminus 0, \mathcal{TOP}_m \times \mathcal{TOP}_n)$-bundle (resp. $((\mathbb{R}^m \setminus 0) * (\mathbb{R}^n \setminus 0), \mathcal{TOP}_m \times \mathcal{TOP}_n)$-bundle) associated with $\xi \times \eta$. We define the map

$$f : (\mathbb{R}^m \setminus 0) * (\mathbb{R}^n \setminus 0) \to (\mathbb{R}^m \times \mathbb{R}^n) \setminus 0,\ f[x, t, y] = \frac{1}{\sqrt{2t^2 - 4t + 4}} (tx, (2 - t)y).$$

(Comment: $t^2 + (2-t)^2 = 2t^2 - 4t + 4$.) Clearly, f is a $\mathcal{TOP}_m \times \mathcal{TOP}_n$-equivariant map, and so it yields a morphism $\varphi : \xi' * \eta' \to (\xi \times \eta)'$ of the associated bundles.

Furthermore, f is a homotopy equivalence. Indeed, it is easy to see that

$$f|S^{m-1} * S^{n-1} : S^{m-1} * S^{n-1} \to S^{m+n-1}$$

is a homeomorphism, where $S^{k-1} := \{x \in \mathbb{R}^k \mid ||x|| = 1\}$.

(iv) We leave it to the reader. □

Let $B\mathcal{V}$ be the telescope of the sequence

$$\cdots \to B\mathcal{V}_{n-1} \to B\mathcal{V}_n \xrightarrow{r_n} B\mathcal{V}_{n+1} \to \cdots .$$

We denote by

$$(4.7) \qquad j_n = j_n^{\mathcal{V}} : B\mathcal{V}_n \to B\mathcal{V}$$

the obvious inclusion $B\mathcal{V}_n = B\mathcal{V}_n \times \{n\} \subset B\mathcal{V}_n \times [n, n+1] \to B\mathcal{V}$.

It is well known that for every finite (and in fact finite dimensional, see 4.27(viii) below) CW-space X the set $[X, B\mathcal{V}]$ can be described as follows. One says that two $\mathcal{V}$-objects ξ and η, $\dim \xi = m$, $\dim \eta = n$, are stably equivalent if $\xi \oplus \theta^{N+n} \simeq \eta \oplus \theta^{N+m}$ for some (large) N. Then, the set of all stable $\mathcal{V}$-objects over X is in a bijective correspondence with $[X, B\mathcal{V}]$.

Based on this, we give the following definition.

4.8. Definition. Given a space $X \in \mathscr{C}$, we define a stable $\mathcal{V}$-object ξ over X to be a map $f : X \to B\mathcal{V}$. In this case we also write $\xi = \{f : X \to B\mathcal{V}\}$ and say (tautologically) that f classifies ξ. We say that two stable $\mathcal{V}$-objects $\alpha = \{f : X \to B\mathcal{V}\}$ and $\beta = \{g : X \to B\mathcal{V}\}$ are *equivalent* if $f \simeq g : X \to B\mathcal{V}$.

(b) Given two stable $\mathcal{V}$-objects $\xi = \{f : X \to B\mathcal{V}\}$ and $\eta = \{g : Y \to B\mathcal{V}\}$, a morphism $q : \xi \to \eta$ is a map $q : X \to Y$ such that $gq = f$.

For every map $h : X \to Y$ and every stable $\mathcal{V}$-object $\xi = \{f : Y \to B\mathcal{V}\}$ we define the induced $\mathcal{V}$-object $h^*\xi := \{fh : X \to B\mathcal{V}\}$. We also have the canonical morphism $\mathfrak{I}_{h,\xi} := h : h^*\xi \to \xi$.

Notice that $1_{B\mathcal{V}}$ is a universal stable $\mathcal{V}$-object.

(c) Let ξ be a $\mathcal{V}_n$-object classified by $f : X \to B\mathcal{V}_n$. We define its stabilization $\xi_{\text{st}} = (\xi, f)_{\text{st}}$ to be the stable $\mathcal{V}$-object $X \xrightarrow{f} B\mathcal{V}_n \xrightarrow{j_n} B\mathcal{V}$.

4.9. Definition. Let $\xi = \{f : X \to B\mathcal{V}\}$ be a stable $\mathcal{V}$-object. Given a map $\varphi : B \to B\mathcal{V}$, a (B, φ)-prestructure on ξ is a pair (a, H) where $a : X \to B$ is a map and $H : X \times I \to B\mathcal{V}$ is a homotopy from φa to f. Two prestructures (a_0, H_0) and (a_1, H_1) are equivalent if there are maps $b : X \times I \to B$ and $J : X \times I \times I \to B\mathcal{V}$ such that $b|X \times \{i\} = a_i, i = 0, 1$ and $J|X \times I \times \{0\} = \varphi b$, $J(x, t, 1) = f(x)$ for every $t \in I$, $J|X \times \{i\} \times I = H_i, i = 0, 1$. A (B, φ)-*structure* on ξ is an equivalence class of prestructures. We denote by $[a, H]$ the equivalence class of (a, H).

Below in this context we sometimes call φ a *structure map*, i.e., we use the term "structure map" when we want to emphasize that φ is not just a map but a map which is used for structuralization of $\mathcal{V}$-objects.

4.10. Proposition. *Let $\xi = \{f : X \to B\mathcal{V}\}$ be a stable $\mathcal{V}$-object.*

(i) If $\varphi : B \to B\mathcal{V}$ and $\psi : C \to B\mathcal{V}$ are homotopy equivalent maps then (B, φ)-structures on ξ are in a bijective correspondence with (C, ψ)-structures on it.

(ii) If $\varphi : B \to B\mathcal{V}$ is a fibration then the set of all (B, φ)-structures on ξ is in a canonical bijective correspondence with the set $[\mathrm{Lift}_\varphi f]$.

Proof. (i) We leave it to the reader.

(ii) Let $g : X \to B$ be a φ-lifting of f. Considering the (stationary) homotopy $H : X \times I \to B\mathcal{V}, H(x,t) := f(x)$, we conclude that (g, H) is a (B, φ)-prestructure on ξ. Clearly, vertically homotopic liftings yield equivalent prestructures, and so we have a correspondence

$$\Phi : [\mathrm{Lift}_\varphi f] \to \{(B, \varphi)\text{-structures on } \xi\}.$$

We prove that Φ is surjective. Consider a (B, φ)-prestructure (a_0, H_0) on ξ where $a_0 : X \to B$ is a map and $H_0 : X \times I \to B\mathcal{V}$ is a homotopy from φa_0 to f. Then there is a homotopy $b : X \times I \to B$ with $b|X \times \{0\} = a_0$ and $\varphi b = H_0$. We set $a_1 := b|X \times \{1\}$ and define $H_1 : X \times I \to B\mathcal{V}, H_1(x,t) := f(x)$. Finally, we define $J : X \times I \times I \to B\mathcal{V}, J(x,s,t) := H_0(x, s+t-st)$, and it is clear that (b, J) yields an equivalence between (a_0, H_0) and (a_1, H_1).

We prove that Φ is injective. Let $g_0, g_1 : X \to B$ be two φ-liftings of f, and let $H_i : \varphi g_i \simeq f, i = 1, 2$, be the stationary homotopies. Suppose that (g_0, H_0) and (g_1, H_1) are equivalent (B, φ)-prestructures on ξ, and consider $b : X \times I \to B$ and $J : X \times I \times I \to B\mathcal{V}$ such that (b, J) yields this equivalence, see 4.9. By 1.17(ii), there is a φ-lifting $\widehat{J} : X \times I \times I \to B$ of J such that $\widehat{J}|X \times I \times \{0\} = b$, $\widehat{J}(x, i, t) = g_i(x)$ for every $x \in X, t \in I, i = 0, 1$. Then $\widehat{J}|X \times I \times \{1\}$ is a vertical homotopy between g_0 and g_1. □

4.11. Proposition-Construction. *Let $\xi = \{f : X \to B\mathcal{V}\}$ and $\eta = \{g : X \to B\mathcal{V}\}$ be two equivalent stable $\mathcal{V}$-objects over X. Then every homotopy $F : f \simeq g$ induces a bijection*

$$\{(B, \varphi)\text{-structures on } \xi\} \longleftrightarrow \{(B, \varphi)\text{-structures on } \eta\}$$

for every structure map $\varphi : B \to B\mathcal{V}$.

Proof. Let (a, H) be a (B, φ)-prestructure on ξ. We define the homotopy $H' : \varphi a \simeq f \simeq g$ where the first homotopy is H and the second one is F. Then, clearly, (a, H') is a (B, φ)-prestructure on η, and it is easy to see that in this way we get a well-defined correspondence

$$L_F : \{(B,\varphi)\text{-structures on } \xi\} \longrightarrow \{(B,\varphi)\text{-structures on } \eta\}.$$

Furthermore, we define a homotopy $G : g \simeq f, G(x,t) := F(x,1-t)$. Similarly to above, we get a correspondence

$$L_G : \{(B,\varphi)\text{-structures on } \eta\} \longrightarrow \{(B,\varphi)\text{-structures on } \xi\}.$$

We leave it to the reader to check that L_G is inverse to L_F. □

4.12. Proposition–Construction. *Let $\omega : \xi \to \gamma^n_{\mathcal{V}}$ be a classifying morphism for a $\mathcal{V}_n$-object ξ, and let $\varphi : B \to B\mathcal{V}$ be an arbitrary fibration.*

(i) *Consider the pull-back diagram*

$$\begin{array}{ccc} B_n & \longrightarrow & B \\ {\scriptstyle \varphi_n}\big\downarrow & & \big\downarrow{\scriptstyle \varphi} \\ B\mathcal{V}_n & \xrightarrow{j_n} & B\mathcal{V}. \end{array}$$

Then ω induces a bijection

$$\Phi_\omega : \{(B,\varphi)\textit{-structures on } \xi_{\mathrm{st}}\} \to \{B_n,\varphi_n)\textit{-structures on } \xi\}$$

(ii) *Every morphism $\sigma : \eta \to \xi$ of $\mathcal{V}_n$-objects induces a function*

$$\sigma^! : \{(B,\varphi)\textit{-structures on } \xi_{\mathrm{st}}\} \to \{(B,\varphi)\textit{-structures on } \eta_{\mathrm{st}}\}.$$

Proof. (i) We let $f := \mathrm{bs}\,\omega$. Because of 4.10(ii) and 4.5(b), we have the bijections

$$\begin{aligned} \{(B,\varphi)\text{-structures on } \xi_{\mathrm{st}}\} \longleftrightarrow & [\mathrm{Lift}_\varphi\, j_n f] \\ = & [\mathrm{Lift}_{\varphi_n} f] \xrightarrow{\Psi_\omega} \{(B_n,\varphi_n)\text{-structures on } \xi\}. \end{aligned}$$

(ii) We equip η with the classifying morphism $\omega\sigma : \eta \to \gamma^n_{\mathcal{V}}$. We define $\sigma^!$ to be a function such that the diagram

$$\begin{array}{ccc} \{(B,\varphi)\text{-structures on } \xi_{\mathrm{st}}\} & \xrightarrow{\Phi_\omega} & \{(B_n,\varphi_n)\text{-structures on } \xi\} \\ {\scriptstyle \sigma^!}\big\downarrow & & \big\downarrow{\scriptstyle \sigma^\flat} \\ \{(B,\varphi)\text{-structures on } \eta_{\mathrm{st}}\} & \xrightarrow{\Phi_{\omega\sigma}} & \{(B_n,\varphi)\text{-structures on } \eta\} \end{array}$$

commutes. Here the horizontal arrows are the bijections from (i) and $\sigma^\flat$ is the function described in 4.5(c) (recall that a (B_n,φ_n)-structure is just a $(\varphi_n^*\gamma^n_{\mathcal{V}})$-structure). □

Recall that θ^k_X denotes the standard trivial $\mathcal{V}_k$-object over a space X.

4.13. Lemma. *Let ξ be a $\mathcal{V}_n$-object classified by a map $f : X \to B\mathcal{V}_n$. Then for every fibration $\varphi : B \to B\mathcal{V}$ there is a canonical bijection*

$$K_f : \{(B,\varphi)\textit{-structures on } \xi_{\rm st}\} \longleftrightarrow \{(B,\varphi)\textit{-structures on } (\xi \oplus \theta^1_X)_{\rm st}\}$$

where we assume that $\xi \oplus \theta^1_X$ *is classified by* $r_n f : X \to B\mathcal{V}_{n+1}$.

Proof. Consider the composition $B\mathcal{V}_n \xrightarrow{r_n} B\mathcal{V}_{n+1} \xrightarrow{j_{n+1}} B\mathcal{V}$. The standard deformation $F : Mr_n \times I \to B\mathcal{V}_{n+1}$, see I.3.16(a), yields canonically a homotopy $H : j_n \simeq j_{n+1}r_n$. Now, by 4.11, the homotopy Hf induces the desired bijection. □

4.14. Constructions. Note that $\operatorname{ts}\theta^1_X = X \times F$ where $F = \mathbb{R}$ for $\mathcal{V} \leq \mathcal{TOP}$ and $F = \{-1, 1\}$ for $\mathcal{V} = \mathcal{G}$. For simplicity, we write γ^n instead of $\gamma^n_{\mathcal{V}}$.

(a) We define a morphism $e = e_X : \theta^1_X \to \theta^1_X$ by setting $(\operatorname{ts} e)(x, f) := (x, -f), x \in X, f \in F$.

(b) For simplicity, we denote $\theta^1_{\rm pt}$ by θ^1. Let $\omega : \xi \to \gamma^n_{\mathcal{V}}$ classify a $\mathcal{V}_n$-object ξ over X. We define

$$\widehat{\omega} : \xi \oplus \theta^1_X = \xi \times \theta^1 \xrightarrow{\omega \times 1} \gamma^n_{\mathcal{V}} \times \theta^1 = \gamma^n_{\mathcal{V}} \oplus \theta^1_{B\mathcal{V}} \xrightarrow{\rho_n} \gamma^{n+1}_{\mathcal{V}}.$$

(c) Given ω as in (b), consider the morphism $\widehat{\omega} : \xi \oplus \theta^1_X \to \gamma^{n+1}$. Now, because of 4.12 and 4.13, we have the bijection

$$\Phi_{\widehat{\omega}} K_{\operatorname{bs}\omega} : \{(B, \varphi)\text{-structures on } \xi_{\rm st}\} \leftrightarrow \{(B_{n+1}, \varphi_{n+1})\text{-structures on } \xi \oplus \theta^1_X\}.$$

4.15. Definition. Let ξ be a $\mathcal{V}_n$-object over X.

(a) Given a $\mathcal{V}_{n+1}$-object λ, let a morphism $a : \xi \oplus \theta \to \lambda$ give a λ-structure on $\xi \oplus \theta$, see 2.1. Then the morphism

$$\xi \oplus \theta \xrightarrow{1 \oplus e} \xi \oplus \theta \xrightarrow{a} \lambda$$

gives us a certain (in general, another) λ-structure on $\xi \oplus \theta$. This structure is called the *opposite λ-structure* to the given one.

We leave it to the reader to prove that the opposite structure is well-defined and that opposite to opposite yields the original structure.

(b) Given a map $\psi : C \to B\mathcal{V}_{n+1}$, we set $\lambda = \psi^*\gamma^{n+1}$ and recall that, by definition,

$$\{(C, \psi)\text{-structures on } \xi \oplus \theta^1_X\} = \{\lambda\text{-structures on } \xi \oplus \theta^1_X\},$$

So, two (C, ψ)-structures on $\xi \oplus \theta^1_X$ are called *opposite* to one another if the corresponding λ-structures on $\xi \oplus \theta$ are opposite, as defined in (a).

(c) Now we assume that ξ is equipped with a classifying morphism $\xi \to \gamma^n_{\mathcal{V}}$. Given a fibration $\varphi : B \to B\mathcal{V}$, consider the bijection as in 4.14(c). Two (B, φ)-structures on $\xi_{\rm st}$ are called *opposite* to one another if the corresponding (B_{n+1}, φ_{n+1})-structures on $\xi \oplus \theta^1_X$ are opposite, as defined in (b).

Now we consider $(S^n, *)$-fibrations. By 1.74, they are classified by a space $B_{(S^n,*)}$. By 1.75(iii), $B_{(S^n,*)} \simeq B\mathcal{F}_n$, where $\mathcal{F}_n := \mathscr{H}(S^n, *)$. We shall use the simpler notation $B\mathcal{F}_n$ rather then $B_{(S^n,*)}$, and we shall use the term "$\mathcal{F}_n$-objects" for $(S^n,*)$-fibrations. The universal $\mathcal{F}_n$-object over $B\mathcal{F}_n$ will be denoted by $\gamma^n_{\mathcal{F}}$.

Let $\theta = \theta^1_{\mathcal{F}}$ be the (standard trivial) $(S^1,*)$-bundle over pt. Let $\rho^{\mathcal{F}}_n : \gamma^n_{\mathcal{F}} \wedge^h \theta \to \gamma^{n+1}_{\mathcal{F}}$ be the classifying morphism for $\gamma^n_{\mathcal{F}} \wedge^h \theta$. We set $r^{\mathcal{F}}_n := \operatorname{bs} \rho^n_{\mathcal{F}}$ and define $B\mathcal{F}$ to be the telescope of the sequence

$$\{\cdots \to B\mathcal{F}_n \xrightarrow{r^{\mathcal{F}}_n} B\mathcal{F}_{n+1} \to \cdots\},$$

and we denote by $j_n = j^{\mathcal{F}}_n : B\mathcal{F}_n \to B\mathcal{F}$ the obvious inclusion, cf. (4.7).

4.16. Definition. Similarly to 4.8, we define a stable $\mathcal{F}$-object α over X to be a map $f : X \to B\mathcal{F}$. Given two stable $\mathcal{F}$-objects $\alpha = \{f : X \to B\mathcal{F}\}$ and $\beta = \{g : Y \to B\mathcal{F}\}$, a morphism $\varphi : \alpha \to \beta$ is a map $a : X \to Y$ with $ga = f$.

Given an $\mathcal{F}_n$-object $\alpha = \{f : X \to B\mathcal{F}_n\}$, we define its *stabilization* $\alpha_{\text{st}} = (\alpha, f)_{\text{st}}$ to be a stable $\mathcal{F}$-object $X \xrightarrow{f} B\mathcal{F}_n \xrightarrow{j_n} B\mathcal{F}$.

4.17. Construction. The $\mathcal{TOP}_n$-action on $\mathbb{R}^n$ can be extended to a $\mathcal{TOP}_n$-action on the one-point compactification $S^n = \mathbb{R}^n \cup \{\infty\}$ of $\mathbb{R}^n$. So, we have a $\mathcal{TOP}_n$-space $(S^n,*)$ where the base point $*$ is ∞. Now, given a $\mathcal{TOP}_n$-object ξ, one can form the $((S^n,*),\mathcal{TOP}_n)$-bundle $\xi^\bullet$ using the fiberwise one-point compactification, where the "infinities" of fibers form the section. More accurately, if ξ is associated with a locally trivial principal $\mathcal{TOP}_n$-bundle λ, then $\xi^\bullet$ is the $((S^n,*),\mathcal{TOP}_n)$-bundle associated with λ. Similarly, for every $\mathcal{V}_n$-object ξ with $\mathcal{V} \le \mathcal{TOP}$ we can construct the $((S^n,*),\mathcal{V}_n)$-object $\xi^\bullet$.

Furthermore, consider $S^0 = \{-1,+1\}$ as the trivial S^0-bundle θ over pt. Given a $\mathcal{G}_n$-object ξ, set $\xi^\bullet = \xi * \theta$. Then the points of $\operatorname{ts}\xi^\bullet$ are suitable equivalence classes $[x,t,y]$ of triples (x,t,y), $x \in \operatorname{ts}\xi, t \in [0,2], y \in \{-1,1\}$. Now, we define

$$s : \operatorname{bs}\xi \to \operatorname{ts}\xi^\bullet,\ s(b) = [x,2,1],\ \text{where } b = p_\xi(x).$$

So, every $\mathcal{V}_n$-object ξ can be naturally converted into an $\mathcal{F}_n$-object $\xi^\bullet$.

Note that the correspondence $\xi \mapsto \xi^\bullet$ yields maps

(4.18) $$a^{\mathcal{V}}_{\mathcal{F}}(n) : B\mathcal{V}_n \to B\mathcal{F}_n \text{ and (as } n \to \infty)\ a^{\mathcal{V}}_{\mathcal{F}} : B\mathcal{V} \to B\mathcal{F}.$$

So, the sequence (4.3) can be elongated as

$$BO \xrightarrow{a^{\mathcal{O}}_{\mathcal{PL}}} B\mathcal{PL} \xrightarrow{a^{\mathcal{PL}}_{\mathcal{TOP}}} B\mathcal{TOP} \xrightarrow{a^{\mathcal{TOP}}_{\mathcal{G}}} B\mathcal{G} \xrightarrow{a^{\mathcal{G}}_{\mathcal{F}}} B\mathcal{F}$$

and we extend the ordering above by setting $\mathcal{O} \le \mathcal{PL} \le \mathcal{TOP} \le \mathcal{G} \le \mathcal{F}$.

4.19. Lemma. *For every $\mathcal{V}_m$-object ξ and $\mathcal{V}_n$-object η over CW-bases,*

$$(\xi \dagger \eta)^\bullet \simeq_B \xi^\bullet \wedge^h \eta^\bullet$$

where $B = \mathrm{bs}\,\xi \times \mathrm{bs}\,\eta$ *and* $\dagger$ *means* $\times$ *for* $\mathcal{V} \leq \mathcal{TOP}$ *and* $*$ *for* $\mathcal{V} = \mathcal{G}$.

Proof. Because of 4.6(iii), it suffices to consider only $\mathcal{V} = \mathcal{G}$. Firstly, some constructions for spaces. Given a space Z, we regard (parametrize) the suspension SZ as $SZ = Z \times [0,1]/(Z \times \{0\} \cup Z \times \{1\})$, and we denote by $[i] \in SZ$ the point given by $Z \times \{i\}$, $i = 0, 1$. As usual, points of the join $X * Y$ are written as triples $[x,t,y], x \in X, y \in Y, t \in [0,2]$.

Given two spaces X, Y, we consider the map $f : SX \times SY \to S(X * Y)$,

$$f([x,s],[y,t]) = \begin{cases} \left[\left[x, \frac{2t}{s+t}, y\right], s+t-st\right] & \text{if } (s,t) \neq (0,0), \\ [0] & \text{otherwise,} \end{cases}$$

where $x \in X, y \in Y, s, t \in [0,1]$.

We regard SZ as a pointed space with base point $[1]$. Now, f maps the wedge

$$SX \vee SY = SX \times \{*\} \cup \{*\} \times SY$$

to the base point of $S(X * Y)$, and so one can pass f through a quotient map $g : SX \wedge^h SY \to S(X * Y)$ such that $g|C(SX \vee SY)$ is a constant map.

We prove that g is a pointed homotopy equivalence if $X = S^{m-1}, Y = S^{n-1}$. Indeed, g can be decomposed as

$$g : S^m \wedge^h S^n \xrightarrow{q} S^m \wedge S^n \xrightarrow{h} S(S^{m-1} * S^{n-1})$$

where the quotient map q is a homotopy equivalence. Note that both spaces $S^m \wedge S^n$ and $S(S^{m-1} * S^{n-1})$ are homeomorphic to S^{m+n}. Now, let

$$U = \left\{[[x,\lambda,y],\mu] \in S(S^{m-1} * S^{n-1}) \mid 0 < \lambda < 1/2, 0 < \mu < 1\right\}.$$

Then U is an open set of $S(S^{m-1} * S^{n-1})$, and $h|h^{-1}U : h^{-1}U \to U$ is a homeomorphism because for every $(\lambda,\mu) \in (0,1/2) \times (0,1)$ the system

$$\begin{cases} \frac{2t}{s+t} = \lambda \\ s+t-st = \mu \end{cases}$$

has just one solution $(s,t) \in (0,1) \times (0,1)$. Thus, $\deg h = 1$, and so, by I.3.29, g is a pointed homotopy equivalence.

Now, the desired sectioned equivalence $\psi : \xi^\bullet \wedge^h \eta^\bullet \to (\xi * \eta)^\bullet$ occurs as a "fiberwise version" of the above g. Let θ be as in 4.17. Since $p_{\xi*\theta}^{-1}(b) = Sp_\xi^{-1}(b)$, we can write points of $\mathrm{ts}(\xi * \theta)$ as suitable equivalence classes $[x,t], x \in \mathrm{ts}\,\xi, t \in I$. Furthermore, the points of $\mathrm{ts}(\xi * \eta)$ will be written as suitable equivalence classes $[x,t,y], x \in \mathrm{ts}\,\xi, y \in \mathrm{ts}\,\eta, t \in [0,2]$. We define a morphism

$$\varphi : \xi^\bullet \times \eta^\bullet = (\xi * \theta) \times (\eta * \theta) \to (\xi * \eta) * \theta = (\xi * \eta)^\bullet$$

by setting

$$\operatorname{ts}\varphi([x,s],[y,t]) = \begin{cases} \left[\left[x, \frac{2t}{s+t}, y\right], s+t-st\right] & \text{if } (s,t) \neq (0,0), \\ [0] & \text{otherwise,} \end{cases}$$

where $x \in \operatorname{ts}\xi, y \in \operatorname{ts}\eta, s, t \in [0,1]$. Since the composition

$$\xi^\bullet \vee^h \eta^\bullet \to \xi^\bullet \times \eta^\bullet \xrightarrow{\varphi} (\xi * \eta)^\bullet$$

maps $\operatorname{ts}(\xi^\bullet \vee^h \eta^\bullet)$ onto the section of $(\xi * \eta)^\bullet$, we can pass φ through a sectioned morphism $\psi : \xi^\bullet \wedge^h \eta^\bullet \to (\xi * \eta)^\bullet$ which maps $\operatorname{ts}(\xi \vee^h \eta) \times I$ to the section. Now, ψ induces a pointed homotopy equivalence of fibers, since its restriction to fibers coincides with g. Thus, by 1.30(ii), ψ is a sectioned equivalence over B. □

So, the homotopy smash product plays the same role for $\mathcal{F}$-objects which the direct product (or join) plays for $\mathcal{V}$-objects. Now, given an $\mathcal{F}_m$-object ξ and $\mathcal{F}_n$-object η, we define the *Whitney sum* $\xi \oplus \eta$ by setting $\xi \oplus \eta := d^*(\xi \wedge^h \eta)$ where $d : X \to X \times X$ is the diagonal.

Let $\mu^{\mathcal{F}}_{m,n} : B\mathcal{F}_m \times B\mathcal{F}_n \to B\mathcal{F}_{m+n}$ classify the $\mathcal{F}$-object $\gamma^m_{\mathcal{F}} \wedge^h \gamma^n_{\mathcal{F}}$.

It follows from 4.19 that the diagram

$$\begin{array}{ccc} B\mathcal{V}_m \times B\mathcal{V}_n & \xrightarrow{\mu^{\mathcal{V}}} & B\mathcal{V}_{m+n} \\ \Big\downarrow{\scriptstyle a^{\mathcal{V}}_{\mathcal{F}} \times a^{\mathcal{V}}_{\mathcal{F}}} & & \Big\downarrow{\scriptstyle a^{\mathcal{V}}_{\mathcal{F}}} \\ B\mathcal{F}_m \times B\mathcal{F}_n & \xrightarrow{\mu^{\mathcal{F}}} & B\mathcal{F}_{m+n} \end{array}$$

commutes up to homotopy. Moreover, an obvious analog of 4.6 holds for $\mu^{\mathcal{F}}$ (and the above diagram is the analog of 4.6(iii)); we leave it to the reader to figure it out.

4.20. Theorem. *Let $\mathcal{Z}$ denote one of the symbols $\mathcal{V}, \mathcal{F}$. There is a map $\mu = \mu^{\mathcal{Z}} : B\mathcal{Z} \times B\mathcal{Z} \to B\mathcal{Z}$ such that, for every m, n, the diagram*

$$\begin{array}{ccc} B\mathcal{Z}_m \times B\mathcal{Z}_n & \xrightarrow{j_m \times j_n} & B\mathcal{Z} \times B\mathcal{Z} \\ \Big\downarrow{\scriptstyle \mu_{m,n}} & & \Big\downarrow{\scriptstyle \mu} \\ B\mathcal{Z}_{m+n} & \xrightarrow{j_{m+n}} & B\mathcal{Z} \end{array}$$

commutes up to homotopy.

Proof. Let B_n be the telescope of the finite sequence $\cdots \to B\mathcal{Z}_m \to \cdots \to B\mathcal{Z}_n$. Then we have a filtration $\{\cdots \subset B_n \subset B_{n+1} \subset \cdots\}$ of $B\mathcal{Z}$, and it is clear that $B\mathcal{Z} \times B\mathcal{Z} = \cup_n(B_n \times B_n)$. So, by III.1.16, we have a surjection $\rho : [B\mathcal{Z} \times B\mathcal{Z}, B\mathcal{Z}] \to \varprojlim\{[B_n \times B_n, B\mathcal{Z}]\}$. We can form a homotopy commutative diagram where u_k is a standard deformation retraction.

$$\begin{array}{ccc} B_m \times B_n & \xrightarrow{\nu_{m,n}} & B_{m+n} \\ {\scriptstyle u_m \times u_n}\downarrow & & \downarrow{\scriptstyle u_{m+n}} \\ B\mathcal{Z}_m \times B\mathcal{Z}_n & \xrightarrow{\mu_{m,n}} & B\mathcal{Z}_{m+n}. \end{array}$$

Now, we define $f : B_n \times B_n \xrightarrow{\nu_{n,n}} B_{2n} \subset B\mathcal{Z}$ and note that, by 4.6(ii), $f_{n+1}|(B_n \times B_n) \simeq f_n$. So, $\{[f_n]\}$ is a string, and we define $\mu : B\mathcal{Z} \times B\mathcal{Z} \to B\mathcal{Z}$ by requiring $\rho[\mu] = \{[f_n]\}$, i.e., $\mu|(B_n \times B_n) \simeq f_n$. The commutativity of the diagram is obvious. □

4.21. Definition. Given two stable $\mathcal{V}$-objects ξ, η, we define their product (or join for $\mathcal{V} = \mathcal{G}$) to be the map $X \times Y \xrightarrow{f \times g} B\mathcal{V} \times B\mathcal{V} \xrightarrow{\mu} B\mathcal{V}$ where f (resp. g) classifies ξ (resp. η). The *Whitney sum* of two stable $\mathcal{V}$-objects ξ, η over X is the $\mathcal{V}$-object $\xi \oplus \eta := d^*(\xi \times \eta)$ where $d : X \to X \times X$ is the diagonal. Similarly, given two stable $\mathcal{F}$-objects α, β, we define

$$\alpha \wedge^h \beta := \{X \times Y \xrightarrow{f \times g} B\mathcal{F} \times B\mathcal{F} \xrightarrow{\mu} B\mathcal{F}\}$$

where f (resp. g) classifies α (resp. β). The *Whitney sum* of two stable $\mathcal{F}$-objects α, β over X is the $\mathcal{F}$-object $\alpha \oplus \beta := d^*(\alpha \wedge^h \beta)$.

Because of 4.20, $(\alpha \times \beta)_{\text{st}} = \alpha_{\text{st}} \times \beta_{\text{st}}$, etc.

For future needs, we give the following definition.

4.22. Definition. A *multiplicative structure map* is a structure map $\varphi : B \to B\mathcal{V}$ equipped with a map $\mu_B : B \times B \to B$ (multiplication) and a homotopy $H : \varphi\mu_B \simeq \mu\circ(\varphi \times \varphi)$.

The space $\Omega^n S^n$ can be interpreted as the space $(S^n, *)^{(S^n,*)}$ of all pointed maps $S^n \to S^n$. Let $\Omega^n_k S^n$ be the subspace of $\Omega^n S^n$ consisting of all maps of degree k. It is clear that $\mathcal{F}_n \simeq \Omega^n_{\pm 1} := \Omega^n_1 \cup \Omega^n_{-1}$.

Every self-equivalence $f : (S^n, *) \to (S^n, *)$ gives a self-equivalence

$$f \wedge 1 : (S^{n+1}, *) = (S^n \wedge S^1, *) \to (S^n \wedge S^1, *) = (S^{n+1}, *).$$

We define $i_n : \mathcal{F}_n \to \mathcal{F}_{n+1}$, $i_n(f) = f \wedge 1$.

4.23. Lemma. *Let $k : S^n \to \Omega S^{n+1}$ be the adjoint map to $1_{S^{n+1}}$. Then $(\Omega^n k)_{\pm 1} : \Omega^n_{\pm 1} S^n \to \Omega^{n+1}_{\pm 1} S^{n+1}$ is homotopic to $i_n : \mathcal{F}_n \to \mathcal{F}_{n+1}$. Furthermore, the homotopy fiber of i_n is $(n-2)$-connected.*

Proof. Only the last assertion needs proof. The map

$$\pi_i(S^n) \xrightarrow{k_*} \pi_i(\Omega S^{n+1}) \cong \pi_{i+1}(S^{n+1})$$

is just the suspension map, and, by the Freudenthal Suspension Theorem, it is an isomorphism for $i < 2n-2$ and an epimorphism for $i = 2n-2$. Thus, the homotopy fiber of k is $(2n-2)$-connected, and therefore the homotopy fiber of $\Omega^n k$ is $(n-2)$-connected. □

Let $p_n : B\mathcal{F}_n \to B\mathcal{G}_{n+1}$ be the forgetful map (regarding $(S^n,*)$-fibrations as S^n-fibrations).

4.24. Proposition. (i) *The homotopy fiber of the forgetful map p_n is S^n. In particular, $B\mathcal{F} \simeq B\mathcal{G}$.*

(ii) *The homotopy fiber of $r_n^{\mathcal{F}} : B\mathcal{F}_n \to B\mathcal{F}_{n+1}$ is $(n-1)$-connected.*

(iii) *The homotopy fiber of $r_n^{\mathcal{G}} : B\mathcal{G}_n \to B\mathcal{G}_{n+1}$ is $(n-2)$-connected.*

Proof. (i) This follows from 1.76.

(ii) This follows from 4.23 and 1.75(v).

(iii) This follows from (i) and (ii), because $r_{n+1}^{\mathcal{G}} p_n \simeq p_{n+1} r_n^{\mathcal{F}}$. □

4.25. Let $\mathcal{U}_n$ be the group of all unitary transformations of the complex vector space $\mathbb{C}^n$. Its classifying space $B\mathcal{U}_n$ also classifies n-dimensional complex vector bundles, see e.g. Husemoller [1], Stong [3]. Let $\gamma_{\mathbb{C}}^n$ be the universal $(\mathbb{C}^n, \mathcal{U}_n)$-bundle over $B\mathcal{U}_n$, and let θ be the product $\mathbb{C}^1$-bundle over pt. Then $\gamma_{\mathbb{C}}^n \times \theta$ is classified by a map $r_n^{\mathcal{U}} : B\mathcal{U}_n \to B\mathcal{U}_{n+1}$, and we define $B\mathcal{U}$ to be the telescope of the sequence $\{r_n^{\mathcal{U}}\}$. Note that $r_n^{\mathcal{U}}$ is homotopic to Bi_n, where the inclusion $i_n : \mathcal{U}_n \to \mathcal{U}_{n+1}$ is given by the splitting $\mathbb{C}^{n+1} = \mathbb{C}^n \oplus \mathbb{C}^1$, cf. 3.4. Furthermore, similarly to (4.7), we define the map $j_n^{\mathcal{U}} : B\mathcal{U}_n \to B\mathcal{U}$ as the inclusion $B\mathcal{U}_n = B\mathcal{U}_n \times \{n\} \subset B\mathcal{U}_n \times [n, n+1] \to B\mathcal{U}$. Finally, there is a map $\mu_{m,n}^{\mathcal{U}} B\mathcal{U}_m \times B\mathcal{U}_n \to B\mathcal{U}_{m+n}$ which classifies $\gamma^m \times \gamma^n$. Based on this, one can construct a map

$$\mu^{\mathcal{U}} : B\mathcal{U} \times B\mathcal{U} \to B\mathcal{U} \tag{4.26}$$

with properties like 4.20.

Similarly to 4.8, we define a stable complex vector bundle to be the homotopy class of a map $X \to B\mathcal{U}$. In particular, there is a universal stable complex vector bundle $\gamma_{\mathbb{C}}$ given by $1_{B\mathcal{U}}$.

Regarding $\gamma_{\mathbb{C}}^n$ as a real vector bundle, we can classify it by a map $R_n : B\mathcal{U}_n \to B\mathcal{O}_{2n}$, called *realification.* Conversely, given a real vector bundle ξ, the vector bundle $\xi \otimes \mathbb{C}$ admits a canonical complex structure and thus can be considered as a complex vector bundle. In particular, $\gamma_{\mathcal{O}}^n \otimes \mathbb{C}$ is classified by a map $C_n : B\mathcal{O}_n \to B\mathcal{U}_n$, called *complexification.* As usual, there are maps $R : B\mathcal{U} \to B\mathcal{O}$ and $C : B\mathcal{O} \to B\mathcal{U}$ as $n \to \infty$.

Now we recall the necessary information on the homotopy of classifying spaces.

4.27. Theorem. (i) $\pi_i(B\mathcal{G}) = \pi_{i-1}(S)$ *for every $i > 1$, where S is the sphere spectrum. In particular, every group $\pi_i(B\mathcal{G})$ is finite.*

(ii) *$\pi_i(BO) = \mathbb{Z}$ for $i \equiv 0, 4 \bmod 8, i > 0$; $\pi_i(BO) = \mathbb{Z}/2$ for $i \equiv 1, 2 \bmod 8$, $\pi_i(BO) = 0$ otherwise. Moreover, $\Omega^8 BO \simeq BO \times \mathbb{Z}$. Furthermore, $\pi_{2i}(BU) = \mathbb{Z}, i > 0$, $\pi_{2i+1}(BU) = 0$, and $\Omega^2 BU \simeq BU \times \mathbb{Z}$.*

(iii) *Each of the maps $C_* : \mathbb{Z} = \pi_{4k}(BO) \to \pi_{4k}(BU) = \mathbb{Z}$ and $R_* : \mathbb{Z} = \pi_{4k+4}(BU) \to \pi_{4k+4}(BO) = \mathbb{Z}$ is multiplication by a_k, where $a_k = 1$ for k even and $a_k = 2$ for k odd.*

(iv) *The groups $\pi_i(\mathcal{PL}/O)$ are finite. Moreover, $\pi_i(\mathcal{PL}/O) = 0$ for $i < 7$ and $\pi_7(\mathcal{PL}/O) = \mathbb{Z}/28$.*

(v) *$\mathcal{TOP}/\mathcal{PL} \simeq K(\mathbb{Z}/2, 3)$, $\pi_4(B\mathcal{TOP}) = \mathbb{Z}$. Furthermore, the homomorphism*

$$\mathbb{Z} = \pi_4(B\mathcal{PL}) \xrightarrow{(\alpha^{\mathcal{PL}}_{\mathcal{TOP}})_*} \pi_4(B\mathcal{TOP}) = \mathbb{Z}$$

is multiplication by 2.

(vi) *$\pi_{4k}(\mathcal{G}/\mathcal{PL}) = \mathbb{Z}$ for $k > 0$, $\pi_{4k+2}(\mathcal{G}/\mathcal{PL}) = \mathbb{Z}/2$, $\pi_{2k+1}(\mathcal{G}/\mathcal{PL}) = 0$.*

(vii) *$\pi_i(\mathcal{G}/\mathcal{TOP}) \cong \pi_i(\mathcal{G}/\mathcal{PL})$ for every i.*

(viii) *For every n there exists $N = N(n)$ such that $r_k : B\mathcal{V}_k \to B\mathcal{V}_{k+1}$, as well as $j_k : B\mathcal{V}_k \to B\mathcal{V}$, is an n-equivalence for every $k > N$.*

(ix) *The spaces $B\mathcal{V}_n$ and $B\mathcal{V}$ are connected, $\pi_1(B\mathcal{V}_n) = \pi_1(B\mathcal{V}) = \mathbb{Z}/2$, and the groups $\pi_i(B\mathcal{V})$ are finitely generated. Furthermore, the space $B\mathcal{V}$ is simple, and the groups $H_i(B\mathcal{V})$ are finitely generated.*

Proof-survey. (i) If $i << N$, then

$$\begin{aligned}\pi_i(B\mathcal{G}) &= \pi_i(B\mathcal{G}_{N+1}) = \pi_{i-1}(\mathcal{G}_{N+1}) = \pi_{i-1}(\mathcal{F}_N)\\ &= \pi_{i-1}(\Omega^N S^N) = \pi_{i+N-1}(S^N) = \pi_{i-1}(S).\end{aligned}$$

(ii), (iii) This is the famous Bott Periodicity Theorem, see e.g. Milnor [6] or Husemoller [1] (or the original paper, Bott [1]).

(iv) Hirsch–Mazur [1] proved that the group $\pi_i(\mathcal{PL}/O)$ is isomorphic to the group Φ_i of smooth structures on a $\mathcal{PL}$ sphere S^i. One can prove that every smooth manifold which is $\mathcal{PL}$ isomorphic to a sphere is a so-called twisted sphere, i.e., it can be constructed by gluing the two standard disks along the boundary. The group (under the connected sum) of twisted n-spheres is denoted by Γ_n. So, $\pi_n(\mathcal{PL}/O) = \Gamma_n$. One can prove that $\Gamma_n = \pi_0(\mathrm{Diff}_n)$, where Diff_n is the group of all diffeomorphisms of S^{n-1}. It is easy to see that $\Gamma_n = 0$ for $n < 3$. Smale [2] and Munkres [1] proved that $\Gamma_3 = 0$, Cerf [1] proved that $\Gamma_4 = 0$. The h-cobordism Theorem of Smale [3] (a good proof can be found in Milnor [8]) implies that $\Gamma_n = \Theta_n$ for $n > 4$, where Θ_n is the group of homotopy n-spheres. Kervaire–Milnor [1] considered these groups and proved that $\Theta_5 = 0 = \Theta_6$, $\Theta_7 = \mathbb{Z}/28$ and Θ_n is finite for $n > 3$.

(v) This is a theorem of Kirby–Siebenmann [1].

(vi) This is a theorem of Sullivan [1]. A good proof can be found in Madsen–Milgram [1].

(vii) Consider the homotopy exact sequence of the $\mathcal{TOP}/\mathcal{PL}$-fibration $\mathcal{G}/\mathcal{PL} \to \mathcal{G}/\mathcal{TOP}$. Since $\mathcal{TOP}/\mathcal{PL} = K(\mathbb{Z}/2, 3)$ and $\pi_3(\mathcal{G}/\mathcal{PL}) = 0$, we conclude that $\pi_i(\mathcal{G}/\mathcal{PL}) \cong \pi_i(\mathcal{G}/\mathcal{TOP})$ for $i \neq 4$. Furthermore, this fibration yields the exact sequence

$$0 \to \pi_4(\mathcal{G}/\mathcal{PL}) \to \pi_4(\mathcal{G}/\mathcal{TOP}) \to \pi_3(\mathcal{TOP}/\mathcal{PL}) \to 0.$$

Kirby–Siebenmann [1] proved that this exact sequence does not split. Thus, $\pi_4(\mathcal{G}/\mathcal{TOP}) = \mathbb{Z}$.

(viii) The case $\mathcal{V} = \mathcal{G}$ follows from 4.24(iii), and for $\mathcal{V} = \mathcal{O}$ it is clear because there is a locally trivial bundle $B\mathcal{O}_n \to B\mathcal{O}_{n+1}$ with fiber S^n. The remaining cases can be found in Kirby–Sibenmann [1].

(ix) The spaces $B\mathcal{V}_n$ are connected because there is just one $\mathcal{V}_n$-object over pt. The connectedness of $B\mathcal{V}$ follows from that of $B\mathcal{V}_n$. Furthermore, $\pi_1(B\mathcal{O}_n) = \pi_0(\mathcal{O}_n) = \mathbb{Z}/2$, $\pi_1(B\mathcal{G}_n) = \pi_0(\mathcal{G}_n) = \mathbb{Z}/2$. Hence, by (viii), $\pi_1(B\mathcal{O}) = \mathbb{Z}/2 = \pi_1(B\mathcal{G})$, and so, by (iv) and (v), $\pi_1(B\mathcal{PL}) = \pi_1(B\mathcal{TOP}) = \mathbb{Z}/2$. The isomorphisms $\pi_1(B\mathcal{PL}_n) = \mathbb{Z}/2 = \pi_1(B\mathcal{TOP}_n)$ are proved in Kirby–Siebenmann [1]. The groups $\pi_i(B\mathcal{G})$ and $\pi_i(B\mathcal{O})$ are finitely generated by (i) and (ii), respectively. The groups $\pi_i(B\mathcal{PL})$ are finitely generated by (iv) or (vi), and $\pi_i(B\mathcal{TOP})$ are finitely generated by (v). Furthermore, $B\mathcal{V}$ is a simple space since, by 4.20 and (viii), the map $B\mathcal{V} \times \mathrm{pt} \to B\mathcal{V} \times B\mathcal{V} \xrightarrow{\mu} B\mathcal{V}$ is weakly homotopic to the identity (the proof can be done just as for H-spaces, see e.g. Hu [1], Whitehead [2]). Thus, $H_i(B\mathcal{V})$ are finitely generated because so are $\pi_i(B\mathcal{V})$ (use the Hurewicz Theorem mod $\mathfrak{C}$ for spaces, where $\mathfrak{C}$ is the Serre class of finitely generated abelian groups, see e.g. Mosher–Tangora [1]). □

4.28. Theorem. *Let $\mathcal{Z}$ denote one of the symbols $\mathcal{V}, \mathcal{F}$. The map $\mu : B\mathcal{Z} \times B\mathcal{Z} \to B\mathcal{Z}$ in* 4.20 *is uniquely determined up to homotopy. Furthermore, the following diagrams commute up to homotopy:*

(i) (*Associativity.*)

$$\begin{array}{ccc} B\mathcal{Z} \times B\mathcal{Z} \times B\mathcal{Z} & \xrightarrow{\mu \times 1} & B\mathcal{Z} \times B\mathcal{Z} \\ {\scriptstyle 1\times\mu}\downarrow & & \downarrow{\scriptstyle \mu} \\ B\mathcal{Z} \times B\mathcal{Z} & \xrightarrow{\mu} & B\mathcal{Z}. \end{array}$$

(ii) (*Commutativity.*)

$$\begin{array}{ccc} B\mathcal{Z} \times B\mathcal{Z} & \xrightarrow{T} & B\mathcal{Z} \times B\mathcal{Z} \\ {\scriptstyle \mu}\downarrow & & \downarrow{\scriptstyle \mu} \\ B\mathcal{Z} & = & B\mathcal{Z} \end{array}$$

where T switches the factors.

(iii)

$$\begin{array}{ccc} B\mathcal{Z}' \times B\mathcal{Z}' & \xrightarrow{\mu'} & B\mathcal{Z}' \\ {\scriptstyle a\times a}\downarrow & & \downarrow{\scriptstyle a} \\ B\mathcal{Z} \times B\mathcal{Z} & \xrightarrow{\mu} & B\mathcal{Z}. \end{array}$$

where $\mathcal{Z}' \leq \mathcal{Z}$.

Similarly, the map $\mu^{\mathcal{U}} : B\mathcal{U} \times B\mathcal{U} \to B\mathcal{U}$ *in* 4.26 *is associative and commutative, and the properties like* 4.20 *determine it uniquely up to homotopy.*

Proof. Since, by 4.24(i), $B\mathcal{F} \simeq B\mathcal{G}$, it suffices to consider the case of the spaces $B\mathcal{V}$. Firstly, we prove that $\mu = \mu^{\mathcal{V}}$ is uniquely determined up to homotopy. Let B_n be the telescope of the finite sequence $B\mathcal{V}_0 \to \cdots \xrightarrow{r_{n-1}} B\mathcal{V}_n$. We prove the homotopy uniqueness of μ if we prove that

$$\rho : [B\mathcal{V} \times B\mathcal{V}, B\mathcal{V}] \to \varprojlim\{[B_n \times B_n, B\mathcal{V}]\}$$

is an injection (and so, by III.1.16, a bijection).

By 4.27(viii), for every n there is $N = N(n)$ such that

$$(j_N \times j_N)_* : [(B\mathcal{V} \times B\mathcal{V})^{(n)}, B_N \times B_N] \to [(B\mathcal{V} \times B\mathcal{V})^{(n)}, B\mathcal{V} \times B\mathcal{V}]$$

is a bijection. Let $h_n : (B\mathcal{V} \times B\mathcal{V})^{(n)} \to B\mathcal{V} \times B\mathcal{V}$ be a map such that $(j_N \times j_N)_*(h_n)$ is the inclusion $(B\mathcal{V} \times B\mathcal{V})^{(n)} \subset B\mathcal{V} \times B\mathcal{V}$. Then the family $\{[h_n]\}$ yields a function

$$h := \{h_n | \varprojlim\} : \varprojlim\{[B_n \times B_n, B\mathcal{V}]\} \to \varprojlim\{[(B\mathcal{V} \times B\mathcal{V})^{(n)}, B\mathcal{V}]\},$$

and we have the commutative diagram

$$\begin{array}{ccc} [B\mathcal{V} \times B\mathcal{V}, B\mathcal{V}] & \xrightarrow{\rho} & \varprojlim\{[B_n \times B_n, B\mathcal{V}]\} \\ \| & & \downarrow{\scriptstyle h} \\ [B\mathcal{V} \times B\mathcal{V}, B\mathcal{V}] & \xrightarrow{\bar\rho} & \varprojlim\{[(B\mathcal{V} \times B\mathcal{V})^{(n)}, B\mathcal{V}]\}. \end{array}$$

So, it suffices to prove that $\bar\rho$ is injective, and now we do it.

By 4.27(ix), $B\mathcal{V}$ is a simple space. So, by III.1.18(ii), it suffices to prove that all the groups $H^{i-1}(B\mathcal{V} \times B\mathcal{V}; \pi_i(B\mathcal{V}))$ are finite. Since, by 4.27(ix), all the groups $\pi_i(B\mathcal{V})$ and $H_i(B\mathcal{V})$ are finitely generated, it suffices to prove that $H^{i-1}(B\mathcal{V} \times B\mathcal{V}; \pi_i(B\mathcal{V})) \otimes \mathbb{Q} = 0$, i.e., that

$$H^{i-1}(B\mathcal{V} \times B\mathcal{V}; \mathbb{Q}) \otimes \pi_i(B\mathcal{V}) \otimes \mathbb{Q} = 0.$$

If $\mathcal{V} = \mathcal{G}$ then, by 4.27(i), $\pi_i(B\mathcal{G}) \otimes \mathbb{Q} = 0$ for every $i > 0$, and the result follows.

If $\mathcal{V} = \mathcal{O}$ then $H^i(B\mathcal{O} \times B\mathcal{O}; \mathbb{Q}) = 0$ for $i \neq 4k$, see e.g. Milnor–Stasheff [1]. But, by 4.27(ii), $\pi_i(B\mathcal{O}) \otimes \mathbb{Q} = 0$ for $i \neq 4k$. So, $H^{i-1}(B\mathcal{O} \times B\mathcal{O}; \mathbb{Q}) \otimes \pi_i(B\mathcal{O}) \otimes \mathbb{Q} = 0$ for every i.

Finally, if $\mathcal{V} = \mathcal{PL}, \mathcal{TOP}$ then, by 4.27(iv,v), the groups $\pi_i(\mathcal{V}/\mathcal{O})$ are finite, and hence $a^{\mathcal{O}}_{\mathcal{V}}[0] : B\mathcal{O}[0] \to B\mathcal{V}[0]$ is a homotopy equivalence, i.e., $B\mathcal{O}$ and $B\mathcal{V}$ have the same rational homotopy type. Thus,

$$H^{i-1}(B\mathcal{V} \times B\mathcal{V}; \mathbb{Q}) \otimes \pi_i(B\mathcal{V}) \otimes \mathbb{Q} = H^{i-1}(B\mathcal{O} \times B\mathcal{O}; \mathbb{Q}) \otimes \pi_i(B\mathcal{O}) \otimes \mathbb{Q} = 0.$$

The commutativity of the diagrams in question can be proved similarly. For example, we prove the associativity of μ. Consider the diagram

$$\begin{array}{ccccc} B\mathcal{V}_m \times B\mathcal{V}_n \times B\mathcal{V}_p & \xrightarrow{j_m \times j_n \times j_p} & B\mathcal{V} \times B\mathcal{V} \times B\mathcal{V} & \xrightarrow{\mu \times 1} & B\mathcal{V} \times B\mathcal{V} \\ & & \downarrow{\scriptstyle 1\times\mu} & & \downarrow{\scriptstyle \mu} \\ & & B\mathcal{V} \times B\mathcal{V} & \xrightarrow{\mu} & B\mathcal{V} \end{array} .$$

By 4.6(i) and 4.20, $\mu(\mu \times 1)(j_m \times j_n \times j_p) \simeq \mu(1 \times \mu)(j_m \times j_n \times j_p)$. So, it suffices to prove that $\rho : [B\mathcal{V} \times B\mathcal{V} \times B\mathcal{V}, B\mathcal{V}] \to \varprojlim\{[B_n \times B_n \times B_n, B\mathcal{V}]\}$ is an injection. This can be done as above; we leave it to the reader. □

Let $\mathcal{SV}_n$ be the submonoid of $\mathcal{V}_n$ consisting of the orientation preserving maps $\mathbb{R}^n \to \mathbb{R}^n$ or $S^{n-1} \to S^{n-1}$ (we leave it to the reader to fix the case $\mathcal{V} = \mathcal{PL}$). By 4.25(ix), $\pi_1(B\mathcal{V}_n) = \mathbb{Z}/2$ for $n \geq 0$, and it is clear that $B\mathcal{SV}_n$ is a 2-sheeted (i.e., the universal) covering of $B\mathcal{V}_n$. Furthermore, the space $B\mathcal{SV} = \lim_{n\to\infty} B\mathcal{SV}_n$ can be defined to be the universal covering space of $B\mathcal{V}$. Finally, there is a hierarchy $B\mathcal{SO} \to B\mathcal{SPL} \to B\mathcal{STOP} \to B\mathcal{SG}$ similar to (and given by) (4.3), and the homotopy fiber, say, of $B\mathcal{SO} \to B\mathcal{SPL}$ is $\mathcal{PL}/\mathcal{O}$.

The (co)homology of $B\mathcal{V}$ has been studied quite extensively, but we use only a small part of the known information. Additional information can be found in Madsen–Milgram [1] and May [4].

4.29. Theorem. (i) $H^*(B\mathcal{O}; \mathbb{Z}/2) = \mathbb{Z}/2\,[w_1, \ldots, w_n, \ldots]$, $\dim w_i = i$. *Furthermore, $H^*(B\mathcal{O}; \mathbb{Z}/2)$ is contained in $H^*(B\mathcal{V}; \mathbb{Z}/2)$ as a subalgebra for every $\mathcal{V}$, and the homomorphism $(a^{\mathcal{O}}_{\mathcal{V}})^* : H^*(B\mathcal{V}; \mathbb{Z}/2) \to H^*(B\mathcal{O}; \mathbb{Z}/2)$ is an epimorphism.*

(ii) *If R is a ring such that $1/2 \in R$, then*

$$H^*(B\mathcal{O}; R) = H^*(B\mathcal{SO}; R) = R[p_1, \ldots, p_k, \ldots],\ \dim p_k = 4k.$$

(iii) $H^*(B\mathcal{U}) = \mathbb{Z}[c_1, \ldots c_n, \ldots]$, $\dim c_i = 2i$.

Proof. See e.g. Milnor–Stasheff [1]. □

We have $\mathcal{U}_1 = \{ z \in \mathbb{C} \mid |z| = 1 \}$. Set $\mathcal{SU}_n := \operatorname{Ker}(\det : \mathcal{U}_n \to \mathcal{U}_1)$, where det maps a matrix to its determinant. So, we have the inclusion $t_n : \mathcal{SU}_n \to \mathcal{U}_n$. Furthermore, the inclusion $i_n : \mathcal{U}_n \to \mathcal{U}_{n+1}$ induces the inclusion $k_n : \mathcal{SU}_n \to \mathcal{SU}_{n+1}$. Since $t_{n+1}k_n = i_n t_n$, we get the homotopy commutative diagram

$$
\begin{array}{ccc}
BSU_n & \xrightarrow{Bk_n} & BSU_{n+1} \\
{\scriptstyle Bt_n}\downarrow & & \downarrow{\scriptstyle Bt_{n+1}} \\
BU_n & \xrightarrow{Bi_n} & BU_{n+1}\,.
\end{array}
\tag{4.30}
$$

We can and shall assume that Bt_n is a fibration.

4.31. Lemma. (i) *The homotopy fiber of Bt_n is S^1.*
(ii) *The fibration $Bt_n : BSU_n \to BU_n$ is a 3-connective covering.*
(iii) *The square (4.30) is a morphism of S^1-fibrations for every n.*

Proof. (i) This follows from 3.4 since $\mathcal{U}_n/\mathcal{SU}_n = S^1$.

(ii) Let $i : \mathcal{U}_1 \to \mathcal{U}_n$ map $z \in \mathcal{U}_1$ to the matrix with $a_{11} = z, a_{ii} = 1$ for $i > 1$, and $a_{ij} = 0$ for $i \neq j$. We have $\pi_1(\mathcal{U}_n) = \mathbb{Z}$, $\pi_2(\mathcal{U}_n) = 0$, see e.g. Milnor [6]. Since $\det \circ i = 1_{\mathcal{U}_1}$, $\det_* : \pi_1(\mathcal{U}_n) \to \pi_1(\mathcal{U}_1)$ is an isomorphism. Considering the homotopy exact sequence of the locally trivial bundle $\mathcal{SU}_n \to \mathcal{U}_n \xrightarrow{\det} S^1$, we conclude that $\pi_i(\mathcal{SU}_n) = 0$ for $i < 3$. Thus, $\pi_i(B\mathcal{SU}_n) = 0$ for $i < 4$. Since Bt_n is an S^1-fibration and $\pi_i(S^1) = 0$ for $i > 1$, (ii) is proved.

(iii) The square (4.30) induces a morphism of the homotopy exact sequences of the vertical fibrations. Since $\pi_i(B\mathcal{SU}_n) = 0$ for $i < 3$, the map Bk_n provides an isomorphism of the fundamental groups of fibers. □

The squares (4.30) can be aggregated in a homotopy commutative diagram

$$
\begin{array}{ccccccc}
\cdots & \longrightarrow & BSU_n & \xrightarrow{Bk_n} & BSU_{n+1} & \longrightarrow & \cdots \\
 & & {\scriptstyle Bt_n}\downarrow & & \downarrow{\scriptstyle Bt_{n+1}} & & \\
\cdots & \longrightarrow & BU_n & \xrightarrow{Bi_n} & BU_{n+1} & \longrightarrow & \cdots,
\end{array}
$$

where every vertical map is an S^1-fibration. We can assume that this ladder commutes (changing Bk_n map by map, using the covering homotopy property). Defining BSU to be the telescope of the top sequence, we have the map $q : BSU \to BU$ (the telescope of the Bt_n's). By 1.41(iii), q is a quasi-fibration, and each fiber is homotopy equivalent to S^1. Passing to a fibrational substitute of q, we have a fibration $F \to BSU \xrightarrow{p} BU$, where F is CW-equivalent to S^1.

4.32. Lemma. (i) *The fibration $p : BSU \to BU$ is a 3-connective covering.*
(ii) $H^*(BSU) = \mathbb{Z}[c_2, \dots, c_n, \dots]$, $\dim c_n = 2n$.

Proof. (i) By 3.4, the homotopy fiber of Bk_n is $\mathcal{SU}_{n+1}/\mathcal{SU}_n = S^{2n+1}$, and so $\pi_i(BSU) = \pi_i(BSU_n)$ for $i < n$. Thus, $\pi_i(BSU) = 0$ for $i < 4$. Since $\pi_i(F) = \pi_i(S^1) = 0$ for $i > 1$, p is a 3-connective covering.

(ii) Consider the cohomology Leray–Serre spectral sequence of the fibration $F \to BSU \xrightarrow{p} BU$. We have $H^*(BU) = \mathbb{Z}[c_1, \dots, c_n, \dots]$ and

$H^*(F) = H^*(S^1)$. Let $x \in H^1(F) = \mathbb{Z}$ be a generator, and let τ denote the transgression. Since $H^i(BS\mathcal{U}) = 0$ for $i < 4$, $\tau(x) = c_1$ (up to sign). Hence, $\tau(c_1^{\alpha_1} \cdots c_r^{\alpha_r} x) = c_1^{\alpha_1+1} \cdots c_r^{\alpha_r}$. Thus, $E_\infty^{i,j} = 0$ for $j > 1$, $E_\infty^{*,0} = \mathbb{Z}[c_2, \dots, c_n, \dots]$. □

§5. Thom Spaces and Thom Spectra

5.1. Definition. (a) Let $\alpha = \{p : Y \to X\}$ be an $\mathcal{F}_n$-object with a section s. We define the *Thom space* $T\alpha$ of α by setting $T\alpha := Y/s(X)$. We set $T\alpha := \mathrm{pt}$ if $\mathrm{bs}\,\alpha = \emptyset$.

(b) Given a $\mathcal{V}_n$-object ξ, define the *Thom space* $T\xi$ of ξ as $T\xi := T(\xi^\bullet)$, where $\xi^\bullet$ is as in 4.17.

Notice that $T\alpha$ has a canonical base point (the image of $s(X)$). Furthermore, $T\alpha = (\mathrm{bs}(\alpha))^+$ for every $\mathcal{F}_0$-object α.

It is easy to see that every morphism $\varphi : \alpha \to \beta$ of $\mathcal{F}_n$-objects induces a map $T\varphi : T\alpha \to T\beta$ of Thom spaces, and in fact we have a *Thom functor* T. Moreover, an $\mathcal{F}$-equivalence of $\mathcal{F}$-objects induces a homotopy equivalence of Thom spaces.

5.2. Examples. (a) The Thom space of the (trivial) $\mathcal{V}_n$-object over a point is S^n.

(b) The open Möbius band fibered over the middle circle can be considered as a line bundle. In greater detail, if we glue (identify) the points $(-1, -x)$ and $(1, x)$ in $[-1, 1] \times \mathbb{R}$, we obtain a space Y homeomorphic to the open Möbius band. Now, the projection $p_1 : [-1, 1] \times \mathbb{R} \to [-1, 1]$ yields the $((\mathbb{R}, 0), \mathcal{O}_1)$-bundle $\zeta = \{p : Y \to S^1 = [-1, 1]/\{-1, 1\}\}$. Then $T\zeta$ is the real projective plane RP^2 (prove this).

(c) More generally, let ξ_n be the canonical line bundle over RP^n. Then $T\xi_n = RP^{n+1}$ (prove this, or see e.g. Stong [3]).

(d) Similarly to (c), let λ_n be the canonical complex line bundle over the complex projective space CP^n. Then $T\lambda_n = CP^{n+1}$.

(e) There is the Thom space $T\gamma_{\mathcal{V}}^n$ of the universal $\mathcal{V}_n$-object $\gamma_{\mathcal{V}}^n$. It is usually denoted by $M\mathcal{V}_n$ or $TB\mathcal{V}_n$.

5.3. Definition. Given a point $x \in X$, the pair $(p^{-1}(x), s(x))$ can be regarded as an $(S^n, *)$-fibration $\theta = \theta_{\mathcal{F}}^n$ over $\{x\}$. The morphism (the inclusion of the fiber) $\theta \to \alpha$ of $\mathcal{F}_n$-objects induces a map $j = j_x : S^n = T\theta \to T\alpha$. We call this map j_x a *root* of $T\alpha$ at x.

If X is connected, then the homotopy class of j_x is uniquely determined up to sign. In this case we write just j (and say just root).

There are also some other models of $T\xi$. For example, given a spherical fibration ξ, one can define $T\xi := C(p_\xi)$. Furthermore, given a $\mathcal{V}$-object with $\mathcal{V} \le \mathcal{TOP}$, one can consider the underlying (see (4.3)) spherical fibration $\overline{\xi}$ and set $T\xi := C(p_{\overline{\xi}})$. These Thom spaces are homotopy equivalent (but not homeomorphic) to ours. Furthermore, given a vector bundle ξ, let $D(\xi)$ (resp. $S(\xi)$) be the unit disk (resp. unit sphere) subbundle of ξ with respect to some Riemannian metric in ξ. Then $T\xi$ is homeomorphic to $\mathrm{ts}(D(\xi))/(\mathrm{ts}(S(\xi)))$. Moreover, if $\mathcal{V} \leq \mathcal{TOP}$ and the base of a $\mathcal{V}$-object ξ is compact (e.g. it is a finite CW-space) then $T\xi$ is (homeomorphic to) the one-point compactification of $\mathrm{ts}(\xi)$.

5.4. Construction–Definition. Let ξ be any $\mathcal{V}_n$-object. Define a section $s' : \mathrm{bs}\,\xi \to \mathrm{ts}\,\xi^\bullet$ of $\xi^\bullet$ as follows. If $\mathcal{V} \leq \mathcal{TOP}$ then s' is the composition

$$\mathrm{bs}\,\xi \xrightarrow{s} \mathrm{ts}\,\xi \subset \mathrm{ts}\,\xi^\bullet$$

where s is the zero section of ξ. If $\mathcal{V} = \mathcal{G}$ then

$$s'(b) := [x, 2, -1] \text{ where } b \in \mathrm{bs}\,\xi, 2 \in [0,2], -1 \in S^0 = \{-1, 1\} \text{ and } p_\xi(x) = b,$$

cf. 4.17. We define the *zero section of* $T\xi$

$$\mathfrak{z} : \mathrm{bs}\,\xi \to T\xi$$

to be the composition $\mathrm{bs}\,\xi \xrightarrow{s'} \mathrm{ts}\,\xi^\bullet \xrightarrow{\text{quotient}} T\xi$. Clearly, $\mathfrak{z}$ is an injective map.

5.5. Proposition. (i) *Given an $\mathcal{F}_m$-object α and an $\mathcal{F}_n$-object β, we have $T(\alpha \wedge^h \beta) \simeq T\alpha \wedge T\beta$.*

(ii) *$T(\xi \times \eta) \simeq T\xi \wedge T\eta$ for all $\mathcal{V}$-objects ξ, η with $\mathcal{V} \leq \mathcal{TOP}$, and $T(\xi * \eta) \simeq T\xi \wedge T\eta$ for all $\mathcal{G}$-objects ξ, η.*

(iii) *$T(\xi \oplus \theta^1) \simeq ST(\xi)$. In particular, $T(\theta^n_X) \simeq S^n X^+$. Furthermore, for every $x \in X$ the root j_x is homotopic (up to sign) to the inclusion $S^n(\{x\}^+) \subset S^n X^+$.*

Proof. Exercise. □

Again, consider an $\mathcal{F}_n$-object $\alpha = \{p : Y \to X\}$. Choose a point $x \in X$ and set $F = p^{-1}(x)$. Given a loop $\omega : [0,1] \to X$ at x, $\omega(0) = x = \omega(1)$, consider a covering homotopy $h_t : F \to Y, t \in [0,1]$ such that $ph_t(a) = \omega(t)$ for every $a \in F, t \in [0,1]$. Since $h_1(F) \subset F$, there is a map $f : F \to F$ such that the composition $F \xrightarrow{f} F \subset Y$ coincides with h_1. Since $F \simeq S^n$, the degree of f is defined, and we set $d(\omega) = \deg f$. It is clear that $d(\omega)$ is well-defined, and $d(\omega) = \pm 1$ because f is a self-equivalence.

5.6. Definition. An $\mathcal{F}_n$-object α over X is called *orientable* if $d(\omega) = 1$ for every $x \in X$ and every loop ω at x. A $\mathcal{V}_n$-object ξ is called orientable if $\xi^\bullet$ is orientable.

It is clear that an $\mathcal{F}_n$-object over a connected base is orientable iff $d(\omega) = 1$ for some single point x and every ω at x.

Given α as above and an abelian group G, there are the homology local system [11] $\{\widetilde{H}_n(F_x;G)\}$ and the cohomology local sytem $\{\widetilde{H}^n(F_x;G)\}$, $x \in X$, over X. Recall that $F_x \simeq S^n$, and so $\widetilde{H}_n(F_x;G) \cong G \cong \widetilde{H}^n(F_x;G)$ for every $x \in X$.

Thom [1], [2] discovered the following important fact.

5.7. Theorem-Definition. *For every abelian group G and every i there are isomorphisms*

$$H_i(X;\{\widetilde{H}_n(F_x;G)\}) \cong \widetilde{H}_{i+n}(T\alpha;G),$$

$$H^i(X;\{\widetilde{H}^n(F_x;G)\}) \cong \widetilde{H}^{i+n}(T\alpha;G).$$

These isomorphisms are called Thom isomorphisms. *They are natural in α and G.*

Proof. We can assume that X is connected. We prove only the homological Thom isomorphism. Let $s : X \to Y$ be the section. Consider the homology Leray–Serre spectral sequence of the *relative* fibration $(Y, s(X)) \to X$ (see e.g. Switzer [1], p.351–352, or, in detail, Prieto [1]). This spectral sequence converges to $H_*(Y, s(X); G) \simeq \widetilde{H}_*(T\alpha; G)$, and $E^2_{p,q} = H_p(X;\{H_q(F_x,*;G)\})$. So, $E^2_{p,q} = 0$ for $q \neq n$. Therefore, $E^2_{p,q} = E^\infty_{p,q}$, and $H_p(X;\{\widetilde{H}_n(F_x;G)\}) = H_p(X;\{H_n(F_x,*;G)\}) = E^2_{p,n} = E^\infty_{p,n} = \widetilde{H}_{p+n}(T\alpha;G)$.

The naturality of the Thom isomorphisms follows from the naturality of the Leray–Serre spectral sequence. □

5.8. Corollary. (i) $\widetilde{H}_i(T\alpha;G) = 0 = \widetilde{H}^i(T\alpha;G)$ *for $i < n$. Furthermore, if the base X of α is connected, then $\widetilde{H}_n(T\alpha;G) = G = \widetilde{H}^n(T\alpha;G)$ for orientable α and $\widetilde{H}_n(T\alpha;G) = G/2G$, $\widetilde{H}^n(T\alpha;G) = \{\, g \in G \mid g = -g \,\}$ for non-orientable α.*

(ii) $\pi_i(T\alpha) = 0$ *for every $i < n$. Furthermore, if the base X of α is connected, then $\pi_n(T\alpha) = \mathbb{Z}$ for orientable α, and $\pi_n(T\alpha) = \mathbb{Z}/2$ for non-orientable α. Finally, the root $j : S^n \to T\alpha$ yields a generator of $\pi_n(T\alpha)$ in both cases.*

Proof. (i) The first assertion is a trivial corollary of 5.7. The last assertion follows from 5.7, because $H_0(X;M) = M/\{tm - m\}$ and $H^0(X;M) = \{\, m \mid tm = m \,\}, m \in M, t \in \pi_1(X)$, for every $\pi_1(X)$-module M.

(ii) Since α has a section, the map $p_* : \pi_k(Y) \to \pi_k(X)$ is onto for every k. Let $q : Y \to Y/s(X) = T\alpha$ be the quotient map. Consider the diagram

[11] See e.g. Hilton–Wiley [1], Hu [1], Spanier [2], Fuks–Rochlin [1] about local systems.

$$\begin{array}{ccccccccc} & & & & \pi_1(X) & = & \pi_1(X) & & \\ & & & & {\scriptstyle s_*}\downarrow & & \| & & \\ 0 & \longrightarrow & \pi_1(S^n) & \xrightarrow{i_*} & \pi_1(Y) & \xrightarrow{p_*} & \pi_1(X) & \longrightarrow & 0 \\ & & \| & & {\scriptstyle q_*}\downarrow & & & & \\ & & \pi_1(S^n) & \xrightarrow{j_*} & \pi_1(T\alpha) & & & & \end{array}$$

with exact row. We prove that j_* is epic. Indeed, by the van Kampen Theorem,

$$\begin{aligned} \pi_1(Y/s(X)) &= \pi_1(Y \cup Cs(X)) \\ &= \pi_1(Y) *_{\pi_1(s(X))} \pi_1(Cs(X)) = \pi_1(Y) *_{\pi_1(s(X))} \{1\}. \end{aligned}$$

Hence, $q: Y \to Y/s(X)$ induces an epimorphism

$$q_*: \pi_1(Y) \to \pi_1(Y) *_{\pi_1(s(X))} \{1\} = \pi_1(Y/s(X)) = \pi_1(T\alpha).$$

So, for every $x \in \pi_1(T\alpha)$ there exists $y \in \pi_1(Y)$ with $q_*(y) = x$. Furthermore, $y - s_*p_*(y) = i_*(z)$ for some $z \in \pi_1(S^n)$. Now, $j_*(z) = q_*i_*(z) = q_*(y - s_*p_*(y)) = q_*(y) = x$, and so j_* is epic.

If $n > 1$, then $\pi_1(T\alpha) = 0$, and so, by the Hurewicz Theorem, $\pi_i(T\alpha) \simeq H_i(T\alpha)$ for $0 < i \leq n$. Thus, $\pi_i(T\alpha) = 0$ for $i < n$, and $\pi_n(T\alpha)$ has the required properties.

The inclusion of a fiber yields a morphism of relative fibrations

$$\begin{array}{ccc} (S^n, *) & \longrightarrow & (Y, s(X)) \\ \downarrow & & \downarrow \\ \mathrm{pt} & \longrightarrow & X. \end{array}$$

Considering the corresponding morphism of the homology Leray-Serre spectral sequences (as in 5.7), we conclude that $j_*: H_n(S^n; \mathbb{Z}/2) \to H_n(T\alpha; \mathbb{Z}/2)$ is an isomorphism; moreover, $j_*: H_n(S^n) \to H_n(T\alpha)$ is an isomorphism for orientable α. Hence, j yields a generator of $\pi_n(T\alpha)$.

Let $n = 1$. Since j_* is epic, $\pi_1(T\alpha)$ is cyclic and j yields a generator. So, again $\pi_1(T\alpha) = H_1(T\alpha)$, etc. □

It makes sense to separate the orientable and non-orientable cases in 5.7. Suppose that $X = \operatorname{bs}\alpha$ is connected. It follows from 5.8 that $H^n(T\alpha; \mathbb{Z}/2) = \mathbb{Z}/2$. The non-trivial element $u_{\mathbb{Z}/2} \in H^n(T\alpha; \mathbb{Z}/2)$ is called the *Thom class* (mod 2) of α. Moreover, if α is orientable, then, by 5.7, $H^n(T\alpha) = \mathbb{Z}$. A generator $u_{\mathbb{Z}}$ (either one) of $H^n(T\alpha)$ is called the *Thom class* (integral) of α. Thus, orientability is equivalent to the existence of the integral Thom class.

5.9. Corollary. (i) *If $2G = 0$ then there are Thom isomorphisms*

$$\varphi_G : H^i(X;G) \xrightarrow{\cong} H^{i+n}(T\alpha;G), \quad \varphi^G : H_i(X;G) \xrightarrow{\cong} H_{i+n}(T\alpha;G).$$

(ii) *If α is orientable then there are Thom isomorphisms*

$$\varphi_G : H^i(X;G) \xrightarrow{\cong} H^{i+n}(T\alpha;G), \qquad \varphi^G : H_i(X;G) \xrightarrow{\cong} H_{i+n}(T\alpha;G).$$

Proof. In both cases the local systems $\{H_n(F_x;G)\}$ and $\{H^n(F_x;G)\}$ are simple (prove this). Besause of this, $H_i(X;\{H_n(F_x;G)\}) = H_i(X;G)$ and $H^i(X;\{H^n(F_x;G)\}) = H^i(X;G)$. □

In the next chapter we discuss the orientability problem, interconnections between orientability and Thom classes, etc. Here we remark that a $\mathcal{V}_n$-object is orientable iff its structure group (monoid) can be reduced to $\mathcal{SV}_n$ (prove this).

The line bundle ζ in 5.2(b) gives us an example of a non-orientable bundle. You can see it immediately, or notice that $T\zeta = RP^2$ and apply 5.8.

It is clear that $\alpha = \{p : Y \to X\}$ is orientable iff $\alpha \oplus \theta^1_X$ is (cf. V.1.10(iii) below). Hence, $\zeta \oplus \theta^k_X$ is non-orientable for every k, i.e., for every n there exists a non-orientable $\mathcal{V}_n$-object. Moreover, if α is orientable then $f^*\alpha$ is orientable for every map $f : Z \to \operatorname{bs}\alpha$. Thus, the universal $\mathcal{V}_n$-object $\gamma^n_{\mathcal{V}}$ is non-orientable.

Consider a pointed space $K = (K,*)$, and let $A \subset \pi_n(K)$ be such that $\pm A = A$.

5.10. Definition. Let $\alpha = \{p : Y \to X\}$ be an $\mathcal{F}_n$-object over a space X, and let $j_x : S^n \to T\alpha$ be a root with respect to a point $x \in X$. We regard j_x as a canonically pointed map. An element $v \in [T\alpha, K]^\bullet$ is called an (A,K)-*marking of* α if $j^*_x(v) \in A$ for all $x \in X$. An (A,K)-marking of a $\mathcal{V}_n$-object ξ is defined to be an (A,K)-marking of $\xi^\bullet$.

It is clear that if the base X of α is connected then v is an (A,K)-marking iff $j^*_{x_0}(v) \in A$ for some single point $x_0 \in X$.

Let $\Omega^n_A(K)$ be the subspace of $\Omega^n K = (K,*)^{(S^n,*)}$ consisting of all maps $\varphi : (S^n,*) \to (K,*)$ such that $[\varphi] \in A \subset \pi_n(K,*)$.

5.11. Theorem. *Let $\mathcal{Z}$ be one of the symbols $\mathcal{V},\mathcal{F}$. There exists a CW-space $B(\mathcal{Z}_n,A,K)$ with the following properties:*

(i) *The set of equivalence classes of (A,K)-marked $\mathcal{Z}_n$-objects over X is in a bijective correspondence with the set $[X, B(\mathcal{Z}_n,A,K)]$, i.e., $B(\mathcal{Z}_n,A,K)$ is a classifying space for (A,K)-marked $\mathcal{Z}$-objects;*

(ii) *The homotopy fiber of the forgetful map $B(\mathcal{Z}_n,A,K) \to B\mathcal{Z}_n$ is $\Omega^n_A(K)$.*

Proof. An (A, K)-marking of any $\mathcal{F}_n$-object α is just an $(\Omega^n_A(K), (K, *))$-structure on α, see 2.12. Thus, the case $\mathcal{Z} = \mathcal{F}$ follows from 2.13. Let $a = a^{\mathcal{V}}_{\mathcal{F}}(n) : B\mathcal{V}_n \to B\mathcal{F}_n$ be as in (4.18), and let $\lambda = \{B(\mathcal{F}_n, A, K) \to B\mathcal{F}_n\}$ be the forgetful $\Omega^n_A(K)$-fibration. Defining $B(\mathcal{V}_n, A, E)$ to be the total space of the fibration $a^*\lambda$, we conclude that (i) and (ii) are true for $\mathcal{Z} = \mathcal{V}$. (Note that, by 1.38, $B(\mathcal{V}_n, A, E)$ has the homotopy type of a CW-space.) □

If $\mathcal{V}$ is $\mathcal{O}$ or $\mathcal{TOP}$, then $B(\mathcal{V}_n, A, K)$ has an explicit geometrical description, see 3.9.

Now we turn to stable objects.

5.12. Constructions, Definitions, Notation. Let $\overline{B\mathcal{F}_n}$ be the telescope of the finite sequence $\{B\mathcal{F}_1 \to \cdots \xrightarrow{r^{\mathcal{F}}_{n-1}} B\mathcal{F}_n\}$. We regard $\overline{B\mathcal{F}_n}$ as a CW-subcomplex of $B\mathcal{F}$, i.e., there is a CW-filtration $\{\overline{B\mathcal{F}_n}\}$ of $B\mathcal{F}$. Recall that $\overline{B\mathcal{F}_n} \simeq B\mathcal{F}_n$, and so we have the universal $\mathcal{F}_n$-object $\gamma^n_{\mathcal{F}}$ over $\overline{B\mathcal{F}_n}$.

(a) Let $\alpha = \{f : X \to B\mathcal{F}\}$ be a stable $\mathcal{F}$-object over a CW-complex X, and let $\mathscr{F} = \{\emptyset = X_{-1} \subset X_0 \subset \cdots \subset X_n \subset \cdots\}$, $\cup X_n = X$ be a CW-filtration such that $f(X_n) \subset \overline{B\mathcal{F}_n}$. We define $f_n : X_n \to \overline{B\mathcal{F}_n}, f_n(x) = f(x)$ and put for simplicity $\zeta^n := f^*_n\gamma^n_{\mathcal{F}}$. Clearly, $i^*_n\zeta^{n+1} = \zeta^n \oplus \theta^1$ where $i_n : X_n \to X_{n+1}$ is the inclusion. Considering the maps $s_n := T\mathfrak{I}_{i_n,\zeta^{n+1}} : ST\zeta^n = T(\zeta^n \oplus \theta) \to T\zeta^{n+1}$, we get the *Thom spectrum*

$$T(\mathscr{F}, \alpha) := \{T\zeta^n, s_n\}.$$

In most applications $T\zeta^n$ is a CW-space. Nevertheless, if not, one can apply II.1.19 in order to get a spectrum $T(\mathscr{F}, \alpha)$.

(b) If X is connected, then the family of roots $j_n : S^n \to T\zeta^n$ yields a morphism $j : S \to T(\mathscr{F}, \alpha)$ of spectra, which we call a *root of* $T(\mathscr{F}, \alpha)$.

(c) Given a stable $\mathcal{F}$-object $\alpha = \{f : X \to B\mathcal{F}\}$ over a CW-complex X, let $X_n(\alpha)$ be the maximal CW-subcomplex which is contained in $f^{-1}(\overline{B\mathcal{F}_n})$. So, we have a canonical filtration $\mathscr{X} = \{X_n(\alpha)\}$ of X, and we set

$$T\alpha := T(\mathscr{X}, \alpha).$$

(d) Given a stable $\mathscr{F}$-object $\alpha = \{f : X \to B\mathcal{F}\}$ and a map $h : Y \to X$, we define a map $h_n : Y_n(h^*\alpha) \to X_n(\alpha), h_n(y) := h(y)$. Then we have the map $T\mathfrak{I}_{h_n,\zeta^n} : Th^*_n\zeta_n \to T\zeta^n$. So, we get a morphism

$$Th := \{T\mathfrak{I}_{h_n,\zeta^n}\} : T(Y, h^*\alpha) \to T\alpha.$$

(e) Given a stable $\mathcal{V}$-object $\xi = \{u : X \to B\mathcal{V}\}$, we set $T\xi := T\xi^\bullet$ where, as usual, $\xi^\bullet := \{X \xrightarrow{u} B\mathcal{V} \xrightarrow{a^{\mathcal{V}}_{\mathcal{F}}} B\mathcal{F}\}$.

(f) Given a structure map $\varphi : B \to B\mathcal{V}$, we can regard it as a stable $\mathcal{V}$-object $\varphi^*\gamma_{\mathcal{V}}$ and construct the spectrum $T(\varphi^*\gamma_{\mathcal{V}})$. However, as usual, we introduce a special notation $T(B, \varphi) := T(\varphi^* \gamma_{\mathcal{V}})$ in order to emphasize that φ is a structure map.

(g) Because of (e), there is the Thom spectrum $T\gamma_{\mathcal{V}}$ of the universal stable $\mathcal{V}$-object $\gamma_{\mathcal{V}}$. This spectrum is usually denoted by $M\mathcal{V}$ (or by $TB\mathcal{V}$, as in Stong [3]). Clearly, its n-th term is (homotopy equivalent to) $M\mathcal{V}_n$. In greater detail, let $\overline{B\mathcal{V}_n}$ be the telescope of the finite sequence $\{B\mathcal{V}_1 \to \cdots \xrightarrow{r_{n-1}^{\mathcal{V}}} B\mathcal{V}_n\}$; so, we have the filtration $\{\overline{B\mathcal{V}_n}\}$ of $B\mathcal{V}$. Then the maps $a_{\mathcal{F}}^{\mathcal{V}}(n) : \overline{B\mathcal{V}_n} \to \overline{B\mathcal{F}_n}$ yield a map $a_{\mathcal{F}}^{\mathcal{V}} : B\mathcal{V} \to B\mathcal{F}$ of filtered spaces, i.e., $(a_{\mathcal{F}}^{\mathcal{V}})^{-1}(\overline{B\mathcal{F}_n}) = \overline{B\mathcal{V}_n}$, and the n-th term of $M\mathcal{V}$ is the Thom space $M\mathcal{V}_n$ of the universal $\mathcal{V}_n$-object over $\overline{B\mathcal{V}_n}$.

5.13. Lemma. *Let $\alpha = \{f : X \to B\mathcal{F}\}$ and $\mathscr{F}$ be as in 5.12(a). Then $T(\mathscr{F}, \alpha) \simeq T\alpha$, i.e., the homotopy type of the Thom spectrum does not depend on filtration.*

Proof. We define $f_n : X_n(\alpha) \to \overline{B\mathcal{F}_n}, f_n(x) = f(x)$. We set $E_n := T(f_n^* \gamma^n)$, $F_n := T(f_n^* \gamma^n | X_n)$. Then $T\alpha = \{E_n\}$ and $T(\mathscr{F}, \alpha) = \{F_n\}$.

According to II.(1.4), we regard $\Sigma^{-n}\Sigma^\infty E_n$ as a subspectrum of $T\alpha$, and $T\alpha = \bigcup_n \Sigma^{-n}\Sigma^\infty E_n$. Similarly, $T(\mathscr{F}, \alpha) = \bigcup_n \Sigma^{-n}\Sigma^\infty F_n$. Clearly, $T(\mathscr{F}, \alpha) \subset T\alpha$. On the other hand, $\Sigma^{-n}\Sigma^\infty E_n \subset \bigcup_{m=1}^{\infty} \Sigma^{-m}\Sigma^\infty F_m$ for every n, and so $T(\mathscr{F}, \alpha)$ is cofinal in $T\alpha$. □

5.14. Lemma. *Let $\alpha = \{f : X \to B\mathcal{F}\}$ and $\beta = \{g : Y \to B\mathcal{F}\}$ be two stable $\mathcal{F}$-objects.*

(i) *Let $u, v : \alpha \to \beta$ be two morphisms of stable $\mathcal{F}$-objects, see 4.16. If $u \simeq_{B\mathcal{F}} v : X \to Y$ then $Tu \simeq Tv : T\alpha \to T\beta$.*

(ii) *Let $a : \alpha \to \beta$ be a morphism of stable $\mathcal{F}$-objects. If $a : X \to Y$ is a homotopy equivalence then $Ta : T\alpha \to T\beta$ is an equivalence.*

(iii) *Here we assume that $Y = X$. If $f \simeq g : X \to B\mathcal{F}$ (i.e., α and β are equivalent stable $\mathcal{F}$-objects) then $T\alpha \simeq T\beta$.*

Proof. (i) Let $U : u \simeq_{B\mathcal{F}} v$ be a homotopy over $B\mathcal{F}$. We have the commutative diagram

$$\begin{array}{ccc} X \times I & \xrightarrow{U} & Y \\ {\scriptstyle p_1}\downarrow & & \downarrow{\scriptstyle g} \\ X & \xrightarrow{f} & B\mathcal{F}. \end{array}$$

Then

$$T(U^*\beta) = T(U^* g^* \gamma_{\mathcal{F}}) = T(p_1^* f^* \gamma_{\mathcal{F}}) = T(p_1^* \alpha) = T\alpha \wedge I^+.$$

Now, $TU : T\alpha \wedge I^+ = T(U^*\beta) \to T\beta$ is a homotopy between Tu and Tv.

(ii) Let $\widehat{f} : \widehat{X} \to B\mathcal{F}$ be the fibrational substitute of f constructed in the proof of 1.35. Given a filtration $\mathscr{X} = \{X_n\}$ of X, we define the filtration $\widehat{\mathscr{X}}$ of $\widehat{X}$ by setting $\widehat{X}_n = \{(x, \omega) | x \in X_n, \omega \in X_n^I, \omega(0) = f(x)\}$. The inclusion $X \to \widehat{X}, x \mapsto (x, \omega_x)$ maps X_n to $\widehat{X}_n$ and so yields a map $T(\mathscr{X}, f) \to T(\widehat{\mathscr{X}}, \widehat{f})$ of Thom (pre)spectra, and this map induces an isomorphism

$$\pi_i(T(\mathscr{X}, f)) \to \pi_i(T(\widehat{\mathscr{X}}, \widehat{f})) = \varinjlim\{\pi_{i+N}(T\zeta^n)\}$$

where ζ^n is as in 5.12(a). So, without loss of generality, we can assume that f and g are fibrations. But then, by 1.27, a is an equivalence over $B\mathcal{F}$, and the result follows from (i).

(iii) Let $F : X \times I \to B\mathcal{F}$ be a homotopy between f and g. Considering the commutative diagram

$$\begin{array}{ccc} X & \xrightarrow{a} & X \times I \\ {\scriptstyle f}\downarrow & & \downarrow{\scriptstyle F} \\ B\mathcal{F} & = & B\mathcal{F} \end{array}$$

where $a(x) = (x, 0)$, we conclude, by (ii), that $Ta : T\alpha \to T(F^*\gamma_{\mathcal{F}})$ is an equivalence. Similarly, $T\beta \simeq T(F^*\gamma_{\mathcal{F}})$, and thus $T\alpha \simeq T\beta$. □

5.15. Construction. Let ξ be a stable $\mathcal{V}$-object classified by $f : X \to B\mathcal{V}$, and let $\varphi : B \to B\mathcal{V}$ be a map. Consider a (B, φ)-(pre)structure (a, H) on ξ as defined in 4.9, i.e., $H : X \times I \to B\mathcal{V}$ is a homotopy between f and φa. The inclusions $i_k : X = X \times \{k\} \to X \times I, k = 0, 1$, yield the morphisms

$$b_0 := Ti_0 : T\xi \to T(H^*\gamma_{\mathcal{V}}),\ b_1 := Ti_1 : T(a^*\varphi^*\gamma_{\mathcal{V}}) \to T(H^*\gamma_{\mathcal{V}}),$$

and each b_i is an equivalence by 5.14(ii). We define a morphism

$$T_H(a) : T\xi \xrightarrow{b_0} T(H^*\gamma_{\mathcal{V}}) \xrightarrow{b_1^{-1}} T(a^*\varphi^*\gamma_{\mathcal{V}}) \xrightarrow{Ta} T(B, \varphi).$$

By 5.14(i), equivalent prestructures yield homotopic morphisms, i.e., the homotopy class of the morphism $T_H(a) : T\xi \to T(B, \varphi)$ depends only on the (B, φ)-structure.

5.16. Remark. Let $\alpha = \{f : X \to B\mathcal{F}_k\}$ be an $\mathcal{F}_k$-object, and let α_{st} be its stabilization. Considering the Thom spectrum $T(\alpha_{\text{st}})$, we see that its n-th term $T_n(\alpha_{\text{st}})$ is $T(\alpha \oplus \theta^{n-k}), n \geq k$, i.e., $T_n(\alpha_{st}) = \Sigma^{n-k}T\alpha$. So, we have an isomorphism in $\mathscr{S}$

$$\mathfrak{e} : T(\alpha_{\text{st}}) \cong \Sigma^{-k}\Sigma^{\infty}T\alpha.$$

5.17. Definition. Given a CW-complex X, we say that a map $f : X \to B\mathcal{F}$ is *regular* if $f(X^{(n-2)}) \subset B\mathcal{F}_n$ for every n. A stable $\mathcal{F}$-object α is called *regular* if it is classified by a regular map.

5.18. Lemma. *Let (X, A) be a CW-pair, and let $h : X \to B\mathcal{F}$ be such that $h|A$ is regular. Then h is homotopic rel A to a regular map $f : X \to B\mathcal{F}$.*

Proof. This follows from 4.24 in a routine way. □

5.19. Construction, Notation. Let α be a stable $\mathcal{F}$-object classified by a regular map $f : X \to B\mathcal{F}$. We define $f_n : X^{(n-2)} \to B\mathcal{F}_n$, $f_n(x) = f(x)$ for every $x \in X^{(n-2)}$ and set

$$\alpha^n := f_n^* \gamma_{\mathcal{F}}^n.$$

Note that we are able to write $T\alpha = \{T\alpha^n\}$. Clearly, $\alpha_{\text{st}}^n = \alpha|X^{(n-2)}$ (where, of course, $\alpha_{\text{st}}^n = (\alpha^n)_{\text{st}}$).

5.20. Lemma. *Let $h : Y \to X$ be a k-connected cellular map.*

(i) *If $n > 1$ then, for every $\mathcal{F}_n$-object α over X, the map $T(\mathfrak{I}_{h,\alpha}) : T(h^*\alpha) \to T\alpha$ is $(n + k)$-connected.*

(ii) *For every stable $\mathcal{F}$-object α over X, the map $Th : T(h^*\alpha) \to T\alpha$ is k-connected.*

Proof. (i) The homomorphism $h_* : H_i(Y; \{H_n(F_y)\}) \to H_i(X; \{H_n(F_x)\})$ is an isomorphism for $i \le k$ and an epimorphism for $i = k + 1$. So, in view of the Thom isomorphism 5.7, $T(\mathfrak{I}_{h,\alpha})_* : H_i(T(h^*\alpha)) \to H_i(T\alpha)$ is an isomorphism for $i \le k+n$ and an epimorphism for $i = k+n+1$. Since $n > 1$, both Thom spaces are simply connected, and thus $T\mathfrak{I}_{h,\alpha}$ is $(n+k)$-connected.

(ii) By 5.14(iii) and 5.18, we can assume that α is classified by a regular map $f : X \to B\mathcal{F}$. Moreover, by 5.13, $T\alpha \simeq T(\mathscr{F}, f)$ where $\mathscr{F}$ is the filtration such that $X_n = X^{(n-2)}$. Since $h : Y^{(N-2)} \to X^{(N-2)}$ is k-connected for $N >> k$, the map $T(\mathfrak{I}_{h,\alpha^N}) : Th^*(\alpha^N) \to T\alpha^N$ is $(k + N)$-connected by (i). Thus, by II.4.5(iii), Th is k-connected. □

5.21. Theorem. (i) *$T(\alpha \wedge^h \beta) \simeq T\alpha \wedge T\beta$ for all stable $\mathcal{F}$-objects α, β, and this equivalence can be chosen naturally with respect to α and β.*

(ii) *$T(\xi \dagger \eta) \simeq T\xi \wedge T\eta$ for all stable $\mathcal{V}$-objects ξ, η, and this equivalence can be chosen naturally with respect to ξ and η. (Here, as in 4.19, $\dagger$ means $\times$ for $\mathcal{V} \le \mathcal{TOP}$ and $*$ for $\mathcal{V} = \mathcal{G}$.)*

Proof. (i) We let $\alpha = \{f : X \to B\mathcal{F}\}$ and $\beta = \{g : Y \to B\mathcal{F}\}$. Without loss of generality we can assume that $\mu^F(\overline{B\mathcal{F}_n} \times \overline{B\mathcal{F}_n}) \subset \overline{B\mathcal{F}_{2n}}$. Then there is an inclusion

$$h_n : X_n(\alpha) \times Y_n(\beta) \subset (X \times Y)_{2n}(\alpha \wedge^h \beta).$$

This map h_n induces a map

$$Th_n : T(f_n^* \gamma_{\mathcal{F}}^n) \wedge T(g_n^* \gamma_{\mathcal{F}}^n) \to T((f \times g)_{2n}^* \gamma_{\mathcal{F}}^{2n}),$$

and these maps form a morphism $Th : T\alpha \wedge T\beta \to T(\alpha \wedge^h \beta)$. Clearly, Th is natural with respect to α and β. We prove that Th is an equivalence.

By 5.14(iii) and 5.18, we can assume that α and β are regular. Then $X^{(n-2)} \subset X_n(\alpha)$ and $Y^{(n-2)} \subset Y_n(\beta)$, and we have the following commutative diagram of inclusions:

$$\begin{array}{ccc} X^{(n-2)} \times Y^{(n-2)} & \xrightarrow{a_1} & X_n(\alpha) \times Y_n(\beta) \\ {\scriptstyle a_2}\uparrow & & \downarrow{\scriptstyle h_n} \\ (X \times Y)^{(n-2)} & \xrightarrow{a_3} & (X \times Y)_{2n}(\alpha \wedge^h \beta). \end{array}$$

Clearly, every $a_i, i = 1, 2, 3$, is an $(n-3)$-equivalence. Hence, h_n is an $(n-3)$-equivalence. So, by 5.20, Th_n is a $(2n-3)$-equivalence, and so, by II.4.5(iii), Th is an $(n-3)$-equivalence for every n. Thus, Th is an equivalence.

(ii) This is an immediate consequence of (i). □

5.22. Corollary. *The spectrum $M\mathcal{V}$ is a commutative ring spectrum.*

Proof. Firstly, we consider $M\mathcal{G}$. Let $\mu = \mu^G : B\mathcal{G} \times B\mathcal{G} \to B\mathcal{G}$ be as in 4.20 and 4.28. For simplicity, let γ denote $\gamma_{\mathcal{G}}$ and $\wedge$ denote $\wedge^h$. So, $\gamma \wedge \gamma = \mu^*\gamma$, and so there is a morphism $\nu := \mathfrak{I}_{\mu,\gamma} : \gamma \wedge \gamma \to \gamma$ which yields a pairing $\overline{\mu} : T\gamma \wedge T\gamma \to T(\gamma \wedge \gamma) \xrightarrow{T\nu} T\gamma$. In order to prove the associativity of $\overline{\mu}$ we must prove that the morphisms $\nu \circ (\nu \wedge 1)$ and $\nu \circ (1 \wedge \nu)$ are bundle homotopic, i.e., that there exists a bundle homotopy $\Phi : \gamma \wedge \gamma \wedge \gamma \wedge 1_I \to \gamma$ such that $\Phi|\gamma \wedge \gamma \wedge \gamma \wedge 1_{\{0\}} = \nu \circ (\nu \wedge 1)$ and $\Phi|\gamma \wedge \gamma \wedge \gamma \wedge 1_{\{1\}} = \nu \circ (1 \wedge \nu)$. But this follows easily from 1.72. Clearly, the root $j : S \to T\gamma$ of $T\gamma$ can play the role of the unit for $\overline{\mu}$.

The commutativity can be proved similarly.

The proof for the spectra $M\mathcal{V}$ with $\mathcal{V} \leq \mathcal{TOP}$ can be done similarly to that for $M\mathcal{G}$, using the universal property 3.2(iii) and the bijective correspondence between principal $\mathcal{V}_n$-bundles and $((\mathbb{R}^n, 0), \mathcal{V}_n)$-bundles. Here $\mathcal{PL}_n$ is the group described in 4.1(c). □

Given a regular stable $\mathcal{F}$-object α, we say that α is *orientable* if α^4 is orientable. Given a stable $\mathcal{F}$-object β, we say that β is orientable if it is equivalent to an orientable regular stable $\mathcal{F}$-object α. Finally, we say that a stable $\mathcal{V}$-object ξ is orientable if $\xi^\bullet$ is.

The following stable version of 5.7–5.9 holds.

5.23. Theorem. *Let α be a stable $\mathcal{F}$-object over a CW-complex X.*

(i) *$\pi_i(T\alpha) = 0$ for $i < 0$. If X is connected then $\pi_0(T\alpha) = \mathbb{Z}$ for orientable α and $\pi_0(T\alpha) = \mathbb{Z}/2$ for non-orientable α, and the root $j : S \to T\alpha$ yields a generator of $\pi_0(T\alpha)$.*

(ii) *Let G be an abelian group. Suppose that either $2G = 0$ or α is orientable. Then there are Thom isomorphisms*

$$\varphi_G : H^i(X;G) \xrightarrow{\cong} H^i(T\alpha;G), \qquad \varphi^G : H_i(X;G) \xrightarrow{\cong} H_i(T\alpha;G).$$

Notice that the (co)homology of a space X *appears as the domain, while the (co)homology of a* spectrum $T\alpha$ *appears as the range.*

Proof. (i) We have $\pi_i(T\alpha) = \varinjlim \pi_{i+n}(T\alpha^n)$. So, by 5.8(ii), $\pi_i(T\alpha) = 0$ for $i < 0$. Furthermore, if $j : S^n \to T\alpha^n$ is a root of $T\alpha^n$ then

$$SS^n \xrightarrow{Sj} ST\alpha^n \to T\alpha^{n+1}$$

is a root of $T\alpha^{n+1}$. Hence, by 5.8(ii), $\pi_n(T\alpha^n) \to \pi_{n+1}(T\alpha^{n+1})$ is an isomorphism, and thus $\pi_0(T\alpha)$ is such as claimed.

(ii) The isomorphism, say, φ_G can be constructed as

$$H^i(X;G) = H^i(X^{(N-2)};G) \cong H^{i+N}(T\alpha^N;G) = H^i(T\alpha;G),$$

where $i << N$. □

In particular, if the base of α is connected, then $H^0(T\alpha;\mathbb{Z}/2) = \mathbb{Z}/2$, and $H^0(T\alpha) = \mathbb{Z}$ for orientable α. Thus, one can define a *stable* Thom class $u_{\mathbb{Z}/2} \in H^0(T\alpha;\mathbb{Z}/2)$ and, for orientable α, $u_{\mathbb{Z}} \in H^0(T\alpha)$ to be a generator of the group.

Frequently we shall write simply u instead of $u_{\mathbb{Z}}$ or $u_{\mathbb{Z}/2}$.

Recall that, for every connected spectrum E, there is a morphism $\tau_0 : E \to H(\pi_0(E))$ as in II.4.12.

5.24. Proposition. *Let α be a stable $\mathcal{F}$-object, and let $u \in H^0(T\alpha;\pi_0(T\alpha))$ be a Thom class. We assume that* $\operatorname{bs}\alpha$ *is connected.*

(i) *If α is orientable then the morphism $u : T\alpha \to H\mathbb{Z}$ coincides (up to sign) with $\tau_0 : T\alpha \to H\mathbb{Z}$.*

(ii) *If α is non-orientable then the morphism $u : T\alpha \to H\mathbb{Z}/2$ coincides with $\tau_0 : T\alpha \to H\mathbb{Z}/2$.*

Proof. We prove (i) only. For simplicity, we denote τ_0 by τ. It suffices to prove that $\tau \in H^0(T\alpha)$ generates $H^0(T\alpha) = \mathbb{Z}$. By II.4.9, the evaluation

$$\text{ev} : \mathbb{Z} = H^0(T\alpha) \to \text{Hom}(H_0(T\alpha), \mathbb{Z}) = \text{Hom}(\mathbb{Z}, \mathbb{Z}) = \mathbb{Z}$$

is an isomorphism, and so we must prove that $\tau_* : H_0(T\alpha) \to H_0(H\mathbb{Z})$ is an isomorphism. But, since $\tau_* : \pi_0(T\alpha) \to \pi_0(H\mathbb{Z})$ is an isomorphism, this follows from the commutativity of the diagram

$$\begin{array}{ccc} \pi_0(T\alpha) & \xrightarrow{\tau_*^\pi} & \pi_0(H\mathbb{Z}) \\ \cong\downarrow & & \downarrow\cong \\ H_0(T\alpha) & \xrightarrow{\tau_*^H} & H_0(H\mathbb{Z}) \end{array}$$

where the vertical isomorphisms are the Hurewicz homomorphisms. □

Now we introduce a stable analog of (A, K)-markings.

5.25. Definition. (a) Let E be a spectrum, let α be a stable $\mathcal{F}$-object over X, and let $A \subset \pi_0(E)$ be such that $\pm A = A$. If X is connected, we define an *(A, E)-marking* of α to be an element $v \in E^0(T\alpha)$ such that $j^*(v) \in A$, where $j : S \to T\alpha$ is a root of $T\alpha$. If $X = \sqcup X_\lambda$ with connected X_λ, we define an (A, E)-marking of α to be a family $\{v_\lambda\}$, where v_λ is an (A, E)-marking of $\alpha|X_\lambda$. Furthermore, an (A, E)-marking of a stable $\mathcal{V}$-object ξ is defined to be an (A, E)-marking of $\xi^\bullet$.

(b) An equivalence of two (A, E)-marked $\mathcal{F}$-, resp. $\mathcal{V}$-objects is an equivalence of $\mathcal{F}$-, resp. $\mathcal{V}$-objects which carries one of the given (A, E)-markings to the other.

Let $t_{(\mathcal{F},A,E)}(X)$ be the set of all equivalence classes of (A, E)-marked $\mathcal{F}$-objects over X. An $\mathcal{F}$-object induced from an (A, E)-marked one gets an obvious (A, E)-marking. So, $t_{(\mathcal{F},A,E)}$ is a functor. I can't prove the representability of $t_{(\mathcal{F},A,E)}$ on $\mathscr{C}$, but this holds on $\mathscr{C}_{\text{fd}}$. We prove this below, but we need some preliminaries.

Let E be an Ω-spectrum $\{E_n\}$, and let $\Omega_A^\infty E$ be the union of all components of $\Omega^\infty E$ belonging to A. Since $\pi_0(E) = \pi_n(E_n)$ for every n, one can regard A as a subset of $\pi_n(E_n)$ and consider (A, E_n)-markings of $\mathcal{F}_n$-objects. For simplicity, let B_n denote the space $B(\mathcal{F}_n, A, E_n)$ as in 5.11, and let ζ^n be the universal (A, E_n)-marked $\mathcal{F}_n$-object over B_n with the universal (A, E_n)-marking $a_n : T\zeta^n \to E_n$. Then the map

$$T(\zeta^n \oplus \theta^1) = ST\zeta^n \xrightarrow{Sa_n} SE_n \to E_{n+1}$$

gives us an (A, E_{n+1})-marking of $\zeta^n \oplus \theta^1$. This marking can be classified by a map $b_n : B_n \to B_{n+1}$. We define $B(\mathcal{F}, A, E)$ to be the telescope of the sequence $\cdots \to B_n \xrightarrow{b_n} B_{n+1} \xrightarrow{b_{n+1}} \cdots$, and we define $\overline{B}_n$ (resp. $\overline{B\mathcal{F}_n}$) to be the telescope of the finite sequence $\cdots \to B_i \xrightarrow{b_i} \cdots \xrightarrow{b_{n-1}} B_n$ (resp. $\cdots \to B\mathcal{F}_i \xrightarrow{r_{i-1}} \cdots \xrightarrow{r_{n-1}} B\mathcal{F}_n$). So, $\{\overline{B}_n\}$ (resp. $\{\overline{B\mathcal{F}_n}\}$) is a filtration of $B(\mathcal{F}, A, E)$ (resp. $B\mathcal{F}$). Recall that there are standard deformation retractions $d_n : \overline{B}_n \to B_n$ and $d_n' : \overline{B\mathcal{F}}_n \to B\mathcal{F}_n$.

Let $l_n : B_n \to B\mathcal{F}_n$ classify ζ_n.

5.26. Lemma. *The map $b_n : B_n \to B_{n+1}$ is $(n-1)$-connected.*

Proof. Let $q_n : C_n \to B\mathcal{F}_n$ be a fibrational substitute of l_n. Then there are maps $c_n : C_n \to C_{n+1}$ homotopy equivalent to b_n and such that the square

$$\begin{array}{ccc} C_n & \xrightarrow{c_n} & C_{n+1} \\ {\scriptstyle q_n}\downarrow & & \downarrow{\scriptstyle q_{n+1}} \\ B\mathcal{F}_n & \xrightarrow{r_n} & B\mathcal{F}_{n+1} \end{array} \tag{5.27}$$

commutes up to homotopy, and we can assume (deforming c_n if necessary, using the covering homotopy property) that it commutes. Recall that $E_0 \simeq \Omega^n E_n \simeq \Omega^\infty E$. Thus, by 5.11(ii), the homotopy fiber of q_n is $\Omega^n_A E_n \simeq \Omega^\infty_A E$. We fix $x \in B\mathcal{F}_n$ and set $\Phi_n = q_n^{-1}(x), \Phi_{n+1} = q_{n+1}^{-1}(r_n x)$. First, we consider the map $\bar{c}_n : \Phi_n \to \Phi_{n+1}$, $\bar{c}_n(a) = c_n(a)$ for every $a \in \Phi_n$, and prove that $\bar{c}_n$ is a homotopy equivalence.

Given a connected CW-space X, let $M_n(X)$ be the set of all (A, E_n)-markings of θ^n_X. Every (A, E_n)-marking v of θ^n_X gives us the (A, E_{n+1})-marking

$$\widehat{v} : T\theta^{n+1}_X = ST\theta^n_X \xrightarrow{v} SE_n \to_{n+1},$$

and we define

$$a : M_n(X) \to M_{n+1}(X), \quad a(v) = \widehat{v},$$

for every (A, E_n)-marking v of θ^n_X. Consider the map $f : X \to \{x\} \subset B\mathcal{F}_n$ and the commutative diagram

$$\begin{array}{ccccc} M_n(X) & = & [\mathrm{Lift}_{q_n} f] & = & [X, \Phi_n] \\ \downarrow a & & \downarrow (c_n)_* & & \downarrow (\bar{c}_n)_* \\ M_{n+1}(X) & = & [\mathrm{Lift}_{q_{n+1}} c_n f] & = & [X, \Phi_{n+1}] \end{array}$$

where $(c_n)_*[g] = [c_n g]$ for every q_n-lifting $g : X \to C_n$ of f. Clearly, for every connected CW-space X, a is a bijection, and so $(\bar{c}_n)_*$ is a bijection, and thus c_n is a homotopy equivalence (since $\Phi_n \simeq \Omega^\infty_A E$ is homotopy equivalent to a CW-space).

Now, the square (5.27) induces a ladder of the homotopy exact sequences of fibrations q_n, q_{n+1}. By the above, c_n yields a homotopy equivalence of fibers. Furthermore, by 4.24(ii), r_n is $(n-1)$-connected, and so c_n is $(n-1)$-connected (by a diagram chase). □

We define $\bar{l}_n : \overline{B}_n \xrightarrow{d_n} B_n \xrightarrow{l_n} B\mathcal{F}_n \subset \overline{B\mathcal{F}}_n$ and consider the ladder

$$\begin{array}{ccccccc} \cdots \longrightarrow & \overline{B}_n & \xrightarrow{\subset} & \overline{B}_{n+1} & \xrightarrow{\subset} & \cdots \\ & \downarrow \bar{l}_n & & \downarrow \bar{l}_{n+1} & & \\ \cdots \longrightarrow & \overline{B\mathcal{F}}_n & \xrightarrow{\subset} & \overline{B\mathcal{F}}_{n+1} & \xrightarrow{\subset} & \cdots . \end{array}$$

This ladder commutes up to homotopy, and we can assume that it is commutative (deforming $\bar{l}_n$ map by map if necessary, using the homotopy extension property). Thus, we get a map $l = \cup \bar{l}_n : B(\mathcal{F}, A, E) = \cup \overline{B}_n \to \cup \overline{B\mathcal{F}}_n = B\mathcal{F}$ of filtered spaces. We set $\eta = l^* \gamma_{\mathcal{F}}$.

Clearly, the universal (A, E_n)-marking a_n on ζ^n yields an (A, E)-marking v_n on ζ^n_{st}. Furthermore, we have $e_n^* \eta = \zeta^n_{\mathrm{st}}$ where $e_n : B_n \to B(\mathcal{F}, A, E)$ is the inclusion. Consider the morphism $T(e_n, \eta) : T\zeta^n_{\mathrm{st}} \to T\eta$.

5.28. Proposition. *The $\mathcal{F}$-object η admits an (A, E)-marking v such that $T(e_n, \eta)^*(v) = v_n$ for every n.*

Proof. We have $l_n^*\zeta^{n+1} = \zeta^n \oplus \theta^1$. Consider the Thom spectrum $T := \{T\zeta^n, s_n\}$, where $s_n = T\mathfrak{I}_{l_n,\zeta^{n+1}} : ST\zeta^n \to T\zeta^{n+1}$. For every $k \geq n$ there is a map $s_{k-1}\circ\cdots\circ S^{k-1}s_n : S^{k-n}T\zeta^n \to T\zeta^k$. These maps form a morphism $\sigma_n : T\zeta^n_{\text{st}} \to T$ of spectra.

The family of the universal (A, E_n)-markings a_n, $n = 1, 2, \dots$, yields an element $a \in E^0(T)$, and it is clear that $\sigma_n^*(a) = v_n$ for every n.

By 5.26, the map $1_{B(\mathcal{F},A,E)}$ is homotopic to a map $f : B(\mathcal{F}, A, E) \to B(\mathcal{F}, A, E)$ such that $f(B(\mathcal{F}, A, E)^{(n-2)}) \subset \overline{B}_n$. Clearly, $f^*\eta \simeq \eta$. We have $g^*\zeta^n \simeq \eta^n$, where g is the composition $g : B(\mathcal{F}, A, E)^{(n-2)} \xrightarrow{f} \overline{B}_n \xrightarrow{d_n} B_n$. Thus, for every n, we get a map $T\eta^n \simeq Tg^*\zeta^n \xrightarrow{T\mathfrak{I}_{g,\zeta^n}} T\zeta^n$, and these maps form a morphism $\tau : T\eta \to T$. Now, since the diagram

$$\begin{array}{ccc} T\zeta^n_{\text{st}} & \xrightarrow{T(e_n,\eta)} & T\eta \\ \| & & \downarrow{\scriptstyle\tau} \\ T\zeta^n_{\text{st}} & \xrightarrow{\sigma_n} & T \end{array}$$

commutes, we are able to set $v := \tau^*(a)$. □

We assume that η is equipped with the (A, E)-marking v, and we define

$$\phi : [X, B(\mathcal{F}, A, E)] \to t_{(\mathcal{F},A,E)}(X), X \in \mathscr{C},$$

by setting $\phi(f) = f^*(\eta)$ for every map $f : X \to B(\mathcal{F}, A, E)$.

5.29. Theorem. *The function ϕ is bijective for every $X \in \mathscr{C}_{\text{fd}}$. In other words, the functor $t_{(\mathcal{F},A,E)} : \mathscr{C}_{\text{fd}} \to \mathscr{E}ns$ can be represented by $B(\mathcal{F}, A, E)$. Furthermore, ϕ is surjective for every $X \in \mathscr{C}$.*

Proof. Let $t_{(\mathcal{F}_n,A,E_n)}(X)$ be the set of all equivalence classes of (A, E_n)-marked $\mathcal{F}_n$-objects over X. For every $\mathcal{F}_n$-object β the equivalence $\mathfrak{e} : T\beta_{\text{st}} \to \Sigma^{-n}\Sigma^\infty T\beta$ as in 5.16 induces an isomorphism $\mathfrak{e}^* : \widetilde{E}^n(T\beta) = E^0(\Sigma^{-n}\Sigma^\infty T\beta) \to E^0(T\beta_{\text{st}})$, which maps (A, E_n)-markings of β to (A, E)-markings of β_{st}. So, we have a function

$$\sigma_n : t_{(\mathcal{F}_n,A,E_n)}(X) \to t_{(\mathcal{F},A,E)}(X), \sigma_n(\beta) = \beta_{\text{st}}.$$

On the other hand, for every stable α there is the morphism $Ti_n : T\alpha^n_{\text{st}} \to T\alpha$, where $i_n : X^{(n)} \to X$ is the inclusion. It induces a homomorphism

$$E^0(T\alpha) \xrightarrow{(Ti_n)^*} E^0(T\alpha^n_{\text{st}}) \xrightarrow{\mathfrak{e}_*^{-1}} E^0(\Sigma^{-n}\Sigma^\infty T\alpha^n) = \widetilde{E}^n(T\alpha^n)$$

which maps (A, E)-markings of α to (A, E_n)-markings of α^n. So, we have a function

$$\tau_n : t_{(\mathcal{F},A,E)}(X) \to t_{(\mathcal{F}_n,A,E_n)}(X^{(n-2)}), \tau_n(\alpha) = \alpha^n.$$

If $\dim X << n$, then τ_n is inverse to σ_n. In particular, σ_n is bijective if $\dim X << n$.

Consider the following diagram, where $\phi_n(f) = f^*\zeta^n$ and $k : B_n \to B(\mathcal{F}, A, E)$ is the obvious inclusion like (4.7):

$$\begin{array}{ccc} [X, B_n] & \xrightarrow{k_*} & [X, B(\mathcal{F}, A, E)] \\ \phi_n \downarrow & & \downarrow \phi \\ t_{(\mathcal{F}_n, A, E_n)}(X) & \xrightarrow{\sigma_n} & t_{(\mathcal{F}, A, E)}(X) \end{array} .$$

By 5.28, this diagram commutes. Now, by 5.11(i), ϕ_n is bijective. Furthermore, if $\dim X << n$ then, by 5.26, k_* is bijective, and, by the above, σ_n is bijective. Thus, ϕ is bijective for every $X \in \mathscr{C}_{\text{fd}}$.

The surjectivity of ϕ follows because, by III.1.16, the map $\rho : [X, Y] \to \varprojlim [X^{(n)}, Y]$ is surjective for all X, Y and, in particular, for $Y = B(\mathcal{F}, A, E)$. □

5.30. Proposition. *The homotopy fiber of $l : B(\mathcal{F}, A, E) \to B\mathcal{F}$ is $\Omega_A^\infty E$.*

Proof. Consider the ladder which is composed of the squares (5.27),

$$(5.31) \qquad \begin{array}{ccccccc} \cdots \longrightarrow & C_n & \xrightarrow{c_n} & C_{n+1} & \xrightarrow{c_{n+1}} & \cdots \\ & q_n \downarrow & & \downarrow q_{n+1} & & \\ \cdots \longrightarrow & B\mathcal{F}_n & \xrightarrow{r_n} & B\mathcal{F}_{n+1} & \xrightarrow{r_{n+1}} & \cdots , \end{array}$$

where the (forgetful) maps q_n are $(\Omega_A^\infty E)$-fibrations. This ladder commutes up to homotopy, and we can assume that it is commutative (deforming c_n map by map, using the covering homotopy property). Defining $\overline{C(A, E)}$ to be the telescope of the top sequence

$$\cdots \to C_n \xrightarrow{c_n} C_{n+1} \xrightarrow{c_{n+1}} \cdots ,$$

we get the map $q : \overline{C(A, E)} \to B\mathcal{F}$, which is the telescope of q_n's. Since q_n is homotopy equivalent to l_n, its homotopy fiber is $\Omega_A^n E_n \simeq \Omega_A^\infty E$, see 5.11(ii). So, by 1.41(i), every fiber of q_n is homotopy equivalent to $\Omega_A^\infty E$. Furthermore, by 1.41(iv), q is a quasi-fibration, and every fiber of q is CW-equivalent to $\Omega_A^\infty E$. So, by 1.41(i), the homotopy fiber of q is $\Omega_A^\infty E$. Now the proposition holds because q is homotopy equivalent to l. □

We have considered Ω-spectra E, but this is not a real restriction. Indeed, if E is an arbitrary spectrum, take an Ω-spectrum E' equivalent to E, see II.1.21, and set $B(\mathcal{F}, A, E) := B(\mathcal{F}, A, E')$. Clearly, 5.29 and 5.30 hold in this case also.

Now we turn to $\mathcal{V}$-objects. Let $\widehat{l} : C(\mathcal{F}, A, E) \to B\mathcal{F}$ be a fibrational substitute of the forgetful map $l : B(\mathcal{F}, A, E) \to B\mathcal{F}$. We define $B(\mathcal{V}, A, E)$ via the pull-back diagram

$$\begin{array}{ccc} B(\mathcal{V}, A, E) & \xrightarrow{h} & C(\mathcal{F}, A, E) \\ \downarrow & & \downarrow \widehat{\imath} \\ B\mathcal{V} & \xrightarrow{a_{\mathcal{F}}^{\mathcal{V}}} & B\mathcal{F} \end{array}$$

By 1.38. $B(\mathcal{V}, A, E)$ has the homotopy type of a CW-space. Now, one has an (A, E)-marked stable $\mathcal{V}$-object $\eta^{\mathcal{V}} := h^*\eta$.

Let $t_{(\mathcal{V},A,E)}(X)$ be the set of all equivalence classes of (A, E)-marked $\mathcal{V}$-objects over X. Define $\phi^{\mathcal{V}} : [X, B(\mathcal{V}, A, E)] \to t_{(\mathcal{F},A,E)}(X), X \in \mathscr{C}$, by setting $\phi^{\mathcal{V}}(f) = f^*(\eta^{\mathcal{V}})$. Now, 5.29 and 5.30 imply the following theorem.

5.32. Theorem. (i) *The map $\phi^{\mathcal{V}}$ is bijective for every $X \in \mathscr{C}_{\text{fd}}$. In other words, the functor $t_{(\mathcal{V},A,E)} : \mathscr{C}_{\text{fd}} \to \mathscr{E}ns$ can be represented by $B(\mathcal{V}, A, E)$. Furthermore, $\phi_{\mathcal{V}}$ is surjective for every $X \in \mathscr{C}$.*

(ii) *The homotopy fiber of the forgetful map $B(\mathcal{V}, A, E) \to B\mathcal{V}$ is $\Omega_A^\infty E$.* □

From here to the end of this section, we choose a natural number N and fix a base point $s_0 \in S^N$.

5.33. Construction. Consider a map $t : X \to \mathcal{F}_N$ of a CW-space X. It yields a map $\tau : X \times S^N \to X \times S^N$, $\tau(x, s) := (x, t(x)(s))$. Let $i : X \subset CX, i(x) = (x, 0)$, be the inclusion of the bottom. We regard $X \times S^N$ as the subspace $i(X) \times S^N$ of $CX \times S^N$ and define

$$p : (CX \times S^N)_{\text{left}} \cup_\tau (CX \times S^N)_{\text{right}} \xrightarrow{p_1 \cup p_1} (CX)_{\text{left}} \cup (CX)_{\text{right}} = SX.$$

One can prove that p is a quasi-fibration, see Dold–Thom [1]. Furthermore, p has a section

$$SX \to (CX \times S^N)_{\text{left}} \cup_\tau (CX \times S^N)_{\text{right}},\ a \mapsto (a, s_0),$$

where $a \in (CX)_{\text{left}}$ or $a \in (CX)_{\text{right}}$.

Let $\xi = \xi_t$ be a fibrational substitute of p. Then, by 1.41(i), ξ is an $(S^N, *)$-fibration (i.e, an $\mathcal{F}_N$-object) over SX.

Alternatively, ξ is classified by a map $h : SX \to B\mathcal{F}_N$ which is the adjoint map to $t : X \to \mathcal{F}_N \simeq \Omega B\mathcal{F}_N$.

We define $\varphi : X \times S^N \xrightarrow{\tau} X \times S^N \xrightarrow{p_2} S^N$. It is clear that φ factors through

$$\overline{\varphi} : \frac{X \times S^N}{X} = \frac{X \times S^N}{X \times \{s_0\}} \to S^N.$$

Consider the diagram

$$\frac{CX \times S^N}{CX} \supset \frac{X \times S^N}{X} \xrightarrow{\overline{\varphi}} S^N.$$

5.34. Lemma. $T\xi \simeq \frac{CX \times S^N}{CX} \cup_{\overline{\varphi}} S^N$.

Proof. Clearly, $T\xi \simeq Cp$. Now, $Cp = \left(\frac{CX \times S^N}{CX}\right)_{\text{left}} \bigcup_{\frac{X \times S^N}{X}} \left(\frac{CX \times S^N}{CX}\right)_{\text{right}}$, where $\frac{X \times S^N}{X} \subset \left(\frac{CX \times S^N}{CX}\right)_{\text{left}}$ is induced by the inclusion $i : X \subset CX$ and the map $\frac{X \times S^N}{X} \to \left(\frac{CX \times S^N}{CX}\right)_{\text{right}}$ is induced by $\theta : X \times S^N \to CX \times S^N$, $\theta(x, s) = (i(x), \overline{\varphi}(x, s))$. Now the lemma follows because $X \times S^N \xrightarrow{\theta} CX \times S^N \xrightarrow{p_2} S^N$ coincides with $\overline{\varphi}$. □

There are two H-space structures on $\Omega^N S^N$. One of them is given via the loop structure, while another one is given via the compositions of maps $S^N \to S^N$. The corresponding multiplications are denoted by $*$ and $\circ$. These H-structures do not coincide: for example, if $x \in \Omega_k^N S^N$, $y \in \Omega_l^N S^N$, then $x * y \in \Omega_{k+l}^N S^N$ while $x \circ y \in \Omega_{kl}^N S^N$. Note that $*1_{S^N} : \Omega_k^N S^N \to \Omega_{k+1}^N S^N$ is a homotopy equivalence for every k.

Let X be a pointed connected space, let $f : S^N X \to S^N$ be a pointed map, and let $g : X \to \Omega^N S^N$ be the adjoint map to f. It is easy to see that $g(X) \subset \Omega_0^N S^N$. Consider the composition

$$t : X \xrightarrow{g} \Omega_0^N S^N \xrightarrow{*1_{S^N}} \Omega_1^N S^N \subset S\mathcal{F}_N.$$

This t gives an $\mathcal{F}_N$-object ξ over SX as in 5.33.

5.35. Theorem (cf. May [3], Ravenel [1]). $T\xi \simeq C(f)$.

Proof. For simplicity, let $q : X \times S^N \to S^N$ denote the projection p_2, and let $\pi : S^N \vee S^N \to S^N$ be the folding map, $\pi|S^N = 1_{S^N}$ for each of the two summands. Consider the map $h : X \times S^N \to X \wedge S^N \xrightarrow{f} S^N$, where the first map collapses the wedge. Since $h(x, s_0) = s_0 = q(x, s_0)$ for every $x \in X$, the map $\pi \circ (h \vee q) : (X \times S^N) \vee (X \times S^N) \to S^N$ factors through a map $k : (X \times S^N) \bigcup_{X \times \{s_0\}} (X \times S^N) \to S^N$. Consider the composition

$$\psi : X \times S^N \to X \times (S^N \vee S^N) = (X \times S^N) \bigcup{}_{X \times \{s_0\}} (X \times S^N) \xrightarrow{k} S^N$$

and denote by $\sigma : X \to \Omega_1^N S^N$ the adjoint map to ψ. It is easy to see that σ is homotopic to $t = g * 1_{S^N} : X \to \Omega_1^N S^N$. Thus, in order to prove the theorem it suffices to prove that

$$Cf \simeq \frac{CX \times S^N}{CX} \bigcup_{\overline{\psi}} S^N,$$

where $\overline{\psi} : \frac{X \times S^N}{X} \to S^N$ is constructed as $\overline{\varphi}$ was in 5.34. Consider the diagram

$$\begin{array}{ccccccc} \frac{CX\times(S^N\vee S^N)}{CX} & \supset & \frac{X\times(S^N\vee S^N)}{X} & & & & \\ \| & & \| & & & & \\ \frac{CX\times S^N}{CX} \vee \frac{CX\times S^N}{CX} & \supset & \frac{X\times S^N}{X} \vee \frac{X\times S^N}{X} & \xrightarrow{\overline{h}\vee\overline{q}} & S^N \vee S^N & \xrightarrow{\pi} & S^N \end{array}$$

where $\overline{h}, \overline{q}$ are as in 5.34. By construction, $\overline{h}$ collapses the factor S^N and therefore induces the map $\widehat{h} : X \wedge S^N \to S^N$. Clearly, $\widehat{h} \simeq f$. There are the diagrams

$$(CX \wedge S^N) \vee \frac{CX \times S^N}{CX} \supset (X \wedge S^N) \vee \frac{X \times S^N}{X} \xrightarrow{\widehat{h}\vee\overline{q}} S^N \vee S^N \xrightarrow{\pi} S^N,$$

and

$$(CX \wedge S^N) \vee S^n \supset S^N X \vee S^N \xrightarrow{\widehat{h}\vee 1} S^N \vee S^N \xrightarrow{\pi} S^N.$$

We set $a = \pi\circ(\overline{h} \vee \overline{q})$, $b = \pi\circ(\widehat{h} \vee \overline{q})$, $c = \pi\circ(\widehat{h} \vee 1_{S^N})$. We have

$$\frac{CX \times S^N}{CX} \bigcup{}_{\overline{\psi}} S^N \simeq \left(\frac{CX \times S^N}{CX} \bigvee \frac{CX \times S^N}{CX}\right) \bigcup{}_a S^N$$
$$\simeq \left((CX \wedge S^N) \bigvee \frac{CX \times S^N}{CX}\right) \bigcup{}_b S^N \simeq ((CX \wedge S^N) \vee S^N) \bigcup{}_c S^N$$
$$\simeq (CX \wedge S^N) \bigcup{}_{\widehat{h}} S^N.$$

Now, the last space is homotopy equivalent to Cf because $\widehat{h} \simeq f$. □

5.36. Construction. Let ξ be an $\mathcal{V}_n$-object over X, $\mathcal{V} \le \mathcal{TOP}$, and let θ^0 be the trivial $\mathcal{V}_0$-object over X. We have $T(\xi \times \theta^0) = T\xi \wedge X^+$, and $d^*(\xi \times \theta^0) = \xi$ where $d : X \to X \times X$ is the diagonal. Thus, one has the map

$$\Delta^n := T\mathfrak{I}_{d,\xi\times\theta^0} : T\xi \to T\xi \wedge X^+.$$

In order to construct Δ^n for $\mathcal{V} = \mathcal{G}$, we do the following. Let $p_1 : X \times X \to X$ be the projection, $p_1(x, y) = x$. Then $T(p_1^*\xi) \simeq T\xi \wedge X^+$ (prove this; note also that $p_1^*\xi = \xi \times \theta^0$ for $\mathcal{V} \le \mathcal{TOP}$). Furthermore, $d^* p_1^* \xi = \xi$, and we define $\Delta^n := T\mathfrak{I}_{d,p_1^*\xi}$.

Similarly, if ξ is a *stable* $\mathcal{V}$-object over X, we have a *morphism* (*of spectra*)

$$\Delta := T(d) : T\xi \to T\xi \wedge X^+.$$

§6. Homotopy Properties of Certain Thom Spectra

Recall that $\mathscr{A}_p$ denotes the mod p Steenrod algebra, see II.6.25.

6.1. Proposition. *The homotopy groups of the Thom spectrum $M\mathcal{V}$ are finitely generated $\mathbb{Z}/2$-vector spaces. In particular, $M\mathcal{V}$ is a $\mathbb{Z}[2]$-local spectrum of finite $\mathbb{Z}[2]$-type.*

Proof. By 4.27(ix), every group $H_i(B\mathcal{V})$ is finitely generated, and so every group $H_i(B\mathcal{V};\mathbb{Z}/2)$ is finite. Thus, by 5.23(ii), every group $H_i(M\mathcal{V};\mathbb{Z}/2)$ is finite.

By 5.23(i), $M\mathcal{V}$ is connected. The universal stable $\mathcal{V}$-object over $B\mathcal{V}$ is non-orientable, and so, by 5.23(i), $\pi_0(M\mathcal{V}) = \mathbb{Z}/2$. Since $M\mathcal{V}$ is a ring spectrum, $\pi_i(M\mathcal{V})$ is a $\mathbb{Z}/2$-vector space for every i. Hence, by II.4.24, $2^k H_i(M\mathcal{V}) = 0$ for some $k = k(i)$ (take $\mathfrak{C}$ to be the Serre class of all abelian groups having 2-primary exponents, see II.4.23(iii)). Furthermore, $H_i(M\mathcal{V})$ is a *finite* 2-primary group because $H_i(M\mathcal{V};\mathbb{Z}/2)$ is finite (use the Universal Coefficient Theorem II.4.9). So, by II.4.24, $\pi_i(M\mathcal{V})$ is a finite 2-primary group for every i (take $\mathfrak{C}$ to be the Serre class of all finite 2-primary abelian groups, see II.4.23(ii)). Thus, $\pi_i(M\mathcal{V})$ is a finite dimensional $\mathbb{Z}/2$-vector space for every i. □

6.2. Theorem (Thom [2]). *The spectrum $M\mathcal{V}$ is a wedge of suspensions over $H\mathbb{Z}/2$, i.e., it is a graded Eilenberg–Mac Lane spectrum, $M\mathcal{V} = H(\pi_*(M\mathcal{V}))$.*

Proof. Let $u \in H^*(M\mathcal{V};\mathbb{Z}/2)$ be the Thom class. By 6.1, $M\mathcal{V}$ is a connected $\mathbb{Z}[2]$-local ring spectrum with $\pi_0(M\mathcal{V}) = \mathbb{Z}/2$, and so, by II.7.24 (for $p = 2$), it suffices to prove that $Q_i(u) \neq 0$ for all $i = 0, 1, \dots$. Since the canonical morphisms $MO \to M\mathcal{V}$ maps the Thom class to the Thom class, it suffices to prove that $Q_i(u) \neq 0$ for the Thom class $u \in H^*(MO;\mathbb{Z}/2)$. In view of universality of MO and stability of Q_i, it suffices to find a vector bundle ξ with $Q_i(u_\xi) \neq 0$.

Let η be the canonical 1-dimensional vector bundle over $RP^\infty = BO_1$. It is well known (see e.g. Stong [3]) that $T\eta = RP^\infty$ and that $x := u_\eta$ is the generator of $H^1(RP^\infty;\mathbb{Z}/2) = \mathbb{Z}/2$. We prove by induction that $Q_i(x) = x^{2^{i+1}}$. Firstly, $Q_0(x) = Sq^1 x = x^2$. Suppose that $Q_{n-1}(x) = x^{2^n}$. Then

$$Q_n(x) = [Q_{n-1}, Sq^{2^n}](x) = Q_{n-1}Sq^{2^n}x + Sq^{2^n}Q_{n-1}x = Sq^{2^n}x^{2^n} = x^{2^{n+1}}.$$

The induction is confirmed. □

Now we consider the Thom spectra $MS\mathcal{V}$ of orientable $\mathcal{V}$-objects.

6.3. Proposition. *Let ξ be a $\mathcal{V}$-object over a connected base, and let $u = u_\xi \in H^n(T\xi;\mathbb{Z}/2)$ be the Thom class. Then ξ is orientable iff $Sq^1(u) = 0$.*

Proof. Recall that Sq^1 is the coboundary homomorphism in the exact sequence

$$\cdots \to H^i(X;\mathbb{Z}/2) \to H^i(X;\mathbb{Z}/4) \to H^i(X;\mathbb{Z}/2) \xrightarrow{Sq^1} H^{i+1}(X;\mathbb{Z}/2) \to \cdots$$

induced by the exact sequence $0 \to \mathbb{Z}/2 \to \mathbb{Z}/4 \to \mathbb{Z}/2 \to 0$, see e.g. Mosher–Tangora [1]. If ξ is not orientable, then $H_i(T\xi) = 0$ for $i < n$, $H_n(T\xi) = \mathbb{Z}/2$, and so $Sq^1 u \neq 0$. If ξ is orientable, then u is the reduction mod 2 of a class $v \in H^n(T\xi)$, and so $Sq^1 u = 0$. □

6.4. Proposition. *Every group $\pi_i(M\mathcal{SV})$ is finitely generated. So, $M\mathcal{SV}$ has finite $\mathbb{Z}$-type.*

Proof. Since $B\mathcal{SV}$ is the universal covering of $B\mathcal{V}$, we conclude that $\pi_1(B\mathcal{SV}) = 0$ and $\pi_i(B\mathcal{V}) = \pi_i(B\mathcal{SV})$ for $i > 1$. By 4.27(ix), the groups $\pi_i(B\mathcal{SV})$ are finitely generated, and so the groups $H_i(B\mathcal{SV})$ are finitely generated (use the mod $\mathfrak{C}$ Hurewicz Theorem for spaces, where $\mathfrak{C}$ is the class of finitely generated abelian groups). So, the groups $H_i(M\mathcal{SV}) \cong H_i(B\mathcal{SV})$ are finitely generated. Thus, by II.4.24, $\pi_i(M\mathcal{SV})$ are finitely generated. □

6.5. Theorem. *$M\mathcal{SO}[2]$ is a wedge of spectra of the form $\Sigma^k H\mathbb{Z}$ and $\Sigma^k H\mathbb{Z}/2$. In particular, $M\mathcal{SO}[2]$ is a graded Eilenberg–Mac Lane spectrum, and every torsion element of $\pi_*(M\mathcal{SO})$ has order 2.*

Proof. See Wall [1] (the original proof) or Stong [3], Ch. IX. □

6.6. Theorem. *$M\mathcal{SV}[2]$ is a graded Eilenberg–Mac Lane spectrum.*

Proof. By II.7.1, the homomorphism $\pi_0(H\mathbb{Z}[2]) \to \pi_0(M\mathcal{SO}[2]), 1 \mapsto 1$, is induced by a morphism $f : H\mathbb{Z}/2 \to M\mathcal{SO}[2]$. Since the composition

$$H\mathbb{Z} \to H\mathbb{Z}[2] \xrightarrow{f} M\mathcal{SO}[2] \xrightarrow{\alpha_{\mathcal{V}}^{\mathcal{O}}[2]} M\mathcal{SV}[2]$$

satisfies II.7.7, the result follows. □

Now we fix an *odd* prime p and discuss p-primary properties of $M\mathcal{SV}$. Let $j : S \to M\mathcal{SV}$ be a root; choose the Thom class $v \in H^0(M\mathcal{SV})$ with $j^*(v) = 1 \in H^0(S)$. Let $u \in H^0(M\mathcal{SV};\mathbb{Z}/p)$ be the reduction mod p of v. We consider the action of $\mathscr{A}_p$ on the class u. Since $j : S \to M\mathcal{SV}$ is the unit of the ring spectrum $M\mathcal{SV}$, the class u is the counit of the coalgebra $H^*(M\mathcal{SV};\mathbb{Z}/p)$, cf. II.7.20.

6.7. Lemma. *The operations $Q_i, i \geq 0$, act trivially on $H^*(M\mathcal{SO};\mathbb{Z}/p)$. In particular, $H^*(M\mathcal{SO};\mathbb{Z}/p)$ is an $\mathscr{A}_p/(Q_0)$-module.*

Proof. By 4.29(ii), $H^k(B\mathcal{SO};\mathbb{Z}/p) = 0$ for odd k, and so, by 5.23, $H^k(M\mathcal{SO};\mathbb{Z}/p) = 0$ for odd k. Hence, Q_i acts on $H^*(M\mathcal{SO};\mathbb{Z}/p)$ trivially because $\dim Q_i$ is odd. Since the left ideal $\mathscr{A}_p(Q_0, \dots, Q_n, \dots)$ coincides with the two-sided ideal (Q_0), $H^*(MSO; \mathbb{Z}/p)$ is an $\mathscr{A}_p/(Q_0)$-module. □

The following theorem was proved by Averbuch [1], Milnor [4], and Novikov [1].

6.8. Theorem. *$H^*(MSO; \mathbb{Z}/p)$ is a free $\mathscr{A}_p/(Q_0)$-module.*

Proof. There is a unique morphism $\overline{\Delta}$ such that the diagram

$$\begin{array}{ccc} \mathscr{A}_p & \xrightarrow{\Delta} & \mathscr{A}_p \otimes \mathscr{A}_p \\ \downarrow & & \downarrow \\ \mathscr{A}_p/(Q_0) & \xrightarrow{\overline{\Delta}} & \mathscr{A}_p/(Q_0) \otimes \mathscr{A}_p/(Q_0) \end{array}$$

commutes. This $\overline{\Delta}$ turns $\mathscr{A}_p/(Q_0)$ into a Hopf algebra. Furthermore, the space of primitives of this Hopf algebra is just $\mathbb{Z}/p\{\mathscr{P}^{\Delta_i} | i = 1, 2, \dots\}$. Indeed, the dual Hopf algebra $(\mathscr{A}_p/(Q_0))^*$ is the subalgebra $\mathbb{Z}/p\,[\xi_1, \dots, \xi_n, \dots]$ of $\mathscr{A}_p^*$. Thus, by II.6.31, it suffices to prove that $\mathscr{P}^{\Delta_i}(u) \neq 0$.

Let η be the canonical 1-dimensional complex vector bundle over CP^∞. It is well known (see e.g. Husemoller [1], Stong [3]) that $T\eta = CP^\infty$ and that $x := u_\eta$ is a generator of $H^2(CP^\infty; \mathbb{Z}/p) = \mathbb{Z}/p$. We prove by induction that $\mathscr{P}^{\Delta_i}(x) = x^{p^i}$ (and thus $\mathscr{P}^{\Delta_i}(x) \neq 0$). This implies immediately that $\mathscr{P}^{\Delta_i}(u) \neq 0$ because of the universality of u. We have $\mathscr{P}^{\Delta_1}(x) = P^1(x) = x^p$. Suppose that $\mathscr{P}^{\Delta_n}(x) = x^{p^n}$. Now,

$$\begin{aligned} \mathscr{P}^{\Delta_{n+1}}(x) &= [P^{p^n}, \mathscr{P}^{\Delta_n}](x) = P^{p^n} \mathscr{P}^{\Delta_n}(x) \pm \mathscr{P}^{\Delta_n} P^{p^n}(x) \\ &= P^{p^n}(x^{p^n}) = x^{p^{n+1}}. \end{aligned}$$

The induction is confirmed. □

Based on 6.8, one can prove that $\pi_*(MSO)$ has no odd torsion. More precisely, we have the following theorem (Averbuch–Milnor–Novikov). A proof can be found in Stong [3].

6.9. Theorem. *For every odd prime p,*

$$\pi_*(MSO[p]) = \mathbb{Z}[p][x_1, \dots, x_k, \dots], \dim x_n = 4n. \qquad \square$$

Now we consider the spectrum MSG.

6.10. Theorem. *$Q_i u \neq 0, i = 1, \dots,$ and $\mathscr{P}^{\Delta_j} u \neq 0, j = 1, 2, \dots$ for the class $u = u^G \in H^*(MSG; \mathbb{Z}/p)$.*

Proof. The inequality $\mathscr{P}^{\Delta_j} u \neq 0$ holds because it holds in MSO. We prove that $Q_1 u \neq 0$. Consider the Postnikov tower of the p-localized sphere $S^N[p]$, N large. Since $H_i(S^N[p]) = 0$ for $i > N$, the first non-trivial Postnikov invariant of $S^N[p]$ is P^1. So, $\pi_i(S^N[p]) = 0$ for $N < i < N + 2p - 3$, $\pi_{N+2p-3}(S^N[p]) = \mathbb{Z}/p$. Let $f : S^{N+2p-3} \to S^N[p]$ be an essential map. Then $H^N(Cf;\mathbb{Z}/p) = H^{N+2p-2}(Cf;\mathbb{Z}/p) = \mathbb{Z}/p$, and generators $x \in H^N(Cf;\mathbb{Z}/p), y \in H^{N+2p-2}(Cf;\mathbb{Z}/p)$ can be chosen such that $P^1(x) = y$. Let X be the cone of a map $S^{N+2p-3} \to S^{N+2p-3}$ of degree p; then the Bockstein homomorphism

$$\beta : H^{N+2p-3}(X;\mathbb{Z}/p) \to H^{N+2p-2}(X;\mathbb{Z}/p)$$

is an isomorphism, cf. II.6.27(b). Since $[f] \in \pi_{n+2p-3}(S^N[p])$ has order p, f can be extended to a map $g : X \to S^N[p]$. So, the group $\widetilde{H}^*(Cg;\mathbb{Z}/p)$ is generated by three elements x, y, z, $\dim x = N, \dim y = N + 2p - 2, \dim z = N + 2p - 1$. Moreover, $\beta P^1 x = \beta y = z$. Since $\beta x = 0$,

$$Q_1 x = P^1 \beta x - \beta P^1 x = -\beta P^1 x = -z \neq 0.$$

By 5.35, the space Cg is the Thom space of a certain S^N-fibration ξ over M, and it is clear that x is the Thom class u_ξ of this fibration. Hence, $Q_1 u_\xi \neq 0$, and so $Q_1 u \neq 0$.

Now we prove that $Q_i u \neq 0$ (following Tsuchia [1], [2]). We use some facts about the stable homotopy groups of spheres and some standard notation for their elements, see Toda [1].

Set $q = 2p - 2$. Consider a map $h : S^{pq-1} \to S^{pq-1}$ of degree p and set $M = Ch$. The cofiber sequence $S^{pq-1} \xrightarrow{h} S^{pq-1} \to M$ induces an exact sequence

$$\pi_{pq+2p-3}(M) \xrightarrow{h_*} \pi_{pq+2p-3}(S^{pq}) \xrightarrow{p} \pi_{pq+2p-3}(S^{pq}).$$

(This sequence is exact since, by the Freudenthal Suspension Theorem, these homotopy groups coincide with the stable ones.) Since $\pi_{pq+2p-3}(S^{pq}) = \mathbb{Z}/p$, the generator of this group has the form $h_*(\alpha)$ for some $\alpha \in \pi_{pq+2p-3}(M)$. Set $L = M \cup_\alpha e^{(p+1)q}$. It is easy to see that $\pi_{(p+1)q}(L) \otimes \mathbb{Q} = \mathbb{Q}$, i.e., $\pi_{(p+1)q}(L)$ contains $\mathbb{Z}$ as a direct summand. Let ι be a generator of this subgroup $\mathbb{Z}$. Set $K = L \cup_{p\iota} e^{(p+1)q+1}$.

Let $b : S^{pq-1} \to BS\mathcal{G}$ yield the element $\beta_1 \in \pi_{pq-2}(S)$ (under the isomorphism $\pi_i(S) \cong \pi_{i+1}(BS\mathcal{G})$). Since $p\beta_1 = 0$, b can be extended to M. Furthermore, since $\pi_{(p+1)q-1}(BS\mathcal{G}) = 0$ and $\pi_{(p+1)q}(BS\mathcal{G}) = \mathbb{Z}/p$, f can be extended to some $c : K \to BS\mathcal{G}$. Thus, there arises an S^N-fibration ξ over K, and

$$T\xi = S^N \cup_{\beta_1} e^{N+pq-1} \cup_p e^{N+pq} \cup_\alpha e^{N+(p+1)q} \cup_p e^{N+(p+1)q+1}.$$

Here α_1 is the N-fold suspension of $\alpha \in \pi_{pq+2p-3}(M)$ and p corresponds to the element $p \in \pi_k(D^k, S^{k-1}) = \mathbb{Z}$ for $k = N + pq - 1, N + (p + 1)q$. Let $s \in H^N(T\xi;\mathbb{Z}/p), e_{pq-1} \in H^{N+pq-1}(T\xi;\mathbb{Z}/p)$, $e_{pq} \in H^{N+pq}(T\xi;\mathbb{Z}/p)$, etc., be the cohomology classes corresponding to the cells above. Then in $T\xi$ we have:

1. $P^p(s) = e_{pq}$.
2. $P^1P^p(s) = P^{p+1}(s) = e_{(p+1)q}; P^pP^1(s) = 0$.
3. $\beta P^{p+1}(s) = e_{(p+1)q+1}; P^{p+1}\beta(s) = 0$.
4. $P^pP^1\beta(s) = \beta P^pP^1(s) = P^p\beta P^1(s) = P^1\beta P^p(s) = 0$.
5. $\beta(e_{pq-1}) = e_{pq}$.
6. $P^1(e_{pq}) = e_{(p+1)q}; \beta P^1(e_{pq}) = e_{(p+1)q+1}$.
7. $\beta(e_{(p+1)q}) = e_{(p+1)q+1}$.

This implies that $Q_1(s) = 0$ and

$$Q_2(s) = P^pQ_1(s) - Q_1P^p(s) = -Q_1(e_{pq}) = \beta P^1e_{pq} = e_{(p+1)q+1} \neq 0.$$

So, $Q_2(s) \neq 0$, and hence $Q_2(u) \neq 0$. Furthermore, $Q_i(s) = 0$ for $i > 2$ because $\dim Q_i(s) > \dim K$.

Let Σ_p be the symmetric group of degree p, and let π be its cyclic subgroup of order p generated by the permutation which sends i to $i+1$ mod p. Let E be a contractible free π-space. Consider the S^{pN-1}-fibration $\xi * \cdots * \xi$ (p times) over K^p with the projection $q : Y \to K^p, Y := \mathrm{ts}(\xi * \cdots * \xi)$. Since π acts on Y and K^p (via permutations), one can construct the S^{pN-1}-fibration η of the form $1 \times_\pi q : E \times_\pi Y \to E \times_\pi K^p$. (Here $A \times_\pi B$ is $(A \times B)/\pi$.) Furthermore,

$$T\eta = (E \ltimes_\pi (T\xi)^{\wedge p} = E \times_\pi (T\xi \wedge \cdots \wedge T\xi))/((E/\pi) \times \mathrm{pt})$$

where $A \ltimes_\pi B := (A \times_\pi B)/A$. Let $P : H^N(T\xi; \mathbb{Z}/p) \to H^{pN}(T\eta; \mathbb{Z}/p)$ be the Steenrod construction, see Steenrod–Epstein [1], Ch. VII, §2. Since $P(s)$ generates $H^{pN}(T\eta; \mathbb{Z}/p)$, it is the Thom class of η. Let $d : T\xi \to (T\xi)^{\wedge p}$ be the diagonal, and let

$$d_1 := 1 \ltimes_\pi d : (E/\pi) \ltimes T\xi = E \ltimes_\pi T\xi \to E \ltimes_\pi (T\xi)^{\wedge p}.$$

Finally, let $x \in H^1(E/\pi; \mathbb{Z}/p) = \mathbb{Z}/p$ and $y \in H^2(E/\pi; \mathbb{Z}/p) = \mathbb{Z}/p$ be the generators. By the definition of the Steenrod operations P^j (see Steenrod–Epstein [1], Definition VII.3.2), we have

$$d_1^*P(s) = \sum_j (-1)^{N+j+m\frac{N(N+1)}{2}}(m!)^N y^{m(N-2j)} \otimes P^j(s) + xy^{m(N-2j)-1}\beta P^j(s)$$

where $2m = p-1$. In view of the properties 1–7 above, every summand with $j \neq 0, p, p+1$ is zero. Hence,

$$\begin{aligned} d_1^*P(s) = (m!)^N(\varepsilon_0 y^{mN} \otimes s + \varepsilon_p y^{mN-p^2+p} \otimes e_{pq} + \varepsilon_{p+1} y^{mN-p^2+1} \otimes e_{(p+1)q}) \\ + (-1)^{N+m(N^2+N)/2} xy^{mN-p^2} \otimes e_{(p+1)q+1}, \end{aligned}$$

where ε_i is 1 or (-1). If $i > 2$, then $Q_i(a) = 0$ for $a = s, e_{pq}, e_{(p+1)q}, e_{(p+1)q+1}$. Furthermore, $Q_i(y) = 0, Q_i(x) = y^{p^i}$. Since Q_i is primitive, we conclude that

$$Q_i d_1^* P(s) = (-1)^{N+m(N^2+N)/2} y^{p^i} y^{mN-p^2} \otimes e_{(p+1)q+1} \neq 0.$$

So, $Q_i P(s) \neq 0$, and thus $Q_i u \neq 0$. □

6.11. Theorem (Peterson–Toda [1]). *$MS\mathcal{G}$ is a graded Eilenberg–Mac Lane spectrum.*

Proof. By II.7.4, it suffices to prove that $MS\mathcal{G}[p]$ is a graded Eilenberg–Mac Lane spectrum for every prime p. For $p = 2$ this follows from 6.6. If $p > 2$, then, by II.7.14, it suffices to prove that $E := MS\mathcal{G} \wedge M(\mathbb{Z}/p)$ is a graded Eilenberg–Mac Lane spectrum. If $p > 3$, then E is a ring spectrum because $M(\mathbb{Z}/p)$ is (cf. the proof of II.7.14). Furthermore, the spectrum $M = M(\mathbb{Z}/3)$ admits a non-associative pairing $M \wedge M \to M$, and so E admits a pairing $E \wedge E \to E$ (possibly non-associative).

Now, $\pi_0(E) = \mathbb{Z}/p$ because $\pi_0(MS\mathcal{G}) = \mathbb{Z}$. Hence, $H^*(E; \mathbb{Z}/p)$ is a connected coalgebra (possibly non-associative for $p = 3$). Let $v \in H^0(E; \mathbb{Z}/p)$ be its counit. Then, by II.7.20, $Q_0(v) \neq 0$ because $\pi_0(E) = \mathbb{Z}/p$. Furthermore, let $\iota : S \to M(\mathbb{Z}/p)$ represent a generator of $\pi_0(M(\mathbb{Z}/p)) = \mathbb{Z}/p$. Considering the map

$$f : MS\mathcal{G} = MS\mathcal{G} \wedge S \xrightarrow{1 \wedge \iota} MS\mathcal{G} \wedge (M(\mathbb{Z}/p)) = E$$

we conclude that $f^*(v) = u \in H^*(MS\mathcal{G}, \mathbb{Z}/p)$ with u as in 6.10. Hence, by 6.10, $Q_i(v) \neq 0$ for $i > 0$ and $\mathscr{P}^{\Delta_j}(v) \neq 0$ for $j > 0$. Thus, by II.7.24 (and II.7.25), E is a graded Eilenberg–Mac Lane spectrum. □

Now we consider the action of the Steenrod algebra $\mathscr{A}_p$ on the Thom class $u_{\mathcal{PL}} \in H^*(MS\mathcal{PL}; \mathbb{Z}/p)$, $p > 2$.

6.12. Theorem. $Q_0(u_{\mathcal{PL}}) = 0$, $Q_1(u_{\mathcal{PL}}) = 0$.

Proof. We have $\pi_0(MS\mathcal{PL}) = \mathbb{Z}$. Hence, $H_0(MS\mathcal{PL}) = \mathbb{Z} = H^0(MS\mathcal{PL})$, and thus $Q_0(u) = 0$.

Furthermore, Sullivan established a splitting

$$BS\mathcal{PL}[p] \simeq BS\mathcal{O}[p] \times B\,\mathrm{Coker}\,J_p,$$

where $B\,\mathrm{Coker}\,J_p$ is some mysterious space (a good proof can be found in Madsen–Milgram [1]). The notation $B\,\mathrm{Coker}\,J_p$ is inspired by the isomorphism

$$\pi_i(B\,\mathrm{Coker}\,J_p) \cong \mathrm{Coker}(J[p] : \pi_{i-1}(S\mathcal{O}[p]) \to \pi_{i-1}(S\mathcal{G}[p])).$$

So, $\pi_i(B\,\mathrm{Coker}\,J_p) = 0$ for $i < 2p^2 - 2p - 1$, see e.g. Toda [1]. This implies that $H^i(B\,\mathrm{Coker}\,J_p; \mathbb{Z}/p) = 0$ for $i < 2p$, and hence

$$H^{2p-1}(BS\mathcal{PL}[p]; \mathbb{Z}/p) = H^{2p-1}(BS\mathcal{O}[p]; \mathbb{Z}/p) = 0.$$

Thus, $Q_1(u) = 0$ since $H^{2p-1}(MS\mathcal{PL}; \mathbb{Z}/p) = 0$. (Another proof of the equality $Q_1(u) = 0$ can be found in VI.3.32 below). □

6.13. Theorem. *If $i > 1$ then $Q_i(u_{\mathcal{PL}}) \neq 0$.*

Proof. We use the notation from the proof of 6.10. We have the exact sequence

$$\pi_{pq-1}(\mathcal{G}/\mathcal{PL}[p]) \to \pi_{pq-1}(B\mathcal{SPL}[p]) \xrightarrow{(\alpha_{\mathcal{G}}^{\mathcal{PL}})_*} \pi_{pq-1}(B\mathcal{SG}[p]) \\ \to \pi_{pq-2}(\mathcal{G}/\mathcal{PL}[p]).$$

By 4.27(vi), $\pi_{pq-1}(\mathcal{G}/\mathcal{PL}[p]) = 0$ and $\pi_{pq-2}(\mathcal{G}/\mathcal{PL}[p]) = \mathbb{Z}[p]$. Hence, $(\alpha_{\mathcal{G}}^{\mathcal{PL}})_*$ is an isomorphism because $\pi_{pq-1}(B\mathcal{SG}[p])$ is finite.

Let $f : S^{pq-1} \to B\mathcal{SG}$ yield the element $\beta_1 \in \pi_{pq-2}(S)$ (under the isomorphism $\pi_i(S) \cong \pi_{i+1}(B\mathcal{SG})$). Since $(\alpha_{\mathcal{G}}^{\mathcal{PL}})_*$ is an isomorphism, there exists $\widetilde{f} : S^{pq-1} \to B\mathcal{SPL}$ with $\alpha_{\mathcal{G}}^{\mathcal{PL}}\widetilde{f} \simeq f$. We prove that $\widetilde{f}$ can be extended to a map $K \to B\mathcal{SPL}$, and then we can follow the proof of 6.10 and prove that $Q_i(u_{\mathcal{PL}}) \neq 0$.

Elementary obstruction theory implies that $\widetilde{f}$ can be extended to some $h : L \to B\mathcal{SPL}$. Given a map $M \to B\mathcal{SPL}$, any two of its extensions $h_1, h_2 : L \to B\mathcal{SPL}$ differ by a certain element $d(h_1, h_2) \in \pi_{(p+1)q}(B\mathcal{SPL})$, and every element of $\pi_{(p+1)q}(B\mathcal{SPL})$ can be realized as $d(h_1, h_2)$ with fixed h_1. So, one can construct $h : L \to B\mathcal{SPL}$ such that h extends $\widetilde{f}$ and $h_*(\iota) = 0$. Thus, $\widetilde{f}$ can be extended to K. □

Tsuchia [2] has proved the following conjecture of Peterson [1].

6.14. Theorem. *The kernel of the homomorphism*

$$\varphi : \mathscr{A}_p \to H^*(M\mathcal{SPL}; \mathbb{Z}/p), \quad \varphi(a) = a(u_{\mathcal{PL}}),$$

is $\mathscr{A}_p(Q_0, Q_1)$.

Proof. It is clear that $\mathscr{A}_p/\mathscr{A}_p(Q_0, Q_1)$ admits a unique structure of a coalgebra such that the quotient map $q : \mathscr{A}_p \to \mathscr{A}_p/\mathscr{A}_p(Q_0, Q_1)$ is a homomorphism of coalgebras. Moreover, the dual homomorphism of algebras

$$q^* : (\mathscr{A}_p/\mathscr{A}_p(Q_0, Q_1))^* = \mathbb{Z}/p\,[\xi_i | i > 0] \otimes \Lambda(\tau_j | j > 1) \subset \mathscr{A}_p^*$$

is monic. Since the vector space of indecomposables of $(\mathscr{A}_p/\mathscr{A}_p(Q_0, Q_1))^*$ has basis $\{\xi_i, \tau_j | i > 0, j > 1\}$, we conclude that the vector space of primitives of $\mathscr{A}_p/\mathscr{A}_p(Q_0, Q_1)$ has basis $\{\mathscr{P}^{\Delta_i}, Q_j | i > 0, j > 1\}$. By 6.12, the coalgebra homomorphism φ factors through a coalgebra homomorphism $\psi : \mathscr{A}_p/\mathscr{A}_p(Q_0, Q_1) \to H^*(M\mathcal{SPL}; \mathbb{Z}/p)$, and we must prove that ψ is monic. By 6.13, $\psi(Q_i) \neq 0$ for $i > 1$. Furthermore, considering the forgetful morphism $M\mathcal{SO} \to M\mathcal{SPL}$, we conclude, by 6.8, that $\psi(\mathscr{P}^{\Delta_i}) \neq 0$. So, ψ is injective on primitives, and thus, by II.6.14, it is monic.

6.15. Theorem. *The morphism*

$$T(a_{\mathcal{TOP}}^{\mathcal{PL}})[1/2] : M\mathcal{SPL}[1/2] \to M\mathcal{STOP}[1/2]$$

is an equivalence. In particular, for every odd prime p the morphism

$$T(a_{\mathcal{TOP}}^{\mathcal{PL}})[p] : M\mathcal{SPL}[p] \to M\mathcal{STOP}[p]$$

is an equivalence. So, theorems 6.12 – 6.14 *hold if we replace* $\mathcal{PL}$ *by* $\mathcal{TOP}$.

Proof. Let a denote

$$a_{\mathcal{TOP}}^{\mathcal{PL}} : B\mathcal{SPL} \to B\mathcal{STOP}.$$

By 4.27(v), $a[1/2] : B\mathcal{SPL}[1/2] \to B\mathcal{STOP}[1/2]$ is an equivalence, and so

$$a^* : H^*(B\mathcal{STOP}; \mathbb{Z}/p) \to H^*(B\mathcal{SPL}; \mathbb{Z}/p)$$

is an isomorphism for every odd prime p. Hence, by 5.23(ii),

$$(Ta)^* : H^*(M\mathcal{STOP}; \mathbb{Z}/p) \to H^*(M\mathcal{SPL}; \mathbb{Z}/p)$$

isomorphism, and so, by II.5.18(ii), $(Ta)[p] : TB\mathcal{STOP}[p] \to TB\mathcal{SPL}[p]$ is an equivalence for every odd prime p. Recall that $X[1/2][p] = X[p]$ for p odd and $X[1/2][2] = X[0]$. Thus, by II.5.19(ii), $T(a)[1/2] : M\mathcal{SPL}[1/2] \to M\mathcal{STOP}[1/2]$ is an equivalence. □

6.16. Remark. Mahowald [1] proved that $H\mathbb{Z}$ is a Thom spectrum of some stable spherical fibration and that $H\mathbb{Z}/2$ is a Thom spectrum of some stable vector bundle, see IX.5.8 below. In fact, it makes sense to state the following problem: how can one recognize whether a given spectrum is a Thom spectrum?

§7. Manifolds and (Co)bordism

Throughout this book the word "manifold" means "metrizable, separable, triangulable topological manifold with a finite number of components". Hence, every manifold belongs to $\mathscr{W}$ since every metrizable space does, see Kelley [1]. The boundary of a manifold M is denoted by ∂M. When we write M^n, it means that the manifold M has dimension n. We consider here topological (in future TOP), piecewise linear (in future PL), and smooth, i.e., C^∞, (in future DIFF) manifolds. The necessary preliminary information can be found in Kirby–Siebenmann [1], Munkres [2], [3].

Every DIFF manifold admits a canonical structure of a PL manifold, see loc. cit., while every PL manifold is a topological manifold for trivial reasons.

Similarly to bundles, we introduce a uniform symbol $\mathcal{T}$ in order to speak about manifolds of these three classes simultaneously. For example, "a $\mathcal{T}$ map of $\mathcal{T}$ manifolds" is a map of topological manifolds, or a smooth map of smooth manifolds, or a PL map of PL manifolds. Furthermore, a $\mathcal{T}$ isomorphism is a homeomorphism of topological manifolds, or a PL isomorphism of PL manifolds, or a diffeomorphism of smooth manifolds.

In view of a well-known connection between manifolds and $((\mathbb{R}^n, 0), \mathcal{V}_n)$-bundles, we introduce a uniform notation $\mathcal{V}^{\mathcal{T}}$, where $\mathcal{V}_n^{\mathcal{T}}$-bundle means $((\mathbb{R}^n, 0), \mathcal{O}_n)$-bundle, $((\mathbb{R}^n, 0), \mathcal{PL}_n)$-bundle, and $((\mathbb{R}^n, 0), \mathcal{TOP}_n)$-bundle if $\mathcal{T}$ is DIFF, PL and TOP respectively.

Clearly, every $((\mathbb{R}^n, 0), \mathcal{O}_n)$-bundle is an $((\mathbb{R}^n, 0), \mathcal{DIFF}_n)$-bundle where $\mathcal{DIFF}_n$ is the group of self-diffeomorphisms of $(\mathbb{R}^n, 0)$. Recall that, conversely, every $((\mathbb{R}^n, 0), \mathcal{DIFF}_n)$-bundle is equivalent to an $((\mathbb{R}^n, 0), \mathcal{O}_n)$-bundle (via assigning to a diffeomorphism its linear part), see loc. cit.

To justify the notation above, we formulate the following obvious fact.

7.1. Proposition. *The total space of any $\mathcal{V}^{\mathcal{T}}$-object over a $\mathcal{T}$ manifold is a $\mathcal{T}$ manifold.* □

7.2. Definition. A $\mathcal{T}$ *embedding* (of $\mathcal{T}$ manifolds) is a $\mathcal{T}$ map $i : M \to V$ such that $i(M)$ is a $\mathcal{T}$ submanifold of V and $i : M \to i(M)$ is a $\mathcal{T}$ isomorphism. A *bordered $\mathcal{T}$ embedding* is a map $i : (M, \partial M) \to (V, \partial V)$ such that the induced maps $M \to V$ and $\partial M \to \partial V$ are $\mathcal{T}$ embeddings and, moreover, $i(m,t) = (i(m), t)$ for some collars $\partial M \times I \subset M, \partial V \times I \subset V$, where $m \in \partial M, t \in I$, and $i(M \setminus (\partial M \times I)) \subset (V \setminus (\partial V \times I))$.

7.3. Definition. Fix any $\mathcal{T}$, and let $\mathcal{V} = \mathcal{V}^{\mathcal{T}}$.

(a) Let $i : (M, \partial M) \to (V, \partial V)$, $\dim V = \dim M + k$, be a bordered $\mathcal{T}$ embedding of $\mathcal{T}$ manifolds. A *$\mathcal{T}$ tubular neighborhood* of i is a triple (U, q, ξ) such that U is a neigborhood of $i(M)$, ξ is a $\mathcal{V}_k$-object over M and $q : (U, i(M)) \to (\operatorname{ts} \xi, s(M))$ is a $\mathcal{T}$ isomorphism where $s : M \to \operatorname{ts} \xi$ is the zero section of ξ.

(b) A bundle ξ which figures in (a) is called a *normal bundle* of the embedding i.

7.4. Construction. Given a tubular neighborhood (U, q, ξ) of a bordered embedding $(M, \partial M) \to (V, \partial V)$, consider the map

$$b : U \xrightarrow{q} \operatorname{ts} \xi \subset \operatorname{ts} \xi^{\bullet} \xrightarrow{\text{quotient}} T\xi.$$

Notice that $T\xi = \operatorname{ts} \xi \cup \{*\}$ where $*$ is the base point of $T\xi$, and define $c_{(U,q,\xi)} : V \to T\xi$ by setting

$$c_{(U,q,\xi)}(x) = \begin{cases} b(x) & \text{if } x \in U, \\ * & \text{otherwise.} \end{cases}$$

Clearly, $c_{(U,q,\xi)}(x)$ can be decomposed as

$$V \xrightarrow{\text{quotient}} V/(V \setminus U) \xrightarrow{h} T\xi.$$

7.5. Proposition. *The map $h : V/(V \setminus U) \to T\xi$ is a homeomorphism.* □

7.6. Definition. (a) An *n-dimensional microbundle* over a space X is a diagram $\xi = \{X \xrightarrow{s} E \xrightarrow{p} X\}$ with the following properties:

(1) $ps = 1_X$;

(2) For every $x \in X$, there are neighborhoods U of x and V of $s(x)$ and a homeomorphism $h_x : U \times \mathbb{R}^n \to V$ such that $ph_x(u, v) = u$ for all $(u, v) \in U \times \mathbb{R}^n$, $h_x(u, 0) = s(u)$ for every $u \in U$.

(b) Two microbundles $\xi_i = \{X \xrightarrow{s_i} E_i \xrightarrow{p_i} X\}$, $i = 1, 2$, over X are *equivalent* if there are neighborhoods V_i of $s_i(X)$ in E_i, $i = 1, 2$, and a homeomorphism $h : V_1 \to V_2$ such that the following diagram commutes:

$$\begin{array}{ccccc} X & \xrightarrow{s_1} & V_1 & \xrightarrow{p_1} & X \\ \| & & \downarrow{\scriptstyle h} & & \| \\ X & \xrightarrow{s_2} & V_2 & \xrightarrow{p_2} & X. \end{array}$$

(c) If X, E are $\mathcal{T}$ manifolds and $\xi = \{X \xrightarrow{s} E \xrightarrow{p} X\}$ is a microbundle, we say that ξ is a *$\mathcal{T}$ microbundle* if s and p are $\mathcal{T}$ maps and h_x is (can be chosen to be) a $\mathcal{T}$ isomorphism for every x.

(d) If X, E_1, E_2 are $\mathcal{T}$ manifolds, we say that two $\mathcal{T}$ microbundles $\xi_i = \{X \xrightarrow{s_i} E_i \xrightarrow{p_i} X\}$, $i = 1, 2$, over X are *$\mathcal{T}$ equivalent* if there are V_1, V_2 and h such as in (b), but the inclusions $V_i \subset E_i$ are required to be $\mathcal{T}$ maps and h is required to be a $\mathcal{T}$ isomorphism.

7.7. Theorem. (i) *Every $\mathcal{V}^{\mathcal{T}}$-bundle over a $\mathcal{T}$ manifold is a $\mathcal{T}$ microbundle.*

(ii) *Every $\mathcal{T}$ microbundle over a $\mathcal{T}$ manifold is equivalent to a $\mathcal{V}^{\mathcal{T}}$-bundle, and this $\mathcal{V}^{\mathcal{T}}$-bundle is unique up to isomorphism.*

Proof. (i) This is trivial.

(ii) This is difficult. We give the references. If $\mathcal{T}$ is TOP, this is proved in Kister [1], Siebenmann–Guillou–Hahl [1]. If $\mathcal{T}$ is PL, resp. DIFF, the proof can be found in Kuiper–Lashof [1], resp. Milnor [7]. All these cases are also considered in Kirby–Siebenmann [1]. □

7.8. Theorem. *Given a bordered $\mathcal{T}$ embedding $i : (M, \partial M) \to (V, \partial V)$, there exists N such that the embedding*

$$(M, \partial M) \xrightarrow{i} (V, \partial V) \xrightarrow{j} (V \times \mathbb{R}^N, \partial V \times \mathbb{R}^N), j(v) = (v, 0),$$

admits a tubular neighborhood.

Proof. We refer the reader to Kirby–Siebenmann [1]. In fact, there the existence of a neighborhood isomorphic to the total space of a microbundle is proved, but, because of 7.7(ii), the required result follows. □

We set $\mathbb{R}^N_+ := \{(x^1, \dots, x^N) \in \mathbb{R}^N | x^N \geq 0\}$.

7.9. Theorem (the Whitney Theorem). *Every $\mathscr{T}$ manifold M^n admits a bordered $\mathscr{T}$ embedding $(M, \partial M) \to (\mathbb{R}^{N+n}_+, \partial\mathbb{R}^{N+n}_+)$ for some $N = N(M)$.*

Proof. The proof can be found e.g. in Munkres [3] or Dubrovin–Novikov–Fomenko [1] for $\mathscr{T} = \mathrm{DIFF}$, but the proof for any $\mathscr{T}$ can be done in a similar way. □

7.10. Definition. Given a $\mathscr{T}$ manifold M, a *tangent bundle* τM of M is a $\mathcal{V}^{\mathscr{T}}$-bundle which is equivalent to the microbundle $M \xrightarrow{d} M \times M \xrightarrow{p_1} M$ where d is the diagonal, $d(m) = (m, m)$.

7.11. Proposition. *Let ξ be a normal $\mathcal{V}^{\mathscr{T}}$-bundle of a bordered $\mathscr{T}$ embedding $i : (M, \partial M) \to (V, \partial V)$. Then the $\mathcal{V}^{\mathscr{T}}$-bundles $i^*\tau V$ and $\tau M \oplus \xi$ over M are equivalent.*

Proof. Do it as an exercise, or see Milnor [7]. □

Of course, if M is a smooth manifold, then τM is equivalent to the usual tangent bundle of M (prove it as an exercise, or see Milnor [7]). Furthermore, if $i : M \to V$ is a smooth embedding of smooth manifolds, then ξ is equivalent to the quotient bundle $(i^*\tau V)/\tau M$.

7.12. Definition. A *normal bundle* of a $\mathscr{T}$ manifold M^n is a pair (ν^N, ω) where ν^N is a normal $\mathcal{V}^{\mathscr{T}}$-bundle of any bordered $\mathscr{T}$ embedding $(M, \partial M) \to (\mathbb{R}^{N+n}_+, \partial\mathbb{R}^{N+n}_+)$ and $\omega : \nu^N \to \gamma^N_{\mathcal{V}^{\mathscr{T}}}$ is a classifying morphism for ν^n. A *stable normal bundle* of M is a stable $\mathcal{V}^{\mathscr{T}}$-object $\nu = \nu_M$ of the form $(\nu^N)_{\mathrm{st}} = (\nu^N, \omega)_{\mathrm{st}}$, where (ν^N, ω) is a normal bundle of M.

Clearly, if M is closed then there is no essential difference between embeddings $M \to \mathbb{R}^N$ and bordered embeddings $M \to \mathbb{R}^N_+$. Because of this, we shall consistently treat a normal bundle of a closed manifold M as a normal bundle of an embedding $M \to \mathbb{R}^N$.

7.13. Theorem. *Let (ν_1^N, ω_1) and $(\nu_2^{N'}, \omega_2)$ be two normal bundles of a $\mathscr{T}$-manifold M. Then the $\mathcal{V}^{\mathscr{T}}$-bundles ν_1^N and $\nu_2^{N'}$ are stably equivalent. In particular, the stable normal bundle of M is uniquely defined up to equivalence.*

Proof. This follows from 7.11. □

7.14. Proposition. *If (ν^N, ω) is a normal bundle of M then $(\nu|\partial M, \omega|\partial M)$ is a normal bundle of ∂M (provided $\partial M \neq \emptyset$).*

Proof. Let (U, q, ν^N) be a tubular neighborhood of a bordered embedding $(M, \partial M) \to (\mathbb{R}_+^{N+n}, \partial\mathbb{R}_+^{N+n})$. Then $q(U \cap \partial\mathbb{R}_+^{N+n}) \subset \operatorname{ts}(\nu^N|\partial M)$, and so we get a map

$$q' : U \cap \partial\mathbb{R}_+^{N+n} \to \operatorname{ts}\nu^N|\partial M, \quad q'(x) := q(x).$$

Clearly, $(U \cap \partial\mathbb{R}_+^{N+n}, q', \nu^N|\partial M)$ is a tubular neighborhood of $i|\partial M : \partial M \to \mathbb{R}^{N+n-1} = \partial\mathbb{R}_+^{N+n}$. Finally, it is obvious that $\omega|\partial M$ classifies $\nu^N|\partial M$. □

7.15. Definition. (a) The following special case of 7.4 turns out to be very important. Let M^n be a closed manifold, and let (U, q, ν^N) be a tubular neighborhood of an embedding $i : M^n \to \mathbb{R}^{N+n}$. We regard S^{N+n} as the one-point compactification of $\mathbb{R}^{N+n}$, and we consider (U, q, ν^N) as a tubular neighborhood of the embedding $M \xrightarrow{i} \mathbb{R}^{N+n} \subset S^{N+n}$. The map

$$c^N := c_{(U,q,\nu^N)} : S^{N+n} \to T\nu^N$$

as in 7.4 is called the *Browder–Novikov map.*

(b) Given data as in (a), let $\nu = (\nu^N)_{\text{st}}$. We define a morphism

$$c : S^n_{\text{spectrum}} = \Sigma^{-N}\Sigma^\infty S^{N+n}_{\text{space}} \xrightarrow{\Sigma^{-N}\Sigma^\infty c_{\nu^N}} \Sigma^{-N}\Sigma^\infty T\nu^N = T\nu$$

of spectra, where the last equality follows from 5.16. We call this morphism c the *Browder–Novikov morphism.*

(c) Let D^k be the standard k-dimensional disk. Similarly to above, given a bordered embedding $i : (M, \partial M) \to (\mathbb{R}_+^{N+n}, \partial\mathbb{R}_+^{N+n})$, we can construct a Browder–Novikov map $(D^{N+n}, \partial D^{N+n}) \to (T\nu_M^N, T(\nu^N|\partial M))$, we leave details to the reader.

7.16. Data. Let M be a compact $\mathscr{T}$ manifold, let $f : M \to Y$ be a map, and let ξ be a $\mathcal{V}^{\mathscr{T}}$-bundle over a CW-space Z. Suppose that $\operatorname{ts}\xi$ is an open subset of the space Y. Considering the inclusion $Z \xrightarrow{s_\xi} \operatorname{ts}\xi \subset Y$, set $N = f^{-1}Z$ and define $g : N \to Z, g(n) := f(n), n \in N$. We set $U := f^{-1}(\operatorname{ts}\xi)$, and we denote by $i : N \to M$ the inclusion.

7.17. Definition. Given data 7.16, the map $f : M \to Y$ is called *transversal to* ξ if the following hold:

(i) N is a $\mathscr{T}$ submanifold of M.

(ii) There is a tubular neighborhood of i of the form $(U, q, g^*\xi)$ such that the following diagram commutes:

$$\begin{array}{ccc} U & \xrightarrow{f|U} & Y \\ {\scriptstyle q}\downarrow & & \uparrow\cup \\ \operatorname{ts}(g^*\xi) & \xrightarrow{\operatorname{ts}(\mathfrak{I}_{g,\xi})} & \operatorname{ts}\xi. \end{array}$$

7.18. Theorem. *We assume data 7.16. Suppose that $\mathscr{T}$ is TOP and $\dim M \neq 4$, or $\mathscr{T}$ is PL, or $\mathscr{T}$ is DIFF. Then every map $f' : M \to Y$ is homotopic to a map $f : M \to Y$ which is transversal to ξ. Moreover, if $A \subset B \subset M$ with A closed and B open and if $f'|B$ is transversal to ξ, then f can be chosen such that $f'|A = f|A$.*

Proof. If $\mathscr{T}$ is DIFF, this can be deduced from the well-known Thom Transversality Theorem. If $\mathscr{T}$ is PL, this was proved by Williamson [1] (for microbundles). Both these cases are also considered in Kirby–Siebenmann [1]. Furthermore, Kirby–Siebenmann [1] proved the theorem when $\mathscr{T}$ is TOP and $\dim M \neq 4 \neq \dim M - \dim \xi$. Scharlemann [1] (cf. also Matsumoto [1]) proved that the theorem holds for $\dim M - \dim \xi = 4$ if there exists a four-dimensional almost parallelizable topological manifold having signature 8, and such a manifold was constructed by Freedman [1]. □

7.19. Definition. Fix any $\mathscr{T}$ and let $\mathcal{V} = \mathcal{V}^{\mathscr{T}}$. Let $\varphi : B \to B\mathcal{V}$ be a (structure) map. Roughly speaking, a (B, φ)-structure on a manifold is a (B, φ)-structure on its stable normal bundle. We pass to a rigorous definition.

(a) A *strict (B, φ)-structure on M* is a tuple $\mathfrak{i} = (i, U, q, \nu^N, \omega, [a, H])$ where $i : (M, \partial M) \to (\mathbb{R}_+^{N+n}, \partial\mathbb{R}_+^{N+n})$ is a $\mathscr{T}$ bordered embedding, (U, q, ν^N) be a $\mathscr{T}$ tubular neighborhood of i, ω is a classifying morphism for ν^N and $[a, H]$ is a (B, φ)-structure on $\nu_{\text{st}}^N = (\nu^N, \omega)_{\text{st}}$, see 4.9.

(b) Given a strict (B, φ)-structure $\mathfrak{i} = (i, U, q, \nu^N, \omega, [a, H])$ on M, we define its *suspension* $\sigma\mathfrak{i}$ to be a strict (B, φ)-structure

$$(i', U', q', \nu^N \oplus \theta^1, \widehat{\omega}, [a'H'])$$

where i' is the embedding

$$\begin{aligned}(M, \partial M) \xrightarrow{i} (\mathbb{R}_+^{N+n}, \partial\mathbb{R}_+^{N+n}) &\subset (\mathbb{R}_+^{N+n} \times \mathbb{R}, \partial\mathbb{R}_+^{N+n} \times \mathbb{R}) \\ &= (\mathbb{R}_+^{N+n+1}, \partial\mathbb{R}_+^{N+n+1}),\end{aligned}$$

$U' := U \times \mathbb{R}$, $q' := q \times 1 : U \times \mathbb{R} \to (\operatorname{ts}\nu^N) \times \mathbb{R} = \operatorname{ts}(\nu^N \oplus \theta^1)$, $\widehat{\omega} : \nu^N \oplus \theta^1 \to \gamma_{\mathcal{V}}^{N+1}$ is the classifying morphism as in 4.14(b), and $[a', H'] = K_{\operatorname{bs}\omega}[a, H]$ as in 4.13.

(c) Let $\mathfrak{i}_k = (i_k, U_k, q_k, \nu_k^N, \omega_k, [a_k, H_k]), k = 0, 1$ be two (B, φ)-structures on M. We say that $\mathfrak{i}_0$ and $\mathfrak{i}_1$ are equivalent if there is a morphism $h : \nu_0^N \to \nu_1^N$ over M and a family $J_t : (\mathbb{R}_+^{N+n}, \partial\mathbb{R}_+^{N+n}) \to (\mathbb{R}_+^{N+n}, \partial\mathbb{R}_+^{N+n}), t \in I$ with the following properties:

1. The map $J : (\mathbb{R}_+^{N+n}, \partial\mathbb{R}_+^{N+n}) \times I \to (\mathbb{R}_+^{N+n}, \partial\mathbb{R}_+^{N+n}) \times I, J(x, t) = (J_t(x), t)$, is a $\mathscr{T}$ isomorphism;
2. $J_1(U_0) = U_1, J_1(i_0(m)) = i_1(m)$ for every $m \in M$;
3. The composition $\operatorname{ts}\nu_0^N \xrightarrow{(q_0)^{-1}} U_0 \xrightarrow{J_1} U_1 \xrightarrow{q_1} \operatorname{ts}\nu_1^N$ coincides with $\operatorname{ts} h$;
4. $\omega_0 = \omega_1 h : \nu_0^N \to \gamma_{\mathcal{V}}^N$;

(5) the (B, φ)-structure $[a_1, H_1]$ on $(\nu_0^N)_{\mathrm{st}}$ is induced by h from the (B, φ)-structure $[a_0, H_0]$ on $(\nu_0^N)_{\mathrm{st}}$, as it is defined in 4.12(ii), i.e., $[a'H'] = h^![a, H]$.

(d) We say that two strict (B, φ)-structures $\mathfrak{i}_0$ and $\mathfrak{i}_1$ are stably equivalent if there are non-negative integers k, l such that the (B, φ)-structures $\sigma^k\mathfrak{i}_0$ and $\sigma^l\mathfrak{i}_1$ are equivalent. (Here, of course, $\sigma^0\mathfrak{i} = \mathfrak{i}$ and $\sigma^k\mathfrak{i} = \sigma\sigma^{k-1}\mathfrak{i}$.) We denote by $[\mathfrak{i}]$ the class of stable equivalence of the strict (B, φ)-structure $\mathfrak{i}$.

(e) A (B, φ)-*structure* on M is a class of stably equivalent strict (B, φ)-structures on it. A (B, φ)-*manifold* is a manifold equipped with a (B, φ)-structure.

7.20. Constructions. (a) Let a (B, φ)-structure on M be represented by a strict (B, φ)-structure $\mathfrak{i} = (i, U, q, \nu^N, \omega, [a, H])$, and let $j : \partial M \to M$ be the inclusion. We define the induced (B, φ)-structure

$$\mathfrak{i}' = (i', U', q', \nu^N|\partial M, \omega', [a', H'])$$

on ∂M as follows: $(U', q', \nu^N|\partial M)$ is the tubular neighborhood defined in the proof of 7.14, $\omega' := \omega|\partial M$, and $[a', H'] := j^![a, H]$, as it is defined in 4.12.

(b) For every strict (B, φ)-structure $\mathfrak{i} = (i, U, q, \nu^N, \omega, [a, H])$ on M, there exists a strict (B, φ)-structure $\mathfrak{i}' = (i', U', q', \nu^N|\partial M, \omega', [a', H'])$ such that $[\mathfrak{i}] = [\mathfrak{i}']$ and $U' \subset \{(x^1, \dots, x^{N+n}) \in \mathbb{R}_+^{N+n} \mid x^{N+n} > 0\}$ (prove this).

(c) Given two (B, φ)-manifolds M and M' of the same dimension n, we equip the disjoint union $M \sqcup M'$ with the following (B, φ)-structure. Let $\mathfrak{i} = (i, U, q, \nu^N, \omega, [a, H])$ (resp. $\mathfrak{i}' = (i', U', q', \nu^{N'}|\partial M, \omega', [a', H'])$) represent a (B, φ)-structure on M (resp. on M'). Without loss of generality we can assume that $N' = N$ and that $U \subset \{(x^1, \dots, x^{N+n}) \in \mathbb{R}_+^{N+n} \mid x^1 > 0\}$, $U' \subset \{(x^1, \dots, x^{N+n}) \in \mathbb{R}_+^{N+n} \mid x^1 < 0\}$. We define $j : M \sqcup M' \to \mathbb{R}_+^{N+n}$ by requiring $j|M = i, j|M' = i'$. Then we get a strict (B, φ)-structure $(j, U \cup U'$, etc.) on $M \sqcup M'$.

The construction 7.20(a) can be generalized. If M^m is a submanifold of V^n with a trivial normal bundle $\nu = \nu_V^M$ then every trivialization of ν yields an isomorphism $\nu_M \cong \nu_V|M \oplus \nu$. (This is more or less clear, but it is not so easy to write it down neatly.) Then, in view of 4.11–4.14, every (B, φ)-structure on V yields a (B, φ)-structure on M. In fact, in 7.20(a) we considered the (equivalence class of a) trivialization given by the inner normal.

7.21. Definition. Let a (B, φ)-structure on M be represented by a strict (B, φ)-structure $\mathfrak{i} := (i, U, q, \nu^N, \omega, [a, H])$. Consider the suspension $\sigma\mathfrak{i} = (i', U', q', \nu^N \oplus \theta, \widehat{\omega}, [a, H])$, see 7.19(b). The morphism $1 \oplus e$ as in 4.14(a) defines the opposite (B, φ)-structure $(1 \oplus e)^![a, H]$ on $\nu^N \oplus \theta$, and we set $-\sigma\mathfrak{i} := (i', U', q', \nu^N \oplus \theta, \widehat{\omega}, (1 \oplus e)^![a, H])$. The (B, φ)-structure on M given by $-\sigma\mathfrak{i}$ is called the *opposite* (B, φ)-*structure* to that given by $\mathfrak{i}$.

The opposite (B,φ)-structure to [i] is denoted by $-[\mathrm{i}]$. Furthermore, given a (B,φ)-manifold M, we denote by $-M$ the (B,φ)-manifold which coincides with M as a manifold but has the opposite (B,φ)-structure.

We leave it to the reader to prove that opposite to opposite is the original (B,φ)-structure, i.e., that "to be opposite" is a symmetric relation.

An example of the opposite structure is the opposite orientation, see V.1.1. We recomend it to the reader to keep it in the mind whenever we discuss opposite structures.

Consider a closed (B,φ)-manifold M and choose a representing strict (B,φ)-structure $\mathrm{i}_0 = (i, U, q, \nu^N, \omega, [a, H])$. Here we assume that i is an embedding $i : M \to \mathbb{R}^{N+n}$ and that $U \subset \{(x^1, \dots, x^{N+n}) \in \mathbb{R}^{N+n} | x^{N+n} > 0\}$. Let $(u^1, \dots, u^{N+n})$ be the coordinates of a point $u \in U$. We define a bordered embedding $j : U \times I \to \{x \in \mathbb{R}^{N+n+1} \mid x^{N+n+1} \geq 0\}$ by setting

$$j(u,t) = (u^1, \dots, u^{N+n-1}, u^{N+n}\cos \pi t, u^{N+n}\sin \pi t), t \in I.$$

Furthermore, we define a tubular neighborhood $(j(U \times I), r, \nu^N \times 1_I)$ of $j|(i(M) \times I)$ by setting

$$r(j(u,t)) := (a(u), t) \in \mathrm{ts}(\nu^N) \times I = \mathrm{ts}(\nu^N \times 1_I).$$

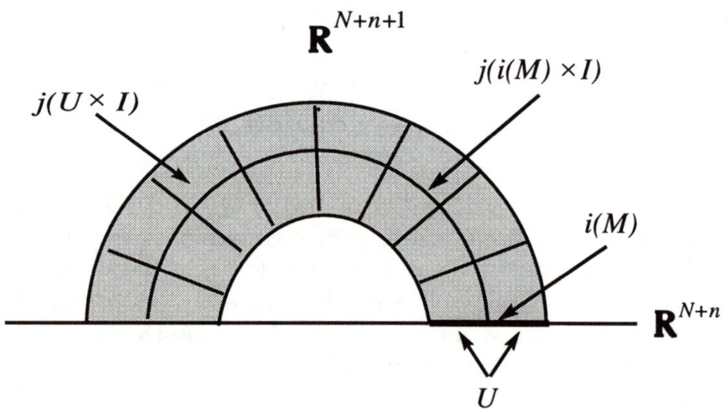

Finally, we define the morphism $\psi : \nu^N \times 1_I \to \nu^N$, $\mathrm{ts}\,\psi(x,t) = x$. Then we get a strict (B,φ)-structure $(j, j(U \times I), r, \nu^N \times 1_I, \omega\psi, \psi^![a, H])$ on $M \times I$. Hence, by 7.20, this (B,φ)-structure yields a strict (B,φ)-structure i_1 on $M = M \times \{1\}$.

7.22. Proposition. *The (B, φ)-structures* $\mathfrak{i}_0$ *and* $\mathfrak{i}_1$ *on M are opposite to one another.*

Proof. We define $J_t : \mathbb{R}^{N+n+1} \to \mathbb{R}^{N+n+1}$ by setting

$$J_t(x^1, \dots, x^{N+n}, x^{N+n+1}) := (x^1, \dots, x^{N+n-1}, x^{N+n} \cos \pi t, x^{N+n+1} \sin \pi t),$$

and we define $h : \nu_0 \oplus \theta^1 \to \nu_1 \oplus \theta^1$ to be the unique morphism with the property 7.19(c,3). It is easy to see that $\{J_t, h\}$ gives an equivalence between $\sigma\mathfrak{i}_1$ and $-\sigma\mathfrak{i}_0$. □

7.23. Construction. Let $\varphi : B \to B\mathcal{V}^{\mathscr{T}}$ be a structure map. Given a (B, φ)-manifold M and a $\mathscr{T}$ isomorphism $f : L \to M$, we can canonically equip L with a (B, φ)-structure. Namely, the embedding $i : M \to \mathbb{R}_+^{N+n}$ yields the embedding $if : L \to \mathbb{R}_+^{N+n}$, etc. We denote this (B, φ)-structure on L by $f^![\mathfrak{i}]$ where $[\mathfrak{i}]$ is the (B, φ)-structure on M.

Furthermore, given a $\mathscr{T}$ isomorphism $f : L \to M$ of (B, φ)-manifolds, we say that f is a (B, φ)-isomorphism of (B, φ)-manifolds if $f^![\mathfrak{i}_M] = [\mathfrak{i}_L]$.

7.24. Construction. Given two compact n-dimensional (B, φ)-manifolds L, M, let

$$f : \partial L \to \partial M$$

be a $\mathscr{T}$ isomorphism such that $f^![\mathfrak{i}_{\partial M}] = -[\mathfrak{i}_{\partial L}]$. Recall that $L \cup_f M$ admits a canonical structure of a $\mathscr{T}$ manifold. We equip $L \cup_f M$ with a (B, φ)-structure as follows. Take strict (B, φ)-structures $\mathfrak{i}_L = (i_0, U_0, q_0, \nu_0^N, [a_0, H_0])$ and $-\mathfrak{i}_M = (i_1, U_1, q_1, \nu_1^N, [a_1, H_1])$. Since $f^![\mathfrak{i}_{\partial L}] = -[\mathfrak{i}_{\partial M}]$, we can assume that $i_0|\partial L = i_1|\partial M$. Furthermore, we assume that, for some collar $\partial L \times I$ of ∂L, we have $i_0(l, t) = (i_0(l), tx^{N+n})$, i.e., L meets $\mathbb{R}^{N+n-1} = \partial\mathbb{R}_+^{N+n}$ orthogonally. Similarly for i_1. We can also assume that $U_0 \cap \partial\mathbb{R}_+^{N+n} = U_1 \cap \partial\mathbb{R}_+^{N+n}$ and $q_0|\partial L : U_0 \cap \partial\mathbb{R}_+^{N+n} \to \mathrm{ts}(\nu_0|\partial L)$ coincides with $q_1|\partial M : U_1 \cap \partial\mathbb{R}_+^{N+n} \to \mathrm{ts}(\nu_1|\partial M)$.

Consider the map

$$\chi : \mathbb{R}^{N+n} \to \mathbb{R}^{N+n}, \quad \chi(x^1, \dots, x^{N+n}) = (x^1, \dots, -x^{N+n})$$

and define the embedding $i : L \cup_f M \to \mathbb{R}^{N+n}$ by setting $i(l) = i_0(l)$ if $l \in L$ and $i(m) = \chi i_1(m)$ if $m \in M$. Now we can regard i as a $\mathscr{T}$ embedding, and the (B, φ)-structures on L and M yield a certain (B, φ)-structure on $L \cup_f M$. In other words, the (B, φ)-structure on $L \cup_f M$ is the result of gluing the ones on L and M, and we are able to make a gluing because $f^![\mathfrak{i}_{\partial M}] = -[\mathfrak{i}_{\partial L}]$.

7.25. Definition. (a) A *singular (B, φ)-manifold (M^n, f) of dimension n in a space X* is a map $f : M^n \to X$, where M^n is a compact (B, φ)-manifold. The singular manifold (M, f) is called *closed* if M is closed. We say that a closed singular (B, φ)-manifold $f : M^n \to X$ *bounds* if there is a singular manifold $F : V^{n+1} \to X$, where V is a (B,φ)-manifold such that $\partial V = M$ as (B,φ)-manifolds and $F|M = f$. In this case we write $\partial(V,F) = (M,f)$. We say that two closed singular (B,φ)-manifolds $e : L \to X$ and $f : M \to X$ are (B,φ)-*bordant* if $e \sqcup (-f) : L \sqcup (-M) \to X$ bounds. Here $(-f) : (-M) \to X$ coincides with f as a map of spaces, but $-M$ is equipped with the opposite (B,φ)-structure to M. A singular manifold (V,F) with $\partial(V,F) = (L\sqcup -M, e\sqcup -f)$ is called a *membrane* or a *bordism* between (L,e) and (M,f).

(b) The relation "to be bordant" is called also the *bordism* relation. This is an equivalence relation (prove this; the reflexivity follows from 7.22). The (B,φ)-bordism class of a closed singular (B,φ)-manifold $f : M \to X$ is denoted by $[M,f]$. When X is a point we write just $[M]$ rather then $[M,f]$. The set of all n-dimensional bordism classes in X is called an n-dimensional (B,φ)-*bordism set of* X and is denoted by $\Omega_n^{(B,\varphi)}(X)$, cf. Atiyah [1].

7.26. Proposition. *The operation $\sqcup$ of disjoint union of singular manifolds induces an abelian group structure on the set $\Omega_n^{(B,\varphi)}(X)$.*

Proof. It is easy to see that $\sqcup$ induces a well-defined associative and commutative operation $+$ on $\Omega_n^{(B,\varphi)}(X)$, $[L,e] + [M,f] := [L \sqcup M, e \sqcup f]$.

Any singular (B,φ)-manifold which bounds can play the role of the neutral element. Indeed, let (A,o) bound. Consider a singular manifold $f : M \to X$ and recall that, by 7.22, $(M,f) \sqcup (-M,-f)$ bounds. So, $(-M,-f) \sqcup (M,f) \sqcup (A,o)$ bounds, i.e., (M,f) is bordant to $(M,f) \sqcup (A,o)$, i.e., $[M,f] + [A,o] = [M,f]$, i.e., $[A,o]$ is the neutral element 0. Furthermore, since $(-M,-f)\sqcup(M,f)\sqcup(A,o)$ bounds, we conclude that $[-M,-f]+[M,f] = 0$, i.e., we find the opposite element for every $[M,f]$.

It remains to find a singular manifold (A,o) which bounds. Consider the disk $D = D^{n+1}$ and define $O : D \to X$ to be a constant map. Since the stable normal bundle of D is classified by a constant map $D \to B\mathcal{V}$, D admits a (B,φ)-structure, and so (D,O) converts into a singular (B,φ)-manifold. Now, set $(A,o) := \partial(D,O) = (S^n$, constant$)$.

Thus, $\Omega_n^{(B,\varphi)}(X)$ gets the structure of an abelian group. □

Frequently one chooses the bordism class of the empty set as the neutral element, but I did not do it since I want to avoid a discussion about structures on the empty manifold.

According to 5.12(f), there is a Thom spectrum $T(B,\varphi)$. The following theorem connects homotopy theory and geometry and plays the pivotal role in algebraic topology.

7.27. Theorem (the Pontrjagin–Thom Theorem). *There is a natural isomorphism $\Omega_n^{(B,\varphi)}(X) \cong T(B,\varphi)_n(X)$.*

Proof. Throughout the proof S^k denotes the space (not the spectrum), while S denotes the sphere spectrum.

We construct a function $\Theta : \Omega_n^{(B,\varphi)}(X) \to T(B,\varphi)_n(X)$ and prove that it is an isomorphism. Recall that

$$T(B,\varphi)_n(X) = \pi_n(T(B,\varphi) \wedge X^+).$$

Let $f : M^n \to X$ be a closed singular (B,φ)-manifold in X, and let the (B,φ)-structure on M be represented by a strict (B,φ)-structure $\mathfrak{i} = (i, U, q, \nu^N, \omega, [a, H])$; here we consider i as an embedding $M^n \subset \mathbb{R}^{N+n}$. By 5.15, the (B,φ)-prestructure (a,H) on $\nu = \nu_{\text{st}}^N$ yields a morphism

$$T_H(a) : T\nu \to T(B,\varphi),$$

and its homotopy class depends only on the (B,φ)-structure $[a,H]$.

Let $\Delta : T\nu \to T\nu \wedge M^+$ be the morphism as in 5.36, and let $c : \Sigma^n S \to T\nu$ be the Browder–Novikov morphism as in 7.15(b). The composition

$$\beta_{(M,f)} : \Sigma^n S \xrightarrow{c} T\nu \xrightarrow{\Delta} T\nu \wedge M^+ \xrightarrow{T_H(a)\wedge f^+} T(B,\varphi) \wedge X^+$$

gives us an element $[\beta_{(M,f)}] \in T(B,\varphi)_n(X)$, and we set $\Theta[M,f] = [\beta_{(M,f)}]$.

We prove that stably equivalent strict (B,φ)-structures on (M,f) give the same element $[\beta_{(M,f)}]$. Clearly, $\beta_{(M,f)}^{\sigma\mathfrak{i}} = \beta_{(M,f)}^{\mathfrak{i}}$. Hence, we must prove that equivalent strict (B,φ)-structures on (M,f) give homotopic morphisms $\Sigma^n S \to T(B,\varphi)$. So, let (J_t, h) as in 7.19(c) give an equivalence between strict (B,φ)-structures $\mathfrak{i}$ and $\mathfrak{i}'$ on M. Define $J : \mathbb{R}^{N+n} \times I \to \mathbb{R}^{N+n}$, $J(x,t) := J_t(x,t)$. Consider the bordered embedding

$$j : M \times I \to \mathbb{R}^{N+n} \times I, j(m,t) = (J_t(m), t)$$

and its tubular neighborhood $(\overline{U}, \overline{q}, \nu^N \times 1_I)$ where

$$\overline{U} := J(U \times I),\ \overline{q}(J(u,t)) = (q(u), t) \in (\operatorname{ts} \nu^n) \times I = \operatorname{ts}(\nu^N \times 1_I).$$

Now, following 7.4, and 7.15(a), construct a map $u'' : S^{N+n} \times I \to T(\nu^N \times 1_I)$ (by collapsing the complement of U). Clearly, u'' can be decomposed as

$$S^{N+n} \times I \xrightarrow{\text{quotient}} S^{N+n} \wedge I^+ \xrightarrow{u'} T(\nu^N \times I).$$

Furthermore, the projection $\nu^N \times 1_I \to \nu^N$ induces a map $T(\nu^N \times I) \to T\nu^N$, and we get the map

$$u : S^{N+n} \wedge I^+ \xrightarrow{u'} T(\nu^N \times I) \to T\nu^N.$$

Now we consider the morphism

$$\Sigma^{-N}\Sigma^\infty u : \Sigma^n S \wedge I^+ = \Sigma^{-N}\Sigma^\infty(S^{N+n} \wedge I^+) \xrightarrow{\Sigma^\infty u} \Sigma^{-N}\Sigma^\infty T\nu^N \cong T\nu,$$

the last isomorphism is given by 5.16. Finally, we define

$$E : \Sigma^n S \wedge I^+ \xrightarrow{\Sigma^{-N}\Sigma^\infty u} T\nu \xrightarrow{\Delta} T\nu \wedge M^+ \xrightarrow{T_H(a)\wedge f^+} T(B,\varphi)\wedge X^+.$$

It is easy to see that E is a homotopy between $\beta^{\mathfrak{i}}_{(M,f)}$ and $\beta^{\mathfrak{i}'}_{(M,f)}$.

We prove that Θ is well-defined, i.e., that $[\beta_{(M,f)}] = [\beta_{(L,e)}]$ if (M,f) and (L,e) are (B,φ)-bordant. Clearly, $[\beta_{(M,f)\sqcup(N,g)}] = [\beta_{(M,f)}] + [\beta_{(N,g)}]$. So, it suffices to prove that $[\beta_{(M,f)}] = 0$ if (M,f) bounds.

Let $(M,f) = \partial(V,f)$ as (B,φ)-manifolds. Choose a representing strict (B,φ)-structure on V and consider the induced (B,φ)-structure on M. Let ν_V^N, resp. ν_M^N, be the corresponding normal bundle of V, resp. M; recall that $\nu_M^N = \nu_V^N|M$. Then, by 7.15(c), we have the commutative diagram

$$\begin{array}{ccc} S^{N+n} & \xrightarrow{c^N} & T(\nu_M^N) \\ \cap\big\downarrow & & l^N\big\downarrow\cap \\ D^{N+n+1} & \longrightarrow & T(\nu_V^N) \end{array}$$

where c^N is the Browder–Novikov map and l^N is induced by the inclusion $\nu_M^N \subset \nu_V^N$. In particular, the map $l^N c^N$ is inessential. Let $\nu_M := (\nu_M^N)_{\mathrm{st}}$ and $\nu_V := (\nu_V^N)_{\mathrm{st}}$. Then the above diagram induces the following homotopy commutative diagram in $\mathscr{S}$:

$$\begin{array}{ccccccc} \Sigma^n S & \xrightarrow{c} & T\nu_M & \longrightarrow & T\nu_M \wedge M^+ & \longrightarrow & T(B,\varphi)\wedge X^+ \\ & & l\big\downarrow\cap & & \big\downarrow\cap & & \big\| \\ & & T\nu_V & \longrightarrow & T\nu_V \wedge V^+ & \longrightarrow & T(B,\varphi)\wedge X^+; \end{array}$$

here $l = \Sigma^{-N}\Sigma^\infty l^N$, the top line is $\beta_{(M,f)}$ and the morphisms in the bottom line are similar to the corresponding morphisms in the top line. Now, we conclude that $\beta_{(M,f)}$ is inessential since $lc = \Sigma^{-N}\Sigma^\infty(l^N c^N)$ is.

So, we have constructed a well-defined map

$$\Theta : \Omega_n^{(B,\varphi)}(X) \to T(B,\varphi)_n(X).$$

It is easy to see that Θ is a homomorphism.

Now we prove that Θ is an isomorphism. Firstly, we prove that Θ is epic.

Let $\overline{B\mathcal{V}}_n$ be the telescope of the finite sequence $\{B\mathcal{V}_1 \to \cdots \xrightarrow{r_{n-1}^{\mathcal{V}}} B\mathcal{V}_n\}$, and let B_k be the maximal CW-complex contained in $\varphi^{-1}(\overline{B\mathcal{V}}_k)$. We set $\zeta^k := \varphi_k^*\gamma^k$ where γ^k is the universal $\mathcal{V}_k$-bundle over $\overline{B\mathcal{V}}_k$, and $\varphi_k : B_k \to \overline{B\mathcal{V}}_k$ is a restriction of φ. Note that $\mathrm{ts}(\zeta^k \times \theta^0)$ is a neighborhood of $B_k \times X$ in $T\zeta^k \wedge X^+$. Clearly, $T(B,\varphi) = \{T\zeta^k\}$, cf 5.12(a,b), 5.15.

Let $x \in T(B,\varphi)_n(X)$ be represented by $h' : S^{N+n} \to T\zeta^N \wedge X^+$, $N = N(x)$. By 7.18, we can deform h' into a map $h : S^{N+n} \to T\zeta^N \wedge X^+$ which is transversal to $\zeta^N \times \theta^0$. We set $M := h^{-1}(B_N \times X)$ and consider the map

$$f = f^h : M \xrightarrow{\widehat{h}} B_N \times X \xrightarrow{p_2} X$$

where $\widehat{h}(m) = h(m)$ for every $m \in M$. Furthermore, we set $g := p_1 f : M \to B_N$. Finally, we notice that $(\zeta^N)_{\text{st}}$ gets a canonical (B,φ)-prestructure $(\varphi v, G)$ where $v : B_N \to B$ is the inclusion and $G(b,t) = \varphi v(b)$ for every $(b,t) \in B_N \times I$.

By 7.17(i), M is a $\mathscr{T}$-manifold. Now we equip M with the following strict (B,φ)-structure $\mathfrak{i} = (i, U, q, g^*\zeta^N, \omega, [a,H])$:

$i : M \subset \mathbb{R}^{N+n} = S^{N+n} \setminus \{\infty\}$ is given by the inclusion $M \subset S^{N+n}$;
$(U, q, g^*\zeta^N)$ is the tubular neighborhood provided by 7.17(ii);
$\omega := \mathfrak{I}_{g,\zeta^N}\mathfrak{I}_{\varphi_N,\gamma^N} : g^*\zeta^N \to \gamma^N$;
$[a,H] := \mathfrak{I}^!_{g,\zeta^N}[\varphi v, G]$, see 4.12(ii).

So, $f : M \to X$ turns into a singular (B,φ)-manifold in X. Clearly, $\Theta[M,f] = x$, i.e., Θ is an epimorphism.

We prove that Θ is monic, but we need some preliminaries. Let (M,f) be a closed singular (B,φ)-manifold in X, and let $\mathfrak{i} = (i, U, q, f^*\zeta^N, \omega, [a,H])$ be a strict (B,φ)-structure on M. Given $k \geq N$, we denote by ν^k the bundle $\nu^N \oplus \theta^{k-N}$, i.e., the normal bundle of M with respect to $\sigma^{k-n}\mathfrak{i}$. Now, given a morphism $\psi = \psi_k : \nu^k \to \zeta^k$ of $\mathcal{V}_k$-bundles, we have the map

$$(7.28) \qquad b_\psi = b_{\psi,\mathfrak{i}} : S^{k+n} \xrightarrow{c^k} T\nu^k \xrightarrow{\Delta^k} T\nu^k \wedge M^+ \xrightarrow{\psi\wedge f^+} T\zeta^k \wedge X^+$$

where c^k is the Browder–Novikov map and Δ^k is a map as in 5.36.

7.29. Lemma. *Given a closed singular (B,φ)-manifold $f : M^n \to X$, there exists a natural number k and a morphism $\psi : \nu^k \to \zeta^k$ such that the morphism $\beta_{(M,f)} : \Sigma^n S \to T(B,\varphi) \wedge X^+$ is homotopic to the morphism*

$$\Sigma^n S = \Sigma^{-k}\Sigma^\infty S^{n+k} \xrightarrow{\Sigma^{-k}\Sigma^\infty b_\psi} \Sigma^{-k}\Sigma^\infty(T\zeta^k \wedge X^+) \xrightarrow{i_k} T(B,\varphi) \wedge X^+$$

where i_k is the morphism II.(1.4). *Moreover, if $\beta_{(M,f)}$ is an inessential morphism then k and ψ can be chosen so that b_ψ is an inessential map.*

Proof. We have the strict (B,φ)-structure $\mathfrak{i} = (i, U, q, \nu^N, \omega, [a,H])$. Given $k \leq N$, we define $\omega_k : \nu^k \to \gamma^k$ to be the composition

$$\nu^k = \nu^N \oplus \theta^{k-N} \xrightarrow{\omega\oplus 1} \gamma^N \oplus \theta^{k-n} \xrightarrow{\rho_N\oplus 1} \gamma^{N+1} \oplus \theta^{k-N-1} \to \cdots$$
$$\to \gamma^{k-1} \oplus \theta^1 \xrightarrow{\rho_{k-1}} \gamma^k,$$

and we set $f_k := \operatorname{bs}\omega_k$. Fix a (B,φ)-structure on $\nu = \nu^N_{\text{st}} = \nu^k_{\text{st}}$.

Since M is compact, and since $\bigcup B_n = B$, there is a number k and maps $a_k : M \to B_k$ and $H_k : M \times I \to \overline{BV}_k$ with the following properties:

(1) The composition $M \xrightarrow{a_k} B_k \subset B$ coincides with a;

(2) The composition $M \times I \xrightarrow{H_k} \overline{B\mathcal{V}}_k \to B\mathcal{V}$ coincides with H;

(3) H is a homotopy between $\varphi_k a_k$ and f_k.

We define $r: M\times I \to M\times I, r(m,t) = (m,1)$. So, $H_k r = f_k \circ p_1 : M \times I \to \overline{B\mathcal{V}}_k$, and hence we have the morphism

$$u: \nu^k \times 1_I \xrightarrow{\mathfrak{F}_{\omega_k}} (f_k \circ p_1)^*(\gamma^k) = r^* H_k^* \gamma^k \xrightarrow{\mathfrak{I}_{r,H_k^*\gamma^k}} H_k^* \gamma^k.$$

Since $H_k|(M \times \{0\}) = \varphi_k a_k$, the restriction of u to the bundles over $M \times \{0\}$ yields the morphism

$$\psi : \nu^k = \nu^k \times \{0\} \to a_k^* \varphi_k^* \gamma^k = a_k^* \zeta^k \xrightarrow{\mathfrak{I}_{a_k,\zeta^k}} \zeta^k.$$

It is clear that $\beta_{(M,f)}$ has the form

$$\Sigma^n S = \Sigma^{-k}\Sigma^\infty S^{k+n} \xrightarrow{\Sigma^{-k}\Sigma^\infty b_\psi} \Sigma^{-k}\Sigma^\infty(T\zeta^k \wedge X^+) \xrightarrow{i_k} T(B,\varphi) \wedge X^+$$

with b_ψ as in (7.28).

Now we prove the last assertion (about inessential morphisms). Recall that $\pi_n(T(B,\varphi)\wedge X^+) = \lim_{r\to\infty} \pi_{n+r}(T\zeta^r)$, and, by the above, the homomorphism

$$\pi_{n+k}(T\zeta^k) \to \lim_{r\to\infty} \pi_{n+r}(T\zeta^r)$$

maps $[b_k] = [b_{\psi,i}] \in \pi_{n+k}(T\zeta^k)$ to $[\beta_{(M,f)}]$. Let $\overline{r}_k : \overline{B\mathcal{V}}_k \to \overline{B\mathcal{V}}_{k+1}$ be the inclusion. Given a_k and H_k as above, we set $a_{k+1} := \overline{r}_k a_k$ and $H_{k+1} := \overline{r}_k H_k$. Then the pair (a_{k+1}, H_{k+1}) yields a new morphism $\psi = \psi_{k+1} : \zeta^{k+1} \to \gamma^{k+1}$, and we get an element b_{k+1}. Clearly, the homomorphism

$$\pi_{n+k}(T\zeta^k) \to \pi_{n+k+1}(T\zeta^{k+1})$$

maps $[b_k]$ to $[b_{k+1}]$. So, we have a sequence

$$[b_k] \longmapsto [b_{k+1}] \longmapsto \cdots \longmapsto [b_m] \longmapsto \cdots$$

where each $[b_m]$ maps to $\beta_{(M,f)}$ under the homomorphism

$$\pi_{n+m}(T\zeta^m) \to \lim_{r\to\infty} \pi_{n+r}(T\zeta^r).$$

Thus, if $[\beta_{(M,f)}] = 0$ then $[b_m] = 0$ for m large enough. ⊞

We continue the proof of the theorem. We prove that Θ is monic, i.e., we suppose that $[\beta_{(M,f)}] = 0$ and prove that (M,f) bounds. Consider any map $b = b_\psi : S^{n+k} \to T\zeta^k$ as in 7.29; by 7.29, we can assume that b is inessential, i.e., b can be extended to a map $g' : D^{n+k+1} \to T\zeta^k$. We regard D^{n+k+1} as $S^{k+n} \times I/S^{k+n} \times \{1\}$, and we let $[x,t] \in D^{n+k+1}$ be the equivalence class of $(x,t) \in S^{k+n} \times I$. Given $\varepsilon \in (0,1)$, we can assume that $g'[x,t] = g'[x,0] = b(x)$ for every $(x,t) \in S^{n+k} \times [0,\varepsilon)$. Then, clearly, $g' \mid S^{k+n} \times [0,\varepsilon)$ is transversal to $\zeta^k \times \theta^0$. Hence, by 7.18, g' is homotopic to a map $g : D^{n+k+1} \to T\zeta^k \wedge X^+$

which is transversal to $\zeta^k \times \theta^0$ and , moreover, $g \mid S^{k+n} = g' \mid S^{k+n} = b$. We set $V := g^{-1}(B_K \times X)$ and define

$$F : V \xrightarrow{\widehat{g}} B_k \times X \xrightarrow{p_2} X, \widehat{g}(v) := g(v) \text{ for every } v \in V.$$

Now, asserting as in the proof of the epimorphicity of Θ, we turn (V, F) into a certain (B, φ)-manifold. Clearly, $\partial(V, F) = (M, f)$ as (B, φ)-manifolds. □

Theorem 7.27 shows that $\Omega_*^{(B,\varphi)}$ can be considered as a homology theory. Namely, we can set $\Omega_n^{(B,\varphi)}(X, A) := \Omega_n^{(B,\varphi)}(X/A) = \pi_n(T(B,\varphi)\wedge(X/A))$, etc. Geometrically, the group $\Omega_n^{(B,\varphi)}(X, A)$ can be described as follows. A closed singular (B, φ)-manifold in a pair (X, A) is a map $f : (M, \partial M) \to (X, A)$ of a compact (B, φ)-manifold M. Of course, $\partial M = \emptyset$ if $A = \emptyset$. We say that a closed singular manifold $f : (M, \partial M) \to (X, A)$ *bounds* if there exists a map $F : V \to X$ such that M is a (B, φ)-submanifold of ∂V, $F|M = f$, and $F(\partial V \setminus M) \subset A$. Now, similarly to 7.25, we can define the bordism classes $[M, f]$, which form a group $\Omega_n^{(B,\varphi)}(X, A)$. Furthermore, we define $\partial : \Omega_n^{(B,\varphi)}(X, A) \to \Omega_{n-1}^{(B,\varphi)}(A)$ by setting $\partial[M, f] = [\partial M, f|\partial M]$. One can prove that $\{\Omega_n^{(B,\varphi)}, \partial\}$ is a homology theory. See the details in Conner [1], Stong [3].

7.30. Definition. The homology theory $T(B, \varphi)_*(-) = \Omega_*^{(B,\varphi)}(-)$ is called a *(B, φ)-bordism theory.* The dual cohomology theory $T(B, \varphi)^*(-)$ is called a *cobordism theory.*

If $\varphi : B \to B\mathcal{V}$ is a multiplicative structure map (see 4.22), then there is the following pairing $T(B, \varphi) \wedge T(B, \varphi) \to T(B, \varphi)$. Set $\gamma = \gamma_{\mathcal{V}}$ and $\lambda := \varphi^*\gamma_{\mathcal{V}}$. Now, the homotopy H yields the following equivalence over B

$$\mu_B^*(\lambda) = \mu_B^*\varphi^*\gamma \simeq_B (\varphi \times \varphi)^*\mu^*(\gamma) = (\varphi \times \varphi)^*(\gamma \wedge^h \gamma) = \lambda \wedge^h \lambda,$$

and we get a pairing

$$\overline{\mu} : T(B, \varphi) \wedge T(B, \varphi) = T\lambda \wedge T\lambda \simeq T(\lambda \wedge^h \lambda) \xrightarrow{T\mu_B} T\lambda = T(B, \varphi),$$

where the equivalence is given by 5.21(i). This pairing induces a pairing

$$T(B, \varphi)_m(X) \otimes T(B, \varphi)_n(Y) \to T(B, \varphi)_{m+n}(X \times Y),$$

i.e., by 7.27, a pairing

$$\Omega_m^{(B,\varphi)}(X) \otimes \Omega_n^{(B,\varphi)}(Y) \to \Omega_{m+n}^{(B,\varphi)}(X \times Y).$$

We leave it to the reader to prove that this pairing has the form

$$\{f : M \to X\} \otimes \{g : N \to Y\} \longmapsto \{f \times g : M \times N \to X \times Y\}.$$

7.31. Examples. (a) (Pontrjagin, 1937, the available publication is Pontrjagin [2]). Let B be a contractible space. Then $T(B, \varphi) \simeq S$, i.e., the corresponding bordism group is just $\Pi_*(X)$ for every space X. Geometrically, a (B, φ)-manifold is just a manifold with an equivalence class of trivializations of its normal bundle. Such manifolds are called *framed manifolds.*

Pontrjagin used (proved) 7.27 with $\mathscr{T} = \mathrm{DIFF}$ in order to compute $\pi_*(S)$, and this was the first application of (co)bordism.

(b) (Milnor [4], Novikov [1]). There is a Thom spectrum $T(B\mathcal{U}, R)$ where $R : B\mathcal{U} \to B\mathcal{SO}$ is the realification. Geometrically, $(B\mathcal{U}, R)$-manifold is a smooth manifold with an equivalence class of complex vector bundle structures on $\nu^N, N >> \dim M$. Such a manifold is called a *stably almost complex manifold.* The corresponding (co)bordism group is called *complex (co)bordism.*

The spectrum $T(B\mathcal{U}, R)$ is usually denoted by $M\mathcal{U}$ and can also be described as follows. Let $M\mathcal{U}_n$ be the Thom space $T\gamma_{\mathbb{C}}^n$ of the universal n-dimensional complex vector bundle $\gamma_{\mathbb{C}}^n$ over $B\mathcal{U}_n$. Let $r_n = r_n^{\mathcal{U}} : B\mathcal{U}_n \to B\mathcal{U}_{n+1}$ classify the bundle $\gamma_{\mathbb{C}}^n \oplus \theta_{\mathbb{C}}^1$. Without loss of generality we can aasume that r_n is a CW-embedding. We consider the map $Tr_n := T\mathfrak{I}_{r_n, \gamma^{n+1}} : S^2 M\mathcal{U}_n \to M\mathcal{U}_{n+1}$. Then $M\mathcal{U} = \{(M\mathcal{U})_n, s_n\}$, where $(M\mathcal{U})_{2n} = M\mathcal{U}_n$, $(M\mathcal{U})_{2n+1} = SM\mathcal{U}_n$ and $s_{2n} = 1_{SM\mathcal{U}_n}$, $s_{2n+1} = Tr_n$.

Following 5.22, one can prove that $M\mathcal{U}$ is a ring spectrum.

(c) Let $\varphi : BS\mathcal{V} \to B\mathcal{V}, \mathcal{V} \leq \mathcal{TOP}$, be the direct limit of the two-sheeted coverings $BS\mathcal{V}_n \to B\mathcal{V}_n$, see the text after 4.28. Then a $\mathcal{V}$-object ξ admits a (B, φ)-structure iff it is orientable in the sense of 5.6. It is easy to see that in this case a (B, φ)-manifold is in fact a manifold which is oriented in the classical sense (see the definition e.g. in Dold [5]), cf. V.2.4 below. In particular, $H_n(M, \partial M) = \mathbb{Z}$ for every connected (B, φ)-manifold M^n, and an orientation $[M, \partial M]$ of M is just a generator (either of two) of the group $H_n(M, \partial M) = \mathbb{Z}$. Furthermore, if $M = \sqcup_{i=1}^k M_i^n$ where every M_i is a connected manifold, then $H_n(M, \partial M) = \oplus H_n(M_i, \partial M_i) = \mathbb{Z}^k$, and an orientation $[M, \partial M]$ of M is an element $[M, \partial M] = ([M_1, \partial M_1], \dots, [M_k, \partial M_k]) \in \mathbb{Z}^k = H_n(M)$, where $[M_i, \partial M_i]$ is an orientation of M_i.

The groups $\pi_*(M\mathcal{SO})$ were computed by Averbuch–Milnor–Novikov–Rochlin–Wall–Thom; the complete information about these groups is contained in Stong [3] (cf. 6.5 and 6.9). The groups $\pi_*(M\mathcal{SPL})$ have not been computed yet.

(d) We can consider the structure $1 : B\mathcal{V} \to B\mathcal{V}$. In this case opposite structure coincides with the original one. In particular, every element of the group $M\mathcal{V}_*(X)$ has order 2. Of course, this follows also from 6.1. Actually, $\pi_*(M\mathcal{O})$ was the first example of the successful complete calculation of the (co)bordism groups, Thom [2]. Namely,

$$\pi_*(M\mathcal{O}) = \mathbb{Z}/2\,[x_i \mid \dim x_i = i,\ i \in \mathbb{N},\ i \neq 2^s - 1].$$

Furthermore, the groups $\pi_*(M\mathcal{PL})$ and $\pi_*(M\mathcal{TOP})$ are computed by Brumfiel–Madsen–Milgram [1]

(e) Many other interesting examples are considered in Stong [3]. I want also to remark that the general concept of (B, φ)-(co)bordism was originated by Lashof [1] and developed by Stong [3].

Pontrjagin used bordism in order to calculate homotopy groups, while in other examples one applies homotopy techniques in order to investigate bordism. It is reflected in the following: Pontrjagin introduced exotic objects (framed manifolds) in order to compute homotopy groups of very natural objects (spheres), while Thom computed homotopy groups of exotic objects (Thom spaces $M\mathcal{SO}_n, M\mathcal{O}_n$) in order to deal with very natural objects (manifolds).

Now we consider the problem of realizability of homology classes. Let M^n be a closed connected manifold and $[[M]] \in H_n(M; \mathbb{Z}/2) = \mathbb{Z}/2$ be its fundamental class mod 2, i.e., $[[M]] \neq 0 \in \mathbb{Z}/2$. If $M = \sqcup_{i=1}^k M_i^n$ where every M_i is a closed connected manifold, set

$$[[M]] = ([[M_1]], \dots , [[M_k]]) \in (\mathbb{Z}/2)^k = H_n(M; \mathbb{Z}/2).$$

Given a space X and a map $f : M \to X$, we get a homology class $f_*[[M]] \in H_n(X; \mathbb{Z}/2)$. We say that a homology class $z \in H_n(X; \mathbb{Z}/2)$ *can be realized* (*by a manifold*) if it can be represented as $f_*[[M]]$ with some $f : M \to X$.

Similarly, we say that a homology class $z \in H_n(X)$ can be realized if it can be represented as $f_*[M]$ with some $f : M \to X$, where M^n is a closed oriented manifold and $[M] \in H_n(M)$ is an orientation as in 7.31(c).

Question: Can every homology class be realized? If not, how can one describe the realizable classes? Moreover, we can restrict this problem, considering $\mathcal{T}$ manifolds with given $\mathcal{T}$, or even (B, φ)-manifolds with some (B, φ).

Define a homomorphism

$$\mathfrak{t}^X : M\mathcal{SV}_n(X) \to H_n(X)$$

as follows. By 7.27, $M\mathcal{SV}_n(X)$ can be interpreted as the bordism group of oriented manifolds. If an element $a \in M\mathcal{SV}_n(X)$ is represented by a singular oriented manifold $f : M \to X$, set $\mathfrak{t}^X(a) := f_*[M] \in H_n(X)$. We leave it to the reader to prove that $\mathfrak{t}^X$ is a well-defined homomorphism. It is clear that the image of $\mathfrak{t}$ consists precisely of all realizable integral homology classes.

Similarly, we define a homomorphism

$$\bar{\mathfrak{t}}^X : M\mathcal{V}_n(X) \to H_n(X; \mathbb{Z}/2)$$

by setting $\bar{\mathfrak{t}}^X[M, f] = f_*[[M]] \in H_n(X; \mathbb{Z}/2)$ for every $f : M^n \to X$. The image of $\bar{\mathfrak{t}}$ consists precisely of all realizable homology classes mod 2.

The homomorphisms $\mathfrak{t}^X, \bar{\mathfrak{t}}^X$ are called the *Steenrod–Thom homomorphisms*.

The Thom class $u \in H^0(MS\mathcal{V})$ gives us a morphism $u : MS\mathcal{V} \to H\mathbb{Z}$. Hence, for every space X we have a homomorphism

$$u^X : MS\mathcal{V}_*(X) \to H_*(X).$$

Furthermore, u generates $H^0(MS\mathcal{V})$, and so, by 5.24(i), the homomorphism

$$u^{\mathrm{pt}} : \mathbb{Z} = MS\mathcal{V}_0(\mathrm{pt}) \to H_0(\mathrm{pt}) = \mathbb{Z}$$

is an isomorphism. We choose u such that $u^{\mathrm{pt}}(1) = 1$.

Similarly, the Thom class $\bar{u} \in H^0(M\mathcal{V}; \mathbb{Z}/2)$ gives us a homomorphism $\bar{u}^X : M\mathcal{V}_*(X) \to H_*(X; \mathbb{Z}/2)$ natural in X.

7.32. Proposition. (i) *The homomorphism u^X coincides with $\mathfrak{t}^X$. Furthermore, u^X coincides with the edge homomorphism*

$$MS\mathcal{V}_n(X) \to \oplus_i E^\infty_{i,n-i}(X) \to E^\infty_{n,0}(X) \subset E^2_{n,0}(X) = H_n(X)$$

in the AHSS $E^*_{*,*}(X) \Longrightarrow MS\mathcal{V}_*(X)$, $E^2_{p,q}(X) = H_p(X; \pi_q(MS\mathcal{V}))$.

(ii) *The homomorphism $\bar{u}^X$ coincides with $\bar{\mathfrak{t}}^X$. Furthermore, $\bar{u}^X$ coincides with the edge homomorphism*

$$M\mathcal{V}_n(X) \to \oplus_i E^\infty_{i,n-i}(X) \to E^\infty_{n,0}(X) \subset E^2_{n,0}(X) = H_n(X; \mathbb{Z}/2)$$

in the AHSS $E^*_{*,*}(X) \Longrightarrow M\mathcal{V}_*(X)$, $E^2_{p,q}(X) = H_p(X; \pi_q(M\mathcal{V}))$.

Proof. We prove only (i). Given a CW-pair (X, A), we define a homomorphism

$$\mathfrak{t}^{(X,A)} : MS\mathcal{V}_n(X, A) \to H_n(X, A)$$

as follows. Let $f : (M, \partial M) \to (X, A)$ be a singular oriented manifold in the pair (X, A), and let $[M, \partial M] \in H_n(M, \partial M)$ be the orientation of M. Then we have a homology class $f_*[M, \partial M] \in H_n(X, A)$. Now, we set $t^{(X,A)}[M, f] = f_*[M, \partial M]$. Clearly, $\mathfrak{t}^{(X,\emptyset)} = \mathfrak{t}^X$.

It is obvious that the family $\mathfrak{t} = \{\mathfrak{t}^{(X,A)} : MS\mathcal{V}_*(X, A) \to H_*(X, A)\}$ is a morphism of homology theories. By III.3.23(ii), this morphism $\mathfrak{t}$ is induced by a morphism $\mathfrak{t} : MS\mathcal{V} \to H\mathbb{Z}$ of spectra. Since $\mathfrak{t}^{\mathrm{pt}} : \mathbb{Z} = MS\mathcal{V}_0(\mathrm{pt}) \to H_*(\mathrm{pt}) = \mathbb{Z}$ is an isomorphism, the element $\mathfrak{t} \in H^0(MS\mathcal{V}) = \mathbb{Z}$ must be a generator, cf. 5.24. So, $\mathfrak{t} = \pm u$. But both morphisms $\mathfrak{t}$ and u map 1 to 1, and so $\mathfrak{t} = u$.

Similarly, one can see that the edge homomorphism is natural, and so it coincides (up to sign) with u_*. □

7.33. Theorem. *The homomorphism $\bar{\mathfrak{t}}^X : M\mathcal{V}_*(X) \to H_*(X; \mathbb{Z}/2)$ is epic for every X and every $\mathcal{V}$. Thus, every homology class* mod 2 *can be realized, and, in particular, by a smooth manifold.*

Proof. By 6.2, $M\mathcal{V}$ is a graded Eilenberg–Mac Lane spectrum. Furthermore, by 5.23(i), $\pi_0(M\mathcal{V}) = \mathbb{Z}/2$. Hence, by II.7.2, there is a morphism

$j : H\mathbb{Z}/2 \to M\mathcal{V}$ such that $H\mathbb{Z}/2 \xrightarrow{j} M\mathcal{V} \xrightarrow{\bar{u}} H\mathbb{Z}/2$ is an equivalence. Thus, $\bar{\mathrm{t}}^X = \bar{u}^X$ is epic. □

On the other hand, t^X is not an epimorphism in general.

Given an odd prime p, consider a morphism

$$\widehat{Q}_n : H\mathbb{Z} \to H\mathbb{Z}/p \xrightarrow{Q_n} \Sigma^{2p^n-1} H\mathbb{Z}/p,$$

where the first morphism is the mod p reduction. Let $(\widehat{Q}_n)_* : H_*(X) \to H_*(X; \mathbb{Z}/p)$ be the induced homomorphism.

7.34. Lemma. (i) *If a homology class $z \in H_*(X)$ can be realized by a DIFF manifold, then $(\widehat{Q}_i)_*(z) = 0$ for every i and every odd prime p.*

(ii) *If a homology class $z \in H_*(X)$ can be realized by a TOP manifold, then $(\widehat{Q}_1)_*(z) = 0$ for every odd prime p.*

Proof. (i) By 6.7, the composition $MSO \xrightarrow{u} H\mathbb{Z} \xrightarrow{\widehat{Q}_i} \Sigma^{2p^i-1} H\mathbb{Z}$ is trivial.

(ii) By 6.15 and 6.14, the composition $MSTOP \xrightarrow{u} H\mathbb{Z} \xrightarrow{\widehat{Q}_1} \Sigma^{2p-1} H\mathbb{Z}$ is trivial. □

7.35. Theorem. *There exists an element $z \in H_7(K(\mathbb{Z}/3, 2))$ such that $(\widehat{Q}_1)_*(z) \neq 0$ (for $p = 3$). In particular, z cannot be realized by a (topological) manifold.*

Proof. Let $\iota \in H^3(K(\mathbb{Z}/3, 2); \mathbb{Z}/3) = \mathbb{Z}/3$ be a generator of this group. Let z be a generator of $H_7(K(\mathbb{Z}/3, 2)) = \mathbb{Z}/3$, and let $\bar{z}$ be the mod 3 reduction of z. Consider the $\mathbb{Z}/3$-basis $\{\beta P^1\iota, Q_1\iota\}$ of $H^7(K(\mathbb{Z}/3, 2); \mathbb{Z}/3)$. We have $\langle \beta P^1\iota, \bar{z}\rangle = 0$, because $\bar{z}$ comes from integral homology. So, $\langle Q_1\iota, \bar{z}\rangle \neq 0$. Now, by II.6.36,

$$0 \neq \langle Q_1\iota, \bar{z}\rangle = \langle \iota, (Q_1)_\bullet \bar{z}\rangle = \langle \iota, (\chi(Q_1))_*(\bar{z})\rangle.$$

But $\chi(Q_1) = -Q_1$. Thus, $0 \neq Q_1(\bar{z}) = (\widehat{Q}_1)_*(z)$. □

7.36. Theorem. *For every $z \in H_n(X)$ there exists k such that $(2k+1)z$ can be realized by a smooth manifold.*

Proof. The proof is similar to that of 7.33. By 6.5, $MSO[2]$ is a graded Eilenberg–Mac Lane spectrum. Furthermore, by 5.23(i), $\pi_0(MSO) = \mathbb{Z}$. Hence, $\mathrm{t}[2] : MSO[2]_*(X) \to H_*(X; \mathbb{Z}[2])$ is epic. □

7.37. Theorem. *Every homology class $z \in H_i(X)$ with $i \leq 6$ can be realized by a smooth manifold.*

Proof. It suffices to prove that the homomorphism

$$\mathfrak{t}[p] : MSO[p]_*(X) \to H_*(X; \mathbb{Z}[p])$$

is epic for every prime p and every $i \le 6$. For $p = 2$ this follows from 7.36. If $p > 2$, then, by 6.9, $MSO[p]_{(3)} = H\mathbb{Z}[p]$, $\pi_4(MSO[p]) = \mathbb{Z}[p]$ and $MSO[p]_{(4)} = MSO[p]_{(7)}$, where the subscript denotes the coskeleton. So, the cofiber sequence

$$MSO[p]_{(4)} \xrightarrow{p_4} MSO[p]_{(3)} \xrightarrow{\kappa} \Sigma^5 H\mathbb{Z}[p],$$

see II.4.19, can be rewritten as

$$MSO[p]_{(7)} \xrightarrow{p_4} H\mathbb{Z}[p] \xrightarrow{\kappa} \Sigma^5 H\mathbb{Z}[p].$$

Furthermore, by 7.32 and 5.24(i), the morphism

$$MSO[p] \xrightarrow{\tau_7} MSO[p]_{(7)} \xrightarrow{p_4} H\mathbb{Z}[p]$$

coincides with $\mathfrak{t}[p]$. Now, by II.4.5(ii), the homomorphism

$$(\tau_7)_* : (MSO[p])_i(X) \to (MSO[p]_{(7)})_i(X)$$

is epic for $i \le 6$ and every CW-space X. So, for $i \le 6$, we have an exact sequence

$$(MSO[p])_i(X) \xrightarrow{\mathfrak{t}[p]} H_i(X; \mathbb{Z}[p]) \xrightarrow{\kappa_*} H_{i-5}(X; \mathbb{Z}[p]).$$

Hence, for $i \le 6$, $\mathfrak{t}[p] : (MSO[p])_i(X) \to H_i(X; \mathbb{Z}[p])$ is epic if $\kappa_* : H_i(X; \mathbb{Z}[p]) \to H_{i-5}(X; \mathbb{Z}[p])$ is a zero homomorphism. So, trivially, $\mathfrak{t}[p]$ is epic if $i \le 4$.

If $p > 3$ then $H^5(H\mathbb{Z}[p]; \mathbb{Z}[p]) = 0$, see e.g. Cartan [1], and so $\kappa_* = 0$. Thus, $\mathfrak{t}[p] : MSO[p]_i(X) \to H_i(X; \mathbb{Z}[p])$ is epic for $p > 3$ and $i \le 6$.

Let $p = 3$. We have

$$H^5(H\mathbb{Z}[3]; \mathbb{Z}[3]) = \mathbb{Z}/3 = \{\delta P^1 \rho\},$$

where $\delta : H\mathbb{Z}/p \to H\mathbb{Z}[p]$ is the integral Bockstein homomorphism and $\rho : H\mathbb{Z}[p] \to H\mathbb{Z}/p$ is the reduction mod p, i.e., $\kappa = \lambda \delta P^1 \rho, \lambda \in \mathbb{Z}/3$. (In fact, it follows from 6.8 that $\lambda \ne 0$, but we do not need it here.) So, it suffices to prove that $(\delta P^1 \rho)_* : H_i(X; \mathbb{Z}[3]) \to H_{i-5}(X; \mathbb{Z}[3])$ is a zero homomorphism for $i \le 6$. This is clear for $i < 5$.

We choose any $z \in H_i(X; \mathbb{Z}[3]), i = 5, 6$, and prove that $(\delta P^1 \rho)_*(z) = 0$. To the contrary, suppose that $(\delta P^1 \rho)_*(z) \ne 0$.

Firstly, let $i = 5$. We have $0 \ne (P^1 \rho)_*(z) \in H_1(X; \mathbb{Z}/3)$. So, there is $y \in H^1(X; \mathbb{Z}/3)$ such that $0 \ne \langle (P^1 \rho)_*(z), y \rangle \in \mathbb{Z}/3$. Since $P^1(x) = 0$ for every $x \in H^1(X; \mathbb{Z}/3)$, we conclude that

$$0 \ne \langle (P^1 \rho)_*(z), y \rangle = \langle z, \rho P^1 y \rangle = 0.$$

This is a contradiction.

Now, let $i = 6$. We can assume that X is a finite CW-space (since z is contained in a finite CW-subspace of X.) In particular, $H_1(X; \mathbb{Z}[3])$ is a finitely generated $\mathbb{Z}[3]$-module. Since $H^1(X; \mathbb{Z}/3^m) = \operatorname{Hom}(H_1(X; \mathbb{Z}[3]), \mathbb{Z}/3^m)$ and since $(\delta P^1 \rho)_*(z) \neq 0$, there exist a natural number m and a class $y \in H^1(X; \mathbb{Z}/3^m)$ such that

$$0 \neq \langle (\delta P^1 \rho)_*(z), y \rangle \in \mathbb{Z}/3^m.$$

Let $f : X \to K(\mathbb{Z}/3^m, 1)$ be a map such that $f^* \iota = y$, where

$$\iota \in H^1(K(\mathbb{Z}/3^m, 1); \mathbb{Z}/3^m) = \mathbb{Z}/3^m$$

is a generator. Note that $f_* z = 0$ because $H_6(K(\mathbb{Z}/3^m, 1)) = 0$. Now

$$0 \neq \langle (\delta P^1 \rho)_*(z), y \rangle = \langle (\delta P^1 \rho)_*(z), f^* \iota \rangle = \langle (\delta P^1 \rho)_*(f_* z), \iota \rangle = 0.$$

This is a contradiction. □

7.38. Theorem. (i) *There exists a class $z \in H_*(X)$ which can be realized by a PL manifold, but cannot be realized by a DIFF manifold.*

(ii) *If a homology class can be realized by a TOP manifold, then it can be realized by a PL manifold.*

Proof. (i) By 7.34, it suffices to find $z \in H_*(X)$ such that $(\widehat{Q}_2)_*(z) \neq 0$ and z can be realized by PL manifolds. Dually, it suffices to find a finite CW–space Y and a class $y \in H^*(Y)$ such that $\widehat{Q}_2(y) \neq 0$ and

$$y \in \operatorname{Im}(u^Y_{\mathcal{PL}} : M\mathcal{SPL}^*(Y) \to H^*(Y)).$$

(Then we set $z = Dy$, where $D : H^i(Y) \to H_{N-i}(X)$ is the duality isomorphism and X is N-dual to Y.) Let Y be the $2N$-skeleton of $M\mathcal{SPL}_N$, and let $y = i^* u_N$, where $N > 2p^2$, $u_N \in H^N(M\mathcal{SPL}_N; \mathbb{Z}/p)$ is the Thom class and $i : Y \to M\mathcal{SPL}_N$ is the inclusion. It is clear that y can be represented by a morphism

$$\Sigma^{-N}\Sigma^\infty Y \subset \Sigma^{-N}\Sigma^\infty \subset M\mathcal{SPL}_N \to M\mathcal{SPL} \xrightarrow{u_{\mathcal{PL}}} H\mathbb{Z}.$$

Hence, $y \in \operatorname{Im}(u^Y_{\mathcal{PL}} : M\mathcal{SPL}^*(Y) \to H^*(Y))$. Finally, by 6.13, $\widehat{Q}_2(y) \neq 0$.

(ii) Suppose that a class z can be realized by a TOP manifold. By 6.15, $M\mathcal{SPL}[1/2] \simeq M\mathcal{STOP}[1/2]$. Hence, there exists k such that $2^k z$ can be realized by a PL manifold. On the other hand, by 7.36, there is n such that $(2n+1)z$ can be realized by a DIFF (and hence a PL) manifold. Taking $a, b \in \mathbb{Z}$ such that $2^k a + (2n+1)b = 1$, we conclude that $z = a(2^k z) + b((2n+1)z)$ can be realized by a PL manifold. □

7.39. Remarks. (a) The problem of realizing homology classes was formulated explicitly by Steenrod, see Eilenberg [1]. However, it is really a much older question in algebraic topology, dating back to Poincaré. In fact,

Poincaré [1] used the term "homology" for what we call bordism, and one can say that Poincaré constructed a bordism theory. Of course, (sub)manifolds are good naive models for cycles, but the correct definition of homology joins geometric and algebraic concepts. So, the problem of realizing homology classes can be considered as an attempt to compare a naive conception of homology with its strict definition.

(b) The problem of realizing homology classes was solved by Thom [2] in principle. Namely, Thom proved 7.32 and hence reduced the problem to a pure homotopy problem. In this way he proved 7.33 and 7.37. Moreover, he proved that a certain class $z \in H_7(K(\mathbb{Z}/3 \oplus \mathbb{Z}/3, 1))$ cannot be realized by a *smooth* manifold. In fact, his proof was like that of 7.35, but at that time there was no information about $BTOP$. We remark that Thom [2] introduced the spaces MO_k and MSO_k in order to attack the realizability problem.

(c) Thom [2] constructed the map $\Theta : \Omega_n^O(X) \to MO_n(X)$ (as in the proof of 7.27) as follows. Let M^n be embedded in S^{N+n}. Then we have a map $U \xrightarrow{q} \operatorname{ts} \nu^N \to \operatorname{ts} \gamma_O^N$ and it can be extended to a map $S^{N+n} \to MO_N = T\gamma_O^N$. (In fact, Thom did not distinguish U and $\operatorname{ts} \xi$.) The collapsing map $c : S^{N+n} \to T\nu$ arose later, in papers of Browder [1] and Novikov [2]. Of course, this construction follows general ideas of Thom, but, I think, it is a certain step further: we have here some universality. For this reason, I named the collapsing map c the Browder–Novikov map.

I want to remark that Browder and Novikov introduced the collapsing map for needs of differential topology, i.e., in some sense, outside of cobordism theory. Namely, Browder [1] described homotopy types containing smooth closed manifolds; Novikov [2,3] also did it and even classified smooth closed manifolds which are homotopy equivalent to a given one. So, we have here another remarkable application of Thom spaces. However, this topic is beyond this book; we refer the reader to Browder [3], Novikov [3].

7.40. Remarks. (a) It follows from 7.35 that for every $i > 6$ there is a class $y \in H_i(X)$ which cannot be realized by a manifold. Namely, given i, consider the suspension isomorphism $s : H_7(K(\mathbb{Z}/3, 2)) = H_i(S^{i-7}K(\mathbb{Z}/3, 2))$ and put $y = sz$ for z as in 7.35. Then $\widehat{Q}_1(z) \neq 0$ (for $p = 3$).

(b) The minimal dimension n in 7.38(i) is $n = 19$, Brumfiel [1]. The proof is similar to that of 7.37. Namely, $6 = \dim Q_1 + 1$, while $18 = \dim Q_2 + 1$, if $p = 3$. Now, similarly to (a), one can see that the class in 7.38(i) exists for every $n > 18$.

Résumé on Realizability

(i) *Every homology class* $z \in H_*(X;\mathbb{Z}/2)$ *can be realized by a smooth manifold.*

(ii) *Every class* $z \in H_i(X)$ *with* $i \leq 6$ *can be realized by a smooth manifold.*

(iii) *For every* $i > 6$, *there is a class* $z \in H_i(X)$ *which cannot be realized by a topological manifold.*

(iv) *Given a class* $z \in H_i(X)$, *there is a natural number* n *such that* $(2n+1)z$ *can be realized by a smooth manifold.*

(v) *If a homology class can be realized by a topological manifold, then it can be realized by a PL manifold.*

(vi) *If a homology class of dimension* ≤ 18 *can be realized by a topological manifold, then it can be realized by a smooth manifold.*

(vii) *For every* $i > 18$, *there is a class* $z \in H_i(X)$ *which can be realized by a PL manifold but cannot be realized by a smooth manifold.*

Chapter V. Orientability and Orientations

It seems that the orientability concept arose implicitly in the infancy of humanity, when people became able to distinguish upward and downward (as well as left and right) directions. Many epochs later we had suitable concepts of the orientation of the line (arrow), the plane (circle arrow) and space (right-left triples of vectors, spiralled arrow, etc.). Finally, in the nineteenth century a satisfactory concept of the orientation of the space $\mathbb{R}^n$ as an equivalence class of frames was formulated.

The orientability concept was developed further by considering families X of spaces $\mathbb{R}^n$, the orientation of X being a family of compatible (in some sense) orientations of the members $\mathbb{R}^n$ of X. For example, an orientation of a manifold is given by a family of compatible orientations of the charts, an orientation of a bundle is given by the family of compatible orientations of fibers.

Later the (co)homological nature of orientability was understood. For instance, an orientation of $\mathbb{R}^n$ can be treated as one of the two generators of the group $H_n(\hat{\mathbb{R}}^n)$ (or $H^n(\hat{\mathbb{R}}^n)$), where $\hat{\mathbb{R}}^n = S^n$ is the one-point compactification of $\mathbb{R}^n$. In this way we can define an orientation of an $\mathbb{R}^n$-bundle ξ to be a compatible family of orientations of the fibers, i.e., (successfully formalizing this naive idea) to be a Thom class $u \in \widetilde{H}^n(T\xi)$. Similarly, an orientation of a (closed) manifold M^n can be defined to be a compatible family of orientations of the charts, i.e., to be a fundamental class $[M] \in H_n(M)$. It is clear that the definitions above are suitable for any (co)homology theory, and in fact this generalization has been made and has turned out to be very fruitful. For example, it makes very lucid such matters as Poincaré duality, integrality phenomena, characteristic classes, etc.

§1. Orientations of Bundles and Fibrations

As in §IV.5, we deal "theoretically" with $\mathcal{F}_n$-objects, i.e., with $(S^n,*)$-fibrations, but the results will be applied to $\mathcal{V}_n$-objects.

Let E be a ring spectrum, and let $\sigma^d \in \widetilde{E}^d(S^d)$ be the image of $1 \in \pi_0(E) = \widetilde{E}^0(S^0)$ via the iterated suspension isomorphism $\widetilde{E}^0(S^0) \cong \widetilde{E}^d(S^d)$.

1.1. Definition (Dold [1]). Let α be an $\mathcal{F}^d$-object over a CW-space X, and let $j_x : S^d \to T\alpha$ be a root with respect to a point $x \in X$. The element $u = u_{\alpha,E} \in \widetilde{E}^d(T\alpha)$ is called an *orientation of* α *with respect to* E, or, briefly, an E*-orientation of* α, if $j_x^*(u) = \pm\sigma^d$ for all $x \in X$. Furthermore, we define an E-orientation of a $\mathcal{V}_n$-object ξ to be an E-orientation of $\xi^\bullet$.

Here the sign before σ^d can depend on x. Note that we are forced to say $j_x^*(u) = \pm\sigma^d$ rather than $j_x^*(u) = \sigma^d$, because the homotopy class of a root j is determined up to sign only. More precisely, the sign arises when we want to fix a homotopy equivalence of the standard sphere S^n with the fiber F_x in order to construct the root $j_x : S^d \to F_x \subset T\alpha$. From another viewpoint, the sphere admits an involution of degree -1. This indeterminacy can be eliminated if we consider rooted $\mathcal{V}$-objects. Maybe, sometimes this makes sense, but for a lot of applications such rigidity is not necessary.

Of course, the sign ± 1 really depends only on the component of the base X. More precisely, for a connected base it is possible to choose homotopy equivalences of the standard sphere S^d with fibers such that all maps $j_x : S^d \to T\alpha$ are homotopic, cf. §IV.5. Thus, for a connected base X an orientation can be characterized by the equality $j_{x_0}^*(u) = \pm\sigma^d$ for some single point $x_0 \in X$.

An $\mathcal{F}$-object with a fixed E-orientation is said to be E*-oriented.* In other words, an E-oriented $\mathcal{F}$-object is a pair (α, u) where u is an E-orientation of the $\mathcal{F}$-object α. An $\mathcal{F}$-object is said to be E*-orientable* if it admits an E-orientation. Similar terminology is used for $\mathcal{V}$-objects (replacing $\mathcal{F}$ by $\mathcal{V}$).

It is obvious that E-orientability is an invariant of the $\mathcal{F}$-equivalence because Thom spaces of $\mathcal{F}$-equivalent $\mathcal{F}$-objects are homotopy equivalent.

Corollary IV.5.8 shows that orientability as defined in IV.5.6 is just $H\mathbb{Z}$-orientability. Here, roughly speaking, a Thom class $u \in H^*(T\alpha)$ enables us to cohere orientations of fibers. Indeed, using the cohomological description of orientability (see the introduction to this chapter) we conclude that the restriction of the Thom class to each fiber induces an orientation of the fiber. In other words, the existence of a Thom class enables us to equip the fibers with compatible orientations. It seems that, rigorously speaking, we must treat orientations of the fibers (i.e., generators of the groups $H^*(\hat{\mathbb{R}}^d_x)$) as being compatible iff there exists a Thom class. Thus, it makes sense *to call* a

Thom class an orientation of an $\mathcal{F}$-object. This argument justifies the term E-orientation of Definition 1.1; it is a generalization of $H\mathbb{Z}$-orientation (Thom class). For this reason an E-orientation is also called a generalized Thom class, or a Thom–Dold class, because Dold [1] introduced this concept for arbitrary E. This generalization turns out to be very fruitful, e.g., in this way it is possible to generalize the Thom Isomorphism Theorem IV.5.9, see 1.3 and/or 1.7 below.

Theorem IV.5.7 can also be generalized to an arbitrary ring spectrum E, see Becker [1], Rudyak [3], but this generalization does not seem completely satisfactory, e.g. I cannot *immediately* deduce from it Theorem 1.3 (or even (1.7)) below (as we deduced IV.5.9 from IV.5.7).

Let α be an $\mathcal{F}_d$-object over X, and let $\Delta^d : T\alpha \to T\alpha \wedge X^+$ be a map as in IV.5.36. Let $F = (F, m)$ be any E-module spectrum with the pairing $m : E \wedge F \to F$. Define

$$\overline{\varphi} : \widetilde{E}^d(T\alpha) \otimes \widetilde{F}^n(X^+) \xrightarrow{m^{T\alpha, X^+}} \widetilde{F}^{n+d}(T\alpha \wedge X^+) \xrightarrow{(\Delta^d)^*} \widetilde{F}^{n+d}(T\alpha),$$

$$\underline{\varphi} : \widetilde{E}^d(T\alpha) \otimes \widetilde{F}_n(T\alpha) \xrightarrow{1\otimes\Delta^d_*} \widetilde{E}^d(T\alpha) \otimes \widetilde{F}_n(T\alpha \wedge X^+) \xrightarrow{m^{T\alpha}_{\bullet, X^+}} \widetilde{F}_{n-d}(X^+).$$

Now suppose that α is equipped with an E-orientation $u \in \widetilde{E}^d(T\alpha)$. Define

$$\varphi_F = \varphi_{F,X,u} : F^n(X) = \widetilde{F}^n(X^+) \to \widetilde{F}^{n+d}(T\alpha), \ \varphi_F(x) = \overline{\varphi}(u \otimes x),$$
$$\varphi^F = \varphi^{F,X,u} : \widetilde{F}_n(T\alpha) \to \widetilde{F}_{n-d}(X^+) = F_{n-d}(X), \ \varphi^F(x) = \underline{\varphi}(u \otimes x).$$

1.2. Proposition. *Let $\omega : \alpha \to \beta$ be a morphism of $\mathcal{F}_d$-objects, and let $T\omega : T\alpha \to T\beta$ be the induced map of the Thom spaces.*

(i) *If $u \in \widetilde{E}^d(T\beta)$ is an E-orientation of β then $(T\omega)^*(u)$ is an orientation of α. In particular, α is E-orientable if β is.*

(ii) *We set $X := \operatorname{bs}\alpha, Y := \operatorname{bs}\beta, f := \operatorname{bs}\omega : X \to Y$. Let u be an E-orientation of β. If we equip α with the orientation $(T\omega)^*(u)$, then the following diagrams commute:*

$$\begin{array}{ccc} F^n(Y) & \xrightarrow{\varphi_F} & \widetilde{F}^{n+d}(T\beta) \\ {\scriptstyle f^*}\downarrow & & \downarrow{\scriptstyle (T\omega)^*} \\ F^n(X) & \xrightarrow{\varphi_F} & \widetilde{F}^{n+d}(T\alpha), \end{array} \qquad \begin{array}{ccc} \widetilde{F}_n(T\beta) & \xrightarrow{\varphi^F} & F_{n-d}(X) \\ {\scriptstyle (T\omega)_*}\uparrow & & \uparrow{\scriptstyle f_*} \\ \widetilde{F}_n(T\alpha) & \xrightarrow{\varphi^F} & F_{n-d}(Y). \end{array}$$

Proof. This is obvious. □

1.3. Theorem-Definition. *For every CW-space X the homomorphisms φ^F and φ_F are isomorphisms. These isomorphisms are called* Thom–Dold isomorphisms.

Proof. The case $X = \emptyset$ is trivial. So, we assume that $X \neq \emptyset$. Firstly, we prove that φ_F is an isomorphism. We need some preliminaries.

For every X the pairing $m^{S^d,X^+} : \widetilde{E}^i(S^d) \otimes \widetilde{F}^j(X^+) \to \widetilde{F}^{i+j}(X^+)$ yields a homomorphism

$$h_F = h_F^X : \widetilde{F}^i(X^+) \to \widetilde{F}^{i+d}(S^d X^+),\ a \mapsto m^{S^d,X^+}(\sigma^d \otimes a).$$

Since m commutes with suspensions, h_F coincides with the iterated suspension isomorphism $\mathfrak{s}^{i+d-1} \cdots \mathfrak{s}^i : \widetilde{F}^i(X^+) \to \widetilde{F}^{i+d}(X^+)$. In particular, h_F is an isomorphism.

Similarly to the above, using the multiplication $\mu : E \wedge E \to E$ instead of the pairing $m : E \wedge F \to F$, we have an isomorphism

$$h_E = h_E^X : \widetilde{E}^i(X^+) \to \widetilde{E}^{i+d}(S^d X^+).$$

Now we start to prove that φ_F is an isomorphism.

Step 1. We prove that φ_F is an isomorphism if α is the standard trivial $\mathcal{F}_d$-object over a connected finite dimensional CW-space X. By IV.5.5(iii), $T\alpha \simeq S^d X^+$, and so $E^n(X) \cong \widetilde{E}^{n+d}(T\alpha)$, but we must prove that φ_F establishes an isomorphism of these groups. Fix a point $x_0 \in X$ and let $i : \{x_0\} \subset X$ be the inclusion. We set $j := S^d(i^+) : S^d(\{x_0\}^+) \to S^d X^+$. By IV.5.5(iii), j is a root of $T\alpha$. Now, we have the commutative diagram

$$\begin{array}{ccc} \widetilde{E}^0(\{x_0\}^+) & \xleftarrow{\ i^*\ } & \widetilde{E}^0(X^+) \\ \cong \big\downarrow h' & & \cong \big\downarrow h'' \\ \widetilde{E}^d(S^d(\{x_0\}^+)) & \xleftarrow{\ j^*\ } & \widetilde{E}^d(S^d X^+) \end{array}$$

where h' and h'' are the isomorphisms h_E for the spaces $\{x_0\}$ and X, respectively. In order to distinguish units of the rings $\widetilde{E}^0(\{x_0\}^+)$ and $\widetilde{E}^0(X^+)$, we use the notation $1'$ for $1 \in \widetilde{E}^0(\{x_0\}^+)$ and $1''$ for $1 \in \widetilde{E}^0(X^+)$.

Let $u \in \widetilde{E}^d(S^d X^+)$ be an E-orientation of α. Firstly, we prove that $u = \pm h''(1'' + a)$ where

$$a \in \operatorname{Ker}\{i^* : E^0(X) \to E^0(\{x_0\})\} = \operatorname{Ker}\{i^* : \widetilde{E}^0(X^+) \to \widetilde{E}^0(\{x_0\}^+)\}.$$

Indeed, let $u = h''(x)$ for some $x \in \widetilde{E}^0(X^+) = E^0(X)$. Without loss of generality we can assume that $j^*u = \sigma^d$. Then $j^*h''x = \sigma^d$, and so $h'i^*x = \sigma^d$. Since $\sigma^d = h'1'$, we conclude that $h'(i^*x - 1') = 0$, i.e., $i^*x - 1' = 0$, i.e., $i^*(x - 1'') = 0$. Thus, $x = 1 + a$ where $a \in \operatorname{Ker} i^*$.

Now, the E-orientation u yields the homomorphism φ_F, and it is easy to see that φ_F is a composition of the isomorphism h_F and multiplication by $\pm(1 + a)$. So, it suffices to prove that $1 + a$ is invertible in $E^0(X)$.

Since X is connected, the reduced iterated diagonal

$$d : (X, x_0) \to \underbrace{(X, x_0) \wedge \cdots \wedge (X, x_0)}_{k \text{ times}} = (X \wedge \cdots \wedge X, *)$$

is inessential if $k > \dim X$. Hence, $a^k = 0$ if $k > \dim X$, and thus $1 + a$ is an invertible element of $E^0(X)$.

Step 2. We prove that φ_F is an isomorphism if α is the standard trivial $\mathcal{F}_d$-object over a finite dimensional CW-space X. Let $X = \sqcup X_\lambda$ with X_λ connected. Then (cf. II.3.16(c))

$$F^n(X) = \prod_\lambda F^n(X_\lambda) \cong \prod_\lambda \widetilde{F}^{n+d}(T\alpha_\lambda) = \widetilde{F}^{n+d}(\vee_\lambda T\alpha_\lambda) = \widetilde{F}^{n+d}(T\alpha)$$

where the second homomorphism is given by Step 1.

Step 3. We prove that φ_F is an isomorphism if α is $\mathcal{F}_d$-equivalent to the standard trivial $\mathcal{F}_d$-object θ^d_X over a finite dimensional CW-space X (i.e., there is a sectioned bundle homotopy equivalence between α and θ^d_X). This follows from Step 2, because the equivalence $\alpha \to \theta^d_{\mathcal{F}}$ yields a homotopy equivalence $T\alpha \to T\theta^n_{\mathcal{F}}$, and the last one commutes with φ_F, cf. 1.2(ii).

Now we pass to the general case. Given a CW-subspace Y of X, let $\varphi_Y : F^n(Y) \to \widetilde{F}^{n+d}(T(\alpha|Y))$ be the restriction of φ_F to Y.

1.4. Lemma. (i) *Let Y be a CW-subspace of X, and let $(Y; A, B)$ be a CW-triad. Set $C = A \cap B$. If φ_A, φ_B and φ_C are isomorphisms then so is φ_Y. In particular, if φ_A is an isomorphism and B is a finite dimensional space such that $\alpha|B$ is $\mathcal{F}_d$-equivalent to a trivial $\mathcal{F}_d$-object, then φ_Y is an isomorphism.*

(ii) *Let $X_0 \subset \cdots \subset X_n \subset \cdots$ be a sequence of CW-subspaces of X such that $X = \bigcup X_n$. If φ_{X_n} is an isomorphism for every n then φ_X is an isomorphism.*

Proof. (i) By 1.2(i), we have the following commutative diagram of the Mayer–Vietoris sequences, where TZ denotes $T(\alpha|Z)$ for $Z = Y, A, B, C$ (the bottom sequence is the Mayer–Vietoris sequence of the triad $(TY; TA, TB)$; notice that $TC = TA \cap TB$, even if $C = \emptyset$):

$$\begin{array}{ccccccc}
\cdots \to & F^{n-1}(C) & \longrightarrow & F^n(Y) & \longrightarrow & F^n(A) \oplus F^n(B) & \to \cdots \\
& \cong \downarrow \varphi_C & & \downarrow \varphi_Y & & \cong \downarrow \varphi_A \oplus \varphi_B & \\
\cdots \to & \widetilde{F}^{n+d-1}(TC) & \longrightarrow & \widetilde{F}^{n+d}(TY) & \longrightarrow & \widetilde{F}^{n+d}(TA) \oplus \widetilde{F}^n(TB) & \to \cdots .
\end{array}$$

Now, by the Five Lemma, φ_Y is an isomorphism.

If $\alpha|B$ is $\mathcal{F}_d$-equivalent to a trivial $\mathcal{F}_d$-object then, by the above, φ_B and φ_C are isomorphisms provided B is finite dimensional.

(ii) By III.4.18, we have the following diagram, where $T_r := T(\alpha|X_r)$ and $\varphi_r := \varphi_{X_r}$:

$$\begin{array}{ccccccc}
0 \longrightarrow & \varprojlim^1\{F^{n-1}(X_r)\} & \longrightarrow & F^n(X) & \longrightarrow & \varprojlim\{F^n(X_r)\} & \longrightarrow 0 \\
& \cong \downarrow \varprojlim^1\{\varphi_r\} & & \downarrow \varphi_X & & \cong \downarrow \varprojlim\{\varphi_r\} & \\
0 \longrightarrow & \varprojlim^1\{\widetilde{F}^{n+d-1}(T_r)\} & \longrightarrow & \widetilde{F}^{n+d}(T_r) & \longrightarrow & \varprojlim\{\widetilde{F}^{n+d}(T_r)\} & \longrightarrow 0.
\end{array}$$

It is commutative by 1.2(ii). Since the left and right vertical arrows are isomorphisms, φ_X is an isomorphism. ⊞

For pedantic persons, we remark that TC, etc. in (i) and T_r in (ii) are not CW-spaces in general. Nevertheless, all used sequences are exact, cf. II.3.31 and III.4.20(a).

We continue the proof of the theorem. Consider the subsets

$$D^n_+ = \{x \in D^n \mid ||x|| \geq 1/2\},\ D^n_- = \{x \in D^n \mid ||x|| \leq 1/2\}$$

of the unit disk $D^n = \{x \in \mathbb{R}^n \mid ||x|| \leq 1\}$. Given an n-dimensional cell $e^n, n > 0$, in X with the characteristic map $\chi : D^n \to X$, we set $e^n_+ = \chi(D^n_+)$ and $e^n_- = \chi(D^n_-)$. Given $n > 0$, let A_n (resp. B_n) be the union of all subsets e^n_+ (resp. e^n_-) where e^n runs over all n-cells of X. Let X_n be the n-skeleton of X. We prove by induction that φ_{X_n} is an isomorphism for every n.

The assertion is trivial for $n = 0$. Suppose that φ_{X_n} is an isomorphism. Set $Y_n = X_n \cup A_{n+1}$. Then φ_{Y_n} is an isomorphism since Y_n is a deformation retract of X_n. Now, $X_{n+1} = Y_n \cup B_{n+1}$, and $\alpha|B_{n+1}$ is $\mathcal{F}_d$-equivalent to a trivial $\mathcal{F}_d$-object since B_{n+1} is a disjoint union of contractible spaces. So, by 1.4(i), $\varphi_F = \varphi_{X_{n+1}}$ is an isomorphism.

Thus, by 1.4(ii), $\varphi = \varphi_X$ is an isomorphism.

The proof of that φ^F is an isomorphism is similar, but it is simpler because we use the direct limit instead of the inverse one. □

1.5. Corollary. *Let α be an E-oriented $\mathcal{F}_d$-object over X.*

(i) *Let A be a CW-subspace of X. Then there are the relative Thom–Dold isomorphisms*

$$\varphi_F : F^n(X, A) \to F^{n+d}(T\alpha, T(\alpha|A)),$$
$$\varphi^F : F_n(T\alpha, T(\alpha|A)) \to F_{n-d}(X, A).$$

(ii) *Let β be an $\mathcal{F}$-object over X. Then there are the Thom–Dold isomorphisms*

$$\varphi_F : \widetilde{F}^n(T\beta) \to \widetilde{F}^{n+d}(T(\alpha \oplus \beta)),$$
$$\varphi^F : \widetilde{F}_{n+d}(T(\alpha \oplus \beta)) \to \widetilde{F}_n(T\beta)).$$

Proof. (i) One can prove this, just following the proof of 1.3. Another way: consider the exact sequences of pairs (X, A) and $(T\alpha, T(\alpha|A))$, map one of them to the other and use 1.3 and the Five Lemma.

(ii) We consider the cohomological case only. We set $\lambda := p^*_\beta(\alpha)$. Then $T(\alpha \oplus \beta) \simeq T(\lambda)/T(\lambda|s(X))$, where $s : X \to \operatorname{ts}\beta$ is the section. Furthermore, by 1.2(i), the canonical morphism $\lambda \to \alpha$ equips λ with an E-orientation, and we obtain an isomorphism

$$\widetilde{F}^n(T\beta) = F^n(\text{ts}(\beta), s(X)) \xrightarrow{\varphi_F} F^{n+d}(T(\lambda), T(\lambda|s(X))) = \widetilde{F}^{n+d}(T(\alpha \oplus \beta))$$

where φ_F is as in (i). □

1.6. Proposition. *Let $\tau : D \to E$ be a ring morphism of ring spectra. Then for every D-orientation u of α the element $\tau(u)$ is an E-orientation of α. Furthermore, let (F, m) be any D-module spectrum, and let (G, n) be any E-module spectrum. Let $\rho : F \to G$ be a morphism such that the diagram*

$$\begin{array}{ccc} D \wedge F & \xrightarrow{\tau \wedge \rho} & E \wedge G \\ {\scriptstyle m}\downarrow & & \downarrow{\scriptstyle n} \\ F & \xrightarrow{\rho} & G \end{array}$$

commutes. Then the diagrams

$$\begin{array}{ccc} F^*(X) & \xrightarrow{\rho_*} & G^*(X) \\ {\scriptstyle \varphi_F}\downarrow & & \downarrow{\scriptstyle \varphi_G} \\ \widetilde{F}^*(T\alpha) & \xrightarrow{\rho_*} & \widetilde{G}^*(T\alpha) \end{array} \qquad \begin{array}{ccc} F_*(X) & \xrightarrow{\rho_*} & G_*(X) \\ {\scriptstyle \varphi^F}\uparrow & & \uparrow{\scriptstyle \varphi^G} \\ \widetilde{F}_*(T\alpha) & \xrightarrow{\rho_*} & \widetilde{G}_*(T\alpha) \end{array}$$

commute. Here φ^F, φ_F are given by u and φ^G, φ_G are given by $\tau(u)$.

Proof. Decode the definitions. □

Clearly, the relative version (with (X, A) instead of X, etc.) of 1.6 holds as well.

Since every ring spectrum E is an E-module spectrum, we can put $F = E$ in 1.5 and obtain Thom–Dold isomorphisms

$$\begin{aligned} \varphi_E &: E^n(X, A) \to E^{n+d}(T\alpha, T(\alpha|A)), \\ \varphi^E &: E_n(T\alpha, T(\alpha|A)) \to E_{n-d}(X, A). \end{aligned} \tag{1.7}$$

Similarly, we can put $F = D$, $G = E$ in 1.6 and get the commutative diagrams

$$\begin{array}{ccc} D^*(X) & \xrightarrow{\tau_*} & E^*(X) \\ {\scriptstyle \varphi_D}\downarrow & & \downarrow{\scriptstyle \varphi_E} \\ \widetilde{D}^*(T\alpha) & \xrightarrow{\tau_*} & \widetilde{E}^*(T\alpha), \end{array} \qquad \begin{array}{ccc} D_*(X) & \xrightarrow{\tau_*} & E_*(X) \\ {\scriptstyle \varphi^D}\uparrow & & \uparrow{\scriptstyle \varphi^E} \\ \widetilde{D}_*(T\alpha) & \xrightarrow{\tau_*} & \widetilde{E}_*(T\alpha). \end{array}$$

Notice a curious consequence of Theorem 1.3.

1.8. Corollary. *Let (X, x_0) be a connected pointed space, and let $i : \{x_0\} \to X$ be the inclusion. We set $\widetilde{E}^*(X) := \operatorname{Ker}\{i^* : E^*(X) \to E^*(\{x_0\})\}$. Then for every $a \in \widetilde{E}^0(X)$ the element $1 + a \in E^0(X)$ is invertible in the ring $E^0(X)$.*

Proof. Clearly, $1 + a$ is an E-orientation of the 0-dimensional $\mathcal{V}$-object over X, and the corresponding Thom–Dold isomorphism coincides with the multiplication by $1 + a : E^0(X) \to E^0(X) = \widetilde{E}^0(X^+)$. □

Now we compute the number of E-orientations of any E-orientable $\mathcal{F}_d$-object α. Let $S(\alpha)$ denote the set of all E-orientations of α, and let X be the base of α.

1.9. Proposition. (i) *Suppose that X is a connected space. Choose a point $x_0 \in X$, let $i : \{x_0\} \subset X$ be the inclusion and regard $\widetilde{E}^*(X) := \widetilde{E}^*(X, x_0)$ as the subset* $\operatorname{Ker}\{i^* : E^*(X) \to E^*(\{x_0\})\}$ *of $E^*(X)$. Then every Thom–Dold isomorphism*

$$\varphi : E^0(X) \to \widetilde{E}^d(X)$$

establishes a bijection between the subset $\{1 + \widetilde{E}^0(X)\} \cup \{-1 + \widetilde{E}^0(X)\}$ of $E^0(X)$ and the set $S(\alpha)$. In other words, $S(\alpha)$ is in a bijective correspondence with $\mathbb{Z}/2 \times \bar{E}^0(X)$ where $\bar{E}^0(X) :=$ $\operatorname{Coker}\{\varepsilon^* : E^0(\mathrm{pt}) \to E^0(X)\}$ and $\varepsilon : X \to \mathrm{pt}$.

(ii) *If X is a disjoint union of its connected components X_λ, then*

$$S(\alpha) = \prod_\lambda S(\alpha|X_\lambda).$$

Proof. (i) Let $j : S^d \to T\alpha$ be a root at x_0. Choose any E-orientation u of α such that $j^*(u) = \sigma^n$ and consider the commutative diagram

$$\begin{array}{ccc} \widetilde{E}^d(T\alpha) & \xrightarrow{j^*} & \widetilde{E}^d(S^d) \\ {\scriptstyle\varphi}\uparrow & & \uparrow{\scriptstyle\varphi'} \\ E^*(X) & \xrightarrow{i^*} & E^*(\{x_0\}) \end{array}$$

where φ (resp φ') is the Thom–Dold isomorphism given by u (resp. by σ^d). Firstly, we prove that φ establishes a bijection between the sets $\{1+\widetilde{E}^0(X)\}$ and $\{v \in \widetilde{E}^d(X) | j^*v = \sigma^d\}$. Indeed, let $b = 1 + a$ for some $a \in \widetilde{E}^0(X)$. Then

$$j^*\varphi(b) = j^*\varphi(1 + a) = \varphi' i^*(1 + a) = \varphi' i^*(1) = \sigma^d.$$

Conversely, let v be an E-orientation of α with $j^*v = \sigma^d$. We set $b := \varphi^{-1}(v)$. Then

$$0 = \sigma^d - \sigma^d = j^*v - j^*u = j^*\varphi(b) - j^*\varphi(1) = j^*\varphi(b - 1) = \varphi' i^*(b - 1).$$

So, $i^*(b - 1) = 0$, i.e., $b = 1 + a$ for some $a \in \widetilde{E}^0(X)$.

Now we prove that φ establishes a bijective correspondence between the sets $\{-1+\widetilde{E}^0(X)\}$ and $\{v \in \widetilde{E}^d(X) | j^*v = -\sigma^d\}$. Indeed,

$$\{b \in \{-1+\widetilde{E}^0(X)\}\} \iff \{-b \in \{1+\widetilde{E}^0(X)\}\} \iff \{j^*\varphi(-b) = \sigma^d\}$$
$$\iff \{j^*\varphi(b) = -\sigma^d\}.$$

Finally, since X is connected, we conclude that $S(\alpha) = \{v | j^*(v) = \pm\sigma^d\}$.

(ii) This is obvious. □

1.10. Proposition. (i) *If α is an E-orientable $\mathcal{F}$-object over Y then $f^*\alpha$ is E-orientable for every map $f : X \to Y$.*

(ii) *If any two of three $\mathcal{F}$-objects $\alpha, \beta, \alpha \wedge^h \beta$ are E-orientable, then so is the third one.*

(iii) *If any two of three $\mathcal{F}$-objects $\alpha, \beta, \alpha \oplus \beta$ over the same base X are E-orientable, then so is the third one.*

(iv) *The standard trivial $\mathcal{F}_n$-object θ^n over any space X is E-orientable for every ring spectrum E.*

Proof. (i) This holds by 1.2(i), since there is a morphism $\mathfrak{I}_{f,\alpha} : f^*\alpha \to \alpha$.

(ii) Let $u_\alpha \in \widetilde{E}^m(T\alpha), u_\beta \in \widetilde{E}^n(T\beta)$ be E-orientations of α, β respectively. Then the image of the class $u_\alpha \otimes u_\beta$ under the pairing

$$\widetilde{E}^m(T\alpha) \otimes \widetilde{E}^n(T\beta) \to \widetilde{E}^{m+n}(T\alpha \wedge T\beta)$$

is an E-orientation of $\alpha \wedge^h \beta$. Conversely, let $\alpha \wedge^h \beta$ and α be E-oriented $\mathcal{F}$-objects. Set $v = \varphi^E(1) \in \widetilde{E}^*(T\alpha)$. Then the image of $u_{\alpha\wedge^h\beta} \otimes v$ under the homomorphism

$$\mu_{T\beta}^{\bullet, T\alpha} : \widetilde{E}^*(T\alpha \wedge^h T\beta) \otimes \widetilde{E}_*(T\alpha) \to \widetilde{E}^*(T\beta)$$

(see II.(3.40)) is an E-orientation of β. Indeed, one can check it by simple verification on fibers, i.e., for the case when the bases are points.

(iii) The E-orientability of $\alpha \oplus \beta$ follows from (i) and (ii). Suppose now that $\alpha \oplus \beta$ and α are E-oriented. Then the isomorphism

$$\widetilde{E}^*(T(\alpha \oplus \beta)) \cong \widetilde{E}^*(T\beta).$$

(as in 1.5(ii)) maps the E-orientation of $\alpha \oplus \beta$ to an E-orientation of β. One can check it by a verification of fibers, i.e., for the case when X is a point.

(iv) Without loss of generality we can assume that X is connected. Choose a point $x_0 \in X$ and consider the maps $i : \{x_0\} \subset X$ and $\varepsilon : X \to \{x_0\}$. Then the composition

$$S^n(\{x_0\}^+) \xrightarrow{S^n(i^+)} S^nX^+ \xrightarrow{S^n(\varepsilon^+)} S^n(\{x_0\}^+)$$

is the identity map, and so $(S^ni^+)^* : E^*(S^nX^+) \to E^*(S^n(\{x_0\}^+))$ is an epimorphism for every spectrum E. By IV.5.5(iii), $T\theta^n \simeq S^nX^+$, and S^ni^+ is a root j at x_0. In other words, $j^* : E^*(T\theta^n) \to E^*(S^n)$ is an epimorphism. Thus, θ^n is E-orientable for every ring spectrum E. □

Note that if E is an Ω-spectrum and if $Y = E_n$ and $A = \{\pm 1\}$, then an (A, Y)-marking is just an E-orientation. Thus, we can apply IV.5.11 to the classification of E-oriented $\mathcal{V}$-objects and prove the following fact.

1.11. Theorem. *There exists a space $B(\mathcal{V}_n, E)$ which classifies E-oriented $\mathcal{V}_n$-objects. Thus, for every X the set of all equivalence classes of E-oriented $\mathcal{V}_n$-objects over X is in a natural bijective correspondence with the set $[X, B(\mathcal{V}_n, E)]$. The construction $B(\mathcal{V}_n, E)$ is natural with respect to morphisms $E \to F$ preserving the elements ± 1 of the coefficient rings. The homotopy fiber of the forgetful map $p : B(\mathcal{V}_n, E) \to B\mathcal{V}_n$ is $\Omega^\infty_{\pm 1} E$, i.e., the union of the components of $\Omega^\infty E$ which correspond to $\{\pm 1\} \subset \pi_0(\Omega^\infty E) = \pi_0(E)$.* □

Now we consider orientation theory for stable objects.

1.12. Definition. Let $T\alpha$ be the Thom spectrum of a stable $\mathcal{F}$-object α over X. (Recall that, according to IV.5.12, X is assumed to be a CW-complex.) If X is connected, define an *E-orientation* of α to be an element $u \in \widetilde{E}^0(T\alpha)$ such that $j^*(u) = \pm 1 \in \pi_0(E)$, where $j : S \to T\alpha$ is a root of $T\alpha$.

If $X = \sqcup X_\lambda$ with connected X_λ, then an E-orientation of α is a family $\{u_\lambda\}$, where u_λ is an E-orientation of $\alpha | X_\lambda$. Furthermore, an E-orientation of a stable $\mathcal{V}$-object ξ is defined to be an E-orientation of $\xi^\bullet$.

For every $\mathcal{F}_n$-object α the isomorphism $\mathfrak{e} : T(\alpha_{\mathrm{st}}) \to \Sigma^{-n}\Sigma^\infty T\alpha$ in IV.5.16 induces an isomorphism $\mathfrak{e}^* : \widetilde{E}^n(T\alpha) = E^0(\Sigma^{-n}\Sigma^\infty T\alpha) \to E^0(T\alpha_{\mathrm{st}})$.

1.13. Proposition. *The isomorphism $\mathfrak{e}^*$ yields a bijective correspondence between E-orientations of α and α_{st}. Hence, α is E-orientable iff α_{st} is.*

Proof. It suffices to consider α over a connected base. Choose roots $j_1 : S \to T(\alpha_{\mathrm{st}})$ and $j_2 : S^n \to T\alpha$. Then $\mathfrak{e} j_1$ and $\Sigma^{-n}\Sigma^\infty j_2$ are homotopic (up to sign), and the result holds because $\mathfrak{e}^*$ is an isomorphism. □

Let α be a stable $\mathcal{F}$-object over X. Consider the morphism $\Delta : T\alpha \to T\alpha \wedge X^+$ as in IV.5.36. Let (F, m) be any E-module spectrum. Define

$$\overline{\varphi} : E^0(T\alpha) \otimes F^i(X) = E^0(T\alpha) \otimes \widetilde{F}^i(X^+) \xrightarrow{m^{T\alpha, X^+}} F^i(T\alpha \wedge X^+) \xrightarrow{\Delta^*} F^i(T\alpha),$$

$$\underline{\varphi} : E^0(T\alpha) \otimes F_i(T\alpha) \xrightarrow{1 \otimes \Delta_*} E^0(T\alpha) \otimes F_i(T\alpha \wedge X^+) \xrightarrow{m^{T\alpha}_{\bullet, X^+}} \widetilde{F}_i(X^+) = F_i(X).$$

Now suppose that α is equipped with an E-orientation $u \in E^0(T\alpha)$. Define

$$\varphi_F : F^i(X) \to F^i(T\alpha),\ \varphi_F(x) = \overline{\varphi}(u \otimes x),$$
$$\varphi^F : F_i(T\alpha) \to F_i(X),\ \varphi^F(x) = \underline{\varphi}(u \otimes x).$$

1.14. Theorem. *The homomorphisms φ_F and φ^F are isomorphisms for every CW-space X.*

Proof. We consider only the cohomology case. Let α^n be as in IV.5.19.

Step 1. Let X be finite dimensional. Then one has the commutative diagram

$$\begin{array}{ccccccc}\cdots \longleftarrow & F^i(X^{(n-2)}) & \longleftarrow & F^i(X^{(n-1)}) & \longleftarrow \cdots \\ & \cong\downarrow & & \downarrow\cong & \\ \cdots \longleftarrow & \widetilde{F}^{i+n}(T\alpha^n) & \longleftarrow & \widetilde{F}^{i+n+1}(T\alpha^{n+1}) & \longleftarrow \cdots,\end{array}$$

which stabilizes as $n \to \infty$ and gives the desired isomorphism.

Step 2. Let X be a disjoint union of finite dimensional spaces, $X = \sqcup X_\lambda$. Then $T\alpha = \vee_\lambda T(\alpha|X_\lambda)$, and so (cf. II.3.16(c))

$$F^i(T\alpha) \cong \prod_\lambda F^i(T(\alpha|X_\lambda)) \cong \prod_\lambda F^i(X_\lambda) \cong F^i(\sqcup X_\lambda) \cong F^i(X).$$

The second isomorphism is given by Step 1.

Step 3. Consider the telescope $T = T_{\text{ev}} \cup T_{\text{od}}$ of the skeletal filtration of X. We have

$$T \simeq X, T_{\text{ev}} \simeq \bigvee_{n=0}^{\infty} X^{(2n)}, T_{\text{od}} \simeq \bigvee_{n=0}^{\infty} X^{(2n+1)}, T_{\text{ev}} \cap T_{\text{od}} \simeq \bigvee_{n=0}^{\infty} X^{(n)}.$$

Let $h : T \to X$ be the canonical homotopy equivalence. Set $\overline{\alpha} = h^*\alpha, \alpha_{\text{ev}} = \overline{\alpha}|T_{\text{ev}}, \alpha_{\text{od}} = \overline{\alpha}|T_{\text{od}}$. It is easy to see that $T\alpha_{\text{ev}} \cap T\alpha_{\text{od}} \simeq T(\overline{\alpha}|(T_{\text{ev}} \cap T_{\text{od}}))$. Consider the following commutative diagram of Mayer–Vietoris exact sequences:

$$\begin{array}{ccccccccc}\cdots \longrightarrow & F^{k-1}(T\alpha_{\text{ev}} \cap T\alpha_{\text{od}}) & \longrightarrow & F^k(T\overline{\alpha}) & \longrightarrow & F^k(T\alpha_{\text{ev}}) \oplus F^k(T\alpha_{\text{od}}) & \longrightarrow \cdots \\ \cong\uparrow\varphi'' & \cong\uparrow\varphi' & & \uparrow\varphi_F & & \cong\uparrow\varphi'' & \cong\uparrow\varphi' \\ \cdots \longrightarrow & F^{k-1}(T_{\text{ev}} \cap T_{\text{od}}) & \longrightarrow & F^k(T) & \longrightarrow & F^k(T_{\text{ev}}) \oplus F^k(T_{\text{od}}) & \longrightarrow \cdots.\end{array}$$

Now, by Step 2, φ' and φ'' are isomorphisms, so, by the Five Lemma, φ_F is also an isomorphism. □

The Thom–Dold isomorphism φ^F can be lifted to a geometric level.

1.15. Theorem (Mahowald–Ray [1]). *Let α be a stable $\mathcal{F}$-object over X equipped with an E-orientation $u : T\alpha \to E$. Then the morphism*

$$T\alpha \wedge F \xrightarrow{\Delta \wedge 1} T\alpha \wedge X^+ \wedge F \xrightarrow{\tau \wedge 1} X^+ \wedge T\alpha \wedge F \xrightarrow{1 \wedge u \wedge 1} X^+ \wedge E \wedge F \xrightarrow{1 \wedge m} X^+ \wedge F$$

is an equivalence.

Proof. This morphism induces a homomorphism of homotopy groups which coincides (up to sign) with the isomorphism φ^F. □

Let $p : \widehat{E} \to E$ be a connective covering of a spectrum E.

1.16. Proposition. *An $\mathcal{F}$-object α is E-orientable iff it is $\widehat{E}$-orientable. Furthermore, for every $\mathcal{F}$-object α the morphism p induces a bijection between $\widehat{E}$-orientations of α and E-orientations of it.*

Proof. Firstly, let α be a stable $\mathcal{F}$-object. By IV.5.8(ii), $T\alpha$ is connected. Hence, by II.4.16, every E-orientation $u : T\alpha \to E$ can be lifted to $\widehat{E}$, and this lifting is an $\widehat{E}$-orientation because $p_* : \pi_0(\widehat{E}) \to \pi_0(E)$ is an isomorphism. Furthermore, again by II.4.16, the morphism p yields the desired bijection. So, the proposition holds for stable $\mathcal{F}$-objects. Finally, it holds for $\mathcal{F}_n$-objects because of 1.13. □

Considering $A = \{\pm 1\} \subset \pi_0(E)$, we conclude that an (A, E)-marking of a stable $\mathcal{V}$-object (or $\mathcal{F}$-object) ξ is just an E-orientation of ξ. Let $t_{(\mathcal{V},E)}(X)$ be the set of all equivalence classes of E-oriented (i.e., (A, E)-marked) stable $\mathcal{V}$-objects over X. Now, IV.5.29 yields the following theorem.

1.17. Theorem. *There exist a CW-space $B(\mathcal{V}, E)$ and an E-oriented stable $\mathcal{V}$-object $\eta = \eta_{\mathcal{V},E}$ over $B(\mathcal{V}, E)$ such that the map $[X, B(\mathcal{V}, E)] \to t_{(\mathcal{V},E)}(X), f \mapsto f^*\eta$ is bijective for every finite dimensional CW-space X. In other words, $B(\mathcal{V}, E)$ classifies E-oriented stable $\mathcal{V}$-objects over finite dimensional CW-spaces. The homotopy fiber of the forgetful map $p : B(\mathcal{V}, E) \to B\mathcal{V}$ is $\Omega^\infty_{\pm 1} E$. Furthermore, for every CW-space X the map $[X, B(\mathcal{V}, E)] \to t_{(\mathcal{V},E)}(X)$ is surjective.* □

We set $M(\mathcal{V}, E) := T\eta$. The E-orientation u_η of η is called a *universal E-orientation* for stable $\mathcal{V}$-objects, and η is called a *universal E-oriented stable $\mathcal{V}$-object.*

1.18. Theorem (naturality with respect to E). *Given a ring morphism $\tau : D \to E$ of ring spectra, there are maps $B(\mathcal{V}, \tau) : B(\mathcal{V}, D) \to B(\mathcal{V}, E)$ and $M(\mathcal{V}, \tau) : M(\mathcal{V}, D) \to M(\mathcal{V}, E)$ such that $B(\mathcal{V}, \tau)^*\eta_E \simeq \eta_D$ and the following diagrams commute up to homotopy:*

$$\begin{array}{ccc} \Omega^\infty_{\pm 1} D & \longrightarrow & \Omega^\infty_{\pm 1} E \\ \downarrow & & \downarrow \\ B(\mathcal{V}, D) & \xrightarrow{B(\mathcal{V},\tau)} & B(\mathcal{V}, E) \\ \downarrow & & \downarrow \\ B\mathcal{V} & = & B\mathcal{V} \end{array} \qquad \begin{array}{ccc} M(\mathcal{V}, D) & \xrightarrow{M(\mathcal{V},\tau)} & M(\mathcal{V}, E) \\ {\scriptstyle u_D}\downarrow & & \downarrow{\scriptstyle u_E} \\ D & \xrightarrow{\tau} & E. \end{array}$$

Proof. The composition $M(\mathcal{V}, D) \xrightarrow{u_D} D \xrightarrow{\tau} E$ is an E-orientation of η_D, and hence, by 1.17, it yields a map $B(\mathcal{V}, \tau) : B(\mathcal{V}, D) \to B(\mathcal{V}, E)$ with $B(\mathcal{V}, \tau)^* \eta_E \simeq \eta_D$. So, we get the map

$$M(\mathcal{V}, \tau) := T\mathfrak{I}_{B(\mathcal{V},\tau),\eta_E} : M(\mathcal{V}, D) \to M(\mathcal{V}, E).$$

Now, the right diagram commutes because $M(\mathcal{V}, \tau)^*(u_E)$ is the E-orientation τu_D of η_D.

Now we prove that the left diagram commutes. It suffices to prove that the maps $[X, \Omega^\infty_{\pm 1} D] \to [X, \Omega^\infty_{\pm 1} E] \to [X, B(\mathcal{V}, E)]$ and $[X, \Omega^\infty_{\pm 1} D] \to [X, B(\mathcal{V}, D)] \to [X, B(\mathcal{V}, E)]$ coincide for every X. But each of these maps treats a trivial E-oriented $\mathcal{V}$-object as a certain E-oriented $\mathcal{V}$-object. □

1.19. Remarks. (a) Dold [1] proved Theorem 1.3.

(b) As we remarked in Ch. IV, the classifying spaces $B(\mathcal{V}_n, E)$ were introduced by May [2].

(c) Sometimes one defines an E-orientation by the condition $j^*(u) = \varepsilon\sigma^n$, where ε is an invertible element of the ring $\pi_0(E)$, see May [3] or Switzer [1]. Certainly, the class of E-orientable $\mathcal{V}$-objects in this case is just the same as in our case. Furthermore, this situation is in some sense a direct sum of ours. For example, in this case the classifying space $B(\mathcal{V}, E)$ is just the disjoint union of copies of ours.

(d) There are some reasons to write simply ab instead of $\overline{\varphi}(a \otimes b)$, $a \in \widetilde{E}^*(T\alpha), b \in F^*(X)$, cf. II.3.43. Then we can write $\varphi_F(x) = u_E x$, and, for instance, the commutativity of the left diagram after (1.7) can be expressed as $\tau(ux) = \tau(u)\tau(x)$.

(e) We have seen above that one can consider an E-orientation as a structure on $\mathcal{V}$-objects. Moreover, we say that a structure map $\varphi : B \to B\mathcal{V}$ is E-orientable if there exists a map $(B, \varphi) \to B(\mathcal{V}, E)$ over $B\mathcal{V}$, i.e., if $\varphi^*\gamma_\mathcal{V}$ is E-oriented. In this case we have a morphism $T(B, \varphi) \to M(\mathcal{V}, E)$, which induces an E-orientation $u \in E^0(T(B, \varphi))$ of $\varphi^*\gamma_\mathcal{V}$. So, in this case every (B, φ)-structured $\mathcal{V}$-object gets a certain E-orientation.

(f) Similarly to 1.15, the cohomological Thom–Dold isomorphism can also be lifted to the spectra level. This "geometric lifting" has the form $\varphi : F(X^+, E) \to F(T\alpha, E)$ where $F(-,-)$ is the functional spectrum. Moreover, an analogous "geometric lifting" also exists for Thom–Dold isomorphisms as in 1.5 (see Lewis–May–Steinberger [1], p. 436).

Now we consider the relations between E- and $E[p]$-orientability, where p runs over all primes and $E[p]$ is the $\mathbb{Z}[p]$-localization of E.

Let q be the order of the element $1 \in \pi_0(E)$ in the additive group $\pi_0(E)$, $0 \le q < \infty$.

1.20. Proposition. *An $\mathcal{F}$-object α is E-orientable iff it is $E[p]$-orientable for all primes p such that $p|q$.*

Proof. By 1.13, we can concentrate our attention on stable $\mathcal{F}$-objects only, and we can assume that $\operatorname{bs}\alpha$ is connected. By 1.6 and II.5.15(i), E-orientability implies $E[p]$-orientability. We prove the converse. Let $j_p^* := E[p](j) : \widetilde{E[p]}^n(T\alpha) \to \widetilde{E[p]}^n(S^n)$ be the homomorphism induced by the root $j : S^n \to T\alpha$.

Firstly, suppose that $q = 0$. Then for every p there exists an element $\widetilde{v}_p \in \widetilde{E[p]}^n(T\alpha)$ such that $j_p^*\widetilde{v}_p = \sigma^n \otimes 1 \in \widetilde{E}^n(S^n)\otimes\mathbb{Z}[p] = \widetilde{E[p]}^n(S^n)$. Hence, for every p there exists an element $v_p \in \widetilde{E}^n(T\alpha)$ such that $j^*v_p = a_p\sigma^n + y_p$, where $a_p \in \mathbb{Z}, (a_p, p) = 1$, and $y_p \in \widetilde{E}^n(T\alpha)$ is such that $m_p y_p = 0$ for some $m_p \in \mathbb{Z}$ with $(m_p, p) = 1$. Set $u_p := m_p v_p$. Then $j^*u_p = a_p m_p \sigma^n = b_p \sigma^n$ with $(b_p, p) = 1$. Since $(b_p, p) = 1$, there exists a finite set $\{p_1, ..., p_k\}$ of primes such that $(b_{p_1}, ..., b_{p_k}) = 1$. Hence, there exists a finite set $\{x_1, ..., x_k\}$ of integers such that $\sum x_i b_{p_i} = 1$. Now, $j^*u = \sigma^n$ if $u = \sum x_i u_{p_i}$, i.e., α is E-orientable.

The case $q \neq 0$ is similar; the only difference is that we have the equality $\sum x_i b_{p_i} = t$, where $(t, q) = 1$. Let $s \in \mathbb{Z}$ be such that $st \equiv 1 \bmod q$. Then $j^*u = \sigma^n$ if $u = s\sum x_i u_{p_i}$. □

1.21. Proposition. *Let E be a ring spectrum, and let $l : E \to E_\Lambda$ be the localization with respect to a subring Λ of $\mathbb{Q}$. Let $u = u_E : M(\mathcal{V}, E) \to E$ be the universal E-orientation, and let $u_{E_\Lambda} : M(\mathcal{V}, E_\Lambda) \to E_\Lambda$ be the universal E_Λ-orientation. Set $\hat{u} = (u_{E_\Lambda})_\Lambda$, and let $l_\# := M(\mathcal{V}, l)$ be as in* 1.18. *Then the diagram*

$$\begin{array}{ccc} M(\mathcal{V}, E)_\Lambda & \xrightarrow{u_\Lambda} & E_\Lambda \\ {\scriptstyle (l_\#)_\Lambda}\big\downarrow & & \big\downarrow{\scriptstyle =} \\ M(\mathcal{V}, E_\Lambda)_\Lambda & \xrightarrow{\hat{u}} & E_\Lambda \end{array}$$

commutes up to homotopy, i.e., $\hat{u}l_\# \simeq u_\Lambda$.

Proof. By 1.18, the following diagram commutes up to homotopy:

$$\begin{array}{ccc} M(\mathcal{V}, E) & \xrightarrow{u} & E \\ {\scriptstyle l_\#}\big\downarrow & & \big\downarrow{\scriptstyle l} \\ M(\mathcal{V}, E_\Lambda) & \xrightarrow{u_{E_\Lambda}} & E_\Lambda. \end{array}$$

The Λ-localization of this diagram is the desired diagram since $l_\Lambda = 1_{E_\Lambda}$. □

1.22. Definition. Let α be an $\mathcal{F}_d$-object over a CW-space X, and let $u \in \widetilde{H}^d(T\alpha; \mathbb{Z}/2)$ be the Thom class of α. We define the *i-th Stiefel–Whitney class of α* by setting

$$w_i(\alpha) := \varphi_{H\mathbb{Z}/2}^{-1} Sq^i u \in H^i(X; \mathbb{Z}/2).$$

Similarly, given a stable $\mathcal{F}$-object α, we set $w_i(\alpha) := \varphi^{-1}Sq^i u$ where u is the stable Thom class and φ is the Thom isomorphism as in IV.5.23(ii). Finally, given a $\mathcal{V}$-object ξ, we set $w_i(\xi) := w_i(\xi^\bullet)$.

1.23. Examples. (a) Every $\mathcal{F}$-object is $H\mathbb{Z}/2$-orientable, see § IV.2. Vice versa, if a ring spectrum E is such that every $\mathcal{F}$-object is E-orientable, then there exists a morphism $MO \to E$ and hence, by IV.6.2, a morphism $H\mathbb{Z}/2 \to E$ compatible with the units. Hence, by II.7.7, E is a graded Eilenberg–Mac Lane spectrum and $2\pi_*(E) = 0$.

(b) By IV.5.8(ii), orientability as defined in IV.5.6 is just $H\mathbb{Z}$-orientability. It is easy to see that $B(\mathcal{V}_n, H\mathbb{Z})$ is just the two-sheeted (universal) covering $BS\mathcal{V}_n$ over $B\mathcal{V}_n$. In particular, a vector bundle is $H\mathbb{Z}$-orientable iff its structure group can be reduced to SO. Observe that $B(S\mathcal{V}_n, H\mathbb{Z})$ is just the disjoint union of two copies of $BS\mathcal{V}_n$. Besides, $H\mathbb{Z}$-orientability of any $\mathcal{F}$-object ξ is equivalent to the equality $w_1(\xi) = 0$. This holds because $BS\mathcal{V}$ can be obtained from $B\mathcal{V}$ just by killing the class w_1. Alternatively, this follows from IV.6.3 since $w_1(\xi) = \varphi^{-1}Sq^1 u_\xi$.

(c) Atiyah–Bott–Shapiro [1] proved that a vector bundle ξ is KO-orientable iff it admits a Spin-structure. This holds, in turn, iff $w_1(\xi) = 0 = w_2(\xi)$. This condition is purely homotopic and can be formulated for every $\mathcal{F}$-object. It is necessary for KO-orientability of any $\mathcal{F}$-object, but it is not sufficient for KO-orientability of $\mathcal{PL}$-bundles, see Ch. VI. One the other hand, Sullivan proved that every $\mathcal{STOP}$-bundle is $KO[1/2]$-orientable, see Madsen–Milgram [1] for a good proof.

(d) The complexification $C : BO \to BU$ induces a ring morphism $KO \to K$, see VI.3.3 below. So, every KO-orientable $\mathcal{F}$-object is K-orientable. Atiyah–Bott–Shapiro [1] proved that a vector bundle ξ is K-orientable iff it admits a $\mathrm{Spin}^{\mathbb{C}}$-structure. The last condition is equivalent to the purely homotopic conditions $w_1(\xi) = 0 = \delta w_2(\xi)$, where δ is the connecting homomorphism in the Bockstein exact sequence

$$\cdots \to H^*(X) \xrightarrow{2} H^*(X) \xrightarrow{\bmod 2} H^*(X;\mathbb{Z}/2) \xrightarrow{\delta} H^*(X) \to \cdots .$$

As in the KO-case, this condition is necessary for K-orientability of any $\mathcal{F}$-object, but it is not sufficient for K-orientability of $\mathcal{PL}$-bundles, see Ch. VI. One the other hand, every $\mathcal{STOP}$-bundle is $K[1/2]$-orientable in view of Sullivan's result mentioned in example (c).

(e) An $\mathcal{F}_d$-object α over a finite CW-space is orientable with respect to the sphere spectrum S iff it has trivial stable fiber homotopy type, i.e., iff there exists N such that $\alpha \oplus \theta^N$ is equivalent to θ^{N+d}. The simple proof (for vector bundles, but this does not matter) can be found in Husemoller [1], Ch. XV, Th. 7.7.

By 1.6, an S-orientable α is E-orientable for every ring spectrum E, cf. also 1.10(iv). So, (a) and (e) appear as two extremal cases.

1.24. Definition. Let (ξ, u) be an E-oriented $\mathcal{V}_n$-object over X. The *Euler class* of ξ is

$$\chi(\xi) = \chi^E(\xi, u) := \varepsilon^* \mathfrak{z}^* u \in \widetilde{E}^n(X^+) = E^n(X),$$

where $\mathfrak{z} : X \to T\xi$ is the zero section as in IV.5.4, and $\varepsilon : X^+ \to X$ is a map such that $\varepsilon | X = 1_X$.

1.25. Theorem (The Gysin exact sequence). *Let $\xi = \{p : Y \to X\}$ be any E-oriented S^{n-1}-fibration (i.e., $\mathcal{G}_n$-object). Then there exists an exact sequence*

$$\cdots \to E^k(X) \xrightarrow{\chi} E^{k+n}(X) \xrightarrow{p^*} E^{k+n}(Y) \to E^{k+1}(X) \to \cdots,$$

where χ denotes the multiplication by the Euler class $\chi = \chi(\xi)$.

Proof. Since $T\xi \simeq C(p)$, we have a long cofiber sequence $Y \xrightarrow{p} X \xrightarrow{\mathfrak{z}} T\xi \to \cdots$. It yields a long cofiber sequence

$$Y^+ \xrightarrow{p^+} X^+ \xrightarrow{\hat{\mathfrak{z}}} T\xi \to \cdots.$$

where $\hat{\mathfrak{z}} = \mathfrak{z}\varepsilon$. This sequence induces an exact sequence

$$\cdots \to \widetilde{E}^k(T\xi) \xrightarrow{\hat{\mathfrak{z}}^*} E^k(X) \xrightarrow{(p^+)^*} E^k(Y) \to \cdots.$$

Now, the composition $E^{k-n}(X) \xrightarrow{\varphi} \widetilde{E}^k(T\xi) \xrightarrow{\hat{\mathfrak{z}}^*} E^k(X)$ coincides with χ, because

$$\hat{\mathfrak{z}}^*(\varphi(x)) = \hat{\mathfrak{z}}^*(ux) = \hat{\mathfrak{z}}^*(u)x = \chi x,$$

the second equality holding because of commutativity of the diagram

$$\begin{array}{ccc} T\xi & \xrightarrow{\Delta^n} & T\xi \wedge X^+ \\ {\scriptstyle \hat{\mathfrak{z}}}\uparrow & & \uparrow{\scriptstyle \hat{\mathfrak{z}}\wedge 1} \\ X^+ & \xrightarrow{d^+} & X^+ \wedge X^+ \end{array}$$

with Δ^n as in IV.5.36.

If we replace $\widetilde{E}^k(T\xi)$ by $E^{k-n}(X)$, we get the desired exact sequence. □

1.26. Proposition. (i) *Let $f : Z \to X$ be a map, and let ξ be any E-oriented $\mathcal{V}_n$-object over X. Then $\chi(f^*\xi) = f^*(\chi(\xi))$ provided $f^*\xi$ is equipped with the induced orientation.*

(ii) *Let ξ be an E-oriented $\mathcal{V}_m$-object over X, and let η be an E-oriented $\mathcal{V}_n$-object over Y. Assume that $\xi \times \eta$ (or $\xi * \eta$ for $\mathcal{V} = \mathcal{G}$) is equipped with the product E-orientation (see* 1.10(ii)*). Then $\chi(\xi \times \eta) = \mu(\chi(x), \chi(\eta))$, where $\mu : E^*(X) \otimes E^*(Y) \to E^*(X \times Y)$ is given by the multiplication in E.*

(iii) *Let ξ, η be two E-oriented $\mathcal{V}$-objects over X. Assume that $\xi \oplus \eta$ is equipped with the sum E-orientation as in* 1.10(iii). *Then $\chi(\xi \oplus \eta) = \chi(\xi)\chi(\eta)$.*

(iv) *Let $\xi = \{p : Y \to X\}$ be an $H\mathbb{Z}$-oriented S^{n-1}-fibration over a connected base X. If $E = H\mathbb{Z}$, then $\chi(\xi)$ coincides up to sign with the characteristic class of ξ, i.e., $\chi(\xi) = \pm\tau\iota$, where $\tau : H^{n-1}(F) \to H^n(X)$ is the transgression, $\iota \in H^{n-1}(F) = \mathbb{Z}$ is a generator and $F = F_x \simeq S^{n-1}$ is the fiber of ξ; here x is an arbitrary point of X.*

Proof. The properties (i)–(iii) are clear. We prove (iv). We denote $\chi(\xi)$ just by χ. Put $\varkappa = \tau\iota$. Consider the following diagram, where the bottom line is the exact sequence of the pair (Y, F):

$$\begin{array}{ccccc} & & H^n(X) & = & H^n(X) \\ & & \downarrow{\scriptstyle \bar{p}^*} & & \downarrow{\scriptstyle p^*} \\ H^{n-1}(F) & \xrightarrow{\delta} & H^n(Y,F) & \xrightarrow{k^*} & H^n(Y). \end{array}$$

We have $\tau = (\bar{p}^*)^{-1}\delta$, i.e $\varkappa = (\bar{p}^*)^{-1}\delta\iota$. By 1.25, the group $\operatorname{Ker} p^*$ is cyclic (because $H^0(X) = \mathbb{Z}$), and χ generates this cyclic group. Since $p^*\varkappa = 0$, $\varkappa = m\chi$ for some $m \in \mathbb{Z}$.

On the other hand, $\bar{p}^*\varkappa$ generates a cyclic group $\operatorname{Im}\delta$. Since $k^*\bar{p}^*\chi = p^*\chi = 0$, $\bar{p}^*\chi \in \operatorname{Im}\delta$, and so $\chi = m'\varkappa$ for some $m' \in \mathbb{Z}$. Thus, $\chi = \pm\varkappa$. □

1.27. Proposition. *Let (ξ, u) be an E-oriented $\mathcal{V}_n$-object over X, and let $\mathfrak{z} : X \to T\xi$ and $\varepsilon : X^+ \to X$ be as in* 1.24. *Then $\varepsilon^*\mathfrak{z}^*(\varphi(x)) = \chi(\xi)x$ for every $x \in E^*(X)$ where $\varphi : E^*(X) \to \widetilde{E}^*(T\xi)$ is a Thom–Dold isomorphism with respect to u.*

Proof. Let $d : X \to X \times X$ be the diagonal, and let $\Delta^n : T\xi \to T\xi \wedge X^+$ be as in IV.5.36. Let $\mathfrak{z}_1 : X \times X \to T\xi \wedge X^+$ be the zero section for $\xi \times \theta^0$ (i.e., for $p_1^*\xi$, see IV.5.36), and let $i : X \to X^+$ be the inclusion. Consider the commutative diagram

$$\begin{array}{ccccc} \widetilde{E}^*(T\xi) \otimes \widetilde{E}^*(X^+) & \xrightarrow{\mu} & \widetilde{E}^*(T\xi \wedge X^+) & \xrightarrow{(\Delta^n)^*} & \widetilde{E}^*(T\xi) \\ \downarrow{\scriptstyle \mathfrak{z}^* \otimes i^*} & & \downarrow{\scriptstyle \mathfrak{z}_1^*} & & \downarrow{\scriptstyle \mathfrak{z}^*} \\ \widetilde{E}^*(X) \otimes \widetilde{E}^*(X) & \xrightarrow{\overline{\mu}} & \widetilde{E}^*(X \times X) & \xrightarrow{d^*} & \widetilde{E}^*(X) \end{array}$$

where $\mu := \mu_E^{T\xi, X^+}$ and $\overline{\mu} := \mu_E^{X,X}$. Given $x \in E^*(X) = \widetilde{E}^*(X^+)$, we have

$$\mathfrak{z}^*\varphi(x) = \mathfrak{z}^*\Delta^*\mu(u \otimes x) = d^*\overline{\mu}(\mathfrak{z}^* \otimes i^*)(u \otimes x) = d^*\overline{\mu}(\mathfrak{z}^*u \otimes i^*x) = (\mathfrak{z}^*u)(i^*x).$$

Hence, $\varepsilon^*\mathfrak{z}^*\varphi(x) = (\varepsilon^*\mathfrak{z}^*u)(\varepsilon^*i^*(x)) = \chi(\xi)x$. □

1.28. Exercises. (a) Prove the following stable version of 1.5(ii). Let α, β be two stable $\mathcal{F}$-objects over X, and let α be equipped with an E-orientation u. Then there are the Thom–Dold isomorphisms

$$F^i(T\beta) \xrightarrow{u^\flat} F^i(T\alpha \wedge T\beta) \xrightarrow{\tau^*} F^i(T(\alpha \oplus \beta)),\ u^\flat(x) = m^{T\alpha, T\beta}(u \otimes x),$$
$$F_i(T(\alpha \oplus \beta)) \xrightarrow{\tau_*} F_i(T\alpha \wedge T\beta) \xrightarrow{u_\flat} F_i(T\beta),\ u_\flat(x) = m^{T\alpha}_{\bullet, T\alpha\wedge T\beta}(u \otimes x),$$

where $\tau = T\mathfrak{I}_{d,\alpha\wedge^h\beta} : T(\alpha \oplus \beta) \to T\alpha \wedge T\beta$ and $d : X \to X \times X$ is the diagonal.

(b) Let α be a stable $\mathcal{F}$-object over X equipped with an E-orientation $u : T\alpha \to E$ where E is a commutative ring spectrum, and let $f : X \to B\mathcal{F}$ classify α. Let X be equipped with a homotopy associative multiplication $\nu : X \times X \to X$, and let $f : X \to B\mathcal{F}$ respect the multiplications. Clearly, the pairing $E_*(X) \otimes E_*(X) \to E_*(X \times X) \xrightarrow{\nu_*} E_*(X)$ turns $E_*(X)$ into a ring. Furthermore, $\nu^*\alpha \simeq \alpha \wedge^h \alpha$, and so, by IV.5.21(i), ν yields a pairing (possibly non-associative) $T\alpha \wedge T\alpha \to T\alpha$. So, similarly to above, $E_*(T\alpha)$ turns out to be a "non-associative ring". Nevertheless, prove that

$$\varphi^E : E_*(T\alpha) \to \widetilde{E}_*(X)$$

is a ring isomorphism. In particular, $E_*(T\alpha)$ is actually a ring. (Hint: consider the morphism in 1.15 and prove that it respects pairings.)

§2. Orientations of Manifolds

Let M be a topological n-dimensional manifold. Consider a point $m \in M \setminus \partial M$ and a disk neighborhood U of m. Let $\varepsilon = \varepsilon^{m,U} : M \to S^n$ be the map which collapses the complement of U. Let E be a ring spectrum, and let $s_n \in E_n(S^n, *)$ be the image of $1 \in \pi_0(E)$ under the homomorphism

$$\pi_0(E) = \widetilde{E}_0(S^0) \cong \widetilde{E}_n(S^n) = E_n(S^n, *).$$

2.1. Definition. Let M be a compact topological manifold. An element $[M, \partial M] = [M, \partial M]_E \in E_n(M, \partial M)$ is called an *orientation of M with respect to E*, or, briefly, an *E-orientation* of M, if $\varepsilon^{m,U}_*[M, \partial M] = \pm s_n$ for every m and every disk neighborhood U of m.

A manifold with a fixed E-orientation is called E-oriented, and a manifold which admits an E-orientation is called E-orientable. So, an E-oriented manifold is in fact a pair $(M, [M]_E)$.

Clearly, a connected manifold M is $H\mathbb{Z}$-orientable iff $H_n(M, \partial M) = \mathbb{Z}$, i.e., iff M is orientable in the classical sense.

Note that s_n is a canonical E-orientation of the sphere S^n.

2.2. Proposition. *Let M be a connected manifold, and let U_0 be a disk neighborhood of a point $m_0 \in M$. If an element $[M, \partial M] \in E_n(M)$ is such that $\varepsilon_*^{m_0, U_0}[M, \partial M] = \pm s_n$, then $[M, \partial M]$ is an E-orientation of M.*

Proof. If a disk neighborhood V of m_0 satisfies $\overline{V} \subset U_0$, then $\varepsilon^{m_0, U_0} \simeq \varepsilon^{m_0, V}$. Hence, $\varepsilon^{m,U} \simeq \varepsilon^{m,U'}$ for every $m \in M$ and every pair of disk neighborhoods U, U' of m. Connect an arbitrary point $m \in M$ with m_0 by some arc (homeomorphic to I) and consider a neighborhood W of this arc such that W is homeomorphic to a disk. Let $V_0 \subset W, V \subset W$ be disk neighborhoods of m_0 and m, respectively. Then $\varepsilon^{m_0, V_0} \simeq \varepsilon^{m,V}$, and hence $\varepsilon^{m_0, U_0} \simeq \varepsilon^{m,U}$ for every pair (m, U). □

Consider an embedding of a closed manifold M^n in $\mathbb{R}^{N+n}$ and a tubular neighborhood (U, q, ν^N) of this embedding. The diagonal $d : M \to M \times M$ induces the map $\Delta^N : T\nu^N \to T\nu^N \wedge M^+$, see IV.5.36. Let $\overline{v} : S^{N+n} \to T\nu^N \wedge M^+$ be the composition

$$\overline{v} : S^{N+n} \xrightarrow{c^N} T\nu^N \xrightarrow{\Delta^N} T\nu^N \wedge M^+$$

where c^N is the Browder–Novikov map as in IV.7.15(a). As in IV.7.12, set $\nu = (\nu^N)_{\mathrm{st}}$. Then, by IV.5.16, $T\nu = \Sigma^{-N}\Sigma^\infty T\nu^N$. For simplicity, denote $\Sigma^{-n}\Sigma^\infty M^+$ by $\widehat{M}$. The map $\overline{v}$ induces a morphism

$$\begin{aligned} v := \Sigma^{-N-n}\Sigma^\infty \overline{v} : S \to \Sigma^{-N-n}\Sigma^\infty(T\nu^N \wedge M^+) \\ = \Sigma^{-N}\Sigma^\infty T\nu^N \wedge \Sigma^{-n}\Sigma^\infty M^+ = T\nu \wedge \widehat{M} \end{aligned}$$

of spectra. Furthermore, the root $j : S^N \to T\nu^N$ yields a stable root $J := \Sigma^{-N}\Sigma^\infty j : S \to T\nu$. Finally, the collapse $\varepsilon : M \to S^n$ yields a pointed map $\overline{\varepsilon} : M^+ \to S^n, \overline{\varepsilon}|M = \varepsilon$, and we set $\mathscr{E} := \Sigma^{-n}\Sigma^\infty \overline{\varepsilon} : \widehat{M} \to S$.

2.3. Theorem. (i) *For every closed manifold M, the map $\overline{v} : S^{N+n} \to T\nu^N \wedge M^+$ is an $(N+n)$-duality map between $T\nu$ and M^+. In other words, $v : S \to T\nu \wedge \widehat{M}$ is a duality morphism.*

(ii) *Let M^n be a connected closed manifold. Then the root $J : S \to T\nu$ is dual (up to sign) to the morphism $\mathscr{E}$.*

Proof. (i) By IV.7.5, $T\nu^N \simeq \mathbb{R}^{N+n}/(\mathbb{R}^{N+n} \setminus U)$. Thus, by II.2.8(a), $\overline{v}$ is an $(N+n)$-duality.

(ii) Let $DJ : \widehat{M} \to S$ be the morphism which is dual to J, see II.2.3(c). Since $[M, S^n] = H^n(M)$, we have $[M, S^n] = \mathbb{Z}$ for $H\mathbb{Z}$-orientable M and $[M, S^n] = \mathbb{Z}/2$ for $H\mathbb{Z}$-non-orientable M, and $[M, S^n]$ is generated by ε in both these cases. Thus, it suffices to prove that $(DJ)_* : H_0(\widehat{M}) = H_n(M) \to \mathbb{Z}$ is an isomorphism for $H\mathbb{Z}$-orientable M and $(DJ)_* : H_0(\widehat{M}; \mathbb{Z}/2) = H_n(M; \mathbb{Z}/2) \to \mathbb{Z}/2$ is an isomorphism for arbitrary

M. Firstly, $(DJ)_*: H_n(M;\mathbb{Z}/2) \to \mathbb{Z}/2$ is an isomorphism for every M because, by IV.5.9(i), $J^*: H^0(T\nu;\mathbb{Z}/2) \to H^0(S;\mathbb{Z}/2)$ is an isomorphism. Furthermore, if M is $H\mathbb{Z}$-orientable, then $\mathbb{Z} = H_n(M) = \widetilde{H}_n(M^+) = \widetilde{H}^N(T\nu)$. Hence, ν is $H\mathbb{Z}$-orientable, and hence, by IV.5.9(ii), $J^*: H^0(T\nu) \to H^0(S)$ is an isomorphism. Thus, $(DJ)_*$ must be an isomorphism. □

2.4. Theorem. *A closed manifold M is E-orientable iff its stable normal bundle is E-orientable.*

Notice that, by 1.13, the stable normal bundle ν is E-orientable iff ν^N is.

Proof. Without loss of generality, we can assume that M is connected. Given a root $j: S^N \to T\nu^N$, set $J := \Sigma^{-N}\Sigma^\infty j: S \to T\nu$. The homeomorphism $\overline{w}: S^{N+n} \to S^N \wedge S^n$ is an $(N+n)$-duality map between S^N and S^n, and $w = \Sigma^{-N-n}\Sigma^\infty\overline{w}$ is just the identification $S \wedge S = S$. By 2.3(i) and II.2.4(i), we have the commutative diagram

$$
(2.5)\qquad
\begin{array}{ccccccc}
\widetilde{E}^0(T\nu) & = & [T\nu, E] & \xrightarrow{v_E} & [S, E\wedge \widehat{M}] & = & E_n(M)\\
 & & \big\downarrow{\scriptstyle J^*} & & \big\downarrow{\scriptstyle (1\wedge DJ)_*} & & \\
 & & [S, E] & \xrightarrow{w_E} & [S, E\wedge S] & &
\end{array}
$$

where all horizontal maps are isomorphisms.

We prove that v_E gives a bijective correspondence between E-orientations of ν and E-orientations of M. (In particular, ν is E-orientable iff M is.) Indeed, E-orientations $u \in \widetilde{E}^n(T\nu)$ are defined by the equality $J^*(u) = \pm 1$, while E-orientations $[M] \in E_n(M) = E_0(\widehat{M})$ of M are characterized by the equality $(\mathscr{E})_*[M] = \pm 1 \in \pi_0(E)$. But, by 2.3(i), DJ is homotopic (up to sign) to $\mathscr{E}: \widehat{M} \to S$, and the result follows. □

2.6. Corollary (of the proof). *Every duality $v: S \to T\nu \wedge \widehat{M}$ yields a bijective correspondence (given by v_E) between E-orientations of M and E-orientations of ν.* □

2.7. Remarks. (a) Milnor–Spanier [1] established an $(N+n)$-duality between $T\nu$ and M^+.

(b) Note that, in fact, we do not need to know a concrete form of the duality morphism $v: S \to T\nu \wedge \widehat{M}$ in order to prove 2.4: it suffices just to know that such a duality exists.

The bijective correspondence noted in 2.6 admits another description. Let $[T\nu]_E \in E_n(T\nu)$ be the image of the unit $1 \in E_0(S)$ under the homomorphism

$$E_0(S) \xrightarrow{\cong} E_n(S^n) \xrightarrow{c_*} E_n(T\nu).$$

Note that $[T\nu]_E$ is completely determined by the Browder–Novikov morphism $c: S^n \to T\nu$. Moreover, we have $\tau_*([T\nu]_E) = [T\nu]_F$ for every ring morphism $\tau: E \to F$ of ring spectra. Now, the following proposition holds.

2.8. Proposition. *Let $u \in E^0(T\nu)$ be an E-orientation of the stable bundle ν of M, and let $\varphi : E_n(M) \to E_n(T\nu)$ be the Thom–Dold isomorphism associated with u. Then $v_E(u) = \varphi^{-1}([T\nu]_E)$ (here we identify $E_n(M)$ with $\widetilde{E}_0(\widehat{M})$), i.e., $[M]_E = \varphi^{-1}([T\nu]_E)$.*

Proof. The class $v_E(u)$, as well as the class $\varphi^{-1}([T\nu]_E)$, is given by the map $S \xrightarrow{v} T\nu \wedge \widehat{M} \xrightarrow{u\wedge 1} E \wedge \widehat{M}$. □

Note that u is not a canonical element, but since φ, as well as $[M]_E$, depends on u, this indeterminacy vanishes for $\varphi([M]_E) = [T\nu]_E$.

Let F be an E-module spectrum. Given a closed E-oriented manifold $(M, [M]_E)$, consider the isomorphism

$$P = P_{[M]_E} : F^i(M) \xrightarrow{\varphi} F^i(T\nu) \cong F_i(\widehat{M}) = \widetilde{F}_{n-i}(M^+) = F_{n-i}(M).$$

Here φ is the Thom–Dold isomorphism given by an E-orientation u of ν, which, in turn, is given by the E-orientation $[M]_E$ of M according to 2.4.

The isomorphism P is called *Poincaré duality* and admits the following alternative description.

2.9. Theorem. *The homomorphism*

$$\cap [M]_E : F^i(M) \to F_{n-i}(M)$$

coincides with P.

Proof. Let $d: M \to M \times M$ be the diagonal, and let $\Delta: T\nu \to T\nu \wedge M^+$ be the morphism as in IV.5.36. We define

$$\mathcal{T} : T\nu \xrightarrow{\Delta} T\nu \wedge M^+ \simeq T\nu \wedge \Sigma^\infty M^+ = T\nu \wedge \Sigma^n \widehat{M}$$

and

$$\nabla : \Sigma^n \widehat{M} = \Sigma^\infty M^+ \xrightarrow{\Sigma^\infty d^+} \Sigma^\infty (M^+ \wedge M^+) = \Sigma^n \widehat{M} \wedge M^+ \simeq \Sigma^n \widehat{M} \wedge \Sigma^n \widehat{M}.$$

Since the maps

$$M \xrightarrow{d} M \times M \xrightarrow{d \times 1} M \times M \times M, \qquad M \xrightarrow{d} M \times M \xrightarrow{1 \times d} M \times M \times M$$

coincide, we conclude that the morphisms

$$T\nu^N \xrightarrow{\Delta^N} T\nu^N \wedge M^+ \xrightarrow{\Delta^N \wedge 1} T\nu^N \wedge M^+ \wedge M^+,$$

$$T\nu^N \xrightarrow{\Delta^N} T\nu^N \wedge M^+ \xrightarrow{1 \wedge (d^+)} T\nu^N \wedge M^+ \wedge M^+$$

are homotopic. Hence, the morphisms

$$T\nu \xrightarrow{\mathcal{T}} T\nu \wedge \Sigma^n \widehat{M} \xrightarrow{\mathcal{T}\wedge 1} T\nu \wedge \Sigma^n \widehat{M} \wedge \Sigma^n \widehat{M}$$

and

$$T\nu \xrightarrow{\mathcal{T}} T\nu \wedge \Sigma^n \widehat{M} \xrightarrow{\nabla} T\nu \wedge \Sigma^n \widehat{M} \wedge \Sigma^n \widehat{M}$$

are homotopic. In particular, in the diagram

$$\begin{array}{ccc} & & [S, T\nu \wedge \Sigma^n \widehat{M}] \\ & & \big\downarrow {\scriptstyle (\mathcal{T}\wedge 1)_*} \\ [S, T\nu \wedge \Sigma^n \widehat{M}] & \xrightarrow{(1\wedge\nabla)_*} & [S, T\nu \wedge \Sigma^n \widehat{M} \wedge \Sigma^n \widehat{M}] \end{array}$$

we have

$$(1 \wedge \nabla)_*(\Sigma^n v) \simeq (\mathcal{T} \wedge 1)_*(\Sigma^n v), \tag{2.10}$$

where $\Sigma^n v : S^n \to \Sigma^n(T\nu \wedge M^+) = T\nu \wedge \Sigma^n \widehat{M}$.

Let $u \in E^0(T\nu)$ be the E-orientation of ν which is dual to $[M]_E$, cf. 2.6. Given $x \in F^i(M) = F^i(\Sigma^n \widehat{M}) = [\Sigma^n \widehat{M}, \Sigma^i F]$, consider the following commutative diagram:

$$\begin{array}{ccc} & & \Sigma^n v \in [S, T\nu \wedge \Sigma^n \widehat{M}] \\ & & {\scriptstyle (\mathcal{T}\wedge 1)_*} \big\downarrow \\ \Sigma^n v \in [S, T\nu \wedge \Sigma^n \widehat{M}] & \xrightarrow{(1\wedge\nabla)_*} & [S, T\nu \wedge \Sigma^n \widehat{M} \wedge \Sigma^n \widehat{M}] \\ \big\downarrow {\scriptstyle (u\wedge 1)_*} & & \big\downarrow {\scriptstyle (u\wedge 1)_*} \\ [M]_E \in [S, E \wedge \Sigma^n \widehat{M}] & \xrightarrow{(1\wedge\nabla)_*} & [S, E \wedge \Sigma^n \widehat{M} \wedge \Sigma^n \widehat{M}] \\ & & \big\downarrow {\scriptstyle (1\wedge x\wedge 1)_*} \\ & & [S, E \wedge \Sigma^i F \wedge \Sigma^n \widehat{M}] \\ & & \big\downarrow {\scriptstyle m\wedge 1)_*} \\ & & [S^n, \Sigma^i F \wedge \Sigma^n \widehat{M}]. \end{array}$$

Now (the second equality follows from (2.10))

$$\begin{aligned} P(x) &= (m \wedge 1)_*(1 \wedge x \wedge 1)_*(u \wedge 1)_*(\mathcal{T} \wedge 1)_*(\Sigma^n v) \\ &= (m \wedge 1)_*(1 \wedge x \wedge 1)_*(u \wedge 1)_*(1 \wedge \nabla)_*(\Sigma^n v) \\ &= (m \wedge 1)_*(1 \wedge x \wedge 1)_*(1 \wedge \nabla)_*(u \wedge 1)_*(\Sigma^n v) = x \cap [M]_E. \end{aligned}$$

This completes the proof. □

2.11. Definition. Let F be a module spectrum over a ring spectrum E. Let $f: M^m \to N^n$ be a map of closed manifolds.

(a) Suppose that both M, N are E-oriented. We define *transfers* (other names: *Umkehrs, Gysin homomorphisms*)

$$f^!: F^i(M) \to F^{n-m+i}(N), \qquad f_!: F_i(N) \to F_{m-n+i}(M)$$

to be the compositions

$$f^!: F^i(M) \cong F_{m-i}(M) \xrightarrow{f_*} F_{m-i}(N) \cong F^{n-m+i}(M), \text{ i.e., } f^! = P_{[N]}^{-1} f_* P_{[M]}$$

$$f_!: F_i(N) \cong F^{n-i}(N) \xrightarrow{f^*} F^{n-i}(M) \cong F_{m-n+i}(N), \text{ i.e., } f_! = P_{[M]} f^* P_{[N]}^{-1}.$$

The reader can find many good properties of transfers in Dold [5], Dyer [1].

(b) More generally, we do not assume that M and/or N is E-oriented, but we suppose that there is a morphism $\omega: \nu_M \oplus \xi \to \nu_N, \operatorname{bs}\omega = f$ where ξ is an E-oriented stable bundle. (In other words, the difference $\nu_N - f^*\nu_M$ is E-orientable.) We define transfers

$$\omega^!: F^i(M) \to F^{n-m+i}(N), \qquad \omega_!: F_i(N) \to F_{m-n+i}(M)$$

to be the compositions

$$\begin{aligned} \omega^!: F^i(M) = F^{i-m}(\widehat{M}) \cong F_{m-i}(T\nu_M) &\xrightarrow{\varphi} F_{m-i}(T(\nu_M \oplus \xi)) \\ &\xrightarrow{(T\omega)_*} F_{m-i}(T\nu_N) \cong F^{i-m}(\widehat{N}) = F^{n-m+i}(N), \\ \omega_!: F_i(N) = F_{i-n}(\widehat{N}) \cong F^{n-i}(T\nu_N) &\xrightarrow{(T\omega)^*} F^{n-i}(T(\xi \oplus \nu_M)) \\ &\xrightarrow{\varphi} F^{n-i}(T\nu_M) \cong F_{i-n}(\widehat{M}) = F_{m-n+i}(M) \end{aligned}$$

where the φ's are the Thom–Dold isomorphisms as in 1.28(a).

If $f: M^n \to N^n$ is a map of closed $H\mathbb{Z}$-oriented manifolds then

$$f_* f_!(x) = (\deg f)x$$

for every $x \in H_*(N)$ (prove this!). In particular, if $\deg f = 1$ then $f_*: H_*(M) \to H_*(N)$ is epic. Similarly, $f^*: H^*(N) \to H^*(M)$ is a monomorphism if $\deg f = 1$. Theorem 2.13 below generalizes this fact.

2.12. Lemma. *Let E be a ring spectrum. Let $f: M^n \to N^n$ be a map of degree ± 1 of closed $H\mathbb{Z}$-orientable manifolds. If $[M]$ is an E-orientation of M then $f_*[M]$ is an E-orientation of N. In particular, N is E-orientable if M is.*

Proof. The map $M \xrightarrow{f} N \xrightarrow{\varepsilon} S^n$ has degree ± 1, and so $\varepsilon_*(f_*[M]) = (\varepsilon f)_*[M] = \pm s_n$. Thus, $f_*[M]$ is an E-orientation of N. □

2.13. Theorem. *Let E be a ring spectrum. Let $f : M^n \to N^n$ be a map of degree ± 1 of closed $H\mathbb{Z}$-orientable manifolds. If M is E-orientable then $f^* : F^*(N) \to F^*(M)$ is monic and $f_* : F_*(M) \to F_*(N)$ is epic for every E-module spectrum F.*

Proof. Let $[M]$ be an E-orientation of M. Given $x \in F^*(N)$, we have $f_*(f^*(x) \cap [M]) = x \cap f_*[M]$. But $f_*[M]$ is an E-orientation of N, and so $x \cap f_*[M] \neq 0$ if $x \neq 0$, and thus $f^*(x) \neq 0$ if $x \neq 0$. Furthermore, since $f_*[M]$ is an E-orientation of N, every $a \in F_*(N)$ has the form

$$a = u \cap f_*[M] = f_*(f^*(u) \cap [M]). \qquad \square$$

2.14. Remarks-Exercises. (a) The results of this section can be generalized for manifolds with boundary. Namely, Atiyah [3] generalized 2.3 and proved that $M/\partial M$ is $(N+n)$-dual to $T\nu^N$, where ν^N is a normal bundle of $(M, \partial M)$ in $(\mathbb{R}^{N+n}_+, \partial\mathbb{R}^{N+n}_+)$. In this way one can generalize 2.4 and 2.6, i.e., the word "closed" can be omitted there. Moreover, there is an isomorphism P of the form

$$P : F^i(M) \xrightarrow{\varphi} \widetilde{F}^{i+N}(T\nu^N) \cong \widetilde{F}_{n-i}(M/\partial M) = F_{n-i}(M, \partial M),$$

and it coincides with the isomorphism

$$\cap[M, \partial M]_E : F^i(M) \to F_{n-i}(M, \partial M).$$

(b) We note the following generalization of 2.5 and 2.6. Consider a spectrum E and a subset $A \subset \pi_0(E)$ with $\pm A = A$. We regard A as a subset of $E_n(S^n, *) = \widetilde{E}_n(S^n) \cong \widetilde{E}_0(S^0)$. We define an (A, E)-marking of a manifold M^n to be an element $V \in E_n(M, \partial M)$ such that $\varepsilon(V) \in A \subset E_n(S^n, *)$, where ε is as in 2.1. Based on 2.3, one can prove that M is (A, E)-markable iff its normal bundle is (see Definition IV.5.25). Moreover, (A, E)-markings of M are in a bijective correspondence with (A, E)-markings of ν.

(c) Interpret 2.11(a) as a special case of 2.11(b).

§3. Orientability and Integrality

Here we establish some interconnections between orientability and some integrality theorems. Let $\tau : D \to E$ be a ring morphism of ring spectra. Let ξ be any D-orientable (and hence E-orientable) $\mathcal{V}$-object over X, and let u_D (resp. u_E) be a D- (resp. E-) orientation of ξ. We do not say that $\tau(u_D) = u_E$. On the contrary, integrality phenomena arise precisely because of incompatibility of the orientations.

The orientation u_E gives rise to the Thom–Dold isomorphism $\varphi_E : E^*(X) \to \widetilde{E}^*(T\xi)$. Set

$$R(\xi) = R_{u_D,u_E}(\xi) := \varphi_E^{-1}\tau(u_D) \in E^0(X). \tag{3.1}$$

Now, let M^n be any D-orientable manifold, and let $[M]_D, [M]_E$ be D-, E-orientations of it respectively. Consider a stable normal bundle ν of M and fix a Browder–Novikov morphism $c : S^n \to T\nu$, Then, according to 2.6, the orientations $[M]_D, [M]_E$ determine certain orientations $u_D(\nu), u_E(\nu)$ in a canonical manner, and so the class $R(\nu)$ is defined.

3.2. Theorem. *For every $x \in D^k(M)$ we have*

$$\langle \tau(x)R(\nu), [M]_E\rangle = \tau\langle x, [M]_D\rangle$$

(where $\langle -, -\rangle$ is the Kronecker pairing).

Proof. Let $[T\nu]_D, [T\nu]_E$ be as in 2.8. We have

$$\begin{aligned}\langle \tau(x)R(\nu), [M]_E\rangle &= \langle \tau(x)\varphi_E^{-1}\tau(u_D), [M]_E\rangle = \langle \varphi_E^{-1}(\tau(x)\tau(u_D)), [M]_E\rangle \\ &= \langle \tau(xu_D), \varphi_E[M]_E\rangle = \langle \tau(xu_D), [T\nu]_E\rangle \\ &= \langle \tau(xu_D), \tau[T\nu]_D\rangle = \tau\langle (xu_D), [T\nu]_D\rangle \\ &= \tau\langle (xu_D), \varphi_D[M]_D\rangle = \tau\langle \varphi_D^{-1}(xu_D), [M]_D\rangle \\ &= \tau\langle x, [M]_D\rangle.\end{aligned}$$

□

3.3. Corollary (the "Integrality" Theorem). *The element $\langle \tau(x)R(\nu), [M]_E\rangle$ of the group $\pi_{n-k}(E)$ belongs to the subgroup* $\operatorname{Im}\{\tau_* : \pi_{n-k}(D) \to \pi_{n-k}(E)\}$. □

Now we give some examples, but to understand them the reader should know something about characteristic classes and vector bundles. However, in future we do not use these examples, they just give a nice illustration of 3.3.

3.4. Examples. (a) Given a complex vector bundle ξ, we define a class

$$\mathscr{T}(\xi) := \prod \frac{t_i}{e^{t_i} - 1}$$

where t_i are the Wu generators, i.e., the Chern class $c_i(\xi)$ is the i-th elementary symmetric polynomial of t_i, see Milnor–Stasheff [1]. We also define the *Todd class*

$$T(\xi) := \prod \frac{t_i}{1 - e^{-t_i}}$$

Let $\operatorname{ch} : K^0(X) \to H^{**}(X; \mathbb{Q}) := \prod_n H^n(X; \mathbb{Q})$ be the classical Chern character.

3.5. Theorem (see Hirzebruch [1], Palais [1]). *Let η be an arbitrary complex vector bundle over an almost complex closed manifold M^{2n}. Then*

$$\langle \operatorname{ch}(\eta)\mathscr{T}(\tau M), [M]\rangle \text{ and } \langle \operatorname{ch}(\eta)T(\tau M), [M]\rangle$$

are integers for every $H\mathbb{Q}$-orientation $[M]$ of M (here τM is the tangent bundle of M).

Proof. Clearly, it suffices to prove the theorem for some single $H\mathbb{Q}$-orientation $[M]$. Put $D = K$, $E = \prod_{n\in\mathbb{Z}} \Sigma^{2n} H\mathbb{Q} = \bigvee_{n\in\mathbb{Z}} \Sigma^{2n} H\mathbb{Q}$, see III.3.26, i.e., $E^i(X) = \prod_{n\in\mathbb{Z}} H^{i+2n}(X;\mathbb{Q})$. Based on II.7.13, define $\tau : D \to E$ to be the composition

$$K \xrightarrow{l[0]} K[0] \xrightarrow{\mathrm{ch}_K} H(\pi_*(K)\otimes\mathbb{Q}) = \bigvee_{n\in\mathbb{Z}} \Sigma^{2n} H\mathbb{Q}.$$

The inclusion $H\mathbb{Q} \subset E$ equips the $H\mathbb{Q}$-oriented manifold M with an E-orientation, and every complex bundle ξ admits a canonical K-orientation u_ξ such that $\varphi_E^{-1}\,\mathrm{ch}_K\, u_\xi = \mathscr{T}(-\xi)$, see e.g. Stong [3], p.294. (Stong considers a family of isomorphisms φ_H, but really this family is an isomorphism φ_E.) By Bott periodicity, we have $\pi_{2i}(K) = \mathbb{Z}, \pi_{2i+1}(K) = 0$. Interpreting η as an element of $K^0(M)$, we conclude that

$$\langle \mathrm{ch}(\eta)\mathscr{T}(\tau M), [M]_E\rangle = \langle \mathrm{ch}(\eta)R(\nu M), [M]_E\rangle \in \mathrm{Im}\{\tau_* : \pi_{2n}(K) \to \pi_{2n}(E)\}.$$

Now, considering the projection $E \to H\mathbb{Q}$, we conclude that

$$\langle \mathrm{ch}(\eta)\mathscr{T}(\tau M), [M]\rangle \in \mathrm{Im}\{\pi_{2n}(K) \to \pi_{2n}(E) \to \pi_{2n}(H\mathbb{Q})\} = \mathbb{Z}.$$

Similarly, there exists a K-orientation v_ξ with $\varphi_E^{-1}\,\mathrm{ch}\, v_\xi = T(-\xi)$. So, we obtain integrality of the second number[12]. □

(b) Given a real vector bundle ξ, we define the class

$$\hat{A}(\xi) := \prod \frac{y_i}{2\sinh(y_i/2)}$$

where the Pontrjagin class $p_i(\xi)$ is the i-th elementary symmetric polynomial of y_i^2.

3.6. Theorem (see Hirzebruch [1], Palais [1]). *Let M^{4n} be a smooth closed manifold with $w_1(M) = 0$ and $w_2(M) = \rho(c)$ for some $c \in H^2(M)$, where $\rho : H^2(M) \to H^2(M;\mathbb{Z}/2)$ is the* mod 2 *reduction. Then*

[12] We indicate a construction of such v_ξ. Let $\gamma = \gamma_{\mathbb{C}}^1$ be the universal complex line bundle over $B\mathcal{U}_1 = CP^\infty$, and let $t := c_1(\gamma) \in H^2(CP^\infty)$. We require that $\varphi^{-1}\,\mathrm{ch}\, v_\gamma = \frac{t}{1-e^{-t}}$, i.e., $\mathrm{ch}\, v_\gamma = \frac{1}{1-e^{-t}}$; here $\varphi = \varphi_H$. Since $\mathrm{ch}\,\gamma = e^t$ and $\gamma \otimes \bar\gamma = \theta_{\mathbb{C}}^1$, we have $\mathrm{ch}\,\bar\gamma = e^{-t}$, i.e., we can put $v_\gamma = \frac{\beta}{1-\bar\gamma} = \beta(1 + \bar\gamma + \bar\gamma^2 + \cdots) \in K^2(CP^\infty)$, where $1 \in K^0(CP^\infty)$ represents $\theta_{\mathbb{C}}^1$ and $\beta \in K^2(\mathrm{pt})$ is the Bott element. Every complex line bundle ξ has the form $\xi = f^*\gamma$ for some $f : \mathrm{bs}\,\xi \to CP^\infty$, and we put $v_\xi = f^* v_\gamma$. Now, using the splitting principle, we can construct v_ξ with $\varphi_E^{-1}\,\mathrm{ch}\, v_\xi = T(-\xi)$ for every complex vector bundle ξ, cf. §VII.2 and Conner–Floyd [1].

$$\langle e^{c/2}\operatorname{ch}(\eta)\hat{A}(\tau M), [M]\rangle$$

is an integer for every complex vector bundle η *over* M *and every* $H\mathbb{Q}$-*orientation* $[M]$ *of* M.

Proof. This theorem also can be deduced from 3.3 with D, E, τ just the same as in Example (a). Namely, M is K-orientable, and the element c (in fact, the Spin$^{\mathbb{C}}$-structure on M) enables us to construct a canonical K-orientation of M, see Stong [3], Ch. XI. (In fact, every K-orientation of a $\mathcal{V}$-object ξ yields some c with $w_2(\xi) = \rho(c)$, see Ch. VI. It can happen that different K-orientations yield the same class c, but there exists a canonical "lifting", i.e., a canonical K-orientation, for every class c.) By 2.4, we get a certain K-orientation u of a normal bundle ν, and $\varphi_E^{-1}\tau u = e^{c/2}\operatorname{ch}(\eta)\hat{A}(\tau M)$, see loc. cit. Now the proof can be finished as in Example (a). □

(c) There is a stronger version of 3.6.

3.7. Theorem (see Hirzebruch [1], Palais [1]). *Let* M^{8n+4} *be a smooth closed manifold with* $w_1(M) = 0, w_2(M) = 0$. *Then for every real vector bundle* η *and for every* $H\mathbb{Q}$-*orientation* $[M]$ *of* M *the number*

$$\langle \operatorname{ph}(\eta)\hat{A}(\tau M), [M]_{H\mathbb{Z}}\rangle$$

is even. Here ph *is the Pontrjagin character,* $\operatorname{ph}\eta = \operatorname{ch}(\eta\otimes_{\mathbb{R}}\mathbb{C})$.

Proof. This theorem can be deduced from 3.3 if we put $E^i(X) = \bigoplus_{n\in\mathbb{Z}} H^{i+4n}(X;\mathbb{Q})$, $\tau = \operatorname{ph} : KO(X) \to E^*(X)$. The complexification $C : BO \to BU$ induces the homomorphism

$$C_* : \mathbb{Z} = \pi_{8n+4}(BO) \to \pi_{8n+4}(BU) = \mathbb{Z}$$

which is multiplication by 2, see IV.4.27(iii), and so the image $\operatorname{Im}\{\operatorname{ph} : \pi_{8n+4}(KO) \to \pi_{8n+4}(E)\}$ consists of even numbers, and the theorem holds. □

(d) Considering the morphisms $Sq^i : H\mathbb{Z}/2 \to \Sigma^i H\mathbb{Z}/2$, we get, by III.3.27, a morphism $Sq : H\mathbb{Z}/2 \to \prod_{n=-\infty}^{\infty} \Sigma^n H\mathbb{Z}/2$ where $Sq(x) = \prod_{i\geq 0} Sq^i(x)$. By the Cartan formula, $Sq : D \to E$ is a ring morphism. Let M^n be a closed connected topological manifold, and let $[[M]] \in H_n(M;\mathbb{Z}/2)$ be its fundamental class mod 2, i.e., the non-zero element of the group $H_n(M;\mathbb{Z}/2) = \mathbb{Z}/2$. Let $W(\xi) = \sum w_i(\xi)$. Since $w_i(\xi) = \varphi^{-1}Sq^i(u)$ where $u \in \widetilde{H}(T\xi;\mathbb{Z}/2)$ is a Thom class of ξ, we deduce from 3.2 that

$$\langle x, [[M]]\rangle = \langle Sq(x)W(\nu), [[M]]\rangle.$$

Indeed, put $D = H\mathbb{Z}/2, E = \prod_{n=-\infty}^{\infty} \Sigma^n H\mathbb{Z}/2 = \bigvee_{n=-\infty}^{\infty} \Sigma^n H\mathbb{Z}/2$ and $\tau = Sq : D \to E$. Then put $[M]_D = [[M]]$ and define $[M]_E$ to be the image of $[M]_D$ under the inclusion $H\mathbb{Z}/2 \to \vee_n \Sigma^n H\mathbb{Z}/2$ of the summand.

This formula is well-known and can also be deduced from the Wu formula $\langle Sq(x), [M] \rangle = \langle (V \cup x), [M] \rangle$, where V is the so-called Wu class of M (see the definition of V e.g. in Stong [3], pp. 98–100).

In particular, if $\dim x \neq n$ then

$$\langle Sq(x)W(\nu), [[M]] \rangle = 0. \tag{3.8}$$

This implies, for example, that $w_n(\nu) = 0$ (put $x = 1 \in H^0(M; \mathbb{Z}/2)$). Besides, it follows from (3.8) that every $H\mathbb{Z}$-orientable 3-dimensional manifold M is parallelizable. Indeed, it suffices to prove that $w_i(\nu) = 0$ for $i = 1, 2, 3$. We have $w_1(\nu) = 0, w_3(\nu) = 0$. If $w_2(\nu) \neq 0$, then there exists (by duality) a class $l \in H^1(M; \mathbb{Z}/2)$ with $\langle l, w_2(\nu) \rangle \neq 0$. But then $\langle Sq(l)W(\nu), [M] \rangle \neq 0$.

One can deduce from (3.8) the following theorem of Massey [1]. *Let $\alpha(n)$ be the number of ones in the dyadic expansion of n. Then $w_i(\nu) = 0$ for $i > n - \alpha(n)$.* In fact, all the relations between the Stiefel–Whitney classes follow from (3.8), see Brown–Peterson [2].

(e) Similarly to Example (d) one can consider the operation $P = \sum P^i$ for an odd prime p.

Analogs of Examples (d), (e) hold also in some other cohomology theories.

(f) Let MU be the complex (co)bordism theory (see Ch. VII), and let $s_\omega : MU \to \Sigma^{|\omega|} MU$ be the Novikov cohomology operation associated with a partition $\omega = (i_1, ..., i_k)$, where $|\omega| = \sum i_k$. Put $D = MU, E = \bigvee_\omega \Sigma^{|\omega|} MU$. Given a finite CW-space X and $x \in MU^k(X)$, set $S(x) := \oplus_\omega s_\omega(x) \in \oplus_\omega MU^{k+2|\omega|}(x) = E^k(X)$. Then, by III.3.23(ii), S is induced by a morphism $S : D \to E$ of spectra, and we put $\tau := S$. Furthermore, we note that the inclusion $MU \to E$ of the summand equips every MU-oriented $\mathcal{V}$-object with an E-orientation. Let ξ be a complex vector bundle with the canonical MU-orientation u. Then $\varphi_{MU}^{-1} s_\omega(u) = c_\omega(\xi)$, where $c_\omega(\xi) \in MU^{2|\omega|}(\mathrm{bs}\,\xi)$ is the Conner–Floyd–Chern class of ξ. So, $\varphi_E^{-1} S(u) = C(\xi)$ where $C(\xi); = \sum_\omega c_\omega(\xi)$. Let M be a closed almost complex manifold with the canonical MU-orientation $[M]$. By 3.3,

$$\langle S(x)C(\nu), [M] \rangle \in \bigoplus_{k \leq n-i} \pi_{2k}(MU)$$

for every $x \in MU^{2i}(M^{2n})$ (the point of this formula is the inequality $k \leq n - i$).

More generally, one can consider an arbitrary set Λ of partitions such that $\omega_1, \omega_2 \in \Lambda$ iff $(\omega_1, \omega_2) \in \Lambda$ and introduce the spectrum $E = \vee_{\omega \in \Lambda} \Sigma^{|\omega|} MU$, etc.

(g) The Adams power operation $\psi^n : K^0(X) \to K^0(X)$ can be extended to a stable operation, but this requires the expense of localization. More precisely, there exists a ring morphism $\psi^n : K^*(X) \to K^*(X)[1/n]$ which coincides with ψ^n on $K^0(X)$. Adams [6] defined a "cannibalistic" characteristic class $\rho_n(\xi) := \varphi^{-1}\psi^n(u) \in K^*(X)[1/n]$ for every K-oriented (for example, complex) vector bundle (ξ, u) over X. Thus, if M is any K-oriented (for example, stably almost complex) manifold, then, by 3.2, for every $x \in K^*(M)$ we have

$$\langle \psi^n(x)\rho_n(\nu), [M]\rangle \in \mathbb{Z} \subset \mathbb{Z}[1/n].$$

Theorem 3.2 is a partial case of the following result. Let M, N be two closed D-orientable manifolds, and let $\tau : D \to E$ be a ring morphism of ring spectra. Choose orientations $[M]_D$, $[M]_E$, $[N]_D$, $[N]_E$.

3.9. Theorem (Dyer [1]). *For every $f : M \to N$ and every $x \in D^*(M)$ we have*

$$f^!_E(\tau(x)R(\nu_M)) = \tau(f^!_D(x)R(\nu_N)).$$

Here $f^!$ denotes the transfers defined in (2.11).

Proof. This can be easily deduced from 3.2, see Dyer [1]. □

Note that 3.2 follows from 3.9, if we take N to be a point.

§4. Obstructions to Orientability

In this and the next section we give an obstruction theory for orientability with respect to spectra of finite type. By 1.16, it suffices to consider connected spectra only. Furthermore, by 1.20, an $\mathcal{F}$-object is E-orientable iff it is $E[p]$-orientable for all primes p which divide the order of $1 \in \pi_0(E)$. Thus, up to the end of this chapter we fix any prime p and consider a connected p-local ring spectrum E of finite $\mathbb{Z}[p]$-type. As usual, $\mathbb{Z}[p]^*$ denotes the set of invertible elements of the ring $\mathbb{Z}[p]$.

In this and the next section the Postnikov n-stage of a spectrum E is denoted simply by E_n. This is not compatible with the notation of Ch. II, where E_n was just the n-th term of E, and the Postnikov n-stage of a spectrum E was denoted by $E_{(n)}$, but this will not confuse us because we shall not consider (nor even mention) terms of spectra in this chapter.

By 1.13, without loss of generality we may and shall restrict our attention to stable $\mathcal{V}$-objects only. So, up to the end of the chapter the term "$\mathcal{V}$-object" means "stable $\mathcal{V}$-object over a connected CW-base", and similarly for $\mathcal{F}$-objects.

4.1. Recollection (See Fomenko–Fuchs–Gutenmacher [1], Hu [1], Mosher–Tangora [1], Spanier [1]).

(a) Recall that an F-fibration $p : Y \to X$ is called *simple* if the $\pi_1(B)$-action on $\pi_n(F)$ is trivial for every n.

(b) Let F be a CW-space with $\pi_i(F) = 0$ for $i < n$ where $n > 1$. Then, by the Hurewicz Theorem, $H_i(F) = 0$ for $i < n$, and so, by II.4.9, there is a canonical isomorphism $a : H^n(F; \pi_n(F)) \xrightarrow{\cong} \operatorname{Hom}(H_n(F), \pi_n(F))$. Let $g : H_n(X) \to \pi_n(X)$ be the inverse isomorphism to the Hurewicz isomorphism $h : \pi_n(F) \to H_n(F)$. The element

$$\iota_n := a^{-1}(g) \in H^n(F; \pi_n(F)) \tag{4.2}$$

is called the *fundamental class* of a space F.

(c) Now, let

$$p : Y \to X \tag{4.3}$$

be a simple F-fibration with F as in (b), let F_x be a fiber over a point $x \in X$, and let $\tau : H^n(F_x; \pi_n(F_x)) \to H^{n+1}(X; \pi_n(F_x))$ be the transgression. We define the *characteristic class* χ *of the fibration* (4.3) by setting

$$\chi := \tau\iota_n \in H^{n+1}(X; \pi_n(F)) \tag{4.4}$$

(d) In particular, if F is an Eilenberg–Mac Lane space $K(\pi, n)$ then there is the fundamental class

$$\iota_n \in H^n(K(\pi, n); \pi_n(K(\pi, n))) = H^n(K(\pi, n); \pi).$$

However, we must be careful with the last (traditionally used) equality. Namely, this equality means that there is a standard group π, and we somehow identify $\pi_n(K(\pi, n))$ with π. A similar problem arises when we consider characteristic classes. In other words, when we say, for instance, "Let $\chi \in H^{n+1}(X; \pi)$ be a characteristic class of the $K(\pi, n)$-fibration over X" it means that χ is an element of the correseponding $\operatorname{Aut}\pi$-orbit in $H^{n+1}(X; \pi)$, cf. Spanier [1], Ch.8, §1. (By the way, cf. II.4.19.)

(e) Recall that a $K(\pi, n)$-fibration has trivial (i.e., =0) characteristic class iff it admits a section, see loc. cit.

4.5. Lemma. *Let p be a prime, and let $\pi = \mathbb{Z}[p]$ or π be a cyclic group of order p^k. Let $q : Y \to B$ be a simple fibration with fiber $K(\pi, n), n \geq 1$, and with non-trivial characteristic class $\chi \in H^{n+1}(B; \pi)$. Furthermore, let Y, B be homotopy equivalent to CW-spaces, and let $H^{n+1}(B; \pi)$ be a finitely generated $\mathbb{Z}[p]$-module. Finally, let*

$$\begin{array}{ccc} Y & \xrightarrow{g} & Y \\ {\scriptstyle q}\downarrow & & \downarrow{\scriptstyle q} \\ B & \xrightarrow[f]{} & B \end{array}$$

be a commutaive diagram such that f is a homotopy equivalence. Then g is a homotopy equivalence.

Proof. Let $f(b_1)=b_2, b_i \in B$, $i=1,2$, and let $F_i := q^{-1}(b_i)$, $i=1,2$. Let $\overline{g}: F_1 \to F_2$ be the restriction of g. In order to prove that g is a homotopy equivalence, it suffices to prove that $\overline{g}$ is a homotopy equivalence. Let $\iota_i \in H^n(F_i;\pi)=\pi$ be the fundamental class of F_i, $i=1,2$. It suffices to prove that $\overline{g}^*(\iota_2)=\lambda\iota_1$ with $\lambda \in \mathbb{Z}[p]^*$, where

$$\overline{g}^*: H^*(F_2;\pi) \to H^*(F_1;\pi)$$

is the induced homomorphism. We have $\chi \neq 0$. Furthermore, $H^{n+1}(B;\pi)$ is a finitely generated $\mathbb{Z}[p]$-module, and so there exists r such that $p^r|\chi, p^{r+1} \nmid \chi$. Set $\chi_i = \tau\iota_i$, where τ is the transgression in the Leray–Serre spectral sequence of q. Since f is a homotopy equivalence, $p^r|f^*(\chi_2), p^{r+1} \nmid f^*\chi_2$. Now, if $\overline{g}^*(\iota_2)=p^s\lambda\iota_1$ with $s>0, \lambda \in \mathbb{Z}[p]$, then

$$f^*(\chi_2)=f^*(\tau\iota_2)=\tau\overline{g}^*(\iota_2)=\tau(p^s\lambda\iota_1)=p^s\lambda\chi_1. \tag{4.6}$$

In particular, $p^{r+1}|f^*(\chi_2)$. This is a contradiction. □

We denote $\pi_i(E)$ by π_i and consider the Postnikov tower of E

$$\begin{array}{ccccccccc} & & E & & & & & & \\ & & \big\downarrow{\scriptstyle \tau_n} & & & & & & \\ \cdots \longrightarrow & & E_{n+1} & \longrightarrow & E_n & \xrightarrow{p_n} & E_{n-1} & \longrightarrow \cdots \longrightarrow & E_0 \\ & & & & & & {\scriptstyle \kappa_n}\big\downarrow & & \big\| \\ & & & & & & \Sigma^{n+1}H(\pi_n) & & H(\pi_0) \end{array} \tag{4.7}$$

Here κ_n is the n-th Postnikov invariant of E (and also the corresponding higher cohomology operation).

By II.4.30(i), every E_n is a ring spectrum.

4.8. Proposition. *Every E-orientable $\mathcal{F}$-object α is E_n-orientable for every n. In particular, it is $H(\pi_0)$-orientable.*

Proof. This holds by 1.6 since, by II.4.30(i), τ_n is a ring morphism. □

So, as a first step, we must clarify when α is $H(\pi_0)$-orientable.

4.9. Proposition. *Let R be a ring (non-graded), and let HR be the corresponding Eilenberg–Mac Lane spectrum.*

(i) *If α is $H\mathbb{Z}$-orientable then α is HR-orientable for every R.*

(ii) *If α is not $H\mathbb{Z}$-orientable, then it is HR-orientable iff $2R=0$.*

Proof. (i) The (unique) ring homomorphism $\mathbb{Z} \to R$ induces a ring morphism $H\mathbb{Z} \to HR$ of spectra. Thus, by 1.6, α is HR-orientable.

(ii) Since every $\mathcal{F}$-object is $H\mathbb{Z}/2$-orientable, we conclude that α is HR-orientable if $2R = 0$. Namely, the ring homomorphism $\mathbb{Z}/2 \to R$ induces a ring morphism $H\mathbb{Z}/2 \to HR$, and we can apply 1.6.

Now, if α is not $H\mathbb{Z}$-orientable then, by IV.5.23(i) and II.4.7(i), $H_0(T\alpha) = \mathbb{Z}/2$, and so, by II.4.9,

$$H^0(T\alpha; R) = \mathrm{Hom}(H_0(T\alpha), R) = \mathrm{Hom}(\mathbb{Z}/2, R).$$

On the other hand, if α is HR-orientable then $H^0(T\alpha; R) \cong H^0(\mathrm{bs}\,\alpha; R) = R$. Thus, $2R = 2\,\mathrm{Hom}(\mathbb{Z}/2, R) = 0$. □

Going further, consider any $H(\pi_0)$-oriented $\mathcal{F}$-object α with an $H(\pi_0)$-orientation $v : T\alpha \to H(\pi_0)$.

4.10. Proposition. *Every lifting $u : T\alpha \to E$ of v is an E-orientation of α.*

Proof. Let $j : S \to T\alpha$ be a root of the spectrum $T\alpha$. The composition $S \xrightarrow{j} T\alpha \xrightarrow{u} E$ yields exactly the same element of $\pi_0(E)$ as the composition $S \xrightarrow{j} T\alpha \xrightarrow{v} H(\pi_0)$ does. □

4.11. Proposition. *A $\mathcal{V}$-object α over a finite dimensional CW-base is E-orientable iff $0 \in \kappa_n(v)$ for all n.*

Proof. I explain here why the base should be finite dimensional. Because of obstruction theory, if $0 \in \kappa_n(v)$ for every n then α is E_n-orientable for every n, and vice versa. However, if $\mathrm{bs}\,\alpha$ is not finite dimensional then we can't guarantee that α is E-orientable, since E_n-orientations can be incompatible. In other words, we have a phenomenon of "phantomic orientability". On the other hand, if $\mathrm{bs}\,\alpha$ is finite dimensional then E-orientability is equivalent to E_N-orientability with $N >> \dim X$ (prove this!), and the result follows. □

Since E is a ring spectrum, π_n is a module over the ring π_0, and so $H(\pi_n)$ is a module spectrum over the ring spectrum $H(\pi_0)$. Hence, for every k, n we have a Thom isomorphism $\varphi : H^k(X; \pi_n) \xrightarrow{\cong} \widetilde{H}^k(T\alpha; \pi_n)$. We introduce the higher characteristic classes $e_n(\alpha) \subset H^{n+1}(X; \pi_n)$ by setting

$$e_n(\alpha) = \varphi^{-1}\kappa_n(v). \tag{4.12}$$

Of course, $e_n(\alpha)$ is defined iff $0 \in e_i(\alpha)$ for all $i < n$. Now, 4.11 can be reformulated as follows.

4.13. Proposition. *An $\mathcal{F}$-object α over a finite dimensional base is E-orientable iff $0 \in e_n(\alpha)$ for all n.* □

As usual, we set

$$e_n(\xi) := e_n(\xi^\bullet) \tag{4.14}$$

for any $\mathcal{V}$-object ξ. We say that the characteristic class e_n can be realized by $\mathcal{V}$-objects if there exists a $\mathcal{V}$-object ξ such that $e_n(\xi)$ is defined and $0 \notin e_n(\xi)$. The realizability problem for characteristic classes seems to be of great interest. We shall see below that for fixed E and different $\mathcal{V}$ this problem has different solutions.

Let $\gamma_n = \gamma_n^{\mathcal{V}}$ be the universal stable E-oriented $\mathcal{V}$-object over $B(\mathcal{V}, E_n)$ (in other words, $\gamma_n^{\mathcal{V}}$ is $\eta_{\mathcal{V},E_n}$ in the notation of 1.17). (Do not confuse $\gamma_n^{\mathcal{V}}$ with the n-dimensional $\mathcal{V}$-object $\gamma_{\mathcal{V}}^n$ from IV.4.2.) Let $u_n \in E_n^0(T\gamma_n) = E_n^0(M(\mathcal{V}, E_n))$ be the universal E_n-orientation of γ_n. Set

$$e_n^{\mathcal{V}} = \varphi^{-1}\kappa_n u_{n-1} \in H^{n+1}(B(\mathcal{V}, E_{n-1}); \pi_n). \tag{4.15}$$

4.16. Proposition (universality of $e_n^{\mathcal{V}}$). *Let ξ be any E_n-oriented $\mathcal{V}$-object over X, and let $a \in e_n(\xi)$. Then $a = f^* e_n^{\mathcal{V}}$ for some $f : X \to B(\mathcal{V}, E_{n-1})$.*

Proof. The element $\varphi(a)$ is given by the composition

$$T\xi \xrightarrow{h} E_{n-1} \xrightarrow{\kappa_n} \Sigma^{n+1} H(\pi_n)$$

where, by 4.10, h must be an E_{n-1}-orientation of ξ. Hence, h yields a map $f : X \to B(\mathcal{V}, E_{n-1})$ such that $(Tf)^* u_{n-1} = h$, where $Tf := T\mathfrak{I}_{f,\gamma_{n-1}} : T\xi \to T\gamma_{n-1}$. Thus,

$$f^* e_n^{\mathcal{V}} = f^*(\varphi^{-1}\kappa_n u_{n-1}) = \varphi^{-1}\kappa_n (Tf)^* u_{n-1} = \varphi^{-1}\kappa_n h = \varphi^{-1}\varphi a = a. \qquad \square$$

Of course, it makes sense to realize classes e_n by the universal objects $\gamma_{n-1}^{\mathcal{V}}$. Since $e_n^{\mathcal{V}} \in e_n(\gamma_{n-1}^{\mathcal{V}})$, the condition $e_n^{\mathcal{V}} \neq 0$ is necessary for the realizability of e_n. We shall see below that this condition is not sufficient for the realizability of e_n. So, it would be useful to find some condition of non-triviality of $e_n^{\mathcal{V}}$ and to find when this non-triviality implies that $0 \notin e_n(\gamma_{n-1}^{\mathcal{V}})$. The first of these will be done in 4.19, and the second one will be done in 5.1, 5.6.

By 1.18, $B(\mathcal{V}, E)$ is a functor of E (on the category of ring spectra and ring morphisms). By II.4.30, each spectrum E_n is a ring spectrum and all the morphisms τ_n, p_n are ring morphisms for every $n \geq 0$. Thus, we can apply the functor $B(\mathcal{V}, -)$ to the tower (4.7) and obtain the tower (4.17), where $\rho_n := B(\mathcal{V}, \tau_n), q_n := B(\mathcal{V}, p_n)$.

(4.17)

$$\begin{array}{ccccccc} & B(\mathcal{V}, E) & & & & & \\ & \downarrow{\scriptstyle \rho_n} & & & & & \\ \cdots \longrightarrow & B(\mathcal{V}, E_n) & \xrightarrow{q_n} & B(\mathcal{V}, E_{n-1}) \longrightarrow \cdots \longrightarrow & B(\mathcal{V}, E_0) & \xrightarrow{q_0} & B\mathcal{V} \end{array}$$

Because of IV.1.35, we can and shall assume that every map q_n is a fibration.

4.18. Theorem. (i) *The fiber of* q_n *is* $K(\pi_n, n)$. *Thus, the tower* (4.17) *is the Postnikov–Moore tower*[13] *of the forgetful map* $q : B(\mathcal{V}, E) \to B\mathcal{V}$.

(ii) *The* $K(\pi_n, n)$*-fibration* q_n *has a section iff* $e_n^{\mathcal{V}} = 0$.

Proof. (i) This is obvious because the fiber of q_n is just the fiber of $\Omega_{\pm 1}^{\infty}(p_n)$, and hence it is the Eilenberg–Mac Lane space $K(\pi_n, n)$.

(ii) Note that $e_n^{\mathcal{V}} = 0$ iff the universal E_{n-1}-orientation of $\gamma_{n-1}^{\mathcal{V}}$ can be extended to an E_n-orientation of $\gamma_{n-1}^{\mathcal{V}}$, i.e., iff q_n admits a section. □

4.19. Corollary. *If* $\Omega^{\infty}\kappa_n \neq 0$, *then* $e_n^{\mathcal{V}} \neq 0$.

Proof. The restriction of the tower (4.17) to a point $b \in B\mathcal{V}$ gives us the Postnikov tower of $\Omega_{\pm 1}^{\infty} E$. One has the pull-back diagram

$$\begin{array}{ccc} \Omega_{\pm 1}^{\infty} E_n & \longrightarrow & B(\mathcal{V}, E_n) \\ {\scriptstyle \Omega^{\infty} p_n}\downarrow & & \downarrow{\scriptstyle q_n} \\ \Omega_{\pm 1}^{\infty} E_{n-1} & \longrightarrow & B(\mathcal{V}, E_{n-1}). \end{array}$$

If $\Omega^{\infty}\kappa_n \neq 0$, then $\Omega^{\infty} p_n$ does not admit a section. So, q_n does not admit a section. Thus, by 4.18(ii), $e_n^{\mathcal{V}} \neq 0$. □

4.20. Definition. We say that a connected ring spectrum E of finite $\mathbb{Z}[p]$-type is *simple* if the fibration $q_{n+1} \cdots q_m : B(\mathcal{V}, E_m) \to B(\mathcal{V}, E_n)$ is simple for every m, n, $m > n$ and every $\mathcal{V}$.

4.21. Lemma. (i) *The spectrum* E *is simple iff every fibration* $q_n : B(\mathcal{V}, E_n) \to B(\mathcal{V}, E_{n-1})$ *is simple for every* n *and every* $\mathcal{V}$.

(ii) *If the space* $B(\mathcal{V}, E_0)$ *is simply connected* (*i.e.*, $B(\mathcal{V}, E_0) \simeq BS\mathcal{V}$), *then* E *is simple.*

(iii) *If* $2\pi_*(E) \neq 0$ *then* E *is simple.*

(iv) *Let* A *be a finitely generated* $\mathbb{Z}[p]$*-module. If a spectrum* E *is simple then the group* $H^i(B(\mathcal{V}, E_n); A)$ *is a finitely generated* $\mathbb{Z}[p]$*-module for every* i, n.

Proof. (i) Let F be a fiber of $q_{n+1} \cdots q_m$. It follows from 4.18(i) that $\pi_i(F) = \pi_i$ for $n + 1 \leq i \leq m$ and $\pi_i(F) = 0$ otherwise. Now, the $\pi_1(B(\mathcal{V}, E_n))$-action on $\pi_i(F), n + 1 \leq i \leq m$, coincides the $\pi_1(B(\mathcal{V}, i-1))$-action on $\pi_i(\text{fiber of } q_i) = \pi_i(F)$.

(ii) The action of $\pi_1(B(\mathcal{V}, E_{n-1})) = \pi_1$ on $\pi_n(\text{fiber of } q_n) = \pi_n$ coincides with the action of $\pi_1(\Omega^{\infty} E) = \pi_1$ on $\pi_n = \pi_n(\Omega^{\infty} E)$; but the last action is trivial because $\Omega^{\infty} E$ is an H-space. Now the result follows from (i).

---

[13] See e.g. Spanier [2] about Postnikov–Moore towers.

(iii) If $2\pi_*(E) \neq 0$ then the fiber of q_0 is homotopy equivalent to $\mathbb{Z}/2$. Since $[\text{pt}, B(\mathcal{V}, H(\pi_0))]$ is the one-point set, $B(\mathcal{V}, E_0) = B(\mathcal{V}, H(\pi_0))$ is a connected space. Considering the homotopy exact sequence of the fibration q_0, we conclude that $B(\mathcal{V}, E_0)$ is simply connected, and the claim follows from (ii).

(iv) Firstly, we prove that every group $H^i(B(\mathcal{V}, E_0); A)$ is a finitely generated $\mathbb{Z}[p]$-module. There are two possibilities: q_0 is the identity map or q_0 is homotopy equivalent to the universal covering (cf. the proof of (iii)). If q_0 is the identity map then, by IV.4.27(ix), every group $H_i(B\mathcal{V})$ is finitely generated, and so $H^i(B\mathcal{V}; A) = H^i(B(\mathcal{V}, E_0); A)$ is a finitely generated $\mathbb{Z}[p]$-module. Furthermore, if q_0 is homotopy equivalent to the universal covering then $B(\mathcal{V}, E_0)$ is simply connected and $\pi_i(B(\mathcal{V}, E_0)) = \pi_i(B\mathcal{V})$ for every $i > 1$. So, $\pi_i(B(\mathcal{V}, E_0))$ are finitely generated abelian groups by IV.4.27(ix). Hence, $H_i(B(\mathcal{V}, E_0))$ are finitely generated abelian groups by the mod $\mathfrak{C}$ Hurewicz Theorem for spaces, where $\mathfrak{C}$ is the Serre class of the finitely generated abelian groups, see e.g. Mosher–Tangora [1]. Thus, $H^i(B(\mathcal{V}, E_0); A)$ is a finitely generated $\mathbb{Z}[p]$-module for every i.

Suppose inductively that $H^i(B(\mathcal{V}, E_{n-1}); \mathbb{Z}[p])$ is a finitely generated $\mathbb{Z}[p]$-module for every i. It is well known (or one can prove this as above) that $H^i(K(\pi, n); A)$ is a finitely generated $\mathbb{Z}[p]$-module for every finitely generated $\mathbb{Z}[p]$-module π and every i, n. Considering the Leray–Serre spectral sequence of the fibration $K(\pi_n, n) \to B(\mathcal{V}, E_{n-1}) \to B(\mathcal{V}, E_n)$, one can see that $H^i(B(\mathcal{V}, E_n); A)$ is a finitely generated $\mathbb{Z}[p]$-module for every i. The induction is confirmed. □

4.22. Theorem. *Suppose that E is a simple spectrum and that the group π_n is cyclic or isomorphic to $\mathbb{Z}[p]$. Then $e_n^{\mathcal{V}}$ is a characteristic class of the $K(\pi, n)$-fibration q_n. (In other words, if χ_n is a characteristic class of the $K(\pi, n)$-fibration q_n then $\chi_n = \varepsilon e_n^{\mathcal{V}}$ for some $\varepsilon \in \mathbb{Z}[p]^*$.)*

Proof. Let $\chi = \chi_n$ be a charactersitic class of q_n. If $e_n^{\mathcal{V}} = 0$ then, by 4.18(ii), q_n admits a section, and so $\chi = 0$. Similarly, the converse holds. So, we can assume that $e_n^{\mathcal{V}} \neq 0 \neq \chi$. Let $r : B \to B(\mathcal{V}, E_{n-1})$ be a simple $K(\pi_n, n)$-fibration with characteristic class $e_n^{\mathcal{V}}$. (For example, take the fibration induced from the standard $K(\pi_n, n)$-fibration $PK(\pi_n, n+1) \to K(\pi_n, n+1)$ (see IV.1.39(a)) by the map $e_n^{\mathcal{V}} : B(\mathcal{V}, E_{n-1}) \to K(\pi_n, n+1)$.) We have

$$q_n^*(e_n^{\mathcal{V}}) = q_n^* \varphi^{-1} \kappa_n u_{n-1} = \varphi^{-1}(Tq_n)^* \kappa_n u_{n-1} \text{ where } Tq_n := T\mathfrak{I}_{q_n, \gamma_{n-1}}.$$

But $T(q_n)^* \kappa_n u_{n-1}$ is given by the composition

$$M(\mathcal{V}, E_n) \xrightarrow{Tq_n} M(\mathcal{V}, E_{n-1}) \xrightarrow{u_{n-1}} E_{n-1} \xrightarrow{\kappa_n} \Sigma^{n+1} H(\pi_n),$$

and, by 1.18,

$$\kappa_n u_{n-1} Tq_n = \kappa_n u_{n-1} M(\mathcal{V}, p_n) = \kappa_n p_n u_{n-1}.$$

So, $q_n^*(e_n^{\mathcal{V}}) = 0$ since $\kappa_n p_n = 0$. Hence, there exists a map $\alpha : B(\mathcal{V}, E_n) \to B$ over $B(\mathcal{V}, E_{n-1})$. Similarly, since $r^*(e_n^{\mathcal{V}}) = 0$, the E_{n-1}-oriented $\mathcal{V}$-object $r^*\gamma_{n-1}^{\mathcal{V}}$ admits an E_n-orientation, and hence there exists a map $\beta : B \to B(\mathcal{V}, E_n)$ over $B(\mathcal{V}, E_{n-1})$.

$$\begin{array}{ccccccc} B(\mathcal{V}, E_n) & \xrightarrow{\alpha} & B & \qquad & B & \xrightarrow{\beta} & B(\mathcal{V}, E_n) \\ \downarrow{\scriptstyle q_n} & & \downarrow{\scriptstyle r} & & \downarrow{\scriptstyle r} & & \downarrow{\scriptstyle q_n} \\ B(\mathcal{V}, E_{n-1}) & = & B(\mathcal{V}, E_{n-1}) & & B(\mathcal{V}, E_{n-1}) & = & B(\mathcal{V}, E_{n-1}) \end{array}$$

Since characteristic classes of the fibrations q_n and r are non-zero, $\alpha\beta$ and $\beta\alpha$ are homotopy equivalences by 4.5. Hence, α is a homotopy equivalence.

Let $K'(\pi, n)$ (resp. $K''(\pi, n)$) be the fiber of r (resp. of q_n), and let $\iota' \in H^n(K'(\pi, n); \pi)$ (resp. $\iota'' \in H^n(K''(\pi, n); \pi)$) be a fundamental class (we can't say "the fundamental class", see 4.1). Furthermore, the homotopy equivalence α induces a homotopy equivalence $\overline{\alpha} : K''(\pi, n) \to K'(\pi, n)$. Since $H^n(K(\pi, n); \pi) = \pi$, we conclude that $\overline{\alpha}^*\iota' = \varepsilon\iota''$ for some $\varepsilon \in \mathbb{Z}[p]^*$. Thus, $\chi = \varepsilon e_n^{\mathcal{V}}$. □

4.23. Remarks. (a) Resuming the above, we have two ways to E-orient a $\mathcal{V}$-object $\xi = \{f : X \to B\mathcal{V}\}$. The first way is to lift a morphism $v : T\xi \to H(\pi_0)$ to E along the tower (4.7), and in this way we meet the obstructions given by the κ_n's. The second way is to lift the map $f : X \to B\mathcal{V}$ to $B(\mathcal{V}, E)$ along the tower (4.17), and in this way we meet the obstructions given by the e_n's. So, the transfer from tower (4.7) to tower (4.17) can be considered as a form of the Thom isomorphism.

(b) Theorems 4.18(ii) and 4.22 show that it makes sense to introduce a class $e_0^{\mathcal{V}} \in H^1(B\mathcal{V}; \pi_0)$ as the characteristic class of the covering $B(\mathcal{V}, E_0) \to B\mathcal{V}$. In fact, $e_0^{\mathcal{V}} = 0$ if $2\pi_0(E) = 0$ and $e_0^{\mathcal{V}} = w_1$ otherwise. Clearly, $e_0^{\mathcal{V}}(\xi)$ is the obstruction to E_0-orientability of ξ, i.e., it is the first obstruction to E-orientability of ξ.

(c) Knapp–Ossa [1] defined the *E-codegree* of an $\mathcal{F}_n$-object α to be the minimal natural number k such that $k\sigma^n \in \operatorname{Im}\{j^* : \widetilde{E}^n(T\alpha) \to \widetilde{E}^n(S^n)\}$. Clearly, the codegree of α is equal to 1 iff α is E-orientable. Knapp–Ossa [1] considered just K- and KO-codegree of vector bundles; however, I think, the general concept of E-codegree is interesting and is able to be considered in some general context related to the results of this and the next section.

§5. Realizability of Obstructions to Orientability

Now we are ready to attack the problem of realizability of the classes e_n. Let $\{\pi_{i_r}\}, 0 = i_0 < i_1 < \cdots$ be the set of all non-trivial homotopy groups of E. In other words, $\pi_i \neq 0$ iff $i = i_r$ for some r. Note that $E_{i_{r-1}} = E_{i_r-1}$ and $\gamma^{\mathcal{V}}_{i_{r-1}} = \gamma^{\mathcal{V}}_{i_r-1}$. Of course, it makes sense to realize the classes e_{i_r} only, because each class e_j with $j \neq i_r$ belongs to the trivial group. For simplicity, denote e_{i_r} by $\varkappa_r$ and κ_{i_r} by σ_r.

Let $X|n \to X$ be the $(n-1)$-connective covering of X (see IV.1.39(b)). Let $(\Omega^\infty \sigma_i)|n$ be the corresponding Postnikov invariants of $(\Omega^\infty E)|n$. (Recall that Ω^∞ transforms the Postnikov tower of a spectrum to that of the space.)

5.1. Theorem. *Let E be a simple spectrum, and let the group π_j be cyclic (possibly trivial) or $\mathbb{Z}[p]$ for every j. Suppose that there exists n such that*

(i) $0 \notin \varkappa_n(\gamma^{\mathcal{V}}_{i_n-1})$,

(ii) $(\Omega^\infty \sigma_r)|i_n \neq 0$ *for all* $r > n$.

Then $0 \notin \varkappa_r(\gamma^{\mathcal{V}}_{i_r-1})$ for $r \geq n$. In other words, all characteristic classes $\varkappa_r, r \geq n$ can be realized by a $\mathcal{V}$-object.

Proof. Firstly, we prove the following lemma.

5.2. Lemma. *Let $r > n$, and let $\omega : \Omega^\infty E_{i_{r-1}} \to \Omega^\infty E_{i_{r-1}}$ be a map such that $\omega_* : \pi_{i_n} \to \pi_{i_n}$ is an isomorphism. Then ω cannot be lifted to $\Omega^\infty E_{i_r}$ (with respect to the projection $\Omega^\infty E_{i_r} \to \Omega^\infty E_{i_{r-1}}$).*

Proof. Because of naturality, ω gives a self-map of the Postnikov–Moore tower

$$\begin{array}{ccccccc}
\Omega^\infty E_{i_{r-1}} & \longrightarrow & \Omega^\infty E_{i_{r-2}} & \longrightarrow & \cdots & \longrightarrow & \Omega^\infty E_{i_n} \\
\downarrow{\scriptstyle\omega} & & \downarrow & & & & \downarrow \\
\Omega^\infty E_{i_{r-1}} & \longrightarrow & \Omega^\infty E_{i_{r-2}} & \longrightarrow & \cdots & \longrightarrow & \Omega^\infty E_{i_n}.
\end{array}$$

Let $X_s \to \Omega^\infty E_{i_s}$ be an $(i_n - 1)$-connective covering of $\Omega^\infty E_{i_s}$. Because of naturality of connective coverings, this diagram induces the diagram

$$\begin{array}{ccccccc}
X_{r-1} & \longrightarrow & X_{r-2} & \longrightarrow & \cdots & \longrightarrow & X_n \\
{\scriptstyle\omega_{r-1}}\downarrow & & \downarrow{\scriptstyle\omega_{r-2}} & & & & \downarrow{\scriptstyle\omega_n} \\
X_{r-1} & \longrightarrow & X_{r-2} & \longrightarrow & \cdots & \longrightarrow & X_n.
\end{array}$$

Note that $X_n = K(\pi_{i_n}, i_n)$. Hence, ω_n induces an isomorphism of homotopy groups, and so it is a homotopy equivalence. Suppose inductively that $\omega_s, n \leq s < r-1$, is a homotopy equivalence. In the commutative diagram

$$\begin{array}{ccc} X_{s+1} & \xrightarrow{\omega_{s+1}} & X_{s+1} \\ \downarrow & & \downarrow \\ X_s & \xrightarrow{\omega_s} & X_s \end{array}$$

the vertical arrows are $K(\pi_{i_s+1}, i_s+1)$-fibrations with the characteristic class $\Omega^\infty \sigma_{s+1}|i_n$, which is non-trivial in view of the condition (ii) of the theorem. Furthermore, ω_s is a homotopy equivalence. Now, $\pi_i(X_s)$ are finitely generated $\mathbb{Z}[p]$-modules, and hence $H^i(X_s; \pi_{i_s+1})$ are finitely generated $\mathbb{Z}[p]$-modules. So, by 4.5, ω_{s+1} is a homotopy equivalence. Thus, inductively, ω_{r-1} is a homotopy equivalence. But the characteristic class of the fibration $X_r \to X_{r-1}$ is non-trivial according to the condition (ii) of the theorem, and hence ω_{r-1} cannot be lifted to X_r. This implies that ω cannot be lifted to $\Omega^\infty E_{i_r}$. ⊞

We continue the proof of the theorem. For simplicity, we denote $B(\mathcal{V}, E_{i_k})$ by B_k. The assertion $0 \notin \varkappa_r(\gamma_{i_r-1})$ is equivalent to the following one: there is no map $B_{r-1} \to B_{r-1}$ over $B\mathcal{V}$ which can be lifted to B_r, i.e., that the diagram (5.3) below cannot be completed:

(5.3)
$$\begin{array}{ccc} B_{r-1} & & B_r \\ \| & & \downarrow \\ B_{r-1} & \longrightarrow & B_{r-1} \\ \downarrow & & \downarrow \\ B\mathcal{V} & = & B\mathcal{V}\,. \end{array}$$

We prove the last assertion. Consider any map $g : B_n \to B_n$ over $B\mathcal{V}$. By naturality of Postnikov–Moore towers, we have a commutative diagram

$$\begin{array}{ccccccc} K(\pi_{i_n}, i_n) & \longrightarrow & B_n & \longrightarrow & B_{n-1} & \longrightarrow & B\mathcal{V} \\ {\scriptstyle \overline{g}}\downarrow & & {\scriptstyle g}\downarrow & & \downarrow{\scriptstyle f} & & \| \\ K(\pi_{i_n}, i_n) & \longrightarrow & B_n & \longrightarrow & B_{n-1} & \longrightarrow & B\mathcal{V}. \end{array}$$

Here $K(\pi_{i_n}, i_n)$ is the fiber of the fibration $B_n \to B_{n-1}$, and $\overline{g}$ is the map of fibers induced by g. Let $\iota \in H^{i_n}(K(\pi_{i_n}, i_n); \pi_{i_n})$ be a fundamental class.

5.4. Lemma. *The fibration $B_n \to B_{n-1}$ is simple, and its characteristic class $\tau\iota$ has finite order.*

Proof. The fibration is simple because E is a simple spectrum. Now, let $u : T\gamma^{\mathcal{V}}_{i_n-1} \to E_{n-1}$ be the universal E_{n-1}-orientation of $\gamma^{\mathcal{V}}_{i_n-1}$. By (4.15), $\varphi(\varkappa^{\mathcal{V}}_n) = \sigma_n u = u^*\sigma_n$, where $\sigma_n \in H^{i_n+1}(E_{i_n-1}; \pi_{i_n})$ is the Postnikov invariant of E. By II.7.12(i), σ_n has finite order, and so $\varphi(\varkappa^{\mathcal{V}}_n)$ has finite order, and so $\varkappa^{\mathcal{V}}_n$ has finite order. But, by 4.22, $\tau\iota = \varepsilon\varkappa^{\mathcal{V}}_n$ with $\varepsilon \in \mathbb{Z}[p]^*$. ⊞

5.5. Lemma. *The map $\overline{g}$ is a homotopy equivalence, i.e., $\overline{g}^*\iota = \lambda\iota$ for some $\lambda \in \mathbb{Z}[p]^*$.*

Proof. By the condition (i) of the theorem, $0 \notin \varkappa_n(\gamma^{\mathcal{V}}_{i_n-1})$, and so $\varkappa^{\mathcal{V}}_n \neq 0$. Hence, by 4.22, $\tau\iota \neq 0$. By 5.4, $p^N\tau\iota = 0$ for some N. Suppose that $\overline{g}^*\iota = p^s\lambda\iota, s > 0, \lambda \in \mathbb{Z}[p]$. Consider the commutative diagram (where f^N denotes $f\circ\cdots\circ f$)

$$\begin{array}{ccc} B_n & \xrightarrow{g^N} & B_n \\ \downarrow & & \downarrow \\ B_{n-1} & \xrightarrow[f^N]{} & B_{n-1}. \end{array}$$

Note that $\gamma^{\mathcal{V}}_{i_n-1}$ is not E_{i_n-1}-orientable because $0 \notin \varkappa_n(\gamma^{\mathcal{V}}_{i_n-1})$. Since f is a map over $B\mathcal{V}$, this non-orientability implies that f^N cannot be lifted to B_n. Hence, $(f^N)^*\tau\iota \neq 0$. But

$$(f^N)^*\tau\iota = \tau((\overline{g}^N)^*\iota) = \tau(p^{sN}\lambda^N\iota) = p^{sN}\lambda^N\tau\iota = 0.$$

This is a contradiction. ⊞

Now we finish the proof of the theorem. Let $r > n$. We must prove that $0 \notin \varkappa_n(\gamma^{\mathcal{V}}_{i_n-1})$, i.e., that any $h : B_{r-1} \to B_{r-1}$ over $B\mathcal{V}$ cannot be lifted to B_r, see (5.3). Suppose that there exists h which can be lifted to B_r. Then it induces a map of the Postnikov–Moore towers, and, in particular, a map $g : B_n \to B_n$ over $B\mathcal{V}$. We have the two diagrams below, where the bottom vertical maps are the forgetful fibrations from 1.17 and the top vertical maps are inclusions of fibers:

$$\begin{array}{ccc} \Omega^\infty_{\pm1}E_{i_r-1} & \xrightarrow{\omega} & \Omega^\infty_{\pm1}E_{i_r-1} \\ \downarrow & & \downarrow \\ B_{r-1} & \xrightarrow{h} & B_{r-1} \\ \downarrow & & \downarrow \\ B\mathcal{V} & = & B\mathcal{V}, \end{array} \qquad \begin{array}{ccc} \Omega^\infty_{\pm1}E_{i_n} & \xrightarrow{\psi} & \Omega^\infty_{\pm1}E_{i_n} \\ \downarrow & & \downarrow \\ B_n & \xrightarrow{g} & B_n \\ \downarrow & & \downarrow \\ B\mathcal{V} & = & B\mathcal{V} \ . \end{array}$$

By 5.5, $\psi_* : \pi_{i_n}(\Omega^\infty E_{i_n}) \to \pi_{i_n}(\Omega^\infty E_{i_n})$ is an isomorphism. So, since ψ is the "Postnikov i_n-stage" of ω, we conclude that $\omega_* : \pi_{i_n}(\Omega^\infty E_{i_r-1}) \to \pi_{i_n}(\Omega^\infty E_{i_r-1})$ is an isomorphism. Now, by 5.2, ω cannot be lifted to $\Omega^\infty E_{i_r}$, and hence h cannot be lifted to B_r. □

Note that in Theorem 5.1 the class $\mathcal{V}$ appears in condition (i) only: condition (ii) is related purely to the spectrum E. It is possible to weaken condition (ii) with simultaneous strengthening of condition (i) to obtain the following theorem.

5.6. Theorem. *Let E be a simple spectrum, and let the group π_j be cyclic or isomorphic to $\mathbb{Z}[p]$ for every j. Suppose that*

(i) $\varkappa_1(\gamma_0^{\mathcal{V}}) \neq 0$,

(ii) $\varkappa_r^{\mathcal{V}} \neq 0$ *for every* $r > 1$.

Then $0 \notin \varkappa_r(\gamma_{i_r-1}^{\mathcal{V}}), r = 1, 2, \ldots$.

Proof. This is similar to the proof of 5.1, but simpler. It suffices to prove that there is no map $B_{r-1} \to B_r$ over $B\mathcal{V}$. By 4.18(ii), non-triviality of the class $\varkappa_r^{\mathcal{V}}$ implies the the fibration $B_r \to B_{r-1}$ does not admit a section, $r = 2, 3, \ldots$. Hence, it suffices to prove that every map $B_{r-1} \to B_{r-1}$ over $B\mathcal{V}$ is a homotopy equivalence for every $r > 0$. For $r = 1$ this follows from the equality $B(\mathcal{V}, E_0) = BS\mathcal{V}$. Consider any r and suppose inductively that every map $B_{r-1} \to B_{r-1}$ over $B\mathcal{V}$ is a homotopy equivalence. Given any map $g : B_r \to B_r$ over $B\mathcal{V}$, we have, by naturality, a diagram

$$\begin{array}{ccc} B_r & \xrightarrow{g} & B_r \\ \downarrow & & \downarrow \\ B_{r-1} & \xrightarrow[f]{} & B_{r-1} \end{array}$$

where f is an equivalence. Thus, by 4.21 (iv) and 4.5, g is a homotopy equivalence. □

5.7. Remark. In Section 4 we discussed $\mathcal{V}$-objects over finite dimensional spaces. The spaces $B(\mathcal{V}, E_n)$ are not finite dimensional, but it is easy to see that the classes $\varkappa_r$ from 5.1, 5.6 can be realized by $\mathcal{V}$-objects over certain skeletons of $B(\mathcal{V}, E_n)$.

5.8. Remark. The results of this and previous sections were obtained by Rudyak [6,8].

Chapter VI. K- and $K\mathcal{O}$-Orientability

In this chapter we apply the results of the previous one to the orientability of $\mathcal{V}$-objects with respect to K and $K\mathcal{O}$. The case $\mathcal{V} = \mathcal{O}$ was considered by Atiyah–Bott–Shapiro [1], the other cases were considered mainly by the author, see Rudyak [6,8,9]. To be convenient, we collect the results as a résumé, see the ends of §§ 3,4. Here K, resp. $K\mathcal{O}$, means complex, resp. real K-theory, see Atiyah [4], Husemoller [1], Karoubi [1], etc.

Set $k := K|0$, $k\mathcal{O} := K\mathcal{O}|0$. In view of V.1.17 K-, resp. $K\mathcal{O}$-orientability is equivalent to k-, resp. $k\mathcal{O}$-orientabilty. So, it suffices to consider the k- and $k\mathcal{O}$-orientability problems.

As usual, given a space X and an abelian group π, we do not distinguish elements of $H^n(X;\pi)$ and maps (homotopy classes) $X \to K(\pi, n)$. For example, we can and shall speak about the map $Sq^k : K(\mathbb{Z}/2, n) \to K(\mathbb{Z}/2, n+k)$; this map corresponds to the element $Sq^k \iota_n$ where $\iota_n \in H^n(K(\mathbb{Z}/2, n); \mathbb{Z}/2)$ is the fundamental class.

Let G be one of the groups $\mathbb{Z}$ or $\mathbb{Z}[2]$. In this chapter $\rho : HG \to H\mathbb{Z}/2$ (as well as $\rho : K(\mathbb{Z}, n) \to K(\mathbb{Z}/2, n)$) denotes the reduction mod 2, and $\delta : H\mathbb{Z}/2 \to \Sigma HG$ denotes the integral Bockstein morphism. We use the same symbols $\rho : H^n(X;G) \to H^n(X;\mathbb{Z}/2)$ and $\delta : H^n(X;\mathbb{Z}/2) \to H^{n+1}(X;\mathbb{G})$ for the corresponding homomorphisms. Thus, there is the Bockstein exact sequence, where 2 over the arrow means multiplication by 2,

$$\cdots \to H^n(X;G) \xrightarrow{2} H^n(X;G) \xrightarrow{\rho} H^n(X;\mathbb{Z}/2) \xrightarrow{\delta} H^{n+1}(X;G) \to \cdots .$$

Finally, as usual, $\mathbb{Z}[2]^*$ denotes the set of invertible elements of $\mathbb{Z}[2]$.

§1. Some Secondary Operations on Thom Classes

The results of this section were obtained by Hegenbarth [1]. In this section H denotes $H\mathbb{Z}/2$, $H_*(A)$ denotes $H_*(A;\mathbb{Z}/2)$ and $H^*(A)$ denotes $H^*(A;\mathbb{Z}/2)$.

The information that we shall need on secondary cohomology operations can be found in Mosher–Tangora [1]. One says that a secondary operation Φ can be realized by $\mathcal{V}$-objects if there exists a $\mathcal{V}$-object ξ such that Φ is defined on the Thom class $u_\xi \in H^*(T\xi)$ and $0 \notin \Phi(u_\xi)$ (recall that usually Φ is a multivalued map).

Consider an Adem relation

$$Sq^a Sq^b = \sum_{c=0}^{[a/2]} \binom{b-c-1}{a-2c} Sq^{a+b-c} Sq^c$$

with $a < 2b$. Let $\Phi(a,b)$ be the secondary operation associated with this relation. The goal of this section is to prove that $\Phi(2,2)$ and $\Phi(3,3)$ can be realized by spherical fibrations.

We need some preliminaries about Kudo–Araki–Dyer–Lashof operations. We just give a brief description of their properties: more detailed information can be found in Cohen–Lada–May [1], Madsen–Milgram [1]. These operations were introduced by Kudo–Araki [1] for $p = 2$, while Dyer–Lashof [1] have given the construction for $p > 2$.

Let X be an n-fold loop space. The product on $H_*(X)$ we denote by $*$. There are operations $Q^i : H_k(X) \to H_{k+i}(X)$ with the following properties:

(1) Q^i is defined for $i - k < n - 1$;

(2) Q^i is natural with respect to n-fold loop maps (i.e., $Q^i(f)$ is defined for $f = \Omega^n g$);

(3) $Q^i(x) = 0$ for $i < \dim x$;

(4) $Q^i(x) = x * x$ for $i = \dim x$;

(5) Let $\sigma : S\Omega X \to X$ be the adjoint map to $1_{\Omega X}$, and let

$$\sigma_* : H_i(\Omega X) \cong H_{i+1}(S\Omega X) \to H_{i+1}(X)$$

be the homological suspension. Then $Q^i(\sigma_* x) = \sigma_*(Q^i x)$;

(6) (Cartan formula) Given $x \in H_*(X), y \in H_*(Y)$, one has

$$Q^i(x \otimes y) = \sum_{j+k=i} Q^j(x) \otimes Q^k(y),$$

where $x \otimes y \in H_*(X \times Y)$ and $X \times Y$ is equipped with the product loop structure. By naturality,

$$Q^i(x * y) = \sum_{j+k=i} Q^j(x) * Q^k(y)$$

for every $x, y \in H_*(X)$, and

$$Q^i(d_* x) = \sum_{j+k=i} Q^j(x') * Q^k(x''),$$

where $d : X \to X \times X$ is the diagonal and $d_*(x) = \sum x' \otimes x''$;

(7) (Adem relations) If $r > 2s$, then

$$Q^r Q^s = \sum_i \binom{2i-r}{r-s-i-1} Q^{r+s-i} Q^i;$$

(8) (Nishida relations) Let $Sq^i_\bullet : H_k(X) \to H_{k-i}(X)$ be the dual operation to Sq^i, i.e., $\langle Sq^i(x), a\rangle = \langle x, Sq^i_\bullet(a)\rangle$, cf. II.6.36. Then

$$Sq^i_\bullet Q^j(a) = \sum_{2k<i} Q^{j-i+k} Sq^k_\bullet(a);$$

(9) If $a \in H_*(\Omega^n_k S^n)$, then $Q^i(a) \in H_*(\Omega^n_{2k} S^n)$, where $\Omega^n_k S^n$ is the component of $\Omega^n S^n$ consisting of all maps of degree k.

Let $0 \neq a_k \in H_0(\Omega^n_k S^n) = \mathbb{Z}/2$, and let $[k]$ be the image of a_k in $H_0(\Omega^n S^n)$. Note that $[k] * [l] = [k+l]$.

Given a connected space A, let ε be the non-trivial element of $H_0(A)$. Let K_n denote $K(\mathbb{Z}/2, n)$, and let $x_n \in H_n(K_n)$, $y_n \in H_n(K(\mathbb{Z}, n))$ be the generators.

Consider the two-stage Postnikov system (n is large, in fact $n > k$)

$$\begin{array}{ccc} K_{n+k-1} & \xrightarrow{j} & E & & \\ & & \downarrow{\scriptstyle p} & & \\ & & K(\mathbb{Z}, n) & \xrightarrow{Sq^k \rho} & K_{n+k}, \end{array}$$

i.e., $p : E \to K(\mathbb{Z}, n)$ is a fibration with fiber K_{n+k-1} and characteristic class $Sq^k \rho$, i.e., E is the homotopy fiber of $Sq^k \rho$. Here j is the inclusion of a fiber. Let $\iota : S^n \to E$ yield the generator $1 \in \pi_n(E) = \mathbb{Z}$. The element $(\Omega^n \iota)[l] \in H_0(\Omega^n E)$ we denote also by $[l]$. The image $(\Omega^n j)(x_{k-1}) \in H_{k-1}(\Omega^n E)$ of x_{k-1} we denote also by x_{k-1}.

1.1. Lemma. *$Q^{k-1}[1] = [2] * x_{k-1}$ in $H_*(\Omega^n E)$.*

Proof. It is clear that $Q^{k-1}[1] = 0$ or $Q^{k-1}[1] = [l] * x_{k-1}$. Moreover, in the latter case $l = 2$ because of the naturality of Q^i and (9). Hence, we must prove that $Q^{k-1}[1] \neq 0$. Let

$$\sigma_*^{k-1} : H_0(\Omega^n E) \to H_{k-1}(\Omega^{n-k+1} E)$$

be the iterated homological suspension. Under the homotopy equivalence (not as loop spaces)

$$\Omega^{n-k+1} E \simeq K(\mathbb{Z}, k-1) \times K_{2k-2}$$

we have $\sigma_*^{k-1}[1] = y_{k-1} \otimes \varepsilon$. Now

$$\sigma_*^{k-1}(Q^{k-1}[1]) = Q^{k-1}(\sigma_*^{k-1}[1]) = Q^{k-1}(y_{k-1} \otimes \varepsilon) = (y_{k-1} \otimes \varepsilon)^2,$$

(where a^2 means $a * a$). Hence, we must prove that $(y_{k-1} \otimes \varepsilon)^2 \neq 0$.

Let $u' \in H^{2k-2}(K_{2k-2})$ and $v' \in H^{k-1}(K(\mathbb{Z}, k-1))$ be the generators. Let

$$u \in H^{2k-2}(\Omega^{n-k+1}E), \quad v \in H^{k-1}(\Omega^{n-k+1}E)$$

be the images of u' and v' under the equivalence $\Omega^{n-k+1}E \simeq K(\mathbb{Z}, k-1) \times K_{2k-2}$. Let

$$\psi : H^*(\Omega^{n-k+1}E) \to H^*(\Omega^{n-k+1}E) \otimes H^*(\Omega^{n-k+1}E)$$

be induced by the product on the loop space. It is clear that the inequality $(y_{k-1} \otimes \varepsilon)^2 \neq 0$ follows from the following fact.

1.2. Sublemma (Milgram [2]). $\psi(u) = u \otimes 1 + v \otimes v + 1 \otimes u.$

Proof. Put $X = \Omega^{n-k}E$ and consider the $\mathbb{Z}/2$-cohomology Leray–Serre spectral sequence of the ΩX-fibration $PX \to X$. This spectral sequence is a spectral sequence of Hopf algebras because of the loop product on X. In particular, there is a family of comultiplications $\psi_r : E_r^{*,*} \to E_r^{*,*} \otimes E_r^{*,*}$ commuting with the differentials.

Let $\bar{\otimes}$ be the multiplication in this spectral sequence. Since $Sq^k(\tau v) = 0$ (τ is the transgression), $\tau v \bar{\otimes} v$ is killed by an element of the fiber. Hence, $\tau v \bar{\otimes} v = d_k u$. Now

$$\begin{aligned} d_k(\psi_k u) &= \psi_{k+1}(d_k u) = \psi_{k+1}(\tau v \bar{\otimes} v) = \psi(\tau v) \bar{\otimes} \psi(v) \\ &= (\tau v \otimes 1 + 1 \otimes \tau v) \bar{\otimes} (v \otimes 1 + 1 \otimes v) \\ &= \tau v \bar{\otimes} v \otimes 1 + v \otimes \tau v + \tau v \otimes v + 1 \otimes \tau v \bar{\otimes} v = \\ & d_k(u \otimes 1 + v \otimes v + 1 \otimes u). \end{aligned}$$

Hence, $\psi(u) = u \otimes 1 + v \otimes v + 1 \otimes u$. Hence, 1.2, and thus 1.1, is proved. □

1.3. Lemma. (i) *If $k-1 \geq r > 0$, then $\Omega^{n-r}E \simeq K(\mathbb{Z}, r) \times K_{k+r-1}$ (not as loop spaces), and $Q^{k-1}(y_r \otimes \varepsilon) = \varepsilon \otimes x_{k+r-1}$.*

(ii) *Consider the Postnikov tower*

$$\begin{array}{ccc} K_{n+k-1} & \xrightarrow{\widetilde{j}} & \widetilde{E} \\ & & \downarrow p \\ & K_n & \xrightarrow{Sq^k} K_{n+k}. \end{array}$$

If $k-1 \geq r > 0$, then $\Omega^{n-r}\widetilde{E} \simeq K_r \times K_{k+r-1}$, and $Q^{k-1}(x_r \otimes \varepsilon) = \varepsilon \otimes x_{k+r-1}$.

Proof. (i) The equivalence $\Omega^{n-r}E \simeq K(\mathbb{Z}, r) \times K_{k+r-1}$ is clear. Now, we have the equality $\sigma_*([j] * x_{k-1}) = \varepsilon \otimes x_k$, where

$$\sigma_* : H_{k-1}(\Omega^n E) \to H_k(\Omega^{n-1}E) \cong H_0(K(\mathbb{Z},1)) \otimes H_k(K_k)$$

is the homological suspension. Furthermore, $\sigma_*(y_p \otimes \varepsilon) = y_{p+1} \otimes \varepsilon$ and $\sigma_*(\varepsilon \otimes x_q) = \varepsilon \otimes x_{q+1}$, where

$$\sigma_* : H_*(K(\mathbb{Z},p) \times K_q) \to H_*(K(\mathbb{Z},p+1) \times K_{q+1}).$$

Thus, by 1.1,

$$\begin{aligned} Q^{k-1}(y_r \otimes \varepsilon) &= Q^{k-1}(\sigma_*^r[1]) = \sigma_*^r(Q^{k-1}[1]) = \sigma_*^r([2] * x_{k-1}) = \sigma_*^{r-1}(\varepsilon \otimes x_k) \\ &= \varepsilon \otimes x_{k+r-1}. \end{aligned}$$

(ii) This follows from (i), because there is a morphism of Postnikov towers

$$\begin{array}{ccccc} E & \longrightarrow & \widetilde{E} & & \\ \downarrow & & \downarrow & & \\ K(\mathbb{Z},n) & \xrightarrow{\rho} & K_n & \xrightarrow{Sq^k} & K_{n+k}. \end{array}$$

□

1.4. Lemma. *Given a Postinkov tower*

$$\begin{array}{ccccc} K_{n+2} & \xrightarrow{j} & E & & \\ & & \downarrow{\scriptstyle p} & & \\ & & K(\mathbb{Z},n) & \xrightarrow{Sq^3\rho} & K_{n+3}, \end{array}$$

there is a map $\Phi : E \to K_{n+5}$ *such that* $\Phi j = Sq^3$.

Proof. There is the Adem relation $Sq^3Sq^3 + Sq^5Sq^1 = 0$. Since $Sq^1\rho = 0$, we have the relation $Sq^3Sq^3\rho = 0$ which holds on integral cohomology classes. Let $\iota_{n+2} \in H^{n+2}(K_{n+2})$ be the fundamental class, and let τ denote the transgression in the Leray–Serre spectral sequence of the K_{n+2}-fibration p. We have

$$\tau(Sq^3\iota_{n+2}) = Sq^3(\tau\iota_{n+2}) = Sq^3Sq^3\rho = 0.$$

Hence,

$$Sq^3\iota_{n+2} \in \mathrm{Im}\{j^* : H^5(E) \to H^5(K_{n+2})\}.$$

Thus, there is $\Phi \in H^5(E)$ with $j^*\Phi = Sq^3\iota_{n+2}$, i.e., $\Phi j = Sq^3$. □

Consider now the 3-stage Postnikov tower

$$\begin{array}{ccccc} E'' & & E' & & \\ {\scriptstyle p''}\downarrow & & \downarrow{\scriptstyle p'} & & \\ K_{n+2} & \xrightarrow{j} & E & \xrightarrow{\Phi} & K_{n+5} \\ & & \downarrow{\scriptstyle p} & & \\ & & K(\mathbb{Z},n) & \xrightarrow{Sq^3\rho} & K_{n+3}, \end{array}$$

where $\Phi j = Sq^3$, $p' : E' \to E$ is the K_{n+4}-fibration with characteristic class φ, and p'' is induced from p' by j. Of course, E'' is the homotopy fiber of pp'. Moreover,

$$\Omega^n E' \simeq K(\mathbb{Z}, 0) \times \Omega^n E''.$$

Furthermore, $\Omega^n E'' \simeq K_2 \times K_4$.

1.5. Lemma. *In the group*

$$H_*(\Omega^n E') = H_*(K(\mathbb{Z}, 0)) \otimes H_*(K_2) \otimes H_*(K_4)$$

we have $Q^2 Q^2[1] = [4] * (\varepsilon \otimes x_4)$.

Proof. Firstly, we note that $Q^i[2] = 0$ for $i = 1, 2$. Indeed, by the Adem relations,

$$Q^i[2] = Q^i Q^0[1] = \sum a_{rs} Q^r Q^s[1]$$

with $0 < s < i \leq 2$. But $H_s(\Omega^n E') = 0$ for $s < 2$, and so $Q^s[1] = 0$.

Now, $\Omega^n E' \simeq K(\mathbb{Z}, 0) \times K_2 \times K_4$, and the map

$$\Omega^n p' : \Omega^n E' \to \Omega^n E \simeq K(\mathbb{Z}, 0) \times K_2$$

is the projection onto the first two factors. By 1.1, $(\Omega^n p')_* : H_*(\Omega^n E') \to H_*(\Omega^n E)$ maps $Q^2[1]$ to $Q^2[1] = 2 * x_2$. It is clear that

$$(\Omega^n p')_*([2] * (x_2 \otimes \varepsilon)) = [2] * x_2.$$

But $(\Omega^n p')_*$ is monic in dimension 2, and so $Q^2[1] = [2] * x_2$ in $H_*(\Omega^n E')$. It follows from the Cartan formula (6) that

$$Q^2 Q^2[1] = Q^2([2] * (x_2 \otimes \varepsilon)) = [4] * Q^2(x_2 \otimes \varepsilon),$$

because $Q^0[2] = [4]$ and $Q^i[2] = 0$ for $i > 0$. In order to compute $Q^2(x_2 \otimes \varepsilon)$ we can compute it in $H_*(\Omega^n E'')$. By 1.3(ii), $Q^2(x_2 \otimes \varepsilon) = \varepsilon \otimes x_4$ in $H_*(\Omega^n E'')$. Thus, this holds in $H_*(\Omega^n E')$ also. □

Consider now any Adem relation

$$Sq^a Sq^b + \sum Sq^{l_i} Sq^{k_i} = 0, \quad a < 2b.$$

This relation implies the relation $Sq^a Sq^b \rho + \sum Sq^{l_i} Sq^{k_i} \rho = 0$, which holds on integral cohomology classes. Moreover, it will hold if we exclude terms with $k_i = 1$ (since $Sq^1 \rho = 0$.) Thus, we get the relation

$$Sq^a Sq^b \rho + \sum Sq^{l_i} Sq^{k_i} \rho = 0, k_i > 1$$

and the (partially defined, multivalued) operation $\Phi(a, b) : H\mathbb{Z} \to H\mathbb{Z}/2$ associated with this relation.

Recall that

$$\Phi = \Phi(a,b) : H^i(X;\mathbb{Z}) \to H^{i+a+b-1}(X)$$

is defined on the subgroup $\operatorname{Ker}(Sq^b\rho \cap (\cap_i \operatorname{Ker} Sq^{k_i}\rho))$, and for i large $\Phi(x)$ is a coset with respect to the indeterminacy subgroup

$$\Phi(0) = \operatorname{Im}(Sq^a \oplus (\oplus_i \operatorname{Im} Sq^{l_i})) \subset H^{i+a+b-1}(X).$$

1.6. Theorem. *There exists an oriented stable spherical fibration ξ such that:*

(i) $w_2(\xi) = 0$,
(ii) $\Phi(3,3)$ *is defined on the Thom class* $u_\xi \in H^0(T\xi;\mathbb{Z})$,
(iii) $\Phi(3,3)$ *has zero indeterminacy on* u_ξ,
(iv) $\Phi(3,3)(u_\xi) \neq 0$.

Proof. Let $p : E \to K(\mathbb{Z}, n)$ be a K_{n+2}-fibration with characteristic class $Sq^3\rho : K(\mathbb{Z}, n) \to K_{n+3}$. Consider the diagram

$$\begin{array}{ccccc}
K_{n+4} & \xrightarrow{j'} & E' & & \\
 & & \downarrow{\scriptstyle p'} & & \\
K_{n+2} & \xrightarrow{j} & E & \xrightarrow{\Phi} & K_{n+5} \\
 & & \downarrow{\scriptstyle p} & & \\
 & & K(\mathbb{Z},n) & \xrightarrow{Sq^3\rho} & K_{n+3}
\end{array}$$

where $\Phi j = Sq^3$ and p' is a K_{n+4}-fibration with characteristic class Φ, cf. 1.4. Here j and j' are the inclusions of fibers. Clearly, this diagram is a defining diagram for the operation $\Phi(3,3)$, i.e., $\Phi(3,3) = \Phi$. (As usual, we use the same symbol for a Postnikov invariant and the corresponding cohomology operation.)

Let ζ be an $S\mathcal{F}_n$-object over X with $w_2(\xi) = 0$. Then Φ is defined on u_ζ because

$$Sq^3\rho(u_\zeta) = w_3(\zeta)u_\zeta = Sq^1 w_2(\zeta)u_\zeta = 0.$$

Furthermore, Φ has the indeterminacy

$$\operatorname{Im}(Sq^3 : H^{n+2}(T\zeta) \to H^{n+5}(T\zeta)).$$

Every $x \in H^{n+2}(T\zeta)$ has the form $x = yu_\zeta$ for some $y \in H^2(X)$. Thus,

$$Sq^3(x) = Sq^3(yu_\zeta) = \sum_{i+j=3} Sq^i(y)Sq^j(u_\zeta).$$

But $Sq^j(u_\zeta) = w_j(\zeta)u_\zeta = 0$ for $0 < j \leq 3$, while $Sq^3(y) = 0$. Hence, Φ has zero indeterminacy on u_ζ.

Let D be the homotopy fiber of

$$Sq^2\rho : K(\mathbb{Z}, n) \to K_{n+2}.$$

Since $Sq^1 Sq^2 = Sq^3$, there exists a map $e : D \to E$ such that the diagram

$$\begin{array}{ccccc} D & \longrightarrow & K(\mathbb{Z}, n) & \xrightarrow{Sq^2\rho} & K_{n+2} \\ {\scriptstyle e}\big\downarrow & & \Big\| & & \big\downarrow{\scriptstyle Sq^1} \\ E & \longrightarrow & K(\mathbb{Z}, n) & \xrightarrow{Sq^3\rho} & K_{n+3} \end{array}$$

commutes. Here every row is a fibration, where $Sq^i\rho, i = 2, 3$ is the projection and the left arrow is the inclusion of a fiber. Consider the diagram

$$\begin{array}{ccccc} K_{n+4} & \xrightarrow{f'} & D' & \xrightarrow{e'} & E' \\ & & {\scriptstyle q'}\big\downarrow & & \big\downarrow{\scriptstyle p'} \\ K_{n+2} & \xrightarrow{f} & D & \xrightarrow{e} & E \\ & & {\scriptstyle q}\big\downarrow & & \big\downarrow{\scriptstyle p} \\ & & K(\mathbb{Z}, n) & = & K(\mathbb{Z}, n), \end{array}$$

where the top square is a pull-back diagram. Here $ef = j, e'f' = j'$. It follows from 1.5 that

(1.7) $$Q^2Q^2[1] * [-4] = (\Omega^n j')_*(x_4) \neq 0 \in H_*(\Omega^n D').$$

Let $\widehat{f}$ be a fibrational substitute of f'. Consider the pull-back diagram

$$\begin{array}{ccccc} M & \xrightarrow{b} & K_{n+4} & & \\ \big\downarrow{\scriptstyle t} & & \big\downarrow{\scriptstyle f'} & & \\ S^n & \xrightarrow{\iota} & D' & \xrightarrow{q'} & D, \end{array}$$

where ι gives $1 \in \pi_0(D') = \mathbb{Z}$. (So, the fibration $t : M \to S^n$ is induced from $\widehat{f}$ by ι.) Clearly, the fibration $t : M \to S^n$ is induced from the standard ΩD-fibration $PD \to D$ by $q'\iota$, and so there arises the K_2-fibration

$$\Omega^n t : \Omega^n M \to \Omega_0^n S^n.$$

Let τ be the transgression in the homology Leray–Serre spectral sequence of this fibration. One has $Sq^1_*(Q^2Q^2[1] * [-4]) = 0$ by the Nishida relations (8) and property (3), and so

$$0 = \tau Sq^1_*(Q^2Q^2[1] * [-4]) = Sq^1_*\tau(Q^2Q^2[1] * [-4]).$$

Since $Sq^1_* : H_3(K_2) \to H_2(K_2)$ is an isomorphism, $\tau(Q^2Q^2[1] * [-4]) = 0$. Hence

(1.8) $$Q^2Q^2[1] * [-4] \in \mathrm{Im}(H_4(\Omega^n M) \to H_4(\Omega_0^n S^n)).$$

Let $a : S^n\Omega^n M \to M$ be the map adjoint to $1_{\Omega^n M}$, and let Cg be the cone of $g := ta$. By IV.5.35, Cg is the Thom space of a certain $\mathcal{SF}_n$-object α over $S\Omega^n M$. Let $u \in H^n(Cg;\mathbb{Z})$ be the Thom class of α. Consider the following diagram where the left vertical sequence is a long cofiber sequence:

$$\begin{array}{ccccccccc}
S^n\Omega^n M & \xrightarrow{a} & M & \xrightarrow{b} & K_{n+4} & & & & \\
{\scriptstyle g}\downarrow & & \downarrow{\scriptstyle t} & & \downarrow{\scriptstyle f'} & & & & \\
S^n & = & S^n & \xrightarrow{\iota} & D' & \xrightarrow{e'} & E' & & \\
{\scriptstyle r}\downarrow & & {\scriptstyle r}\downarrow & & \downarrow{\scriptstyle q'} & & \downarrow{\scriptstyle p'} & & \\
Cg & = & Cg & \xrightarrow{v} & D & \xrightarrow{e} & E & \xrightarrow{\Phi} & K_{n+5} \\
 & & {\scriptstyle u}\downarrow & & \downarrow{\scriptstyle q} & & \downarrow{\scriptstyle p} & & \\
 & & K(\mathbb{Z},n) & = & K(\mathbb{Z},n) & = & K(\mathbb{Z},n) & & \\
 & & & & \downarrow{\scriptstyle Sq^2\rho} & & \downarrow{\scriptstyle Sq^3\rho} & & \\
 & & & & K_{n+2} & & K_{n+3}. & &
\end{array}$$

Since $q'\iota t$ is inessential, there exists $v : Cg \to D$ with $q'\iota = vr$. Furthermore, $qvr = ur$, and $r^* : H^n(Cg;\mathbb{Z}) \to H^n(S^n;\mathbb{Z})$ is an isomorphism. Hence, $qv = u$. So, Φ is defined on u.

We prove that it has zero indeterminacy on u. By the above, it suffices to prove that $w_2(\alpha) = 0$. But $w_2(\alpha)\rho u = Sq^2\rho u = Sq^2\rho qv = 0$.

We prove that $\Phi(u) \neq 0$. If $\Phi(u) = 0$, then Φev is inessential. (Indeed, Φev is one of the values of $\Phi(u)$, but Φ has zero indeterminacy on u.) But then there exists $h : Cg \to D'$ with $q'h = v$. Since $q'h = v$, one has $\iota = hr$. Since rg is inessential, hrg is. Hence $\iota g = \iota ta$ is. But ιta is adjoint to

$$\Omega^n M \xrightarrow{1} \Omega^n M \xrightarrow{\Omega^n t} \Omega^n S^n \xrightarrow{\Omega^n \iota} \Omega^n D'.$$

Thus, if $\Phi(u) = 0$, then $(\Omega^n\iota)(\Omega^n t)$ is inessential. Consider the diagram

$$\begin{array}{ccc}
H_*(\Omega^n M) & \xrightarrow{(\Omega^n b)_*} & H_*(K_4) \\
{\scriptstyle (\Omega^n t)_*}\downarrow & & \downarrow{\scriptstyle (\Omega^n f')_*} \\
H_*(\Omega_0^n S^n) & \xrightarrow{(\Omega^n \iota)_*} & H_*(\Omega_0^n D').
\end{array}$$

By (1.7), $Q^2Q^2[1] * [-4] \neq 0 \in H_*(\Omega^n D')$, while

$$Q^2Q^2[1] * [-4] \in \mathrm{Im}((\Omega^n\iota)_*(\Omega^n t)_*)$$

by (1.8). Hence, $(\Omega^n\iota)(\Omega^n t)$ is essential, and hence $\Phi(u) \neq 0$.

Recall that $B\mathcal{F} \simeq B\mathcal{G}$. Now, the stabilization of α is the desired stable spherical fibration ξ. □

1.9. Theorem. *There exists an oriented stable spherical fibration ξ such that:*

(i) $w_2(\xi) = 0$,

(ii) $\Phi(2,2)$ *is defined on the Thom class* $u \in H^n(T\xi; \mathbb{Z})$,

(iii) $\Phi(2,2)$ *has zero indeterminacy on* u,

(iv) $\Phi(2,2)(u) \neq 0$.

Proof. This can be proved as was 1.6, but we show another way. For n large we have $\pi_n(S^n) = \mathbb{Z}, \pi_{n+1}(S^n) = \mathbb{Z}/2 = \pi_{n+2}(S^n)$. Furthermore, the first Postnikov invariant of S^n is $Sq^2\rho$, and the second one is $\Phi(2,2)$, see e.g. Mosher–Tangora [1]. In other words, the defining tower for $\Phi(2,2)$ is just the $(n+2)$-coskeleton of S^n.

Regarding $w_2 : BS\mathcal{G} \to K_2$ as a fibration (passing to a fibrational substitute if necessary), let $i : B \to BS\mathcal{G}$ be an inclusion of a fiber. We set $\eta := i^*\gamma_{S\mathcal{G}}$ and $\xi := \eta|B^{(3)}$ (replacing B by a cellular substitute). If $\Phi(u_\xi) = 0$, then ξ must be S-orientable and so trivial, cf. V.1.23(e). But ξ is non-trivial because the homomorphism $i^* : H^3(BS\mathcal{G}) \to H^3(B)$ is non-zero. Indeed,

$$H^3(BS\mathcal{G}) = \mathbb{Z}/2 \oplus \mathbb{Z}/2 = \{w_3, e_3\},$$

where e_3 is the Gitler–Stasheff class, see e.g. Madsen–Milgram [1]. Considering the Leray–Serre spectral sequence of the fibration $B \xrightarrow{i} BS\mathcal{G} \xrightarrow{w_2} K_2$, we conclude that $i^*e_3 \neq 0$. □

1.10. Remark. Hegenbarth [1] proved that every operation $\Phi(a,b), 1 < b \leq a < 2b$, can be realized by spherical fibrations. In fact, we followed this proof in 1.6. It is based on ideas of Peterson [2] and Ravenel [1].

Here we considered the operations $\Phi(a,b)$ with $a \geq b$. What about $a < b$? In Rudyak–Khokhlov [1] it was asserted that every such an operation can be realized by vector bundles. Unfortunately, there is a gap in that paper, and now I can only prove that some such operations can be realized. Namely, let $B(i_1, \dots, i_k)$ be the homotopy fiber of the map

$$\prod_{j=1}^{k} w_{i_j} : BO \to \prod_{j=1}^{k} K_{i_j}.$$

We say that the sequence $\{i_1, \dots, i_s\}, i_r < i_{r+1}$ is *apt* if $0 \neq w_j \in H^*(B(i_1, \dots, i_{j-1}))$ for every $j = 1, \dots, s$.

1.11. Theorem. *Consider the Adem relation*

$$Sq^a Sq^b = \sum_{i=1}^{s} Sq^{l_i} Sq^{k_i},\ l_i \geq 2k_i,$$

where $a < b$ and $k_i < k_{i+1}$. If $1 < k_1$ and the sequence $\{k_1, \dots, k_s, b\}$ is apt, then $\Phi(a,b)$ can be realized by vector bundles.

Proof. It is based on the ideas from Ch.V, §5. Set $l_{s+1} := a$, $k_{s+1} := b$, $q := k_i + l_i$. Let

$$\theta : H \to \bigvee_{i=1}^{s+1} \Sigma^{k_i} H$$

be the morphism which corresponds to $\{Sq^{k_1}, \cdots, Sq^{k_{s+1}}\} \in \oplus[H, \Sigma^{k_i} H]$ under the isomorphism

$$[H, \bigvee_{i=1}^{s+1} \Sigma^{k_i} H] \cong \oplus[H, \Sigma^{k_i} H].$$

Consider the diagram

$$\begin{array}{ccccc}
 & & E' & & \\
 & & \downarrow{\scriptstyle p'} & & \\
\bigvee_{i=1}^{s+1} \Sigma^{k_i - 1} H & \xrightarrow{j} & E & \xrightarrow{\Phi} & \Sigma^{q-1} H \\
 & & \downarrow{\scriptstyle p} & & \\
 & & H & \xrightarrow{\theta} & \bigvee_{i=1}^{s+1} \Sigma^{k_i} H,
\end{array}$$

where

$$\bigvee \Sigma^{k_i - 1} H \xrightarrow{j} E \xrightarrow{p} H \xrightarrow{\theta} \bigvee \Sigma^{k_i} H$$

and

$$E' \xrightarrow{p'} E \xrightarrow{\Phi} \Sigma^{q-1} H$$

are long cofiber sequences and

$$\Phi j | \Sigma^{k_i - 1} H = Sq^{l_i}.$$

This diagram is just the defining diagram for the secondary operation Φ, see e.g. Mosher–Tangora [1]. The Postnikov tower of E has the form

$$\begin{array}{ccccccccccc}
E & & & & & & & & & & \\
= \downarrow {\scriptstyle \tau_{s+1}} & & & & & & & & & & \\
E_{s+1} & \xrightarrow{p_{s+1}} & E_s & \longrightarrow & \cdots & \xrightarrow{p_n} & E_{n-1} & \xrightarrow{p_{n-1}} & \cdots & \xrightarrow{p_1} & H \\
 & & {\scriptstyle \sigma_{s+1}} \downarrow & & & & {\scriptstyle \sigma_n} \downarrow & & & & \downarrow {\scriptstyle \sigma_1} \\
 & & \Sigma^{k_{s+1}} H & & & & \Sigma^{k_n} H & & & & \Sigma^{k_1} H
\end{array}$$

where $\bullet \xrightarrow{p_i} \bullet \xrightarrow{\sigma_i} \bullet$ is a cofiber sequence and

$$(1.12) \qquad (\sigma_i)_*(x) = Sq^{d+k_i}((p_1 \circ \cdots \circ p_i)_*(x)) \in H^{k_i}(X)$$

for every $x \in (E_i)^d(X)$, where E_i denotes the coskeleton $E_{(k_1+\cdots+k_i-i)}$ of E. Since $k_1 > 1$, we conclude that $E_{(0)} = H$ and $\sigma_0 = Sq^{k_1}$.

We prove that the map $\Omega^\infty p' : \Omega^\infty E' \to \Omega^\infty E$ does not admit a section. Indeed,

$$\Omega^\infty E = \mathbb{Z}/2 \times \prod K_{k_i-1},$$

and $\Omega^\infty E'$ is the homotopy fiber of the map

$$\Omega^\infty \Phi : \mathbb{Z}/2 \times \prod K_{k_i-1} \to K_{q-1}.$$

Furthermore, $\Omega^\infty\Phi(\iota_{k_r-1}) = Sq^{l_r}(\iota_{k_r-1})$, where $\iota_r \in H^r(K_r)$ is the fundamental class. Since $l_0 < k_0$, $Sq^{l_0}(\iota_{k_0-1}) \neq 0$. Hence $\Omega^\infty\Phi$ is an essential map. Hence $\Omega^\infty p'$ does not admit a section.

Since $k_i > 1$, $\pi_0(E) \simeq \mathbb{Z}/2$. Let ξ be the universal stable $(1, E)$-marked vector bundle (see IV.5.25)[14] , and let $u \in H^0(T\xi)$ be the Thom class. Let $B(\mathcal{O}, 1, E)$ be the classifying space for $(1, E)$-marked stable vector bundles, see IV.5.32(i). For simplicity, we denote $B(\mathcal{O}, 1, E)$ by $B(\mathcal{O}, E)$. Suppose that $0 \in \Phi(u)$. Then there exists a map $M(\mathcal{O}, E) \to E'$ such that the left hand diagram of (1.13) below commutes. Hence ξ admits a $(1, E')$-marking, and hence there exists a map f such that the right hand diagram of (1.13) commutes, cf. IV.5.32.

$$(1.13) \qquad \begin{array}{ccc} M(\mathcal{O},E) & \longrightarrow & E' \\ \| & & \downarrow{\scriptstyle p'} \\ M(\mathcal{O},E) & & E \\ \| & & \downarrow{\scriptstyle p} \\ M(\mathcal{O},E) & \xrightarrow{u} & H \end{array} \qquad \begin{array}{ccc} B(\mathcal{O},E) & \xrightarrow{f} & B(\mathcal{O},E') \\ \| & & \downarrow{\scriptstyle q} \\ B(\mathcal{O},E) & & B(\mathcal{O},E) \\ \| & & \downarrow{\scriptstyle \pi} \\ B(\mathcal{O},E) & \xrightarrow{\pi} & B\mathcal{O}. \end{array}$$

Here π is the forgetful map and $q := B(\mathcal{O}, p')$. Furthermore, q does not admit a section because $\Omega^\infty p'$ does not admit a section, cf. V.4.19. Hence, it suffices to prove that qf is a homotopy equivalence for every map $f : B(\mathcal{O}, E) \to B(\mathcal{O}, E')$ over $B\mathcal{O}$. (Indeed, if qf is a homotopy equivalence, then q admits a section. This is a contradiction.)

We prove that every map $g : B(\mathcal{O}, E) \to B(\mathcal{O}, E)$ over $B\mathcal{O}$ is a homotopy equivalence. Consider any such a map g. It induces the following self-map of the Postnikov–Moore tower, where $q_n := B(\mathcal{O}, p_n)$ is a K_{k_n-1}-fibration and $g_{s+1} := g$:

[14] We do not want to speak about E-orientations because E is not a ring spectrum.

$$\begin{array}{ccc}
B(\mathcal{O},E) & \xrightarrow{g_{s+1}} & B(\mathcal{O},E) \\
{\scriptstyle q_{s+1}}\downarrow & & \downarrow{\scriptstyle q_{s+1}} \\
B(\mathcal{O},E_s) & \xrightarrow{g_s} & B(\mathcal{O},E_s) \\
{\scriptstyle q_s}\downarrow & & \downarrow{\scriptstyle q_s} \\
\vdots & & \vdots \\
{\scriptstyle q_{n+1}}\downarrow & & \downarrow{\scriptstyle q_{n+1}} \\
B(\mathcal{O},E_n) & \xrightarrow{g_n} & B(\mathcal{O},E_n) \\
{\scriptstyle q_n}\downarrow & & \downarrow{\scriptstyle q_n} \\
B(\mathcal{O},E_{n-1}) & \xrightarrow{g_{n-1}} & B(\mathcal{O},E_{n-1}) \\
{\scriptstyle q_{n-1}}\downarrow & & \downarrow{\scriptstyle q_{n-1}} \\
\vdots & & \vdots \\
{\scriptstyle q_1}\downarrow & & \downarrow{\scriptstyle q_1} \\
B\mathcal{O} & = & B\mathcal{O}.
\end{array}$$

Set $\xi_n = (q_n \circ \cdots \circ q_1)^* \gamma_{\mathcal{O}}$. Let $u_n \in H^0(M(\mathcal{O}, E_n))$ be the Thom class of ξ_n, and let $v \in (E_n)^0(M(\mathcal{O}, E_n))$ be the universal $(1, E_n)$-marking of ξ. Let χ_n be the characteristic class of q_n. Following V.4.22, one can prove that $\chi_n = \varphi^{-1}\sigma_n v_n$. Hence,

$$\chi_n = \varphi^{-1}\sigma_n v_n = \varphi^{-1} Sq^{k_n} u_n = w_{k_n}(\xi_n)$$

(the second equality follows from (1.12)). So, $B(\mathcal{O}, E_n) = B(k_1, \dots, k_{n-1})$.

By setting $g_0 = 1_{B\mathcal{O}}$, suppose inductively that g_{n-1} is a homotopy equivalence. Since the sequence $\{k_1, \dots, k_s\}$ is apt, we have $\chi_n = w_{k_n}(\xi_n) \neq 0$. Hence, by 4.5, g_n is a homotopy equivalence. Thus, g is a homotopy equivalence. □

§2. Some Calculations with Classifying Spaces

Recall that K and $K\mathcal{O}$ are ring spectra. Hence, by II.4.28, k and $k\mathcal{O}$ are ring spectra. Furthermore,

$$\Omega^\infty K \simeq B\mathcal{U} \times \mathbb{Z} \simeq \Omega^\infty k; \; \Omega^\infty K\mathcal{O} \simeq B\mathcal{O} \times \mathbb{Z} \simeq \Omega^\infty k\mathcal{O}.$$

Finally, by Bott periodicity, IV.4.27(ii), $\Sigma^2 K \simeq K$, $\Sigma^8 K\mathcal{O} \simeq K\mathcal{O}$.

2.1. Lemma (cf. Adams [2]). *If $n > 2$, then the first non-trivial Postnikov invariant of $BU|(2n)$ is $\delta Sq^2\rho\iota_{2n} \in H^{2n+3}(K(\mathbb{Z},2n))$. Here $\iota_{2n} \in H^{2n}(K(\mathbb{Z},2n))$ is a fundamental class.*

Proof. By IV.4.27(ii), $\Omega^2 BU \simeq BU \times \mathbb{Z}$. So,

$$\Omega^{2n-4}(BU|(2n)) \simeq BU|4 \simeq BSU.$$

Since $H^{2n+3}(K(\mathbb{Z},2n)) = \mathbb{Z}/2 = \{\delta Sq^2\rho\iota_{2n}\}$, it suffices to prove the non-triviality of the Postnikov invariant $\psi \in H^5(K(\mathbb{Z},4))$ of BSU. The Postnikov tower of BSU has the form

$$\begin{array}{ccc} & & \vdots \\ & & \downarrow \\ BSU & \xrightarrow{\tau} & X \\ & & \downarrow \\ & & K(\mathbb{Z},4) \xrightarrow{\psi} K(\mathbb{Z},7). \end{array}$$

It is clear that $\tau_* : H_i(BSU) \to H_i(X)$ is an isomorphism for $i \leq 6$. In particular, $H_6(X)$ is torsion free, see IV.4.32. If $\psi = 0$, then $X = K(\mathbb{Z},4) \times K(\mathbb{Z},6)$, i.e., $H_6(X) = \mathbb{Z}/2$. This is a contradiction. □

By Bott periodicity, $\pi_*(K) = \mathbb{Z}[t,t^{-1}]$, $\dim t = 2$. Thus, $\pi_*(k) = \mathbb{Z}[t]$, $\dim t = 2$.

The multiplication by t in $k^*(X)$ is given by a morphism

$$t_\# : k = S \wedge k \xrightarrow{t\wedge 1} \Sigma^{-2}k \wedge k \xrightarrow{\mu} \Sigma^{-2}k.$$

For simplicity, we also denote the suspension $\Sigma^n t_\# : \Sigma^n k \to \Sigma^{n-2}k, n \in \mathbb{Z}$, of this morphism by $t_\#$.

Consider the morphism $t_\#^{r+1} : \Sigma^{2r+2}k \to k$ and denote its cone by k^r. We have the exact sequence

$$\cdots \to \pi_i(k) \to \pi_{i+2r+2}(k) \to \pi_{i+2r+2}(k^r) \to \pi_{i-1}(k) \to \cdots,$$

and hence $\pi_*(k^r) = \mathbb{Z}[t]/(t^{r+1})$. Moreover, $k^0 = H\mathbb{Z}$.

2.2. Proposition. *If $r > 0$, then $(\Sigma^{-2}k^r)|0 \simeq k^{r-1}$.*

Proof. There exists f such that the diagram

$$\begin{array}{ccccc} \Sigma^{2r}k & \xrightarrow{t_\#^r} & k & \longrightarrow & k^{r-1} \\ \| & & \downarrow{\scriptstyle t_\#} & & \downarrow{\scriptstyle f} \\ \Sigma^{2r}k & \xrightarrow{t_\#^{r+1}} & \Sigma^{-2}k & \longrightarrow & \Sigma^{-2}k^r \end{array}$$

commutes. It is clear that $f_*(t^s) = t^s$ for $s < r$, where $f_* : \pi_*(k^{r-1}) \to \pi_*(\Sigma^{-2}k^r)$. Let $\varphi : \widehat{\Sigma^{-2}k^r} \to \Sigma^{-2}k^r$ be a connective covering of $\Sigma^{-2}k^r$. Since k^{r-1} is connected, there is a φ-lifting $\hat{f} : k^{r-1} \to \widehat{\Sigma^{-2}k^r}$ of f. Clearly, $\hat{f}$ is a homotopy equivalence. □

2.3. Proposition. *There exists a commutative diagram*

$$\begin{array}{ccccc}
\Sigma^{2r+2}k & \xrightarrow{t_\#^{r+1}} & k & \longrightarrow & k^r \\
\downarrow{\scriptstyle t_\#} & & \| & & \downarrow{\scriptstyle p_r} \\
\Sigma^{2r}k & \xrightarrow{t_\#^{r}} & k & \longrightarrow & k^{r-1}
\end{array}$$

such that $(p_r)_*(t^s) = t^s$ for $s < r$ and $(p_r)_*(t^r) = 0$, where $(p_r)_* : \pi_*(k^r) \to \pi_*(k^{r-1})$. Furthermore, the cone of p_r is $\Sigma^{2r+1}H\mathbb{Z}$, and so there is the long cofiber sequence

$$\cdots \to \Sigma^{2r}H\mathbb{Z} \xrightarrow{j_r} k^r \xrightarrow{p_r} k^{r-1} \xrightarrow{\sigma_r} \Sigma^{2r+1}H\mathbb{Z} \to \cdots .$$

Moreover, $\Sigma^2(\sigma_r j_{r-1}) = \sigma_{r+1}j_r$.

Proof. The existence of p_r follows from the commutativity of the left square of the diagram. The properties of p_r follow from the commutativity of the right square of the diagram. The equality $\Sigma^2(\sigma_r j_{r-1}) = \sigma_{r+1}j_r$ follows from 2.2. □

Consider the diagram

(2.5)

$$\begin{array}{ccccccccc}
 & & k & & & & & & H\mathbb{Z} \\
 & & \downarrow{\scriptstyle \tau_{2r}} & & & & & & \| \\
\cdots \longrightarrow & k^{r+1} \longrightarrow & k^r & \xrightarrow{p_r} & k^{r-1} & \longrightarrow & \cdots & \longrightarrow & k^0 \\
 & & & & \downarrow{\scriptstyle \sigma_r} & & & & \downarrow{\scriptstyle \sigma_1} \\
 & & & & \Sigma^{2r+1}H\mathbb{Z} & & & & \Sigma^3 H\mathbb{Z}
\end{array}$$

2.6. Theorem (cf. Adams-Priddy [1]). *The diagram (2.5) is the Postnikov tower of k, and for the Postnikov invariants σ_r we have:*

(i) *If $i > n > 1$, then $((\Omega^\infty\sigma_i)|2n)[2] \neq 0$.*

(ii) *$\Omega^\infty\sigma_1 = 0 = \Omega^\infty\sigma_2$, $(\Omega^\infty\sigma_i)[2] \neq 0$ for every $i \geq 3$.*

(iii) *The morphism $\sigma_r j_{r-1} : \Sigma^{2r-2}H\mathbb{Z} \to \Sigma^{2r+1}H\mathbb{Z}$ is $\delta Sq^2\rho$. In particular, $\sigma_1 = \delta Sq^2\rho$. Furthermore, $(\delta Sq^2\rho)\sigma_n = 0$ for every $n \geq 1$, and the higher operation σ_{n+1} is associated with this relation.*

Proof. The tower (2.5) is the Postnikov tower of k because of 2.3. Now we prove the properties (i)–(iii).

(i) Since $\Omega^\infty k = BU \times \mathbb{Z}$, we deduce from 2.1 that $(\Omega^\infty \sigma_{n+1})|2n = \delta Sq^2 \rho\iota_{2n} \neq 0$. Now, $(\Omega^\infty \sigma_{n+1})|2n[2] \neq 0$ since the element $\delta Sq^2 \rho\iota_{2n}$ has order 2.

(ii) We have $\Omega^\infty k^0 = \Omega^\infty H\mathbb{Z} = \mathbb{Z}, \Omega^\infty k^1 = \mathbb{C}P^\infty \times \mathbb{Z}$. Hence $\Omega^\infty \sigma_1 = 0 = \Omega^\infty \sigma_2$. Moreover, $(\Omega^\infty \sigma_i)[2] \neq 0$ for every $i \geq 3$ by (i).

(iii) Consider the diagram

$$\begin{array}{ccccccc}
\Sigma^{2r-2}H\mathbb{Z} & \xrightarrow{j_{r-1}} & k^{r-1} & \xrightarrow{\sigma_r} & \Sigma^{2r+1}H\mathbb{Z} & & \\
 & & \downarrow{\scriptstyle p_{r-1}} & & & & \\
 & & k^{r-2} & \xrightarrow{\sigma_{r-1}} & \Sigma^{2r-1}H\mathbb{Z} & \xrightarrow{\psi} & \Sigma^{2r+2}H\mathbb{Z}
\end{array}$$

where $\psi = \Sigma(\sigma_r j_{r-1})$. Note that $\sigma_r j_{r-1} \in H^{2r+1}(\Sigma^{2r-2}H\mathbb{Z}) = \mathbb{Z}/2 = \{\delta Sq^2 \rho\}$.

It follows from 2.1 that $\Omega^\infty(\sigma_r j_{r-1}) = \delta Sq^2 \rho\iota_{2r}$. So $\sigma_r j_{r-1} \neq 0$, and hence $\sigma_r j_{r-1} = \delta Sq^2 \rho$. Hence, $\psi = \delta Sq^2 \rho$. Furthermore, the equality $\psi = \Sigma \sigma_r j_{r-1}$ means that σ_r is associated with the relation $\psi \sigma_{r-1} = 0$.[15] But, by the above, $\psi = \delta Sq^2 \rho$. □

2.7. Lemma. *We have $H^5(K(\mathbb{Z}/2,2);\mathbb{Z}[2]) = \mathbb{Z}/4$. Furthermore, the element $\delta Sq^2 \iota \in H^5(K(\mathbb{Z}/2,2);\mathbb{Z}[2])$ has order 2; here $\iota \in H^2(K(\mathbb{Z}/2,2);\mathbb{Z}/2)$ is the fundamental class.*

Proof(Sketch). Firstly, because of the Serre class theory, every group $H^i(K(\mathbb{Z}/2,n);\mathbb{Z}[2])$ is a finite 2-primary group. Now, using information about the ring $H^*(K(\mathbb{Z}/2,2);\mathbb{Z}/2)$ (see e.g. Mosher–Tangora[1]), and applying the Universal Coefficient Theorem, one can prove that the group $H^5(K(\mathbb{Z}/2,2);\mathbb{Z}[2])$ is cyclic. Finally, considering the $\mathbb{Z}[2]$-cohomology Leray–Serre spectral sequence of the fibration

$$K(\mathbb{Z}/2,1) \to PK(\mathbb{Z}/2,2) \to K(\mathbb{Z}/2,2),$$

one can conclude that $H^5(K(\mathbb{Z}/2,2))$ has order 4. Thus, it is $\mathbb{Z}/4$.

Let $a \in H^1(RP^5;\mathbb{Z}/2), a \neq 0$. Since $\delta Sq^2 a^2 = \delta a^4 \neq 0$, we conclude that $\delta Sq^2 \iota \neq 0$. Furthermore, $\rho\delta Sq^2 \iota = Sq^1 Sq^2 \iota = Sq^3 \iota = 0$. Thus, $\delta Sq^2 \iota$ has order 2. □

Let Y be the homotopy fiber of the map $\delta Sq^2 : K(\mathbb{Z}/2,2) \to K(\mathbb{Z}[2],5)$. So, we have a fibration

$$K(\mathbb{Z}[2],4) \xrightarrow{j} Y \xrightarrow{p} K(\mathbb{Z}/2,2). \tag{2.8}$$

with characteristic class δSq^2.

---

[15] This relation holds because $j_{r-1}(\Sigma^{-1}\sigma_{r-1}) = 0$. Actually, this is a general argument in the theory of higher order cohomology operations.

2.9. Lemma. (i) $H^4(Y;\mathbb{Z}[2])=\mathbb{Z}[2]$.

(ii) *The homomorphism* $j^*:\mathbb{Z}[2]=H^4(Y;\mathbb{Z}[2])\to H^4(K(\mathbb{Z}[2],4);\mathbb{Z}[2])=\mathbb{Z}[2]$ *is multiplication by* 2ε *for some* $\varepsilon\in\mathbb{Z}[2]^*$.

(iii) *A map* $h:Y\to K(\mathbb{Z}[2],4)$ *yields a generator (i.e., an element of* $\mathbb{Z}[2]^*$*) of* $H^4(Y;\mathbb{Z}[2])$ *iff* $h_*:\mathbb{Z}[2]=\pi_4(Y)\to\pi_4(K(\mathbb{Z}[2],4))=\mathbb{Z}[2]$ *is multiplication by* 2ε *for some* $\varepsilon\in\mathbb{Z}[2]^*$.

Proof. By 2.7, $\delta Sq^2\iota$ has order 2. Now the assertions (i) and (ii) can be proved by routine calculations with the spectral sequence of the fibration (2.8). Furthermore, (ii) implies that $h:Y\to K(\mathbb{Z}[2],4)$ yields a generator iff $(hj)^*(u)=2u$ where $u\in H^4(K(\mathbb{Z}[2],4);\mathbb{Z}[2])=\mathbb{Z}[2]$ is a generator (fundamental class). Thus, the assertion (iii) follows because $j_*:\pi_4(K(\mathbb{Z}[2],4))\to\pi_4(Y)$ is an isomorphism. □

Since $BSU=\Omega^\infty k|4$, the 8-stage of the Postnikov tower of $BSU[2]$ has the form (2.10) below, where we write σ_i instead of $(\Omega^\infty\sigma_i|4)[2]$. Moreover, by 2.6(iii), $\sigma_3 i=\delta Sq^2\rho$.

$$
\begin{array}{ccccc}
 & & BSU[2]_{(8)} & & \\
 & & \downarrow & & \\
K(\mathbb{Z}[2],6) & \xrightarrow{i} & BSU[2]_{(6)} & \xrightarrow{\sigma_3} & K(\mathbb{Z}[2],9) \\
 & & \downarrow{\scriptstyle\pi} & & \\
 & & K(\mathbb{Z}[2],4) & \xrightarrow{\sigma_2=\delta Sq^2\rho} & K(\mathbb{Z}[2],7)
\end{array}
\tag{2.10}
$$

2.11. Lemma. *Let* $h:Y\to K(\mathbb{Z}[2],4)$ *be a map such that* $\sigma_2h=0$. *(In fact this is true for all h, but we do not use it.) Then for every two* π*-liftings* $g_1,g_2:Y\to BSU[2]_{(6)}$ *of h we have* $\sigma_3g_1=\sigma_3g_2$.

Proof. The difference $g_1-g_2:Y\to BSU[2]_{(6)}$ is homotopic to $i\varphi:Y\to BSU[2]_{(6)}$ for some $\varphi:Y\to K(\mathbb{Z}[2],6)$. Hence, $\sigma_3(g_1-g_2)=\sigma_3 i\varphi=\delta Sq^2\rho\varphi$. But $H^6(Y;\mathbb{Z}[2])=\mathbb{Z}/2=\{p^*(\delta\iota)^2\}$ where p is as in (2.8). So, it suffices to prove that $\delta Sq^2\rho(p^*(\delta\iota)^2)=0$. But $Sq^2\rho((\delta\iota)^2)=Sq^2(Sq^1\iota)^2=0$. □

2.12. Theorem (cf. Adams–Priddy [1]). *The 8-stage of the Postnikov tower of* $BSO[2]$ *has the form*

$$
\begin{array}{ccccc}
 & & BSO[2]_{(8)} & & \\
 & & \downarrow & & \\
 & & BSO[2]_{(4)} & \xrightarrow{\psi} & K(\mathbb{Z}[2],9) \\
 & & \downarrow & & \\
BSO[2]_{(2)} & = & K(\mathbb{Z}/2,2) & \xrightarrow{\sigma} & K(\mathbb{Z}[2],5),
\end{array}
$$

where the class $\psi \in H^9(BSO[2]_{(4)}; \mathbb{Z}[2])$ *is non-zero. Furthermore,* $\sigma = \delta Sq^2$, *and hence* $BSO[2]_{(4)} = Y$.

Proof. For simplicity, denote $BSO[2]_{(n)}$ by B_n. The groups $\pi_i(BSO)$ are well known in view of Bott periodicity, see IV.4.27(ii). In order to prove the non-triviality of σ and ψ we must prove that B_n is not equivalent to $B_{n-1} \times K(\mathbb{Z}[2], n)$, where $n = 4$ (for σ) and $n = 8$ (for ψ). We prove this for both values of n simultaneously.

Let $Q^*(X)$ be the indecomposable quotient of $H^*(X; \mathbb{Z}/2)$. Suppose that $B_n \simeq B_{n-1} \times K(\mathbb{Z}[2], n)$. Then $Q^n(B_n) = Q^n(B_{n-1}) \oplus Q^n(K(\mathbb{Z}[2], n))$ by the Künneth formula. Since $Q^n(B_n) = \mathbb{Z}/2$, we have $Q^n(B_{n-1}) = 0$. Thus, for every $x \in H^n(B_n; \mathbb{Z}/2)$ we have (under the Künneth isomorphism) $x = a\iota_n + d$, where d is decomposable in $H^n(B_n)$ and $\iota_n \in H^n(K(\mathbb{Z}[2], n); \mathbb{Z}/2) = \mathbb{Z}/2$ is a generator. Since $Sq^1 \iota_n = 0$, $Sq^1 x$ is decomposable in $H^{n+1}(B_n)$ for every $x \in H^n(B_n)$.

Since $H^i(B_n; \mathbb{Z}/2) = H^i(BSO[2]; \mathbb{Z}/2)$ for $i \le n$, the image of the homomorphism $Sq^1 : H^n(BSO[2]; \mathbb{Z}/2) \to H^{n+1}(BSO[2]; \mathbb{Z}/2)$ consists of decomposables. But this contradicts the equality $Sq^1 w_n = w_{n+1}$. Thus, σ and ψ are non-trivial.

By 2.7, $H^5(K(\mathbb{Z}/2, 2); \mathbb{Z}[2]) = \mathbb{Z}/4 = \{x\}$ with $2x = \delta Sq^2 \iota$. Hence, σ must be equal to one of the elements $x, 3x$ or $\delta Sq^2 \iota$. We have

$$H^5(BSO[2]_{(4)}; \mathbb{Z}/2) = H^5(BSO[2]_{(7)}; \mathbb{Z}/2) = \mathbb{Z}/2 \oplus \mathbb{Z}/2 = \{w_5, w_2 w_3\}.$$

If $\sigma = x$ or $\sigma = 3x$, then $H^5(BSO[2]_{(4)}; \mathbb{Z}/2)$ would be $\mathbb{Z}/2$. (To see this, consider the spectral sequence of the fibration $K(\mathbb{Z}[2], 4) \to E \to K(\mathbb{Z}/2, 2)$ with characteristic class σ and use the fact that $\rho(\sigma)$ is non-zero.) Hence, $\sigma = \delta Sq^2 \iota$, and thus $BSO[2]_{(4)} = Y$. □

2.13. Lemma. *Consider the following diagram:*

$$\begin{array}{ccccc}
 & & BSU[2]_{(8)} & & \\
 & & \downarrow & & \\
Y & \xrightarrow{g} & BSU[2]_{(6)} & \xrightarrow{\sigma_3} & K(\mathbb{Z}[2], 9) \\
\| & & \downarrow \pi & & \\
Y & \xrightarrow{h} & K(\mathbb{Z}[2], 4) & \xrightarrow{\sigma_2 = \delta Sq^2 \rho} & K(\mathbb{Z}[2], 7).
\end{array}$$

Let the map h *give a generator of* $H^4(Y; \mathbb{Z}[2]) = \mathbb{Z}[2]$. *Then for every lifting* $g : Y \to BSU[2]_{(6)}$ *of* h *we have* $\sigma_3 g \neq 0$. *In particular,* h *cannot be lifted to* $BSU[2]_{(8)}$.

Proof. In view of 2.11, it suffices to prove the lemma for only one such a lifting g. Let $C : BO \to BU$ be the complexification. Since $H^2(BSO) = 0$, there is a commutative diagram

$$\begin{array}{ccc} BSO & \xrightarrow{c} & BSU \\ {\scriptstyle k}\downarrow & & \downarrow{\scriptstyle l} \\ BO & \xrightarrow{C} & BU \end{array}$$

where k (resp. l) is a 1-connected (resp. 3-connected) covering. Then, passing to Postnikov towers, we have the commutative diagram

$$\begin{array}{ccccc} BSO[2]_{(8)} & \xrightarrow{c_8} & BSU[2]_{(8)} & & \\ {\scriptstyle p_1}\downarrow & & \downarrow{\scriptstyle p_2} & & \\ BSO[2]_{(7)} & \xrightarrow{c_7} & BSU[2]_{(6)} & \xrightarrow{\sigma_3} & K(\mathbb{Z}[2],9) \\ \| & & \downarrow{\scriptstyle \pi} & & \\ Y & \xrightarrow{h} & K(\mathbb{Z}[2],4), & & \end{array}$$

where, by 2.12, $BSO[2]_{(7)} = Y$. By IV.4.27(iii), $h_* : \pi_4(Y) \to \pi_4(K(\mathbb{Z}[2],4))$ is multiplication by 2, and so, by 2.9(iii), h is a generator of $H^4(Y;\mathbb{Z}[2])$. Hence, the last diagram with $c_7 = g$ coincides with the diagram of the lemma. Thus, by 2.11, it suffices to prove that $\sigma_3 c_7 \neq 0$.

Let $K_i = K_i(\mathbb{Z}[2],8)$ be the fiber of p_i, $i = 1,2$, and let $u_i \in H^8(K_i;\mathbb{Z}[2])$ be a fundamental class. By IV.4.27(iii), $(c_8)_* : \pi_8(BSO) \to \pi_8(BSU)$ is an isomorphism, and so the map $\bar{c}_8 : K_1 \to K_2$ of fibers is a homotopy equivalence. In particular, $(\bar{c}_8)^*(u_2) = au_1$ for some $a \in \mathbb{Z}[2]^*$. Let τ_i be the transgressions in the $\mathbb{Z}[2]$-cohomological spectral sequences of the fibrations $p_i, i = 1,2$. By 2.12, $\tau_1(u_1) = \psi \neq 0$, and so

$$0 \neq \tau_1(au_1) = \tau_1((\bar{c}_8)^* u_2) = (c_7)^*(\tau_2 u_2) = c_7^*(\sigma_3). \qquad \square$$

Sullivan [1] established the homotopy equivalence

$$\mathcal{G}/\mathcal{PL}[2] \simeq Y \times \prod_{i>1} K(\mathbb{Z}[2],4i) \times K(\mathbb{Z}/2,4i-2) \tag{2.14}$$

(for a good proof see Madsen–Milgram [1]). Basing on this and using IV.4.27(v), one can prove (see loc. cit.) that

$$\mathcal{G}/\mathcal{TOP}[2] \simeq \prod_{i\geq 1} K(\mathbb{Z}[2],4i) \times K(\mathbb{Z}/2,4i-2). \tag{2.15}$$

We define $j_{\mathcal{PL}} : K(\mathbb{Z}[2],4) \xrightarrow{j} Y \xrightarrow{a} \mathcal{G}/\mathcal{PL}[2]$, where a is the inclusion of the factor in (2.14). Similarly, let $j_{\mathcal{TOP}} : K(\mathbb{Z}[2],4) \to \mathcal{G}/\mathcal{TOP}[2]$ be the inclusion of the factor in (2.15). Lemma 2.9 yields the following proposition.

2.16. Proposition. (i) $H^4(\mathcal{G}/\mathcal{PL}[2];\mathbb{Z}[2]) = \mathbb{Z}[2] = H^4(\mathcal{G}/\mathcal{TOP}[2];\mathbb{Z}[2])$.

(ii) $j^*_{\mathcal{PL}} : H^4(\mathcal{G}/\mathcal{PL}[2]; \mathbb{Z}[2]) \to H^4(K(\mathbb{Z}[2],4); \mathbb{Z}[2])$ *is multiplication by* 2ε *for some* $\varepsilon \in \mathbb{Z}[2]^*$.

(iii) $j^*_{\mathcal{TOP}} : H^4(\mathcal{G}/\mathcal{TOP}[2]; \mathbb{Z}[2]) \to H^4(K(\mathbb{Z}[2],4); \mathbb{Z}[2])$ *is an isomorphism.* □

Let $p : \widetilde{X} \to X$ be a 3-connective covering of X for $X = \mathcal{G}/\mathcal{PL}[2]$ or $X = \mathcal{G}/\mathcal{TOP}[2]$.

2.17. Proposition. (i) *The map*

$$p^* : \mathbb{Z}[2] = H^4(\mathcal{G}/\mathcal{PL}[2]; \mathbb{Z}[2]) \to H^4\left(\mathcal{G}/\widetilde{\mathcal{PL}}[2]; \mathbb{Z}[2]\right) = \mathbb{Z}[2]$$

is multiplication by 2ε *for some* $\varepsilon \in \mathbb{Z}[2]^*$.

(ii) *The map*

$$p^* : \mathbb{Z}[2] = H^4(\mathcal{G}/\mathcal{TOP}[2]; \mathbb{Z}[2]) \to H^4\left(\mathcal{G}/\widetilde{\mathcal{TOP}}[2]; \mathbb{Z}[2]\right) = \mathbb{Z}[2]$$

is an isomorphism.

Proof. This follows immediately from (2.14), (2.15) and 2.16. □

Consider the diagram

$$\begin{array}{ccccc} \mathcal{G}/\mathcal{PL}[2] & \longrightarrow & B\mathcal{SPL}[2] & \longrightarrow & B\mathcal{SG}[2] \\ \varphi\downarrow & & \downarrow a^{\mathcal{PL}}_{\mathcal{TOP}} & & \| \\ \mathcal{G}/\mathcal{TOP}[2] & \longrightarrow & B\mathcal{STOP}[2] & \longrightarrow & B\mathcal{SG}[2] \end{array}$$

with any φ such that the diagram commutes. Such a map φ exists, but it is not unique. It follows from (2.14) and (2.15) that $\varphi_* : \pi_i(\mathcal{G}/\mathcal{PL}[2]) \to \pi_i(\mathcal{G}/\mathcal{TOP}[2])$ is an isomorphism for $i \neq 4$ and that

$$\varphi_* : \mathbb{Z}[2] = \pi_4(\mathcal{G}/\mathcal{PL}[2]) \to \pi_4(\mathcal{G}/\mathcal{TOP}[2]) = \mathbb{Z}[2].$$

is multiplication by 2ε for some $\varepsilon \in \mathbb{Z}[2]^*$. Fix one such map φ. It is easy to see that φ admits a 3-connective covering (unique up to homotopy equivalence) $\widetilde{\varphi} : \mathcal{G}/\widetilde{\mathcal{PL}}[2] \to \mathcal{G}/\widetilde{\mathcal{TOP}}[2]$.

2.18. Proposition. (i) *The map*

$$\varphi^* : \mathbb{Z}[2] = H^4(\mathcal{G}/\mathcal{TOP}[2]; \mathbb{Z}[2]) \to H^4(\mathcal{G}/\mathcal{PL}[2]; \mathbb{Z}[2]) = \mathbb{Z}[2]$$

is an isomorphism.

(ii) *The map*

$$\widetilde{\varphi}^* : \mathbb{Z}[2] = H^4\left(\mathcal{G}/\widetilde{\mathcal{TOP}}[2]; \mathbb{Z}[2]\right) \to H^4\left(\mathcal{G}/\widetilde{\mathcal{PL}}[2]; \mathbb{Z}[2]\right) = \mathbb{Z}[2]$$

is multiplication by 2ε *for some* $\varepsilon \in \mathbb{Z}[2]^*$.

Proof. (i) Consider the composition (where a is the inclusion of the factor, see (2.14))

$$b : Y \xrightarrow{a} \mathcal{G}/\mathcal{PL}\,[2] \xrightarrow{\varphi} \mathcal{G}/\mathcal{TOP}\,[2] \xrightarrow{\text{proj}} K(\mathbb{Z}[2], 4).$$

Since the map $\varphi_* : \pi_4\,(\mathcal{G}/\mathcal{PL}\,[2]) \to \pi_4(\mathcal{G}/\mathcal{TOP}\,[2])$ is multiplication by 2ε, so is the map $b_* : \pi_4(Y) \to \pi_4\,(K(\mathbb{Z}[2], 4))$. By 2.9,

$$j^* : H^*(Y; \mathbb{Z}[2]) = \mathbb{Z}[2] \to \mathbb{Z}[2] = H^*(K(\mathbb{Z}[2], 4); \mathbb{Z}[2])$$

is also multiplication by 2ε. Hence $b^* : H^4(K(\mathbb{Z}[2], 4); \mathbb{Z}[2]) \to H^4(Y; \mathbb{Z}[2])$ is an isomorphism, and hence φ^* is an isomorphism.

(ii) By 2.17, in the diagram

$$\begin{array}{ccc} H^4(\widetilde{\mathcal{G}/\mathcal{TOP}}\,[2]; \mathbb{Z}[2]) & \xrightarrow{\widetilde{\varphi}} & H^4(\widetilde{\mathcal{G}/\mathcal{PL}}\,[2]; \mathbb{Z}[2]) \\ {\scriptstyle p_1^*}\uparrow & & \uparrow{\scriptstyle p_2^*} \\ H^4\,(\mathcal{G}/\mathcal{TOP}\,[2]; \mathbb{Z}[2]) & \xrightarrow{\varphi} & H^4\,(\mathcal{G}/\mathcal{PL}\,[2]; \mathbb{Z}[2]) \end{array}$$

p_1^* is an isomorphism, while p_2^* is multiplication by 2ε, and (ii) follows. □

§3. *k*-Orientability

Let $\sigma_r : k^{r-1} \to \Sigma^{2r+1} H\mathbb{Z}$ be just the same as in §2. Following V.4.23(b) and the beginning of section V.5, we set $\varkappa_0(\xi) = e_0(\xi) = w_1(\xi)$ for every stable $\mathcal{V}$-object ξ. Furthermore, following V.(4.6) and the beginning of section V.5, for every stable $H\mathbb{Z}$-oriented $\mathcal{V}$-object (ξ, u_ξ) we set $\varkappa_r(\xi) = \varphi^{-1}\sigma_r u_\xi \subset H^{2r+1}(\operatorname{bs}\xi)$ where φ is the Thom isomorphism. By V.4.13, a $\mathcal{V}$-object ξ over a finite dimensional space X is k-orientable (and therefore K-orientable) iff $0 \in \varkappa_r(\xi) \subset H^{2r+1}(X), r = 0, 1, \ldots$.

Throughout this section $\gamma_{2n}^{\mathcal{V}} = \gamma_{2n+1}^{\mathcal{V}}$ means the universal k^n-oriented $\mathcal{V}$-object over $B(\mathcal{V}, k^n)$.

Notice that k, as well as $k[2]$, is a simple spectrum by V.4.21.

In order to use the Realizability Theorems V.5.1, V.5.6, we set $\widehat{\sigma}_r = \sigma_r[2]$ and consider also the classes $\widehat{\varkappa}_r = \varphi^{-1}(\widehat{\sigma}_r \widehat{u}_\xi) \subset H^{2r+1}(X; \mathbb{Z}[2])$, i.e., the higher characteristic classes corresponding to $k[2]$; here $\widehat{u}_\xi \in H^0(T\xi; \mathbb{Z}[2])$ is the $\mathbb{Z}[2]$-localization of u_ξ.

3.1. Proposition. (i) $\varkappa_0(\xi) = 0$ *iff* X *is* $H\mathbb{Z}$*-orientable.*

(ii) $\varkappa_1(\xi) = \delta w_2(\xi)$ *provided that* $\varkappa_1(\xi)$ *is defined, i.e., if* $\varkappa_0(\xi) = 0$.

Proof. (i) See V.1.23(b).

(ii) Let v be an $H\mathbb{Z}$-orientation of ξ. Then

$$\varkappa_1(\xi) = \varphi^{-1}\sigma_1 v = \varphi^{-1}\delta Sq^2 \rho v = \delta\varphi^{-1}Sq^2\rho v = \delta w_2(\xi). \qquad \square$$

Atiyah–Bott–Shapiro [1] proved that a stable *vector* bundle ξ is k-orientable iff it admits a Spin$^{\mathbb{C}}$-structure. This holds, in turn, iff $w_1(\xi) = 0$ and $w_2(\xi)$ is the reduction mod 2 of some integral class, i.e., $\delta w_2(\xi) = 0$, see e.g. Stong [3], Ch XI. In other words, we have the following fact:

3.2. Theorem. *If $\varkappa_0(\xi) = 0, \varkappa_1(\xi) = 0$ for some vector bundle ξ, then $0 \in \varkappa_r(\xi)$ for all r. In other words, none of the classes $\varkappa_r, r > 1$, can be realized by vector bundles.* □

One should clarify the situation. Note that we cannot apply V.5.1 (for $n = 1$) because $\Omega^\infty\sigma_2 = 0$ by 2.6(ii), and hence $(\Omega^\infty\sigma_2)|4 = 0$. Furthermore, we cannot apply V.5.6 because $\varkappa_2^{\mathcal{O}} = 0$.

3.3. Lemma. *The complexification $C : B\mathcal{O} \to B\mathcal{U}$ can be lifted to a morphism of spectra $\widehat{C} : K\mathcal{O} \to K$.*

Proof. In view of Bott periodicity it suffices to prove that the diagram

$$\begin{array}{ccc} \Omega^8 B\mathcal{O} & \xrightarrow{\Omega^8 C} & \Omega^8 B\mathcal{U} \\ \beta\downarrow & & \downarrow\beta' \\ B\mathcal{O} & \xrightarrow{C} & B\mathcal{U} \end{array}$$

commutes up to homotopy (where β, β' are the homotopy equivalences given by Bott periodicity). But this follows immediately from the commutativity of the diagram

$$\begin{array}{ccc} K\mathcal{O}^0(X) & \xrightarrow{\lambda\otimes} & K\mathcal{O}^0(S^8X) \\ c\downarrow & & \downarrow c \\ K^0(X) & \xrightarrow{\mu^4\otimes_{\mathbb{C}}} & K^0(S^8X) \end{array}$$

where $\lambda \in K\mathcal{O}^0(S^8) = \mathbb{Z}$ and $\mu \in K^0(S^2) = \mathbb{Z}$ are suitable generators. This diagram commutes, in turn, because of IV.4.27(iii) (for $n = 8$). □

Sullivan [1] proved that every $\mathcal{SPL}$-bundle is $K\mathcal{O}[1/2]$-orientable, a good proof can be found in Madsen–Milgram [1]. Hence, by IV.4.27(v), every $\mathcal{STOP}$-bundle is $K\mathcal{O}[1/2]$-orientable. Since the complexification $\widehat{C} : K\mathcal{O} \to K$ preserves the units, every $\mathcal{STOP}$-bundle is $K[1/2]$-orientable. Hence, every $\mathcal{STOP}$-bundle is $k[1/2]$-orientable by V.1.16. Thus, every $\mathcal{SPL}$- and/or $\mathcal{STOP}$-bundle is $k^r[1/2]$-orientable for every r, $0 \le r \le \infty$ (where k^∞ means k). By V.1.20, we have the following fact:

3.4. Theorem. *Given r, $0 \le r \le \infty$, an $\mathcal{STOP}$-bundle (as well as an $\mathcal{SPL}$-bundle) is k^r-orientable iff it is $k^r[2]$-orientable. In particular, the class $\varkappa_r$ can be realized by $\mathcal{PL}$- or $\mathcal{TOP}$-bundles iff $\widehat{\varkappa}_r$ can.* □

Put $BR\mathcal{V}{=}B(\mathcal{V}, k^1)$, i.e., $BR\mathcal{V}$ is the homotopy fiber of $\delta w_2 : BS\mathcal{V} \to K(\mathbb{Z}, 3)$. By the way, note that $BR\mathcal{O} = B\,\mathrm{Spin}$. Firstly, we compute the order of the class $\widehat{\varkappa}_2^{\mathcal{G}} \in H^5(BR\mathcal{G}; \mathbb{Z}[2])$. Recall that $\widehat{\varkappa}_2^{\mathcal{G}} = \varphi^{-1}\widehat{\sigma}_2 u_1$ where u_1 is the universal $k^1[2]$-orientation. We have $4H^*(k^1[2]; \mathbb{Z}[2]) = 0$, because every group $H^i(H\mathbb{Z}[2]; \mathbb{Z}[2]), i > 0$ has exponent 2. Hence, the order of $\widehat{\sigma}_2 \in H^5(k^1[2]; \mathbb{Z}[2])$ is 2 or 4. Hence, the order of $\widehat{\varkappa}_2^{\mathcal{G}}$ is 2 or 4.

3.5. Lemma. $\rho_* \widehat{\varkappa}_2^{\mathcal{G}} \neq 0 \in H^5(BR\mathcal{G}; \mathbb{Z}/2)$.

Proof. Let E be the homotopy fiber of $Sq^3\rho : H\mathbb{Z}[2] \to \Sigma^3 H\mathbb{Z}/2$. It is easy to see that there is a morphism a such that the diagram below commutes.

$$\begin{array}{ccc}
k^1[2] & \xrightarrow{\ a\ } & E \\
{\scriptstyle p}\big\downarrow & & \big\downarrow{\scriptstyle p'} \\
H\mathbb{Z}[2] & = & H\mathbb{Z}[2] \\
{\scriptstyle \delta Sq^2\rho}\big\downarrow & & {\scriptstyle Sq^3\rho}\big\downarrow \\
\Sigma^3 H\mathbb{Z}[2] & \xrightarrow[\rho]{} & \Sigma^3 H\mathbb{Z}/2
\end{array}$$

Consider the cofiber sequences $\Sigma^2 H\mathbb{Z}[2] \xrightarrow{j} k^1[2] \xrightarrow{p} H\mathbb{Z}[2], \Sigma^2 H\mathbb{Z}/2 \xrightarrow{j'} E \xrightarrow{p'} H\mathbb{Z}[2]$ and $\Sigma^2 H\mathbb{Z}[2] \xrightarrow{i} k^1[2] \xrightarrow{a} E$. We have the diagram

$$\begin{array}{ccccc}
\Sigma^2 H\mathbb{Z}[2] & = & \Sigma^2 H\mathbb{Z}[2] & & \\
{\scriptstyle \times 2}\big\downarrow & & \big\downarrow{\scriptstyle i} & & \\
\Sigma^2 H\mathbb{Z}[2] & \xrightarrow{\ j\ } & k^1[2] & \xrightarrow{\ \widehat{\sigma}_2\ } & \Sigma^5 H\mathbb{Z}[2] \\
{\scriptstyle \rho}\big\downarrow & & \big\downarrow{\scriptstyle a} & & \big\downarrow{\scriptstyle \rho} \\
\Sigma^2 H\mathbb{Z}/2 & \xrightarrow{\ j'\ } & E & \xrightarrow{\ \Phi\ } & \Sigma^5 H\mathbb{Z}/2.
\end{array} \tag{3.6}$$

Clearly, the left bottom square commutes. It is easy to see that

$$\mathrm{Im}\{i_* : \mathbb{Z}[2] = \pi_2(\Sigma^2 H\mathbb{Z}[2]) \to \pi_2(k^1[2]) = \mathbb{Z}[2]\} = 2\mathbb{Z}[2].$$

One can prove that $k^1[2]^*(\Sigma^2 H\mathbb{Z}[2]) = \mathbb{Z}[2]$ and that the homotopy class of any morphism $f : \Sigma^2 H\mathbb{Z}[2] \to k^1[2]$ is determined by the homomorphism

$$f_* : \mathbb{Z}[2] = \pi_2(\Sigma^2 H\mathbb{Z}[2]) \to \pi_2(k^1[2]) = \mathbb{Z}[2].$$

Hence, without loss of generality we can assume that $i = 2j$. So, the left top square commutes. Furthermore,

$$\rho\widehat{\sigma}_2 i = \rho\widehat{\sigma}_2(2j) = 2\rho\widehat{\sigma}_2 j = 0.$$

Hence, there exists $\Phi : E \to \Sigma^5 H\mathbb{Z}/2$ such that the diagram commutes. It is clear that

$$\Phi j'\rho = \rho\widehat{\sigma}_2 j = \rho\delta Sq^2\rho = Sq^3\rho,$$

and so $\Phi j' = Sq^3$ or $\Phi j' = Sq^3 + Sq^2Sq^1$. We want to prove that there exists Φ with $\Phi j' = Sq^3$.

3.7. Sublemma. *There is* $b : E \to \Sigma^5 H\mathbb{Z}[2]$ *such that* $bj' = Sq^2Sq^1$, $ba = 0$.

$$\begin{array}{ccccc}
\Sigma^2 H\mathbb{Z}[2] & \xrightarrow{j} & k^1[2] & & \\
\downarrow{\scriptstyle \rho} & & \downarrow{\scriptstyle a} & & \\
\Sigma^2 H\mathbb{Z}/2 & \xrightarrow{j'} & E & \xrightarrow{b} & \Sigma^5 H\mathbb{Z}/2 \\
 & & \downarrow{\scriptstyle p'} & & \\
 & & H\mathbb{Z}[2] & &
\end{array}$$

Proof. Consider an arbitrary morphism b with $bj' = Sq^2Sq^1$. It exists because $(Sq^2Sq^1)(Sq^3\rho) = 0$. By 2.6(iii), the cofiber sequence

$$\Sigma^2 H\mathbb{Z}[2] \xrightarrow{j} k^1[2] \xrightarrow{p} H\mathbb{Z}[2]$$

induces an exact sequence

$$H^5(\Sigma^2 H\mathbb{Z}[2];\mathbb{Z}/2) \xleftarrow{j^*} H^5(k^1[2];\mathbb{Z}/2) \xleftarrow{p^*} H^5(H\mathbb{Z}[2];\mathbb{Z}/2) \xleftarrow{(\delta Sq^2\rho)^*} \cdots,$$

where

$$(\delta Sq^2\rho)^* : \{Sq^2\rho\} = H^2(H\mathbb{Z}[2];\mathbb{Z}/2) \to H^5(H\mathbb{Z}[2];\mathbb{Z}/2) = \{Sq^5\rho\}.$$

We have

$$(\delta Sq^2\rho)^*(Sq^2\rho) = Sq^2\rho\delta Sq^2\rho = Sq^2Sq^3\rho = Sq^5\rho + Sq^4Sq^1\rho = Sq^5\rho,$$

and therefore $(\delta Sq^2\rho)^* : H^2(H\mathbb{Z}[2];\mathbb{Z}/2) \to H^5(H\mathbb{Z}[2];\mathbb{Z}/2)$ is an isomorphism. Hence, $H^5(k^1[2];\mathbb{Z}/2) = \mathbb{Z}/2 = \{\rho\widehat{\sigma}_2\}$ and $j^*\rho\widehat{\sigma}_2 = Sq^3\rho$. Now, if $ba \neq 0$ then $ba = \rho\widehat{\sigma}_2$ and so $baj = Sq^3\rho$. Thus, $bj'\rho = Sq^3\rho$, which contradicts $bj' = Sq^2Sq^1$. ⊞

Now let Φ be any morphism such that (3.6) commutes. If $\Phi j' = Sq^3 + Sq^2Sq^1$, then for every morphism b as in 3.7 we have $(\Phi + b)j' = Sq^3$. But the replacement of Φ by $\Phi' = \Phi + b$ keeps the commutativity of the diagram (3.6) because $ba = 0$. Hence, we may assume that $\Phi j' = Sq^3$ in (3.6).

Since E is the fiber of

$$Sq^3\rho : H\mathbb{Z}[2] \to \Sigma^3 H\mathbb{Z}/2,$$

the morphism $\Phi : E \to \Sigma^5 H\mathbb{Z}/2$ yields the secondary cohomology operation related to $Sq^3(Sq^3\rho) = 0$, i.e., the operation $\Phi = \Phi(3,3)$ in the notation of §1. Recall that $\widehat{\varkappa}_2^{\mathcal{G}} = \varphi^{-1}\widehat{\sigma}_2 u$ for some $k^1[2]$-orientation u of $\gamma_3^{\mathcal{G}}$. Let $MR\mathcal{G}$ denote the Thom spectrum of $\gamma_3^{\mathcal{G}}$. Consider the diagram

$$\begin{array}{ccccc} MR\mathcal{G} & \xrightarrow{u} & k^1[2] & \xrightarrow{\widehat{\sigma}_2} & \Sigma^5 H\mathbb{Z}[2] \\ \| & & \downarrow{\scriptstyle a} & & \downarrow{\scriptstyle \rho} \\ MR\mathcal{G} & \xrightarrow{v} & E & \xrightarrow{\Phi} & \Sigma^5 H\mathbb{Z}/2 \end{array}$$

where $v := au$. Let $x \in H^0(MR\mathcal{G})$ be the Thom class of $\gamma_3^{\mathcal{G}}$. The stable spherical fibration ξ as in 1.6 is such that $w_1(\xi) = 0 = w_2(\xi)$. Thus, ξ can be induced from the universal fibration $\gamma_3^{\mathcal{G}}$ over $BR\mathcal{G}$. By 1.6, $\Phi(u_\xi) \neq 0$, and so $0 \notin \Phi(x)$. Since $x = (p')_*(v)$, we conclude that $\Phi v \neq 0$, and so $\rho\widehat{\sigma}_2 u \neq 0$. Finally, $\widehat{\varkappa}_2^{\mathcal{G}} = \varphi^{-1}\widehat{\sigma}_2 u$, and hence $\rho_*\widehat{\varkappa}_2^{\mathcal{G}} \neq 0$. □

Madsen [1] computed the Bockstein spectral sequence for the 2-torsion of $BS\mathcal{G}$, but we only need the following fact.

3.8. Lemma. $H^5(BS\mathcal{G}; \mathbb{Z}[2]) = \mathbb{Z}/8 \oplus \mathbb{Z}/2$.

Proof. Throughout the proof $H_*(A)$ means $H_*(A; \mathbb{Z}[2])$. Similarly for H^*. Given a finite 2-primary group G, let $c(G)$ denote the dimension of the $\mathbb{Z}/2$-vector space $G \otimes \mathbb{Z}/2$.

Information about $H^*(BS\mathcal{G}; \mathbb{Z}/2)$ can be found in Madsen–Milgram [1], Theorem 3.35 or May [4]. By IV.4.27(i), the groups $\pi_i(BS\mathcal{G})$ are finite, and so the groups $H_i(BS\mathcal{G})$ are finite by the Hurewicz–Serre Theorem. Thus, $H_i(BS\mathcal{G}) \cong H^{i+1}(BS\mathcal{G})$ for every $i > 0$. Let

$$\begin{array}{ccccccc} & & K(\mathbb{Z}/8,4) & & K(\mathbb{Z}/2,3) & & \\ & & \downarrow & & \downarrow & & \\ \cdots & \longrightarrow & X & \longrightarrow & E & \longrightarrow & K(\mathbb{Z}/2,2) \end{array}$$

be the Postnikov tower of $BS\mathcal{G}[2]$, so that $E = BS\mathcal{G}[2]_{(3)}$, $X = BS\mathcal{G}[2]_{(4)}$. We have $H^3(BS\mathcal{G}; \mathbb{Z}/2) = (\mathbb{Z}/2)^2$. Thus, $H^3(E; \mathbb{Z}/2) = (\mathbb{Z}/2)^2$. Hence in the $\mathbb{Z}/2$-cohomology spectral sequence of the fibration $K(\mathbb{Z}/2,3) \to E \to K(\mathbb{Z}/2,2)$ the fundamental class $\iota \in H^2(K(\mathbb{Z}/2,2))$ transgresses to zero, and so this fibration is trivial, $E \simeq K(\mathbb{Z}/2,2) \times K(\mathbb{Z}/2,3)$. Thus, $H_4(E) = H_4(K(\mathbb{Z}/2,2)) = \mathbb{Z}/4$, $H_5(E) = (\mathbb{Z}/2)^3$.

The element $\delta w_4 \in H^5(BS\mathcal{G})$ has order 2, and it is not divisible by 2 because $w_5 = \rho\delta w_4$ is not divisible by 2. Since $H_4(X) = H^5(X)$, we conclude that $H_4(X)$ contains $\mathbb{Z}/2$ as a direct summand.

Consider now the $\mathbb{Z}[2]$-homology spectral sequence of the fibration

$$K(\mathbb{Z}/8, 4) \to X \to E.$$

We have $E^2_{0,4} = \mathbb{Z}/8, E^2_{4,0} = \mathbb{Z}/4, E^2_{i,4-i} = 0$ otherwise; $E^2_{5,0} = (\mathbb{Z}/2)^3$, $E^2_{i,5-i} = 0$ otherwise. We prove that $d^5 : (\mathbb{Z}/2)^3 = E^2_{5,0} \to E^2_{0,4} = \mathbb{Z}/8$ is non-zero. Indeed, suppose $d^5 = 0$. Then $H_4(X)$ is an extension of $\mathbb{Z}/4$ by $\mathbb{Z}/8$. Since $H_4(X)$ contains $\mathbb{Z}/2$ as a direct summand, we conclude that $c(H_4(X)) = 2$. Furthermore, if $d^5 = 0$ then $H_5(X) = H_5(E) = (\mathbb{Z}/2)^3$. On the other hand, $H_5(X;\mathbb{Z}/2) = H_5(BS\mathcal{G};\mathbb{Z}/2) = (\mathbb{Z}/2)^4$ (the first equality holds because $\pi_5(BS\mathcal{G}) = 0$). Now we have

$$4 = c(H_5(X;\mathbb{Z}/2)) = c(H_4(X)) + c(H_5(X)) = 2 + 3 = 5.$$

This is a contradiction. Hence, $d^5 \neq 0$. Thus, $E^\infty_{0,4} = \mathbb{Z}/4$, and $H_4(X)$ is an extension of $\mathbb{Z}/4$ by $\mathbb{Z}/4$. Since $H_4(X)$ contains $\mathbb{Z}/2$ as a direct summand, we conclude that $H_4(X) = \mathbb{Z}/8 \oplus \mathbb{Z}/2$. Thus, $H^5(BS\mathcal{G}) = \mathbb{Z}/8 \oplus \mathbb{Z}/2$. □

3.9. Theorem. *$H^5(BR\mathcal{G};\mathbb{Z}[2]) = \mathbb{Z}/4$, and the class $\widehat{\varkappa}_2^{\mathcal{G}}$ generates this group.*

Proof. By 3.8, $H^5(BS\mathcal{G};\mathbb{Z}[2]) = \mathbb{Z}/8 \oplus \mathbb{Z}/2 = \{x, \delta w_4\}$ where $\operatorname{ord} x = 8$. Let $u \in H^2(K(\mathbb{Z}[2], 2);\mathbb{Z}[2])$ be a fundamental class. Consider the $\mathbb{Z}[2]$-cohomology spectral sequence of the fibration

$$K(\mathbb{Z}[2], 2) \to BR\mathcal{G}[2] \to BS\mathcal{G}[2].$$

It is easy to see that $\tau(u^2) = \delta w_4 + \alpha x$, where $\alpha \in 2(\mathbb{Z}/8)$. Thus, $H^5(BR\mathcal{G};\mathbb{Z}[2]) = \mathbb{Z}/8$ (for $\alpha = 0, 4$) or $H^5(BR\mathcal{G};\mathbb{Z}[2]) = \mathbb{Z}/4$ (for $\alpha = 2, 6$). Hence, $H^5(BR\mathcal{G};\mathbb{Z}[2])$ is cyclic, and, by 3.5, the element $\widehat{\varkappa}_2^{\mathcal{G}}$ generates this group. But, as we noted before 3.5, the order of $\widehat{\varkappa}_2^{\mathcal{G}}$ is not more than 4, and hence $H^5(BR\mathcal{G};\mathbb{Z}[2]) = \mathbb{Z}/4 = \{\widehat{\varkappa}_2^{\mathcal{G}}\}$. □

3.10. Lemma. $\pi_2(BR\mathcal{V}[2]) = \mathbb{Z}[2]$.

Proof. By IV.4.27(i,ii), $\pi_2(BS\mathcal{V}) = \mathbb{Z}/2$ and $\pi_3(BS\mathcal{V})$ is finite for $\mathcal{V} = \mathcal{O}, \mathcal{G}$. Because of (2.14), (2.15) and IV.4.27(iv,v), the same is valid for $\mathcal{V} = \mathcal{PL}, \mathcal{TOP}$ also. Hence, the fibration

$$K(\mathbb{Z}[2], 2) \to BR\mathcal{V}[2] \to BS\mathcal{V}[2]$$

induces the exact sequence

$$0 \to \mathbb{Z}[2] \to \pi_2(BR\mathcal{V}[2]) \to \mathbb{Z}/2 \to 0.$$

Furthermore, $H^3(BS\mathcal{V};\mathbb{Z}[2]) = \mathbb{Z}/2 = \{\delta w_2\}$ (for $\mathcal{V} = \mathcal{O}$ see e.g. Milnor–Stasheff [1], for $\mathcal{V} = \mathcal{G}$ see e.g. Madsen–Milgram [1], for $\mathcal{V} = \mathcal{PL}$, $\mathcal{TOP}$ use (2.14) and (2.15), or IV.4.27(iv,v)), and the Leray–Serre spectral sequence of the fibration above implies that $H^3(BR\mathcal{V};\mathbb{Z}[2]) = 0$. Thus, $H_2(BR\mathcal{V};\mathbb{Z}[2])$ is torsion free, and hence $\pi_2(BR\mathcal{V})$ is. Thus, $\pi_2(BR\mathcal{V}) = \mathbb{Z}[2]$. □

Let $q = q^{\mathcal{V}} : BR\mathcal{V} = B(\mathcal{V}, k^1) \to B(\mathcal{V}, k^0) = BS\mathcal{V}$ be the map as in V.(4.17). The map $BR\mathcal{V}[2] \xrightarrow{q[2]} BS\mathcal{V}[2] \to BS\mathcal{G}[2]$ turns $BR\mathcal{V}[2]$ into a bundle over $BS\mathcal{G}[2]$.

3.11. Lemma. *The homotopy fiber of any map $g_{\mathcal{PL}} : BR\mathcal{PL}[2] \to BR\mathcal{G}[2]$ over $BS\mathcal{G}[2]$ is $\mathcal{G}/\mathcal{PL}[2]$. The homotopy fiber of any map $g_{\mathcal{TOP}} : BR\mathcal{TOP}[2] \to BR\mathcal{G}[2]$ over $BS\mathcal{G}[2]$ is $\mathcal{G}/\mathcal{TOP}[2]$.*

Proof. We consider the $\mathcal{PL}$ case only; the $\mathcal{TOP}$ case can be considered similarly. Consider the following diagram, where the rows are fibrations:

$$\begin{array}{ccccc} F & \longrightarrow & BR\mathcal{PL}[2] & \xrightarrow{g} & BR\mathcal{G}[2] \\ \downarrow{\scriptstyle h} & & \downarrow{\scriptstyle q^{\mathcal{PL}}[2]} & & \downarrow{\scriptstyle q^{\mathcal{G}}[2]} \\ \mathcal{G}/\mathcal{PL}[2] & \longrightarrow & BS\mathcal{PL}[2] & \longrightarrow & BS\mathcal{G}[2]. \end{array}$$

By the Five Lemma, h induces an isomorphism of homotopy groups $\pi_i, i \geq 3$. Applying π_2 to the right hand square of the diagram, we obtain (by 3.10) the square

$$\begin{array}{ccc} \mathbb{Z}[2] & \xrightarrow{g_*} & \mathbb{Z}[2] \\ \downarrow & & \downarrow \\ \mathbb{Z}/2 & \xrightarrow{\cong} & \mathbb{Z}/2, \end{array}$$

where the vertical arrows are epic. Furthermore, the bottom arrow is an isomorphism because $\pi_1(\mathcal{G}/\mathcal{PL}) = 0$. Thus, g_* is an isomorphism. Hence, by diagram chasing, $h_* : \pi_i(F) \to \pi_i(\mathcal{G}/\mathcal{PL})$ is an isomorphism for $i = 1, 2$. □

Note that the $\mathbb{Z}[2]$-localization in 3.11 is essential. Namely, there is a certain map $BR\mathcal{PL} \to BR\mathcal{G}$ over $BR\mathcal{G}$ with homotopy fiber X such that $\pi_1(X)$ is an abelian group of odd order.

3.12. Lemma. $H^5(BR\mathcal{PL};\mathbb{Z}[2]) = 0$.

Proof. By IV.4.27(iv),

$$H^5(BR\mathcal{PL};\mathbb{Z}/2) = H^5(BR\mathcal{O};\mathbb{Z}/2) = H^5(B\,\mathrm{Spin}^{\mathbb{C}};\mathbb{Z}/2).$$

But (see e.g. Stong [3], Ch. XI)

$$H^*(B\,\mathrm{Spin}^{\mathbb{C}};\mathbb{Z}/2) = H^*(BS\mathcal{O};\mathbb{Z}/2)/(\mathscr{A}_2 w_3).$$

So, $H^5(B\,\mathrm{Spin}^{\mathbb{C}};\mathbb{Z}/2) = 0$ since $Sq^2 w_3 = w_2 w_3 + w_5$ in $H^*(BSO;\mathbb{Z}/2)$. Hence, $H^5(BR\mathcal{PL};\mathbb{Z}/2) = 0$, and thus $H^5(BR\mathcal{PL};\mathbb{Z}[2]) = 0$. □

3.13. Theorem. *If $\varkappa_1(\xi) = 0$ for some $S\mathcal{PL}$-bundle ξ then $0 \in \varkappa_2(\xi)$ and $0 \in \varkappa_3(\xi)$. Thus, the classes $\varkappa_2$ and $\varkappa_3$ cannot be realized by $\mathcal{PL}$-bundles.*

Proof. By IV.4.27(iv), the map $a^{\mathcal{O}}_{\mathcal{PL}} : B\mathcal{O} \to B\mathcal{PL}$ is a 6-equivalence, i.e, informally speaking, there is no difference between $\mathcal{O}$-bundles and $\mathcal{PL}$-bundles over 6-dimensional CW-complexes. Now, the result follows from 3.2 since $\varkappa_3 \subset H^5(\mathrm{bs}\,\xi)$. □

3.14. Theorem. $0 \notin \varkappa_4(\gamma_7^{\mathcal{PL}})$. *Furthermore, all the classes $\varkappa_n, n \geq 4$, as well as the classes $\widehat{\varkappa}_n, n \geq 4$, can be realized by $\mathcal{PL}$-bundles.*

Proof. Let $q' : B(\mathcal{G}, k^4) \to BR\mathcal{G} = B(\mathcal{G}, k^1)$ be a map (fibration) induced by a projection $k^4 \to k^1$ in the Postnikov tower of k, cf. V.4.17. By 3.13, every k^1-orientable $\mathcal{PL}$-bundle is k^3-orientable. Hence, if $0 \in \varkappa_4(\gamma_7^{\mathcal{PL}})$, then every k^1-orientable $\mathcal{PL}$-bundle is k^4-orientable. In particular, the universal $\mathcal{PL}$-bundle $\gamma_3^{\mathcal{PL}}$ over $BR\mathcal{PL}$ is k^4-orientable, i.e., there exists a map f' : $BR\mathcal{PL} \to B(\mathcal{G}, k^4)$ over $BS\mathcal{G}$. Thus, we have a commutative diagram

$$\begin{array}{ccc}
BR\mathcal{PL} & \xrightarrow{f'} & B(\mathcal{G}, k^4) \\
\| & & \downarrow{\scriptstyle q'} \\
BR\mathcal{PL} & \xrightarrow{g'} & BR\mathcal{G} \\
\downarrow & & \downarrow \\
BS\mathcal{PL} & \longrightarrow & BS\mathcal{G},
\end{array}$$

where $g' := q'f'$. It induces a commutative diagram

$$\begin{array}{ccc}
BR\mathcal{PL}[2] & \xrightarrow{f} & B(\mathcal{G}, k^4)[2] \\
\| & & \downarrow{\scriptstyle q} \\
BR\mathcal{PL}[2] & \xrightarrow{g} & BR\mathcal{G}[2] \\
\downarrow & & \downarrow \\
BS\mathcal{PL}[2] & \longrightarrow & BS\mathcal{G}[2],
\end{array}$$

where $q = q'[2], f = f'[2]$, and $g := qf$ is a map over $BS\mathcal{G}[2]$. Without loss of generality, we can assume q and f to be fibrations. By 3.11, the homotopy fiber of g is $\mathcal{G}/\mathcal{PL}[2]$, and, by V.4.18(i), the homotopy fiber of q is $(\Omega^\infty k^4)|4 = BS\mathcal{U}[2]_{(8)}$. Thus, f induces a map of fibers

$$\bar{f} : \mathcal{G}/\mathcal{PL}[2] \to BS\mathcal{U}[2]_{(8)}.$$

Let $u \in H^4(B\mathcal{SU}[2]_{(8)}; \mathbb{Z}[2]) = \mathbb{Z}[2]$ and $v \in H^4(\mathcal{G}/\mathcal{PL}[2]; \mathbb{Z}[2]) = \mathbb{Z}[2]$ be generators, i.e., $u, v \in \mathbb{Z}[2]^*$. Consider the $\mathbb{Z}[2]$ -cohomology spectral sequences of the fibrations g and q. By V.4.22, we have $\tau u = \varepsilon \overline{\varkappa}_2^{\mathcal{G}}$ where $\varepsilon \in \mathbb{Z}[2]^*$. Furthermore, by 3.12, $H^5(BR\mathcal{PL}; \mathbb{Z}[2]) = 0$. Hence, $\tau v = \pm \overline{\varkappa}_2^{\mathcal{G}}$ since, by 3.9, $\widehat{\varkappa}_2^{\mathcal{G}}$ generates the group $H^5(BR\mathcal{G}; \mathbb{Z}[2]) = \mathbb{Z}/4$. Thus, $\bar{f}^* u = \varepsilon v$ with $\varepsilon \in \mathbb{Z}[2]$.

We define the map

$$h : Y \xrightarrow{a} \mathcal{G}/\mathcal{PL}[2] \xrightarrow{\bar{f}} B\mathcal{SU}[2]_{(8)} \xrightarrow{p} B\mathcal{SU}[2]_{(4)} = K(\mathbb{Z}[2], 4),$$

where a is the inclusion of the factor and p is the projection in the Postnikov tower of $B\mathcal{SU}[2]$. Since $\bar{f}^* u = \varepsilon v$, the map $h : Y \to K(\mathbb{Z}[2], 4)$ yields a generator of $H^4(Y; \mathbb{Z}[2]) = \mathbb{Z}[2]$. Hence, by 2.11, h cannot be lifted to $B\mathcal{SU}[2]_{(8)}$. This is a contradiction. Thus, $0 \notin \varkappa_4(\gamma_7^{\mathcal{PL}})$.

Hence, by 3.4, $0 \notin \widehat{\varkappa}_4(\gamma_7^{\mathcal{PL}})$. By V.5.1, all the classes $\widehat{\varkappa}_n$, $n \geq 4$, can be realized by $\mathcal{PL}$-bundles, and so (by 3.4) all the classes $\varkappa_n$, $n \geq 4$, can. □

Now we pass to $\mathcal{TOP}$-bundles.

Consider a commutative diagram

$$(3.15) \qquad \begin{array}{ccccc} \mathcal{G}/\mathcal{PL}[2] & \xrightarrow{i_{\mathcal{PL}}} & BR\mathcal{PL}[2] & \xrightarrow{g_{\mathcal{PL}}} & BR\mathcal{G}[2] \\ \downarrow{\scriptstyle \bar{h}} & & \downarrow{\scriptstyle h} & & \| \\ \mathcal{G}/\mathcal{TOP}[2] & \xrightarrow{i_{\mathcal{TOP}}} & BR\mathcal{TOP}[2] & \xrightarrow{g_{\mathcal{TOP}}} & BR\mathcal{G}[2] \end{array}$$

where the rows are fibrations, $g_{\mathcal{TOP}}$ is any map as in 3.11, h is a forgetful map over $BS\mathcal{G}[2]$, $g_{\mathcal{PL}} = g_{\mathcal{TOP}} h$ and $\bar{h}$ is the induced map of the fibers.

3.16. Lemma. *The homomorphism*

$$\bar{h}_* : \mathbb{Z}/2 = \pi_2(\mathcal{G}/\mathcal{PL}[2]) \to \pi_2(\mathcal{G}/\mathcal{TOP}[2]) = \mathbb{Z}/2$$

is an isomorphism.

Proof. Consider the fibration $\mathcal{G}/\mathcal{V}[2] \to BR\mathcal{V}[2] \to BR\mathcal{G}[2]$. By 3.10,

$$\pi_2(BR\mathcal{PL}[2]) = \mathbb{Z}[2] = \pi_2(BR\mathcal{TOP}[2]).$$

Hence, if $\mathcal{V} = \mathcal{PL}$ or $\mathcal{TOP}$, then the boundary homomorphism

$$\mathbb{Z}/2 = \pi_3(BR\mathcal{G}[2]) \to \pi_2(\mathcal{G}/\mathcal{V}[2]) = \mathbb{Z}/2$$

for the above fibration is an epimorphism, and hence an isomorphism. □

We fix maps $g_{\mathcal{TOP}}$ and $g_{\mathcal{PL}}$ as in (3.15).

3.17. Proposition. *Consider the map* $i = i_{\mathcal{PL}} : \mathcal{G}/\mathcal{PL}[2] \to BR\mathcal{PL}[2]$ *in* (3.15). *We have*

$$\operatorname{Im}\{i^* : H^4(BR\mathcal{PL}[2];\mathbb{Z}[2]) \to H^4(\mathcal{G}/\mathcal{PL}[2];\mathbb{Z}[2])\} = 4H^4(\mathcal{G}/\mathcal{PL}[2];\mathbb{Z}[2]).$$

Proof. Consider the fibration

$$\mathcal{G}/\mathcal{PL}[2] \xrightarrow{i} BR\mathcal{PL}[2] \xrightarrow{g} BR\mathcal{G}[2]$$

as in (3.15).By 3.12, $H^5(BR\mathcal{PL}[2];\mathbb{Z}[2]) = 0$. Hence, $g^*\widehat{\varkappa}_2^{\mathcal{G}} = 0$. So, there exists $\iota \in H^4(\mathcal{G}/\mathcal{PL}[2];\mathbb{Z}[2]) = \mathbb{Z}[2]$ such that $\tau\iota = \widehat{\varkappa}_2^{\mathcal{G}}$ in the $\mathbb{Z}[2]$-cohomology spectral sequence of the fibration. By 3.5, $\iota \notin 2\mathbb{Z}[2]$. Furthermore, by 3.9, $\tau(2\iota) \neq 0, \tau(4\iota) = 0$, and hence $\operatorname{Im} j^* = 4H^4(\mathcal{G}/\mathcal{PL}[2];\mathbb{Z}[2])$. □

Consider the Postnikov–Moore tower of the fibration

$$\mathcal{G}/\mathcal{PL}[2] \xrightarrow{i_{\mathcal{PL}}} BR\mathcal{PL}[2] \xrightarrow{g_{\mathcal{PL}}} BR\mathcal{G}[2]$$

as in (3.15). Its first term gives us a $K(\mathbb{Z}[2],2)$-fibration

(3.18) $$K(\mathbb{Z}[2],2) \to B \xrightarrow{f} BR\mathcal{G}[2],$$

and all remaining terms form a $\mathcal{G}/\widetilde{\mathcal{PL}}[2]$-fibration

(3.19) $$\mathcal{G}/\widetilde{\mathcal{PL}}[2] \to BR\mathcal{PL}[2] \to B.$$

Similarly, consider the Postnikov–Moore tower of the fibration

$$\mathcal{G}/\mathcal{TOP}[2] \xrightarrow{i_{\mathcal{TOP}}} BR\mathcal{TOP}[2] \xrightarrow{g_{\mathcal{TOP}}} BR\mathcal{G}[2]$$

as in (3.15). Its first term gives us a $K(\mathbb{Z}[2],2)$-fibration which, by 3.16, is equivalent to (3.18), and all remaining terms form a $\mathcal{G}/\widetilde{\mathcal{TOP}}[2]$-fibration

(3.20) $$\mathcal{G}/\widetilde{\mathcal{TOP}}[2] \to BR\mathcal{TOP}[2] \xrightarrow{p} B.$$

3.21. Proposition. $H^5(B;\mathbb{Z}[2]) = \mathbb{Z}/8$, *and*

$$f^* : H^5(BR\mathcal{G}[2];\mathbb{Z}[2]) \to H^5(B;\mathbb{Z}[2])$$

is a monomorphism (onto the subgroup of index 2 *since* $H^5(BR\mathcal{G};\mathbb{Z}[2]) = \mathbb{Z}/4$).

Proof. Firstly, $H^i(BR\mathcal{G};\mathbb{Q}) = 0 = H^i(K(\mathbb{Z}[2],2);\mathbb{Q})$ for $i \neq 0,2$. So, $H^5(B;\mathbb{Q}) = 0$, and hence $H^5(B;\mathbb{Z}[2])$ is finite. Furthermore, by 3.12, $H^5(BR\mathcal{PL}[2];\mathbb{Z}[2]) = 0$. Hence in the $\mathbb{Z}[2]$-cohomology spectral sequence of the fibration (3.19) the transgression

$$\tau : \mathbb{Z}[2] = H^4(\mathcal{G}/\widetilde{\mathcal{PL}}[2];\mathbb{Z}[2]) \to H^5(B;\mathbb{Z}[2])$$

is epic. Hence $H^5(B;\mathbb{Z}[2])$ is cyclic. By 3.17 and 2.17(i),

$$\operatorname{Im}\{H^4(BR\mathcal{PL}[2];\mathbb{Z}[2]) \to H^4(\mathcal{G}/\mathcal{PL}[2];\mathbb{Z}[2]) \to H^4(\mathcal{G}/\widetilde{\mathcal{PL}}[2];\mathbb{Z}[2])\}$$

is $8H^4(\mathcal{G}/\widetilde{\mathcal{PL}}[2];\mathbb{Z}[2])$. Hence $H^5(B;\mathbb{Z}[2]) = \mathbb{Z}/8$. Now it is easy to prove that f^* is monic by considering the Leray-Serre spectral sequence of (3.18). □

Let e be a generator of $H^5(B;\mathbb{Z}[2]) = \mathbb{Z}/8$. Let $\iota_P \in H^4(\mathcal{G}/\widetilde{\mathcal{PL}}[2];\mathbb{Z}[2]) = \mathbb{Z}[2]$ and $\iota_T \in H^4(\mathcal{G}/\widetilde{\mathcal{TOP}}[2];\mathbb{Z}[2]) = \mathbb{Z}[2]$ be generators of the corresponding groups.

3.22. Lemma. *In the $\mathbb{Z}[2]$-cohomology spectral sequence of (3.19) we have $\tau\iota_P = \varepsilon e$, and in that of (3.20) we have $\tau\iota_T = 2\varepsilon' e$, where τ is the transgression and $\varepsilon, \varepsilon' \in \mathbb{Z}[2]^*$.*

Proof. By 3.12, $H^5(BR\mathcal{PL};\mathbb{Z}[2]) = 0$, and so $\tau\iota_P = \varepsilon e$. The diagram (3.15) yields the diagram

$$\begin{array}{ccccc} \mathcal{G}/\widetilde{\mathcal{PL}}[2] & \longrightarrow & BR\mathcal{PL}[2] & \longrightarrow & B \\ \downarrow{\scriptstyle\widetilde{\varphi}} & & \downarrow & & \| \\ \mathcal{G}/\widetilde{\mathcal{TOP}}[2] & \longrightarrow & BR\mathcal{TOP}[2] & \longrightarrow & B, \end{array}$$

where the top line is (3.19), the bottom line is (3.20) and $\widetilde{\varphi}$ is as in 2.18. By 2.18(ii), $\widetilde{\varphi}(\iota_T) = 2\varepsilon''\iota_P$. Hence, $\tau\iota_T = 2\varepsilon' e$ because $\tau\iota_P = \varepsilon e$. □

The projection $k^2 \to k^1$ in the Postnikov tower of k induces the $K(\mathbb{Z}[2],4)$-fibration

(3.23) $$K(\mathbb{Z}[2],4) \to B(\mathcal{G},k^2)[2] \xrightarrow{q} B(\mathcal{G},k^1)[2] = BR\mathcal{G}[2].$$

3.24. Lemma. *There is a map F such that the diagram*

$$\begin{array}{ccccc} \mathcal{G}/\widetilde{\mathcal{TOP}}[2] & \longrightarrow & BR\mathcal{TOP}[2] & \xrightarrow{p} & B \\ & & \downarrow{\scriptstyle F} & & \downarrow{\scriptstyle f} \\ K(\mathbb{Z}[2],4) & \longrightarrow & B(\mathcal{G},k^2)[2] & \xrightarrow{q} & BR\mathcal{G}[2] \end{array}$$

commutes. (Here the top line is (3.20), the bottom line is (3.23), and f is as in (3.18).)

Proof. By 3.22, $\tau\iota_T = 2\varepsilon e$, $\varepsilon \in \mathbb{Z}[2]^*$. Hence, $p^*(2e) = 0$. Furthermore, $f^*(\widehat{\varkappa}_2^{\mathcal{G}}) = \pm 2e$ because of 3.21 and 3.9. Hence, $p^* f^*(\widehat{\varkappa}_2^{\mathcal{G}}) = 0$. Finally, by V.4.22, $\widehat{\varkappa}_2^{\mathcal{G}}$ is a characteristic class of (3.23), and hence there exists a q-lifting F of pf. □

3.25. Lemma. *Let F be a map as in 3.24. Then F cannot be lifted to $B(\mathcal{G},k^3)[2]$, i.e., there is no commutative diagram*

$$\begin{array}{ccc} BR\mathcal{TOP}[2] & \longrightarrow & B(\mathcal{G},k^3)[2] \\ \| & & \downarrow{\scriptstyle q'} \\ BR\mathcal{TOP}[2] & \xrightarrow{F} & B(\mathcal{G},k^2)[2] \\ {\scriptstyle p}\downarrow & & \downarrow{\scriptstyle q} \\ B & \xrightarrow{f} & BR\mathcal{G}[2] \end{array} .$$

Proof. Suppose that there is a commutative diagram as above. Passing to the fibers of p, q, and qq', we have a commutative diagram of fibrations

$$\begin{array}{ccc} \mathcal{G}/\widetilde{\mathcal{TOP}}[2] & \xrightarrow{s} & E \\ \| & & \downarrow{\scriptstyle \pi} \\ \mathcal{G}/\widetilde{\mathcal{TOP}}[2] & \xrightarrow{\overline{F}} & (K(\mathbb{Z}[2],4)) \end{array}$$

where $\overline{F}$ is the restriction of F, s is the restriction of a hypothetical lifting of F, E is the homotopy fiber of

$$\delta Sq^2\rho : K(\mathbb{Z}[2],4) \to K(\mathbb{Z}[2],7),$$

and the projection π is homotopic to the inclusion of this fiber. In particular, π does not admit a section.

Let $\iota \in H^4(K(\mathbb{Z}[2],4);\mathbb{Z}[2])$ be a fundamental class such that $\tau\iota = \widehat{\varkappa}_2^{\mathcal{G}}$ in the $\mathbb{Z}[2]$-cohomology spectral sequence of q. Hence,

$$f^*(\tau\iota) = f^*(\widehat{\varkappa}_2^{\mathcal{G}}) = \pm 2e = \varepsilon'\tau\iota_T, \quad \varepsilon' \in \mathbb{Z}[2]^*.$$

Therefore, $\overline{F}^*\iota = \varepsilon\iota_T, \varepsilon \in \mathbb{Z}[2]^*$. Hence,

$$\overline{F}^* : \mathbb{Z}[2] = H^4(K(\mathbb{Z}[2],4);\mathbb{Z}[2]) \xrightarrow{\cong} H^4(\mathcal{G}/\widetilde{\mathcal{TOP}}[2];\mathbb{Z}[2]) = \mathbb{Z}[2]$$

is an isomorphism. Thus, $\overline{F}$ can be considered as the projection in (2.15), and hence $\overline{F}$ admits a section t. Hence π admits a section st. This is a contradiction. □

3.26. Theorem. *The canonical bundle $\gamma_3^{\mathcal{TOP}}$ over $BR\mathcal{TOP}$ is k^2-orientable. In particular, if $\varkappa_1(\xi) = 0$ for any $S\mathcal{TOP}$-bundle ξ, then $0 \in \varkappa_2(\xi)$; in other words, $\varkappa_2$ cannot be realized by $\mathcal{TOP}$-bundles. On the other hand, for every $r \geq 3$ we have $0 \notin \varkappa_r(\gamma_{2r-1}^{\mathcal{TOP}})$. Thus, all the classes $\varkappa_r, r \geq 3$, as well as the classes $\widehat{\varkappa}_r, r \geq 3$, can be realized by $S\mathcal{TOP}$-bundles.*

Proof. The canonical bundle $\gamma_3^{\mathcal{TOP}}$ is k^2-orientable because of 3.24. Indeed, the map F in 3.24 yields a k^2-orientation of $\gamma_3^{\mathcal{TOP}}$. In order to prove the realizability of all the classes $\varkappa_r, r \geq 3$, it suffices to prove that $0 \notin \varkappa_3(\gamma_3^{\mathcal{TOP}})$.

Indeed, this implies that $0 \notin \varkappa_3(\gamma_5^{\mathcal{TOP}})$, because otherwise $0 \in \varkappa_3(\xi)$ for every k^2-oriented ξ and in particular for $\gamma_3^{\mathcal{TOP}}$, see 3.24. Hence $0 \notin \varkappa_r(\gamma_{2r-1}^{\mathcal{TOP}})$ in view of V.5.1.

Suppose that $0 \in \varkappa_3(\gamma_3^{\mathcal{TOP}})$, i.e., that the bundle $\gamma_3^{\mathcal{TOP}}$ is k^3-orientable, i.e., that there exists

$$h' : BR\mathcal{TOP} \to B(\mathcal{G}, k^3)$$

over $BS\mathcal{G}$. Let F' be the composition of h' with the projection $B(\mathcal{G}, k^3) \to B(\mathcal{G}, k^2)$. After $\mathbb{Z}[2]$-localization we have the following diagram, where $F := F'[2], h := h'[2]$, and $g := qF$:

$$\begin{array}{ccc} BR\mathcal{TOP}[2] & \xrightarrow{h} & B(\mathcal{G}, k^3)[2] \\ \| & & {\scriptstyle q'}\downarrow \\ BR\mathcal{TOP}[2] & \xrightarrow{F} & B(\mathcal{G}, k^2)[2] \\ \| & & {\scriptstyle q}\downarrow \\ BR\mathcal{TOP}[2] & \xrightarrow{g} & BR\mathcal{G}[2] \\ \| & & \downarrow \\ BR\mathcal{TOP}[2] & \longrightarrow & BS\mathcal{G}[2] \end{array} .$$

In particular, g is a map over $BS\mathcal{G}[2]$, i.e., g can be regarded as $g_{\mathcal{TOP}}$ described in 3.11.

The Postnikov–Moore tower for g yields the diagram

$$\begin{array}{ccc} BR\mathcal{TOP}[2] & \xrightarrow{F} & B(\mathcal{G}, k^2)[2] \\ \| & & \downarrow{\scriptstyle q} \\ BR\mathcal{TOP}[2] & \xrightarrow{g} & BR\mathcal{G}[2] \\ {\scriptstyle p}\downarrow & & \| \\ B & \xrightarrow{f} & BR\mathcal{G}[2] \end{array}$$

where p and f are as in 3.24. Thus, F has the lifting $h : BR\mathcal{TOP}[2] \to B(\mathcal{G}, k^3)[2]$. But this contradicts 3.25. □

3.27. Theorem. *All the classes $\varkappa_r, r \geq 1$, as well as the classes $\widehat{\varkappa}_r, r \geq 1$, can be realized by spherical fibrations.*

Proof. By 3.9, $\widehat{\varkappa}_2^{\mathcal{G}} \neq 0$. Furthermore, by V.4.9 and 2.6(ii), $\widehat{\varkappa}_r^{\mathcal{G}} \neq 0$ for every $r \geq 3$. Hence, by V.5.6, all the classes $\widehat{\varkappa}_n$ can be realized by spherical fibrations. Finally, the realizability of $\widehat{\varkappa}_2$ implies easily the realizability of $\varkappa_2$ (by exactly the same fibration). □

Recall that we have considered above the $\mathbb{Z}[2]$-local case. What about $\mathbb{Z}[p]$-localization with odd prime p?

In the beginning of the section (see the text before 3.4) we discussed $k[1/2]$- and hence $k[p]$-orientability of $\mathcal{STOP}$-objects. Thus, it makes sense to consider the realizability problem for spherical fibrations only.

The following well-known fact was proved by Adams [6], see also IX.4.16 below.

3.28. Theorem. *For every odd prime p there exists a ring spectrum L such that*

$$K[p] \simeq \vee_{i=0}^{p-2} \Sigma^{2i} L.$$

Furthermore, the inclusion $L \to K[p]$ of the direct summand is a ring morphism. □

3.29. Corollary. *For every odd prime p there exists a ring spectrum ℓ such that*

$$k\,[p] \simeq \vee_{i=0}^{p-2} \Sigma^{2i} \ell.$$

Furthermore, the inclusion $\ell \to k\,[p]$ of the direct summand is a ring morphism. Finally, $k\,[p]$-orientability of any object is equivalent to its l-orientability.

Proof. Set $\ell = L|0$. The existence of the required splitting follows directly from 3.28 and naturality of connective coverings. Furthermore, since $\ell \to k\,[p]$ is a ring morphism, ℓ-orientability implies $k\,[p]$-orientability. Conversely, if $S \to k\,[p]$ is the unit of $k\,[p]$, then

$$S \to k\,[p] \xrightarrow{\text{proj}} \ell$$

is the unit of ℓ. Hence, $k\,[p]$-orientability implies ℓ-orientability. □

The Postnikov tower of ℓ looks similar to that of k. Clearly,

$$\pi_*(\ell) = \mathbb{Z}[p][x], \deg x = 2(p-1).$$

Let ℓ^r be the cone of

$$x_{\#}^{r+1} : \Sigma^{2(r+1)(p-1)} \ell \to \ell,$$

and compare this with k^r. Similarly to 2.3, there is a cofiber sequence

$$\ell^r \xrightarrow{p_r} \ell^{r-1} \xrightarrow{\sigma_r} \Sigma^{2r(p-1)+1} H\mathbb{Z}[p].$$

Moreover, $(\Sigma^{-2(p-1)} \ell^r)|0 = \ell^{r-1}$.

Let $\rho_p : H\mathbb{Z}[p] \to H\mathbb{Z}/p$ and $\delta_p : H\mathbb{Z}/p \to H\mathbb{Z}[p]$ be the reduction mod p and the Bockstein morphism, respectively.

3.30. Theorem. *The diagram*

$$\begin{array}{ccccccc}
 & \ell & & & & & H\mathbb{Z} \\
 & \big\downarrow{\scriptstyle \tau_{2r(p-1)}} & & & & & \big\| \\
\cdots \longrightarrow & \ell^r & \xrightarrow{p_r} & \ell^{r-1} & \longrightarrow \cdots \longrightarrow & & \ell^0 \\
 & & & {\scriptstyle \sigma_r}\big\downarrow & & & \big\downarrow{\scriptstyle \sigma_1} \\
 & & & \Sigma^{2r(p-1)+1}H\mathbb{Z} & & & \Sigma^{2p-1}H\mathbb{Z}
\end{array}$$

is the Postnikov tower of ℓ, *and for the Postnikov invariants* σ_r *we have:*

(i) $(\Omega^\infty \sigma_n)|(2p-2) \neq 0$ *for every* $n > 2$.

(ii) $\sigma_1 \neq 0$, *i.e.*, $\sigma_1 = \lambda\delta_p P^1 \rho_p \in H^{2p-1}(H\mathbb{Z}[p]; \mathbb{Z}[p]) = \mathbb{Z}/p$, *where* $\lambda \in \mathbb{Z}/p, \lambda \neq 0$.

(iii) $\delta_p P_1 \rho_p \sigma_n = 0$ *for every* $n \geq 1$, *and the higher operation* σ_{n+1} *is associated with this relation.*

Proof. (i) Set $Y = \Omega^\infty \ell$, and let X be any connected component of Y. Then $\pi_{2n(p-2)}(X) = \mathbb{Z}[p]$ for $n > 0$ and $\pi_i(X) = 0$ otherwise. Furthermore,

$$\Omega^{2p-2} X = \Omega^{2p-2} Y \simeq Y.$$

Now, X is a factor of $B\mathcal{U}[p]$, and so $H^*(X; \mathbb{Z}[p])$ is torsion free. Consider the Postnikov invariant $\kappa \in H^{4p-3}(K(\mathbb{Z}[p], 2p-2); \mathbb{Z}[p])$ of X. If $\kappa = 0$, then

$$H^{4p-3}(X) = H^{4p-3}(K(\mathbb{Z}[p], 2p-2) \times K(\mathbb{Z}[p], 4p-4))$$

has a non-zero torsion subgroup. This is a contradiction, and so $\kappa \neq 0$. Clearly, κ in fact coincides with $\Omega^\infty \sigma_2$, and so $\Omega^\infty \sigma_2 \neq 0$. This implies that $(\Omega^\infty \sigma_n)|(2n(p-1)) \neq 0$, because

$$(\Omega^{2n(p-1)}(\Omega^\infty \ell))|(2n(p-1)) \simeq \Omega^\infty \ell.$$

Hence, $(\Omega^\infty \sigma_n)|(2p-2) \neq 0$ for $n > 2$. Thus, we have proved (i).

The remaining part of the proof can be done in just the same way as in 2.6. □

As above, we introduce higher characteristic classes $\varkappa_r(\xi) = \varphi^{-1}\sigma_r u_\xi$ related to ℓ.

3.31. Theorem. *All the characteristic classes* $\varkappa_r$ *can be realized by spherical fibrations.*

Proof. Let $u \in H^0(MS\mathcal{G}; \mathbb{Z}[p])$ be the universal Thom class. By IV.6.10, $\delta_p P^1 \rho_p(u) \neq 0$. Now the theorem follows from V.5.6. and 3.30(ii). □

3.32. Remark. Let $v \in H^0(k; \mathbb{Z}/p) = \mathbb{Z}/p$ be a generator. It follows from 3.30(ii) that $Q_1(v) = 0$. Since every $\mathcal{SPL}$-bundle is $k[p]$-orientable, see the text before 3.4, $Q_1(u_{\mathcal{PL}}) = 0$ for the universal Thom class $u_{\mathcal{PL}} \in H^0(MS\mathcal{PL}; \mathbb{Z}/p)$, cf. IV.6.12.

Résumé on k-orientability

The conditions $w_1(\xi) = 0, \delta w_2(\xi) = 0$ are necessary for k-orientability of any $\mathcal{V}$-object ξ. In particular, it makes sense to consider only $\mathcal{V}$-objects ξ with $w_1(\xi) = 0$. We have $\varkappa_1 = \delta w_2$.

Theorem. (i) *The class $\varkappa_1$ can be realized by vector bundles (namely, by the universal oriented vector bundle). If for some $H\mathbb{Z}$-oriented vector bundle ξ we have $\varkappa_1(\xi) = 0$, then $0 \in \varkappa_r(\xi)$ for all r, i.e., ξ is k-orientable. In other words, none of the classes $\varkappa_r, r > 1$, can be realized by vector bundles.*

(ii) *If for some $S\mathcal{PL}$-bundle ξ we have $\varkappa_1(\xi) = 0$, then $0 \in \varkappa_2(\xi)$ and $0 \in \varkappa_3(\xi)$. Thus, the classes $\varkappa_2$ and $\varkappa_3$ cannot be realized by $\mathcal{PL}$-bundles. However, all the classes $\varkappa_r, r \geq 4$, (and $\varkappa_1$, of course) can be realized by $\mathcal{PL}$-bundles.*

(iii) *If for some $S\mathcal{TOP}$-bundle ξ we have $\varkappa_1(\xi) = 0$, then $0 \in \varkappa_2(\xi)$, and hence $\varkappa_2$ cannot be realized by $\mathcal{TOP}$-bundles. However, all the classes $\varkappa_r, r \geq 3$, (and $\varkappa_1$, of course) can be realized by $\mathcal{TOP}$-bundles.*

(iv) *For every $n \geq 1$ we have $0 \notin \varkappa_n(\gamma^{\mathcal{G}}_{2n-1})$. In other words, all the classes $\varkappa_r, r \geq 1$, can be realized by spherical fibrations.*

Thus, we have a remarkable contrast among all the four classes above.

Also, the following observation looks interesting. Every vector bundle over a 3-connected space is k-orientable, but there is no universal n such that every PL bundle over an n-connected space is k-orientable. Another interpretation: a vector bundle ξ over a (finitely dimensional) CW-complex X is k-orientable iff $\xi|X^{(3)}$ is. However, there is no universal n such that, for every PL bundle ξ, k-orientability of $\xi|X^{(n)}$ guarantees k-orientability of ξ.

§4. $k\mathcal{O}$-Orientability

Here we consider the $k\mathcal{O}$-orientability (which is equivalent to the $K\mathcal{O}$-orientability) problem. In particular, we show that it has mutually different solutions for all the four classes of $\mathcal{V}$-objects, as the k-orientability problem does. Since most of the arguments are similar to the arguments of the previous section, we will not be very detailed in the exposition.

Note that the homotopy groups of $k\mathcal{O}$ are well known in view of Bott periodicity $\Omega^8 B\mathcal{O} = B\mathcal{O} \times \mathbb{Z}$, see IV.4.27(ii) (recall that $\Omega^\infty k\mathcal{O} \simeq B\mathcal{O} \times \mathbb{Z}$, and so $\pi_i(k\mathcal{O}) = \pi_i(B\mathcal{O})$).

Let $k\mathcal{O}_n$ denote the Postnikov n-stage of $k\mathcal{O}$.

4.1. Theorem (cf. Adams–Priddy [1], Stong [1]). (i) *The Postnikov tower of $k\mathcal{O}$ has the form*

$$\begin{array}{ccccc}
 & & \vdots & & \\
 & & \downarrow & & \\
\Sigma^8 H\mathbb{Z} & \xrightarrow{j_4} & k\mathcal{O}_8 & \xrightarrow{\sigma_5} & \Sigma^{10} H\mathbb{Z}/2 \\
 & & {\scriptstyle p_4}\downarrow & & \\
\Sigma^4 H\mathbb{Z} & \xrightarrow{j_3} & k\mathcal{O}_4 & \xrightarrow{\sigma_4} & \Sigma^9 H\mathbb{Z} \\
 & & {\scriptstyle p_3}\downarrow & & \\
\Sigma^2 H\mathbb{Z}/2 & \xrightarrow{j_2} & k\mathcal{O}_2 & \xrightarrow{\sigma_3} & \Sigma^5 H\mathbb{Z} \\
 & & {\scriptstyle p_2}\downarrow & & \\
\Sigma H\mathbb{Z}/2 & \xrightarrow{j_1} & k\mathcal{O}_1 & \xrightarrow{\sigma_2} & \Sigma^3 H\mathbb{Z}/2 \\
 & & {\scriptstyle p_1}\downarrow & & \\
H\mathbb{Z} & = & k\mathcal{O}_0 & \xrightarrow{\sigma_1} & \Sigma^2 H\mathbb{Z}/2.
\end{array}$$

Here σ_r are the Postnikov invariants and

$$\bullet \xrightarrow{j_r} \bullet \xrightarrow{p_r} \bullet \xrightarrow{\sigma_r} \bullet$$

are long cofiber sequences.

(ii) $\sigma_1 = Sq^2\rho, \sigma_2 j_1 = Sq^2$, *and so* $\Omega^\infty \sigma_1 = 0 = \Omega^\infty \sigma_2$.

(iii) $((\Omega^\infty \sigma_n)|2)[2] \neq 0$ *for* $n > 2$.

Proof. (i) This follows from IV.4.27(ii).

(ii) By Bott peridicity, $(\Sigma^{-8} k\mathcal{O})|0 \simeq k\mathcal{O}$. Thus, it suffices to prove that $\sigma_5 j_4 = Sq^2\rho$, $\sigma_6 j_5 = Sq^2$, i.e., that $\sigma_5 j_4$ and $\sigma_6 j_5$ are essential morphisms. To prove this, it suffices to prove that $\Omega^4(\Omega^\infty(\sigma_5 j_4)) : K(\mathbb{Z},4) \to K(\mathbb{Z}/2,6)$ and $\Omega^4(\Omega^\infty(\sigma_6 j_5)) : K(\mathbb{Z}/2,5) \to K(\mathbb{Z}/2,7)$ are essential maps. Let BSp be the classifying space for the infinite dimensional symplectic group $Sp = \lim Sp(n)$. (Alternatively, BSp is the homotopy direct limit of the sequence $\cdots \to BSp_n \to BSp_{n+1} \to \cdots$.) We have $\Omega^4 BO \simeq BSp$, see e.g. Milnor [6]. Hence, the Postnikov tower for BSp has the form

$$\begin{array}{ccc}
\vdots & & \\
\downarrow & & \\
BSp_{(6)} & \xrightarrow{a_2} & K(\mathbb{Z}/2,7) \\
\downarrow & & \\
K(\mathbb{Z},4) & \xrightarrow{a_1} & K(\mathbb{Z}/2,6),
\end{array}$$

where $a_i = \Omega^4(\Omega^\infty \sigma_{4-i})$. Now, $H_*(BSp)$ is torsion free, see e.g. Switzer [1], 16.17. This implies easily that a_1 and a_2 are essential maps.

(iii) We have $(\Omega^\infty k\mathcal{O})|2 = B\mathcal{SO}$. Hence, by 2.12,

$$((\Omega^\infty \sigma_3)|2)[2] = \delta Sq^2 \iota_2 \neq 0, \quad ((\Omega^\infty \sigma_4)|2)[2] = \psi \neq 0.$$

Furthermore, by Bott periodicity, $k\mathcal{O}|8 = \Sigma^8 k\mathcal{O}$, and hence

$$(\Omega^\infty \sigma_5)|8 = Sq^2\rho : K(\mathbb{Z}, 8) \to K(\mathbb{Z}/2, 10),$$
$$(\Omega^\infty \sigma_6)|9 = Sq^2 : K(\mathbb{Z}/2, 9) \to K(\mathbb{Z}/2, 11).$$

So $((\Omega^\infty \sigma_r)|2)[2] \neq 0$ for $r = 5, 6$. Finally, the inequality $((\Omega^\infty \sigma_n)|2)[2] \neq 0$ for $n > 6$ follows from the above and Bott periodicity. □

Following our program, now we consider the $k\mathcal{O}$-characteristic classes $\varkappa_r$. Firstly, we set $\varkappa_0(\xi) := w_1(\xi)$. Furthermore, for every stable $H\mathbb{Z}$-oriented $\mathcal{V}$-object (ξ, u_ξ) we set $\varkappa_r(\xi) := \varphi^{-1}\sigma_r u_\xi$. Finally, it will be convenient to introduce the classes $\widehat{\varkappa}_r := \varphi^{-1}\widehat{\sigma}_r u_\xi$ corresponding to $k\mathcal{O}[2]$, i.e., $\widehat{\sigma}_r$ is the $\mathbb{Z}[2]$-localization of σ_r. By 4.1(ii), we have

(4.2) $$\varkappa_1 = w_2.$$

Hence, $B(\mathcal{V}, k\mathcal{O}_1) = B\mathcal{V}|3$.

Atiyah–Bott–Shapiro [1] proved that a stable *vector* bundle ξ is $k\mathcal{O}$-orientable iff it admits a Spin-structure, i.e., iff $w_1(\xi) = 0 = w_2(\xi)$. Thus, we have the following fact:

4.3. Theorem. *If for some vector bundle ξ we have $\varkappa_0(\xi) = 0, \varkappa_1(\xi) = 0$, then $0 \in \varkappa_r(\xi)$ for all $r \geq 1$. Thus, none of the classes $\varkappa_r, r > 1$, can be realized by vector bundles.* □

Because of $k\mathcal{O}[1/2]$-orientability of $\mathcal{STOP}$-bundles (see the text before 3.4), the following analog of 3.4 holds.

4.4. Theorem. *Given r, $0 \leq r \leq \infty$, an $\mathcal{STOP}$-bundle (as well as an $\mathcal{SPL}$-bundle) is $k\mathcal{O}_r$-orientable iff it is $(k\mathcal{O}_r)[2]$-orientable. In particular, the class $\varkappa_r$ can be realized by $\mathcal{PL}$- or $\mathcal{TOP}$-bundles iff $\widehat{\varkappa}_r$ can.* □

As usual, we set $\varkappa_2^{\mathcal{G}} = \varphi^{-1}\sigma_2 u \in H^3(B(G, k\mathcal{O}^1); \mathbb{Z}/2)$ where $u \in k\mathcal{O}^1(M(\mathcal{G}, k\mathcal{O}^1))$ is the universal $k\mathcal{O}^1$-orientation.

4.5. Lemma. *$H^3(B(\mathcal{G}, k\mathcal{O}_1); \mathbb{Z}/2) = \mathbb{Z}/2$, and the class $\varkappa_2^{\mathcal{G}}$ generates this group.*

Proof. Considering the $\mathbb{Z}/2$-cohomology spectral sequence of the fibration

$$K(\mathbb{Z}/2, 2) \to B(\mathcal{G}, k\mathcal{O}_1) \to B\mathcal{SG},$$

one obtains that $H^3(B(\mathcal{G}, k\mathcal{O}_1); \mathbb{Z}/2) = \mathbb{Z}/2$. Let ξ be the spherical fibration in 1.9. By 4.1(ii), the secondary operation σ_2 is just $\Phi(2,2)$. Hence, $\varkappa_2$ can be realized (by ξ). Thus, $\varkappa_2^{\mathcal{G}} \neq 0$. □

4.6. Remark. By 4.5, the set $\varkappa_2(\gamma_1^{\mathcal{G}})$ contains just one element $\varkappa_2^{\mathcal{G}}$, which in fact coincides with $p^* e_3$, where $e_3 \in H^3(B\mathcal{G}; \mathbb{Z}/2)$ is the Gitler–Stasheff class and $p : B(\mathcal{G}, k\mathcal{O}_1) \to B\mathcal{G}$.

Set $BM\mathcal{V} := B(\mathcal{V}, k\mathcal{O}_1) = B\mathcal{V}|3$, i.e., $BM\mathcal{V}$ is the homotopy fiber of $w_2 : BS\mathcal{V} \to K(\mathbb{Z}/2, 2)$. (Note that $BMO = B\,\mathrm{Spin}$.) The map $BM\mathcal{PL} \to BS\mathcal{PL} \to BS\mathcal{G}$ turns $BM\mathcal{PL}$ into a bundle over $BS\mathcal{G}$.

4.7. Lemma. (i) *If $\mathcal{V} \leq \mathcal{TOP}$, then $BM\mathcal{V}$ is 3-connected. Furthermore, $\pi_3(BM\mathcal{G}) = \mathbb{Z}/2$.*

(ii) *The homotopy fiber of any map $BM\mathcal{PL} \to BM\mathcal{G}$ over $BS\mathcal{G}$ is $\mathcal{G}/\mathcal{PL}$. The homotopy fiber of any map $BM\mathcal{TOP} \to BM\mathcal{G}$ over $BS\mathcal{G}$ is $\mathcal{G}/\mathcal{TOP}$.*

Proof. (i) By IV.4.27(ii), $\pi_3(BS\mathcal{O}) = 0$. So, by IV.4.27(iv), $\pi_3(BS\mathcal{PL}) = 0$, and so, by IV.4.27(v), $\pi_3(BS\mathcal{TOP}) = 0$. Hence, $\pi_3(BS\mathcal{V}) = 0$ for $\mathcal{V} \leq \mathcal{TOP}$, and so $\pi_3(BM\mathcal{V}) = 0$. Furthermore, $\pi_3(B\mathcal{G}) = \mathbb{Z}/2$, and so $\pi_3(BM\mathcal{G}) = \mathbb{Z}/2$.

(ii) This is similar to the proof of 3.11. □

4.8. Theorem. *If $\varkappa_1(\xi) = 0$ for some $S\mathcal{PL}$-bundle ξ, then $0 \in \varkappa_2(\xi)$ and $0 \in \varkappa_3(\xi)$. Furthermore, $0 \notin \varkappa_4(\gamma_3^{\mathcal{PL}}) \subset H^9(BM\mathcal{PL}; \mathbb{Z})$. Thus, all the classes $\varkappa_r, r \geq 4$, as well as $\widehat{\varkappa}_r, r \geq 4$, can be realized by $\mathcal{PL}$-bundles.*

Proof. The triviality of $\varkappa_2(\xi), \varkappa_3(\xi)$ follows from 4.3 and IV.4.27(iv), cf. 3.13. As in 3.14, if $0 \in \varkappa_4(\gamma_3^{\mathcal{PL}})$ then we have the diagram

$$\begin{array}{ccc} BM\mathcal{PL}[2] & \xrightarrow{f} & B(\mathcal{G}, k\mathcal{O}_8)[2] \\ \| & & \downarrow p \\ BM\mathcal{PL}[2] & \xrightarrow{g} & BM\mathcal{G}[2] \\ \downarrow & & \downarrow \\ BS\mathcal{PL}[2] & \longrightarrow & BS\mathcal{G}[2]. \end{array}$$

Passing to the homotopy fibers of g and p, we have, by 4.7(ii), a map

$$\bar{f} : \mathcal{G}/\mathcal{PL}[2] \to BSO[2]_{(8)}.$$

Since in the $\mathbb{Z}/2$-cohomology spectral sequences of the fibrations g and p the class $\varkappa_2^{\mathcal{G}} \in H^3(BM\mathcal{G}[2]; \mathbb{Z}/2)$ does not survive (for g this holds because $H^3(BM\mathcal{PL}; \mathbb{Z}/2) = 0$), the map

$$\bar{f} : \mathcal{G}/\mathcal{PL}[2] \to BSO[2]_{(8)}$$

is a 4-equivalence. (In greater detail, it induces an isomorphism of π_2's, and hence, by V.4.5, it induces an isomorphism of π_4's.) Hence, the map

$$\mathcal{BSO}[2]_{(4)} = Y \xrightarrow{a} \mathcal{G}/\mathcal{PL}[2] \xrightarrow{\bar{f}} \mathcal{BSO}[2]_{(8)}$$

is a 4-equivalence (here a is the inclusion of the factor, see (2.14)). In other words, the projection

$$\mathcal{BSO}[2]_{(8)} \to \mathcal{BSO}[2]_{(4)}$$

in the Postnikov tower of $\mathcal{BSO}[2]$ has a section. But this contradicts 2.12. Thus, $0 \notin \varkappa_4(\gamma_3^{\mathcal{PL}})$. The realizability of all the classes $\varkappa_r$ and $\widehat{\varkappa}_r$, $r \geq 4$, now follows from V.5.1 and 4.1(iii), cf. 3.14. □

4.9. Theorem. *If $\varkappa_1(\xi) = 0$ for some $\mathcal{STOP}$-bundle ξ, then $\varkappa_2(\xi) = 0$. Furthermore, $0 \notin \varkappa_3(\gamma_2^{\mathcal{TOP}})$, and hence all the classes $\varkappa_r, r \geq 3$, as well as $\widehat{\varkappa}_r, r \geq 3$, can be realized by $\mathcal{TOP}$-bundles.*

Proof. By 4.7(i), we have $H^3(BM\mathcal{TOP}; \mathbb{Z}/2) = 0$, and hence $\varkappa_2$ cannot be realized by $\mathcal{TOP}$-bundles. Furthermore, if $\gamma_2^{\mathcal{TOP}}$ is $k\mathcal{O}_2$-orientable, then we have the following commutative diagram over $B\mathcal{G}$:

$$\begin{array}{ccc} BM\mathcal{TOP} & \longrightarrow & B(\mathcal{G}, k\mathcal{O}_2) \\ \downarrow & & \downarrow \\ BM\mathcal{G} & = & BM\mathcal{G}. \end{array}$$

Let us localize this diagram and pass to the homotopy fibers of the vertical arrows. Then, by 4.7(ii), we get a map $\mathcal{G}/\mathcal{TOP}[2] \to \mathcal{BSO}[2]_{(4)}$ of the homotopy fibers, and this map is a 2-equivalence. In view of (2.15), this contradicts the non-triviality of the Postnikov invariant $\delta Sq^2\iota$ of $\mathcal{BSO}[2]$, see 2.12. Thus, $0 \notin \varkappa_3(\gamma_2^{\mathcal{TOP}})$. Again, the realizability of all the classes $\varkappa_r$ and $\widehat{\varkappa}_r$, $r \geq 3$, follows from V.5.1 and 4.1(iii). □

4.10. Theorem. *All the classes $\varkappa_r, r \geq 1$, as well as $\widehat{\varkappa}_r, r \geq 1$, can be realized by spherical fibrations.*

Proof. This follows from (4.2) for $r = 1$, from 4.5 for $r = 2$, and from 4.9 for $r \geq 3$. □

4.11. Remark. In fact, Adams [4, IV, Theorem 1.2] proved that all the classes $\varkappa_r$ of dimensions $8n+2$, $8n+3$ can be realized by spherical fibrations over spheres.

Again, one can ask about $k\mathcal{O}[p]$-orientability with odd prime p. We just remark that complexification induces a ring morphism $k\mathcal{O}[p] \to k[p]$ *onto a direct summand* in view of 3.3 and IV.4.27(iii). Thus, $k\mathcal{O}[p]$-orientability is equivalent to $k[p]$-orientability.

Résumé on kO-orientability

The conditions $w_1(\xi) = 0$, $w_2(\xi) = 0$ are necessary for kO-orientability of any $\mathcal{V}$-object ξ. As in §3, we consider $\mathcal{V}$-objects with trivial w_1. We have $\varkappa_1 = w_2$.

Theorem. (i) *The class $\varkappa_1$ can be realized by vector bundles (namely, by the universal oriented vector bundle). If for some vector bundle ξ we have $\varkappa_0(\xi) = 0$ and $\varkappa_1(\xi) = 0$, then $0 \in \varkappa_r(\xi)$ for all $r \geq 1$. Thus, none of the classes $\varkappa_r, r > 1$, can be realized by vector bundles.*

(ii) *If for some $S\mathcal{PL}$-bundle ξ we have $\varkappa_1(\xi) = 0$, then $0 \in \varkappa_2(\xi)$ and $0 \in \varkappa_3(\xi)$. On the other hand, all the classes $\varkappa_r, r \geq 4$, can be realized by $\mathcal{PL}$-bundles.*

(iii) *If for some $S\mathcal{TOP}$-bundle ξ we have $\varkappa_1(\xi) = 0$, then $\varkappa_2(\xi) = 0$. On the other hand, all the classes $\varkappa_r, r \geq 3$, can be realized by $\mathcal{TOP}$-bundles.*

(iv) *All the classes $\varkappa_r, r \geq 1$, can be realized by spherical fibrations.*

§5. A Few Geometric Observations

Here we give some results connected with k- and kO-orientability, but not situated on the main line of this chapter.

5.1. Theorem. *For every $n \geq 23$ there exists a simply connected topological manifold V^n with the following properties:*

(i) *$\delta w_2(V) = 0$;*

(ii) *No odd multiple of a generator $[V]_H \in H_n(V) = \mathbb{Z}$ can be realized by a PL manifold with $\delta w_2(M) = 0$. In particular, V does not admit any $\mathcal{PL}$-structure.*

Proof. Treating $BR\mathcal{TOP}$ as a CW-complex, take its 7-skeleton A and embed it in $\mathbb{R}^m, m \geq 15$. Consider a regular neighborhood X of A, and let $p: X \to A$ be the standard deformation retraction. Consider an $(\mathbb{R}^7, \mathcal{TOP}_7)$-bundle ξ such that $\xi \oplus (\gamma_3^{\mathcal{TOP}})|A$ is stably trivial, and set $\zeta := p^*\xi$. Note that $\zeta \oplus \theta^1$ admits a 8-disk subbundle. (Indeed, one can take the cylinder of the projection of $\zeta^\bullet$.) Let Y be the total space of this disk bundle; clearly, it is a topological manifold with boundary. Consider the embedding $j: A \to X \to Y$ (the last map is given by the zero section). One has $j^*\nu \simeq \gamma_3^{\mathcal{TOP}}|A$, where ν is the stable normal bundle of Y. Let V be the double of Y, $V := Y \cup_{\partial Y} Y$. Then $\varkappa_1(\nu) = 0, 0 \in \varkappa_2(\nu), 0 \notin \varkappa_3(\nu)$, where $\varkappa_i$ are the characteristic classes with respect to k-theory. We prove that V has properties (i) and (ii).

(i) Since V is simply connected, $w_1(V) = 0$. Thus, $w_2(V) = w_2(\nu)$, and so $\delta w_2(V) = 0$ because $\delta w_2(\nu) = \varkappa_2(\nu) = 0$.

(ii) Suppose that there is a map $f : M \to V$ of odd degree and such that M is an $H\mathbb{Z}$-oriented PL manifold with $\delta w_2(M) = 0$. Then $\delta w_2(\nu_M) = 0$, and so, by 3.2, M is k^3-orientable. Let $[M]$ be a k^3-orientation of M. Then, by V.2.12, $f_*([M]) \in k^3_*(V)$ gives us a $k^3[2]$-orientation of V. Thus, V is $k^3[2]$-orientable, and so it is k^3-orientable by 3.4 (and V.2.4). But $\varkappa_3(\nu) \neq 0$. This is a contradiction. □

An analogous theorem holds for $k\mathcal{O}$. We just formulate it; the proof is similar.

5.2. Theorem. *For every $n \geq 17$ there exists a simply connected topological manifold V^n with the following properties:*

(i) $w_2(V) = 0$;

(ii) *No odd multiple of a generator $[V]_H \in H_n(V) = \mathbb{Z}$ can be realized by a PL manifold M with $w_2(M) = 0$. In particular, V does not admit any $\mathcal{PL}$-structure.* □

5.3. Remark. For every homology class z there exists an odd number N such that Nz can be realized by a smooth manifold, see IV.36. That contrasts with 5.1 and 5.2 (recall that every smooth manifold is a PL manifold in a canonical way).

Let $BS\mathcal{PL}(8)$ be the classifying space for $\mathcal{PL}_8$-bundles. Let $BR\mathcal{PL}(8)$ be the homotopy fiber of $\delta w_2 : BS\mathcal{PL}(8) \to K(\mathbb{Z}/2, 3)$, and let $j : BRS\mathcal{PL}(8) \to BS\mathcal{PL}(8)$ be the inclusion of the homotopy fiber. Set $X := BR\mathcal{PL}(8)^{(8)}$, and set $\eta := j^*(\gamma^8_{\mathcal{PL}})|X$. By setting $Y := (B\,\mathrm{Spin}^{\mathbb{C}}(8))^{(8)}$, we have the map $f : Y \to X$, which is induced by the forgetful map $B\,\mathrm{Spin}^{\mathbb{C}}(8) \to BR\mathcal{PL}(8)$. It is clear that $f^*(\eta)$ is the restriction of the canonical $\mathrm{Spin}^{\mathbb{C}}(8)$-bundle. Let $u \in \widetilde{K}^8(T(f^*\eta))$ be the K-orientation of the $\mathrm{Spin}^{\mathbb{C}}$-bundle $f^*\eta$ constructed by Atiyah–Bott–Shapiro [1].

5.4. Theorem. *The bundle η is K-orientable, but the orientation u cannot be extended to X, i.e., $u \notin \operatorname{Im} f^*$.*

Proof. We have $\varkappa_1(\eta) = 0$ by the construction of η. So, by 3.13, $0 \in \varkappa_2(\eta)$, $0 \in \varkappa_3(\eta)$. Finally, $0 \in \varkappa_n(\eta), n > 3$, because $\dim X = 8$. Thus, η is k-orientable and so K-orientable.

Suppose now that $u = f^*v$ for some $v \in K^8(T\eta)$. Since $f^* : H^*(X; \mathbb{Q}) \to H^*(Y; \mathbb{Q})$ is an isomorphism, we have $\varphi^{-1}\,\mathrm{ch}\,v = e^{z/2}\hat{A}(\eta)$ for some $z \in H^2(X)$, (see V.3.4 (b), V.3.6). This implies easily that $\hat{A}(M)$ is an integer for every PL manifold M with $w_1(M) = 0 = w_2(M)$ and $\dim M \leq 8$.

Let M, $\dim M = 8$, be a closed almost parallelizable PL manifold of signature 8. Such manifolds were constructed by Milnor, see e.g. Kervaire–Milnor [1] or Browder [3]. The theorem will be proved if we prove that $\hat{A}(M)$ is not an integer. Now,

$$\hat{A}(M) = (7p_1^2 - 4p_2)/(2^7 \cdot 45),$$

where p_i is the i-th Pontrjagin class of M, see e.g. Hirzebruch [1]. Here $p_1(M) = 0$ because M is almost parallelizable. We have $8 = \sigma(M) = (7p_2 - p_1^2)/45$, where σ is the signature, see e.g. Hirzebruch [1], and so $7p_2 = 8 \cdot 45$. Thus, $\hat{A}(M) = -1/28$. □

5.5. Remarks. (a) Theorem 5.4 means that in the diagram

$$\begin{array}{ccccc} Y & \xrightarrow{f} & X & \longrightarrow & BR\mathcal{PL} \\ {\scriptstyle u}\downarrow & & \vdots\downarrow{\scriptstyle v} & & \| \\ B(\mathcal{O},K) & \longrightarrow & B(\mathcal{PL},K) & \longrightarrow & BR\mathcal{PL} \end{array}$$

there exists a map v such that the right square commutes, but there is no map v such that the left square commutes.

(b) Milnor constructed certain closed almost parallelizable PL manifolds M^{4k} of signature 8, see e.g. Kervaire–Milnor [1] or Browder [3]. Every such manifold M^{4k} is K-orientable because $H^{2i+1}(M) = 0$ for each i. Hence, one can try to find a K-orientation of $\mathrm{Spin}^{\mathbb{C}}$-bundles with the corresponding genus φ such that $\varphi(M)$ is integral for every Milnor manifold M.

(c) We leave it to the reader to formulate and prove a $K\mathcal{O}$-analog of 5.4.

Chapter VII. Complex (Co)bordism

In order to work with complex (co)bordism with singularities we need some preliminaries on complex (co)bordism. Therefore we collect here some facts which will be used below. A standard reference on complex (co)bordism is the book of Ravenel [1], see also Stong [3], Ch. VI.

In this chapter "cohomology theory" means "additive cohomology theory".

§1. Homotopy and Homology Properties of the Spectrum $M\mathcal{U}$

Let $M\mathcal{U} = T(B\mathcal{U}, R)$ be the ring spectrum defined in IV.7.31(b). The complex (co)bordism theory is the (co)homology theory given by $M\mathcal{U}$. Since $B\mathcal{U}$ is simply connected, the $\mathcal{F}$-object

$$B\mathcal{U} \xrightarrow{R} B\mathcal{O} \xrightarrow{a^{\mathcal{O}}_{\mathcal{F}}} B\mathcal{F}$$

is $H\mathbb{Z}$-orientable, and hence, by IV.5.23, $H^0(T(B\mathcal{U}, R)) = \mathbb{Z}$. Therefore, $H^0(M\mathcal{U}) = \mathbb{Z}$. Consider a Thom class $u \in H^0(M\mathcal{U})$ (i.e., either of two generators of $H^0(M\mathcal{U}) = \mathbb{Z}$) and the corresponding morphism $u : M\mathcal{U} \to H\mathbb{Z}$. By IV.5.26, the induced homomorphism

$$u_* : \mathbb{Z} = \pi_0(M\mathcal{U}) \to \pi_0(H\mathbb{Z}) = \mathbb{Z}$$

is an isomorphism. We choose the Thom class $u \in H^0(M\mathcal{U})$ such that u_* maps 1 to 1.

We do not care about any concrete form of the universal bundles $\gamma^n_{\mathbb{C}}$. They can be canonical bundles, as in Milnor–Stasheff [1], or conjugated to these ones, etc. Fixing γ^n's, we fix certain maps (homotopy classes) $B\mathcal{U}_n \to B\mathcal{U}_{n+1}$ and $B\mathcal{U}_m \times B\mathcal{U}_n \to B\mathcal{U}_{m+n}$, but in any case $M\mathcal{U}$ turns out to be a ring spectrum. Furthermore, the ring equivalence class of the spectrum $M\mathcal{U}$ does not depend on the choice of $\gamma^n_{\mathbb{C}}$'s. However, if you want, we can agree that $\gamma^n_{\mathbb{C}}$ is a canonical complex vector bundle.

According to the Pontrjagin–Thom Theorem IV.7.27, $MU_*(-)$ can be interpreted as the (BU, R)-bordism theory, i.e., the bordism theory based on stably almost complex manifold. By the way, every complex analytic manifold is a stably almost complex manifold.

Throughout this chapter the word "bordant" means "(BU, R)-bordant". As usual, given a closed stably almost complex manifold M, the bordism class of a singular manifold $f : M \to X$ will be denoted by $[M, f]$, and the bordism class of M will be denoted by $[M]$. Furthermore, $[M]_{MU}$ denotes the bordism class $[M, 1_M] \in MU_*(M)$.

It is easy to see that the universal $H\mathbb{Z}$-orientation u yields a certain $H\mathbb{Z}$-orientation on every complex vector bundle, cf. 2.8 below. (In fact, this orientaion coincides with the one described in Milnor–Stasheff [1].) Therefore, by V.2.4 and V.2.14, every stably almost complex manifold M^n gets a certain $H\mathbb{Z}$-orientation $[M]_H \in H_n(M, \partial M)$. Now one can define the Steenrod–Thom homomorphism $MU_*(X) \to H_*(X), [M, f] \mapsto f_*([M]_H)$ and prove an analog of IV.7.32.

1.1. Proposition. *$u : MU \to H\mathbb{Z}$ is a ring morphism.*

Proof. Let H denote $H\mathbb{Z}$. We must prove that the diagram

$$\begin{array}{ccc} MU \wedge MU & \xrightarrow{\mu} & MU \\ {\scriptstyle u\wedge u}\downarrow & & \downarrow{\scriptstyle u} \\ H \wedge H & \xrightarrow{\mu_H} & H \end{array}$$

commutes (up to homotopy). Let $\iota : S \to MU$ be the unit of MU. The morphism

$$S = S \wedge S \xrightarrow{\iota\wedge\iota} MU \wedge MU \xrightarrow{\mu} MU \xrightarrow{u} H$$

is homotopic to $u\iota$, while the morphism

$$S = S \wedge S \xrightarrow{\iota\wedge\iota} MU \wedge MU \xrightarrow{u\wedge u} H \wedge H \xrightarrow{\mu_H} H$$

coincides with the unit ι_H of H. Furthermore, $\iota_H \simeq u\iota$ because $\iota^*(u) = 1 \in H^0(S)$ and $u_* : \pi_0(MU) \to \pi_0(H)$ maps 1 to 1. Hence,

$$u \circ \mu \circ (\iota \wedge \iota) \simeq u\iota \simeq \iota_H \simeq \mu_H \circ (u \wedge u) \circ (\iota \wedge \iota).$$

Since $(\iota \wedge \iota)^* : H^0(MU \wedge MU) \to H^0(S \wedge S)$ is an isomorphism, $u \circ \mu \simeq \mu_H \circ (u \wedge u)$. □

The group $\mathcal{U}_1 = \{z \in \mathbb{C} \mid |z| = 1\}$ acts on $\mathbb{C}^{n+1}$ (via the map $a \mapsto za, a \in \mathbb{C}^{n+1}, z \in \mathcal{U}_1$) and hence on its unit sphere S^{2n+1}. The quotient space $S^{2n+1}/\mathcal{U}_1$ is just the complex projective space CP^n. Considering the homotopy exact sequence of the locally trivial principal $\mathcal{U}_1$-bundle

$$\mathcal{U}_1 \to S^{2n+1} \to CP^n,$$

we conclude that $\pi_1(CP^n) = 0, \pi_2(CP^n) = \mathbb{Z}$ and $\pi_i(CP^n) = 0$ for $i \leq 2n$.

The inclusion $\mathbb{C}^n \to \mathbb{C}^{n+1}$ of the hyperspace $z_{n+1} = 0$ induces an inclusion $l_n : CP^n \to CP^{n+1}$. Set $CP^\infty := \cup CP^n$. By the above, $\pi_2(CP^\infty) = \mathbb{Z}$ and $\pi_i(CP^\infty) = 0$ for $i \neq 2$. Thus, $CP^\infty = K(\mathbb{Z}, 2)$.

1.2. Lemma. $B\mathcal{U}_1 \simeq CP^\infty$.

Proof. By IV.3.2(ii), $B\mathcal{U}_1$ is the base of a principal $\mathcal{U}_1$-bundle with a contractible total space, and so $\pi_i(B\mathcal{U}_1) = \pi_{i-1}(\mathcal{U}_1) = \pi_{i-1}S^1$. Hence, $B\mathcal{U}_1 = K(\mathbb{Z}, 2) = CP^\infty$. □

1.3. Notation. (a) We denote by $l_n^m : CP^m \subset CP^n, m \leq n \leq \infty$, the inclusion $l_{n-1} \circ \cdots \circ l_m : CP^m \subset CP^n$. So, $l_{n+1}^n = l_n$. We also use the specific notation $j_n := l_\infty^n : CP^n \to CP^\infty$.

(b) We denote by $r_n^m : B\mathcal{U}_m \to B\mathcal{U}_n, m \leq n < \infty$, the composition $r_m \circ \cdots \circ r_{n-1} : B\mathcal{U}_m \to B\mathcal{U}_n$ where $r_n : B\mathcal{U}_n \to B\mathcal{U}_{n+1}$ classifies $\gamma_\mathbb{C}^n \oplus \theta_\mathbb{C}^1$ as in IV.4.25. So, $r_{n+1}^n = r_n$. Also, we recall the map $j_n^\mathcal{U} : B\mathcal{U}_n \to B\mathcal{U}$, see IV.4.25.

(c) Given a complex vector bundle ξ, dim ξ denotes its complex dimension.

(d) Given a complex vector bundle ξ, we denote by $\bar{\xi}$ the conjugated complex vector bundle.

(e) In this chapter we denote $\gamma_\mathbb{C}^n$ by γ^n, $\gamma_\mathbb{C}$ by γ and $\theta_\mathbb{C}^n$ by θ^n.

(f) Let λ be the canonical complex line bundle over CP^∞. (It is well-known that λ is a universal complex line bundle, but, as I said before, I do not insist that $\gamma^1 = \lambda$.) We set $\eta := \bar{\lambda}$. [16]

(g) Let $e_n : (CP^\infty)^n \to B\mathcal{U}_n$ classify $\eta \times \cdots \times \eta$, and let $p_i : (CP^\infty)^n \to CP^\infty$ be the projection onto the i-th factor.

Let $c_i(\xi)$ denote the i-th Chern class of a complex vector bundle ξ. Recall that $c_i(\xi \times \eta) = \sum_{j+k=i} c_j(\xi)c_k(\eta)$ for every pair of complex vector bundles ξ, η (where $c_0(\xi) = 1 = c_0(\eta)$) and $c_i(\xi) = 0$ for $i > \dim \xi$.

1.4. Theorem. (i) $H^*(CP^\infty) = \mathbb{Z}[t]$ *where* $t = c_1(\eta), \dim t = 2$.

(ii) $H^*((CP^\infty)^n) = \mathbb{Z}[t_1, \ldots, t_n]$, *where* $t_i = p_i^* t$.

(iii) $H^*(B\mathcal{U}_n) = \mathbb{Z}[c_{1,n}, \ldots, c_{n,n}], \dim c_{i,n} = 2i$, *where $c_{i,n}$ is the i-th Chern class of* γ^n, $c_{i,n} = c_i(\gamma^n)$. *Furthermore,* $r_n^*(c_{i,n+1}) = c_{i,n}$ *for* $i \leq n$, *and* $r_n^*(c_{n+1,n+1}) = 0$. *Finally,* $e_n^*(c_{i,n})$ *is the i-th elementary symmetric polynomial* $\sigma_i(t_1, \ldots, t_n)$.

Proof. See Milnor–Stasheff [1], §14. □

Let $\mathrm{Vect}_n(X)$ denote the set of all equivalence classes of n-dimensional complex vector bundles over a space X. Note that $\xi \otimes \theta^1 = \xi$ and $\xi \otimes \bar{\xi} = \theta^1$. Hence, $\mathrm{Vect}_1(X)$ is an abelian group with respect to the tensor product.

---

[16] Note that this notation differs from that in Milnor–Stasheff [1]. They denoted the canonical line bundle over CP^∞ by η.

1.5. Lemma. (i) *For every CW-space X, the function* $c_1 : \mathrm{Vect}_1(X) \to H^2(X), \xi \mapsto c_1(\xi)$, *is a homomorphism of groups.*

(ii) *The homomorphism* $c_1 : \mathrm{Vect}_1(X) \to H^2(X)$ *is an isomorphism for every CW-space X.*

Proof. (i) See e.g. Karoubi [1], V.3.10.

(ii) Since $t = c_1(\eta)$ generates the group $H^2(CP^\infty) = H^2(K(\mathbb{Z},2)) = \mathbb{Z}$, we conclude that t is a fundamental class of the Eilenberg-Mac Lane space $K(\mathbb{Z},2)$.

We prove that c_1 is epic. Given $x \in H^2(X)$, consider $f : X \to CP^\infty$ with $f^*t = x$. Now, $c_1(f^*\eta) = f^*(c_1(\eta)) = x$.

We prove that c_1 is monic. Consider the map $e_1 : CP^\infty \to B\mathcal{U}_1$. We have $t = c_1(\eta) = c_1(e_1^*\gamma^1) = e_1^*(c_1(\gamma^1))$. Hence, $c_1(\gamma^1)$ generates the group $H^2(B\mathcal{U}_1) = \mathbb{Z}$, i.e., $c_1(\gamma^1)$ is a fundamental class of $B\mathcal{U}_1 = K(\mathbb{Z},2)$. Let a complex line bundle ξ be classified by a map $f : X \to CP^\infty$. We have

$$\begin{aligned} \xi \simeq \theta^1 &\iff f \text{ is inessential} \iff f^*c_1(\gamma^1) = 0 \\ &\iff c_1(f^*\gamma^1) = 0 \iff c_1(\xi) = 0. \end{aligned}$$ □

1.6. Corollary. *Let* $m : CP^\infty \times CP^\infty \to CP^\infty$ *be the multiplication in the H-space* $CP^\infty = K(\mathbb{Z},2)$, *and let* $p_i : CP^\infty \times CP^\infty \to CP^\infty, i = 1,2$, *be the projection onto the i-th factor. Then*

$$m^*(\xi) = p_1^*\xi \otimes p_2^*\xi$$

for every complex line bundle ξ *over* CP^∞.

Proof. Note that $m^*x = p_1^*x + p_2^*x$ for every $x \in H^2(CP^\infty)$. Now,

$$c_1(m^*\xi) = m^*c_1(\xi) = p_1^*c_1(\xi) + p_2^*c_1(\xi) = c_1(p_1^*\xi) + c_1(p_2^*\xi) = c_1(p_1^*\xi \otimes p_2^*\xi)$$

□

Given a partition $\omega = (i_1, \dots, i_k)$ (see Milnor–Stasheff [1], §16), set $|\omega| = \sum i_r$, $l(\omega) = k$. We define the universal Chern classes

$$c_\omega \in H^{2|\omega|}(B\mathcal{U}) = H^{2|\omega|}(B\mathcal{U}_N), \; |\omega| << N,$$

via the formula $e_N^*(c_\omega) = t_\omega \in \mathbb{Z}[t_1, \dots, t_N] = H^*((CP^\infty)^n)$ where t_ω is the smallest symmetric polynomial which contains $t_1^{i_1} \cdots t_k^{i_k}$. (Here "smallest" means "with minimal number of summands".) Finally, $c_{(0)} := 1$.

Given a complex vector bundle ξ, $\dim \xi = n$, we define $c_\omega(\xi)$ by setting $c_\omega(\xi) := g^*c_\omega$, where g is the composition $\mathrm{bs}\,\xi \xrightarrow{f} B\mathcal{U}_n \xrightarrow{j_n^{\mathcal{U}}} B\mathcal{U}$ and f classifies ξ. We have $c_\omega(\xi) = 0$ for $l(\omega) > \dim \xi$ and

$$c_\omega(\xi \times \eta) = \sum_{(\omega_1,\omega_2)=\omega} c_{\omega_1}(\xi)c_{\omega_2}(\eta),$$

see Milnor–Stasheff [1].

For every $x \in \pi_{2|\omega|}(MU)$ we define its characteristic numbers $s_\omega(x)$ as

$$s_\omega(x) := f^*\varphi(c_\omega) \in H^{2|\omega|}\left(S^{2|\omega|}\right) = \mathbb{Z}, \tag{1.7}$$

where $f : S^{2|\omega|} \to MU$ represents x and $\varphi : H^*(BU) \to H^*(MU)$ is the Thom isomorphism. Moreover, if x is represented by a stably almost complex manifold M, then

$$s_\omega(x) = \langle c_\omega(\nu_M), [M]_H \rangle,$$

see e.g. Stong [3] or 2.22 below.

We set

$$\lambda_n := \begin{cases} p & \text{if } n = p^k - 1 \text{ for a prime } p, \\ 1 & \text{otherwise.} \end{cases} \tag{1.8}$$

1.9. Theorem. (i) $\pi_*(MU) = \mathbb{Z}[x_1, \dots, x_n, \dots], \dim x_n = 2n$.

(ii) $H_*(MU) = \mathbb{Z}[y_1, \dots, y_n, \dots], \dim y_n = 2n$. *In particular,* $H_*(MU)$ *and* $H^*(MU)$ *are torsion free.*

(iii) *The Hurewicz homomorphism* $h : \pi_*(MU) \to H_*(MU)$ *is monic. In particular, the bordism class of every stably almost complex manifold is completely determined by its Chern numbers.*

(iv) *A family* $\{x_n\}$, $n = 1, 2, \dots$, *of elements of* $\pi_*(MU)$ *is a system of free polynomial generators of it iff* $s_{(n)}(x_n) = \pm\lambda_n$ *for every* n.

(v) *There exists a system* $\{x_n\}$ *of free polynomial generators of* $\pi_*(MU)$ *such that for every prime number* p *and natural number* k *all Chern numbers of* x_{p^k-1} *are divisible by* p. *In particular,* $h(x_{p^k-1}) \in pH_*(MU)$.

(vi) $H^*(MU; \mathbb{Z}/p)$ *is a free* $\mathscr{A}_p/(Q_0)$*-module.*

(vii) *The AHSS for* $MU^*(MU)$ *and* $MU_*(MU)$ *are trivial.*

(viii) *The Hurewicz homomorphism* $\pi_*(MU) \to K_*(MU)$ *is a monomorphism on a direct summand.*

Here in (vi) (Q_0) is the two-sided ideal generated by Q_0. It coincides with the left ideal $\mathscr{A}_p(Q_0, \dots Q_n, \dots)$.

Proof. A proof of (i), (iii)-(vi), (viii) can be found in Stong [3], see also Ravenel [2]. A proof of (ii) can be found in Switzer [1]. The assertion (vii) follows from (i), (ii) and II.7.12(ii). □

1.10. Remark. The assertions (i)–(iv), (vi) were proved by Milnor [4] and Novikov [1]. The assertion (v) was remarked by Conner and Floyd, see Conner [1]; they proved the existence of generators with all characteristic numbers divisible by p (and called them Milnor's generators); you can see such a system in 6.14 below. The assertion (viii) is the well-known Stong–Hattory Theorem, see Stong [2], Hattory [1].

For every spectrum X the morphism $u : M\mathcal{U}_*(X) \to H_*(X)$ induces a homomorphism

$$\overline{u} : M\mathcal{U}_*(X) \otimes_\varepsilon \mathbb{Z} \to H_*(X), \ \overline{u}(x \otimes a) = au(x),$$

where $\varepsilon = u^S : M\mathcal{U}_*(S) \to H_*(S) = \mathbb{Z}$. Note that $\varepsilon(x) = 0$ if $\dim x > 0$.

1.11. Proposition. *The homomorphism*

$$\overline{u} \otimes 1 : M\mathcal{U}_*(X) \otimes_\varepsilon \mathbb{Q} \to H_*(X) \otimes \mathbb{Q} = H_*(X; \mathbb{Q})$$

is an isomorphism of homology theories on $\mathscr{S}$.

Proof. By II.7.13, there is a natural ring (and hence $M\mathcal{U}_*(S)$-module) isomorphism

$$M\mathcal{U}_*(X) \otimes \mathbb{Q} \cong H_*(X) \otimes M\mathcal{U}_*(S) \otimes \mathbb{Q}.$$

Hence, $M\mathcal{U}_*(X) \otimes_\varepsilon \mathbb{Q} \cong H_*(X; \mathbb{Q})$. So, $\overline{u} \otimes 1$ is a morphism of homology theories. It is an isomorphism for $X = S$ and thus, by II.3.19(iii), for every X. □

1.12. Definition. A module M over a commutative ring is called *coherent* if it is finitely generated and every finitely generated submodule of M is finitely presented. A commutative ring R is called *coherent* if the R-module R is coherent.

1.13. Proposition. (i) *The ring $k[x_1, \dots, x_n, \dots]$ is coherent for every commutative Noetherian ring k. In particular, $\pi_*(MU)$ is a coherent ring.*

(ii) *If in an exact triangle*

$$\begin{array}{ccc} M_1 & \longrightarrow & M_2 \\ \nwarrow & & \swarrow \\ & M_3 & \end{array}$$

any two of three modules $M_i, i = 1, 2, 3$, are coherent, then so is the third.

Proof. Do this as an exercise; or see Bourbaki [3], L. Smith [1]. □

1.14. Theorem (L. Smith [1], cf. also Novikov [4])**.** *Let E be a ring spectrum such that $\pi_*(E)$ is a coherent ring. Then $E_*(X)$ is a coherent (and so finitely generated) $E_*(S)$-module for every finite spectrum X. In particular, $M\mathcal{U}_*(X)$ is a coherent $M\mathcal{U}_*(S)$-module for every finite spectrum X.*

Similarly, $E^*(X)$ is a coherent $E^*(S)$-module.

Proof. We consider only the homological case, the cohomological case can be proved similarly. Since $\pi_*(E)$ is coherent, $E_*(S)$ is a coherent $E_*(S)$-module. Now the proposition follows from 1.13(ii) by induction on the number of cells of X. □

1.15. Theorem (cf. Conner–Smith [1]). *For every spectrum X bounded below the following conditions are equivalent:*

(i) *$H_*(X)$ is a free abelian group;*

(ii) *$M\mathcal{U}_*(X)$ is a free $M\mathcal{U}_*(S)$-module.*

Furthermore, $\overline{u} : M\mathcal{U}_*(X) \otimes_\varepsilon \mathbb{Z} \to H_*(X)$ is an isomorphism under these conditions.

Proof. (i) $\Longrightarrow$ (ii). Since $H_*(X)$ is torsion free, the E^2-term of the AHSS

$$E^r_{**}(X) \Rightarrow M\mathcal{U}_*(X), \; E^2_{**}(X) = H_*(X) \otimes M\mathcal{U}_*(S)$$

is torsion free. So, by II.7.12(ii), all its differentials are trivial. Hence, a free abelian basis of $H_*(X)$ yields a free $M\mathcal{U}_*(S)$-basis of $E^\infty_{**}(X)$. Hence, $E^\infty_{**}(X)$ is a free graded $M\mathcal{U}_*(S)$-module. Thus, $M\mathcal{U}_*(X)$ is a free $M\mathcal{U}_*(S)$-module.

(ii) $\Longrightarrow$ (i). Firstly, we prove that $H_*(X)$ is torsion free. Let $x \in H_*(X)$ be a torsion element of minimal dimension. Then the element

$$x \otimes 1 \in E^2_{|x|,0}(X) = H_{|x|}(X) \otimes \pi_0(M\mathcal{U})$$

is a permanent cycle, and so $x \in \operatorname{Im} u$. On the other hand, $M\mathcal{U}_*(X)$ is a free $M\mathcal{U}_*(S)$-module, and hence $M\mathcal{U}_*(X) \otimes_\varepsilon \mathbb{Z}$ is a free abelian group. Hence, by 1.11, $\overline{u} : M\mathcal{U}_*(X) \otimes_\varepsilon \mathbb{Z} \to H_*(X)$ is monic, and so $\operatorname{Im} \overline{u}$ is torsion free. Thus, $x \notin \operatorname{Im} \overline{u}$. This is a contradiction. Hence, $H_*(X)$ is torsion free. In particular, all differentials in the AHSS for $M\mathcal{U}_*(X)$ are trivial.

We have already proved that $\overline{u}$ is monic. On the other hand, $\overline{u}$ is epic because all differentials in the AHSS are trivial. So, $H_*(X) \cong M\mathcal{U}_*(X) \otimes_\varepsilon \mathbb{Z}$, i.e., $H_*(X)$ is a free abelian group.

The last claim has already been proved. □

1.16. Corollary. *For every finite spectrum X the following conditions are equivalent:*

(i) *$H^*(X)$ is a free abelian group;*

(ii) *$M\mathcal{U}^*(X)$ is a free $M\mathcal{U}^*(S)$-module.*

Proof. Let Y be a spectrum dual to X, i.e., $Y = X^\perp$. Then $H^*(X) = H_*(Y)$, $M\mathcal{U}^*(X) = M\mathcal{U}_*(Y)$. Now apply 1.15. □

1.17. Theorem (Conner–Smith [1], Landweber [2], Yosimura [1]). *Given a k-connected spectrum X, there exists a morphism $f : W \to X$ such that W is a k-connected spectrum, $M\mathcal{U}_*(W)$ is a free $\pi_*(M\mathcal{U})$-module and $f_* : M\mathcal{U}_*(W) \to M\mathcal{U}_*(X)$ is an epimorphism.*

Furthermore, if every group $H_i(X)$ is finitely generated then there exists W as above such that every group $H_i(W)$ is finitely generated.

Proof. Without loss of generality we assume that $k = -1$, i.e., that X is connected. Choose an element $x \in M\mathcal{U}_d(X)$. Then x can be represented by a morphism

$$S^d \xrightarrow{h} MU^{(N)} \wedge X \subset MU \wedge X$$

for N large enough. Without loss of generality we can assume that $H_*(MU^{(N)})$ is a free abelian group. Let Y be a dual spectrum to $MU^{(N)}$, and let $u : S \to MU^{(N)} \wedge Y$ be a duality morphism. Then there is the duality isomorphism

$$u^X : [Y, X] \to [S, MU^{(N)} \wedge X], \quad u^X(\varphi) = (1_{MU^{(N)}} \wedge \varphi)u,$$

and hence the isomorphism $\Sigma^d u^X : [\Sigma^d Y, X] \to [S^d, MU^{(N)} \wedge X]$. Let $g : \Sigma^d Y \to X$ be a morphism such that $\Sigma^d u^X[g] = [h]$. Consider the homomorphism $g_* : MU_d(\Sigma^d Y) \to MU_d(X)$. We have $g_*[\Sigma^d u] = x$ since the morphism

$$S^d \xrightarrow{\Sigma^d u} MU^{(N)} \wedge \Sigma^d Y \xrightarrow{1 \wedge g} MU^{(N)} \wedge X \tag{1.18}$$

is homotopic to h. In particular, $x \in \operatorname{Im}\{g_* : MU_*(\Sigma^d Y) \to MU_*(X)\}$. Furthermore, $H_*(Y)$ is a free abelian group, and hence, by 1.15, $MU_*(\Sigma^d Y)$ is a free $\pi_*(MU)$-module.

Now, let $\{x_\alpha\}$ be a family of $\pi_*(MU)$-generators of $MU_*(X)$. We use the above arguments and construct maps $g_\alpha : \Sigma^{|x_\alpha|} Y_\alpha \to X$ such that

$$x_\alpha \in \operatorname{Im}\{(g_\alpha)_* : MU_*(\Sigma^{|x_\alpha|} Y_\alpha) \to MU_*(X)\}.$$

We set $W := \bigvee_\alpha \Sigma^{|x_\alpha|} Y_\alpha$ and define $f : W \to X$ by requiring $f|\Sigma^{|x_\alpha|} Y_\alpha \simeq g_\alpha$. Clearly, the spectrum W is connected, the homomorphism $f_* : MU_*(W) \to MU_*(X)$ is epic, and $MU_*(W) = \oplus_\alpha MU_*(\Sigma^{|x_\alpha|} Y_\alpha)$ is a free $\pi_*(MU)$-module.

Furthermore, suppose every group $H_i(X)$ is finitely generated. Given n, consider the AHSS

$$E^r_{**} \Rightarrow MU_*(X^{(n)}), \quad E^2_{**}(X) = H_*(X; \pi_*(MU)).$$

Without loss of generality we can assume that every group $H_i(X^{(n)})$ is finitely generated. Then E^2_{**} turns out to be a coherent $\pi_*(MU)$-module. Hence, by 1.13(ii), E^∞_{**} is a coherent $\pi_*(MU)$-module, and thus $MU_*(X^{(n)})$ is.

Now, $MU_*(X) = \varinjlim_n MU_*(X^{(n)})$. Hence, $MU_*(X)$ admits a family $\{x_\alpha\}$ of $\pi_*(MU)$-generators such that, for every n, the set $\{x_\alpha \mid \dim x_\alpha \leq n\}$ is finite. Thus, constructing W as above, we conclude that the group $H_i(W)$ is finite for every i. □

1.19. Lemma. *Let E be any MU-module spectrum with the pairing $m : MU \wedge E \to E$. Let A be a finite spectrum such that $H^*(A)$ is torsion free. Then the homomorphism $m^{A,B} : MU^*(A) \otimes_{MU^*(S)} E^*(B) \to E^*(A \wedge B)$ is an isomorphism for every finite spectrum B.*

Proof. Because of 1.16, $MU^*(A)$ is a free $MU^*(S)$-module. So, fixing A, we can consider $m^{A,B}$ as a morphism of cohomology theories on $\mathscr{S}_\mathrm{f}$. Since it is an isomorphism for $B = S$, it is an isomorphism for every $B \in \mathscr{S}_\text{f}$, see II.3.19(iii). □

1.20. Theorem. *Let E be as* 1.19, *and suppose that $\pi_i(E)$ is a finite group for every i. Let X be a spectrum of finite type such that $H^*(X)$ is torsion free. Then the homomorphism $m^{X,Y} : MU^*(X)\widehat{\otimes}_{MU^*(S)}E^*(Y) \to E^*(X \wedge Y)$ is an isomorphism for every spectrum Y. (Here $\widehat{\otimes}_{MU^*(S)}$ is the profinitely completed tensor product defined in* III.1.23.)

Proof. Let $\{X_\lambda\}$ (resp. $\{Y_{\lambda'}\}$) be the direct system of all finite subspectra of X (resp. Y). By III.4.17, $E^*(X \wedge Y) = \varprojlim\{E^*(X_\lambda \wedge Y_{\lambda'})\}$. Furthermore, the system $\{X_\lambda\}$ of all finite subspectra of X has a cofinal subsystem $\{X_\alpha\}$ with X_α finite and such that $H^*(X_\alpha)$ is torsion free for every α. Note that, by III.5.7(iii), $MU^*(X) = \varprojlim\{MU^*(X_\alpha)\}$. Thus,

$$\begin{aligned} E^*(X \wedge Y) &= \varprojlim\{E^*(X_\alpha \wedge Y_{\lambda'})\} = \varprojlim\{MU^*(X_\alpha) \otimes_{MU^*(S)} E^*(Y_{\lambda'})\} \\ &= MU^*(X)\widehat{\otimes}_{MU^*(S)}E^*(Y). \end{aligned}$$ □

1.21. Corollary. *Let E be as in* 1.20. *Then the homomorphism*

$$m : MU^*(MU)\widehat{\otimes}_{MU^*(S)}E^*(X) \to E^*(MU \wedge X)$$

is an isomorphism for every spectrum X. □

1.22. Remark. Lemma 1.19 enables us to construct a spectral sequence

$$\operatorname{Tor}_{p,q}^{MU^*(S)}(MU^*(X), E^*(Y)) \Rightarrow E^*(X \wedge Y)$$

for every pair of finite spectra X, Y, see Novikov [4], Conner–Smith [1].

In order to proceed, we need more information about complex vector bundles over CP^n. Note that $e_1 : CP^\infty \to BU_1$ classifies η.

1.23. Proposition. *The map $e_1 : CP^\infty \to BU_1$ is a homotopy equivalence. So, η is a universal complex line bundle.*

Proof. Because of 1.2 and 1.4(i), it suffices to prove that $e_1^*(c_1(\gamma^1))$ generates the group $H^2(CP^\infty) = \mathbb{Z}$. But

$$e_1^*(c_1(\gamma^1)) = c_1(\eta) = t.$$ □

We denote the bundle $\eta | CP^n = j_n^*\eta$ by η_n.

1.24. Lemma. *Let $\tau(CP^n)$ be the tangent bundle of CP^n. Then there is an equivalence $\tau(CP^n) \oplus \theta^1 \simeq (n+1)\eta_n$ of complex vector bundles.*

Proof. See Milnor–Stasheff [1]. □

Let ξ_n be a normal bundle of the inclusion $l_n : CP^n \to CP^{n+1}$. By IV.7.11, $\tau(CP^n) \oplus \xi_n \cong l_n^* \tau(CP^{n+1})$. Since both $\tau(CP^n)$ and $\tau(CP^{n+1})$ are complex vector bundles, we conclude that ξ_n gets a canonical structure of a complex vector bundle.

1.25. Lemma. *The complex normal bundle ξ_n of the inclusion l_n is isomorphic to η_n.*

Proof. By 1.5(ii), it suffices to prove that $c_1(\xi_n) = c_1(\eta_n)$. Notice that $l_n^* \eta_{n+1} = \eta_n$. We have $l_n^* \tau(CP^{n+1}) = \tau(CP^n) \oplus \xi_n$, and hence, by 1.24, $(n+2)\eta_n \oplus \theta^1 = (n+1)\eta_n \oplus \xi_n \oplus \theta^1$. Since $c_1(\xi \oplus \zeta) = c_1(\xi) + c_1(\zeta)$, we conclude that $c_1(\xi_n) = c_1(\eta_n)$. □

1.26. Proposition. *The zero section $\mathfrak{z} : B\mathcal{U}_1 \to M\mathcal{U}_1$ as in* IV.5.4 *is a homotopy equivalence.*

Proof. Let $D(\gamma^1)$ (resp. $S(\gamma^1)$) be the unit disk (resp. unit sphere) bundle associated with γ^1. Then $S(\gamma^1)$ is just the locally trivial principal $\mathcal{U}_1$-bundle associated with γ^1, i.e., $S(\gamma^1)$ is the universal principal $\mathcal{U}_1$-bundle. Hence, by IV.3.2(ii), $\operatorname{ts} S(\gamma^1)$ is a contractible space. Now, $\mathfrak{z}$ has the form

$$\mathfrak{z} : B\mathcal{U}_1 \xrightarrow{s} \operatorname{ts} D(\gamma^1) \xrightarrow{p} (\operatorname{ts} D(\gamma^1))/\operatorname{ts} S(\gamma^1) = M\mathcal{U}_1$$

where the section s and the projection p are homotopy equivalences. □

We define a map

$$\mathfrak{h} : CP^\infty \xrightarrow{e_1} B\mathcal{U}_1 \xrightarrow{\mathfrak{z}} M\mathcal{U}_1. \tag{1.27}$$

By 1.23 and 1.26, $\mathfrak{h}$ is a homotopy equivalence. Since η_n is a normal bundle of the inclusion $CP^n \subset CP^{n+1}$, there is a collapsing map $c : CP^{n+1} \to T\eta_n$; it collapses the complement of a tubular neighborhood of CP^n. (In fact, $CP^{n+1} \simeq T\eta_n$, but we do not use it here.) Let $g := e_1 j_n : CP^n \to B\mathcal{U}_1$. We define a map $f : CP^{n+1} \to M\mathcal{U}_1$ to be the composition

$$CP^{n+1} \xrightarrow{c} T\eta_n \xrightarrow{Tg} M\mathcal{U}_1$$

where $Tg := T\mathfrak{I}_{g,\gamma^1}$. Let $\mathfrak{h}$ be as in (1.27), and let $h : M\mathcal{U}_1 \to B\mathcal{U}_1$ be a homotopy equivalence inverse to $\mathfrak{z}$.

1.28. Lemma. *The map $CP^{n+1} \xrightarrow{f} M\mathcal{U}_1 \xrightarrow{h} B\mathcal{U}_1$ classifies η_{n+1}. In other words, f is homotopic to $CP^{n+1} \xrightarrow{j_{n+1}} CP^\infty \xrightarrow{\mathfrak{h}} M\mathcal{U}_1$.*

Proof. By 1.5(ii), it suffices to prove that $f^* h^* t = c_1(\eta_{n+1})$, i.e., that $l_n^* f^* h^* t = c_1(\eta_n)$. But this follows immediately from the commutativity of the diagram

$$\begin{array}{ccccc} CP^n & \xrightarrow{e_1 j_n} & BU_1 & = & BU_1 \\ \downarrow{\scriptstyle i_n} & & \downarrow{\scriptstyle \mathfrak{z}} & & \| \\ CP^{n+1} & \xrightarrow{f} & MU_1 & \xrightarrow{h} & BU_1. \end{array}$$

□

1.29. Proposition. *The zero section $\mathfrak{z} : CP^\infty \to T\eta$ is a homotopy equivalence.*

Proof. This follows from 1.26 and 1.23. □

§2. ℂ-oriented Spectra

Recall the inclusion $j_1 : S^2 = CP^1 \subset CP^\infty$.

2.1. Definition (Adams [8]). Let $E = (E, \mu, \iota)$ be a commutative ring spectrum. An element $t = t^E \in \widetilde{E}^2(CP^\infty)$ is called a ℂ*-orientation* of E if $j_1^* t = \mathfrak{s}^2(1) \in \widetilde{E}^2(S^2)$. (Here $\mathfrak{s}^2 : \widetilde{E}^0(S^0) \cong \widetilde{E}^2(S^2)$ is the twofold suspension.) A ℂ*-oriented spectrum* is a spectrum with a fixed ℂ-orientatation, i.e., a pair (E, t) (or, if you want, a quadruple (E, μ, ι, t)). A *morphism* $f : (E, t) \to (E', t')$ *of* ℂ*-oriented spectra* is a ring morphism $E \to E'$ which maps t to t'.

The image of t under the inclusion $\widetilde{E}^*(CP^\infty) \subset E^*(CP^\infty)$ we also denote by t.

2.2. Theorem. *Let (E, t) be any* ℂ*-oriented spectrum. Then*

(i) $E^*(CP^n) = E^*(\mathrm{pt})[t]/(t^{n+1})$.

(ii) $E^*(CP^\infty) = E^*(\mathrm{pt})[[t]]$.

(iii) $E^*((CP^\infty)^n) = E^*(\mathrm{pt})[[t_1, \dots, t_n]]$, *where* $t_i = p_i^*(t)$.

(iv) $E^*(BU_n) = E^*(\mathrm{pt})[[c_{1,n}, \dots, c_{n,n}]], \dim c_{i,n} = 2i$. *Furthermore, $e_n^*(c_{k,n})$ is the elementary symmetric polynomial $\sigma_k(t_1, \dots, t_n)$. So, e_n^* is a monomorphism, and its image just consists of the invariants of the Σ_n-action. (Here Σ_n is the symmetric group of degree n and the action is given by permutation of t_i's.) Finally, $r_n^*(c_{i,n+1}) = c_{i,n}$ for $i \le n$ and $r_n^*(c_{n+1,n+1}) = 0$.*

(v) $E^*(BU) = E^*(\mathrm{pt})[[c_1, \dots, c_n, \dots]]$ *where $(j_n^{\mathcal{U}})^* c_i = c_{i,n}$ for $i \le n$ and $(j_n^{\mathcal{U}})^* c_k = 0$ for $k > n$.*

Proof. See Adams [8], Ch. II, Dold [4], Switzer [1]. □

2.3. Construction-Definition. (a) The classes $c_n = c_n^{E,t}$ as in 2.2(v) are called the *universal Chern–Conner–Floyd classes.*

(b) We introduce universal characteristic classes $c_\omega = c_\omega^{E,t} \in E^{2|\omega|}(BU)$, as we did for the ordinary cohomology. Given $\omega = \{i_1, \dots, i_k\}$, $k \le N$, let

$t_\omega \in \pi_*(E)[[t_1, \dots, t_N]] = E^*((CP^\infty)^N)$ be the smallest symmetric polynomial which contains $t_1^{i_1} \cdots t_k^{i_k}$. Let $c_{\omega,N} \in E^{2|\omega|}(B\mathcal{U}_N)$ be the unique element such that $e_N^*(c_{\omega,N}) = t_\omega$. Since $r_n^* c_{\omega,n+1} = c_{\omega,n}$ for $|\omega| < n+1$, there is a unique element $c_\omega \in E^*(B\mathcal{U})$ such that $(j_n^{\mathcal{U}})^* c_\omega = c_{\omega,n}$ for every $n \geq |\omega|$. Finally, we set $c_{(0)} := 1$. Clearly, $c_n = c_{(1,\dots,1)}$.

Given a complex vector bundle ξ classified by $f : X \to B\mathcal{U}_n$, we define its characterictic class $c_\omega(\xi)$ as $c_\omega(\xi) := (j_n^{\mathcal{U}} f)^* c_\omega \in E^{2|\omega|}(X)$. It is clear that

$$c_\omega(\xi \times \zeta) = \sum_{(\omega_1,\omega_2)=\omega} c_{\omega_1}(\xi) c_{\omega_2}(\zeta).$$

In other words, $(\mu^{\mathcal{U}})^*(c_\omega) = c_{\omega_1} c_{\omega_2}$ where $\mu^{\mathcal{U}}$ is as in IV.(4.26).

Note that $c_{i,n} = c_i(\gamma^n)$ and $c_1(\eta) = t$.

Let

$$i_{2n} : \Sigma^{-2n}\Sigma^\infty M\mathcal{U}_n \to M\mathcal{U} \tag{2.4}$$

be the morphism as in II.(1.4). Define $T \in \widetilde{M\mathcal{U}}^2(CP^\infty)$ via the composition

$$\Sigma^{-2}\Sigma^\infty(CP^\infty) \xrightarrow{\Sigma^{-2}\Sigma^\infty \mathfrak{h}} \Sigma^{-2}\Sigma^\infty M\mathcal{U}_1 \xrightarrow{i_2} M\mathcal{U}$$

with $\mathfrak{h}$ as in (1.27). Clearly, T is a $\mathbb{C}$-orientation of $M\mathcal{U}$. So, by V.1.18, every ring morphism $\tau : M\mathcal{U} \to E$ yields a $\mathbb{C}$-orientation $\tau_*(T)$ of E. Hence, we have a correspondence

$$\{\text{ring morphisms } M\mathcal{U} \to E\} \longrightarrow \{\mathbb{C}\text{-orientations of } E\}.$$

2.5. Theorem. *This correspondence is bijective. Thus, $(M\mathcal{U}, T)$ is the universal $\mathbb{C}$-oriented spectrum.*

Proof. See Adams [8], II.4.6, or Stong [3], Ch. V. □

2.6. Corollary. *Let E be a $\mathbb{C}$-orientable spectrum. Then ring morphisms $M\mathcal{U} \to E$ are in a bijective correspondence with formal power series $\varphi(t) = t + \sum_{i>0} a_i t^{i+1}$, $a_i \in \pi_{2i-2}(E)$.*

Proof. By 2.2(ii), every $\mathbb{C}$-orientation of E has the form

$$t + \sum_{i>0} a_i t^{i+1},\ a_i \in \pi_{2i-2}(E).$$ □

By 2.5, every $\mathbb{C}$-orientation t of E yields a ring morphism

$$u = u^{E,t} : M\mathcal{U} \to E \tag{2.7}$$

with $u_*(T) = t$, and this morphism is unique up to homotopy. We denote $u^{MU,T}$ by U, and it is clear that $U = 1_{MU}$.

2.8. Proposition-Definition. *Let $i_{2n} : \Sigma^{-2n}\Sigma^\infty MU_n \to MU$ be as in (2.4). Let (E,t) be a ℂ-oriented spectrum, and let ξ be a complex vector bundle classified by $f : \mathrm{bs}\,\xi \to BU_n$. We set $g = T\mathfrak{I}_{f,\gamma^n}$ and define $u_\xi \in \widetilde{E}^{2n}(T\xi)$ to be the composition*

$$\text{(2.9)} \qquad \Sigma^{-2n}\Sigma^\infty T\xi \xrightarrow{\Sigma^{-2n}\Sigma^\infty g} \Sigma^{-2n}\Sigma^\infty MU_n \xrightarrow{i_{2n}} MU \xrightarrow{u} E.$$

Then u_ξ is an E-orientation of ξ. We call u_ξ the (E,t)-orientation *of ξ.*

Proof. Since $u = u^{E,t} : MU \to E$ is a ring morphism, $u \in \widetilde{E}^0(MU)$ is an E-orientation of the universal stable complex vector bundle γ. Considering the isomorphism $\mathfrak{e} : T(\xi_{\mathrm{st}}) \to \Sigma^{-2n}\Sigma^\infty T\xi$ as in IV.5.16, we conclude that $i_{2n}\circ\Sigma^{-2n}\Sigma^\infty g\circ\mathfrak{e}$ preserves roots, and so $u\circ i_{2n}\circ\Sigma^{-2n}\Sigma^\infty g\circ\mathfrak{e}$ is an E-orientation of ξ_{st}. Hence, by V.1.13, u_ξ is an E-orientation of ξ. □

2.10. Corollary. *Let E be a commutative ring spectrum. The following three conditions are equivalent:*

(i) *The vector bundle η is E-orientable;*
(ii) *E is a ℂ-orientable spectrum;*
(iii) *Every complex vector bundle is E-orientable.*

Proof. We prove that (i) ⇒ (ii) ⇒ (iii). By 1.29, the zero section $\mathfrak{z} : CP^\infty \to T\eta$ is a homotopy equivalence. Since $S^2 \xrightarrow{j_1} CP^\infty \xrightarrow{\mathfrak{z}} T\eta$ yields a generator of $\pi_2(T\eta) = \mathbb{Z}$, we conclude that $\mathfrak{z}j_1$ can be considered as a root of $T\eta$. Hence, $j_1^*\mathfrak{z}^*v = \pm\mathfrak{s}^2(1) \in \widetilde{E}^2(S^2)$ for every E-orientation v of η, and so $\mathfrak{z}^*v$ or $-\mathfrak{z}^*v$ is a ℂ-orientation of E. Thus, (i) ⇒ (ii). Finally, by 2.8, (ii) ⇒ (iii). □

We denote the (MU,T)-orientation of ξ by U_ξ, and we set

$$\text{(2.11)} \qquad u_n := u_{\gamma^n} \in \widetilde{E}^{2n}(MU_n),\ U_n := U_{\gamma^n} \in \widetilde{MU}^{2n}(MU_n).$$

Let $a_{m,n} : BU_m \times BU_n \to BU_{m+n}$ classify $\gamma^m \times \gamma^n$. We have the map

$$Ta_{m,n} := T\mathfrak{I}_{a_{m,n},\gamma^{m+n}} : MU_m \wedge MU_n \to MU_{m+n}$$

and the morphism

$$\tau_{m,n} : \Sigma^{-2m}\Sigma^\infty MU_m \wedge \Sigma^{-2n}\Sigma^\infty MU_n \simeq \Sigma^{-2m-2n}\Sigma^\infty(MU_m \wedge MU_n) \xrightarrow{\Sigma^{-2m-2n}\Sigma^\infty Ta_{m,n}} \Sigma^{-2(m+n)}\Sigma^\infty MU_{m+n}$$

such that the diagram

$$\begin{array}{ccc}
\Sigma^{-2m}\Sigma^\infty MU_m \wedge \Sigma^{-2n}\Sigma^\infty MU_n & \xrightarrow{\tau_{m,n}} & \Sigma^{-2(m+n)}\Sigma^\infty MU_{m+n} \\
\downarrow{\scriptstyle i_{2m}\wedge i_{2n}} & & \downarrow{\scriptstyle i_{2(m+n)}} \\
MU \wedge MU & \xrightarrow{\mu_{MU}} & MU
\end{array}$$

commutes. Since $u : MU \to E$ as in (2.7) is a ring morphism, we conclude that

(2.12) $$(Ta_{m,n})(u_m \otimes u_n) = u_{m+n}.$$

2.13. Corollary. *Let ξ, ζ be two complex vector bundles. Then the (E,t)-orientation of $\xi \times \zeta$ coincides with the product E-orientation (i.e., the one defined in the proof of* V.1.10(ii)) *of the (E,t)-orientations of ξ and ζ.* □

2.14. Proposition. *Let t be a $\mathbb{C}$-orientation of E. Let ξ be an n-dimensional (E,t)-oriented complex vector bundle. Then $\chi^E(\xi) = c_n^{E,t}(\xi)$, where χ is the Euler class.*

Proof. Firstly, we prove that $\chi(\eta) = c_1(\eta)$. Because of the equality $E^*(CP^\infty) = \varprojlim\{E^*(CP^n)\}$, it suffices to prove that $\chi(\eta_n) = c_1(\eta_n)$ for every n. We define

$$g : CP^n \xrightarrow{j_n} CP^\infty \xrightarrow{e_1} BU_1$$

and set $Tg := T\mathfrak{I}_{g,\gamma^1}$. Because of 1.28 and since $j_{n+1}l_n = j_n$, there is the commutative diagram

$$\begin{array}{ccccc}
CP^n & \xrightarrow{j_n} & CP^\infty & \xrightarrow{e_1} & BU_1 \\
\downarrow{\scriptstyle l_n} & & & & \downarrow{\scriptstyle \mathfrak{z}} \\
CP^{n+1} & \xrightarrow{c} & T\eta_n & \xrightarrow{Tg} & MU_1
\end{array}$$

where c is the collapsing map. Since $T \in \widetilde{MU}^2(CP^\infty)$ is given by the morphism

$$\Sigma^{-2}\Sigma^\infty(CP^\infty) \xrightarrow{\Sigma^{-2}\Sigma^\infty e_1} \Sigma^{-2}\Sigma^\infty BU_1 \xrightarrow{\Sigma^{-2}\Sigma^\infty \mathfrak{z}} \Sigma^{-2}\Sigma^\infty MU_1 \xrightarrow{i_2} MU,$$

we conclude that $T = e_1^*\mathfrak{z}^*U_1$. Hence, $t = e_1^*\mathfrak{z}^*u_1 \in \widetilde{E}^2(CP^\infty)$. Now

$$j_n^*t = j_n^*e_1^*\mathfrak{z}^*u_1 = l_n^*c^*(Tg)^*u_1 = (cl_n)^*(Tg)^*u_1 = (cl_n)^*u_{\eta_n}.$$

But $cl_n : CP^n \to T\eta_n$ is the zero section of $T\eta_n$, and so $\chi(\eta_n) = \varepsilon^*(cl_n)^*u_{\eta_n}$ where $\varepsilon : (CP^n)^+ \to CP^n$ is as in V.1.24. On the other hand, $c_1(\eta_n) = \varepsilon^*j_n^*t \in E^2(CP^\infty)$. So, $\chi(\eta_n) = c_1(\eta_n)$ for every n, and thus $\chi(\eta) = c_1(\eta)$.

Now, by 2.13 and V.1.26(ii),

$$c_n(\eta \times \ldots \times \eta) = t_1 \cdots t_n = \chi(\eta \times \ldots \times \eta) \in E^*((CP^\infty)^n).$$

By 2.2(iv), $e_n^* : E^*(BU_n) \to E^*((CP^\infty)^n)$ is monic. Hence, the proposition holds for γ^n. Hence, it holds for every ξ. □

2.15. Examples. (a) Since $j_1^* : H^2(CP^\infty) \to H^2(CP^1)$ is an isomorphism, there is a unique element $t \in H^2(CP^\infty)$ such that $j_1^*(t) = \mathfrak{s}^2(1)$. So, $(H\mathbb{Z}, t)$ is a $\mathbb{C}$-oriented spectrum. Moreover, the classes $c_i^{H,t}$ coincide with the classical Chern classes, and the element $u^{H,t}$ coincides with the Thom class u defined in §1.

(b) Let (E, t) be a $\mathbb{C}$-oriented spectrum, and let $\tau : E \to F$ be a ring morphism of commutative ring spectra. Then $(F, \tau(t))$ is a $\mathbb{C}$-oriented spectrum.

(c) Let R be a graded commutative ring. Then HR is a $\mathbb{C}$-orientable spectrum since there is a ring morphism $H\mathbb{Z} \to HR$.

(d) Consider complex K-theory. We have $K^*(\text{pt}) = \mathbb{Z}[s, s^{-1}], \deg s = 2$. (We wrote $K^*(\text{pt}) = \mathbb{Z}[t, t^{-1}]$ in Ch. VI, but here the letter t is reserved for $\mathbb{C}$-orientations.) Let $1 \in K^0(CP^\infty)$ represent θ^1. Then $s(\eta - 1) \in \widetilde{K}^2(CP^\infty)$ is a $\mathbb{C}$-orientation of K.

(e) The spectrum KO is not $\mathbb{C}$-orientable. Indeed, $w_2(\eta) \neq 0$, and hence η is not KO-orientable.

(f) The sphere spectrum S is not $\mathbb{C}$-orientable because otherwise every commutative ring spectrum would be $\mathbb{C}$-orientable.

(g) There is the universal $\mathbb{C}$-oriented spectrum (MU, T).

Let $T\eta^{(i)}, i = 1, \dots, n$ be a copy of $T\eta$, and let $d_i \in \widetilde{E}^2(T\eta^{(i)})$ be a copy of u_η. Similarly, D_i is a copy of U_η. The map e_n induces a map

$$Te_n := T\mathfrak{J}_{e_n,\gamma^n} : (T\eta)^{\wedge n} := T\eta^{(1)} \wedge \cdots \wedge T\eta^{(n)} = T(\eta^{(1)} \times \cdots \times \eta^{(n)}) \to MU_n$$

of Thom spaces. Let $d_1 \cdots d_n \in \widetilde{E}^{2n}((T\eta)^{\wedge n})$ be the image of $d_1 \otimes \cdots \otimes d_n$ under the homomorphism $\widetilde{E}^*(T\eta) \otimes \cdots \otimes \widetilde{E}^*(T\eta) \to \widetilde{E}^*((T\eta)^{\wedge n}))$. Clearly, it is an E-orientation of $\eta \times \cdots \times \eta$.

2.16. Proposition. *We have $(Te_n)^*(u_n) = d_1 \cdots d_n$. Furthermore, the map $(Te_n)^* : \widetilde{E}^*(MU_n) \to \widetilde{E}^*((T\eta)^{\wedge n})$ is monic, and its image just consists of the invariants of the Σ_n-action.*

Proof. The first assertion follows from (2.12) inductively. To prove the second one, consider the commutative diagram

$$\begin{array}{ccc} \widetilde{E}^*(MU_n) & \xrightarrow{(Te_n)^*} & \widetilde{E}^*((T\eta)^{\wedge n}) \\ \varphi\uparrow & & \uparrow\overline{\varphi} \\ E^*(BU_n) & \xrightarrow{(e_n)^*} & E^*((CP^\infty)^n), \end{array}$$

where φ and $\overline{\varphi}$ are the Thom–Dold isomorphisms given by u_n and $d_1 \cdots d_n$. Since $d_1 \cdots d_n$ is Σ_n-invariant, the isomorphism $\overline{\varphi}$ is Σ_n-equivariant. Thus, by 2.2(iv), $\mathrm{Im}(Te_n)^*$ just consists of the invariants of the Σ_n-action. □

The element $u \in E^0(MU)$ in (2.7) yields a Thom–Dold isomorphism $\varphi : E^*(BU) \cong E^*(MU)$. We set

$$s_\omega = s_\omega^{E,t} := \varphi(c_\omega) \in E^{2|\omega|}(MU).$$

So, we have a morphism $s_\omega : MU \to \Sigma^{2|\omega|}E$ of spectra, which yields a morphism $s_\omega : MU^*(-) \to E^*(-)$ of cohomology theories.

Let $U_n \in \widetilde{MU}^{2n}(MU_{2n})$ be the Thom–Dold classes as in (2.11). We have (in Notation V.1.19(d)) $s_\omega(U) = c_\omega u$, and so $s_\omega(U_n) = c_\omega(\gamma^n)u_n$. This implies that $s_\omega(U_\xi) = c_\omega(\xi)u_\xi$ for every complex vector bundle ξ, where $U_\xi \in MU^*(T\xi)$ (resp. $u_\xi \in E^*(T\xi)$) is the (MU,T)-orientation (resp. (E,t)-orientation) of ξ.

2.17. Lemma. *Let ξ be a complex line bundle over X, and let u_ξ be the (E,t)-orientation of ξ. Then the following hold:*

(i)
$$c_\omega(\xi) = \begin{cases} (c_1(\xi))^n & \text{if } \omega = (n) \\ 0 & \text{otherwise.} \end{cases}$$

(ii) *Let φ be the Thom–Dold isomorphism given by the orientation u_ξ. Then*

$$\varphi(c_1(\xi)^k) = u_\xi^{k+1}$$

for every k. In other words, $u_\xi c_1(\xi)^k = u_\xi^{k+1}$.

(iii)
$$s_\omega(U_\xi) = \begin{cases} u_\xi^{n+1} & \text{if } \omega = (n) \\ 0 & \text{otherwise.} \end{cases}$$

In particular,

$$s_\omega(T) = \begin{cases} t^{n+1} & \text{if } \omega = (n) \\ 0 & \text{otherwise.} \end{cases}$$

Proof. (i) Clearly, this holds for $\xi = \eta$. Furthermore, this holds for γ^1 since, by 1.23, $e_1 : CP^\infty \to BU$ is a homotopy equivalence. Thus, this holds for every ξ.

(ii) It suffices to consider $\xi = \eta$ (cf. (i)). Let $\mathfrak{z} : CP^\infty \to T\eta$ be the zero section as in IV.5.4. By 2.14 and V.1.27,

$$\varepsilon^*\mathfrak{z}^*u_\eta^{k+1} = (\varepsilon^*\mathfrak{z}^*u_\eta)^{k+1} = c_1(\eta)^{k+1} = \chi(\eta)c_1(\eta)^k = \varepsilon^*\mathfrak{z}^*(\varphi(c_1(\eta)^k)).$$

But, by 1.29, $\varepsilon^*\mathfrak{z}^*$ is a monomorphism.

(iii) Since $s_\omega(U_\xi) = c_\omega(\xi)u_\xi$, the result follows from (i). □

Let μ_{MU} (resp. μ_E) be the multiplication on MU (resp. on E).

2.18. Proposition. *Set*

$$\mu := \mu_{MU}^{MU,MU} : MU^*(MU) \otimes MU^*(MU) \to MU^*(MU \wedge MU)$$

and $\mu' := \mu_E^{MU,MU} : E^*(MU) \otimes E^*(MU) \to E^*(MU \wedge MU)$ *and consider the the diagram*

$$E^*(MU) \otimes E^*(MU) \xrightarrow{\mu'} E^*(MU \wedge MU) \xleftarrow{\mu_{MU}^*} E^*(MU).$$

Then $\mu_{MU}^*(s_\omega) = \sum_{(\omega_1,\omega_2)=\omega} \mu'(s_{\omega_1} \otimes s_{\omega_2})$.

Proof. Let $\mu^{\mathcal{U}} : B\mathcal{U} \times B\mathcal{U} \to B\mathcal{U}$ be as in IV.(4.26). It is easy to see that there is a commutative diagram (a stable analog of V.1.2)

$$\begin{array}{ccc} E^*(MU) & \xrightarrow{\mu_{MU}^*} & E^*(MU \wedge MU) \\ \varphi \uparrow & & \uparrow \varphi' \\ E^*(B\mathcal{U}) & \xrightarrow{(\mu^{\mathcal{U}})^*} & E^*(B\mathcal{U} \times B\mathcal{U}) \end{array}$$

where φ is the Thom isomorphism given by u and φ' is the one given by $\mu'(u \otimes u)$. Now,

$$\begin{aligned} \mu_{MU}^*(s_\omega) &= \mu_{MU}^*(\varphi(c_\omega)) = \varphi'((\mu^{\mathcal{U}})^*(c_\omega)) = \varphi'\left(\sum_{(\omega_1,\omega_2)=\omega} c_{\omega_1} c_{\omega_2}\right) \\ &= \sum_{(\omega_1,\omega_2)=\omega} c_{\omega_1} c_{\omega_2} \mu'(u \otimes u) = \sum_{(\omega_1,\omega_2)=\omega} \mu'(c_{\omega_1} u \otimes c_{\omega_2} u) \\ &= \sum_{(\omega_1,\omega_2)=\omega} \mu'(s_{\omega_1}(U) \otimes s_{\omega_2}(U)) = \sum_{(\omega_1,\omega_2)=\omega} \mu'(s_{\omega_1} \otimes s_{\omega_2}). \qquad \square \end{aligned}$$

Because of the universality of the class U, we have the following corollary.

2.19. Corollary. (i) *For every space* X *and every* $x, y \in MU^*(X)$ *we have*

$$s_\omega(xy) = \sum_{(\omega_1,\omega_2)=\omega} s_{\omega_1}(x) s_{\omega_2}(y).$$

(ii) *For every space* X *and every* $x \in MU^*(X)$, $y \in MU_*(X)$ *we have*

$$s_\omega(x \cap y) = \sum_{(\omega_1,\omega_2)=\omega} s_{\omega_1}(x) \cap s_{\omega_2}(y),$$

$$s_\omega\langle x, y\rangle = \sum_{(\omega_1,\omega_2)=\omega} \langle s_{\omega_1}(x), s_{\omega_2}(y)\rangle.$$

Here $s_{\omega_2} = (s_{\omega_2})_* : MU_*(X) \to E_*(X)$.

(iii) *For every* $a, b \in E_*(MU)$ *and every* $s_\omega \in E^*(MU)$ *we have*

$$\langle s_\omega, ab\rangle = \sum_{(\omega_1,\omega_2)=\omega} \langle s_{\omega_1}, a\rangle \, \langle s_{\omega_2}, b\rangle. \qquad \square$$

2.20. Lemma. *Let* E *be a* $\mathbb{C}$*-oriented spectrum, and let* $h : \pi_*(MU) \to E_*(MU)$ *be the Hurewicz homomorphism (with respect to* $\iota : S \to E$*). Then* $s_\omega(x) = \langle s_\omega, h(x)\rangle$ *for every* $x \in \pi_*(MU)$.

Proof. Given $x : S^k \to MU$, consider $s_\omega(x) : S^k \xrightarrow{x} MU \xrightarrow{s_\omega} \Sigma^{2|\omega|}E$. Now the commutativity of the diagram

$$\begin{array}{ccccc} S^k \wedge S & \xrightarrow{x\wedge\iota} & MU \wedge E & & \\ \Vert & & \downarrow{\scriptstyle s_\omega \wedge 1} & & \\ S^k \wedge S & \xrightarrow{s_\omega(x)\wedge\iota} & \Sigma^{2|\omega|}E \wedge E & \xrightarrow{\mu_E} & E \end{array}$$

implies that

$$\langle s_\omega, h(x)\rangle = \mu_E(s_\omega \wedge 1)(x \wedge \iota) = \mu_E(s_\omega(x) \wedge \iota) = s_\omega(x). \qquad \square$$

By 1.9(i,ii), $\pi_*(MU) = \mathbb{Z}[x_i | \dim x_i = 2i]$, $H_*(MU) = \mathbb{Z}[y_i | \dim y_i = 2i]$.

2.21. Corollary. *Let* $h : \pi_*(MU) \to H_*(MU)$ *be the Hurewicz homomorphism. Then* $h(x_i) \equiv \pm\lambda_i y_i \bmod \mathrm{Dec}\,(H_*(MU))$ *where* λ_i *is the number defined in* (1.8).

Proof. Here s_ω denotes $s_\omega^{H\mathbb{Z}}$. Let $h(x_i) = a_i y_i + d$, where $d \in \mathrm{Dec}$. Since $\{s_\omega\}$ is a free basis of the abelian group $H^*(MU)$, we conclude that $\langle s_{(i)}, y_i\rangle = \pm 1$ or $\langle s_{(i)}, y_i\rangle = 0$. Furthermore, by 2.19(ii), $\langle s_{(i)}, d\rangle = 0$. Now, by 1.9(iv),

$$\lambda_i = s_{(i)}(x_i) = \langle s_{(i)}, h(x_i)\rangle = \langle s_{(i)}, a_i y_i\rangle + \langle s_{(i)}, d\rangle = a_i\langle s_{(i)}, y_i\rangle.$$

Hence, $\langle s_{(i)}, y_i\rangle \neq 0$, and so $\langle s_{(i)}, y_i\rangle = \pm 1$, and thus $\lambda_i = \pm a_i$. $\square$

2.22. Lemma. *Let* V *be a stably almost complex closed manifold of dimension* n. *Let* (E, t) *be a* $\mathbb{C}$*-oriented spectrum, and let* $[V]_E$ *be the image of* $[V]_{MU}$ *under the homomorphism* $u_*^{E,t} : MU_*(V) \to E_*(V)$. *Then*

$$s_\omega^{E,t}[V]_{MU} = c_\omega^{E,t}(\nu) \cap [V]_E \in E_*(V),$$

where ν *is a normal complex bundle of* V. *Furthermore,*

$$s_\omega^{E,t}[V] = \langle c_\omega^{E,t}(\nu), [V]_E\rangle \in E_*(\mathrm{pt}).$$

Proof. Let $N = \dim \nu$, let $U_\nu \in \widetilde{MU}^{2N}(T\nu)$ be the (MU, T)-orientation of ν, and let $[T\nu] = [T\nu]_{MU}$ be as in V.2.8. If $\omega \neq (0)$, then $s_\omega(1) = 0$ for $1 \in \pi_0(MU)$ and so $s_\omega[S^{2N+2n}]_{MU} = 0$. So, $s_\omega[T\nu] = 0$ for $\omega \neq (0)$, $s_{(0)}[T\nu] = [T\nu]_E$. Considering the pairing

$$\Cap : \widetilde{E}^*(T\nu) \otimes \widetilde{E}_*(T\nu) \xrightarrow{\cap} \widetilde{E}_*(T\nu) \xrightarrow{\varphi^E} E_*(V),$$

we conclude that

$$\begin{aligned} s_\omega[V]_{MU} &= \sum_{(\omega_1,\omega_2)=\omega} s_{\omega_1}(U_\nu) \Cap s_{\omega_2}(T\nu) \\ &= s_\omega(U_\nu) \Cap [T\nu]_E = c_\omega(\nu) \cap (u_\nu^E \Cap [T\nu]_E) = c_\omega(\nu) \cap [V]_E. \end{aligned}$$

The last assertion holds since $\langle x, [V]\rangle$ is the image of $x \cap [V]_E$ under the map $V \to \mathrm{pt}$. □

The elements (morphisms) s_ω can be considered as the universal characteristic numbers. Namely, the homomorphism $s_\omega : MU^*(S) \to E^*(S)$ assigns an element of the coefficient ring to a stably almost complex manifold. In particular, for $E = H\mathbb{Z}$ this element (some integer) coincides with the classical characteristic number described in (1.7).

What happens if one changes the ℂ-orientation of E? Given a ℂ-orientation t of E, let $z = t + \sum_{i>0} a_i t^{i+1}$ be another ℂ-orientation of E, and let $u^{E,z}$ be as in (2.7). Then $u^{E,z}(T) = z = t + \sum a_i t^{i+1}$. Let $\mathfrak{z} : CP^\infty \to T\eta$ be the zero section as in 1.29, and let $h : T\eta \to CP^\infty$ be a homotopy equivalence inverse to $\mathfrak{z}$. Then $\mathfrak{z}^* U_\eta = T$ (e.g., by 2.14), and so $h^*T = U_\eta$. Now,

$$u^{E,z}(U_\eta) = u^{E,z}(h^*T) = h^*\left(t + \sum_{i>0} a_i t^{i+1}\right) = u_\eta + \sum_{i>0} a_i u_\eta^{i+1}. \tag{2.23}$$

Considering the element $D_1 \cdots D_N \in \widetilde{MU}^2((T\eta)^N)$ defined before 2.16, and using 2.17(iii), we conclude that

$$\begin{aligned} s_\omega^{E,t}(D_1 \cdots D_N) &= \sum_{(\alpha_1,\dots,\alpha_k)} s_{i_1}^{E,t}(D_{\alpha_1}) \cdots s_{i_k}^{E,t}(D_{\alpha_k}) \\ &= \sum_{(\alpha_1,\dots,\alpha_k)} d_{\alpha_1}^{i_1+1} \cdots d_{\alpha_k}^{i_k+1} \end{aligned}$$

where $\omega = (i_1, \dots, i_k)$. By (2.23), $u^{E,z}(D_j) = \sum_{i>0} (d_j + a_i d_j^{i+1})$, and so

$$\begin{aligned} u^{E,z}(D_1 \cdots D_N) = \prod_{j=1}^N \left(\sum_{i>0} (d_j + a_i d_j^{i+1})\right) &= d_1 \cdots d_N \\ + \sum_{|\omega|>0} a_\omega s_\omega^{E,t}(D_1 \cdots D_N) &= \sum_\omega a_\omega s_\omega^{E,t}(D_1 \cdots D_N), \end{aligned}$$

where $a_\omega = a_{i_1} \cdots a_{i_k}$, $a_{(0)} = 1$. Hence, because of 2.16,

$$u^{E,z}(U_N) = \sum_\omega a_\omega s_\omega^{E,t}(U_N). \tag{2.24}$$

By 1.9(vii) and III.5.7(iii), we have $\varprojlim^1\{MU^*(MU)^{(n)}\} = 0$. This implies easily that $\varprojlim^1\{MU^*(MU_n)\} = 0$. Hence, by III.4.18, the homomorphism

$$\rho : MU^*(MU) \to \varprojlim\{\widetilde{MU}^*(MU_n)\}$$

is an isomorphism. Furthermore, ρ maps the element U to the string $\{U_n\}$. So, we can pass to $\varprojlim$ in (2.24) and replace U_N by U. Hence,

$$u^{E,z}(U) = \sum_\omega a_\omega s_\omega^{E,t}(U), \quad a_\omega = a_{i_1} \cdots a_{i_k},\ a_{(0)} = 1.$$

Thus, because of the universality of U,

$$u^{E,z}(x) = \sum_\omega a_\omega s_\omega^{E,t}(x), \quad a_\omega = a_{i_1} \cdots a_{i_k},\ a_{(0)} = 1 \tag{2.25}$$

for every spectrum X and every $x \in MU^*(X)$, cf. Buhštaber [3].

We finish this section with the remark that $c_\omega^{E,t}$ and $s_\omega^{E,t}$ are natural with respect to morphisms of $\mathbb{C}$-oriented spectra.

§3. Operations on MU. Idempotents. The Brown–Peterson Spectrum

From here to the end of the chapter, S_ω means $s_\omega^{MU,T}$ and s_ω means $s_\omega^{H\mathbb{Z},t}$. Similarly, C_ω means $c_\omega^{MU,T}$ and c_ω means $c_\omega^{H\mathbb{Z},t}$.

Let $MU^*(MU)$ be the ring of MU-operations, see II.3.47. It is easy to see that the $MU^*(S)$-module structure on $MU^*(MU)$ turns $MU^*(MU)$ into an $MU^*(S)$-algebra.

Now we describe this algebra. Firstly, every scalar $a \in MU^*(S)$ is an operation. Furthermore, $S_\omega \in MU^{2|\omega|}(MU)$ can also be considered as an operation. Similarly, every finite homogeneous sum $\sum a_\omega S_\omega$ can be considered as an operation, where $a_\omega S_\omega$ is the composition of operations a_ω and S_ω.

3.1. Theorem (Novikov [4], cf. also Landweber [1]). *Let $\mathbf{S} \subset MU^*(MU)$ be the subalgebra generated by $\{S_\omega\}$.*

(i) *$S_{(n)}(T) = T^{n+1}$, and $S_\omega(T) = 0$ if $l(\omega) > 1$.*

(ii) *For every space X and every $x, y \in MU^*(X)$ we have*

$$S_\omega(xy) = \sum_{(\omega_1,\omega_2)=\omega} S_{\omega_1}(x)S_{\omega_2}(y).$$

Thus, the diagonal $\Delta\colon \mathcal{S} \to \mathcal{S} \otimes \mathcal{S}$ *turns* $\mathcal{S}$ *into a Hopf algebra over* $\mathbb{Z}$*, and this Hopf algebra structure is compatible with the ring strucure on* $M\mathcal{U}$*.*

(iii) *For every pair of partitions* ω', ω''*, the composition* $S_{\omega'} \circ S_{\omega''}$ *is an integral linear combination of elements* S_ω*. Thus,* $\{S_\omega\}$ *is an additive basis of* $\mathcal{S}$*.*

(iv) *For every* $a \in M\mathcal{U}^*(S) \subset M\mathcal{U}^*(M\mathcal{U})$ *we have*

$$S_\omega \circ a = \sum_{(\omega_1,\omega_2)=\omega} S_{\omega_1}(a)S_{\omega_2}.$$

(v) $M\mathcal{U}^*(M\mathcal{U}) = M\mathcal{U}^*(S) \otimes^{\mathrm{grad}} \mathcal{S}$*. (Here* $\otimes^{\mathrm{grad}}$ *is a completed graded tensor product, defined in* III.1.23.*)*

Proof. (i) This is proved in 2.17(iii).

(ii) See 2.19(i).

(iii) Similarly to 2.19(i), we have

$$\begin{aligned}(S_{\omega'} \circ S_{\omega''})(U) &= S_{\omega'}(C_{\omega''}U)\\ &= \sum_{(\omega_1,\omega_2)=\omega'} S_{\omega_1}(C_{\omega''})S_{\omega_2}(U) = \sum S_{\omega_1}(C_{\omega''})C_{\omega_2}U.\end{aligned}$$

So, it remains to prove that $S_{\omega_1}(C_{\omega''})$ is an integral linear combination of classes C_ω. Because of 2.2(iv) and 2.3, it suffices to prove that $S_{\omega_1}(T_{\omega''})$ is an integral linear combination of elements T_ω. But this follows from (i) and (ii).

(iv) We have $(S_\omega \circ a)(x) = S_\omega(ax) = \sum_{(\omega_1,\omega_2)=\omega} S_{\omega_1}(a)S_{\omega_2}(x)$.

(v) By 1.9(vii), the AHSS for $M\mathcal{U}_*(M\mathcal{U})$ collapses. Hence, by II.3.45,

$$M\mathcal{U}^*(M\mathcal{U}) = \mathrm{Hom}_{\pi_*(M\mathcal{U})}(M\mathcal{U}_*(M\mathcal{U}), \pi_*(M\mathcal{U})).$$

This implies that one can also consider homogeneous *infinite* sums (series) $\Sigma a_\omega S_\omega$. In particular, $M\mathcal{U}^*(S) \otimes^{\mathrm{grad}} \mathcal{S} \subset M\mathcal{U}^*(M\mathcal{U})$.

By 1.2, the AHSS for $M\mathcal{U}^*(M\mathcal{U})$ collapses. The morphism $u : M\mathcal{U} \to H\mathbb{Z}$ maps S_ω to s_ω. Since the elements s_ω generate the abelian group $H^*(M\mathcal{U})$, the elements S_ω generate the $M\mathcal{U}^*(S)$-module E_{**}^∞ of this spectral sequence. By III.5.7(iii), $M\mathcal{U}^*(M\mathcal{U})$ does not contain phantoms. Thus, every element of $M\mathcal{U}^*(M\mathcal{U})$ can be represented as a series $\sum a_\omega S_\omega$. □

3.2. Lemma. *Let* Λ *be a subring of* $\mathbb{Q}$*, and let* E*,* F *be two spectra of finite* Λ*-type. Suppose that* $H_*(E)$ *and* $\pi_*(F)$ *are torsion free abelian groups. Then the homomorphism*

(3.3) $$F^*(E) \to \mathrm{Hom}^*(\pi_*(E), \pi_*(F))$$

is monic. In particular, the homomorphism $\mathcal{MU}^*_\Lambda(\mathcal{MU}_\Lambda) \to \operatorname{End}\pi_*(\mathcal{MU}_\Lambda)$ is monic.

Proof. Let $j_X : X \to X[0]$ be the $\mathbb{Q}$-localization. By II.7.12(ii), the AHSS for $F^*(E)$ is trivial, and so $F^*(E)$ is a torsion free abelian group. So, $(j_F)_* : F^*(E) \to F[0]^*(E)$ is monic. Furthermore, by II.5.8(iii), $j_E^* : F[0]^*(E[0]) \to F[0]^*(E)$ is an isomorphism. Thus, the homomorphism

$$h := j_E^*(j_F)_* : F^*(E) \to F[0]^*(E[0])$$

is monic. Now, the diagram

$$\begin{array}{ccc} F^*(E) & \longrightarrow & \operatorname{Hom}^*(\pi_*(E), \pi_*(F)) \\ h\downarrow & & q\downarrow \\ F[0]^*(E[0]) & \longrightarrow & \operatorname{Hom}^*(\pi_*(E)\otimes\mathbb{Q}, \pi_*(F)\otimes\mathbb{Q}), \end{array}$$

where $q(f) = f \otimes 1_{\mathbb{Q}}$, commutes since $h(f) = f \wedge 1_{M(\mathbb{Q})}$. Furthermore, by II.7.11(ii,iii), the bottom map is an isomorphism. Since h is monic, the top homomorphism is monic as well. □

3.4. Lemma. *Let E, F be two spectra as in 3.2.*

(i) *Let $f : E \to F$ be a morphism such that the homomorphism $f_* : \pi_i(E) \to \pi_i(F)$ is zero for every $i \le n$. Then for every CW-space (connected spectrum) X the homomorphism $f_* : E_i(X) \to F_i(X)$ is zero for every $i \le n$.*

(ii) *Let $f, g : E \to F$ be two morphisms such that $f_* = g_* : \pi_i(E) \to \pi_i(F)$ for $i \le n$. Then $f_* = g_* : E_i(X) \to F_i(X)$ for every CW-space (connected spectrum) X and for every $i \le n$.*

In particular, this holds if $E = F = MU_\Lambda$ for some subring Λ of $\mathbb{Q}$.

Proof. (i) Let $F = F_{(n)}$ be the Postnikov n-stage of F, and let $\tau = \tau_n : F \to G$ be the canonical morphism as in II.4.12. By 3.2, the morphism $E \xrightarrow{f} F \xrightarrow{\tau} G$ is trivial. Thus, the composition $E_i(X) \to F_i(X) \to G_i(X)$ is trivial for every i. But, by II.4.5(ii), $\tau_* : F_i(X) \to G_i(X)$ is an isomorphism for every $i \le n$.

(ii) This follows from (i), if we consider a morphism $\varphi : E \to F$ with $\varphi_* = f_* - g_*$. □

Now we want to describe the action of $\mathcal{MU}^*(\mathcal{MU})$ on the coefficient ring $\pi_*(\mathcal{MU})$. We have $c_{(n)}(CP^n) = -(n+1)$, see e.g. Stong [3], and therefore

$$\pi_*(\mathcal{MU}) \otimes \mathbb{Q} = \mathbb{Q}[[CP^1], \dots, [CP^n], \dots].$$

Since $\pi_*(\mathcal{MU})$ is torsion free, it suffices to compute $S_\omega[CP^n]$, $|\omega| \le n$. By the way, $S_\omega(a) = s_\omega(a) \in \mathbb{Z}$ for every $a \in \pi_{2|\omega|}(\mathcal{MU})$.

3.5. Lemma. *$C_\omega(-\eta) = b_\omega T^{|\omega|}$ for some $b_\omega \in \mathbb{Z}$.*

Proof. We prove this by induction on $l(\omega)$. If $l(\omega) = 1$, i.e., $\omega = (k)$, then $0 = C_{(k)}(\eta) + C_{(k)}(-\eta)$, i.e., $C_{(k)}(-\eta) = -T^k$. Assume that the lemma holds for every ω with $l(\omega) < n$. Given any ω with $l(\omega) = n$, by 2.17(i) we have

$$C_\omega(-\eta) + \sum_{(\omega_k, i_k)=\omega} C_{\omega_k}(-\eta) C_{(i_k)}(\eta) = C_\omega(-\eta \oplus \eta) = 0.$$

Since $C_{\omega_k}(-\eta) = b_{\omega_k} T^{|\omega_k|}$ and $C_{(i_k)}(\eta) = T^{i_k}$, the lemma is proved. □

Let $\{CP^d\} \in MU_{2d}(CP^\infty)$ be the bordism class of $j_d : CP^d \subset CP^\infty$.

3.6. Lemma. $T \cap \{CP^n\} = \{CP^{n-1}\}$.

Proof. Remark 1. Consider the map

$$a' : (CP^n)^+ \xrightarrow{\varepsilon} CP^n \xrightarrow{j_n} CP^\infty \xrightarrow{\mathfrak{h}} MU_1$$

where $\varepsilon(x) = x$ for every $x \in CP^n$. Then a' is homotopic to a map $a : (CP^n)^+ \to MU_1$ such that a is transversal to γ^1 and $a^{-1}(BU_1)$ is the subspace CP^{n-1} of $(CP^n)^+$. This follows from 1.28.

Remark 2. Let $\nu = \nu^N$ be a complex normal bundle of an embedding $i : CP^n \to \mathbb{R}^{2N+2n} \subset S^{2N+2n}$. Note that, by 1.25, $\nu \oplus \eta_{n-1}$ is a normal bundle of the embedding $CP^{n-1} \subset CP^n \xrightarrow{i} S^{2N+2n}$. Let $c : S^{2N+2n} \to T\nu$ be the Browder–Novikov map. We leave it to the reader to check that c is transversal to $\nu \oplus \eta$ and $c^{-1}(CP^{n-1}) = CP^{n-1}$, where CP^{n-1} at the left is $cil_{n-1}(CP^{n-1})$ and CP^{n-1} at the right is $il_{n-1}(CP^{n-1})$.

Let $[CP^n]_{MU}$ be the bordism class of $(CP^n, 1_{CP^n})$, and let $[[CP^{n-1}]] \in MU_{2n-2}(CP^n)$ be the bordism class of (CP^{n-1}, l_{n-1}). It suffices to prove that $(j_n^* T) \cap [CP^n]_{MU} = [[CP^{n-1}]]$. Indeed, then

$$\begin{aligned}\{CP^{n-1}\} &= (j_{n-1})_*[[CP^{n-1}]] = (j_{n-1})_*((j_n^* T) \cap [CP^n]_{MU}) \\ &= T \cap (j_n)_*[CP^n]_{MU} = T \cap \{CP^n\}.\end{aligned}$$

According to the Pontrjagin–Thom Theorem IV.7.27, $[CP^n]_{MU}$ is given by the composition

$$f : S^{2N+2n} \xrightarrow{c} T\nu \xrightarrow{\Delta^{2N}} T\nu \wedge (CP^n)^+ \xrightarrow{Tk \wedge 1} MU_N \wedge (CP^n)^+$$

where c is the Browder–Novikov map in Remark 2, $k : \nu \to \gamma^N$ is the classifying morphism for ν, and Δ^{2N} is a map as in IV.5.36. Furthermore, the diagonal $d : CP^n \to CP^n \times CP^n$ yields the map

$$d^+ : (CP^n)^+ \to (CP^n)^+ \wedge (CP^n)^+,$$

and the map $\mu^{\mathcal{U}}_{N,1} : BU_N \times BU_1 \to BU_{N+1}$ (described before IV.(4.26)) yields a map

$$v := T\mathfrak{I}_{\mu_{N,1}, \gamma^{N+1}} : MU_N \wedge MU_1 \to MU_{N+1}.$$

Now, the element $(j_n^*T)\cap[CP^n]_{MU}$ is given by the map

$$h': S^{2N+2n} \xrightarrow{f} MU_N\wedge(CP^n)^+ \xrightarrow{1\wedge d^+} MU_N\wedge(CP^n)^+\wedge(CP^n)^+ \\ \xrightarrow{1\wedge a'\wedge 1} MU_N\wedge MU_1\wedge(CP^n)^+ \xrightarrow{v\wedge 1} MU_{N+1}\wedge(CP^n)^+,$$

where the map a' is described in Remark 1. So, it suffices to prove that h' is homotopic to a map

$$h: S^{2N+2n}\to MU_{N+1}\wedge(CP^n)^+ = T(\gamma^{N+1}\times\theta^0)$$

transversal to $\gamma^{N+1}\times\theta^0$ and such that $h^{-1}(BU_{N+1}\times CP^n)=il_{n-1}(CP^{n-1})$, where $i: CP^n\to S^{2N+2n}$ is the embedding from Remark 2.

Step 1. $v\wedge 1$ is transversal to $\gamma^{N+1}\times\theta^0$,

$$(v\wedge 1)^{-1}(BU_{N+1}\times CP^n)=BU_N\times BU_1\times CP^n,$$

and $(v\wedge 1)^*(\gamma^{N+1}\times\theta^0)=\gamma^N\times\gamma^1\times\theta^0$.

Step 2. By Remark 1, the map $1\wedge a'\wedge 1$ is homotopic to a map

$$1\wedge a\wedge 1: MU_N\wedge(CP^n)^+\wedge(CP^n)^+\to MU_N\wedge MU_1\wedge(CP^n)^+$$

such that $1\wedge a\wedge 1$ is transversal to $\gamma^N\times\gamma^1\times\theta^0$,

$$(1\wedge a\wedge 1)^{-1}(BU_N\times BU_1\times CP^n)=BU_N\times CP^{n-1}\times CP^n,$$

and $(1\wedge a\wedge 1)^*(\gamma^N\times\gamma^1\times\theta^0)=\gamma^N\times\eta_{n-1}\times\theta^0$.

Step 3. We have $d^{-1}(CP^{n-1}\times CP^n)=CP^{n-1}$, and d is transversal to the normal bundle of the inclusion $CP^{n-1}\times CP^n\subset CP^n\times CP^n$. Hence, the map $1\wedge d^+$ is transversal to $\gamma^N\times\eta_{n-1}\times\theta^0$,

$$(1\wedge d^+)^{-1}(BU_N\times CP^{n-1}\times CP^n)=BU_N\times CP^{n-1},$$

and, obviously, $(1\wedge d^+)^*(\gamma^N\times\eta_{n-1}\times\theta^0)=\gamma^N\times\eta_{n-1}$.

Step 4. Clearly, f is transversal to $\gamma^N\times\eta_{n-1}$, and, by Remark 2, $f^{-1}(BU_N\times CP^{n-1})=il_{n-1}CP^{n-1}$.

Now, set $h:=(v\wedge 1)(1\wedge a\wedge 1)(1\wedge d^+)f$. □

3.7. Lemma. *$T^k\cap\{CP^n\}=\{CP^{n-k}\}$ and $\langle T^k,\{CP^n\}\rangle=[CP^{n-k}]$.*

Proof. The first equality follows from 3.6 inductively. The second equality holds because $\langle T^k,\{CP^n\}\rangle$ is the image of the element $T^k\cap\{CP^n\}$ under the map $CP^n\to\text{pt}$. □

3.8. Theorem. *For every ω and n, we have $S_\omega[CP^n]=\lambda_\omega[CP^{n-|\omega|}]$ for some $\lambda_\omega\in\mathbb{Z}$.*

Proof. By 1.24, $\tau(CP^n)\oplus\theta^1=(n+1)\eta_n$. Let ν be a normal bundle (stable or not) of CP^n. By 2.22,

$$S_\omega[CP^n] = \langle C_\omega(\nu), [CP^n]_{MU}\rangle = \langle C_\omega(-\eta_n - \cdots - \eta_n), [CP^n]_{MU}\rangle.$$

By 3.5, $C_\omega(-\eta_n - \cdots - \eta_n) = \lambda_\omega T^{|\omega|}$ for some $\lambda_\omega \in \mathbb{Z}$. Now the theorem follows from 3.7. □

We need some lemmas about binomial coefficients. Given an integer a and a prime p, let $\nu_p(a)$ be the exponent of p in the primary decomposition of a, i.e., $a = p^{\nu_p(a)}b$ with $(b,p) = 1$. The invariant ν_p makes sense for $a \in \mathbb{Q}$ also: we can write a as m/n with $m, n \in \mathbb{Z}$ and set $\nu_p(a) := \nu_p(m) - \nu_p(n)$.

3.9. Lemma. *If $m > n$, then* $\nu_p\binom{p^m}{p^n} = m - n.$

Proof. This follows from the formula

$$\nu_p(n!) = \sum_{k>0}\left[\frac{n}{p^k}\right],$$

where $[m]$ means the "integer part" of m (i.e., $[-]$ is the entire function). □

3.10. Lemma. *If $r < p^n \le p^m$, then* $\nu_p\binom{p^m}{r} > m - n.$

Proof. We have

$$\binom{p^m}{r} = \frac{p^m}{r}\binom{p^m - 1}{r-1}.$$

But $\nu_p\left(\frac{p^m}{r}\right) > m - n$, while $\binom{p^m-1}{r-1} \in \mathbb{Z}$. □

3.11. Notation. Fix a prime p. Given natural numbers m, n with $m > n$, let $\omega(m,n)$ be the partition $(p^{m-n} - 1, \dots, p^{m-n} - 1)$ with $l(\omega) = p^n$. So, $|\omega| = p^m - p^n$.

3.12. Proposition. *Let $\omega = (p^{k_1} - 1, \dots, p^{k_r} - 1)$ with $k_i \ge m - n$ and $|\omega| = p^m - p^n$. Then $S_\omega[CP^{p^m-1}] = \alpha p^{m-n}[CP^{p^n-1}]$, where $\alpha \not\equiv 0 \bmod p$ for $\omega = \omega(m,n)$ and $\alpha \equiv 0 \bmod p$ for all other ω.*

Proof. Let τ (resp. ν) denote the tangent (resp. the normal) bundle of CP^{p^m-1}. Firstly, let $\omega = \omega(m,n)$. Set

$$\omega_r(m,n) = (p^{m-n} - 1, \dots, p^{m-n} - 1), \quad l(\omega_r(m,n)) = r,$$

and set $C^r = C_{\omega_r(m,n)}$. Consider the characteristic class $C := 1 + C^1 + \cdots + C^r + \cdots$. Then for every pair of complex vector bundles ξ, ξ' over the same base we have $C^k(\xi \oplus \xi') = \sum_i C^i(\xi)C^{k-i}(\xi')$, i.e., $C(\xi \oplus \xi') = C(\xi)C(\xi')$.

By 2.17(i), $C(\eta) = 1 + T^{p^{m-n}-1}$. Hence, $C(\tau) = (1 + T^{p^{m-n}-1})^{p^m}$, and therefore

$$C^i(\tau) = \binom{p^m}{i} T^{i(p^{m-n}-1)}.$$

Moreover, $C^k(\nu) + \sum_{i=1}^{k-1} C^i(\nu)C^{k-i}(\tau) + C^k(\tau) = 0$. Hence, by induction, we conclude that $C^k(\nu) = a_k T^{k(p^{m-n}-1)}$ for some $a_k \in \mathbb{Z}$. In particular, for $k = p^n$ we have

$$C^{p^n}(\nu) + \sum_{i=1}^{p^n-1} a_i \binom{p^m}{i} T^{p^m-p^n} + \binom{p^m}{p^n} T^{p^m-p^n} = 0.$$

By 3.9 and 3.10, $\nu_p\binom{p^m}{i} > m-n$ for $1 \le i \le p^n - 1$ and $\nu_p\binom{p^m}{p^n} = m-n$. So, $\nu_p(a_{p^n}) = m - n$. Thus, by 2.22 and 3.7,

$$\begin{aligned} S_{\omega(m,n)}[CP^{p^m-1}] &= \langle C^{p^n}(\nu), \{CP^{p^m-1}\}\rangle = \langle a_{p^n} T^{p^m-p^n}, \{CP^{p^m-1}\}\rangle \\ &= a_{p^n}[CP^{p^n-1}], \end{aligned}$$

where $\nu_p(a_{p^n}) = m - n$.

Now, let $\omega \ne \omega(m,n)$. Since $l(\omega) < p^n$, $\nu_p\binom{p^m}{s} > m-n$ for $s \le l(\omega)$. By 3.5,

$$C_\omega(\nu) = C_\omega(-\eta - \cdots - \eta) = \sum_{s=1}^{l(\omega)} b_s \binom{p^m}{s} T^{p^m-p^n}$$

for some $b_s \in \mathbb{Z}$, i.e., $C_\omega(\nu) = xp^{m-n+1}T^{p^m-p^n}$ for some $x \in \mathbb{Z}$. Thus, $S_\omega[CP^{p^m-1}] = xp^{m-n+1}[CP^{p^n-1}]$, as asserted. □

Let $\{x_n\}$ be a family of free polynomial generators of $\pi_*(MU)$, cf. 1.9(i). By 1.9(iv), we can assume that $s_{(p^n-1)}(x_{p^n-1}) = p$. Since $s_{(p^n-1)}[CP^{p^n-1}] = p^n$, we conclude that

$$[CP^{p^n-1}] \equiv p^{n-1}x_{p^n-1} \bmod \mathrm{Dec}\,(\pi_*(MU)).$$

Based on 3.12, this implies the following fact.

3.13. Corollary. *Let ω be as in* 3.12. *Then*

$$S_\omega(x_{p^m-1}) \equiv \alpha x_{p^n-1} \bmod \mathrm{Dec}\,(\pi_*(MU)),$$

where $\alpha \not\equiv 0 \bmod p$ for $\omega = \omega(m,n)$ and $\alpha \equiv 0 \bmod p$ for $\omega \ne \omega(m,n)$. □

Now, let $Z = \varphi(T) = T + \sum \varphi_i T^{i+1}$ be any $\mathbb{C}$-orientation of MU. By 2.5, it yields a multiplicative operation $v = u^{MU,Z} : MU \to MU$. By (2.25),

(3.14) $$v(x) = \sum_{|\omega|\geq 0} \varphi_\omega S_\omega(x) = x + \sum_{|\omega|>0} \varphi_\omega S_\omega(x)$$

for every Y and every $x \in MU^*(Y)$.

3.15 Theorem. *Choose a system $\{x_i\}$ of polynomial generators of $\pi_*(MU) = \mathbb{Z}[x_i] = MU^*(S)$. Fix any n and set $x = x_n$, $\lambda = \lambda_n$ (see (1.8)). There exists an operation $\Phi : MU[\lambda^{-1}] \to MU[\lambda^{-1}]$ with the following properties:*

(i) *For every spectrum X and every $a \in MU^*(X)[\lambda^{-1}]$, $\Phi(a) = a + xb$ for some $b \in MU^*(X)[\lambda^{-1}]$;*

(ii) *$\Phi(x) = 0$;*

(iii) *$\Phi^2 = \Phi$, i.e., Φ is an idempotent;*

(iv) *$\Phi(x_i) = x_i$ for $i < n$, and $\Phi(x_i) \equiv x_i \mod \mathrm{Dec}$ for $i > n$;*

(v) *$\mathrm{Im}(\Phi : MU^*(X)[\lambda^{-1}] \to MU^*(X)[\lambda^{-1}])$ is a cohomology theory with the coefficient ring $\mathbb{Z}[\lambda^{-1}][x_j | j \neq n, \dim x_j = 2j]$, and this theory is a direct summand of $MU^*(X)[\lambda^{-1}]$.*

Proof. Consider a formal power series $\varphi(z) = z + x\sum_{k\geq n} d_k z^{k+1}$ over the ring $\pi_*(MU)[\lambda^{-1}]$ with $d_n = -\lambda - 1$ and $d_k \in \pi_{2k-2}(MU)[\lambda^{-1}]$. Take the $\mathbb{C}$-orientation $\varphi(T)$ of $MU[\lambda^{-1}]$. It yields a ring morphism $\mu : MU \to MU[\lambda^{-1}]$ and hence a ring operation $\Phi = \mu[\lambda^{-1}] : MU[\lambda^{-1}] \to MU[\lambda^{-1}]$. By (3.14),

$$\Phi(a) = a + \sum_{|\omega|>0} x^{l(\omega)} d_\omega S_\omega(a)$$

for every $a \in MU^*(X)[\lambda^{-1}]$, and so (i) is proved.

Observe that $d_\omega = 0$ for $\omega = (i_1, \dots, i_k)$ with $i_1 \leq \dots \leq i_k < n$. Furthermore, $S_\omega(x) = 0$ for $|\omega| > n$. Thus,

$$\Phi(x) = x + \sum x^{l(\omega)} d_\omega S_\omega(x) = x - \lambda^{-1} x S_{(n)}(x) = 0.$$

Now,

$$\begin{aligned}\Phi(\Phi(a)) &= \Phi(a + \sum_{|\omega|>0} x^{l(\omega)} d_\omega S_\omega(a)) \\ &= \Phi(a) + \sum \Phi(x^{l(\omega)}) \Phi(d_\omega) \Phi(S_\omega(a)) = \Phi(a),\end{aligned}$$

i.e., $\Phi^2 = \Phi$. Thus, we have proved that Φ satisfies (ii) and (iii). Furthermore, by (i), $\Phi(x_i) \equiv x_i \mod \mathrm{Dec}$. Moreover, if $i < n$ then b in (i) must be 0, and so $\Phi(x_i) = x_i$ for $i < n$. Thus, (iv) is proved.

Now we prove (v). Since $\Phi : MU^*(X)[\lambda^{-1}] \to MU^*(X)[\lambda^{-1}]$ is an idempotent, $\mathrm{Im}\,\Phi$ is a direct summand of $MU^*(X)[\lambda^{-1}]$. So, $\mathrm{Im}\,\Phi$ is a cohomology theory. The claim about its coefficient ring follows from (iii) and (iv). □

3.16. Corollary. *If $n \neq p^k - 1$ for any p (i.e., $\lambda_n = 1$), then there exists a multiplicative idempotent $\Phi : MU \to MU$ with $\Phi(x_n) = 0$, $\Phi(x_i) = x_i$ for $i < n$ and $\Phi(x_i) \equiv x_i \bmod \mathrm{Dec}$ for $i > n$. Moreover, for every $a \in MU^*(X)$ we have $\Phi(a) = a + bx_n$, $b \in MU^*(X)$.* □

3.17. Corollary. *Given a prime p, let $n \neq p^k - 1$. Then there exists a multiplicative idempotent $\Phi : MU[p] \to MU[p]$ with $\Phi(x_n) = 0$, $\Phi(x_i) = x_i$ for $i < n$ and $\Phi(x_i) \equiv x_i \bmod \mathrm{Dec}$ for $i > n$. Moreover, for every $a \in MU^*(X)[p]$ we have $\Phi(a) = a + bx_n$, $b \in MU^*(X)[p]$.* □

3.18. Theorem. *Given a prime p, there exists a multiplicative idempotent $\Phi : MU[p] \to MU[p]$ such that $\Phi(x_i) \equiv x_i \bmod \mathrm{Dec}$ for $i = p^k - 1$, $k = 1, 2, \dots$, and $\Phi(x_i) = 0$ otherwise.*

Proof. Consider the generators x_i of $\pi_*(MU)$ with $i \neq p^k - 1$, $k = 1, 2, \dots$, and order them with increasing indices: $\{x_{i_1}, \dots, x_{i_k}, \dots\}$. Let

$$\Phi_k : MU[p] \to MU[p]$$

be an idempotent as in 3.17 related to x_{i_k}. Given k, define $\Phi_k^k := \Phi_k$, $\Phi_i^k := \Phi_{i+1}^k \Phi_i \Phi_{i+1}^k$ for $1 \leq i \leq k-1$, and set $\Phi_{[k]} := \Phi_1^k$. It is easy to see that $\Phi_{[k]}(x_{i_r}) = 0$ for $r \leq k$. Furthermore, since $\Phi_k(x_n) = x_n + bx_{i_k}$ with $b \in MU^*(S)[p]$, we conclude that

$$\Phi_{[k]}(x_n) = x_n + \sum_{r \leq k} b_r x_{i_r},\ b_r \in MU^*(S)[p] \text{ for every } n.$$

Clearly, $\Phi_{[k]}$ is a multiplicative operation. We prove that $\Phi_{[k]}\Phi_{[k]} = \Phi_{[k]}$. By 3.2, it suffices to prove that $\Phi_{[k]}\Phi_{[k]}(x_n) = \Phi_{[k]}(x_n)$ for every n. But

$$\Phi_{[k]}\Phi_{[k]}(x_n) = \Phi_{[k]}(x_n) + \Phi_{[k]}\left(\sum_{r \leq k} b_r x_{i_r}\right) = \Phi_{[k]}(x_n),$$

since $\Phi_{[k]}(x_{i_r}) = 0$. So, $\Phi_{[k]}$ is a multiplicative idempotent, and it is clear that

$$\Phi_{[k]}(x_i) = \begin{cases} \tilde{x}_i & \text{if } i = p^s - 1 \text{ for some } s, \\ 0 & \text{if } i \neq p^s - 1 \text{ and } i \leq i_k, \end{cases}$$

where $\tilde{x}_i \equiv x_i \bmod \mathrm{Dec}$. Moreover, $\Phi_{[k]}(x_i)$ does not depend on k for $k > i$.

By 3.4, for every $a \in MU_*(X)[p]$ there exists k such that $\Phi_{[k]}(a) = \Phi_{[k+r]}(a)$ for every $r \geq 0$. Hence, one can define

$$\Phi := \lim_{k \to \infty} (\Phi_{[k]}) : MU_*(X)[p] \to MU_*(X)[p].$$

By III.3.23(ii), this morphism of homology theories is induced by a morphism $\Phi : MU[p] \to MU[p]$ of spectra. Clearly,

$$\Phi(x_i) = \begin{cases} \tilde{x}_i & \text{if } i = p^s - 1 \text{ for some } s, \\ 0 & \text{otherwise.} \end{cases}$$

This morphism Φ is uniquely determined because of 3.2, and $\Phi^2 = \Phi$, also by 3.2. Furthermore, Φ is a ring morphism (multiplicative operation). Indeed, one needs to check commutativity (up to homotopy) of the diagram

$$\begin{array}{ccc} \mathcal{MU}[p] \wedge \mathcal{MU}[p] & \xrightarrow{\Phi\wedge\Phi} & \mathcal{MU}[p] \wedge \mathcal{MU}[p] \\ \downarrow & & \downarrow \\ \mathcal{MU}[p] & \xrightarrow{\Phi} & \mathcal{MU}[p] \end{array} .$$

By 3.2, it suffices to prove that the diagram of the homotopy groups commutes. But this holds because it holds for $\Phi_{[k]}$. □

3.19. Theorem. *For every prime p there exists a spectrum $BP = BP(p)$ with the following properties:*

(i) *BP is a commutative ring spectrum, and there are morphisms $\varkappa : BP \to \mathcal{MU}[p]$ and $\rho : \mathcal{MU}[p] \to BP$ with $\rho\varkappa = 1_{BP}$. In particular, BP is a direct summand of $\mathcal{MU}[p]$.*

(ii) $\pi_*(BP) = \mathbb{Z}[p][v_1, \dots, v_n, \dots]$, $\dim v_n = 2(p^n - 1)$.

(iii) $H_*(BP) = \mathbb{Z}[p][m_1, \dots, m_n, \dots]$, $\dim m_n = 2(p^n - 1)$.

(iv) $H^*(BP; \mathbb{Z}/p) = \mathscr{A}_p/(Q_0)$.

(v) *Let $h : \pi_*(BP) \to H_*(BP)$ be the Hurewicz homomorphism. Then $p|h(x)$ whenever* $\dim x > 0$.

Proof. Let $\Phi : \mathcal{MU}[p]^*(X) \to \mathcal{MU}[p]^*(X)$ be an idempotent as in 3.18. Since $\Phi^2 = \Phi$, $\operatorname{Im}\Phi$ is a direct summand of $\mathcal{MU}[p]^*(X)$. So, $\operatorname{Im}\Phi$ is a cohomology theory, which we denote by $BP^*(X)$. Let BP be the spectrum of this cohomology theory. Now we prove that it has properties (i)–(v).

(i) Since the cohomology theory $BP^*(X)$ is a direct summand of the cohomology theory $\mathcal{MU}^*[p](X)$, BP is a direct summand of $\mathcal{MU}[p]$. Hence, we have morphisms $\varkappa : BP \to \mathcal{MU}[p]$ and $\rho : \mathcal{MU}[p] \to BP$ with $\rho\varkappa = 1_{BP}$ and $\varkappa\rho = \Phi$.

Let $\mu : \mathcal{MU}[p] \wedge \mathcal{MU}[p] \to \mathcal{MU}[p]$ be the multiplication in $\mathcal{MU}[p]$. We define

$$\mu' : BP \wedge BP \xrightarrow{\varkappa\wedge\varkappa} \mathcal{MU}[p] \wedge \mathcal{MU}[p] \xrightarrow{\mu} \mathcal{MU}[p] \xrightarrow{\rho} BP.$$

We prove that ρ is a multiplicative morphism, i.e., that the diagram

$$\begin{array}{ccc} \mathcal{MU}[p] \wedge \mathcal{MU}[p] & \xrightarrow{\rho\wedge\rho} & BP \wedge BP \\ \mu\downarrow & & \downarrow\mu' \\ \mathcal{MU}[p] & \xrightarrow{\rho} & BP \end{array}$$

commutes (up to homotopy). Decoding the definition of μ', we see that it suffices to prove that the diagram

$$\begin{array}{ccccccc}
MU[p] \wedge MU[p] & \xrightarrow{\rho\wedge\rho} & BP \wedge BP & \xrightarrow{\varkappa\wedge\varkappa} & MU[p] \wedge MU[p] & \xrightarrow{\mu} & MU[p] \\
\Big\downarrow{\scriptstyle\mu} & & & & & & \Big\downarrow{\scriptstyle\rho} \\
MU[p] & = & MU[p] & = & MU[p] & \xrightarrow{\rho} & BP
\end{array}$$

commutes. Since $\rho\Phi = \rho(\varkappa\rho) = (\rho\varkappa)\rho = \rho$, the commutativity of the above diagram follows from the commutativity of the diagram

$$\begin{array}{ccccccc}
MU[p] \wedge MU[p] & \xrightarrow{\rho\wedge\rho} & BP \wedge BP & \xrightarrow{\varkappa\wedge\varkappa} & MU[p] \wedge MU[p] & \xrightarrow{\mu} & MU[p] \\
\Big\downarrow{\scriptstyle\mu} & & & & \Big\downarrow{\scriptstyle\mu} & & \Big\downarrow{\scriptstyle\rho} \\
MU[p] & \xrightarrow{\Phi} & MU[p] & = & MU[p] & \xrightarrow{\rho} & BP
\end{array}$$

But this holds because Φ is a ring morphism.

Similarly, one can prove that $\varkappa$ is a multiplicative morphism, i.e., that $\mu(\varkappa \wedge \varkappa) = \varkappa\mu'$.

Now we prove that μ' is associative. Since BP is a direct summand of $MU[p]$, we conclude that the groups $\pi_*(BP)$ and $H_*(BP; \mathbb{Z}[p])$ are torsion free. So, by 3.2, it suffices to prove that the diagram

$$\begin{array}{ccc}
\pi_*(BP \wedge BP \wedge BP) & \xrightarrow{(\mu'\wedge 1)_*} & \pi_*(BP \wedge BP) \\
\Big\downarrow{\scriptstyle(1\wedge\mu')_*} & & \Big\downarrow{\scriptstyle\mu'_*} \\
\pi_*(BP \wedge BP) & \xrightarrow{\mu'_*} & \pi_*(BP)
\end{array}$$

commutes. Consider the following diagram:

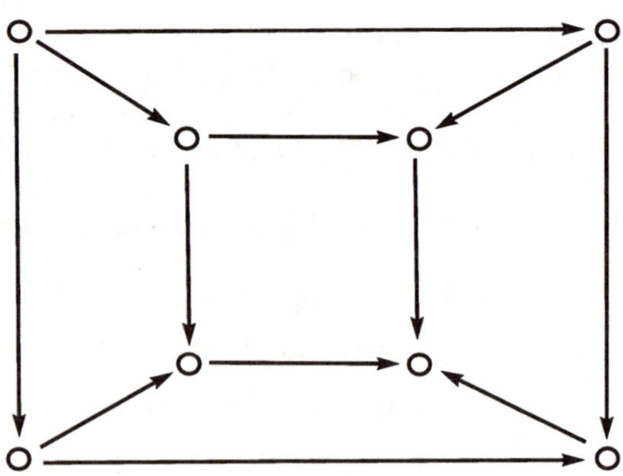

Here the inner rectangle is just the diagram above, the outer rectangle is the similar diagram for $MU[p]$ instead of BP, and the skew arrows are induced by the morphism ρ. The outer rectangle is commutative because $MU[p]$ is a ring spectrum, and all trapezoids are commutative since ρ is a multiplicative morphism. Hence, the inner rectangle is commutative because all the skew arrows are epimorphisms. Thus, we have proved the associativity of μ'.

The commutativity of μ' can be proved similarly.

We define $\iota' : S \xrightarrow{\iota} MU[p] \xrightarrow{\rho} BP$ where ι is the unit of $MU[p]$. Then (BP, μ', ι') is a commutative ring spectrum, and both $\rho, \varkappa$ are ring morphisms.

(ii) Set $v_i := \rho_*(x_{p^i-1})$, where $\rho_* : \pi_*(MU[p]) \to \pi_*(BP)$ is the induced homomorphism. Then $\pi_*(BP) = \mathbb{Z}[p][v_1, \dots, v_n, \dots], \dim v_n = 2(p^n - 1)$. Indeed, the elements v_i generate the ring $\pi_*(BP)$ because ρ_* is epic, and they are algebraically independent because $\varkappa(v_i) \equiv x_{p^i-1} \bmod \mathrm{Dec}$.

(iii) Let $\{y_i\}$ be a family of free polynomial generators of $H_*(MU)$ as in 1.9(ii). Define a ring homomorphism

$$a : \mathbb{Z}[p][m_1, \dots, m_n, \dots] \to H_*(BP), \dim m_i = 2(p^i - 1)$$

by setting $a(m_i) = \rho_H(y_{p^i-1})$, where $\rho_H = H_*(\rho) : H_*(MU[p]) \to H_*(BP)$. We prove that a is an isomorphism.

Firstly, we prove that a is epic. Since the family $\{\rho_H(y_i)\}$ generates $H_*(BP)$, it suffices to prove that $\rho_H(y_j) \in \mathrm{Dec}\,(H_*(BP))$ for $j \neq p^s - 1$. By 2.21, $h(x_i) \equiv \pm\lambda_i y_i \bmod \mathrm{Dec}\,(H_*(MU))$. If $j \neq p^s - 1$ then

$$\Phi(x_j) \equiv 0 \bmod \mathrm{Dec}\,(\pi_*(MU[p])).$$

Note that $\rho = \rho\varkappa\rho = \rho\Phi$. Hence, $\rho_*(x_j) \in \mathrm{Dec}\,(\pi_*(BP))$, and hence $\lambda_j \rho_H y_j \in \mathrm{Dec}\,(H_*(BP))$, and so $\rho_H(y_j) \in \mathrm{Dec}\,(H_*(BP))$ because $(\lambda_j, p) = 1$.

We prove that a is monic. The elements $\Phi(x_{p^k-1})$ are algebraically independent in $\pi_*(MU) \otimes \mathbb{Q}$. By II.7.11(i), the Hurewicz homomorphism $h : \pi_*(MU) \otimes \mathbb{Q} \to H_*(MU) \otimes \mathbb{Q}$ is an isomorphism. Since $h(x_i) \equiv \lambda_i y_i \bmod \mathrm{Dec}$, the elements $\Phi(y_{p^k-1})$ are algebraically independent in $H_*(MU) \otimes \mathbb{Q}$. Hence, these elements are algebraically independent in $H_*(MU[p])$. Since $\Phi = \varkappa\rho$, the elements m_i are algebraically independent in $H_*(BP)$. Thus, a is a monomorphism.

(iv) Choose $u \in H^0(BP; \mathbb{Z}/p) = \mathbb{Z}/p, u \neq 0$. Since $H^*(BP; \mathbb{Z}/p)$ is a direct summand of the free $\mathscr{A}_p/(Q_0)$-module $H^*(MU; \mathbb{Z}/p)$, we have an isomorphism $\mathscr{A}_p u \cong \mathscr{A}_p/(Q_0)$. Computing the dimensions of the $\mathbb{Z}/p$-vector spaces $\mathscr{A}_p/(Q_0)$ and $H^*(BP; \mathbb{Z}/p)$, we conclude that the inclusion $\mathscr{A}_p/(Q_0) \cong \mathscr{A}_p u \subset H^*(BP; \mathbb{Z}/p)$ is an isomorphism.

(v) Choose a system $\{x_k\}$ as in 1.9(v) and construct BP with respect to this system. Then $p | h(v_k)$ for every $k = 1, 2, \dots$. But every $x \in \pi_*(BP), \dim x > 0$, is a polynomial in the v_i. □

Below (in 3.22) we prove that the conditions from 3.19 determine BP uniquely up to equivalence.

3.20. Definition. The spectrum BP is called the *Brown–Peterson spectrum* (with respect to a given p).

3.21. Theorem (Boardman [2]). *Let F be a p-local spectrum of finite $\mathbb{Z}[p]$-type and such that $\pi_*(F)$ and $H_*(F)$ are torsion free groups. Then F is homotopy equivalent to a wedge of suspensions of BP, $F \simeq \vee_{a \in A} \Sigma^a BP$.*

Proof. We can assume that F is connected and that $\pi_0(F) \neq 0$. Let $w \in H^0(F; \pi_0(F))$ be the fundamental class, i.e., the homotopy class of the morphism $\tau_0 : F \to F_{(0)} = H(\pi_0(F))$. Consider any projection $\varepsilon : \pi_0(F) = \mathbb{Z}[p] \oplus \cdots \oplus \mathbb{Z}[p] \to \mathbb{Z}[p]$ and put $x = \varepsilon^{-1}(1) \in \pi_0(F)$. Let $u \in H^0(BP; \mathbb{Z}[p]) = \mathbb{Z}[p]$ be an invertible element of $\mathbb{Z}[p]$. Consider the AHSS $E_r^{**} \Rightarrow F^*(BP), E_2^{**} = H^*(BP; \pi_*(F))$. Note that E_2^{**} is torsion free, and so, by II.7.12(ii), all its differentials are trivial. Furthermore, by III.5.6, this spectral sequence converges. Thus, there exists a morphism $\alpha : BP \to F$ such that $\alpha^*(w) = u \otimes x \in E_\infty^{0,0} = E_2^{0,0}$, where

$$\alpha^* : H^*(F; \pi_*(F)) \to H^*(BP; \pi_*(F)) = H^*(BP; \mathbb{Z}[p]) \otimes \pi_*(F).$$

In the commutative diagram

$$\begin{array}{ccc} H^0(F; \pi_0(F)) & \xrightarrow{\alpha^*} & H^0(BP; \pi_0(F)) \\ \varepsilon_* \downarrow & & \downarrow \varepsilon_* \\ H^0(F; \mathbb{Z}[p]) & \xrightarrow{\alpha^*} & H^0(BP; \mathbb{Z}[p]) \end{array}$$

we have $\varepsilon_*(u \otimes x) = u$, and so $u = \alpha^* v$ for some $v \in H^0(F; \mathbb{Z}[p])$ (in fact, for $v = \varepsilon_* w$).

Similarly, there is the AHSS $E_r^{**} \Rightarrow BP^*(F)$, $E_2^{**} = H^*(F; \pi_*(BP))$. It collapses also, and so we have a morphism $\beta : F \to BP$ with $\beta^* x = v$ for some $x \in H^0(BP; \mathbb{Z}[p])$, where $\beta^* : H^*(BP; \mathbb{Z}[p]) \to H^*(F; \mathbb{Z}[p])$.

We have the composition $BP \xrightarrow{\alpha} F \xrightarrow{\beta} BP$, where

$$(\beta\alpha)^*(x) = u \in H^0(BP; \mathbb{Z}[p]).$$

Let $\overline{u} \in H^0(BP; \mathbb{Z}/p) = \mathbb{Z}/p$ be the reduction of u. Since $(\beta\alpha)^* \neq 0$, we conclude that $(\beta\alpha)^*(\overline{u}) = \lambda\overline{u}, \lambda \neq 0 \in \mathbb{Z}/p$, and so $(\beta\alpha)^*(a\overline{u}) = a(\beta\alpha)^*(\overline{u}) = \lambda a\overline{u}$ for every $a \in \mathscr{A}_p$. Hence, $(\beta\alpha)^* : H^*(BP; \mathbb{Z}/p) \to H^*(BP; \mathbb{Z}/p)$ is an isomorphism since $H^*(BP; \mathbb{Z}/p) = \mathscr{A}_p\overline{u}$. So, by II.5.18(ii), $\beta\alpha$ is an equivalence. Thus, BP splits off F, i.e., $F \simeq BP \vee E$, and E satisfies the conditions for F in the theorem. Iterating the above arguments, and using that F has finite $\mathbb{Z}[p]$-type, one can prove that $F \simeq \vee_{a \in A} \Sigma^a BP$, cf. the proof of II.7.16. □

3.22. Corollary. *Let F be a spectrum such that $\pi_*(F) = \mathbb{Z}[p][v_1, \ldots, v_n, \ldots]$, $\dim v_n = 2(p^n - 2)$, and $H_*(BP)$ is torsion free. Then $F \simeq BP$.* □

By 3.21, $\mathcal{MU}[p]$ splits in a wedge of suspensions of BP. This claim can be refined as follows. Set $A := \operatorname{Ker}(\rho_* : \pi_*(\mathcal{MU}[p]) \to \pi_*(BP))$, i.e.,

$$A = \mathbb{Z}[p][x_i | \dim x_i = 2i],$$

where i runs over all natural numbers different from $p^k - 1$, $k = 1, 2, \ldots$.

3.23. Proposition. *There is an isomorphism of homology theories on* $\mathcal{S}$

$$f : A \otimes BP_*(X) \to \mathcal{MU}_*(X) \otimes \mathbb{Z}[p], \quad f(a \otimes x) = a\varkappa(x), a \in A, x \in BP_*(X).$$

Proof. Clearly, f is natural with respect to X. Furthermore, $A \otimes BP_*(X)$ is a homology theory because A is a free abelian group. Finally, f is an isomorphism for $X = S$ and therefore, by II.3.19(iii), for every X. □

The homomorphism $\pi_*(\mathcal{MU}) \to \pi_*(\mathcal{MU}[p]) \xrightarrow{\rho_*} \pi_*(BP)$ turns $\pi_*(BP)$ into a $\pi_*(\mathcal{MU})$-module.

3.24. Corollary. *There is an isomorphism* $\mathcal{MU}_*(X) \otimes_{\pi_*(\mathcal{MU})} \pi_*(BP) \cong BP_*(X)$ *of homology theories.*

Proof. We have (the third isomorphism holds because $A = \operatorname{Ker} \rho_*$)

$$\begin{aligned} \mathcal{MU}_*(X) \otimes_{\pi_*(\mathcal{MU})} \pi_*(BP) &\cong (A \otimes BP_*(X)) \otimes_{\pi_*(\mathcal{MU})} \pi_*(BP) \\ &\cong (A \otimes BP_*(X)) \otimes_{A \otimes \pi_*(BP)} \pi_*(BP) \\ &\cong BP_*(X) \otimes_{\pi_*(BP)} \pi_*(BP) \cong BP_*(X). \end{aligned}$$ □

The unit $\iota : S \to BP$ yields the morphisms

$$\iota_L = 1 \wedge \iota : BP = BP \wedge S \to BP \wedge BP, \ \iota_R = \iota \wedge 1 : BP = S \wedge BP \to BP \wedge BP.$$

Let $u \in H^0(BP; \mathbb{Z}[p])$ be such that $u^*(\iota) = 1 \in H^0(S; \mathbb{Z}[p]) = \mathbb{Z}[p]$.

3.25. Proposition. *Let* $h_H : \pi_*(BP) \to H_*(BP)$ *and* $h_{BP} : \pi_*(BP) \to BP_*(BP)$ *be the Hurewicz homomorphisms.*

(i) *There is an isomorphism of* $\pi_*(BP)$*-modules*

$$BP_*(BP) \cong \pi_*(BP)[y_1, \ldots, y_n, \ldots], \dim y_n = 2(p^n - 1).$$

Furthermore, $\{y_i\}$ *can be chosen so that the elements* $u_*(y_i)$ *generate the ring* $H_*(BP; \mathbb{Z}[p])$.

(ii) *The homomorphism* $(\iota_R)_* : \pi_*(BP) \to BP_*(BP)$ *coincides with the Hurewicz homomorphism* h_{BP}.

(iii) *The homomorphism* $u_*(\iota_L)_* : \pi_n(BP) \to BP_n(BP) \to H_n(BP)$ *is zero for* $n > 0$ *and an isomorphism for* $n = 0$.

(iv) *The homomorphism* $u_*(\iota_R)_* : \pi_*(BP) \to BP_*(BP) \to H_*(BP)$ *coincides with the Hurewicz homomorphism* h_H.

(v) $h_{BP}(v_n) \equiv v_n \bmod (p, v_1, \dots, v_{n-1})$ *in* $BP_*(BP)$.

Proof. (i) The AHSS for $BP_*(BP)$ is trivial for dimensional reasons, and the result follows.

(ii) This is obvious.

(iii) The morphism $BP \wedge S \xrightarrow{\iota_L} BP \wedge BP \xrightarrow{u\wedge 1} H\mathbb{Z} \wedge BP$ coincides with the morphism $BP \wedge S \xrightarrow{u\wedge 1} H\mathbb{Z} \wedge S \xrightarrow{1\wedge\iota} BP \wedge H\mathbb{Z}$. But $\pi_i(H\mathbb{Z} \wedge S) = 0$ for $i > 0$.

(iv) This is obvious.

(v) By (i), we have $h_{BP}(v_n) = ay_n + bv_n + c$, where $a, b \in \mathbb{Z}[p]$, $c \in (v_1, \dots, v_{n-1}) \subset BP_*(BP)$. By (ii)–(iv), $h_H(v_n) = au_*(y_n)$, and so, by 3.19(v), $p|a$. So, $h_{BP}(v_n) = bv_n + c'$ with $c' \in (p, v_1, \dots, v_{n-1}) \subset BP_*(BP)$. Let $\mu : BP \wedge BP \to BP$ be the multiplication on BP, and let $\mu_* : BP_*(BP) \to \pi_*(BP)$ be the induced homomorphism. Since $\mu\iota_R = 1$, we have

$$v_n = \mu_* h_{BP}(v_n) = \mu_*(bv_n + c') = bv_n + \mu_* c',$$

where $\mu_*(c') \in (p, v_1, \dots, v_{n-1}) \subset \pi_*(BP)$. Thus, $b = 1$. □

For future references, we consider the AHSS E^r_{**} for $BP_*(H\mathbb{Z}/p)$, i.e.,

$$(3.26) \qquad E^r_{**} \Rightarrow BP_*(H\mathbb{Z}/p), \quad E^2_{**} = H_*(H\mathbb{Z}/p; \pi_*(BP)).$$

Firstly, let $p > 2$. By II.6.25,

$$H_*(H\mathbb{Z}/p; \mathbb{Z}/p) = \mathscr{A}_p^* = \mathbb{Z}/p[\xi_1, \dots, \xi_n, \dots] \otimes \Lambda(\tau_0, \dots, \tau_m, \dots),$$
$$\dim \xi_i = 2p^i - 2, \dim \tau_i = 2p^i - 1,$$

and so

$$H_*(H\mathbb{Z}/p; \mathbb{Z}[p]) = H_*(H\mathbb{Z}[p]; \mathbb{Z}/p)$$
$$= \mathbb{Z}/p[\xi_1, \dots, \xi_n, \dots] \otimes \Lambda(\tau_1, \dots, \tau_m, \dots).$$

Thus,

$$E^2_{**} = \mathbb{Z}/p[\xi_1, \dots, \xi_n, \dots] \otimes \Lambda(\tau_1, \dots, \tau_m, \dots) \otimes \mathbb{Z}[p][v_1, \dots, v_l, \dots],$$
$$\operatorname{bideg} \xi_i = (2p^i - 2, 0), \operatorname{bideg} \tau_i = (2p^i - 1, 0), \operatorname{bideg} v_i = (0, 2p^i - 2).$$

3.27. Theorem (cf. J. Cohen [1]). *There are elements $\widehat{\xi}_i$ and $\widehat{\tau}_i$ with the following properties:*

(i) $\widehat{\tau}_i \equiv \tau_i \bmod \operatorname{Dec}(E^2_{*,0})$, $\widehat{\xi}_i \equiv \xi_i \bmod \operatorname{Dec}(E^2_{*,0})$;

(ii) *Every element* $\widehat{\xi}^a := \widehat{\xi}_1^{a_1} \cdots \widehat{\xi}_n^{a_n}$, $a_i \geq 0$, *is a permanent cycle in the spectral sequence* (3.26);

(iii) $d_{2p^i-1}\widehat{\tau}_i = \lambda_i v_i$, $0 \neq \lambda_i \in \mathbb{Z}/p$.

Proof. Throughout the proof $\equiv$ means $\equiv \bmod \operatorname{Dec}(E^2_{*,0})$.

It is easy to see that we can assume that $\beta\tau_i \equiv \xi_i$ where

$$\beta : H_*(H\mathbb{Z};\mathbb{Z}/p) \to H_{*-1}(H\mathbb{Z};\mathbb{Z}/p)$$

is the Bockstein homomorphism.

We have an isomorphism of $\mathbb{Z}/p$-vector spaces

$$E^\infty_{**} = BP_*(H\mathbb{Z}/p) = H_*(BP;\mathbb{Z}/p) = \mathbb{Z}/p\,[m_1,\dots,m_n,\dots],\ |m_n| = 2p^n - 2.$$

The elements τ_i are odd-dimensional, and so they cannot survive. Hence, $d_{2p-1}\tau_1 = \lambda_1 v_1, 0 \neq \lambda_1 \in \mathbb{Z}/p$.

Let $d_{2p-1}\tau_2 = \lambda v_1 \xi_1^a, \lambda \in \mathbb{Z}/p$. Then $d_{2p-1}(\tau_2 - \lambda\tau_1\xi_1^a) = 0$, and we set $\overline{\tau}_2 := \tau_2 - \lambda\tau_1\xi_1^a$, $\overline{\xi}_2 = \beta\overline{\tau}_2$. Then $d_{2p-1}\overline{\tau}_2 = 0 = d_{2p-1}\overline{\xi}_2$, and $\overline{\tau}_2 \equiv \tau_2, \overline{\xi}_2 \equiv \xi_2$. Suppose inductively that we find $\overline{\tau}_i \equiv \tau_i$ and $\overline{\xi}_i \equiv \xi_i$ with $d_{2p-1}\overline{\tau}_i = 0 = d_{2p-1}\overline{\xi}_i$ and $\beta\overline{\tau}_i \equiv \overline{\xi}_i$ for $i = 1,\dots,k$. Let $d_{2p-1}\tau_{k+1} = v_1 x$ for some $x \in \mathbb{Z}/p\,[\overline{\xi}_1,\dots,\overline{\xi}_k] \otimes \Lambda(\overline{\tau}_1,\dots,\overline{\tau}_k)$. Then we set $\overline{\tau}_{k+1} := \tau_1 x$ and $\overline{\xi}_{k+1} := \beta\overline{\tau}_{k+1}$. Clearly, $d_{2p-1}\overline{\tau}_{k+1} = 0 = d_{2p-1}\overline{\xi}_{k+1}$. The induction is confirmed.

Furthermore, $d_{2p-1}(\tau_1 x) = \lambda_1 v_1 x$ for every $x \in E^2_{**}$. In particular, $E^{2p}_{r,q} = 0$ whenever $r < 2p^2 - 2$ and $q > 0$.

Now, $\overline{\tau}_2$ does not survive, and so $d_{2p^2-1}\overline{\tau}_2 = \lambda v_2, 0 \neq \lambda \in \mathbb{Z}/p$. Asserting as above, we find $\overline{\overline{\tau}}_i, \overline{\overline{\xi}}_i$ such that $\overline{\overline{\tau}}_i \equiv \tau_i, \overline{\overline{\xi}}_i \equiv \xi_i$ and $d_{2p^2-1}\overline{\overline{\tau}}_i = 0 = d_{2p^2-1}\overline{\overline{\xi}}_i$ for $i \geq 2$. Moreover $E^{2p^2}_{r,q} = 0$ if $r < 2p^3 - 2$ and $q > 0$.

Now, we can proceed by induction and find the required $\widehat{\tau}_i$ and $\widehat{\xi}_i$. □

By the way, we get another proof of 3.19(v) here. Indeed, $E^\infty_{*,q} = 0$ for $q > 0$, and so the Hurewicz homomorphism $h : \pi_k(BP) \to H_k(BP)$ is zero for $k > 0$.

The case $p = 2$ can be considered similarly. Here

$$\begin{aligned} E^2_{**} =&\mathbb{Z}/2\,[\zeta_1^2, \zeta_2, \dots, \zeta_n, \dots] \otimes \mathbb{Z}[2][v_1,\dots,v_n,\dots],\\ &\dim \zeta_n = 2^n - 1, \dim v_n = 2^{n+1} - 2. \end{aligned}$$

Moreover, ζ_i^2 plays the role of ξ_i, and ζ_i plays the role of τ_{i-1}. Similarly to 3.27, we have the following theorem.

3.28. Theorem. *There are elements $\widehat{\zeta}_i \in E^2_{0,*}$ such that $d_{2^{i+1}-1}\widehat{\zeta}_{i+1} = v_i$ and $\widehat{\zeta}_i \equiv \zeta_i$ mod Dec. Furthermore, all elements $\widehat{\zeta}_i^2$ and their products are permanent cycles.* □

3.29. Theorem. (i) *If X is a finite spectrum, then $BP_*(X)$ is a coherent and hence finitely generated $\pi_*(BP)$-module.*

(ii) *Let X be a spectrum bounded below. Then $H_*(X;\mathbb{Z}[p])$ is a free $\mathbb{Z}[p]$-module iff $BP_*(X)$ is a free $\pi_*(BP)$-module. Furthermore, given $u \in H^0(BP;\mathbb{Z}[p])$ as in 3.25, the homomorphism $u_* : BP_*(X) \to H_*(X;\mathbb{Z}[p])$ is epic if $H_*(X;\mathbb{Z}[p])$ is a free $\mathbb{Z}[p]$-module.*

(iii) *Given a spectrum X bounded below, there exists a morphism $f : W \to X$ such that W is a spectrum bounded below, $BP_*(W)$ is a free $\pi_*(BP)$-module,* *and* $f_* : BP_*(W) \to BP_*(X)$ *is epic. Furthermore, if every* $\mathbb{Z}[p]$*-module* $H_i(X;\mathbb{Z}[p])$ *is finitely generated then there exists* W *as above such that every* $\mathbb{Z}[p]$*-module* $H_i(W;\mathbb{Z}[p])$ *is finitely generated.*

(iv) *Let* E *be a* BP*-module spectrum. Let* X *be a finite spectrum such that* $H^*(X;\mathbb{Z}[p])$ *is a free* $\mathbb{Z}[p]$*-module. Then for every finite spectrum* Y *the pairing* $BP \wedge E \to E$ *induces an isomorphism*

$$E^*(X \wedge Y) \cong BP^*(X)\widehat{\otimes}_{BP^*(S)} E^*(Y).$$

(v) *Let* E *be a* BP*-module spectrum such that every group* $\pi_i(E)$ *is finite. Let* X *be a spectrum of finite* $\mathbb{Z}[p]$*-type such that* $H^*(X;\mathbb{Z}[p])$ *is a free* $\mathbb{Z}[p]$*-module. Then for every spectrum* Y *the pairing* $BP \wedge E \to E$ *induces an isomorphism*

$$E^*(X \wedge Y) \cong BP^*(X)\widehat{\otimes}_{BP^*(S)} E^*(Y).$$

Proof. Since BP is a direct summand of $M\mathcal{U}[p]$, this can be deduced from 1.14, 1.15, 1.17, 1.19 and 1.20. □

3.30. Remarks. (a) The algebra $M\mathcal{U}^*(M\mathcal{U})$ was described by Novikov [4] and Landweber [1].

(b) Theorem 1.9(vi) stimulated a search of spectra having $\mathbb{Z}/p$-cohomology $\mathscr{A}_p/(Q_0)$. Brown–Peterson [1] constructed such a spectrum using Postnikov towers. Novikov [4] proved Theorem 3.15 and its Corollaries 3.16–3.19. This gave a new proof of the Brown–Peterson result. Quillen [1] gave another proof of 3.18. In the proof of 3.15 we followed Buhštaber [3].

(c) There is an integral version of 3.23. Let

$$I = \{n \in \mathbb{N} \mid n = p^k \text{ for some prime } p \text{ and integer } k\}.$$

Basing on 3.15 and following the proof of 3.19, one can construct a spectrum V with the following properties:

$$\pi_*(V) = \mathbb{Z}[x_i, i \in I. \dim x_i = 2i], \quad H_*(V) = \mathbb{Z}[y_i, i \in I, \dim y_i = 2i].$$

Furthermore, V is a direct summand of $M\mathcal{U}$, and

$$MU_*(X) = V_*(X) \otimes \mathbb{Z}[x_i | i \notin I].$$

It is remarkable that V does not split multiplicatively but splits additively, Boardman [2].

§4. Invariant Prime Ideals. The Filtration Theorem

In this section we fix a prime p and denote the ring

$$\pi_*(BP) = \mathbb{Z}[p][v_1, \dots, v_n, \dots], \quad \dim v_i = 2(p^i - 1),$$

by Ω. We set $v_0 := p$ and consider the ideals $I_n := (v_0, v_1, \dots, v_{n-1})$ of Ω. In particular, $I_0 = (0), I_1 = (p)$. Finally, we set $I_\infty := (v_0, v_1, \dots, v_n, \dots)$. Clearly, I_n does not depend on the choice of the system $\{v_n\}$.

Every operation $\theta \in MU^d(MU)$ yields an operation $BP \to \Sigma^d BP$ of the form

$$BP \xrightarrow{\varkappa} MU[p] \xrightarrow{\theta[p]} \Sigma^d MU[p] \xrightarrow{\Sigma^d \rho} \Sigma^d BP. \tag{4.1}$$

On the other hand, every operation $\varphi \in BP^d(BP)$ yields an operation

$$MU[p] \xrightarrow{\rho} BP \xrightarrow{\varphi} \Sigma^d BP \xrightarrow{\Sigma^d \varkappa} \Sigma^d MU[p].$$

Furthermore, by II.5.8 and II.5.3, $(MU[p])^*(MU[p]) = MU^*(MU) \otimes \mathbb{Z}[p]$.

Finally, similarly to 3.1(v) one can prove that

$$BP^*(BP) = BP^*(S) \otimes^{\text{grad}} H^*(BP; \mathbb{Z}[p]) \tag{4.2}$$

as abelian groups. Now, $\rho(\varkappa\varphi\rho)\varkappa = \varphi$, i.e., every operation on BP is induced (as in (4.1)) by some (non-unique) operation on $MU[p]$. Below we write just θ instead of $\rho\theta\varkappa$. In particular, every operation on BP can be expanded (non-uniquely) as a series $\sum a_\omega S_\omega, a_\omega \in \pi_*(MU[p])$.

This information about $BP^*(BP)$ is sufficient for us. Additional information about $BP^*(BP)$ can be found in Adams [8] or Ravenel [1].

4.3. Definition. We say that a graded ideal $I \subset \Omega, I \neq \Omega$ is $BP^*(BP)$-*invariant*, or simply *invariant*, if $\theta(I) \subset I$ for every operation $\theta \in BP^*(BP)$.

4.4. Proposition. *The ideal $I_n, 0 \leq n \leq \infty$, is invariant.*

Proof. Since every operation on BP has the form $\sum a_\omega S_\omega$, it suffices to prove that $S_\omega(v_k) \in I_n$ for every $k < n$. The only non-trivial case is $|\omega| = p^k - 1$, i.e., $S_\omega(v_k) \in \pi_0(BP)$. Consider the commutative diagram

$$\begin{array}{ccc} \pi_{2|\omega|}(BP) & \xrightarrow{S_\omega} & \pi_0(BP) \\ {\scriptstyle h}\big\downarrow & & \big\downarrow{\scriptstyle h}\ \cong \\ H_{2|\omega|}(BP) & \xrightarrow{S_\omega} & H_0(BP). \end{array}$$

By 3.19(v), $p|h(v_k)$. Hence, $p|hS_\omega(v_k)$, and so $p|S_\omega(v_k)$, i.e., $S_\omega(v_k) \in I_n$. □

Fix integers m, n with $m > n \geq 0$. Let $\omega(m,n)$ be as in 3.11.

4.5. Lemma. *Let* $\omega = (p^{k_1}-1, \dots, p^{k_r}-1)$ *with* $k_i \geq m-n$ *and* $|\omega| = p^m - p^n$. *Then*

$$S_\omega(v_m) \equiv \begin{cases} \alpha v_n \bmod I_n,\ \alpha \not\equiv 0 \bmod p & \text{if } \omega = \omega(m,n), \\ 0 \bmod I_n & \text{otherwise.} \end{cases}$$

Proof. This follows from 3.13 because $v_m = \rho(x_{p^m-1})$. □

4.6. Lemma. *If* $|\omega| > p^m - p^n$, *then* $S_\omega(v_m) \equiv 0 \bmod I_n$.

Proof. This holds because $S_\omega(v_m) \in I_m$ and $\dim S_\omega(v_m) < \dim v_n$. □

Consider a sequence $E = (e_1, \dots, e_k, \dots)$ of non-negative integers, where all but finitely many e_i's are 0. We order the set of these sequences lexicographically, by setting $E < F$ if there is some $k \geq 1$ with $e_i = f_i$ for $i < k$ and $e_k < f_k$. Set $v^E = v_{n+1}^{e_1} \cdots v_{n+k}^{e_k} \cdots$. Let $k\omega$ be the partition $(\omega, \dots, \omega)$ (k times); set

$$\omega(E) := (e_1\omega(n+1,n), \dots, e_k\omega(n+k,n), \dots).$$

4.7. Lemma. *Let* $E = (e_1, \dots, e_k, \dots)$ *and* $F = (f_1, \dots, f_k, \dots)$ *be such that* $\dim v^E = \dim v^F$ *and* $E \leq F$. *Set* $t = e_1 + \dots + e_k + \cdots$. *Then*

$$S_{\omega(E)}(v^F) \equiv \begin{cases} \alpha v_n^t \bmod I_n,\ \alpha \not\equiv 0 \bmod p & \text{if } E = F, \\ 0 \bmod I_n & \text{if } E < F. \end{cases}$$

Proof. We have
(4.8)
$$S_{\omega(E)}(v^F) = \sum S_{\omega_1}(v_{n+1}) \cdots S_{\omega_{f_1}}(v_{n+1}) S_{\omega_{f_1+1}}(v_{n+2}) \cdots S_{\omega_?}(v_{n+k}) \cdots .$$

where the summation index runs all the sequences

$$\{\omega_1, \dots, \omega_{f_1}, \omega_{f_1+1}, \dots, \omega_?, \dots\}$$

with $(\omega_1, \dots, \omega_{f_1}, \omega_{f_1+1}, \dots, \omega_?, \dots) = \omega$. Firstly, let $E = F$. Consider any summand. If the operation on every v_{n+k} is $S_{\omega(n+k,k)}$ in this summand, then, by 4.5, this summand is $\alpha v_n^t \bmod I_n$, $\alpha \not\equiv 0 \bmod p$. We prove that every other summand belongs to I_n. Indeed, consider any summand which is not in I_n. Then, by 4.6, for every k every factor $S_\omega(v_{n+k})$ must be such that $|\omega| \leq p^{n+k} - p^n$. Moreover, $|\omega| = p^{n+k} - p^n$ (because if $|\omega| < p^{n+k} - p^n$ somewhere, then $|\omega'| > p^{n+l} - p^n$ for some $S_{\omega'}(v_{n+l})$). Hence, by 4.5, $\omega_1 = \omega(n+1,n)$ and in this way we exhaust all e_1 partitions $\omega(n+1,n)$. Now, considering the factor $S_\omega(v_{n+2})$, we have $\omega = (p^{k_1}-1, \dots, p^{k_r}-1)$ with $k_s \geq 2$, $s = 1, \dots, r$. Hence, by 4.5, $\omega = \omega(n+2,n)$, etc. Thus, $S_{\omega(E)}(v^E) \equiv \alpha v_n^t \bmod I_n$, $\alpha \not\equiv 0 \bmod p$.

Now, let $E < F$. Firstly, suppose that $e_1 < f_1$. Consider in (4.8) any summand which is not in I_n. Reasoning as above, we conclude that $\omega = \omega(n+1, n)$ for every factor $S_\omega(v_{n+1})$. But this is impossible because $e_1 < f_1$. Thus, $S_{\omega(E)}(v^F) \in I_n$. If $e_1 = f_1$, but $e_2 < f_2$, then all factors v_{n+1} of v^F find their partners (i.e., partitions $\omega(n+1, n)$), but it is impossible to serve all factors v_{n+2} of v^F, etc. □

4.9. Lemma. *For every $y \notin I_n$ there exists an operation $\theta \in BP^*(BP)$ such that $\theta(y) \equiv \alpha v_n^t \bmod I_n$ with $\alpha \not\equiv 0 \bmod p$.*

Proof. Let $y = \sum_{F \in \{F\}} a_F v^F$, $a_F \in \mathbb{Z}[p], a_F \not\equiv 0 \bmod p$ for every $F \in \{F\}$. Choose the minimal sequence E in $\{F\}$. Then, by 4.7, $S_{\omega(E)}(y) = \sum a_E S_{\omega(E)}(v^E) \equiv \alpha v_n^t \bmod I_n$. □

4.10. Corollary. *Let J be an invariant ideal such that $I_n \subset J$. If $I_n \neq J$, then $v_n^t \in J$ for some t.* □

4.11. Theorem. *If I is an invariant prime ideal, then $I = I_n$, $0 \le n \le \infty$.*

Proof. If $I \neq I_\infty$ then there is $n \ge 0$ such that $I_n \subset I$ and $I_{n+1} \not\subset I$. Suppose $I \neq I_n$. Then, by 4.10, $v_n^t \in I$ for some t, and $t > 0$ because $I \neq \Omega$. Hence, $v_n \in I$ because I is prime. Thus, $I_{n+1} \subset I$. This is a contradiction. □

Let $\mathscr{M}$ be the following category. Its objects are coherent graded Ω-modules M equipped with a $BP^*(BP)$-action $BP^*(BP) \otimes M \to M$ such that:

1. $\dim \theta(m) = \dim m - \dim \theta$ for every $\theta \in BP^*(BP), m \in M$.
2. $S_\omega(\lambda m) = \sum_{(\omega_1, \omega_2) = \omega} S_{\omega_1}(\lambda) S_{\omega_2}(m), \lambda \in \Omega, m \in M$.

Morphisms of $\mathscr{M}$ are $BP^*(BP)$-equivariant Ω-module homomorphisms.

Note that $BP_*(X)$ is an object of $\mathscr{M}$ for every finite spectrum X. Indeed, $BP_*(X)$ is a coherent Ω-module by 3.29(i), and $BP^*(BP)$ operates on $BP_*(X)$ for general reasons. Now, given $\lambda \in \Omega$ and $x \in BP_*(X)$, we have

$$\begin{aligned} S_\omega(\lambda x) &= \rho S_\omega \varkappa(\lambda x) = \rho S_\omega((\varkappa\lambda)(\varkappa x)) = \rho\left(\sum S_{\omega_1}(\varkappa\lambda) S_{\omega_2}(\varkappa x)\right) \\ &= \sum \rho(S_{\omega_1}(\varkappa\lambda))\rho(S_{\omega_2}(\varkappa x)) = \sum S_{\omega_1}(\lambda) S_{\omega_2}(x), \end{aligned}$$

where the third equality follows from 2.18 or 2.19(i). We also used that ρ and $\varkappa$ are the ring morphisms.

4.12. Definition and Notation. (a) Let R be a graded ring (e.g., $R = \Omega$). If two graded R-modules M, N are isomorphic up to dimension shift, we say that M and N are *stably isomorphic* and write $M \approx N$. Clearly, we are also able to say that two objects M, N of $\mathscr{M}$ are stably isomorphic in $\mathscr{M}$.

(b) Given an R-module M and an ideal $J \subset R$, we set

$$M(0 : J) := \{x \in M \mid Jx = 0\}.$$

Furthermore, given $x \in M$, the *annihilator* of x is the subset (ideal) $\operatorname{Ann} x := \{a \in R | ax = 0\}$.

(c) Given $M \in \mathscr{M}$ and an Ω-submodule N of M, we say that N is *invariant* if the inclusion $N \subset M$ is an inclusion in $\mathscr{M}$. In other words, $s_\omega(x) \in N$ for every $x \in N$ and all ω. So, N appears to be an object of $\mathscr{M}$.

(d) Similarly to II.6.12(b), we say that an element $x \in M \in \mathscr{M}$ is *simple* if $s_\omega(x) = 0$ for every $\omega \neq (0)$. Clearly, every non-zero element of least dimension is simple.

4.13. Lemma. *Let $M \in \mathscr{M}$.*

(i) *Let $N \subset M$ be an inclusion in $\mathscr{M}$, and let $N \subset P \subset M$ be inclusions of Ω-modules. If P/N is an invariant submodule of M/N then P is an invariant submodule of M.*

(ii) *Let J be an invariant ideal of Ω. Then $M(0 : J)$ is an invariant submodule of M.*

Proof. (i) This is obvious.

(ii) We must prove that $s_\omega(x) \in M(0 : J)$ for every $x \in M(0 : J)$ and all ω. We do this by induction on $l(\omega)$. So, let $\omega = (k)$ where $k > 0$. Given $a \in J$, we have

$$0 = s_{(k)}(ax) = s_{(k)}(a)x + as_{(k)}(x) = as_{(k)}(x),$$

i.e., $s_{(k)}(x) \in M(0 : J)$. Suppose that $s_\omega(x) \in M(0 : J)$ for every ω with $l(\omega) < n$. Given $a \in J$ and any ω with $l(\omega) = n$, we have

$$0 = s_\omega(ax) = s_\omega(a)x + as_\omega(x) + \sum_{(\omega_1,\omega_2)=\omega, \omega_i \neq \omega} s_{\omega_1}(a)s_{\omega_2}(x).$$

In particular, $l(\omega_2) < n$. So, $s_{\omega_2}(x) \in M(0 : J)$, and so

$$\sum_{(\omega_1,\omega_2)=\omega, \omega_i \neq \omega} s_{\omega_1}(a)s_{\omega_2}(x) = 0,$$

and hence $as_\omega(x) = 0$, and thus $s_\omega(x) \in M(0 : J)$. □

4.14. Lemma. *Let x be a simple element of $M \in \mathscr{M}$. Then:*

(i) *$\operatorname{Ann} x$ is an invariant ideal;*

(ii) *If $I_r \subset \operatorname{Ann} x$ then $v_r^k x$ is simple for every k;*

(iii) *Ωx is an invariant submodule of M, and Ωx is stably isomorphic in $\mathscr{M}$ to $\Omega / \operatorname{Ann} x$.*

Proof. (i) If $a \in \operatorname{Ann} x$ then $0 = s_\omega(ax) = s_\omega(a)x$.

(ii) We have $s_\omega(v_r^k) \in \operatorname{Ann} x$ if $\omega \neq (0)$. So, $s_\omega(v_r^k x) = v_r^k s_\omega(x) = 0$ if $\omega \neq (0)$.

(iii) The invariance of Ωx follows since $s_\omega(ax) = s_\omega(a)x$ for every $a \in \Omega$. Furthermore, the homomorphism $f : \Omega/\operatorname{Ann} x \to \Omega x, f(a) = ax$, establishes the desired stable isomorphism. □

Let $\Omega\langle n\rangle$ be the subring $\mathbb{Z}[p][v_1, \dots, v_n]$ of Ω. So, Ω is an $\Omega\langle n\rangle$-module. Conversely, there is a ring homomorphism $h : \Omega \to \Omega\langle n\rangle, h(v_i) = v_i$ for $i \leq n$ and $h(v_i) = 0$ for $i > n$, which turns $\Omega\langle n\rangle$ into an Ω-module.

Given an Ω-module M, we define an $\Omega\langle n\rangle$-module $M\langle n\rangle := M \otimes_\Omega \Omega\langle n\rangle$. On the other hand, given an $\Omega\langle n\rangle$-module N, we consider $N \otimes_{\Omega\langle n\rangle} \Omega$ and equip it with an Ω-module structure by setting $a(n \otimes b) = n \otimes ab,\ a, b \in \Omega, n \in N$.

Note that we have $(V \otimes_{\Omega\langle n\rangle} \Omega)\langle n\rangle = V$ for every $\Omega\langle n\rangle$-module V.

Let $\{f_\alpha\}$ be a family of free generators of a free Ω-module F. We define a homomorphism

$$\varphi_n : F\langle n\rangle \otimes_{\Omega\langle n\rangle} \Omega \to F,\ \varphi_n(f_\alpha \otimes \omega) = \omega f_\alpha.$$

Clearly, φ_n is a homomorphism of Ω-modules.

4.15. Lemma. *Let $M \in \mathscr{M}$.*

(i) *There are a finitely generated free Ω-module F, an Ω-epimorphism $f : F \to M$, a natural number n and an Ω-isomorphism $\psi_n : M\langle n\rangle \otimes_{\Omega\langle n\rangle} \Omega \to M$ such that the following diagram commutes:*

$$\begin{array}{ccc} F\langle n\rangle \otimes_{\Omega\langle n\rangle} \Omega & \xrightarrow{\varphi_n} & F \\ {\scriptstyle f\otimes 1}\downarrow & & \downarrow{\scriptstyle f} \\ M\langle n\rangle \otimes_{\Omega\langle n\rangle} \Omega & \xrightarrow{\psi_n} & M. \end{array}$$

(ii) *Let n be as in* (i). *Then the homomorphism $v_r : M \to M, x \mapsto v_r x$, is monic for every $r > n$.*

(iii) *There is m such that $I_m = \operatorname{Ann} x$ for some $x \in M$ but $I_{m+1} \not\subset \operatorname{Ann} y$ for every $y \in M, y \neq 0$.*

Proof. (i) Since M is coherent and so finitely presented, there is an exact sequence of Ω-modules

$$R \xrightarrow{a} F \xrightarrow{f} M \to 0$$

where R and F are free finitely generated Ω-modules. Let $\{r_\beta\}$ be a family of free generators of R. We define $i_n : F\langle n\rangle \to F, i_n(x) = \varphi_n(x \otimes 1)$. Since the set $\{r_\beta\}$ is finite, we can choose n such that $a(r_\beta) \in i_n(F\langle n\rangle)$ for every β. Clearly, there exists a unique homomorphism $\psi_n : M\langle n\rangle \otimes_{\Omega\langle n\rangle} \Omega \to M$ such that the above diagram commutes, and this ψ_n is an isomorphism.

(ii) Consider the exact sequence

$$0 \to \Omega \xrightarrow{v_r} \Omega \to \Omega/(v_r) \to 0.$$

If $r > n$ then $\operatorname{Tor}_1^{\Omega\langle n\rangle}(M\langle n\rangle, \Omega/(v_r)) = 0$ because $\Omega/(v_r)$ is a free $\Omega\langle n\rangle$-module. Hence, we get an exact sequence

$$0 \to M\langle n\rangle \otimes_{\Omega\langle n\rangle} \Omega \xrightarrow{v_r} M\langle n\rangle \otimes_{\Omega\langle n\rangle} \Omega \to M\langle n\rangle \otimes_{\Omega\langle n\rangle} \Omega/(v_r) \to 0,$$

i.e, $v_r : M \to M$ is a monomorphism.

(iii) By (ii), there is a maximal m such that $I_m = \operatorname{Ann} x$ for some $x \in M$. Suppose that there is y such that $I_{m+1} \subset \operatorname{Ann} y$. Now we shall find some $r > 0$ such that I_{m+r} is an annihilator of an element of M, and this will be a contradiction, and the claim will be proved.

Note that $M(0 : I_{m+1}) \neq 0$ since $y \in M(0 : I_{m+1})$. Let w be an element of least dimension in $M(0 : I_{m+1})$. Then w is simple, and so $J_1 := \operatorname{Ann} w$ is invariant. Clearly, $J_1 \supset I_{m+1}$. If $J_1 = I_{m+1}$ then we have the desired contradiction. If not, then, by 4.10, there is t such that $v_{m+1}^t \in J_1$, and so $J_2 := \operatorname{Ann}(v_{m+1}^{t-1} w) \supset I_{m+2}$. If $J_2 = I_{m+2}$ then we are done. If not, then, by 4.14(ii) and 4.14(i), J_2 is invariant, and so $v_{m+2}^s \in J_2$, and so $J_3 := \operatorname{Ann}(v_{m+2}^{s-1} v_{m+1}^{t-1} w) \supset I_{m+3}$. Iterating this process, we get a sequence $J_1 \subset J_2 \subset J_3 \subset \cdots$, where $J_r \supset I_{m+r}$ and J_r is an annihilator of some element. By (ii), this process must stop, and so $J_k = I_{m+k}$ for some k. This is the desired contradiction. □

4.16. Lemma. *Let $M \in \mathscr{M}$. Fix an isomorphism $\psi_n : M\langle n\rangle \otimes_{\Omega\langle n\rangle} \Omega \to M$ as in* 4.15.

(i) *Let N be an Ω-submodule of M such that the obvious homomorphism*

$$j : N\langle n\rangle \otimes_{\Omega\langle n\rangle} \Omega \to M\langle n\rangle \otimes_{\Omega\langle n\rangle} \Omega \xrightarrow{\psi_n} M$$

is an monomorphism with $\operatorname{Im} j = N$. *Then there is a unique morphism $(M/N)\langle n\rangle \otimes_{\Omega\langle n\rangle} \Omega \to M/N$ such that the diagram*

$$\begin{array}{ccccc} N\langle n\rangle \otimes_{\Omega\langle n\rangle} \Omega & \longrightarrow & M\langle n\rangle \otimes_{\Omega\langle n\rangle} \Omega & \longrightarrow & ((M/N)\langle n\rangle) \otimes_{\Omega\langle n\rangle} \Omega \\ \downarrow \cong & & \downarrow \cong & & \downarrow \\ N & \longrightarrow & M & \longrightarrow & M/N \end{array}$$

commutes, and this morphism is an isomorphism.

(ii) *For every $x \in M\langle n\rangle$, $\operatorname{Ann}_{\Omega\langle n\rangle} x = I_k\langle n\rangle$ iff $\operatorname{Ann}_\Omega(x \otimes_{\Omega\langle n\rangle} 1) = I_k$.*

(iii) *$M(0 : I_k) \neq 0$ in M iff $M\langle n\rangle(0 : I_k\langle n\rangle) \neq 0$ in $M\langle n\rangle$.*

(iv) *If x is an element of least dimension in $M\langle n\rangle(0 : I_k\langle n\rangle)$ then $\psi_n(x \otimes_{\Omega\langle n\rangle} 1)$ is an element of least dimension in $M(0 : I_k)$.*

Proof. (i) This follows since the rows of the diagram are exact.

(ii) This is obvious.

(iii) We treat $\psi_n : M\langle n\rangle \otimes_{\Omega\langle n\rangle} \Omega \to M$ as the identity map 1_M. Clearly, if $x \in M\langle n\rangle(0 : I_k\langle n\rangle)$ then $x \otimes_{\Omega\langle n\rangle} 1 \in M(0 : I_k)$. Conversely, suppose that $M(0 : I_k) \neq 0$ and take $y \in M(0 : I_k), y \neq 0$. Then $y = \sum_\alpha y_\alpha \otimes_{\Omega\langle n\rangle} \omega_\alpha$, where $y_\alpha \in M\langle n\rangle$ and each ω_α is a polynomial in $v_{n+1}, \dots, v_{n+k}, \dots$. We have $v_r y = 0$ for every $r \leq k$, and so $\sum_\alpha v_r y_\alpha \otimes_{\Omega\langle n\rangle} \omega_\alpha = 0$ for every $r \leq k$. Now, using the isomorphism

$$M\langle n\rangle \otimes_{\Omega\langle n\rangle} \Omega \cong M\langle n\rangle \otimes_{\mathbb{Z}} \mathbb{Z}[p]\,[v_{n+1}, \dots, v_{n+k}, \dots],$$

we conclude that $v_r y_\alpha = 0$ for every α, i.e., $y_\alpha \in M\langle n\rangle(0 : I_k\langle n\rangle)$.

(iv) Suppose that there is $y \in M(0 : I_k)$ such that $\dim y < \dim(x \otimes_{\Omega\langle n\rangle} 1)$. We have $y = \sum_\alpha y_\alpha \otimes_{\Omega\langle n\rangle} \omega_\alpha = 0$ where $y_\alpha \in M\langle n\rangle$ and each ω_α is a polynomial in $v_{n+1}, \dots, v_{n+k}, \dots$. Reasoning as in (iii), we conclude that $y_\alpha \in M\langle n\rangle(0 : I_k\langle n\rangle)$ for every α, i.e., x is not an element of least dimension in $M\langle n\rangle(0 : I_k\langle n\rangle)$. This is a contradiction. □

4.17. Lemma. *Let $M \in \mathscr{M}$, and let m be as in 4.15(iii). Then every element x of least dimension of $M(0 : I_m)$ is simple, and $\operatorname{Ann} x = I_m$.*

Proof. By 4.4 and 4.13(ii), $M(0 : I_m)$ is an invariant submodule, and so x is simple, and so, by 4.14(i), $\operatorname{Ann} x$ is an invariant ideal. We prove that $\operatorname{Ann} x = I_m$. Indeed, if not, then, by 4.10, $v_m^t \in \operatorname{Ann} x$ for some t, and so $\operatorname{Ann}(v_m^{t-1} x) \supset I_{m+1}$. But this contradicts our choice of m. □

4.18. Theorem. *Every object M of $\mathscr{M}$ admits a filtration in $\mathscr{M}$*

$$0 = M_0 \subset M_1 \subset \dots \subset M_k = M,$$

such that M_i/M_{i-1} is stably isomorphic in $\mathscr{M}$ to Ω/I_{r_i} for every $i = 1, \dots, k$. In particular, this holds for $M = BP_*(X)$, where X is a finite spectrum.

Proof. Choose n as in 4.15(i) and take m as in 4.15(iii). We treat $\psi_n : M\langle n\rangle \otimes_{\Omega\langle n\rangle} \Omega \to M$ as the identity map 1_M. Let $x \in M\langle n\rangle$ be an element of least dimension in $M\langle n\rangle(0 : I_m\langle n\rangle)$. Then, by 4.16(iv), $\widehat{x} := x \otimes_{\Omega\langle n\rangle} 1$ is an element of least dimension in $M(0 : I_m)$, and so, by 4.17, $\operatorname{Ann} \widehat{x} = I_m$, and so, by 4.16(ii), $\operatorname{Ann} x = I_m\langle n\rangle$. Set $\overline{M}_1 := \Omega\langle n\rangle x \in M\langle n\rangle$ and note that

$$\overline{M}_1 = \Omega\langle n\rangle x \approx \Omega\langle n\rangle / I_m\langle n\rangle.$$

We set $M_1 := \overline{M}_1 \otimes_{\Omega\langle n\rangle} \Omega$. Since Ω is a flat $\Omega\langle n\rangle$-module, the inclusion $\overline{M}_1 \subset M\langle n\rangle$ induces an inclusion $M_1 \subset M\langle n\rangle \otimes_{\Omega\langle n\rangle} \Omega = M$. (More pedantically: this inclusion is the composition

$$M_1 \xrightarrow{a} M\langle n\rangle \otimes_{\Omega\langle n\rangle} \Omega \xrightarrow{\psi_n} M$$

where a is a monomorphism induced by the inclusion $M_1 \subset M\langle n\rangle$.)

Note that $M_1 = \Omega\widehat{x}$. By 4.16(iv) and 4.17, $\widehat{x}$ is simple, and so, by 4.14(iii), M_1 is an invariant submodule of M. Now, by 4.16(i), we can apply the same process to M/M_1 with the same n and to find a submodule V of $(M/M_1)\langle n\rangle = M\langle n\rangle/M_1\langle n\rangle$ such that $V \approx \Omega\langle n\rangle/I_{r_2}\langle n\rangle$. We define $\overline{M}_2$ to be an inverse image of V under the canonical projection $M\langle n\rangle \to M\langle n\rangle/M_1\langle n\rangle$. So, we get an $\Omega\langle n\rangle$-module $\overline{M}_2$ where $\overline{M}_2/\overline{M}_1 \approx \Omega\langle n\rangle/I_{r_2}\langle n\rangle$. By induction, we get a filtration

$$0 = \overline{M}_0 \subset \overline{M}_1 \subset \cdots \overline{M}_r \subset \cdots$$

of $M\langle n\rangle$ such that $\overline{M}_i/\overline{M}_{i-1} \approx \Omega\langle n\rangle/I_{r_i}\langle n\rangle$ where $r_1 = m$. Since $M\langle n\rangle$ is a finitely generated module over a Noetherian ring $\Omega\langle n\rangle$, this filtration stabilizes, i.e., $\overline{M}_k = M\langle n\rangle$ for some k. So, we get a finite filtration

$$0 = M_0 \subset M_1 \subset \cdots M_r \subset M_k = M$$

where $M_i = \overline{M}_i \otimes_{\Omega\langle n\rangle} \Omega$.

We have already proved that M_1 is an invariant submodule in M. Now, by 4.13(i) and an obvious induction, every M_i is an invariant submodule of M, i.e., this filtration is a filtration in $\mathscr{M}$. Finally, by 4.16(i),

$$M_i/M_{i-1} = (\overline{M}_i/\overline{M}_{i-1}) \otimes_{\Omega\langle n\rangle} \Omega \approx \Omega\langle n\rangle/I_{r_i}\langle n\rangle \otimes_{\Omega\langle n\rangle} \Omega = \Omega/I_{r_i}. \qquad \square$$

Now we give an integral version of the results above. We say that a graded ideal $I \subset \pi_*(MU), I \neq \pi_*(MU)$, is *$MU^*(MU)$-invariant* if $\theta(I) \subset I$ for every $\theta \in MU^*(MU)$. Let $\{x_i\}, \dim x_i = 2i$, be a system of polynomial generators of $\pi_*(MU)$ as in 1.9(v).

Given a prime p and a natural number n, we set

$$(4.19) \qquad I(p,n) := (p, x_{p-1}, \ldots, x_{p^{n-1}-1}) \subset \pi_*(MU).$$

Furthermore, $I(p,\infty) := (p, x_{p-1}, \ldots, x_{p^{n-1}-1}, \ldots) \subset \pi_*(MU)$. Let $I(p) \subset \pi_*(MU)$ be the ideal such that all Chern numbers of every element of $I(p)$ are divisible by p. Let A_n denote the subset of $\pi_*(MU)$ consisting of all elements of dimension $\leq 2n$.

4.20. Proposition. *$I(p,\infty) = I(p)$. Furthermore, $I(p,n)$ coincides with the ideal generated by $I(p) \cap A_{p^{n-1}-1}$, i.e., $I(p,n) = (I(p) \cap A_{p^{n-1}-1})$. In particular, $I(p,n)$ depends only on p, n.*

Proof. It is clear that $I(p,\infty) \subset I(p)$. Conversely, given $x \in I(p)$, we prove that $x \in I(p,\infty)$.

Represent the set of natural numbers $\mathbb{N}$ as the disjoint union $\mathbb{N} = A \sqcup B$, where $A = \{p-1, \ldots, p^k-1, \ldots\}$. Then x can be expanded as $x = x' + f(x_i | i \in B)$, where $x' \in I(p,\infty)$ and f is a polynomial. Then $p | s_\omega(x)$ and $p | s_\omega(x')$ for every ω. We prove that $f = 0$. Let us order monomials $x_{i_1} \cdots x_{i_k}, i_k \leq i_{k+1}$ by setting $x_{i_1} \cdots x_{i_k} < x_{j_1} \cdots x_{j_l}$ iff there exists s such that $i_r = j_r$ for $r < s$ and $i_s < j_s$. Let $x_{i_1} \cdots x_{i_k}$ be the maximal monomial in f. Then $s_\omega(f) = s_\omega(x_{i_1} \cdots x_{i_k})$ for $\omega = (i_1, \dots, i_k)$. But $s_\omega(x_{i_1} \cdots x_{i_k}) = s_{i_1}(x_{i_1}) \cdots s_{i_k}(x_{i_k})$ is not divisible by p in view of 1.9(iv). Thus, $f = 0$, i.e., $x \in I(p, \infty)$.

Finally, $I(p,n) = (I(p) \cap A_{p^{n-1}-1})$ since $I(p,n) = (I(p,\infty) \cap A_{p^{n-1}-1})$. □

4.21. Proposition. *$I(p,n)$ is an $\mathcal{MU}^*(\mathcal{MU})$-invariant ideal for every (p,n).*

Proof. Firstly, we prove that $I(p)$ is $\mathcal{MU}^*(\mathcal{MU})$-invariant. In other words, we must prove that $p|s_{\omega'}(S_\omega(x))$ for every ω' such that $2(|\omega| + |\omega'|) = \dim x$. But in this case $s_{\omega'}(S_\omega(x)) = S_{\omega'}(S_\omega(x)) \in \mathbb{Z}$ is a Chern number of x, and thus $p|s_{\omega'}(S_\omega(x))$.

Now we prove that $I(p,n)$ is $\mathcal{MU}^*(\mathcal{MU})$-invariant. Because of 2.19(i), it suffices to prove that $S_\omega(x_{p^k-1}) \in I(p,n)$ for every $k < n$. Since $I(p,n) \subset I(p)$ and $I(p)$ is invariant, $S_\omega(x_{p^k-1}) \in I(p)$, i.e., by 4.20, $S_\omega(x_{p^k-1}) \in I(p,\infty)$. But $\dim S_\omega(x_{p^k-1}) \leq 2p^k - 2$, and so $S_\omega(x_{p^k-1}) = py + \sum_{i=1}^k x_{p^i-1} y_i$ for some $y, y_i \in \pi_*(\mathcal{MU})$. Thus, $S_\omega(x_{p^k-1}) \in I(p,n)$. □

We leave it to the reader to prove the following results, similar to 4.11 and 4.18 above.

4.22. Theorem. *If I is an $\mathcal{MU}^*(\mathcal{MU})$-invariant prime ideal of $\pi_*(\mathcal{MU})$, then $I = I(p,n)$, where p is a prime and $0 \leq n \leq \infty$.* □

Let $\mathscr{N}$ be the following category. Its objects are coherent graded $\pi_*(\mathcal{MU})$-modules N equipped with an $\mathcal{MU}^*(\mathcal{MU})$-action $\mathcal{MU}^*(\mathcal{MU}) \otimes N \to N$ such that:

1. $\dim \theta(n) = \dim n - \dim \theta$ for every $\theta \in \mathcal{MU}^*(\mathcal{MU}), n \in N$.
2. $S_\omega(\lambda n) = \sum_{(\omega_1,\omega_2)=\omega} S_{\omega_1}(\lambda) S_{\omega_2}(n), \lambda \in \pi_*(\mathcal{MU}), n \in N$, where $S_{\omega_1}(\lambda)$ is the result of the $\mathcal{MU}^*(\mathcal{MU})$-action on $\pi_*(\mathcal{MU})$.

Morphisms of $\mathscr{N}$ are $\mathcal{MU}^*(\mathcal{MU})$-equivariant $\pi_*(\mathcal{MU})$-module homomorphisms.

Note that $\mathcal{MU}_*(X)$ is an object of $\mathscr{N}$ for every finite spectrum X.

4.23. Theorem. *Every object N of $\mathscr{N}$ admits a filtration in $\mathscr{N}$*

$$N = N_0 \supset N_1 \supset \cdots \supset N_k = 0,$$

where N_i/N_{i+1} is stably isomorphic in $\mathscr{N}$ to $\pi_*(\mathcal{MU})/I(p_i, r_i), i = 1, \dots, k$. In particular, this holds for $N = \mathcal{MU}_*(X)$, where X is a finite spectrum. □

4.24. Remark. Theorems 4.11, 4.22 were proved by Landweber [3], see also Morava [2]. In the proof of 4.11 we followed mainly Johnson–Wilson [1]. Landweber [4] proved the Filtration Theorems 4.18, 4.23.

§5. Formal Groups

Given a formal power series $f(x)$ over a commutative ring R, let $f^{-1}(x)$ denote a formal power series such that $f(f^{-1}(x)) = x = f^{-1}(f(x))$. It is easy to see that $f^{-1}(x)$ exists iff $f(x) = \sum_{i\geq 1} a_i x^i$ with $a_1 \in R^*$, and $f^{-1}(x)$ is unique. As usual, we write $f(x) = g(x) + o(x^n)$, if $f(x) = g(x) + x^{n+1}\varphi(x)$ for some $\varphi(x) \in R[[x]]$.

5.1. Definition (a) A *formal group* (more precisely, a *one dimensional commutative formal group law*) over a commutative ring R is a formal power series $x + y + \sum_{i,j\geq 1} a_{ij}x^i y^j \in R[[x,y]]$ with the following properties:

(1) (commutativity) $F(x,y) = F(y,x)$;
(2) (unitarity) $F(x,0) = x$;
(3) (associativity) $F(F(x,y),z) = F(x,F(y,z))$.

(b) Given two formal groups F, G over R, a *homomorphism* $f : F \to G$ is a formal power series $f(x) = \sum_{i\geq 1} a_i x^i$ such that $f(F(x,y)) = G(f(x), f(y))$. A homomorphism is called an *isomorphism* if $a_1 \in R^*$, and it is called an *equivalence* if $a_1 = 1$. We use the notation $F \simeq G$ for equivalent formal groups.

Notice that $G(f(x), g(x))$ is a homomorphism $F \to G$ if f and g are. In particular, the set $\mathrm{Hom}(F,G)$ of all homomorphisms $F \to G$ is closed under the operation $+_G$ where $(f +_G g)(x) := G(f(x), g(x)$. Moreover, the operation $+_G$ converts $\mathrm{Hom}(F,G)$ into an abelian group: the inversion is established by Proposition 5.9 below.

If $f : F \to G$ is an isomorphism, then f^{-1} exists, and it is an isomorphism $G \to F$.

It is clear that in this way we have a category $\mathscr{F}(R)$ of formal groups over R and their homomorphisms. Furthermore, if $\varphi : R \to S$ is a ring homomorphism and $F(x,y) = x + y + \sum_{i,j\geq 1} a_{ij}x^i y^j$ is a formal group over R, then the formal power series $(\varphi_* F)(x,y) := x + y + \sum_{i,j\geq 1} \varphi(a_{ij})x^i y^j$ is a formal group over S. So, φ yields a functor $\varphi_* : \mathscr{F}(R) \to \mathscr{F}(S)$.

5.2. Definition. A formal group $\mathcal{F}(x,y)$ over a commutative ring L is called *universal* if for every formal group F over every ring R there exists a homomorphism $\varphi : L \to R$ with $\varphi_*(\mathcal{F}) = F$ and this homomorphism is unique. In this case we say that φ *classifies* F.

5.3. Proposition. *There exists a universal formal group.*

Proof (Adams [8]). Let $\Lambda = \mathbb{Z}[a_{ij}]$ be the polynomial ring generated by symbols $a_{ij}, i, j \geq 1$. Consider the formal power series $h(x,y) = x + y + \sum_{i,j\geq 1} a_{ij}x^i y^j$ over Λ and set $h(h(x,y),z) - h(x,h(y,z)) = \sum b_{ijk}x^i y^j z^k$. Let I be the ideal in Λ generated by the elements b_{ijk} and $a_{ij} - a_{ji}$. Set $L = \Lambda/I$, and let $\bar{a}_{ij}$ be the image of a_{ij} in L. It is clear that the formal power series $x + y + \sum_{i,j\geq 1} \bar{a}_{ij}x^i y^j$ over $L := \Lambda/I$ is a universal formal group. □

Clearly, the universal formal group is unique in the following sense: if $(\mathcal{F}, L)$ and $(\mathcal{F}', L')$ are two universal formal groups, then there exists a ring isomorphism $\varphi : L \to L'$ with $\varphi_*\mathcal{F} = \mathcal{F}'$.

The following fact is more complicated. It was proved by Lazard [1].

5.4. Theorem. *The underlying ring L of the universal formal group is a polynomial ring on a countable set of variables, $L = \mathbb{Z}[x_1, \dots, x_n, \dots]$.*

Proof. See Adams [8], Buhštaber [1,2], Fröhlich [1], Ravenel [1]. □

The universal formal group can be also described as follows. Consider the polynomial ring $A = \mathbb{Z}[b_1, \dots, b_n, \dots]$ and the formal power series $g(x) = x + \sum b_i x^{i+1}$. Set $f(x,y) = g^{-1}(g(x) + g(y))$. It is clear that f is a formal group. Let $L \subset A$ be the subring generated by the coefficients of f. Then (L, f) is the universal formal group (see Buhštaber [1,2], cf. also 6.17(a) below).

5.5. Examples. (a) The so-called additive formal group $A(x,y) = x + y$.

(b) The so-called multiplicative formal group $M(x,y) = x+y+axy, a \in R$.

(c) The universal formal group.

(d) Let $f(x)$ be a functionally invertible formal power series, i.e., such that f^{-1} exists. Then $F(x,y) = f^{-1}(f(x) + f(y))$ is a formal group.

5.6. Definition. Given a formal group $F(x,y)$ over R, a formal power series $g(x) = x + o(x) \in R[[x]]$ is called a *logarithm* of F if

$$F(x,y) = g^{-1}(g(x) + g(y)).$$

In other words, g is an equivalence between F and the additive formal group.

5.7. Proposition. *Every formal group $F(x,y)$ over a $\mathbb{Q}$-algebra R has a logarithm, and it is unique. In particular, every two formal groups over a $\mathbb{Q}$-algebra are equivalent.*

Proof (Honda [1]). Consider the formal power series $\omega(x) = \partial_2 F(x,0)$. Because of 5.1(a), we have

$$\omega(F(x,y)) = \partial_2 F(F(x,y),0) = \partial_1 F(F(x,0),y) \cdot \partial_2 F(x,0) = \partial_1 F(x,y)\omega(x).$$

Hence, $\dfrac{dx}{\omega(x)} = \dfrac{dF(x,y)}{\omega(F(x,y))}$. Set $g(x) := \displaystyle\int_0^x \frac{dt}{\omega(t)}$. Then $dg(x) = dg(F(x,y))$ and hence

$$g(F(x,y)) = g(x) + C.$$

Since $F(0,y) = y$ and $g(0) = 0$, $C = g(y)$. Thus, $g(F(x,y)) = g(x) + g(y)$.

We prove the uniqueness of the logarithm. Let

$$g^{-1}(g(x) + g(y)) = h^{-1}(h(x) + h(y)), \quad h(x) = x + o(x).$$

Set $f(x) = g(h^{-1}(x))$. Then $f(x+y) = f(x) + f(y)$. Since R is a $\mathbb{Q}$-algebra and $f(x) = x + o(x)$, we conclude that $f(x) = x$. □

5.8. Examples. (a) The multiplicative formal group $m(x,y) = x + y - xy$ over $\mathbb{Z}$ has no logarithm. Indeed, let $i : \mathbb{Z} \to \mathbb{Q}$ be the inclusion. The logarithm of i_*m is $x + x^2/2 + \dots + x^n/n + \dots = -\ln(1-x)$. If $m(x,y)$ has a logarithm $f(x)$, then $f(x)$ is a logarithm of i_*m also, and so $f(x) = -\ln(1-x)$. But $\ln(1-x) \notin \mathbb{Z}[[x]]$.

(b) It is clear that $g(x) = x$ is a logarithm of the additive formal group $A(x,y)$ over any R. We prove that $f(x) = x + x^p$ is a logarithm of $a(x,y)$ over $\mathbb{Z}/p$. Indeed, we have $x + x^p + y + y^p = x + y + (x+y)^p$, i.e., $f(x+y) = f(x) + f(y)$, i.e., $a(x,y) = f^{-1}(f(x) + f(y))$. (In fact, both f and f^{-1} are automorphisms of $a(x,y)$.) Thus, the uniqueness of the logarithm is false in general.

5.9. Proposition. *Given a formal power series* $F(x,y) = x + y + \sum_{i,j\geq 1} a_{ij}x^iy^j$, *there exists a formal power series* $\theta(x)$ *such that* $F(x,\theta(x)) = 0$, *and it is unique.*

Proof. Firstly, we construct a family $\theta_n(x)$ such that $F(x,\theta_n(x)) = o(x^n)$ and $\theta_{n+1}(x) = \theta_n(x) + o(x^{n+1})$. Set $\theta_0(x) = -x$. Then $F(x,\theta_0(x)) = o(x)$. Assume that θ_n is constructed. Let b_{n+1} be the coefficient of x^{n+1} in $F(x,\theta_n(x))$. Set $\theta_{n+1}(x) = \theta_n(x) - b_{n+1}x^{n+1}$. Then

$$F(x,\theta_{n+1}(x)) = F(x,\theta_n(x)) - b_{n+1}x^{n+1} + o(x^{n+1}) = o(x^{n+1}).$$

Now, set $\theta(x) = -x - \sum_{i=2}^{\infty} b_ix^i$, where b_i is the coefficient of x^i in $\theta_i(x) - \theta_{i-1}(x)$. Then $\theta(x) = \theta_n(x) + o(x^n)$ for every n. Thus, $F(x,\theta(x)) = o(x^n)$ for every n, i.e., $F(x,\theta(x)) = 0$.

We leave it to the reader to prove that θ is unique. □

5.10. Definition. Given a formal group F over R and $n \in \mathbb{Z}$, define inductively a formal power series $[n]_F(x) \in R[[x]]$ as follows. Let $\theta(x)$ be as in 5.9. Set $[-1]_F(x) = \theta(x)$. Furthermore, if $n \geq 0$, then $[n]_F(x) := F(x,[n-1]_F(x))$. If $n < 0$, then $[n]_F(x) := F([-1]_F(x),[n+1]_F(x))$.

Clearly, $[0]_F(x) = 0$, $[1]_F(x) = x$, and $F([m]_F(x), [n]_F(x)) = [m+n]_F(x)$. Furthermore, if F admits a logarithm g, then $[n]_F(x) = g^{-1}(ng(x))$. Finally, $[n]_F$ is an endomorphism of F (i.e., a homomorphism $F \to F$) for every $n \in \mathbb{Z}$ (prove this!).

The formal power series $[n]_F$ can also be described as follows. Let $\operatorname{End} F$ be the set of all endomorphisms of F. Set

$$f(x) +_F g(x) := F(f(x), g(x)),\ (g \cdot f)(x) = g(f(x))$$

for every $f, g \in \operatorname{End} F$. Then $(\operatorname{End} F, +_F, \cdot)$ is a ring. Consider the unique ring homomorphism $\mathbb{Z} \to \operatorname{End} F$. Then $[n]_F$ is just the image of $n \in \mathbb{Z}$.

Let R be a commutative torsion free ring, and let F be a formal group over R. Define $i : R \to R \otimes \mathbb{Q}$ by setting $i(r) = r \otimes 1$. Then $i(R)$ is a subring of $R \otimes \mathbb{Q}$. For simplicity, we say that the logarithm of i_*F is a *logarithm of F over $R \otimes \mathbb{Q}$.*

5.11. Proposition. (i) *If F is equivalent to the additive formal group, then for every $n > 0$ all coefficients of $[n]_F(x)$ are divisible by n. In particular, if $nR = 0$ then $[n]_F(x) = 0$.*

(ii) *Let p be a prime, and let R be a torsion free $\mathbb{Z}[p]$-algebra. Let F be a formal group over R such that all coefficients of $[p]_F(x)$ are divisible by p. Let $g(x) \in R[[x]] \otimes \mathbb{Q}$ be the logarithm of F over $R \otimes \mathbb{Q}$. Then $g(x) \in R[[x]]$, i.e., $g(x)$ is a logarithm of F. In particular, F is equivalent over R to the additive formal group.*

Proof. (i) Let $h : F \to A$ be an equivalence, $F(x,y) = h^{-1}(h(x) + h(y))$. Let $h^{-1}(x) = x + \sum a_i x^{i+1}$. Then $[n]_F(x) = h^{-1}(nh(x)) = nh(x) + \sum n^{i+1} a_i (h(x))^{i+1}$.

(ii) Let $g(x) = x + \sum g_i x^{i+1}, g_i \in R \otimes \mathbb{Q}$. Let $g^{-1}(x) = x + \sum a_i x^{i+1}, a_i \in R \otimes \mathbb{Q}$. Since $x = g(x) + \sum a_i (g(x))^{i+1}$, we conclude (equating coefficients of equal powers of x) that $g_1 + a_1 = 0$ and

$$g_n + f_n(a_1, \dots, a_{n-1}, g_1, \dots, g_{n-1}) + a_n = 0$$

for $n > 1$. Here f_n is a polynomial over $\mathbb{Z}$ such that every monomial in f_n contains some a_i. This implies (by induction) that if $g_i \in R$ for every $i < n$, then $a_i \in R$ for every $i < n$ and $a_n = r_n - g_n$ with $r_n \in R$.

We prove by induction that $g_i \in R$. Since $[p]_F(x) = g^{-1}(pg(x))$ then, by the hypothesis, all coefficients of $g^{-1}(pg(x))$ belong to pR. The coefficient of x^2 in $g^{-1}(pg(x))$ is $pg_1 + p^2 a_1 = pg_1 - p^2 g_1 = pg_1(1-p)$. Since it is divisible by p in R, $g_1 \in R$ because $1 - p$ is invertible in R.

We suppose that $g_i \in R$ for every $i < n$ and prove that $g_n \in R$. The coefficient of x^{n+1} in $g^{-1}(pg(x))$ is $pg_n + f_n(p^2 a_1, \dots, p^n a_{n-1}, g_1, \dots, g_{n-1}) + p^{n+1} a_n$. Since $g_i \in R$ for $i < n$, we have $a_i \in R$ for $i < n$ (as we said above). Furthermore, every monomial in f_n contains some a_i, and hence

$f_n(p^2 a_1, \dots, p^n a_{n-1}, g_1, \dots, g_{n-1})$ is divisible by p^2 in R, i.e., it is $p^2 r$ for some $r \in R$. Hence, the coefficient of x^{n+1} in $g^{-1}(pg(x))$ is

$$pg_n + p^2 r + p^{n+1} a_n = pg_n + p^2 r + p^{n+1}(r_n - g_n) = pg_n(1 - p^n) + p^2 r + p^{n+1} r_n,$$

where $r_n, r \in R$. But this coefficient belongs to pR. Thus, $g_n \in R$ because $1 - p^n$ is invertible in R. □

For future reference note the following obvious fact.

5.12. Proposition. *Let $r : R \to S$ be a ring homomorphism, and let F be a formal group over R.*

(i) *If $[n]_F(x) = \sum a_i x^i$, then $[n]_{r_* F}(x) = \sum r(a_i) x^i$.*

(ii) *If R and S are torsion free and $g(x) = x + \sum g_i x^{i+1}$ is a logarithm of F over $R \otimes \mathbb{Q}$, then $x + \sum r(g_i) x^{i+1}$ is a logarithm of $r_*(F)$ over $S \otimes \mathbb{Q}$.* □

5.13. Proposition. *Let p be a prime, and let R be a commutative ring with $pR = 0$. Let $f : F \to G$ be a homomorphism of formal groups over R. If $f(x) \neq 0$, then $f(x) = \varphi(x^{p^h})$ for some $\varphi(x) \in R[[x]]$ with $\varphi(x) = ax + o(x), a \neq 0$.*

Proof (Fröhlich [1]). Recall that $f(x) = \sum_{i=1}^{\infty} a_i x^i$, and so $f(0) = 0$. We have $f(F(x,y)) = G(f(x), f(y))$. By differentiating this equation with respect to y and putting $y = 0$, we have

$$f'(x)\partial_2 F(x,0) = \partial_2 G(f(x), 0) f'(0).$$

Note that $f'(0) = a_1$. If $a_1 \neq 0$, then we can put $\varphi(x) = f(x)$. If $a_1 = 0$ then $f'(x) = 0$, because $\partial_2 F(x,0) = 1 + \sum_{i>0} b_i x^i$. So $f(x) = g(x^p)$. Now we want to proceed by induction, but we must first show that $g(x)$ is a homomorphism of formal groups over R. If $F(x,y) = \sum a_{ij} x^i y^j$, set $F^{(p)}(x,y) := \sum a_{ij}^p x^i y^j$. Since $a \mapsto a^p$ is an endomorphism of R, $F^{(p)}$ is a formal group over R. We prove that g is a homomorphism $F^{(p)} \to G$. Indeed

$$\begin{aligned} g(F^{(p)}(x^p, y^p)) &= g((F(x,y)^p)) = f(F(x,y)) \\ &= G(f(x), f(y)) = G(g(x^p), g(y^p)). \end{aligned}$$

Thus, $g(F^{(p)}(x,y)) = G(g(x), g(y))$. □

5.14. Definition. (a) The number h in 5.13 is called *height* of the homomorphism f and denoted by $\mathrm{ht}(f)$. If $f = 0$, then $\mathrm{ht}(f) = \infty$.

(b) Given a formal group F over a commutative ring R with $pR = 0$ for a prime p, define the *height* of F as the height of $[p]_F$, $\mathrm{ht}(F) := \mathrm{ht}([p]_F)$.

Since $[p]_F(x) = px + o(x)$ for every formal group F, we have $\mathrm{ht}(F) > 0$. It is easy to see that isomorphic formal groups have equal heights. For every natural number n there exists a formal group of height n, see 6.15(iii) below.

5.15. Theorem (Lazard [2]). *Let F, G be two formal groups over an algebraically closed field of characteristic $p > 0$. If $\mathrm{ht}(F) = \mathrm{ht}(G)$, then F and G are isomorphic.*

Proof. See e.g. Fröhlich [1]. □

5.16. Example. Consider the formal groups $U(x,y) = x+y+xy$, $V(x,y) = \tan(\arctan(x)+\arctan(y)) = \frac{x+y}{1-xy} = (x+y)(1+xy+\ldots+x^ny^n+\ldots)$ over $\mathbb{Z}/3$. We have $[3]_U(x) = x^3, [3]_V(x) = -x^3$. We want to show that U and V are not isomorphic, while $\mathrm{ht}(U) = \mathrm{ht}(V) = 1$. It suffices to prove that $[p]_F = [p]_G$ whenever F and G are isomorphic over $\mathbb{Z}/p, p$ prime. Firstly, $[p]_G(x) = \varphi(x^{p^h})$ with $h > 0$. Furthermore, $(u+v)^{p^h} = u^{p^h} + v^{p^h}$ for every $u, v \in \mathbb{Z}/p\,[[x,y]]$, and $a^{p^h} = a$ for every $a \in \mathbb{Z}/p$. Hence, $[p]_G(f(x)) = f([p]_G(x))$ for every formal power series f. Now, let $f : F \to G$ be an isomorphism. It is easy to see that $f^{-1}([p]_G(f(x))) = [p]_F(x)$, and thus $[p]_F(x) = [p]_G(x)$ by the above.

Now we consider the graded version of the notions discussed above. Let R be a graded commutative ring. We treat $R[[x, y, \ldots, z]]$ as a graded ring with $\deg x = \deg y = \cdots = \deg z = 2$.

5.17. Definition. A *graded formal group* over a graded commutative ring R is a formal group $F(x,y) = x + y + \sum a_{ij}x^iy^j$ which at the same time is a homogeneous element of degree 2, i.e., $\deg a_{ij} = 2-2i-2j$. A *homomorphism* of graded formal groups is a homomorphism $f(x) \in R[[x]]$ of formal groups such that $f(x)$ is a homogeneous element of degree 2.

The concept of the universal formal group makes sense in the graded case also, and the following analog of 5.4 holds. The proof is similar to any proof of 5.4.

5.18. Theorem. *There is a universal graded formal group over a graded commutative ring L where $L = \mathbb{Z}[x_1, \ldots, x_n, \ldots]$, $\deg x_n = -2n$.* □

The obvious analog of 5.15 holds in the graded case also.

§6. Formal Groups Input

Passing to topology, consider any $\mathbb{C}$-oriented spectrum (E, t). By 2.2(ii,iii),

$$E^*(CP^\infty) = E^*(\mathrm{pt})[[t]],\ E^*(CP^\infty \times CP^\infty) = E^*(\mathrm{pt})[[x,y]]$$

where we introduce the notation $x = t_1, y = t_2$. Let

(6.1) $$m : CP^\infty \times CP^\infty \to CP^\infty$$

be the multiplication in the H-space $CP^\infty = K(\mathbb{Z}, 2)$. Considering the induced homomorphism $m^* : E^*(CP^\infty) \to E^*(CP^\infty \times CP^\infty)$, we have

$$m^*(t) = F(x, y) \in E^*(\mathrm{pt})[[x, y]].$$

6.2. Proposition. *$F(x, y)$ is a graded formal group.*

Proof. Since $CP^\infty = CP^\infty \times \mathrm{pt} \xrightarrow{1\times \mathrm{incl}} CP^\infty \times CP^\infty \xrightarrow{m} CP^\infty$ is homotopic to 1_{CP^∞}, we have $F(x, 0) = x$. The properties (1) and (3) from 5.1(a) hold because m is commutative and associative up to homotopy. □

Thus, we have associated a graded formal group to any $\mathbb{C}$-oriented spectrum.

We can also do it in the following way. Let $CP_i^\infty, i = 1, 2$, be a copy of CP^∞, and let $\eta_{(i)}, i = 1, 2$ be a copy of η over CP_i^∞. Let

$$p_i : CP_1^\infty \times CP_2^\infty \to CP_i^\infty, i = 1, 2$$

be the projection. By 1.6, $m^*(\eta) = p_1^*(\eta) \otimes p_2^*(\eta)$. Since $c_1^{E,t}(\eta) = t$, we conclude that

(6.3) $$F(x, y) = c_1^{E,t}(p_1^*(\eta) \otimes p_2^*(\eta)).$$

Because of the universality of η, for every pair of complex line bundles ξ, ζ over X, we have

(6.4) $$c_1^{E,t}(\xi \otimes \zeta) = F(c_1^{E,t}(\xi), c_1^{E,t}(\zeta)).$$

6.5. Lemma. *Let $\varphi : (E, t) \to (E', t')$ be a morphism of $\mathbb{C}$-oriented spectra, and let $\varphi_* : E^*(\mathrm{pt}) \to (E')^*(\mathrm{pt})$ be the induced homomorphism of coefficients. Let F (resp. F') denote the formal group of (E, t) (resp. (E', t')). Then $(\varphi_*)_*(F) = F'$.* □

6.6. Lemma. *Let F be the formal group of a $\mathbb{C}$-oriented spectrum (E, t). Let $z = f(t) = t + \sum a_i t^{i+1}$ be another $\mathbb{C}$-orientation of E, and let G be the formal group of (E, z). Then $f(F(x, y)) = G(f(x), f(y))$. In particular, F and G are equivalent formal groups. Furthermore, if H is any graded formal group equivalent to F, then there exists a $\mathbb{C}$-orientation v of E such that H is the formal group of (E, v).*

Proof. We have

$$m^*(z) = m^*\left(t + \sum a_i t^{i+1}\right) = m^*(t) + \sum a_i (m^*(t))^{i+1} = f(F(x, y)).$$

On the other hand, if $z_i = p_i^*(z)$ for $p_i : CP^\infty \times CP^\infty \to CP^\infty$, then

$$m^*(z) = G(z_1, z_2) = G(f(t_1), f(t_2)) = G(f(x), f(y)).$$

Furthermore, let $g : F \to H$ be an equivalence, $g(x) = x + \sum b_i x^{i+1}$. Set $v = t + \sum b_i t^{i+1}$. As above, one can prove that H is the formal group of (E, v). □

Thus, we have correspondences

$$\{\mathbb{C}\text{-oriented spectra}\} \longrightarrow \{\text{graded formal groups}\},$$
$$\{\mathbb{C}\text{-orientable spectra}\} \longrightarrow \{\text{equivalence classes of graded formal groups}\}.$$

6.7. Examples (cf. 2.15). (a) Let $R = \{R_i\}$ be a graded commutative ring such that $R_i = 0$ for $i > 0$. Then the formal group of HR is the additive formal group over R (because $HR^i(\mathrm{pt}) = 0$ for $i < 0$).

(b) Let R be a graded commutative ring, and let $E = HR$. The inclusion $HR_0 \to HR$ yields a ring morphism $HR_0 \to HR$. Hence, by (a) and 6.5, HR admits a $\mathbb{C}$-orientation t such that (E, t) has the additive formal group. Thus, a formal group of any $\mathbb{C}$-oriented spectrum (HR, s) is equivalent to the additive formal group.

(c) Let (E, t) be a $\mathbb{C}$-oriented spectrum such that $\pi_*(E)$ is a $\mathbb{Q}$-algebra. Then, by II.7.11(ii), $E \simeq H(\pi_*(E))$. Hence, by (b), the formal group of (E, t) is equivalent to the additive formal group. On the other hand, this follows from 5.7.

(d) Let E be complex K-theory. We have $E^*(\mathrm{pt}) = \mathbb{Z}[s, s^{-1}], \deg s = 2$. Consider the $\mathbb{C}$-orientation $t = s(\eta - 1) \in \widetilde{K}^2(CP^\infty)$, i.e., $c_1^{K,t}(\eta) = s(\eta - 1)$. Here $1 \in K^0(CP^\infty)$ represents θ^1. Let $F(x, y)$ be the formal group of (K, t). Then, by (6.3),

$$\begin{aligned} F(x,y) =& c_1^{K,t}(\eta_{(1)} \otimes \eta_{(2)-1}) = s(\eta_{(1)} \otimes \eta_{(2)}) \\ =& s(\eta_{(1)} - 1) + s(\eta_{(2)} - 1) + s(\eta_{(1)} - 1)(\eta_{(2)} - 1) = x + y + s^{-1}xy, \end{aligned}$$

where we write $\eta_{(i)}$ instead of $p_i^*(\eta)$. Thus, F is a multiplicative formal group.

(e) Let $(E, t) = (MU, T)$. Observe that $MU^*(\mathrm{pt}) \cong L$, where L is the underlying ring of the universal graded formal group from 5.18. On the other hand, by 2.5, (MU, T) is the universal $\mathbb{C}$-oriented spectrum. This hints that the formal group of (MU, T) coincides with the universal formal group. This is really true and will be proved below.

Let $v_{ij} \in MU_{2i+2j}(CP^\infty \times CP^\infty)$ be the bordism class of the inclusion $j_i \times j_j : V_{ij} := CP^i \times CP^j \subset CP^\infty \times CP^\infty$. Let $m : CP^\infty \times CP^\infty \to CP^\infty$ be as in (6.1). Consider $m^* : MU^2(CP^\infty) \to MU^2(CP^\infty \times CP^\infty)$ and set

$$h_{ij} := \langle m^*T, v_{ij} \rangle \in \pi_{2i+2j-2}(MU) = MU^{2-2i-2j}(\mathrm{pt}).$$

The elements h_{ij} admit the following geometrical description. Let

$$f : CP^i \times CP^j \to CP^N$$

be the restriction of m. By 1.25, η_{N-1} is a normal bundle of $l_{N-1} : CP^{N-1} \subset CP^N$. Assuming f to be transversal to η_{N-1}, set $H_{ij} := f^{-1}(CP^{N-1})$. Then $h_{ij} = [H_{ij}]$.

6.8. Lemma. (i) $s_{(i+j-1)}(h_{ij}) = \binom{i+j}{i}$ *for* $i, j > 1$, $s_{(j)}(h_{1,j}) = 0$ *for* $j > 1$.
(ii) $h_{0,n} = [CP^{n-1}]$, $h_{1,n} = [CP^1] \times [CP^{n-1}]$.
(iii) *The elements* h_{ij} *generate the ring* $\pi_*(MU)$.

Proof. (i) Fix i, j and set $V = V_{ij}, v = v_{ij}$. Let $t \in H^2(CP^\infty)$ be a $\mathbb{C}$-orientation of $H\mathbb{Z}$, and let $u = u^{H\mathbb{Z},t} : MU \to H\mathbb{Z}$ be the Thom class as in (2.7). We have $u_*T = t$ and $m^*(t) = t_1 + t_2$. Set

$$\overline{v} := u_*(v) \in H_{2i+2j}(CP^\infty \times CP^\infty).$$

By 2.19(ii) and 2.17(iii),

$$\begin{aligned} s_{(i+j-1)}(h_{ij}) = s_{(i+j-1)}\langle m^*T, v\rangle &= \langle m^* s_{(i+j-1)}(T), \overline{v}\rangle + \langle m^*t, s_{(i+j-1)}(v)\rangle \\ &= \langle (t_1+t_2)^{i+j}, \overline{v}\rangle + \langle t_1 + t_2, s_{(i+j-1)}(v)\rangle. \end{aligned}$$

Here the left hand summand is

$$\langle (t_1+t_2)^{i+j}, \overline{v}\rangle = \sum \left\langle \binom{i+j}{k} t_1^k t_2^{i+j-k}, \overline{v} \right\rangle = \left\langle \binom{i+j}{i} t_1^i t_2^j, \overline{v} \right\rangle = \binom{i+j}{i}.$$

We compute the right hand summand. We can compute $s_{(i+j-1)}(v)$ in $CP^i \times CP^j$. Let $\eta_{(1)}$ (resp. $\eta_{(2)}$) be the canonical complex line bundle over CP^i (resp. CP^j). By 1.24, $\tau(CP^i) \oplus \theta^1 = (i+1)\eta_i$, $\tau(CP^j) \oplus \theta^1 = (j+1)\eta_j$. Set $\xi_1 := p_1^*\eta_i$, $\xi_2 := p_2^*\eta_j$ where $p_1 : CP^i \times CP^j \to CP^i$, $p_2 : CP^i \times CP^j \to CP^j$ are the projections. Then $\tau(V) \oplus \theta^2 = (i+1)\xi_1 \oplus (j+1)\xi_2$. We have $c_1(\xi_i) = t_i, i = 1, 2$, and so $c_{(k)}(\xi_i) = t_i^k$. By 2.22, $s_{(i+j-1)}(v) = c_{(i+j-1)}(\nu_V) \cap \overline{v}$. But

$$\begin{aligned} c_{(i+j-1)}(\nu_V) = -c_{(i+j-1)}(\tau_V) &= -c_{(i+j-1)}((i+1)\xi_1 \oplus (j+1)\xi_2) \\ &= -(i+1)t_1^{i+j-1} - (j+1)t_2^{i+j-1}. \end{aligned}$$

If $i, j > 1$, then $t_1^{i+j-1} = 0 = t_2^{i+j-1}$, and hence $s_{(i+j-1)}(v) = 0$. Thus,

$$s_{(i+j-1)}(h_{ij}) = \binom{i+j}{i}.$$

If $i = 1, j > 1$, then $t_1^{i+j-1} = t_1^j = 0$, but $t_2^{i+j-1} \neq 0$, Hence,

$$\begin{aligned} \langle t_1 + t_2, s_{(i+j-1)}(v)\rangle &= -\langle t_1 + t_2, (j+1)t_2^j \cap \overline{v}\rangle \\ &= -\langle t_1 + t_2, (j+1)u_*\{CP^1\}\rangle = -(j+1). \end{aligned}$$

Thus, $s_{(j)}(h_{1,j}) = j+1-(j+1) = 0$.

(ii) The equality $h_{0,n} = [CP^{n-1}]$ follows from 3.7. By 1.9(iii), in order to prove that $h_{1,n} = [CP^1 \times CP^{n-1}] = [CP^1] \times [CP^{n-1}]$ it suffices to prove that $s_\omega(h_{1,n}) = s_\omega([CP^1 \times CP^n])$ for every ω with $|\omega| = n$, i.e, that $S_\omega(h_{1,n}) = S_\omega([CP^1 \times CP^n])$ for every $\omega = (i_1, \dots, i_m)$ with $\sum i_k = n$.

Firstly, if $m = 1$, i.e., $\omega = (n)$, then, by (i), $S_\omega(h_{1,n}) = 0$. Furthermore, $S_\omega[CP^1 \times CP^n] = 0$. Hence, we can and shall assume that $k > 1$.

Now suppose that $m > 1$ and that $i_k > 1$ for every k. It is clear that $S_\omega[CP^1 \times CP^n] = 0$. Now, by 2.19(ii),

$$S_\omega(h_{1,n}) = \sum \langle S_{\omega_1}(m^*T), S_{\omega_2}[CP^1 \times CP^n]\rangle.$$

If $\omega_2 \neq (0)$, then $S_{\omega_2}[CP^1 \times CP^n] = 0$ since $i_k > 1$ for every k. If $\omega_2 = (0)$, then $\omega_1 = \omega$. By 2.17(iii), $S_\omega(T) = 0$ for $k > 1$, and so $S_\omega m^*T = m^*(S_\omega(T)) = 0$.

Finally, let $\omega = (1, \overline{\omega})$. Then we have $S_\omega(v_{1,n}) = 2S_{\overline{\omega}}\{CP^{n-1}\}$ and $S_\omega[CP^1 \times CP^{n-1}] = 2S_{\overline{\omega}}[CP^{n-1}]$. Now,

$$\begin{aligned} S_\omega(h_{1,n}) &= \sum \langle S_{\omega_1}(m^*T), S_{\omega_2}(v_{1,n})\rangle = \sum_{(\omega_1, \overline{\omega}_2) = \overline{\omega}} \langle S_{\omega_1}(m^*T), 2S_{\overline{\omega}_2}\{CP^n\}\rangle \\ &= 2S_{\overline{\omega}}\langle m^*T, \{CP^n\}\rangle = 2S_{\overline{\omega}}[CP^{n-1}] = S_\omega[CP^1 \times CP^n]. \end{aligned}$$

(iii) The GCD of the numbers $\binom{n}{i}$, $i = 1, \dots, n-1$, is just λ_n (defined in (1.8)). Using (i) and the equality

$$\binom{i+j}{1} = i+j = -s_{(i+j-1)}[CP^{i+j-1}] = -s_{(i+j-1)}(h_{0,i+j}),$$

we conclude that the GCD of $\{s_{(n)}(h_{i,n+1-i})\}_{i=0}^n$ is λ_n. Hence,

$$s_{(n)}\left(\sum_{i=0}^n a_i h_{i,n+1-i}\right) = \lambda_n$$

for suitable $a_i \in \mathbb{Z}$. Thus, by 1.9(iv), h_{ij} generate $\pi_*(MU)$. □

6.9. Theorem. *The formal group $f(x,y)$ of (MU,T) is*

$$f(x,y) = \frac{\sum_{i,j\geq 0; (i,j)\neq(0,0)} h_{ij}x^i y^j}{CP(x)CP(y)},$$

where $CP(u) = 1 + \sum_{n=1}^{\infty} [CP^n]u^n$.

Proof (cf. Adams [8], Buhštaber [2]). Let $f(x,y) = m^*T = x + y + \sum a_{kl}x^k y^l$ where $x = T_1, y = T_2$. Then in $CP^i \times CP^j$ we have

$$\begin{aligned} h_{ij} &= \langle m^*T, v_{ij}\rangle = \langle x + y + \sum a_{kl}x^k y^l, v_{ij}\rangle \\ &= [CP^{i-1}][CP^j] + [CP^i][CP^{j-1}] + \sum a_{kl}[CP^{i-k}][CP^{j-l}]. \end{aligned}$$

Hence,

$$\begin{aligned} h_{ij}x^i y^j = {} & (x[CP^{i-1}]\,x^{i-1}[CP^j]\,y^j) + (y[CP^i]\,x^i[CP^{j-1}]\,y^{j-1}) \\ & + \sum a_{kl}x^k y^l([CP^{i-k}]\,x^{i-k}[CP^{j-l}]\,y^{j-l}). \end{aligned}$$

Thus,

$$\begin{aligned} \sum_{i,j} h_{ij}x^i y^j = {} & x\left(\sum_{i,j}[CP^{i-1}]\,x^{i-1}[CP^j]\,y^j\right) + y\left(\sum_{i,j}[CP^i]\,x^i[CP^{j-1}]\,y^{j-1}\right) \\ & + \sum_{k,l} a_{kl}x^k y^l\left(\sum_{i,j}[CP^{i-k}]\,x^{i-k}[CP^{j-l}]\,y^{j-l}\right) \\ = {} & \left(x + y + \sum_{k,l} a_{kl}x^k y^l\right) CP(x)CP(y). \end{aligned}$$

□

6.10. Corollary. *The coefficients a_{ij} of the formal group $f(x,y)$ of $(\mathcal{MU},T)$ generate the ring $\mathcal{MU}^*(\mathrm{pt})$.*

Proof. By 6.9, $a_{ij} \equiv h_{ij} \bmod \mathrm{Dec}\,(\pi_*(\mathcal{MU}))$ for $i,j > 1$. If $n > 1$ then $a_{1,n} + [CP^{n-1}] = h_{1,n} = [CP^1][CP^{n-1}]$. Moreover, $a_{11} = -[CP^1]$. Hence, $a_{1,n} \equiv -[CP^{n-1}] \bmod \mathrm{Dec}$. Now apply 6.8(iii) (and use that $\pi_i(\mathcal{MU}) = \mathcal{MU}^{-i}(\mathrm{pt})$)). □

6.11. Corollary (Quillen [1]). *The formal group $f(x,y)$ of $(\mathcal{MU},T)$ coincides with the universal formal group.*

Proof. Let $\varphi : L \to \mathcal{MU}^*(\mathrm{pt})$ classify the formal group $f(x,y)$. By 6.10, φ is epic. Thus, φ is an isomorphism because L_n and $\mathcal{MU}^{-n}(\mathrm{pt})$ are isomorphic finitely generated free abelian groups. □

6.12. Corollary (Miščenko, an Addendum to Novikov [4]). *The logarithm $g(x)$ of the formal group $f(x,y)$ over $\mathcal{MU}^*(\mathrm{pt}) \otimes \mathbb{Q}$ is*

$$g(x) = x + \sum_{n\geq 1} \frac{[CP^n]}{n+1}\,x^{n+1}.$$

Proof. By 5.7, $g(x) = \int_0^x \frac{dt}{\omega(t)}$, where $\omega(x) = \partial_2 f(x, 0)$. Now, because of 6.9 and 6.8(ii),

$$\omega(x) = \frac{\sum_{i\geq 0} h_{i,1}x^i - \sum_{i\geq 1} h_{i,0}x^i[CP^1]}{CP(x)} = \frac{1 + \sum_{i\geq 1}(h_{i,1} - [CP^1]h_{i,0})\, x^i}{CP(x)}$$
$$= \frac{1}{CP(x)} .$$

Thus,

$$g(x) = \int_0^x CP(t)dt = x + \sum \frac{[CP^n]}{n+1} x^{n+1} . \qquad \square$$

Let $\pi = \pi_n : CP^\infty \to CP^\infty$ be a map such that $\pi^*(t) = nt$ for $t \in H^2(CP^\infty)$. So, $\pi^*(T) = [n]_f(T)$. Set $\beta_k(n) := \langle \pi^* T, \{CP^k\}\rangle \in \pi_{2k-2}(MU)$. Note that $\beta_1(n) = n$.

6.13. Theorem. $[n]_f(x) = \frac{\sum_{k\geq 0} \beta_k(n) x^k}{CP(x)}$.

Proof. Let $[n]_f(x) = \sum a_i x^i$. By 3.7, we have

$$\beta_k(n) = \langle \pi^* T, \{CP^k\}\rangle = \left\langle \sum a_i T^i, \{CP^k\} \right\rangle = \sum a_i [CP^{k-i}].$$

Hence,

$$\beta_k(n)x^k = \sum a_i [CP^{k-i}]\, x^k = \sum a_i x^i ([CP^{k-i}]\, x^{k-i}).$$

Thus,

$$\sum \beta_k(n)x^k = \sum_{k,i} a_i x^i (CP^{k-i}\, x^{k-i}) = \left(\sum a_i x^i\right) CP(x). \qquad \square$$

6.14. Lemma. *For every prime p, the following hold:*
(i) *$p | s_\omega(\beta_k(p))$ for every k and every ω.;*
(ii) $s_{(p^k-1)}(\beta_{p^k}(p)) = p^{p^k} - p^{k+1} - p$;
(iii) *Let $f(x, y)$ be the formal group in 6.9. Define $a_i \in \pi_*(MU)$ via the equality $[p]_f(x) = \sum a_i x^i$. Then $p | s_\omega(a_i)$ for every i and every ω.*

Proof. For simplicity, we denote $\beta_k(p)$ by β_k.

(i) We have

$$s_\omega(\beta_k) = s_\omega\langle \pi^*T, \{CP^k\}\rangle = \sum_{(\omega_1,\omega_2)=\omega} \langle s_{\omega_1}\pi^*T, s_{\omega_2}\{CP^k\}\rangle$$
$$= \sum_{(\omega_1,\omega_2)=\omega} \langle \pi^* s_{\omega_1}(T), s_{\omega_2}\{CP^k\}\rangle.$$

If $\omega_1 = (r)$ for some $r \geq 0$, then $s_{\omega_1}(T) = t^{r+1}$. Hence,

$$\langle \pi^* s_{\omega_1}(T), s_{\omega_2}\{CP^k\}\rangle = \langle \pi_*(t^{r+1}), s_\omega\{CP^k\}\rangle = \langle (pt)^{r+1}, s_\omega\{CP^k\}\rangle.$$

If $l(\omega_1) > 1$ then $s_{\omega_1}(T) = 0$, and so $\langle \pi^* s_{\omega_1}(T), s_{\omega_2}\{CP^k\}\rangle = 0$.

Thus, p divides each summand $\langle \pi^* s_{\omega_1}(T), s_{\omega_2}\{CP^k\}\rangle$ of $s_\omega(\beta_k)$.

(ii) Let $[CP^i]_H \in H_{2i}(CP^{p^k})$ be the homology class given by the inclusion $l^i_{p^k} : CP^i \subset CP^{p^k}$, i.e., $[CP^i]_H = (l^i_{p^k})_*\left(u_*[CP^i]_{MU}\right)$ where $u = u^{H\mathbb{Z},t} : MU \to H\mathbb{Z}$ is as in (2.7). Let ν be a complex normal bundle of CP^{p^k}. Then

$$s_{(p^k-1)}\{CP^{p^k}\} \overset{2.22}{=\!=\!=} c_{p^k-1}(\nu) \cap [CP^{p^k}]_H = -(p^k+1)t^{p^k-1} \cap [CP^{p^k}]$$
$$= -(p^k+1)[CP^1]_H.$$

Now,

$$s_{(p^k-1)}(\beta_{p^k}) = s_{(p^k-1)}\langle \pi^*T, \{CP^{p^k}\}\rangle$$
$$= \langle \pi^* s_{(p^k-1)}(T), \{CP^{p^k}\}\rangle + \langle pt, s_{(p^k-1)}\{CP^{p^k}\}\rangle$$
$$= \langle (pt)^{p^k}, \{CP^{p^k}\}\rangle + \langle pt, -(p^k+1)[CP^1]_H\rangle = p^{p^k} - p^{k+1} - p.$$

(iii) We prove this by induction on k. We have

$$\beta_k = \left\langle \sum a_i T^i, \{CP^k\}\right\rangle = \sum_{i=1}^{k} a_i[CP^{k-i}] = a_k + \sum_{i=1}^{k-1} a_i[CP^{k-i}].$$

Hence, $s_\omega(a_k) = s_\omega(\beta_k) - \sum s_{\omega_1}(a_i)s_{\omega_2}[CP^{k-i}]$. Note that $p|s_\omega(a_1)$ since $a_1 = p$. Assume that $p|s_\omega(a_i)$ for every ω and every $i < k$. Then, by (i), $p|s_\omega(a_k)$. The induction is confirmed. □

6.15. Corollary. *Let p be a prime. Let $I(p,n)$ be the ideal defined in* (4.19).

(i) $I(p,n) = (p, \beta_p(p), \ldots, \beta_{p^{n-1}}(p))$.

(ii) *Let $\alpha_n = \alpha_{n,p} \in MU^{2-p^n}(\mathrm{pt})$ be the coefficient of x^{p^n} in the formal power series $[p]_f(x)$. Then $s_{(p^n-1)}(\alpha_n) = p^{p^n} - p$. Moreover, $I(p,n) = (p, \alpha_1, \ldots, \alpha_{n-1})$.*

(iii) *For every $h > 0$ there exists a graded formal group of height h over the ring $\mathbb{Z}/p[t]$, $\dim t = 2(1-p^h)$.*

Proof. (i) This holds because $s_{(p^k-1)}(\beta_{p^k}(p)) \equiv p \bmod p^2$ and $p|s_\omega(\beta_{p^k}(p))$ for every ω. In detail, let $x_{p^k-1} \in \pi_{2p^k-2}(MU)$ be such that $s_{(p^k-1)}(x_{p^k-1}) = p$ and $p|s_\omega(x_{p^k-1})$ for every ω. Then $I(p,n) = (p, x_{p-1}, \dots, x_{p^{n-1}-1})$. Set

$$y_{p^k-1} = (p^{p^k-1} - p^k)x_{p^k-1} - \beta_{p^k}(p).$$

By 6.14(iii), $s_{(p^k-1)}(y_{p^k-1}) = -p$ and $p|s_\omega(y)$ for every ω. So, $I(p,n) = (p, y_{p-1}, \dots, y_{p^{n-1}-1})$. But $-\beta_{p^k}(p) \equiv y_{p^k-1} \bmod p$. Thus,

$$I(p,n) = (p, \beta_p(p), \dots, \beta_{p^{n-1}}(p)).$$

(ii) We have $\alpha_0 = p$. Furthermore, by 6.13,

$$\alpha_n \equiv \beta_{p^n}(p) - p[CP^{p^n-1}] \bmod \operatorname{Dec}(\pi_*(MU)).$$

So, $s_{(p^n-1)}(\alpha_n) = p^{p^n} - p$. Finally, by 6.14(iii), $p|s_\omega(\alpha_n)$ for every ω. Now the equality $I(p,n) = (p, \alpha_1, \dots, \alpha_{n-1})$ can be proved as the equality $I(p,n) = (p, \beta_p(p), \dots, \beta_{p^{n-1}}(p))$ from (i) was.

(iii) Choose a system $\{x_n\}$ of polynomial generators of $MU^*(\mathrm{pt})$ as in 1.9(v). Consider a homomorphism $\rho : MU^*(\mathrm{pt}) \to \mathbb{Z}/p[t]$ such that $\rho(x_i) = 0$ for $i < p^h - 1$ and $\rho(x_{p^h-1}) \neq 0$. Then $\rho(\alpha_h) \neq 0$ because otherwise $\rho(I(p, h+1)) = 0$. So, $\operatorname{ht} \rho_* f = h$. □

Since there is a commutative diagram

$$\begin{array}{ccc} \pi_*(MU) & \xrightarrow{\subset} & \pi_*(MU) \otimes \mathbb{Q} \\ {\scriptstyle h}\downarrow & & {\scriptstyle h}\downarrow{\scriptstyle =} \\ H_*(MU) & \xrightarrow{\subset} & H_*(MU) \otimes \mathbb{Q}, \end{array}$$

we are able to consider the inclusions $\pi_*(MU) \subset H_*(MU) \subset \pi_*(MU) \otimes \mathbb{Q}$. Moreover, the ungraded formal group $f(x,y)$ over $MU^*(\mathrm{pt})$ can be regarded as a formal group over $\pi_*(MU)$.

6.16. Proposition. *The formal group $h_* f(x,y)$ is equivalent to the additive formal group over $H_*(MU)$. In other words, the logarithm of $f(x,y)$ over $\pi_*(MU) \otimes \mathbb{Q}$ is a formal power series over $H_*(MU)$, i.e.,*

$$\frac{[CP^n]}{n+1} \in H_*(MU),$$

i.e., $(n+1)|h([CP^n])$ in $H_*(MU)$. Furthermore, the elements $\dfrac{[CP^n]}{n+1}$ generate the ring $H_*(MU)$.

Proof. Let $\iota_H : S \to H\mathbb{Z}$ be the unit. Consider the morphism

$$\iota_H \wedge 1 : MU = S \wedge MU \to H\mathbb{Z} \wedge MU.$$

Then $(\iota_H \wedge 1)_* : \pi_*(MU) \to \pi_*(H\mathbb{Z} \wedge MU) = H_*(MU)$ coincides with the Hurewicz homomorphism h. Furthermore, $H\mathbb{Z} \wedge MU$ is a commutative ring spectrum (as the smash product of commutative ring spectra), and so it can be $\mathbb{C}$-oriented via $\iota_H \wedge 1$. The formal group of this $\mathbb{C}$-oriented spectrum is $h_* f(x, y)$. On the other hand, there is a ring morphism

$$1 \wedge \iota_{MU} : H\mathbb{Z} \wedge S \to H\mathbb{Z} \wedge MU.$$

So, $H\mathbb{Z} \wedge MU$ admits a $\mathbb{C}$-orientation such that the associated formal group is the additive formal group. Now, by 6.6, $h_* f(x, y)$ is equivalent to the additive formal group over $H_*(MU)$.

The characteristic numbers $S_\omega(U) \in H^{2|\omega|}(MU)$ form a basis of the group $H^*(MU)$. Now,

$$S_\omega\left(\frac{[CP^{i_1}]}{i_1+1} \times \cdots \times \frac{[CP^{i_k}]}{i_k+1}\right) = \begin{cases} 1 & \text{for } \omega = (i_1, \dots, i_k) \\ 0 & \text{for other } \omega \text{ with } |\omega| = \sum i_k. \end{cases}$$

Hence, the products $\frac{[CP^{i_1}]}{i_1+1} \times \cdots \times \frac{[CP^{i_k}]}{i_k+1}$ form a basis of $H_*(MU)$. □

6.17. Remarks. (a) By 6.16, we have

$$H_*(MU) = \mathbb{Z}[b_1, \dots, b_n, \dots], \quad b_n = \frac{[CP^n]}{n+1},$$

and $g(x) = x + \sum b_i x^{i+1}$. This gives us the description of the universal formal group mentioned after 5.4.

(b) We have $[p]_f(x) = g^{-1}(pg(x))$. Since every coefficient of $g(x)$ belongs to $H_*(MU)$, every coefficient of $[p]_f(x)$ belongs to $pH_*(MU)$. This yields another proof (and an explanation) of 6.14(iii).

Let (E, t) be a $\mathbb{C}$-oriented spectrum, let $(E[0], t)$ be its $\mathbb{Q}$-localization, and let $g(x)$ be the logarithm of the formal group of $(E[0], t)$. Let

$$\mathrm{ch} = \mathrm{ch}_E : E^*(X) \otimes \mathbb{Q} \to H^*(X; E^*(\mathrm{pt}) \otimes \mathbb{Q})$$

be the Chern–Dold character, see II.7.13. Let $t^H \in H^2(CP^\infty)$ be a $\mathbb{C}$-orientation of $H\mathbb{Z}$. The ring homomorphism $\mathbb{Z} \to E^0(\mathrm{pt}) \to E^0(\mathrm{pt}) \otimes \mathbb{Q}$ yields a morphism $a : H^2(CP^\infty) \to H^2(CP^\infty; E^*(\mathrm{pt}) \otimes \mathbb{Q})$, and $s := a(t^H)$ is a $\mathbb{C}$-orientation of $H(E^*(\mathrm{pt}) \otimes \mathbb{Q})$. Thus, $\mathrm{ch}(t)$ is a formal power series $\varphi(s)$ over $E^*(\mathrm{pt}) \otimes \mathbb{Q}$.

6.18. Theorem (Buhštaber [3]). *We have $\varphi(s) = g^{-1}(s)$. In other words,* $\mathrm{ch}(g(t)) = s$.

Proof. Firstly, we prove that $\varphi(s) = s + o(s)$. Let $\varphi(s) = \sum \varphi_i s^i$. Consider the inclusion $j_1 : S^2 = CP^1 \subset CP^\infty$ and the suspension isomorphisms

$$\mathfrak{s}^2 : \widetilde{H}^0(S^0; E^*(\mathrm{pt}) \otimes \mathbb{Q}) \to \widetilde{H}^2(S^2; E^*(\mathrm{pt}) \otimes \mathbb{Q}),$$
$$\bar{\mathfrak{s}}^2 : \widetilde{E}^0(S) \otimes \mathbb{Q} \to \widetilde{E}^2(S^2) \otimes \mathbb{Q}.$$

Then $j_1^*(s) = \mathfrak{s}^2(1)$ where $1 \in \widetilde{H}^0(S^0; E^*(\mathrm{pt}) \otimes \mathbb{Q})$, and $j_1^*(t) = \bar{\mathfrak{s}}^2(1)$ where $1 \in \widetilde{E}^0(S) \otimes \mathbb{Q}$. Since ch preserves the units and is compatible with suspensions, we have $\mathrm{ch}(j_1^* t) = j_1^*(s)$. It is clear that $j_1^* \varphi(s) = \varphi_1 j_1^*(s)$. Hence, $\varphi_1 = 1$ because

$$j_1^*(s) = \mathrm{ch}(j_1^*(t)) = j_1^*(\mathrm{ch}(t)) = j_1^*(\varphi(s)) = \varphi_1 j_1^*(s).$$

Let $p_i : CP^\infty \times CP^\infty \to CP^\infty, i = 1, 2$, be the projections. Set $s_i = p_i^*(s), i = 1, 2$. Consider the commutative diagram

$$\begin{array}{ccc} E^*(CP^\infty) \otimes \mathbb{Q} & \xrightarrow{m^*} & E^*(CP^\infty \times CP^\infty) \otimes \mathbb{Q} \\ \Big\downarrow{\scriptstyle \mathrm{ch}} & & \Big\downarrow{\scriptstyle \mathrm{ch}} \\ H^*(CP^\infty; E^*(\mathrm{pt}) \otimes \mathbb{Q}) & \xrightarrow{m^*} & H^*(CP^\infty \times CP^\infty; E^*(\mathrm{pt}) \otimes \mathbb{Q}). \end{array}$$

We have $m^*(t) = F(x, y) = x + y + \sum a_{ij} x^i y^j$, $m^*(s) = s_1 + s_2$. Hence,

$$\begin{aligned} \mathrm{ch}(m^*(t)) &= \mathrm{ch}(x) + \mathrm{ch}(y) + \sum a_{ij} (\mathrm{ch}(x))^i (\mathrm{ch}(y))^j \\ &= \varphi(s_1) + \varphi(s_2) + \sum a_{ij} (\varphi(s_1))^i (\varphi(s_2))^j = F(\varphi(s_1), \varphi(s_2)) \end{aligned}$$

(we use that ch is a homomorphism of $E^*(\mathrm{pt}) \otimes \mathbb{Q}$-algebras). On the other hand,

$$m^* \,\mathrm{ch}(t) = m^* \varphi(s) = m^* \left(\sum \varphi_i s^i \right) = \sum \varphi_i (m^*(s)^i) = \varphi(s_1 + s_2)$$

Thus, $\varphi(s_1 + s_2) = F(\varphi(s_1), \varphi(s_2))$, i.e., $\varphi^{-1}(s) = g(s)$.

In order to prove that $\mathrm{ch}(g(t)) = s$, set $g(t) = \sum g_i t^i$. Then

$$\mathrm{ch}(g(t)) = \mathrm{ch} \left(\sum g_i t^i \right) = \sum g_i (\mathrm{ch}(t))^i = \sum g_i (\varphi(s))^i = g(\varphi(s)) = s. \quad \square$$

It is interesting to ask whether the correspondence before 6.7 is surjective (or injective), i.e., whether every formal group can be realized as the formal group of a $\mathbb{C}$-oriented spectrum. The answer is *negative.* Consider the formal group $\rho_* f$ over $M\mathcal{U}^*(\mathrm{pt}) \otimes \mathbb{Z}/2$, where $\rho : M\mathcal{U}^*(\mathrm{pt}) \to M\mathcal{U}^*(\mathrm{pt}) \otimes \mathbb{Z}/2$ is the modulo 2 reduction. We claim that $\rho_* f$ cannot be realized. Indeed, let E be a $\mathbb{C}$-oriented spectrum whose formal group is $\rho_* f$. According to 2.1, E is a commutative ring spectrum, and $2\pi_*(E) = 0$. But then E is a graded Eilenberg–Mac Lane spectrum, see IX.5.5 below. Hence, $\rho_* f$ is equivalent to the additive formal group. But this is wrong because $\mathrm{ht}\, f = 1$. This is a contradiction.

On the other hand, consider the spectrum $MU \wedge M(\mathbb{Z}/2)$. It yields a cohomology theory

$$MU^*(X; \mathbb{Z}/2) := (MU \wedge M(\mathbb{Z}/2))^*(X).$$

Let $a : S \to M(\mathbb{Z}/2)$ represent the generator of $\pi_0(M(\mathbb{Z}/2)) = \mathbb{Z}/2$. The morphism $1 \wedge a : MU \to MU \wedge M(\mathbb{Z}/2)$ induces the homomorphism

$$(1 \wedge a)_* : MU^*(\text{pt}) \to MU^*(\text{pt}) \otimes \mathbb{Z}/2, \tag{6.19}$$

and $(1 \wedge a)_* = \rho$. So, we can agree that $MU^*(X; \mathbb{Z}/2)$ realizes $\rho_* f$. This hints that it makes sense to extend the class of $\mathbb{C}$-oriented spectra in order to realize formal groups. We suggest such a class below.

6.20. Definition. (a) Let E be an MU-module spectrum, and suppose that $E^*(\text{pt})$ is a commutative ring. An MU-module morphism $u : MU \to E$ is called a $\mathbb{C}$*-marking* of E if $u_* : MU^*(\text{pt}) \to E^*(\text{pt})$ is a ring homomorphism. A $\mathbb{C}$*-marked spectrum* is a pair (E, u) where u is a $\mathbb{C}$-marking of E.

(b) A $\mathbb{C}$*-marked ring spectrum* is a pair (E, u) where E is a ring spectrum and $u : MU \to E$ is a ring morphism. In this case u is called a *ring* $\mathbb{C}$*-marking*. Clearly, every $\mathbb{C}$-marked ring spectrum is a $\mathbb{C}$-marked spectrum since u turns E into an MU-module spectrum.

By 2.5, every $\mathbb{C}$-oriented spectrum is a $\mathbb{C}$-marked spectrum. Moreover, a commutative $\mathbb{C}$-marked ring spectrum is just a $\mathbb{C}$-oriented spectrum.

It is clear that $MU^*(X; \mathbb{Z}/2) = MU^*(X) \otimes \mathbb{Z}/2$ for every $X \in \mathscr{C}_T$. So, the map

$$a \wedge 1 : MU = MU \wedge S \to MU \wedge M(\mathbb{Z}/2)$$

as above turns $MU \wedge M(\mathbb{Z}/2)$ into a $\mathbb{C}$-marked spectrum. (In fact, the spectrum $MU \wedge M(\mathbb{Z}/2)$ admits a non-commutative multiplication, but this fact is non-trivial, see Araki–Toda [1] and/or VIII.2.4 below.) “

By 1.19, $E^*(X) \cong MU^*(X) \otimes_{u_*} E^*(\text{pt})$ for every finite CW-space X and every $\mathbb{C}$-marked spectrum (E, u). In particular, $E^*(CP^n) = E^*(\text{pt})[t]/(t^{n+1})$ where $t := u_*(T)$. Moreover, the following analog of 2.2 holds and can be proved as 2.2 (i.e., following Adams [8]).

6.21. Proposition. *Let (E, u) be a $\mathbb{C}$-marked spectrum and $t := u_*(T)$. Then the spectrum E satisfies conclusions* (i)–(v) *of* 2.2. □

We define the formal group of a $\mathbb{C}$-marked spectrum (E, u) to be the formal group $u_* f(T_1, T_2) \in E^*(CP^\infty \times CP^\infty) = E^*(\text{pt})[[x, y]]$. It easy to see that this formal group coincides with $m^*(t)$ where $m : CP^\infty \times CP^\infty \to CP^\infty$ is the multiplication 6.1.

6.22. Proposition. *Let (E, u) be a $\mathbb{C}$-marked spectrum with a formal group F. Let $v : MU \to E$ be another $\mathbb{C}$-marking of E, and let G be the formal group* *of (E, v). Then $F \simeq G$. Furthermore, if H is any formal group equivalent to F, then there exists a $\mathbb{C}$-marking w of E such that the formal group of (E, w) is H.*

Proof. Set $t = u_*(T) \in E^2(CP^\infty)$. Then $v_*(T) = t + \sum a_i t^{i+1} = f(t)$. Now we have $fF(x, y) = G(f(x), f(y))$, cf. the proof of 6.6.

Let $H \simeq F$, i.e., $fF(x, y) = H(f(x), f(y))$ for some formal power series $f(x) = x + \sum a_i x^{i+1}$. By setting $t_i = u_*(T_i) \in E^2(CP^\infty \times \cdots \times CP^\infty)$, we can define the characteristic classes c_i^u and the characteristic numbers $s_\omega^u \in E^{2|\omega|}(MU)$. Following (2.25), we *define* $w := \sum_\omega a_\omega s_\omega(U)$. Then $w_*(T) = f(u_*(T))$, and thus the formal group of (E, w) is H. □

6.23. Remark. It is still unknown whether every formal group can be realized by a $\mathbb{C}$-marked spectrum. There are several approaches to attack this problem. For example, if a graded formal group F is classified by a homomorphism $\rho : MU^*(S) \to R$ then one can consider the functor $MU^*(-) \otimes \rho R$. Generally speaking, it is *not* a homology theory (the exactness axiom fails). However, sometimes (see Ch. IX,§4) it is a cohomology theory (at least, on $\mathscr{C}_\mathrm{f}$) which, therefore, realizes F.

Another way is to use (co)bordism with singularities, see Ch. VIII, espesially VIII.4.14.

Also, probably, the following program can partially help to attack the realizabilty problem. For simplicity, we denote $MU^*(S)$ by L. Consider a graded commutative ring R and define a spectrum $M(R) := \vee \Sigma^i M(R_i)$; so, $\pi_*(M(R)) = R$. Set $E := MU \wedge M(R)$; then

$$E^*(S) = L \otimes R = R[x_1 \otimes 1, \dots, x_n \otimes 1, \dots]$$

where $L = \mathbb{Z}[x_1, \dots, x_n, \dots]$. Now, let F be a graded formal group over R, and let $\varphi : L \to R$ classify F. We set $y_i := x_i \otimes 1 - 1 \otimes \varphi(x_i)$. Clearly, $L \otimes R = R[y_1, \dots, y_n, \dots]$. Furthermore, we define a homomorphism $\psi : L \otimes R \to R, \psi(a \otimes b) = \varphi(a)b$. Note that $\operatorname{Ker} \psi = (y_1, \dots, y_n, \dots)$.

The obvious morphism $S \to M(R)$ (given by the unit 1 of R) induces a morphism $MU \to E$ which turns E in a $\mathbb{C}$-marked spectrum. Clearly, $F = \psi_*(G)$ where G is the formal group of E.

Now, suppose that E is a ring spectrum. We define $E(1)$ to be the cone of the morphism $y_1 : \Sigma^2 E \to E$. Clearly, $E(1)$ is a $\mathbb{C}$-marked spectrum, and $E(1)^*(S) = R[y_2, \dots, y_n, \dots]$. Furthermore, if(!) $E(1)$ is an E-module spectrum then we define a spectrum $E(2)$ to be the cone of the morphism $y_2 : \Sigma^4 E(1) \to E(1)$. So, we get a $\mathbb{C}$-marked spectrum $E(2)$ with $E(2)^*(S) = R[y_3, \dots, y_n, \dots]$. Now, if $E(2)$ is an E-module spectrum we can define $E(3)$. So, if we are able to proceed we get a tower

$$E \to E(1) \to \cdots \to E(n) \to \cdots$$

of $\mathbb{C}$-marked spectra with $E(n)^*(S) = R[y_{n+1}, \dots, y_n, \dots]$. Thus, passing to the direct limit, we get a $\mathbb{C}$-marked spectrum $E(\infty)$ whose formal group is F.

So, the problem is whether the spectra $E(n)$ are (turn out to be) E-module spectra. It seems that it holds if E is an E_∞ ring spectrum, see Elmendorf–Kriz–Mandell–May [1]. On the other hand, I do not know any counterexample where the above program does not work.

6.24. Remark. It was Novikov who discovered the connection between complex cobordism and formal groups. Namely, in Novikov [4], §5, he said explicitly that $f(u, v)$ is a formal group over $M\mathcal{U}^*(\mathrm{pt})$. Also, Miščenko computed the logarithm of this formal group, see Addendum 1 to Novikov [4]. Two years later Quillen [1] found that this formal group coincides with the universal formal group constructed by Lazard. Moreover, he gave some important applications of this fact. For example, he constructed the idempotent $\Phi : M\mathcal{U}[p] \to M\mathcal{U}[p]$ such that

$$\Phi[CP^n] = \begin{cases} [CP^n] & \text{if } n = p^k - 1 , \\ 0 & \text{otherwise.} \end{cases}$$

After this paper the intensive expansion of formal groups in algebraic topology started. Furthermore, the excellent surveys of Adams [8], Ch. II and Buhštaber–Miščenko–Novikov [1] contributed to this development.

§7. The Steenrod–tom Dieck Operations

Recall that S_ω (resp. C_ω) means $s_\omega^{M\mathcal{U},T}$ (resp. $c_\omega^{M\mathcal{U},T}$).

T. tom Dieck [1] constructed certain operations in cobordism theory. He called them Steenrod operations. Therefore we use the term Steenrod–tom Dieck operation. We expose them for the particular case of complex cobordism and $p = 2$. Here we follow mainly Buhštaber [2], cf. also Quillen [2]. At the end of the section (starting from 7.17) we discuss results of Mironov [2] describing Steenrod–tom Dieck operations on $\pi_*(M\mathcal{U})$. We need some preliminaries.

Given a space X, consider the involution $a : S^n \times X \times X$, $a(s, x, y) = (-s, y, x)$, where $-s$ is the antipode of s. Hence, we have a $\mathbb{Z}/2$-action on $S^n \times X \times X$. Set

(7.1) $$\Gamma_n(X) := S^n \times_{\mathbb{Z}/2} X \times X := (S^n \times X \times X)/(\mathbb{Z}/2).$$

Similarly, given a pointed space X, set

(7.2) $$\Gamma_n^+(X) := (S^n)^+ \wedge_{\mathbb{Z}/2} X \wedge X := ((S^n)^+ \wedge X \wedge X)/(\mathbb{Z}/2),$$

where the $\mathbb{Z}/2$-action interchanges antipodes in $(S^n)^+$ and switches factors in $X \wedge X$.

Given a complex vector bundle ξ over X, we have a complex vector bundle $p^*(\xi \times \xi)$ over $S^n \times X \times X$, where $p : S^n \times X \times X \to X \times X$ is the projection. The free involution a on $S^n \times X \times X$ yields a free fiberwise involution on the total space of $p^*(\xi \times \xi)$. Passing to quotient spaces, we have the map $\operatorname{ts}(p^*(\xi \times \xi)/(\mathbb{Z}/2)) \to \Gamma_n(X)$. This map is a complex vector bundle, which we denote by $\xi(2)$, $\dim \xi(2) = 2 \dim \xi$. It is easy to see that

$$T(\xi(2)) = \Gamma_n^+(T\xi). \tag{7.3}$$

7.4. Definition. Define the *external Steenrod–tom Dieck operation*

$$EP_n^{2r} : \widetilde{MU}^{2r}(X) \to \widetilde{MU}^{4r}(\Gamma_n^+(X))$$

as follows. Let $a \in \widetilde{MU}^{2r}(X)$ be represented by a map $f : S^{2l}X \to MU_{l+r}$. There is the Thom–Dold class $U_{\gamma(2)} \in \widetilde{MU}^{4l+4r}(\Gamma_n^+(MU_{l+r}))$ where $\gamma = \gamma^{l+r}$. Consider the homomorphism

$$h : \widetilde{MU}^{4l+4r}(\Gamma_n^+(MU_{l+r})) \xrightarrow{f^*} \widetilde{MU}^{4l+4r}\Gamma_n^+(S^{2l}X) \cong \widetilde{MU}^{4r}(\Gamma_n^+(X))$$

and define $EP_n^{2r}(a) := h(U_{\gamma(2)})$. It is easy to see that EP_n^{2r} is well-defined for every $r \in \mathbb{Z}$ and $n \geq 0$.

7.5. Theorem. *The operations EP_n^{2r} have the following properties:*

(i) *They are natural with respect to X.*

(ii) *Let $i : S^{n-1} \to S^n$ be the equatorial inclusion. Then $i^* EP_n^{2r}(a) = EP_{n-1}^{2r}(a)$ for every $a \in \widetilde{MU}^{2r}(X)$.*

(iii) *Define $j : X \wedge X \to (S^n)^+ \wedge X \wedge X$ by setting $j(x, y) := (e_0, x, y)$, where $e_0 = (1, 0, \dots, 0) \in S^n \subset \mathbb{R}^{n+1}$. Then $j^* EP_n^{2r}(a) = a \otimes a \in \widetilde{MU}^{4r}(X)$ for every $a \in \widetilde{MU}^{2r}(X)$.*

(iv) *Define $\Delta : \Gamma_n^+(X \wedge Y) \to \Gamma_n^+(X) \wedge \Gamma_n^+(Y)$ by setting*

$$\Delta(s, x_1, y_1, x_2, y_2) = (s, x_1, x_2, s, y_1, y_2)$$

where $s \in (S^n)^+, x_i \in X, y_i \in Y$. Then the diagram

$$\begin{array}{ccc}
\widetilde{MU}^{2k}(X) \otimes \widetilde{MU}^{2l}(Y) & \xrightarrow{\mu_*} & \widetilde{MU}^{2k+2l}(X \wedge Y) \\
\Big\downarrow{\scriptstyle EP_n^{2k} \otimes EP_n^{2l}} & & \Big\downarrow{\scriptstyle EP_n^{2k+2l}} \\
\widetilde{MU}^{4k}(\Gamma_n^+(X)) \otimes \widetilde{MU}^{4l}(\Gamma_n^+(Y)) & & \widetilde{MU}^{4k+4l}(\Gamma_n^+(X \wedge Y)) \\
\Big\downarrow{\scriptstyle \mu_*} & & \Big\| \\
\widetilde{MU}^{4k+4l}(\Gamma_n^+(X) \wedge \Gamma_n^+(Y)) & \xrightarrow{\Delta^*} & \widetilde{MU}^{4k+4l}(\Gamma_n^+(X \wedge Y))
\end{array}$$

commutes. (Here μ is the multiplication in MU.)

(v) $EP_n^{2r}(U_{\gamma^r}) = U_{\gamma^r(2)}$.

Proof. Decode the definitions. □

The reduced diagonal $d : X \to X \wedge X$ yields a $\mathbb{Z}/2$-equivariant map $1 \wedge d : (S^n)^+ \wedge X \to (S^n)^+ \wedge X \wedge X$, which induces, in turn, the embedding of quotients

$$\lambda : (RP^n)^+ \wedge X \to \Gamma_n^+(X). \tag{7.6}$$

7.7. Definition. Define the *Steenrod–tom Dieck operation*

$$P_n^{2r} : \widetilde{MU}^{2r}(X) \to \widetilde{MU}^{4r}((RP^n)^+ \wedge X)$$

as $P_n^{2r}(x) := \lambda^* EP_n^{2r}(x)$.

7.8. Theorem. *The Steenrod–tom Dieck operations P_n^{2r} have the following properties:*

(i) *They are natural with respect to X.*

(ii) *Let $i_n : RP^{n-1} \to RP^n$ be the canonical inclusion. Then $i_n^* P_n^{2r}(a) = P_{n-1}^{2r}(a)$ for every $a \in \widetilde{MU}^{2r}(X)$.*

(iii) *$P_n^{2r+2s}(xy) = P_n^{2r}(x) P_n^{2s}(y)$ for every $x \in \widetilde{MU}^{2r}(X)$, $y \in \widetilde{MU}^{2s}(X)$.*

(iv) *Let $f : \mathrm{pt} \to RP^n$ be any map. Consider the composition $j : X = S^0 \wedge X \xrightarrow{f^+ \wedge 1} (RP^n)^+ \wedge X$. Then $j^* P_n^{2r}(x) = x^2$.*

Proof. This follows from 7.5. □

We have $[RP^n, B\mathcal{U}_1] = [RP^n, CP^\infty] = H^2(RP^n) = \mathbb{Z}/2$, $2 \le n \le \infty$. Hence, there is only one non-trivial complex line bundle over RP^n. We denote it by ζ_n. We have $c_1(\zeta_n) \ne 0$, and so $C_1(\zeta_n) \ne 0$. Set $z_n := C_1(\zeta_n) \in MU^2(RP^n)$.

7.9. Theorem. *Set $z = z_\infty$. Let f be the (universal) formal group of (MU, T). Then $MU^*(RP^\infty) = MU^*(\mathrm{pt})[[z]]/([2]_f(z))$. (Here $[2]_f(z)$ means that we substitute z in the formal power series $[2]_f$.)*

Proof. Consider the complex line bundle $\xi := \eta^2$ over CP^∞, and let $\lambda := \{h : E \to CP^\infty\}$ be the principal $\mathcal{U}_1$-bundle associated with ξ. Recall that $\mathcal{U}_1 = S^1$, and so λ is a fibration $S^1 \to E \to CP^\infty$. Considering the homotopy exact sequence of this fibration, we conclude that $\pi_i(E) = 0$ for $i \ne 1$. Since $\chi^{H\mathbb{Z},t}(\xi) = c_1(\xi) = 2t$, we conclude (e.g., considering the Leray–Serre spectral sequence of λ and using V.1.26(iv)) that $\pi_1(E) = \mathbb{Z}/2$. Hence, $E \simeq RP^\infty$. Now, $0 \ne h^*(T) \in H^2(RP^\infty) = \mathbb{Z}/2$, and hence h classifies ζ_∞, and so $h^*T = z$.

By 2.14 and (6.4),

$$\chi^{MU,T}(\xi) = C_1(\xi) = C_1(\eta^2) = [2]_f(C_1\eta) = [2]_f(T).$$

Recall that $MU^*(CP^\infty) = MU^*(\mathrm{pt})[[T]]$. So, the multiplication by $[2]_f(T)$ is a monic endomorphism of $MU^*(CP^\infty)$. Considering the Gysin sequence V.1.25 of λ and using the equality $h^*T = z$, we complete the proof. □

This theorem can be generalized. Consider any $\mathbb{C}$-oriented spectrum (E, t) with a formal group F. If $[2]_F(z)$ is not a zero divisor in $E^*(\mathrm{pt})[[z]]$, then $E^*(RP^\infty) = E^*(\mathrm{pt})[[z]]/([2]_F(z))$, where $z = z^E := h^*t \in E^2(RP^\infty)$. Similarly, we can replace $RP^\infty = K(\mathbb{Z}/2, 1)$ by $K(\mathbb{Z}/p, 1)$ and prove that $E^*(K(\mathbb{Z}/p, 1)) = E^*(\mathrm{pt})[[z]]/([p]_F(z))$ for suitable $z \in E^2(K(\mathbb{Z}/p, 1))$ provided that $[p]_F(z)$ is not a zero divisor in $E^*(\mathrm{pt})[[z]]$.

It is easy to see that ζ_n is given by the projection

$$S^n \times_{\mathbb{Z}/2} \mathbb{C} \to S^n \times_{\mathbb{Z}/2} \mathrm{pt} = RP^n,$$

where $\mathbb{Z}/2$ acts antipodally on S^n and via multiplication by -1 on $\mathbb{C}$.

Choose a basis $(\bar{e}_1, \bar{e}_2)$ of $\mathbb{C}^2$. Let $\mathbb{Z}/2 = \{a\}$ act on $\mathbb{C}^2$ as a linear map such that $a(\bar{e}_1, \bar{e}_2) = (\bar{e}_2, \bar{e}_1)$. Consider the complex vector bundle α given by the projection $S^n \times_{\mathbb{Z}/2} \mathbb{C}^2 \to S^n \times_{\mathbb{Z}/2} \mathrm{pt} = RP^n$. Since $a(\bar{f}_1, \bar{f}_2) = (\bar{f}_1, -\bar{f}_2)$ for $\bar{f}_1 = \bar{e}_1 + \bar{e}_2$, $\bar{f}_2 = \bar{e}_1 - \bar{e}_2$, we conclude that $\alpha = \theta \oplus \zeta$.

The diagonal $d : X \to X \times X$ induces a $\mathbb{Z}/2$-equivariant map $1 \times d : S^n \times X \to S^n \times X \times X$. Passing to quotients, we get a map

$$\ell : RP^n \times X \to S^n \times_{\mathbb{Z}/2} X \times X = \Gamma_n(X).$$

Let $\pi_1 : RP^n \times X \to RP^n$, $\pi_2 : RP^n \times X \to X$ be the projections. Let ξ be a complex vector bundle over X.

7.10. Lemma. $\ell^*\xi(2) = \pi_2^*\xi \otimes \pi_1^*(\alpha) = \pi_2^*\xi \otimes \pi_1^*(\theta \oplus \zeta) = \pi_2^*\xi \oplus (\pi_1^*\zeta \otimes \pi_2^*\xi)$.

Proof. Let $p : S^n \times X \times X \to X \times X$, $p_2 : S^n \times X \to X$ be the projections. We have $(1 \times d)^* p^*(\xi \times \xi) = p_2^*\xi \otimes \theta^2$. Passing to quotients, we get the desired formula. □

Consider the composition

$$RP^n \times X \xrightarrow{k} RP^n \times X \xrightarrow{\ell} \Gamma_n(X),$$

where $k(a, x) = (*, x)$. Passing to Thom spaces, we get the diagram

$$\begin{array}{ccccc}
T(k^*\ell^*\xi(2)) & \xrightarrow{\bar{k}} & T(\ell^*\xi(2)) & \xrightarrow{\bar{\ell}} & T\xi(2) \\
\| & & \| & & \| \\
(RP^n)^+ \wedge T\xi & \xrightarrow{\bar{k}} & T(\ell^*\xi(2)) & \xrightarrow{\bar{\ell}} & \Gamma_n^+(T\xi).
\end{array}$$

It is easy to see that the bottom map $\bar{\ell}\bar{k}$ is just the map λ in (7.6) for $X = T\xi$.

7.11. Lemma. *We have $C_2(\ell^*\xi(2)) = C_1(\xi)(z + C_1(\xi) + \sum a_{ij}z^iC_1(\xi)^j)$ in $M\mathcal{U}^4(RP^n \times X)$.*

Here a_{ij} are the coefficients of the universal formal group $f(x, y)$, and we consider the elements $z \in M\mathcal{U}^2(RP^n)$ and $C_1(\xi) \in M\mathcal{U}^2(X)$ as elements of $M\mathcal{U}^2(RP^n \times X)$.

Proof. We have (omitting π_1^* and π_2^*)

$$\begin{aligned} C_2(\ell^*\xi(2)) &= C_2(\xi \oplus (\zeta \otimes \xi)) = C_2(\xi) + C_2(\zeta \otimes \xi) + C_1(\xi)C_1(\zeta \otimes \xi) \\ &= C_1(\xi)f(C_1(\zeta), C_1(\xi)) \\ &= C_1(\xi)(z + C_1(\xi) + \sum a_{ij}z^i(C_1(\xi))^j). \end{aligned}$$

□

7.12. Theorem. *$P_n^2(T) = T(z + T + \sum a_{ij}z^iT^j) \in \widetilde{M\mathcal{U}}^4((RP^n)^+ \wedge T\eta)$ where $T \in \widetilde{M\mathcal{U}}^2(CP^\infty)$ is the universal $\mathbb{C}$-orientation.*

Proof. Consider a map $\varepsilon : (RP^n \times CP^\infty)^+ \to RP^n \times CP^\infty, \varepsilon(x) = x$ for every $x \in RP^n \times CP^\infty$. Let

$$\begin{aligned} \mathfrak{z}_1 &: CP^\infty \to T\eta \\ \mathfrak{z}_2 &: RP^n \times CP^\infty \to (RP^n)^+ \wedge T\eta = T(\pi_2^*\eta) \\ \mathfrak{z}_3 &: RP^n \times CP^\infty \to T(\ell^*\eta(2)) \end{aligned}$$

be the zero sections as in IV.5.4. Note that $\bar{k}\mathfrak{z}_2 = \mathfrak{z}_3$. We have $C_1(\eta) = \varepsilon^*T$ and $\mathfrak{z}_1^*U_\eta = T$. Furthermore, $P_n^2(U_\eta) = (\bar{\ell}\bar{k})^*EP_n^2(U_\eta) = \bar{k}^*U_{\ell^*\eta(2)}$. So,

$$\begin{aligned} \varepsilon^*P_n^2(T) &= \varepsilon^*P_n^2(\mathfrak{z}_1^*U_\eta) = \varepsilon^*\mathfrak{z}_2^*P_n^2(U_\eta) = \varepsilon^*\mathfrak{z}_2^*\bar{k}^*U_{\ell^*\eta(2)} = \varepsilon^*\mathfrak{z}_3^*U_{\ell^*\eta(2)} \\ &= C_2(\ell^*\eta(2)) = C_1(\eta)(z + C_1(\eta) + \sum a_{ij}z^i(C_1(\eta))^j) \\ &= \varepsilon^*(T(z + T + \sum a_{ij}z^iT^j)). \end{aligned}$$

Since ε^* is monic, the theorem is proved. □

7.13. Corollary. *Let $U_N \in \widetilde{M\mathcal{U}}^{2N}(M\mathcal{U}_N)$ be the universal Thom–Dold class. Then*

$$P_n^{2N}(U_N) = \sum_\omega z^{N-|\omega|}a_\omega(z)S_\omega(U_N) \in \widetilde{M\mathcal{U}}^{2N}((RP^n)^+ \wedge M\mathcal{U}_N)$$

for some $a_\omega(z) \in M\mathcal{U}^0(\mathrm{pt})[z], a_{(0)} = 1$.

Proof. By 1.29, $\mathfrak{z} : CP^\infty \to T\eta$ is a homotopy equivalence. Furthermore, $\mathfrak{z}^*U_\eta = T$, and so, by 7.12,

$$P_n^2(U_\eta) = U_\eta(z + U_\eta + \sum a_{ij} z^i (U_\eta)^j) = \sum_{|\omega| \le 1} z^{1-|\omega|} a_\omega(z) S_\omega(U_\eta).$$

Let $D_1 \cdots D_N \in \widetilde{MU}^{2N}(T\eta \wedge \cdots \wedge T\eta)$ be as in 2.16. By 7.8(iii),

$$P_n^{2N}(D_1 \cdots D_N) = \sum z^{N-|\omega|} a_\omega(z) S_\omega(D_1 \cdots D_N)$$

in $\widetilde{MU}^{2N}((RP^n)^+ \wedge T\eta \wedge \cdots \wedge T\eta)$. Furthermore, by 2.16 and 1.21, the homomorphism

$$(1_{(RP^n)^+} \wedge Te_N)^* : \widetilde{MU}^*((RP^n)^+ \wedge MU_N) \to \widetilde{MU}^*((RP^n)^+ \wedge T\eta \wedge \cdots \wedge T\eta)$$

is monic, and $(Te_N)^*(U_N) = D_1 \cdots D_N$. Thus, the result follows because of the naturality of the operations. □

7.14. Corollary. *Suppose that an element $x \in \widetilde{MU}^{2q}(X)$ is represented by a map $f : \Sigma^{2m} X \to MU_{q+m}$. Then*

$$z^m P_n^{2q}(x) = \sum z^{q+m-|\omega|} a_\omega S_\omega(x) \in \widetilde{MU}^{4q+2m}((RP^n)^+ \wedge X).$$

Proof. Let $\sigma^{2m} \in \widetilde{MU}^{2m}(S^{2m})$ be the image of the unit $1 \in \widetilde{MU}^0(S^0)$ under the suspension isomorphism. Then $f^* U_{q+m} = \sigma^{2m} x$. Hence, by 7.13,

$$P_n^{2(q+m)}(\sigma^{2m} x) = \sum z^{q+m-|\omega|} a_\omega \sigma^{2m} S_\omega(x)$$

because $S_\omega(\sigma^{2m}) = 0$ for $|\omega| > 0$. On the other hand, by 7.13, $P_n^{2m}\sigma^{2m} = z^m \sigma^{2m}$ because σ^{2m} can be represented by the root $S^{2m} \to MU_m$. Hence, by 7.8(iii),

$$P_n^{2(q+m)}(\sigma^{2m} x) = P_n^{2m}(\sigma^{2m}) P_n^{2q}(x) = z^m \sigma^{2m} P_n^{2q}(x).$$

Equating the right hand sides of these equalities, we conclude that

$$z^m \sigma^{2m} P_n^{2q}(x) = \sum z^{q+m-|\omega|} a_\omega \sigma^{2m} S_\omega(x).$$

But multiplication by σ^{2m} is the (suspension) isomorphism

$$\widetilde{MU}^*((RP^n)^+ \wedge X) \to \widetilde{MU}^{*+2m}(S^{2m} \wedge (RP^n)^+ \wedge X). \qquad \square$$

In order to apply 7.14, we need the following technical lemma based on 7.9. Consider the formal power series

$$\phi(x) := \frac{[2]_f(x)}{x} \in MU^*(\mathrm{pt})[[x]].$$

7.15. Lemma. *Let $b \in \widetilde{MU}^q((RP^{2n})^+ \wedge X)$ be such that $z_{2n}b = 0$. Then there exists $\lambda \in \widetilde{MU}^q(X)$ such that $k_n^* b = \lambda\phi(z_{2n-2})$, where $k_n : (RP^{2n-2})^+ \wedge X \to (RP^{2n})^+ \wedge X$ is induced by the canonical inclusion $RP^{2n-2} \to RP^{2n}$.*

Proof. Firstly, $T\zeta_{2n-2} = RP^{2n}/RP^1 = RP^{2n}/S^1$, cf. Stong [3], Ch. VIII, Lemma 1. Hence, we have a Thom–Dold isomorphism

$$\varphi : MU^i(RP^{2n-2}) \to \widetilde{MU}^{i+2}(RP^{2n}/S^1).$$

Let $q : RP^{2n} \to RP^{2n}/S^1$ be the quotient map. It is easy to see that

$$q^*\varphi : MU^i(RP^{2n-2}) \to \widetilde{MU}^{i+2}(RP^{2n})$$

maps $1 \in MU^0(RP^{2n-2})$ to $z_{2n} \in \widetilde{MU}^2(RP^{2n})$.

Consider the commutative diagram

$$\begin{array}{ccccc}
\widetilde{MU}^1(S^1) & \xrightarrow{\delta} & \widetilde{MU}^2(RP^{2n}, S^1) & \xrightarrow{q^*_{MU}} & \widetilde{MU}^2(RP^{2n}) \\
\downarrow u & & \downarrow u & & \downarrow u \\
H^1(S^1) & \xrightarrow{\delta_H} & H^2(RP^{2n}, S^1) & \xrightarrow{q^*_H} & H^2(RP^{2n}) \\
\| & & \| & & \| \\
\mathbb{Z} & \longrightarrow & \mathbb{Z} & \longrightarrow & \mathbb{Z}/2
\end{array} .$$

Here the rows are the exact sequences of the pair (RP^{2n}, S^1) and u is given by a Thom class $u \in H^0(MU)$.

Let $\iota \in \widetilde{MU}^1(S^1) = \mathbb{Z}$ be a generator. Since q_H^* is epic, $\delta_H(u\iota) = 2\alpha$ for some α with $q_H^*(\alpha) = uz_{2n}$. Let $\widehat{\zeta}_{2n}$ be a complex line bundle over RP^{2n}/S^1 with $c_1(\widehat{\zeta}_{2n}) = \alpha$. Since $q_H^*(\alpha) = uz_{2n}$, $q^*\widehat{\zeta}_{2n} = \zeta_{2n}$. Since ζ_{2n}^2 is trivial, $C_1(\zeta_{2n}^2) = 0 = q_{MU}^*(C_1(\widehat{\zeta_{2n}^2}))$, and so $C_1(\widehat{\zeta_{2n}^2}) = \delta(m\iota)$ for some $m \in \mathbb{Z}$. Now,

$$2m\alpha = \delta_H(mu\iota) = uC_1(\widehat{\zeta_{2n}^2}) = u\delta(m\iota) = m\delta_H(u\iota).$$

So, $m = 1$, and hence $\delta\iota = C_1(\widehat{\zeta_{2n}^2})$.

We have $C_1(\widehat{\zeta_{2n}^2}) = \varphi(x_n)$ for some $x_n \in MU^0(RP^{2n-2})$. Since the map $i = qk_n : RP^{2n-2} \to RP^{2n} \to RP^{2n}/S^1$ coincides with the zero section of $T\zeta_{2n-2}$, we conclude, by V.1.26(i), that $i^*\varphi(y) = z_{2n-2}y$ for every $y \in MU^*(RP^{2n-2})$. Hence,

$$z_{2n-2}x_n = i^*\varphi(x) = i^*C_1(\widehat{\zeta_{2n}^2}) = C_1(i^*(\widehat{\zeta}_{2n})^2) = C_1(\zeta_{2n-2}^2) = 0.$$

In particular, for $n = \infty$ we have $z_\infty x_\infty = 0$. Consider the epimorphism $\pi : MU^*(\mathrm{pt})[[z]] \to MU^*(RP^\infty), \pi(z) = z_\infty$, as in 7.9. By 7.9, there is $x \in MU^*(\mathrm{pt})[[z]]$ such that $\pi(x) = x_\infty$, and

$$xz = [2]_f(z)\left(\sum_{i=0}^{\infty} a_i x^{i+1}\right), a_i \in M\mathcal{U}^{-2i}(\mathrm{pt})$$

in $M\mathcal{U}^*(\mathrm{pt})[[z]]$. Hence,

$$x_\infty = \phi(z_\infty)\left(a_0 + \sum a_i z_\infty^{i+1}\right) = a_0\phi(z_\infty).$$

Let $b_n : RP^{2n} \to RP^\infty$ be the inclusion. Considering the commutative diagram

$$\begin{array}{ccccc}
\widetilde{M\mathcal{U}}^1(S^1) & \xrightarrow{\delta} & \widetilde{M\mathcal{U}}^2(RP^\infty/S^1) & \xleftarrow{\varphi} & M\mathcal{U}^0(RP^\infty) \\
\| & & \downarrow e_n^* & & \downarrow b_{n-1}^* \\
\widetilde{M\mathcal{U}}^1(S^1) & \xrightarrow{\delta} & \widetilde{M\mathcal{U}}^2(RP^{2n}/S^1) & \xleftarrow{\varphi} & M\mathcal{U}^0(RP^{2n-2})
\end{array}$$

and using that $b_{n-1}^* x_\infty = x_n$, we conclude that $x_n = a\phi(z_{2n-2}), a \in M\mathcal{U}^0(\mathrm{pt})$ for every n, i.e.,

$$\varphi^{-1}\delta\iota = a\phi(z_{2n-2}) \tag{7.16}$$

for every n and some $a \in M\mathcal{U}^0(\mathrm{pt})$.

Let $\varepsilon : (RP^{2n})^+ \to S^0$ collapse RP^{2n}. The composition

$$\widetilde{M\mathcal{U}}^{i-1}((RP^{2n})^+ \wedge X) \xrightarrow{z_{2n}} \widetilde{M\mathcal{U}}^{i+1}((RP^{2n})^+ \wedge X) \xrightarrow{\varepsilon_*} \widetilde{M\mathcal{U}}^{i+1}(X)$$

is trivial, i.e., $z_{2n}x \in \widetilde{M\mathcal{U}}^{i+1}(RP^{2n} \wedge X) \subset \widetilde{M\mathcal{U}}^{i+1}((RP^{2n})^+ \wedge X)$. So, we can consider the homomorphism

$$z_{2n} : \widetilde{M\mathcal{U}}^{i-1}((RP^{2n})^+ \wedge X) \to \widetilde{M\mathcal{U}}^{i+1}(RP^{2n} \wedge X).$$

Now we prove the lemma when $X = Y^+$ for some Y. The projection $p : RP^{2n-2} \times Y \to RP^{2n-2}$ induces a bundle $p^*\zeta_{2n-2}$ over $RP^{2n-2} \times Y$, and $T(p^*\zeta_{2n-2}) = RP^{2n}/S^1 \wedge Y^+$. So, we have a Thom–Dold isomorphism

$$\begin{aligned}
\psi : M\mathcal{U}^{i-1}(RP^{2n-2} \times Y) \to \widetilde{M\mathcal{U}}^{i+1}&((RP^{2n}/S^1) \wedge Y^+)| \\
&= \widetilde{M\mathcal{U}}^{i+1}(RP^{2n} \wedge Y^+, S^1 \wedge Y^+).
\end{aligned}$$

Consider the diagram

$$\begin{array}{ccc}
\widetilde{M\mathcal{U}}^{i-1}(Y^+) & \xrightarrow{\mathfrak{s},\cong} & \widetilde{M\mathcal{U}}^{i}(S^1 \wedge Y^+) \\
\tilde{\delta}\downarrow & & \delta\downarrow \\
\widetilde{M\mathcal{U}}^{i-1}((RP^{2n-2})^+ \wedge Y^+) & \xrightarrow{\psi,\cong} & \widetilde{M\mathcal{U}}^{i+1}(RP^{2n} \wedge Y^+, S^1 \wedge Y^+) \\
k_n^*\uparrow & & \gamma\downarrow \\
\widetilde{M\mathcal{U}}^{i-1}((RP^{2n})^+ \wedge Y^+) & \xrightarrow{z_{2n}} & \widetilde{M\mathcal{U}}^{i+1}(RP^{2n} \wedge Y^+)
\end{array}$$

Here the right line is the exact sequence of the pair $(RP^{2n}\wedge Y^+, S^1\wedge Y^+)$, $\mathfrak{s}$ is the suspension isomorphism, and $\tilde{\delta} := \psi^{-1}\delta\mathfrak{s}$. This diagram commutes. Indeed, it is a diagram of $\widetilde{MU}^*(Y)$-modules and homomorphisms, so, it suffices to check its commutativity for $Y = \mathrm{pt}$. We leave this to the reader. (In fact, this follows from the equality $q^*\varphi(1) = z_{2n}$, see the very beginning of the proof.) Furthermore, since this diagram is a diagram of $\widetilde{MU}^*(Y)$-modules, $\tilde{\delta}(x) = xa\phi(z_{2n-2})$ for every $x \in \widetilde{MU}^{i-1}(Y^+)$ by (7.16). Now,

$$\begin{aligned} z_{2n}b = 0 &\Longrightarrow \gamma\psi k_n^*(b) = 0 \Longrightarrow \psi k_n^* b = \delta(\mu) \\ &\Longrightarrow k_n^*(b) = \tilde{\delta}(\mathfrak{s}^{-1}\mu) = \mathfrak{s}^{-1}\mu a\phi(z_{2n-2}) = \lambda\phi(z_{2n-2}) \end{aligned}$$

for some $\lambda \in \widetilde{MU}^{i-1}(Y^+)$.

Finally, for an arbitrary CW-space X consider the diagram $X \xrightarrow{i} X^+ \xrightarrow{r} X$ with $ri = 1_X$. Given $b \in \widetilde{MU}^*((RP^{2n})^+ \wedge X)$ with $z_{2n}b = 0$, we have $z_{2n}r^*b = 0$. Hence, $k_n^*r^*b = \lambda\phi(z_{2n-2})$ for some $\lambda \in \widetilde{MU}^*(X^+)$. Thus,

$$k_n^*b = k_n^*i^*r^*b = i^*k_n^*r^*b = i^*(\lambda\phi(z_{2n-2})) = (i^*\lambda)\phi(z_{2n-2}). \qquad \square$$

7.17. Proposition. *Let X be a finite CW-space. If $\widetilde{MU}^*(X)$ has no 2-torsion, then*

$$P_{2n}^{2r}(x+y) = P_{2n}^{2r}(x) + P_{2n}^{2r}(y) + xy\phi(z_{2n})$$

for every $x, y \in \widetilde{MU}^{2r}(X)$.

Proof. Note that $S_\omega(x+y) = S_\omega(x) + S_\omega(y)$. Hence, by 7.14,

$$z_{2k}^m(P_{2k}^{2r}(x+y) - P_{2k}^{2r}(x) - P_{2k}^{2r}(y)) = 0$$

for every k and suitable m (here we use that X is finite). We set $k = m+n$. Therefore, by 7.15,

$$z_{2k-2}^{m-1}(P_{2k-2}^{2r}(x+y) - P_{2k-2}^{2r}(x) - P_{2k-2}^{2r}(y)) = \lambda\phi(z_{2k-2}).$$

Let $j : X \to (RP^{k-1})^+ \wedge X$ be as in 7.8(iv). Since $j^*z_{2k-2} = 0$, we conclude that $j^*(\lambda\phi(z_{2k-2})) = 0$, i.e., $j^*(\lambda)j^*(\phi(z_{2k-2})) = 0$. But $j^*\lambda = \lambda$, while $j^*(\phi(z_{2k-2})) = 2$. So, $2\lambda = 0$, and so $\lambda = 0$. Therefore,

$$z_{2k-2}^{m-1}(P_{2k-2}^{2r}(x+y) - P_{2k-2}^{2r}(x) - P_{2k-2}^{2r}(y)) = 0.$$

After iteration, we get $z_{2k-2m}(P_{2k-2m}^{2r}(x+y) - P_{2k-2m}^{2r}(x) - P_{2k-2m}^{2r}(y)) = 0$. Hence, by 7.15,

$$P_{2k-2m}^{2r}(x+y) - P_{2k-2m}^{2r}(x) - P_{2k-2m}^{2r}(y) = \lambda'\phi(z_{2k-2m-2}),$$

i.e.,

$$P_{2n}^{2r}(x+y) - P_{2n}^{2r}(x) - P_{2n}^{2r}(y) = \lambda'\phi(z_{2n}).$$

By 7.8(iv), $j^*P^{2r}_{2n}(a) = a^2$ for every $a \in \widetilde{MU}^{2r}(X)$. Hence,

$$\begin{aligned} 2xy = (x+y)^2 - x^2 - y^2 &= j^*(P^{2r}_{2n}(x+y) - P^{2r}_{2n}(x) - P^{2r}_{2n}(y)) \\ &= j^*(\lambda'\phi(z_{2n})) = 2\lambda'. \end{aligned}$$

Thus, $\lambda' = xy$. □

Let I denote the ideal $(2, \operatorname{Dec} MU^*(\mathrm{pt}))$ of $MU^*(\mathrm{pt})$.

7.18. Theorem. *Set* $[2]_f(x) = 2x + \sum a_k x^k$, $\deg a_k = 2 - 2k$. *Then*

$$P^{2-2k}_2(a_k) \equiv z_2 a_{2k} \bmod I.$$

Proof. Let $h : CP^\infty \to CP^\infty$ be the composition

$$CP^\infty \xrightarrow{d} CP^\infty \times CP^\infty \xrightarrow{m} CP^\infty,$$

i.e., $h^*(T) = [2]_f(T)$. Hence, by 7.12, $P^2_2(T) = T(z + T + \sum a_{ij} z^i T^j)$, where $z = z_2$. Now, by 7.17, ($\equiv$ means $\equiv \bmod I$)

$$\begin{aligned} P^2_2(h^*(T)) = P^2_2\left(2T + \sum a_k T^k\right) &\equiv \sum P^{2-2k}_2(a_k)(P^2_2(T))^k \\ &= \sum P^{2-2k}_2(a_k)\left(Tz + T^2 + \sum a_{ij} z^i T^{j+1}\right)^k \\ &\equiv \sum P^{2-2k}_2(a_k)(T^{2k} + kzT^{2k-1}). \end{aligned}$$

(To be rigorous, we remark that we applied 7.17 to the space CP^∞ and considered infinite sums; but one can do the same calculations in CP^N with N large enough.)

On the other hand,

$$\begin{aligned} h^*(P^2_2(T)) &= h^*\left(T(z + T + \sum a_{ij} z^i T^j)\right) \\ &= h^*(T)\left(z + h^*(T) + \sum a_{ij} z^i (h^*(T))^j\right) \equiv z h^*(T) = z \sum a_k T^k. \end{aligned}$$

So, $P^{2-2k}_2(a_k) \equiv z_2 a_{2k} \bmod I$. □

7.19. Corollary. *Let* $\{2, x_1, \dots, x_k, \dots\}, x_i \in MU^{2-2^{i+1}}(\mathrm{pt})$ *be a sequence such that* $I(2,n) = (2, \dots, x_{n-1})$ *for every* n. *Then* $P^{2-2^k}_2(x_{k-1}) \equiv z_2 x_k \bmod I$.

Proof. By 6.15(ii), $I(2,n) = (2, a_1, \dots a_{2^{n-1}-1})$. Hence, $x_k \equiv a_{2^k-1} \bmod I$, and the corollary follows. □

7.20. Remark. Using Steenrod–tom Dieck operations (for all primes p), Quillen [2] proved 6.11. The only necessary information about $\pi_*(MU)$ is that they are finitely generated! So, in view of 5.4, this yields an *alternative way* (without the Adams spectral sequence) of calculating $\pi_*(MU)$. Lemma 7.15 plays a key role in this calculation.

Chapter VIII. (Co)bordism with Singularities

(Co)bordism with singularities have many applications. However, in this book we mainly consider only a few aspects of this theory: namely, we want to demonstrate that (co)bordism with singularities establishes a big source of interesting (co)homology theories and, in particular, enables us to construct cohomology theories with prescribed properties (e.g., realizing certain formal groups, etc.)

§1. Definitions and Basic Properties

Let $\varphi : B \to B\mathcal{V}$ be a multiplicative structure map, where $\mathcal{V}$ is $\mathcal{O}$ or $\mathcal{PL}$, see IV.4.22. Let $\mathscr{K}$ be the class of all compact (B, φ)-manifolds (smooth for $\mathcal{V} = \mathcal{O}$ and PL for $\mathcal{V} = \mathcal{PL}$). For simplicity of notation, we say "$\mathscr{K}$-manifold" instead of "compact (B, φ)-manifold". Because of the multiplicativity of (B, φ), the product of two $\mathscr{K}$-manifolds is a $\mathscr{K}$-manifold in a canonical way. We also require that $M \times N$ and $(-1)^{\dim M \dim N} N \times M$ are isomorphic (B, φ)-manifolds for every $M, N \in \mathscr{K}$.

Let $L = L_{\mathscr{K}}$ be the (co)bordism theory based on $\mathscr{K}$, i.e., L is the (B, φ)-(co)bordism theory.

Fix a closed manifold $P \in \mathscr{K}$, $\dim P = d$. Consider a manifold $M \in \mathscr{K}$ with $\partial M = P \times Z$ where Z is a closed $\mathscr{K}$-manifold, and form a polyhedron

$$K := Z \times C(P) \cup_{\varphi} M. \tag{1.1}$$

Here $C(P)$ is the cone over P and $\varphi : Z \times P \to \partial M$ is a $\mathscr{K}$-isomorphism (e.g. an oriented diffeomorphism if $\mathscr{K}$ is the class of oriented smooth manifolds); also, the inclusion $Z \times P \subset Z \times C(P)$ is given by the inclusion of the bottom $P \subset C(P)$. Clearly, every closed manifold $N \in \mathscr{K}$ has this form if we put $Z = \emptyset$ and $N = M$. On the other hand, K turns out to be a manifold if we delete the set $Z \times \{*\}$ from it.

1.2. Examples. (a) The wedge $S^1 \vee S^1$ has the form (1.1) with $P = \{4 \text{ points}\}$ and $Z = \text{pt}$ since a neighborhood of the singular point is the cone over 4 points. More generally, the wedge $M_1^n \vee \ldots \vee M_k^n$ of closed n-manifields has the form (1.1) with $P = S^{n-1} \sqcup \ldots \sqcup S^{n-1}$ (k times).

(b) Consider the union of two intersecting circles in $\mathbb{R}^2$. It has the form (1.1) with $P = \{4 \text{ points}\}$ and $Z = \{2 \text{ points}\}$. More generally, we can consider the union of two intersected n-spheres in $\mathbb{R}^{n+1}$ which has the form (1.1) with $P = \{4 \text{ points}\}$ and $Z = S^{n-1}$.

(c) If $P = \{m \text{ points}\}$ and $\dim K = n$ then $H_n(K; \mathbb{Z}/m) = \mathbb{Z}/m$ (provided that K is connected). So, such polyhedra give us good models for homology classes mod m, as closed manifolds give us models for integral homology classes. Novikov and Rochlin (1965, unpublished, mentioned in Novikov [5]) used such objects when they considered topological invariance of Pontrjagin classes mod m.

(d) Let W^{n+1} be a smooth manifold such that its boundary ∂W is a nonstandard homotopy sphere, i.e., ∂W is PL isomorphic but not diffeomorphic to the standard sphere S^n, see Kervaire–Milnor [1]. Then $W \cup C(\partial W)$ is a PL manifold, but it turns out to be a smooth manifold with a singularity; it has the form (1.1) with $P = S^n$ and $Z = \text{pt}$.

Sullivan [1] suggested considering bordism theories based on polyhedra of the form (1.1) as "closed manifolds". (The corresponding concept of a "manifold with boundary" can be introduced as well.) Furthermore, this construction can be iterated: we can consider a family $\{P_1, \dots, P_m\}$ of $\mathscr{K}$-manifolds, not only single P.

Baas [1] formalized these ideas successfully. Now we expose his approach. Related material is also contained in Botvinnik [1], Vershinin [1].

1.3. Definition. (a) A *k-dimensional $\mathscr{K}$-manifold with Sullivan–Baas P-singularity* is a quintuple $(V, \partial_0 V, \partial_1 V, \delta V, \varphi)$, where

(1) V is a $\mathscr{K}$-manifold, $\dim V = k$, with $\partial V = \partial_0 V \cup \partial_1 V$ where $\partial_i V, i = 1, 2$ is a $\mathscr{K}$-manifold.

(2) $\partial\partial_0 V = \partial_0 V \cap \partial_1 V = \partial\partial_1 V$.

(3) δV is a certain $\mathscr{K}$-manifold and $\varphi : \delta V \times P \to \partial_1 V$ is a $\mathscr{K}$-isomorphism.

For simplicity, below we say just "$\mathscr{K}^P$-manifold" or "Sullivan–Baas $\mathscr{K}^P$-manifold" instead of "manifold with Sullivan–Baas P-singularity" and "$\mathscr{K}^P$-manifold V" instead of "$\mathscr{K}^P$-manifold $(V, \partial_0 V, \partial_1 V, \delta V, \varphi)$".

(b) Define the *$\mathscr{K}^P$-boundary* ∂^P of a $\mathscr{K}^P$-manifold $(V, \partial_0 V, \partial_1 V, \delta V, \varphi)$ to be the $\mathscr{K}^P$-manifold

$$\partial^P(V, \partial_0 V, \partial_1 V, \delta V, \varphi) := (\partial_0 V, \emptyset, \partial_0 V \cap \partial_1 V, \partial\delta V, \varphi|\partial\delta V).$$

Note that $\partial^P \partial^P V = 0$ for every V.

A *closed $\mathscr{K}^P$-manifold* is a $\mathscr{K}^P$-manifold M such that $\partial^P M = \emptyset$, i.e., there is a fixed $\mathscr{K}$-isomorphism $\varphi : \delta M \times P \to \partial M$, where δM is a closed $\mathscr{K}$-manifold.

Denoting by $\mathscr{K}^P$ the class of $\mathscr{K}^P$-manifolds, we have in fact a *cobordism category* $(\mathscr{K}^P, \partial^P)$, see Stong [3].

It is clear that every $\mathscr{K}$-manifold V can be considered as a $\mathscr{K}^P$-manifold V with $\delta V = \emptyset$. Furthermore, a closed $\mathscr{K}^{\mathrm{pt}}$-manifold is just a $\mathscr{K}$-manifold with boundary.

1.4. Definition. A k-dimensional *singular $\mathscr{K}^P$-manifold* in a pair (X, A) is a map $f : (V, \partial^P V) \to (X, A)$ of a k-dimensional $\mathscr{K}^P$-manifold V such that there exists a commutative diagram

$$\begin{array}{ccc} V & \xrightarrow{f} & X \\ \uparrow & & \uparrow{\scriptstyle f_0} \\ \delta V \times P & \xrightarrow{p_1} & \delta V, \end{array}$$

where the left map $\delta V \times P \to V$ is the embedding $\delta V \times P \xrightarrow{\varphi} \partial_1 V \subset \partial V \subset V$. In other words, $f|\partial_1 V$ passes through $p_1 : \delta V \times P \to \delta V$, i.e., $f\varphi(b, p) = f_0(b)$ for every $b \in \delta V, p \in P$. The commutativity of the diagram above formalizes the gluing of the cone in the Sullivan construction (1.1). Note that f_0 is unique if it exists.

Of course, $\partial_0 V = \emptyset$ if $A = \emptyset$. As usual, a singular manifold in $(X, \emptyset)$ is called a singular manifold in X.

We say that a singular closed $\mathscr{K}^P$-manifold $f : M \to X$ bounds if there exists a singular $\mathscr{K}^P$-manifold $g : V \to X$ with $\partial^P V = M$ and $g|M = f$. In this case we also write $\partial^P(V, g) = (M, f)$. Now, we can define a *$\mathscr{K}^P$-bordism relation* on the class of closed singular $\mathscr{K}^P$-manifolds: two closed singular $\mathscr{K}^P$-manifolds (M, f) and (N, g) are bordant if $(M, f) \sqcup (-N, -g) = \partial^P(V, h)$ for some (V, h). Here $(-g) : (-N) \to X$ coincides with g as a map of spaces, but N is equipped with the opposite (B, φ)-structure (cf. IV.7.25). In this case we say that (V, h) is a *$\mathscr{K}^P$-membrane* (or *$\mathscr{K}^P$-bordism*) between (M, f) and (N, g).

The $\mathscr{K}^P$-bordism class of a closed singular manifold $f : M \to X$ is denoted by $[M, f]$, as usual. Similarly to IV.7.26, for every $n \in \mathbb{Z}$ we have the n-dimensional $\mathscr{K}^P$-bordism group of X: its elements are $\mathscr{K}^P$-bordism classes of n-dimensional $\mathscr{K}^P$-manifolds, and the operation is induced by the disjoint union of $\mathscr{K}^P$-manifolds. We denote this group by $L_n^P(X)$ since we denote by L_n the n-dimensional $\mathscr{K}$-bordism group. For example, MU^{CP^1} is the complex bordism theory with CP^1-singularity.

1.5. Remark. Consider the cobordism category $\mathscr{K}(X)$ generated by singular $\mathscr{K}$-manifolds in X, see Stong [3], Ch. IV, Example 6. There is a functor $P : \mathscr{K}(X) \to \mathscr{K}(X)$, which transforms a map $f : M \to X$ to the map $M \times P \xrightarrow{p_1} M \xrightarrow{f} X$. It is easy to see that $L_*^P(X)$ is the relative bordism group constructed by P, as defined in Stong [3], Ch. I.

We have the following homomorphisms (natural with respect to X):

$$\begin{aligned} P :& L_i(X) \to L_{i+d}(X),\ \{f : M \to X\} \mapsto \{M \times P \xrightarrow{p_1} M \xrightarrow{f} X\}, \\ r = r^X :& L_i(X) \to L_i^P(X),\ \text{a manifold } M \in \mathscr{K} \text{ is considered as} \\ & \text{the } \mathscr{K}^P\text{-manifold } M \text{ with } \delta M = \emptyset = \partial_1 M, \\ \delta = \delta^X :& L_i^P(X) \to L_{i-d-1}(X),\ \delta[M, f] := [\delta M, f_0] \text{ with } f_0 \text{ as in 1.4.} \end{aligned}$$

1.6. Theorem–Definition. *For every space X the sequence*

$$\cdots \to L_n(X) \xrightarrow{P} L_{n+d}(X) \xrightarrow{r} L_{n+d}^P(X) \xrightarrow{\delta} L_{n-1}(X) \to \cdots \tag{1.7}$$

is exact. This sequence is called the Bockstein–Sullivan–Baas *exact sequence.*

Proof. We prove that $rP = 0$. Consider a singular manifold

$$A \times P \xrightarrow{p_1} A \xrightarrow{f} X.$$

Set $V := A \times P \times I, \partial_i V := A \times P \times \{i\}, i = 0, 1$. Set $F : V \to X, F(a, p, t) = f(a)$. Then $\partial^P(V, F) = (A \times P, fp_1)$. The equalities $\delta r = 0$ and $P\delta = 0$ can be proved similarly.

We prove that the kernels are contained in the images.

1. $\operatorname{Ker} r \subset \operatorname{Im} P$. If $r[M, f] = 0$ then $(M, f) = \partial^P(V, g)$. According to 1.4, we have $g|(\delta V \times P) = g_0 p_1$. Furthermore, $\partial\delta V = \delta M = \emptyset$, $\partial_0 V = M$, $\partial_1 V = \delta V \times P$. Since $\partial_0 V \cap \partial_1 V = \emptyset$, V gives a membrane between M and $\delta V \times P$. Now it is clear that (V, g) is a membrane between (M, f) and $(\delta V \times P, g_0 p_1)$. However, $[\delta V \times P, g_0 p_1] \in \operatorname{Im} P$.

2. $\operatorname{Ker} \delta \subset \operatorname{Im} r$. Let $\delta[M, f] = 0$. Then $(\delta M, f|\delta M) = \partial^P(B, g)$ for some $g : B \to X$, and there is an isomorphism $\varphi : \delta B \times P \to \partial M$. It is clear that $U := B \times P \cup_\varphi M$ is a $\mathscr{K}$-manifold. Set $(U, h) := (B \times P, gp_1) \cup_\varphi (M, f)$. We prove that $r[U, h] = [M, f]$. Consider a manifold $V := B \times P \cup_\varphi M \times I$, where φ glues $B \times P$ and $M \times \{0\} \subset M \times I$. By setting $\partial_0 V = M \times \{0, 1\}, \partial_1 V = B \times P$, we get that V is a $\mathscr{K}^P$-membrane between M and U. Define $F : V \to X$ as follows: $F(m, t) = f(m)$ for $m \in M, t \in I$, and $F|U = h$. Then (V, F) is a $\mathscr{K}^P$-membrane between $r(U, h)$ and (M, f).

3. $\operatorname{Ker} P \subset \operatorname{Im} \delta$. Let $P[M, f] = 0$, i.e., $(M \times P, fp_1) = \partial(V, g)$ for some singular $\mathscr{K}$-manifold $g : V \to X$. Since $\partial V = M \times P$, we can consider V as a closed $\mathscr{K}^P$-manifold. Now we have $\delta[V, g] = [M, f]$. □

1.8. Definition. Define the *Bockstein homomorphism*

$$\beta^P : L_i^P(X) \to L_{i-d-1}^P(X)$$

as $\beta^P := r\delta$.

1.9. Remark. The Bockstein exact sequence

$$\cdots \to H_i(X) \xrightarrow{m} H_i(X) \to H_i(X; \mathbb{Z}/m) \to H_{i-1}(X) \to \cdots$$

looks like (1.7). We shall see below that (1.7) is not only an analog but a generalization of the Bockstein exact sequence. Because of this, we call (1.7) the Bockstein–Sullivan–Baas exact sequence. Furthermore, β^P is an analog (and a generalization) of the classical Bockstein homomorphism β. Moreover, based on (1.7), one can construct an analog of the Bockstein exact couple and spectral sequence (about the latter see Mosher–Tangora [1]).

Now we want to treat L_*^P as a homology theory. To achieve this goal we have to introduce relative groups $L_*^P(X, A)$ and check axioms II.3.1. The groups $L_*^P(X, A)$ can be defined in the usual way, cf. Ch. IV, § 7. Namely, a singular $\mathscr{K}^P$-manifold $f : (M, \partial^P M) \to (X, A)$ in (X, A) bounds if there exists a singular $\mathscr{K}^P$-manifold $g : (V, \partial^P V) \to (X, A)$ such that $\partial_0 V = \partial_0' V \cup \partial_0'' V$ with $\partial_0' V = M, g|\partial_0' V = f$ and $g(\partial_0'' V) \subset A$. Now, the corresponding bordism classes form the bordism group $L_*^P(X, A)$. Moreover, there is an analog of 1.6 with (X, A) instead of X (prove it). Define homomorphisms $\partial_n : L_n^P(X, A) \to L_{n-1}^P(A), \partial_n[M, f] := [\partial^P M, f|\partial^P M]$.

1.10. Theorem. *The family* $\{L_*^P(X, A), \partial_*\}$ *is an additive homology theory.*

Proof. The exactness and homotopy axioms (see II.3.1) can be confirmed in a routine way, see e.g. Conner [1]. We prove the collapse axiom. The collapse $p : (X, A) \to (X/A, *)$ induces the following commutative diagram, where the rows are the exact sequences (1.7):

$$\begin{array}{ccccccc}
\cdots \xrightarrow{P} & L_{n+d}(X,A) & \xrightarrow{r} & L_{n+d}^P(X,A) & \xrightarrow{\delta} & L_{n-1}(X,A) & \longrightarrow \cdots \\
& \cong \downarrow p_* & & \downarrow p_* & & \cong \downarrow p_* & \\
\cdots \xrightarrow{P} & L_{n+d}(X/A,*) & \xrightarrow{r} & L_{n+d}^P(X/A,*) & \xrightarrow{\delta} & L_{n-1}(X/A,*) & \longrightarrow \cdots .
\end{array}$$

Now the Five Lemma implies that $p_* : L_*^P(X, A) \to L_*^P(X/A, *)$ is an isomorphism.

We leave it to the reader to prove additivity. □

Thus, by III.3.23(i), L_*^P can be represented by a spectrum L^P, and this spectrum is unique up to equivalence.

The spectrum L^P yields a cohomology theory $(L_P)^*$. Moreover, if X and Y are n-dual then $\widetilde{L}_P^i(X) \cong \widetilde{L}_{n-i}^P(Y)$. In particular, for every *finite CW-space* X we have the exact sequence (dual to (1.7))

$$(1.11) \qquad \cdots \to L^n(X) \xrightarrow{P} L^{n-d}(X) \xrightarrow{r} L_P^{n-d}(X) \xrightarrow{\delta} L^{n+1}(X) \to \cdots$$

and the Bockstein homomorphism (dual to 1.8)

$$\beta_P = r\delta : L_P^n(X) \to L_P^{n+d+1}(X).$$

By III.3.23(ii), the morphisms $P : L_i(X) \to L_{i+d}(X), r : L_i(X) \to L_i^P(X)$ and $\delta : L_i^P(X) \to L_{i-d-1}(X)$ of the homology theories are induced by morphisms $P : \Sigma^d L \to L$, $r : L \to L^P$ and $\delta : L^P \to \Sigma^{d+1} L$ of spectra. So, we have the sequence of spectra

$$\cdots \to \Sigma^d L \xrightarrow{P} L \xrightarrow{r} L^P \xrightarrow{\delta} \Sigma^{d+1} L \to \cdots . \tag{1.12}$$

Given a space X, we can apply the functor $\pi_*(- \wedge X)$ to this sequence. This yields the sequence (1.7).

On the other hand, we can apply the functor $\pi_*(- \wedge X)$ to the long cofiber sequence

$$\cdots \to \Sigma^d L \xrightarrow{P} L \to C(P) \xrightarrow{k} \Sigma^{d+1} L \to \cdots .$$

This yields an exact sequence

$$\cdots \to L_n(X) \xrightarrow{P} L_{n+d}(X) \to C(P)_{n+d}(X) \xrightarrow{k_*} L_{n-1}(X) \to \cdots ,$$

which looks like (1.7). Thus, it makes sense to compare L^P and $C(P)$.

1.13. Theorem. *Let L be a spectrum of finite $\mathbb{Z}$-type. Suppose that the homomorphism $P : \pi_*(L) \to \pi_*(L)$ is monic. Then the following hold:*

(i) *The spectra L^P and $C(P)$ are almost equivalent;*

(ii) *If the group $(L^P)^0(L)$ has no phantoms (i.e., $\varprojlim^1 (L^P)^0(L^{(n)}) = 0$), then L and $C(P)$ are equivalent.*

Proof. Since L has finite $\mathbb{Z}$-type, this follows from 1.6 and III.6.7. □

1.14. Proposition. *If P and Q are bordant $\mathscr{K}$-manifolds then the homology theories L_*^P and L_*^Q are isomorphic.*

Proof. Let A be a membrane between $\mathscr{K}$-manifolds P and Q. Given a $\mathscr{K}^P$-manifold $V = (V, \partial_0 V, \partial_1 V, \delta V, \varphi)$, define a $\mathscr{K}^Q$-manifold $V' = (V', \partial_0 V', \partial_1 V', \delta V', \varphi')$ as follows:

1. $V' := \delta V \times A \cup_\varphi V$, where $\delta V \times A \supset \delta V \times P \xrightarrow{\varphi} \partial_1 V \subset V$,
2. $\partial_0 V' := (\partial \delta V) \times A \cup_\varphi \partial_0 V$,
3. $\delta V' := \delta V$,
4. $\partial_1 V' := \delta V \times Q \subset \delta V \times A \subset V'$,
5. $\varphi' := 1_{\delta V \times Q}$.

The correspondence $V \mapsto V'$ yields a morphism $h : L_*^P(X) \to L_*^Q(X)$ of homology theories. Considering the ladder of the corresponding sequences (1.7), and using the Five Lemma, we conclude that h is an isomorphism. □

Because of 1.14, sometimes we write $L^{[P]}$ instead of L^P. For example, we can consider a spectrum MU^{x_i}, where x_i is a polynomial generator of $\pi_*(MU) = \mathbb{Z}[x_1, \dots , x_n, \dots]$. By 1.6, $\pi_*(MU^{x_i}) = \pi_*(MU)/(x_i)$.

As we remarked above, one can iterate singularities. Namely, if P' is a closed manifold of $\mathscr{K}$ and $V = (V, \partial_0 V, \partial_1 V, \delta V, \varphi)$ is a $\mathscr{K}^P$-manifold, then

$$P' \times V := (P' \times V, P' \times \partial_0 V, P' \times \partial_1 V, P' \times \delta V, 1_{P'} \times \varphi) \tag{1.15}$$

is a $\mathscr{K}^P$-manifold also. Hence, we can consider $(\mathscr{K}^P)^{P'}$-manifolds and form a class $\mathscr{K}^{P,P'}$, etc. This sequential approach is good in order to guess certain facts about cobordism with singularities. However, from a technical point of view, it is better to consider all the singularities simultaneously. The definition below follows Baas [1], with some small modifications; I am grateful to Haynes Miller for a useful discussion about it.

1.16. Definition. (a) Let $\Sigma = \{P_i\}_{i\in S}$ be a finite set of closed manifolds of $\mathscr{K}$, and let $\dim P_i = d_i$. A *closed k-dimensional $\mathscr{K}$-manifold* with Sullivan–Baas Σ-singularity, or briefly a *closed $\mathscr{K}^\Sigma$-manifold*, is a tuple $(M, \delta_I M, \varphi_{I,i})$, where I runs over all subsets of S and $i \in I$, and, moreover, the following hold:

(1) M is a $\mathscr{K}$-manifold, $\dim M = k$.

(2) $\delta_I M$ is a $\mathscr{K}$-manifold, and $\varphi_{I,i} : \delta_I M \times P_i \to \partial(\delta_{I\setminus\{i\}} M)$ is a $\mathscr{K}$-embedding. Furthermore, $\dim \delta_I M + d_i + 1 = \dim \delta_{I\setminus\{i\}} M$, and

$$\partial(\delta_J M) = \bigcup_{i \notin J} \varphi_{J\cup\{i\},i}(\delta_{J\cup\{i\}} M \times P_i).$$

Also, $\delta_\emptyset M = M$, and so $\partial M = \bigcup_i \varphi_{\{i\},i}(\delta_{\{i\}} M \times P_i)$.

(3) For every I and every $i, j \in I, i \neq j$ the diagram

$$\begin{array}{ccccc}
\delta_I M \times P_i \times P_j & \xrightarrow{\varphi_{I,i}\times 1} & \partial(\delta_{I\setminus\{i\}} M) \times P_j & \xrightarrow{\subset} & \delta_{I\setminus\{i\}} M \times P_j \\
\downarrow{\scriptstyle (\varphi_{I,j}\times 1)(1\times T)} & & & & \downarrow{\scriptstyle \varphi_{I\setminus\{i\},j}} \\
\partial(\delta_{I\setminus\{j\}} M) \times P_i & \xrightarrow{\subset} & \delta_{I\setminus\{j\}} M \times P_i & \xrightarrow{\varphi_{I\setminus\{j\},i}} & \delta_{I\setminus\{i,j\}} M
\end{array}$$

is commutative, and

$$\operatorname{Im} \varphi_{I\setminus\{i\},j} \cap \operatorname{Im} \varphi_{I\setminus\{j\},i} = \operatorname{Im}((\varphi_{I\setminus\{i\},j}) \circ (\varphi_{I,i} \times 1)).$$

Similarly to the above, we say "$\mathscr{K}^\Sigma$-manifold M" instead of "$\mathscr{K}^\Sigma$-manifold $(M, \delta_I M, \varphi_{I,i})$".

(b) Choose $i_0 \in S$, put $P = P_{i_0}$ and set $\widetilde{\Sigma} = \Sigma \setminus \{P\}$. Given a $\mathscr{K}^\Sigma$-manifold $M = (M, \delta_I M, \varphi^M_{I,i})$, define a $\mathscr{K}^{\widetilde{\Sigma}}$-manifold $N = \delta_P M := (N, \delta_J N, \varphi^N_{J,j})$ by setting $\delta_J N = \delta_{J\cup\{i_0\}} M$ and

$$\varphi^N_{J,j} : \delta_J N \times P_j = \delta_{J\cup\{i_0\}} M \times P_j \xrightarrow{\varphi^M_{J\cup\{i_0\},j}} \partial(\delta_{J\cup\{i_0\}\setminus\{j\}} M) = \partial(\delta_{J\setminus\{j\}} N).$$

Note that $N = \delta_{\{i_0\}}M$: this fact justifies the notation δ_P. Furthermore, $\dim M = \dim N + \dim P + 1$.

In particular, given a family $\Sigma = \{P_i\}$ and a closed $\mathscr{K}$-manifold P, set $\Sigma' = \Sigma \cup P$. Then we can assign a $\mathscr{K}^\Sigma$-manifold $\delta_P M$ to a $\mathscr{K}^{\Sigma'}$-manifold M.

(c) Given $\Sigma = \{P_i\}$, set $\Sigma' = \Sigma \cup \mathrm{pt}$ and define a *$\mathscr{K}^\Sigma$-manifold with boundary* to be a closed $\mathscr{K}^{\Sigma'}$-manifold. Given a $\mathscr{K}^\Sigma$-manifold V with boundary, set $\partial^\Sigma V := \delta_{\mathrm{pt}} V$. We call $\partial^\Sigma V$ the *$\mathscr{K}^\Sigma$-boundary* of V. It is clear that the $\mathscr{K}^\Sigma$-boundary of a $\mathscr{K}^\Sigma$-manifold with boundary is a closed $\mathscr{K}^\Sigma$-manifold.

(d) Given a closed $\mathscr{K}^{\Sigma'}$-manifold M and a closed $\mathscr{K}$-manifold P, define a closed $\mathscr{K}^{\Sigma'}$-manifold $M \times P$, where $\delta_I(M \times P) = \delta_I(M) \times P$ and $\varphi_{I,i}^{M\times P} = \varphi_{I,i}^M \times 1_P$.

(e) Again, let $\Sigma = \{P_i\}_{i\in S}$. Given a subset $T \subset S$, consider the family $\widetilde{\Sigma} = \{P_i | i \in T\}$. Now, every closed $\mathscr{K}^{\widetilde{\Sigma}}$-manifold M can be considered as a closed $\mathscr{K}^\Sigma$-manifold if we put $\delta_I M = \emptyset$ for every I with $I \not\subset T$.

It is clear that $\partial^\Sigma \partial^\Sigma V = \emptyset$. (This equality makes sense because of 1.16(e).) So, we have a new cobordism category $(\mathscr{K}^\Sigma, \partial^\Sigma)$ where $\mathscr{K}^\Sigma$ is the class of all $\mathscr{K}^\Sigma$-manifolds.

Iterations of the inclusions $\delta_I M \times P_i \subset \partial(\delta_{I\setminus\{i\}} M) \subset \delta_{I\setminus\{i\}} M$ yield an inclusion $\delta_I M \times \prod_{i\in I\setminus J} P_i \to \delta_J M$ for every $J \subset I$. In particular, there is an inclusion $\delta_I M \times \prod_{i\in I} P_i \subset \delta_\emptyset M = M$.

1.17. Definition. A *singular $\mathscr{K}^\Sigma$-manifold in a pair* (X, A) is a map $f : (V, \partial^\Sigma V) \to (X, A)$ where V is a $\mathscr{K}^\Sigma$-manifold with boundary and f is such that for every I there exists a commutative diagram

$$\begin{array}{ccc} V & \xrightarrow{f} & X \\ \uparrow & & \uparrow{\scriptstyle f_I} \\ \delta_I V \times \prod_{i\in I} P_i & \xrightarrow{p_1} & \delta_I V, \end{array}$$

where the left map is the inclusion as above and the bottom map is the projection.

As above, one can define a bordism theory $L_*^\Sigma(X, A)$. Furthermore, if $\Sigma' = \Sigma \cup \{P\}$, $\dim P = d$, then there is an exact sequence

$$(1.18) \quad \cdots \to L_i^\Sigma(X, A) \xrightarrow{P} L_{i+d}^\Sigma(X, A) \xrightarrow{r} L_{i+d}^{\Sigma'}(X, A) \xrightarrow{\delta} L_{i-1}^\Sigma(X, A) \to \cdots$$

(an analog of (1.7)). It is not unexpected, because, informally speaking, $L^{\Sigma'} = (L^\Sigma)^P$, cf. (1.15). Here P is described in 1.16(d), r is described in 1.16(e), and $\delta[V, f] = [\delta_P(V), f|\delta_P(V)]$. We leave the formal proof to the reader.

As above, one can prove that $L_*^\Sigma(X, A)$ is a homology theory. So, it can be represented by a spectrum L^Σ, and this spectrum is unique up to equivalence.

1.19. Definition. The homology (resp. cohomology) theory $L_*^\Sigma(-)$ (resp. $(L^\Sigma)^*(-)$) is called a bordism (resp. cobordism) theory with Σ-singlarities. For brevity, we just say (co)bordism with Σ-singularities.

Observe that L^Σ is *not* determined by the ideal generated by $\{P_i\}$. For example, if $[P] = 0 \in \pi_*(L)$, then the spectra L^P and $L^{\{P,P\}}$ have different coefficients.

Again, let $\Sigma' = \Sigma \cup \{P\}$, $\dim P = d$. By III.3.23(ii), the morphism $P : L_i^\Sigma(X, A) \to L_{i+d}^\Sigma(X, A)$ of homology theories is induced by a morphism $P : \Sigma^d L^\Sigma \to L^\Sigma$ of spectra. Let $C(P)$ be the cone of the morphism P. The following generalization of 1.13 holds.

1.20. Theorem. *Let L be a spectrum of finite $\mathbb{Z}$-type. Suppose that the homomorphism $P_* : \pi_*(L^\Sigma) \to \pi_*(L^\Sigma)$ is monic. Then the following hold:*

(i) *The spectra $L^{\Sigma'}$ and $C(P)$ are almost equivalent;*

(ii) *If the group $(L^{\Sigma'})^0(L^\Sigma)$ has no phantoms then $L^{\Sigma'}$ and $C(P)$ are equivalent.* □

Now, let $\Sigma = \{P_1, \dots, P_n, \dots\}$ be a countable set. Set $\Sigma_n = \{P_1, \dots, P_n\}$ and define $L_*^\Sigma(X, A) := \varinjlim L_*^{\Sigma_n}(X, A)$. Since $\varinjlim$ preserves exactness, L_*^Σ is a homology theory. Moreover, one can see that L_*^Σ is a bordism theory based on manifolds of the class $\mathscr{K}^\Sigma := \cup_n \mathscr{K}^{\Sigma_n}$.

As above, there is a forgetful morphism $r : L_*(X, A) \to L_*^\Sigma(X, A)$ of homology theories. By III.3.23(ii), it is induced by a morphism of spectra

$$r = r^\Sigma : L \to L^\Sigma. \tag{1.21}$$

1.22. Definition. Let R be a commutative ring, and let M be an R-module. A sequence $\{x_1, \dots, x_n, \dots, \}, x_i \in R$ (finite or infinite) is called *proper with respect to M*, or just *M-proper*, if multiplication by $x_1 : M \to M$ is monic and multiplication by $x_i : M/(x_1, \dots, x_{i-1})M \to M/(x_1, \dots, x_{i-1})M$ is monic for every i. An R-proper sequence is called just *proper*.

1.23. Remark. There is a closely related concept of regular sequence, see e.g. Lang [1]; namely, a proper sequence is called *regular* if $(x_1, \dots, x_k) \neq R$ for every k. For example, the sequence $\{1, 0\}$ is proper but not regular in $\mathbb{Z}$. Note that if a finite sequence is regular then it remains regular after any permutation, while this is not true for a proper sequence. So, we preferred to introduce a new term (proper) and not to talk about "weak regularity", etc.

1.24. Proposition. *Let $\Sigma = \{x_1, \dots, x_n, \dots\}$ be a proper sequence (finite or not) in $\pi_*(L)$. Then there is a $\pi_*(L)$-module isomorphism $\pi_*(L^\Sigma) = \pi_*(L)/(x_1, \dots, x_n, \dots)$.*

Proof. Put $\Sigma_n = \{x_1, \dots, x_n\}$. By 1.6, $\pi_*(L^{\Sigma_1}) = \pi_*(L)/(x_1)$. Now, considering the exact sequence (1.18) and using an obvious induction, we conclude that $\pi_*(L^{\Sigma_n}) = \pi_*(L)/(x_1, \dots, x_n)$ for every n. So, the proposition holds for every finite proper Σ. Thus, it is valid for infinite Σ also since $\pi_*(L)/(x_1, \dots, x_n, \dots) = \varinjlim \pi_*(L)/(x_1, \dots, x_n)$. □

1.25. Examples. (a) Let $\mathscr{K}$ be the class of stably almost complex compact manifolds, i.e., $L_{\mathscr{K}} = MU$. By VII.1.9(i), $\pi_*(MU) = \mathbb{Z}[x_1, \dots, x_n, \dots]$. Hence, by 1.24, if $\Sigma = \{x_1, \dots, x_n, \dots\}$ then $\pi_*(MU^\Sigma) = \mathbb{Z}$. So, by the Eilenberg–Steenrod Theorem,

$$MU^\Sigma = H\mathbb{Z},$$

i.e., classical homology can be interpreted as bordism with singularities! Furthermore, the morphism $r : MU \to MU^\Sigma = H\mathbb{Z}$ as in (1.21) coincides with the Thom class, i.e., with the Steenrod–Thom morphism. In particular, every homology class can be realized by a $\mathscr{K}^\Sigma$-manifold with this Σ.

(b) Let m be a natural number, let $\mathfrak{m}$ be the manifold $\{m \text{ points}\}$, and let $\Sigma' = \Sigma \cup \{\mathfrak{m}\}$ with Σ as in (a). Then $MU^{\Sigma'} = H\mathbb{Z}/m$, and in this case sequence (1.18) is just the Bockstein exact sequence.

(c) The sphere spectrum S yields framed bordism theory. Considering $\mathfrak{m}$ as in (b), one can prove that $S^\mathfrak{m}$ is a Moore spectrum $M(\mathbb{Z}/m)$. Indeed, the sequence (1.7) is exact for every spectrum X (because, by III.4.22, $E_i(X) = \varinjlim E_{i+n}(X_n)$ for all spectra E, X). Hence, $S_i^\mathfrak{m}(H\mathbb{Z}) = 0$ for $i \neq 0$ and $S_0^\mathfrak{m}(H\mathbb{Z}) = \mathbb{Z}/m$. Therefore,

$$H_*(S^\mathfrak{m}) = S_*^\mathfrak{m}(H\mathbb{Z}) = H_*(M(\mathbb{Z}/m)).$$

Thus, by II.4.32, $S^\mathfrak{m} = M(\mathbb{Z}/m)$.

§2. Multiplicative Structures

Here we assume that the above spectrum L is a commutative ring spectrum. We discuss multiplicative structures in L_*^Σ with $\Sigma = \{P_1, \dots, P_n\}$. *From here to the end of the Chapter we assume that every P_i is an even-dimensional manifold.*

The main results of this section were proved by Mironov [1], [2]. This material is also exposed in Botvinnik [1]. In order to have neater notation, we shall consider pairings (quasi-multiplications) $L_*^\Sigma(X) \otimes L_*^\Sigma(Y) \to L_*^\Sigma(X \times Y)$ rather than $L_*^\Sigma(X, A) \otimes L_*^\Sigma(Y, B) \to L_*^\Sigma(X \times Y, A \times Y \cup X \times B)$.

If M is a $\mathscr{K}$-manifold and N is a $\mathscr{K}^\Sigma$-manifold, then $M \times N$ is a $\mathscr{K}^\Sigma$-manifold in the canonical way. Namely, we set $\partial_i(M \times N) := M \times \partial_i N$, etc. (cf. (1.15)). Thus, we have pairings

$$m_L : L_*(X) \otimes L_*^\Sigma(Y) \to L_*^\Sigma(X \times Y), m_L(f \otimes g) = f \times g : M \times N \to X \times Y,$$
$$m_R : L_*^\Sigma(Y) \otimes L_*(X) \to L_*^\Sigma(Y \times X), m_R(g \otimes f) = g \times f : N \times M \to Y \times X,$$

where $f : M \to X$ is a singular $\mathscr{K}$-manifold and $g : N \to Y$ is a singular $\mathscr{K}^\Sigma$-manifold. These pairings turn L^Σ into a (left and right) quasi-module spectrum over L.

2.1. Definition. A quasi-multiplication μ in L_*^Σ is called *admissible* if it is compatible with the pairings m_L, m_R above, i.e., if the diagram

$$\begin{array}{ccc} L_*(X) \otimes L_*^\Sigma(Y) & \xrightarrow{m_L} & L_*^\Sigma(X \times Y) \\ {\scriptstyle r\otimes 1}\downarrow & & \| \\ L_*^\Sigma(X) \otimes L_*^\Sigma(Y) & \xrightarrow{\mu} & L_*^\Sigma(X \times Y), \end{array}$$

and the similar diagram for m_R commute.

In this case the forgetful morphism $r : L_*(-) \to L_*^\Sigma(-)$ is a quasi-ring morphism of homology theories.

It is difficult to introduce a quasi-multiplication in $L_*^P(X)$ because the product of two $\mathscr{K}^P$-manifolds is not a $\mathscr{K}^P$-manifold in general. So, one must find some bypasses. Firstly, we consider a geometric situation which makes this difficulty clear.

Fix any closed manifold $P \in \mathscr{K}$, $\dim P = d$. Clearly, $r[P] = 0$, where $r : L_*(\mathrm{pt}) \to L_*^P(\mathrm{pt})$ is the forgetful homomorphism. Hence, if L_* admits a quasi-multiplication μ then $\mu([M] \otimes [P]) = 0$ for every closed $\mathscr{K}^P$-manifold M. So, if this quasi-multiplication is admissible then $[M \times P] = 0 \in L_*^P(\mathrm{pt})$.

Consider a $\mathscr{K}^P$-manifold M^m with $\partial M = \delta M \times P$ and try to prove that $[M \times P]$ is $\mathscr{K}^P$-bordant to zero. We have $\partial(M \times P) = \delta M \times P' \times P''$, where P', P'' are copies of P, and $\delta(M \times P) = \delta M \times P''$. Try to find a membrane for $M \times P$. The most natural way is (see the proof of 1.6): take $M \times P \times I$ and set $\partial_0(M \times P \times I) = M \times P \times \{0\}$. Then we must put

$$\begin{aligned} \partial_1(M \times P \times I) &= \partial(M \times P) \times I \cup (-1)^m M \times P \times \{1\} \\ &= \delta M \times P' \times P'' \times I \cup (-1)^m M \times P \times \{1\}. \end{aligned}$$

In order to get $M \times P \times I$ as a $\mathscr{K}^P$-manifold, we must put

$$\delta(M \times P \times I) = \delta M \times P' \times I \cup (-1)^m M \times \{1\}.$$

But this contradicts the equality $\delta(M \times P \times \{0\}) = \delta M \times P''$. This contradiction, i.e., the difference between P' and P'', is the main obstruction to the existence of a quasi-multiplication. In order to avoid this obstruction we need to fit $\delta M \times P'$ with $\delta M \times P''$ canonically.

Look at this from another point of view. Consider two closed $\mathscr{K}^P$-manifolds M^m, N^n and try to treat $M \times N$ as a $\mathscr{K}^P$-manifold. We have

$$\begin{aligned} \partial(M \times N) &= \partial M \times N \cup_\gamma (-1)^m M \times \partial N \\ &= \delta M \times P \times N \cup_\gamma (-1)^m M \times \delta N \times P. \end{aligned} \tag{2.2}$$

It would be good to write this as $(\delta M \times N \cup_\gamma (-1)^m M \times \delta N) \times P$. But we cannot do this because γ is in fact the identity map of $\delta M \times P' \times \delta N \times P''$, while in the last term of (2.2) the first P is P' but the second P is P''. So, again, we need to fit P' with P''.

Consider a manifold $\overline{P} := P' \times P'' \times I$, where P', P'' are copies of P. Recal that $\dim P$ is even. Turn $\overline{P}$ into a closed $\mathscr{K}^P$-manifold by setting $\partial_0 \overline{P} = \emptyset$, $\partial_1 \overline{P} = \partial \overline{P} = \delta \overline{P} \times P$, where $\delta \overline{P} = P'' \times \{1\} \cup (-P' \times \{0\})$ and $\varphi : \delta \overline{P} \times P \to \partial \overline{P}$ has the form

$$\varphi : \delta \overline{P} \times P \cong P'' \times \{1\} \times P) \cup (-P' \times \{0\} \times P) \xrightarrow{=} \partial \overline{P}.$$

Let $\mathfrak{a}(P) \in L_*^P(\text{pt})$ be the $\mathscr{K}^P$-bordism class of $\overline{P}$.

2.3. Proposition. *Let M be an arbitrary closed $\mathscr{K}^P$-manifold.*
(i) *The $\mathscr{K}^P$-manifolds $M \times P$ and $\delta M \times \overline{P}$ are $\mathscr{K}^P$-bordant.*
(ii) *If L_*^P has an admissible quasi-multiplication μ, then*

$$\mu_{\text{pt,pt}}([\delta M] \otimes \mathfrak{a}(P)) = 0 \in L_*^P(\text{pt}),$$

Proof. (i) We have $\partial M = \delta M \times P$. Hence (omitting signs),

$$\begin{aligned} \partial(M \times P \times I) &= \delta M \times P \times P \times I \cup M \times P \times \partial I \\ &= \delta M \times \overline{P} \cup M \times P \times \{1\} \cup M \times P \times \{0\}. \end{aligned}$$

By setting $\partial_0(M \times P \times I) = \delta M \times \overline{P} \cup M \times P \times \{0\}$, we obtain the proof. (It makes sense to remark that the equality $\partial_0(M \times P \times I) = \delta M \times P \times P \times I$ forces us to regard $P \times P \times I$ as the $\mathscr{K}^P$-manifold $\overline{P}$.)

(ii) Notice that $[P] = 0 \in L_*^P(\text{pt})$. Now

$$\begin{aligned} \mu_{\text{pt,pt}}([\delta M] \otimes \mathfrak{a}(P)) &= [\delta M \times \overline{P}] = [M \times P] = \mu_{\text{pt,pt}}([M] \otimes [P]) \\ &= \mu_{\text{pt,pt}}([M] \otimes 0) = 0; \end{aligned}$$

the first and third equalities hold since μ is admissible, the second equality holds by (i). □

This hints that it is impossible to find an admissible quasi-multiplication if $\mathfrak{a}(P) \neq 0$.

2.4. Theorem. *If $\mathfrak{a}(P) = 0 \in L_*^P(\text{pt})$, then there exists an admissible quasi-multiplication in L_*^P. Thus, $\mathfrak{a}(P)$ plays the role of an obstruction to the existence of an admissible quasi-multiplication.*

Proof. Let $\overline{P} = \partial^P Q$, $\dim Q = 2d + 2$. The idea of the proof is that Q enables us to cohere P' and P'' and so to avoid the difficulty discussed above.

Given two closed $\mathscr{K}^P$-manifolds M, N, we have (omitting signs)

$$\partial(M \times N) = \partial M \times N \cup M \times \partial N.$$

We put

$$\partial_0(M \times N) = \partial_0 M \times N \cup M \times \partial_0 N, \quad \partial_1(M \times N) = \partial_1 M \times N \cup M \times \partial_1 N.$$

Consider the manifold $\widehat{N} = N \cup \partial N \times I$, i.e., we attach a collar to N. Clearly, $\widehat{N} \cong N$. Hence,

$$\partial(M \times N) \cong \partial(M \times \widehat{N}) = \partial M \times N \cup \partial M \times \partial N \times I \cup M \times \partial N \times \{1\}.$$

So,

$$\partial_1(M \times N) = \delta M \times P' \times N \cup \delta M \times P' \times \delta N \times P'' \times I \cup M \times \delta N \times P'' \times \{1\}.$$

But

$$\delta M \times P' \times \delta N \times P'' \times I \cong \delta M \times \delta N \times \overline{P} \subset \delta M \times \delta N \times \partial Q \subset \partial(\delta M \times \delta N \times Q).$$

We set

$$M * N = M \times N \cup_{\psi} \delta M \times \delta N \times Q, \tag{2.5}$$

where ψ identifies $\delta M \times P' \times \delta N \times P'' \times I \subset M \times N$ with $\delta M \times \delta N \times \overline{P} \subset \partial(\delta M \times \delta N \times Q)$. Then

$$\partial(M * N) = \delta M \times P' \times N \cup \delta M \times \delta N \times \partial_1 Q \cup M \times \delta N \times P''.$$

We turn $M * N$ into a $\mathscr{K}^P$-manifold by setting

$$\delta(M * N) = \delta M \times N \cup \delta M \times \delta N \times \delta Q \cup M \times \delta N.$$

Given two singular $\mathscr{K}^P$-manifolds $f : M \to X$, $g : N \to Y$, we have a map $f \times g : M \times N \to X \times Y$. Let $\partial N \times I$ be a collar of ∂N in N. We can assume that $g|(\partial N \times I) = gp$, where $p : \partial N \times I \to \partial N$ is the projection. Define

$$h : \delta M \times \delta N \times Q \xrightarrow{\text{proj}} \delta M \times \delta N \xrightarrow{f \times g} X \times Y.$$

According to Definition 1.4, $h|(\delta M \times \delta N \times \partial^P Q) = h|(\delta M \times \delta N \times \overline{P})$ coincides with $f \times g|(\delta M \times \delta N \times \overline{P}) = f \times g|(\delta M \times P' \times \delta N \times P'' \times I)$.

Define $f * g : M * N \to X \times Y$ by setting

$$f * g|M \times N = f \times g, \; f * g|\delta M \times \delta N \times Q = h.$$

Clearly, $(f * g)(\partial^P(M * N)) \subset X \times B \cup A \times Y$. We leave it to the reader to prove that in this way we have a well-defined pairing

$$\mu = \mu^Q_{X,Y} : L^P_*(X) \otimes L^P_*(Y) \to L^P_*(X \times Y), \tag{2.6}$$
$$\mu([M, f] \otimes [N, g]) := [M * N, f * g]$$

and that the family $\{\mu^Q_{X,Y}\}$ is a quasi-multiplication in L^P_*.

Finally, this quasi-multiplication is admissible. Indeed, if M and/or N is a $\mathscr{K}$-manifold, then $M * N = M \times N$. □

2.7. Corollary. *Suppose that $\mathfrak{a}(P) = 0 \in L_*(\mathrm{pt})$ and that every group $L^P_i(\mathrm{pt})$ is finite. Then the spectrum L^P admits a ring structure such that $r : L \to L^P$ is a ring morphism.*

Proof. This follows from 2.4 and III.7.3, III.7.5. □

The quasi-multiplication (2.6) depends on Q. We clarify this dependence. Let Q_1, Q_2 be two $\mathscr{K}^P$-manifolds with $\partial^P Q_1 = \overline{P} = \partial^P Q_2$, and let μ_1, μ_2 be the corresponding quasi-multiplications in $L^P_*(-)$. We set $V := Q_1 \cup_f Q_2$ where

$$f : \partial_0 Q_1 = \overline{P} \to -\overline{P} = \partial_0(-Q_1), \quad f(p) = p \text{ for every } p \in \overline{P}.$$

Following IV.7.24, one can prove that V is a (B, φ)-manifold. Furthermore, $\partial V = \partial_1 Q_1 \cup_f (-\partial_1 Q_2)$, and we regard V as a closed $\mathscr{K}$-manifold by setting $\partial_1 V = \partial V$. Let $\beta = \beta^P : L^P_*(-) \to L^P_*(-)$ be the Bockstein homomorphism from 1.8.

2.8. Theorem. *Let $b = b(Q_1, Q_2)$ be the $\mathscr{K}^P$-bordism class of V. Then*

$$\mu_1(x \otimes y) - \mu_2(x \otimes y) = \pm m_L(\beta(x) \otimes m_L(\beta(y) \otimes b))$$

for every $x \in L^P_*(X), y \in L^P_*(Y)$. (Here $\pm$ means that we do not care about the sign.)

Proof. Given two closed $\mathscr{K}^P$-manifolds M, N, we have (see the proof of 2.4)

$$\partial(M \times N) \cong \partial(M \times \widehat{N}) = \partial M \times N \cup \partial M \times \partial N \times I \cup M \times \partial N \times \{1\}.$$

We set $\Delta := \delta M \times P' \times N \cup (-1)^m M \times \delta N \times P''$. So, $\partial(M \times N) = \Delta \cup A$ where $A \cong \delta M \times \delta N \times \overline{P} = \delta M \times \delta N \times \partial^P Q$. We define

$$W := I \times M \times N \cup_{\psi_1} (\delta M \times \delta N \times Q_1) \cup_{\psi_2} (\delta M \times \delta N \times (-Q_2))$$

where $\psi_1 : \{1\} \times A \to \delta M \times \delta N \times \overline{P}$ and $\psi_2 : \{0\} \times A \to \delta M \times \delta N \times \overline{P}$ are the isomorphisms described before (2.5). We have

$$\partial(I \times M \times N) = M \times N \times \{1\} \cup (-(M \times N \times \{0\})) \cup (\pm(I \times \partial(M \times N))).$$

Hence,

$$\partial W = M *_1 N \cup (-(M *_2 N)) \cup (\pm(I \times \Delta)) \cup$$
$$(\pm(I \times A \cup_{\psi_1} (\delta M \times \delta N \times Q_1) \cup I \times A \cup_{\psi_2} (\delta M \times \delta N \times (-Q_2))))$$

We set $\partial_1 W := I \times \Delta$. Thus

$$\partial^P W = M *_1 N \cup (-(M *_2 N)) \cup (\pm(\delta M \times \delta N \times (Q_1 \cup (-Q_2)))). \qquad \square$$

2.9. Remarks. (a) Consider a $\mathscr{K}$-manifold $\hat{P} := P \times P \times I/\sim$, where $(x, y, 0) \sim (y, x, 1)$, i.e $\hat{P} = \Gamma_1(P)$. It is easy to see that $\hat{P}$ and $\overline{P}$ are $\mathscr{K}^P$-bordant. In particular, $\mathfrak{a}(P) \in \operatorname{Im}(r : L_*(\mathrm{pt}) \to L_*^P(\mathrm{pt}))$.

(b) As usual, we shall write ab instead of $\mu(a \otimes b)$, where $a \in L_*^P(X), b \in L_*^P(Y), ab \in L_*^P(X \times Y)$.

(c) We know that $L_*^{P_1}$ and $L_*^{P_2}$ are isomorphic if P_1 and P_2 are bordant. Moreover, if $\mathfrak{a}(P_1) = 0 \in L_*^{P_1}(\mathrm{pt})$, then $L_*^{P_1}$ and $L_*^{P_2}$ are isomorphic quasi-ring homology theories (by choosing suitable quasi-multiplications). We do not need this fact and leave the proof to the reader.

We discuss commutativity and associativity of the quasi-multiplication μ in L_*^P. Fix a $\mathscr{K}^P$-manifold Q with $\partial^P Q = \overline{P}$ and consider the quasi-multiplication $\mu = \mu^Q$ as in (2.6). Set $D = D(P) = Q \cup_\chi Q$, where $\chi : \partial^P Q \to \partial^P Q$, $\chi(p_1, p_2, t) = (p_2, p_1, 1-t)$ for $(p_1, p_2, t) \in P \times P \times I$. Notice that χ inverts the (B, φ)-structure on $\overline{P}$, and so D is a (B, φ)-manifold. It is clear that $\partial D = \partial_1 Q \cup_\chi \partial_1 Q$. We turn D into a closed $\mathscr{K}^P$-manifold by setting $\partial_0 D = \emptyset$. Let $\mathfrak{b}(P) = \mathfrak{b}_Q(P)$ be the $\mathscr{K}^P$-bordism class of D.

2.10. Theorem. *For every $x \in L_*^P(X), y \in L_*^P(Y)$ we have*

$$xy - (-1)^{|x||y|}\tau_*(yx) = \pm\mathfrak{b}(P)\beta(x)\beta(y) \in L_*^P(X \times Y),$$

where $\tau : X \times Y \to Y \times X$ switches factors. In particular, μ is commutative if $\mathfrak{b}(P) = 0$.

Proof. Consider two closed $\mathscr{K}^P$-manifolds M^m, N^n. Following the proof of 2.8, we have $\partial(M \times N) = \Delta_1 \cup A_1, \partial(N \times M) = \Delta_2 \cup A_2$. Furthermore, $M * N = M \times N \cup_{\psi_1} \delta M \times \delta N \times Q$ and $N * M = N \times M \cup_{\psi_2} \delta N \times \delta M \times Q$, where $\psi_1 : A_1 \to \delta M \times \delta N \times \overline{P}$ and $\psi_2 : A_2 \to \delta N \times \delta N \times Q$ are identifications as in (2.5). We set

$$W := I \times M \times N \cup_{\psi_1} (\delta M \times \delta N \times Q) \cup_{\psi_2} (\delta N \times \delta M \times Q)$$

where $\psi_1 : \{1\} \times A_1 \to \delta M \times \delta N \times \overline{P}$ and $\psi_2 : \{0\} \times A_2 \to \delta N \times \delta M \times \overline{P}$. Now, arguing as in 2.8, we set $\partial_1 W = I \times \Delta$, and so

$$\partial^P W = M * N \cup (-(-1)^{mn} N * M) \cup (\pm(\delta M \times \delta N \times D)). \qquad \square$$

Now consider another manifold Q' with $\partial^P Q' = \overline{P}$ and form a closed manifold $N = Q \cup_{\overline{P}} (-Q')$. Then $\mathfrak{b}_Q(P) = \mathfrak{b}_{Q'}(P) + 2[N] \in L_*^P(\mathrm{pt})$. In particular, the mod 2 reduction $\varkappa(P) \in L_*^P(\mathrm{pt}) \otimes \mathbb{Z}/2$ of $\mathfrak{b}_Q(P)$ does not depend on Q.

2.11. Corollary. *If $\varkappa(P) = 0$, then L_*^P can be equipped with an admissible commutative quasi-multiplication. In particular, if $L_*^P(\mathrm{pt})$ contains $1/2$, then L_*^P can be equipped with an admissible commutative quasi-multiplication.*

Proof. We have $\mathfrak{b}_Q(P) = 2[N]$ for some Q with $\partial^P Q = \overline{P}$ and some closed $\mathscr{K}^P$-manifold N. Set $Q' = Q \cup (-N)$. Then $\partial^P Q' = \overline{P}$ and $\mathfrak{b}_{Q'}(P) = 0$. □

This corollary shows that $\varkappa$ can be considered as an obstruction to commutativity.

Passing to associativity, take Q with $\partial^P Q = \overline{P}$ and construct a manifold $C(P) = C_Q(P)$ as follows. Consider the arcs

$$A_k = \left\{ z \in S^1 \mid \frac{2k-2}{3}\pi \leq \arg z \leq \frac{2k-1}{3}\pi \right\}, \quad k = 1, 2, 3,$$

where $S^1 = \{ z \in \mathbb{C} \mid |z| = 1 \}$. We turn the manifold $P \times P \times P \times D^2$ into a $\mathscr{K}^P$-manifold by setting $\partial^P(P \times P \times P \times D^2) = P \times P \times P \times (\cup A_k)$. Let P', P'', P''' be copies of P. We write $P \times P \times P$ as $P' \times P'' \times P'''$ in order to distinguish the factors in $P \times P \times P$. There are inclusions

$$\begin{aligned}
j' \ : \partial^P(P' \ \times Q) &= P' \ \times P \times P \times I \\
&= P' \times P'' \times P''' \times A_1 \subset \partial(P \times P \times P \times D^2); \\
j'' \ : \partial^P(P'' \ \times Q) &= P'' \ \times P \times P \times I \\
&= P'' \times P''' \times P' \times A_2 \subset \partial(P \times P \times P \times D^2); \\
j''' : \partial^P(P''' \times Q) &= P''' \times P \times P \times I \\
&= P''' \times P' \times P'' \times A_3 \subset \partial(P \times P \times P \times D^2).
\end{aligned}$$

Set $\Delta := P \times P \times P \times D^2 \cup_{j' \cup j'' \cup j'''} ((P' \times Q) \cup (P'' \times Q) \cup (P''' \times Q))$. Note that Δ is a $\mathscr{K}^P$-manifold, since it is the result of gluing two $\mathscr{K}^P$-manifolds.

Consider the isomorphisms (identity maps)

$$P' \times Q \xrightarrow{i_1} P'' \times Q \xrightarrow{i_2} P''' \times Q \xrightarrow{i_3} P' \times Q,$$

and set

$$i = i_1 \sqcup i_2 \sqcup i_3 : P' \times Q \sqcup P'' \times Q \sqcup P''' \times Q \to P'' \times Q \sqcup P''' \times Q \sqcup P' \times Q.$$

Define $\psi : \Delta \to \Delta$ as follows:

$$\psi(p', p'', p''', \rho, \varphi) = \left(p'', p''', p', \rho, \varphi + \frac{2\pi}{3} \right),$$

where (ρ, φ) are polar coordinates on D^2 and $(p', p'', p''', \rho, \varphi) \in P' \times P'' \times P''' \times D^2$, and

$$\psi|(P' \times Q \sqcup P'' \times Q \sqcup P''' \times Q) = i|(P' \times Q \sqcup P'' \times Q \sqcup P''' \times Q).$$

Set

(2.12) $$C_Q(P) = (\Delta \times I)/\sim \text{ where } (\psi(x), 0) \sim (x, 1).$$

Let $\mathfrak{c}(P) = \mathfrak{c}_Q(P)$ be the $\mathscr{K}^P$-bordism class of $C_Q(P)$.

2.13. Theorem. *Suppose that $\overline{P} = \partial Q$ for some Q. Let $L_*^P(-)$ be equipped with the admissible quasi-multiplication* (2.6). *Then*

$$(xy)z - x(yz) = \pm(\mathfrak{c}(Q) - [\delta(Q)]\mathfrak{b}_Q(P))\beta(x)\beta(y)\beta(z) \in L_*^P(X \times Y \times Z)$$

for every $x \in L_*^P(X), y \in L_*^P(Y), z \in L_*^P(Z)$.

Proof. This is similar to proofs above, but tedious, see Mironov [1], [2]. □

2.14. Corollary. *The Moore spectrum $M(\mathbb{Z}/p), p > 3$, p prime, admits the structure of a commutative ring spectrum.*

Proof. For simplicity, denote $M(\mathbb{Z}/p)$ by M. By 1.25(c), the homology theory M_* is framed bordism theory with the singularity $\{p \text{ points}\}$. Since $\pi_i(S) \otimes \mathbb{Z}/p = 0$ for $0 < i < 5$, we conclude that $\pi_2(M) = 0 = \pi_3(M)$. So, by 2.4, 2.10 and 2.13, M can be equipped with a commutative and associative quasi-multiplication. By III.7.3, this quasi-multiplication is induced by a multiplication. Finally, the admissibility means that the map $r : S \to M$ can be treated as the unit. □

2.15. Remark. The spectrum $M(\mathbb{Z}/3)$ admits a non-associative commutative pairing $M(\mathbb{Z}/3) \wedge M(\mathbb{Z}/3) \to M(\mathbb{Z}/3)$ by 2.4 and 2.10. In particular, $M(\mathbb{Z}/3)_*(X)$ is a $\mathbb{Z}/3$-module. The spectrum $M(\mathbb{Z}/2)$ does not admit any pairing because $\pi_3(M(\mathbb{Z}/2)) = \mathbb{Z}/4$, see Araki–Toda [1]. This means that the obstruction $\mathfrak{a}(P) \in \pi_1(M(\mathbb{Z}/2))$ for $P = \{2 \text{ points}\}$ is non-trivial.

Let $\{P_1, \dots, P_n, \dots\}, \dim P_n = d_n$, be a sequence of $\mathscr{K}$-manifolds. Recall that every d_n is even. We set $\Sigma_n = \{P_1, \dots, P_n\}$. Hence, there is a tower

(2.16) $$L \to L^{\Sigma_1} \to \cdots \to L^{\Sigma_n} \xrightarrow{r_n} L^{\Sigma_{n+1}} \to \cdots,$$

We have also the Bockstein morphisms $\beta_n : L^{\Sigma_n} \to \Sigma^{-1-d_n} L^{\Sigma_n}$.

2.17. Theorem. *Let $\mathfrak{a}(P_i) = 0 \in L_*^{P_i}(\text{pt})$. Assume that there are chosen Q_i with $\partial^{P_i} Q_i = \overline{P}_i$ for every $i = 1, \dots, n, \dots$. Then the following hold:*

(i) *Every homology theory $L_*^{\Sigma_n}$ admits a quasi-multiplication μ_n such that every morphism in* (2.16) *is a quasi-ring morphism. Furthermore, if the groups $L_i^{\Sigma_n}(\text{pt})$ are finite for some n and every i, then the quasi-multiplication μ_n can be induced by a multiplication (which is unique up to homotopy)*

$L^{\Sigma_n} \wedge L^{\Sigma_n} \to L^{\Sigma_n}$, *i.e.,* L^{Σ_n} *becomes a ring spectrum. Moreover, if the groups* $L_i^{\Sigma_n}(\mathrm{pt})$ *and* $L_i^{\Sigma_{n+1}}(\mathrm{pt})$ *are finite for some* n *and every* i, *then the morphism* $r_n : L^{\Sigma_n} \to L^{\Sigma_{n+1}}$ *is a ring morphism.*

(ii) *Suppose that* $r_i(\mathfrak{b}_{q_i}(P_i)) = 0 \in L_*^{\Sigma_{i+1}}(\mathrm{pt})$ *for every* $i < n$. *Then for every* $x \in L_*^{\Sigma_n}(X), y \in L_*^{\Sigma_n}(Y)$ *we have*

$$xy - (-1)^{|x||y|}\tau_*(yx) = \pm \mathfrak{b}_{Q_n}(P_n)\beta_n(x)\beta_n(y).$$

(iii) *Suppose that* $[\delta Q_i] = 0 \in L_*^{P_i}(\mathrm{pt})$ *and* $\mathfrak{c}(Q_i) = 0 \in L_*^{P_i}(\mathrm{pt})$ *for every* $i \le n$. *Then for every* $x \in L_*^{\Sigma_n}(X), y \in L_*^{\Sigma_n}(Y), z \in L_*^{\Sigma_n}(Z)$ *we have*

$$(xy)z - x(yz) = 0,$$

i.e., the quasi-multiplication μ_n *is associative.*

Proof. This can be proved by induction based on 2.4, 2.10 and 2.13. See Botvinnik [1], Mironov [1], [2]. □

2.18. Remark. The obstructions $\mathfrak{a}$, $\mathfrak{b}$, $\mathfrak{c}$ are natural in the following sense. Consider two classes $\mathscr{K}_1, \mathscr{K}_2$, where $\mathscr{K}_2$ is an underlying class for $\mathscr{K}_1$, e.g., $\mathscr{K}_1$ consists of stably almost complex manifolds and $\mathscr{K}_2$ consists of PL manifolds. Consider the corresponding morphism $r : (L_1)_*^P \to (L_2)_*^P$ of bordism theories. Given a $\mathscr{K}_1$-manifold P, we have the obstruction $\mathfrak{a}_1(P) \in (L_1)_*^P(\mathrm{pt})$. Regarding P as a $\mathscr{K}_2$-manifold, we have the obstruction $\mathfrak{a}_2(P) \in (L_2)_*^P(\mathrm{pt})$. By construction, $r(\mathfrak{a}_1) = \mathfrak{a}_2$, i.e., the obstruction $\mathfrak{a}$ is natural with respect to morphisms of bordism theories. Similarly for $\mathfrak{b}$, $\mathfrak{c}$.

§3. Obstructions and Steenrod–tom Dieck Operations

Mironov [2] found that the obstructions $\mathfrak{a}(P)$ and $\mathfrak{b}(P)$ can be expressed in terms of Steenrod–tom Dieck operations. Below we explain this for the obstruction $\mathfrak{b}(P)$, see 3.9. (If $L = MU$, then $\mathfrak{a}(P) = 0$ for dimensional reasons, while we did not define Steenrod–tom Dieck operations for other L.)

In this section γ^n denotes $\gamma_{\mathbb{C}}^n$.

Firstly, we give a geometric description of Steenrod–tom Dieck operations for $X = S^0$, i.e., we describe the homomorphism

$$P_n^{2r} : MU^{2r}(\mathrm{pt}) = \widetilde{MU}^{2r}(S^0) \to \widetilde{MU}^{4r}((RP^n)^+) = MU^{4r}(RP^n)$$

in terms of manifolds. Here we follow Quillen [1].

3.1. Definition. (a) Let X^n be a smooth manifold without boundary and let V^k be a smooth manifold (possibly with a boundary). (We do not assume X and V to be compact.) Roughly speaking, a *stable almost complex structure* *on a map* $f : V \to X$ is a complex structure on a stable normal bundle of a smooth embedding $g : V^k \to X \times \mathbb{R}^{2N+k+n}$ with $p_1 g = f$. We give a detailed definition.

(a) A *strict almost complex structure on* f is a tuple $\mathfrak{g} := (g, U, q, \nu^{2n}, \omega)$ where $g : V^k \to X \times \mathbb{R}^{2N+k+n}$ is a smooth embedding with $p_1 g = f$, (U, q, ν^{2N}) is a smooth tubular neighborhood of g, and $\omega : \nu^{2N} \to \gamma^N$ is a morphism of vector bundles.

(b) Given a strict almost complex structure $\mathfrak{g} := (g, U, q, \nu^{2N}, \omega)$, we define the suspension $\sigma\mathfrak{g} = (g', U', q', \nu^{2N} \oplus \theta^2, \omega')$ of $\mathfrak{g}$, where g' is the embedding

$$V \xrightarrow{g} X \times \mathbb{R}^{2N+k+n} \times \mathbb{R}^2 = X \times \mathbb{R}^{2N+k+n+2}, \quad g'(v) = (v, 0),$$

$U' := U \times \mathbb{R}^2$, $q' = q \times 1 : U \times \mathbb{R}^2 \to \operatorname{ts}(\nu^{2N} \oplus \theta^2) = \operatorname{ts}(\nu^{2N}) \times \mathbb{R}^2$, and $\omega' : \nu^{2N} \oplus \theta^2 \to \gamma^{N+1}$ looks like $\widehat{\omega}$ in IV.4.14(b).

(c) We asy that two strict structures

$$\mathfrak{g}_0 = (g_0, U_0, q_0, \nu_0^{2N}, \omega_0), \ \mathfrak{g}_1 = (g_1, U_1, q_1, \nu_1^{2N}, \omega_1)$$

are *equivalent* if there is a family $J_t : X \times \mathbb{R}^{2N+k+n} \to X \times \mathbb{R}^{2N+k+n}, t \in I$ with the following properties:

(1) $p_1 J_t = f$ for every $t \in I$;
(2) The map $J : X \times \mathbb{R}^{2N+k+n} \times I \to X \times \mathbb{R}^{2N+k+n} \times I, J(a,t) = (J_t(a), t)$ for every $a \in X \times \mathbb{R}^{2N+k+n}, t \in I$, is a diffeomorphism;
(3) $J_1(U_0) = U_1$, $J_1(g_0(v)) = g_1(v)$ for every $v \in V$;
(4) There is an isomorphism $h : \nu_0 \to \nu_1$ of vector bundles such that $\omega_0 = \omega_1 h : \nu_0^{2N} \to \gamma^N$.

(d) We say that two strict almost complex structures $\mathfrak{g}_0$ and $\mathfrak{g}_1$ are stably equivalent if there are non-negative integers k, l such that the strict almost complex structures $\sigma^k \mathfrak{g}_0$ and $\sigma^l \mathfrak{g}_1$ are equivalent.

(e) A stable equivalence class of strict stably almost complex structures on $f : V \to X$ is called a *stable almost complex structure on* f. A *stably almost complex map* $f : V \to X$ is a map f with a fixed stable almost complex structure.

Clearly, this definition can be generalized. Namely, given a structure map $\varphi : B \to B\mathcal{V}^{\mathcal{T}}$, you can define a (B, φ)*-structure on a* $\mathcal{T}$ *map* $f : V \to X, \partial X = \emptyset$, *of* $\mathcal{T}$ *manifolds*. Moreover, every (B, φ)-structure on a manifold V can be regarded as a (B, φ)-structure on the map $V \to \mathrm{pt}$.

Consider a stably almost complex map $f : V^k \to X^n$ of a closed manifold V. Let $\mathfrak{g}$ be as in 3.1(a). Since $\operatorname{ts} \nu^{2N} \subset \operatorname{ts} \theta_X^{2N+k+n} = X \times \mathbb{R}^{2N+k+n}$, there is a collapsing map

$$c : S^{2N+k+n}(X^+) = T(\theta^{2N+k+n}) \to T\nu^{2N} = T(\theta^{2N+k+n})/U,$$

where U is a tubular neighborhood of $g(V)$ in $\mathbb{R}^{2N+k+n}$. So, we have a map $T\omega : T\nu \to MU_{N+n}$. The map $T\omega \circ c : \Sigma^{2N+k+n}(X^+) \to MU_{N+n}$ yields an element $m \in \widetilde{MU}^{n-k}(X^+)$, and we set

$$q_X(V, f) := m \in \widetilde{MU}^{n-k}(X^+).$$

It is easy to see that the element $q_X(V, f)$ is well-defined, i.e., it does not depend on g, N for N large. Besides, every $x \in MU^*(X)$ has the form $q_X(V, f)$ for some $f : V \to X$. Indeed, let $x \in MU^k(X)$ be represented by a map $a : \Sigma^{2N-k}X^+ \to MU_N$ such that a is transversal to γ^N. If we set $V := a^{-1}(BU_N)$, we get a stably almost complex map $f : V \to X$ with $q_X(V, f) = x$.

3.2. Lemma. *Let $f_i : V_i \to X, i = 0, 1$, be two singular closed smooth manifolds. Let $F : W \to X \times \mathbb{R}$ be a stably almost complex map with $\partial W = V_0 \sqcup V_1$ and such that each $F|V_i, i = 0, 1$ has the form $V_i \xrightarrow{f_i} X = X \times \{i\} \subset X \times \mathbb{R}$. Let us equip each f_i with the stable almost complex structure induced from F. Then $q_X(V_0, f_0) = q_X(V_1, f_1)$. (Roughly speaking, q_X is invariant with respect to bordism of stably almost complex maps.)* □

For simplicity, we denote q_{RP^n} by q_n.

3.3. Proposition. *Let $x \in MU^{-2r}(\mathrm{pt}) = \pi_{2r}(MU)$ be represented by a manifold M^{2r}. Consider a singular manifold $f : V^{n+2r} \to RP^n$ of the form*

$$V = S^n \times_{\mathbb{Z}/2} M \times M \xrightarrow{f} S^n \times_{\mathbb{Z}/2} \mathrm{pt} \times \mathrm{pt} = RP^n, \quad f(s, m, m) = (s, *, *).$$

In other words, $f = \Gamma_2(\varepsilon)$, where $\varepsilon : M \to \mathrm{pt}$. Then $f : V \to RP^n$ turns out to be a stably almost complex map such that

$$q_n(V, f) = P_n^{-2r}(x) \in \widetilde{MU}^{-4r}((RP^n)^+).$$

Proof. Routine arguments, based on the Pontrjagin–Thom construction and the definition of Steenrod–tom Dieck operations. □

3.4. Construction. Let X be a stably almost complex manifold, and let $f : V^k \to X^n$ be a stably almost complex map. Consider the composition

$$V \xrightarrow{g} \mathbb{R}^{2N+k+n} \times X \xrightarrow{1 \times i} \mathbb{R}^{2N+k+n} \times \mathbb{R}^{2M+n}$$

where g is as in 3.1 and $i : X \to \mathbb{R}^{2M+n}$ be a smooth embedding. Clearly, the complex structures on ν_g and $\nu_{1\times i}$ yield a complex structure on $\nu_{(1\times i)g}$. Thus, V gets a canonical stable almost complex structure.

Now, put $X = \mathrm{pt}$ in 3.4. Then, by 3.4, every stable almost complex structure on a map $\varepsilon : V \to \mathrm{pt}$ yields a stable almost complex structure on the manifold V (as described in IV.7.13(b)). We leave it to the reader to check that in this way we get a bijective correspondence between stable almost complex structures on ε and stable almost complex structures on V.

3.5. Proposition–Definition. *Let $h : X^k \to Y^l$ be a stably almost complex map such that $\partial X = \emptyset$. We define $h^! : M\mathcal{U}^i(X) \to M\mathcal{U}^{i+l-k}(Y)$ by setting $h^! q_X(V, f) = q_Y(V, hf)$ for every stably almost complex map $f : V \to X$. This homomorphism is well-defined and called the* Gysin homomorphism.[17]

Proof. Routine, based on the Pontrjagin–Thom Theorem. □

3.6. Lemma. (i) *If V^k is a closed stably almost complex manifold then $q(V, \varepsilon) = [V] \in M\mathcal{U}^{-k}(\mathrm{pt})$ where $\varepsilon : V \to \mathrm{pt}$ collapses V.*

(ii) *If X is a stably almost complex manifold and $f : V^k \to X^n$ is a stably almost complex map then $\varepsilon^! q_{\mathrm{pt}}(V, f) = [V] \in M\mathcal{U}^{-k}(\mathrm{pt})$, where $[V]$ is equipped with the stable almost complex structure in* 3.4 *and $\varepsilon : X \to \mathrm{pt}$ collapses X.*

(iii) *Let X be a stably almost complex manifold. If $q(V, f) = q(V', f')$ for some $f : V \to X$ and $f' : V' \to X$ then $[V] = [V']$ where V and V' are equipped with the stable almost complex structures in* 3.4.

Proof. (i) This is obvious.

(ii) This follows from (i).

(iii) This follows from (ii). □

Given a pointed CW-space Y, consider the pointed inclusion $e_Y : Y \subset Y^+$ and the homomorphism $e_Y^* : M\mathcal{U}^*(Y) = \widetilde{M\mathcal{U}}^*(Y^+) \to \widetilde{M\mathcal{U}}^*(Y)$. Let $\sigma^n \in \widetilde{M\mathcal{U}}^n(S^n)$ be the image of $1 \in \widetilde{M\mathcal{U}}^0(S^0)$ under the suspension isomorphism $\widetilde{M\mathcal{U}}^*(S^0) \to \widetilde{M\mathcal{U}}^{*+n}(S^n)$. Since this isomorphism is multiplication by σ^n, we denote it (as usual) by $\sigma^n : \widetilde{M\mathcal{U}}^*(S^0) \to \widetilde{M\mathcal{U}}^{*+n}(S^n)$.

3.7. Lemma. *Let $f : V^k \to S^n$ be a stably almost complex map. Then $e^* q_{S^n}(V, f) = \sigma^n[V] \in \widetilde{M\mathcal{U}}^{n-k}(S^n)$ where V is equipped with the stable almost complex structure as* 3.4.

Proof. Let q denote q_{S^n}. Let $q(V, f)$ be represented by a map $h : S^{2N-n+k}((S^n)^+) = S^{2N+k} \vee S^{2N-n+k} \to M\mathcal{U}_N$. Note that

$$S^{2N-n+k}((S^n)^+) \setminus * \cong S^n \times \mathbb{R}^{2N-n+k}.$$

We can assume that h is transversal to γ^N and that

$$h^{-1}(B\mathcal{U}_n) \subset S^{2N-n+k}((S^n)^+) \setminus *.$$

Then $q(V, f) = q(V', f')$, where $V' = h^{-1}(B\mathcal{U}_N)$ and

[17] Cf. V.2.11.

$$f' : V' \subset S^{2N-n+k}((S^n)^+) \setminus * \cong S^n \times \mathbb{R}^{2N-n+k} \xrightarrow{p_1} S^n$$

is the map equipped with an obvious stable almost complex structure. Furthermore, $e^*q(V', f')$ is represented by $h|S^{2N+k} : S^{2N+k} \to MU_N$, and so $e^*q(V', f') = \sigma^n[V']$. But, by 3.6(iii), $[V] = [V']$ since $q(V, f) = q(V', f')$. □

Let $z_2 \in MU^2(RP^2)$ be as defined before VII.7.9, and let $\widetilde{z}_2 = e^*_{RP^2} z_2$.

3.8. Lemma. *Let $c : RP^2 \to S^2$ be an essential map (unique up to homotopy). Then $c^*\sigma^2 = \widetilde{z}_2$.*

Proof. This is valid since $\sigma^2 = e^*_{S^2} C_1(\eta_1)$ (in the notaion of Ch. VII). □

Given $x \in MU^{2r}(\mathrm{pt})$, set $\widetilde{P}^{2r}_n(x) := e^*_{RP^n} P^{2r}_n(x) \in \widetilde{MU}^{4r}(RP^n)$, where P^{2r}_n is the Steenrod–tom Dieck operation.

3.9. Theorem. *Let $P = P^d$, d even, be a closed stably almost complex manifold. Let $r : MU \to MU^P$ be the forgetful morphism as in (1.12). Then*

$$r\widetilde{P}^{-d}_2([P]) = \widetilde{z}_2 \mathfrak{b}_Q(P) \in (\widetilde{MU^P})^{-2d}(RP^2)$$

for some Q. In particular,

$$\rho r\widetilde{P}^{-d}_2([P]) = \widetilde{z}_2 \varkappa(P) \in ((\widetilde{MU^P})^{-2d}(RP^2))/(2),$$

where ρ is the modulo 2 reduction.

Proof. We construct a stably almost complex map $f : M^{2d+2} \to S^2$ such that $r[M] = \mathfrak{b}_Q(P)$ and $c^*q_{S^2}(M, f) = P^{-d}_2[P]$ with c as in 3.8. Then the theorem will be proved. Indeed, by 3.3, 3.7 and 3.8 we have (where $e^* = e^*_{S^2}, q = q_{S^2}$)

$$\begin{aligned}\widetilde{P}^{-d}_2([P]) &= e^*_{RP^2} c^* q(M, f) = c^* e^* q(M, f) = c^*(\sigma^2[M]) \\ &= c^*(\sigma^2)[M] = \widetilde{z}_2[M].\end{aligned}$$

We interpret S^2 as the Riemannian sphere $\overline{\mathbb{C}} = \mathbb{C} \cup \{\infty\}$. Consider the following subsets of S^2:

$$\begin{aligned}
D_1 &:= \{z \mid |z| \le 1/2\},\ D_2 = \{z \mid |z| \ge 2\},\\
K_1 &:= \{z \mid 1/2 \le |z| \le 2, 0 \le |\arg z| \le \pi/2\},\\
K_2 &:= \{z \mid 1/2 \le |z| \le 2, \pi/2 \le |\arg z| \le \pi\},\\
S^1 &:= \{z \mid |z| = 1\},\\
J &:= (1/2, 2) \subset \mathbb{R} \subset \mathbb{C},\\
A &:= \{z \mid 1/2 < |z| < 2\} = S^1 \times J.
\end{aligned}$$

For simplicity, denote $\Gamma_1(P)$ (i.e., $S^1 \times_{\mathbb{Z}/2} (P \times P)$) by Γ.

Let $g = \Gamma_1(\varepsilon) : \Gamma \to \Gamma_1(\mathrm{pt}) = S^1$ be the map induced by $\varepsilon : P \to \mathrm{pt}$. Consider the map

$$g \times 1 : \Gamma \times J \to S^1 \times J = A$$

and set $y = q_A(\Gamma \times J, g \times 1) \in M\mathcal{U}^{-2d}(A)$. Let $i : A \to A \cup D_1$ be the inclusion. Since $[\Gamma] \in M\mathcal{U}^{-2d-1}(\mathrm{pt}) = 0$, $y = i^*(x)$ for some $x \in M\mathcal{U}^{-2d}(A \cup D_1)$. We have $x = q_{A\cup D_1}(V, f_1)$, where $f_1 : V \to A \cup D_1$ is such that

$$f_1^{-1}(A) = \Gamma \times J \text{ and } f_1|\Gamma \times J = g \times 1.$$

(In fact, V is the result of attaching a membrane to the bottom of $\Gamma \times J$, and f_1 maps this membrane to D_1.)

We define an involution ω on $\Gamma \times J$ by setting

$$\omega(z, p_1, p_2, s) = (\bar{z}, p_2, p_1, 1/s), \quad z \in S^1, (p_1, p_2) \in P \times P, s \in J.$$

We have $(g \times 1)(\omega(v)) = 1/(g \times 1)(v), v \in \Gamma \times J$. Define $f_2 : V \to S^2$ by setting $f_2(v) = 1/f_1(v)$ for every $v \in V$. Gluing f_1 and f_2, we get a map

$$f = f_1 \cup f_2 : M := V \cup_\omega V \to S^2.$$

We prove that $r[M] = \mathfrak{b}_Q(P)$ for some Q. Firstly, Γ can be represented as $P \times P \times I/\sim$, where $(x, y, 0) \sim (y, x, 1)$. We define

$$\varphi : P' \times P'' \times \{0,1\} \to P \times P \times \{0,1\}, \varphi(x,y,0) = (y,x,1), \varphi(x,y,1) = (y,x,0).$$

Consider a $\mathscr{K}^P$-manifold $W = \overline{P} \times I \cup_\varphi \Gamma$, where φ glues $\overline{P} \times \{1\}$ and Γ. It is clear that W is a $\mathscr{K}^P$-membrane between Γ and $\overline{P}$.

Set $N := f_1^{-1}(D_1), \partial N = \Gamma$. Set $Q := \overline{P} \cup_\varphi N$. Then

$$D_Q(P) = N \cup_\varphi \overline{P} \times [-1, 0] \cup_\chi \overline{P} \times [0, 1] \cup_\varphi N,$$

where $\chi : \overline{P} \times \{0\} \to \overline{P} \times \{0\}$ is the involution defined before 2.10. This manifold $D_Q(P)$ is $\mathscr{K}^P$-bordant to the $\mathscr{K}^P$-manifold

$$N \cup \Gamma \times [-1, 0] \cup_\omega \Gamma \times [0, 1] \cup N$$

which, in turn, is $\mathscr{K}^P$-bordant to M. Thus, $r[M] = \mathfrak{b}_Q(P)$.

We prove that $c^* q_{S^2}(M, f) = P_2^{-d}(P)$. Consider the principal $\mathbb{Z}/2$-bundle (two-sheeted covering) $S^1 \times P \times P \to S^1 \times_{\mathbb{Z}/2} P \times P = \Gamma$. Let $\mathbb{Z}/2$ act on $I = [0, 1]$ as the linear map $t \mapsto 1 - t$, and let $X \to \Gamma$ be the associated $(I, \mathbb{Z}/2)$-bundle. So, X is a manifold with $\partial X = S^1 \times P \times P$. It is easy to see that

(3.10) $$\Gamma_2(P) = S^2 \times_{\mathbb{Z}/2} P \times P = D^2 \times P \times P \cup_{S^1 \times P \times P} X,$$

where D^2 is a disk.

Considering the inclusion $\mathbb{R}^2 \subset RP^2$, we have $RP^2 = \overline{\mathbb{R}^2}$, where bar means the closure. Consider $RP^1 = \overline{\mathbb{R}^1} \subset \overline{\mathbb{R}^2} = RP^2$ and set

$$T = \overline{\mathbb{R}^1 \times [-1,1]} \subset RP^2.$$

So, T is a Möbius band. Let $\pi : T \to RP^1$ be the projection in the normal bundle of RP^1 in RP^2. We have $RP^2 = T \cup D^2$.

We choose $c : (RP^2, RP^1) \to (S^2, 1)$ such that:

1. c is smooth, $c|(RP^2 \setminus RP^1) : RP^2 \setminus RP^1 \to S^2 \setminus \{1\} \subset \overline{\mathbb{C}} \setminus \{1\}$ is bijective.

2. $c(D^2) \subset K_2$, $c(-t) = 1/c(t)$ for every $t \in T \subset RP^2 \setminus RP^1$.

3. The set $\pi^{-1}(a) \cap c^{-1}(D_1)$ is connected for every $a \in RP^1$.

Consider the pull-back diagram

$$\begin{array}{ccc} Z & \xrightarrow{\hat{f}} & RP^2 \\ \downarrow & & \downarrow{\scriptstyle c} \\ M & \xrightarrow{f} & S^2. \end{array}$$

(In fact, $Z = (M \setminus (D \times \Gamma)) \cup_{S^1 \times \Gamma} X$, where D is a small open disk.) Clearly,

$$q_2(Z, \hat{f}) = c^* q_{S^2}(M, f).$$

Set $h = c\hat{f} : Z \to S^2$. Switching the two copies of V in M, we get a diffeomorphism $\theta : h^{-1}(D_1) \to h^{-1}(D_2)$ such that $\hat{f}(\theta(m)) = -\hat{f}(m)$ for every $m \in h^{-1}(D_1)$ (recall that $\hat{f}(m) \in T$).

Given $s \in T \setminus RP^1$, $s \in \overline{\mathbb{R}^1} \times (0,1]$, let F_s be the unique fiber of $\pi : T \to RP^1$ such that $s \in F_s$, and let I_s be a unique segment which joins s and $-s$ in F_s. Let $l_s : I \to I_s$ be the linear homeomorphism, $l(0) = s$.

We have $h^{-1}(D_1) = f^{-1}(D_1) = N$. Define a map

$$\psi : N \times \{0\} \cup h^{-1}(D_1 \cap K_1) \cup N \times \{1\} \to Z$$

of the subset of $\partial(N \times I)$ as follows:

$$\psi|N \times \{0\} = 1_{N \times \{0\}},\ \psi|N \times \{1\} = \theta|(N \times \{0\}), \text{ and}$$
$$\psi|(h^{-1}(a) \times I) = 1_P \times 1_P \times l_{c^{-1}(a)} : P \times P \times I \to P \times P \times I_{c^{-1}(a)}$$
$$\text{where } a \in D_1 \cap K_1.$$

Define $\Phi : N \times I \to RP^2$ by setting $\Phi|(n \times I) = l_{c(n)}$ for every $n \in N$. The stably almost complex map $\hat{f}$ extends to a stably almost complex map

$$\hat{\Phi} = \hat{f} \circ \mathrm{proj} \cup \Phi : \hat{Z} = Z \times I \cup_\psi N \times I \to RP^2.$$

Now, $\partial \hat{Z} = Z \sqcup X \cup D^2 \times P \times P$. Thus, by 3.2, and 3.3,

$$q_2(\hat{Z}, \hat{f}) = q_2(X \cup_{S^1 \times P \times P} D^2 \times P \times P, \hat{\Phi}|X \cup_{S^1 \times P \times P} D^2 \times P \times P)$$
$$= q_2(\Gamma_2(P), \hat{\Phi}|\Gamma_2(P)) = q_2(\Gamma_2(P), \Gamma_2(\varepsilon)) = P_2^{-d}(P). \qquad \square$$

Let $\{V_0, \dots, V_n, \dots\}, \dim V_n = 2^{n+1} - 2$, be a family of stably almost complex manifolds such that $([V_0], \dots, [V_{n-1}]) = I(2,n)$ for every n. Set $\Sigma_n = \{V_0, \dots, V_n\}$. Let $r_n : MU \to MU^{\Sigma_n}$ be the forgetful morphism. Since $\pi_{\text{odd}}(MU) = 0$, the obstructions $\mathfrak{b}(V_n) \in \pi_*(MU^{\Sigma_n})$ are defined. (We use the same symbol for the obstruction $\mathfrak{b}(V_n) \in \pi_*(MU^{V_n})$ and its image in $\pi_*(MU^{\Sigma_n})$.) Furthermore, $\mathfrak{b}(V_n) = \varkappa(V_n)$ because $2\pi_*(MU^{\Sigma_n}) = 0$. Let $I = (2, \text{Dec}) \subset \pi_*(MU)$ be the ideal as in VII.7.18.

3.11. Corollary. $\mathfrak{b}(V_n) \equiv r_n[V_{n+1}] \bmod I$.

Proof. By 3.9 and VII.7.19,

$$\widetilde{z}_2 \mathfrak{b}(V_n) = r_n(P_2^{2-2^{n+1}}[V_n]) \equiv \widetilde{z}_2 r_n[V_{n+1}] \bmod I$$

in $(MU^{\Sigma_n})^*(\mathbb{R}P^2)$. Hence, it suffices to prove that $\widetilde{(MU^{\Sigma_n})}^*(RP^2)$ is a free $(MU^{\Sigma_n})^*(\text{pt})$-module. Now, $E^*(S^1)$ is a free $E^*(\text{pt})$-module with one generator of dimension 1 for every ring spectrum E. If, in addition, $2\pi_*(E) = 0$ then, considering the cofiber sequence $S^1 \xrightarrow{2} S^1 \to RP^2$, we conclude that $E^*(RP^2)$ is a free $E^*(\text{pt})$-module with generators $x_1, x_2, \dim x_i = i$. $\square$

3.12. Remarks. (a) This corollary leads to the following description of $I(2,n)$. Set $V_0 = \{2 \text{ points}\}$. Since $\mathfrak{a}(V_0) = 0$, we can construct a V_0-manifold $D(V_0)$ (described before 2.10). Since $\pi_*(MU) \to \pi_*(MU^{V_0})$ is epic, there is a stably almost complex manifold V_1 which is V_0-bordant to $D(V_0)$. Now, $\mathfrak{a}(V_1) = 0$, and we can construct $D(V_1)$ and V_2 as above, and so on. Because of 3.11, $I(2,n) = ([V_0], \dots, [V_{n-1}])$.

(b) The following picture looks interesting. Consider MU^{V_0}. It contains $[V_1] := D(V_0)$ as the obstruction to commutativity. Killing V_1, i.e., constructing MU^{Σ_2}, we get $[V_2] := D(V_1)$ as the obstruction to commutativity. And so on. So, every killing of the obstruction produces a new obstruction, i.e., we have a *Hydra* here. And we can relax after killing all V_n's only, obtaining a commutative spectrum. By the way, in this way we obtain an ordinary (co)homology theory, see IX.5.5 below.

§4. A Universality Theorem for MU with Singularities

This section is an extract from the paper Würgler [1].

Let (E, u) be a $\mathbb{C}$-marked spectrum such that every group $\pi_i(E)$ is finite. Set $\overline{s}_\omega := u_* s_\omega \in E^{2|\omega|}(MU)$. By VII.1.20, we have

$$E^*(MU) = MU^*(MU)\widehat{\otimes}_{MU^*(S)}E^*(S).$$

Thus, every element of $E^*(MU)$ can be represented as a formal series $\sum a_\omega \overline{s}_\omega, a_\omega \in E^*(S)$.

Define $E^*(S)$-homomorphisms

$$\Delta^E : E^*(MU) \to E^*(MU)\widehat{\otimes}_{E^*(S)}E^*(MU), \ \Delta^E(\overline{s}_\omega) = \sum_{(\omega_1,\omega_2)=\omega} \overline{s}_{\omega_1}\widehat{\otimes}\overline{s}_{\omega_2}$$

(see III.1.22–1.23 concerning the notation $\overline{s}_{\omega_1}\widehat{\otimes}\overline{s}_{\omega_2}$) and

$$\varepsilon : E^*(MU) \to E^*(S), \quad \varepsilon(\overline{s}_\omega) = 0 \text{ for } \omega \neq 0.$$

We say that $(E^*(MU), \Delta^E, \varepsilon)$ is an $E^*(S)$-*coalgebra*, because it satisfies II.6.7(a) if we replace $\otimes$ there by $\widehat{\otimes}$.

4.1. Definition. (a) A *profinite $E^*(S)$-module* is any $E^*(S)$-module M of the form $M = \varprojlim\{M_\lambda\}$, where $\{M_\lambda\}$ is any inverse system of finitely generated $E^*(S)$-modules. The category of profinite $E^*(S)$-modules is denoted by Mod.

(b) Given two profinite $E^*(S)$-modules M, N, we set

$$M \boxtimes N := \varprojlim\{M_\lambda \otimes_{E^*(S)} N_{\lambda'}\}.$$

(c) An $E^*(MU)$-*comodule* is a profinite $E^*(S)$-module M equipped with a coaction $\psi = \psi_M : M \to E^*(MU) \boxtimes M$ such that the diagrams like II.6.7 commute. A *homomorphism* of $E^*(MU)$-comodules is a homomorphism $f : M \to N$ of $E^*(S)$-modules which commutes with the coactions, i.e., $\psi_N f = (f \boxtimes 1)\psi_M$. The category of $E^*(MU)$-comodules is denoted by Com.

Note that, by III.4.17, $E^*(X)$ is a profinite $E^*(S)$-module for every spectrum X. Moreover, $E^*(X)\widehat{\otimes}_{E^*(S)}E^*(Y) = E^*(X)\boxtimes E^*(Y)$ for every two spectra X, Y. Furthermore, $E^*(F)$ is an $E^*(MU)$-comodule for every E-module spectrum F. Indeed, the module structure $m : MU \to F$ induces the action

$$\begin{aligned} m^* : E^*(F) \to E^*(MU \wedge F) &\cong MU^*(MU)\widehat{\otimes}_{MU^*(S)}E^*(F) \\ &= E^*(MU)\widehat{\otimes}_{E^*(S)}E^*(F) = E^*(MU) \boxtimes E^*(F). \end{aligned}$$

Let $S : Com \to Mod$ be the forgetful functor, and let $F : Mod \to Com$ assign to each $M \in Mod$ the extended comodule $F(M) := E^*(MU) \boxtimes M$ with the coaction

$$\begin{aligned} \psi_{F(M)} := \Delta^E \boxtimes 1 : E^*(MU) \boxtimes M &\to E^*(MU)\widehat{\otimes}_{E^*(S)}E^*(MU) \boxtimes M \\ &= E^*(MU) \boxtimes E^*(MU) \boxtimes M. \end{aligned}$$

4.2. Lemma. *There is a natural isomorphism*

$$e : Com(A, F(B)) \to Mod(S(A), B)$$

for every $A \in Com, B \in Mod$. In other words, F and S are adjoint functors.

Proof. Given $f : A \to F(B)$, define $e(f)$ as the composition

$$A \xrightarrow{f} E^*(MU) \boxtimes B \xrightarrow{\varepsilon \boxtimes 1} E^*(S) \boxtimes B = B.$$

We leave it to the reader to prove that e is an isomorphism. □

The following lemma is a standard result of relative homological algebra.

4.3. Lemma. *Let $M \in Mod$.*

(i) *(The relative injectivity of $F(M)$.) Consider a diagram*

$$\begin{array}{ccc} A & \xrightarrow{\varkappa} & B \\ \downarrow{\scriptstyle f} & & \\ F(M) & & \end{array}$$

in Com. If $\varkappa$ is a split monomorphism in Mod, then there exists a morphism $g : B \to F(M)$ of $E^*(MU)$-comodules with $g\varkappa = f$.

(ii) *Let $0 \to F(M) \to B \to C \to 0$ be an exact sequence in Com. If it splits in Mod then it splits in Com.*

Proof. (i). Let $a : B \to A$ split $\varkappa$ in Mod. By 4.2, we have a commutative diagram, where t is adjoint to f and $b = tS(a)$:

$$\begin{array}{ccc} S(A) & \xleftarrow{S(a)} & S(B) \\ {\scriptstyle t}\downarrow & & \downarrow{\scriptstyle b} \\ M & = & M \end{array}.$$

Hence $bS(\varkappa) = t$. By 4.2, b is adjoint to some $g : B \to F(M)$, and $g\varkappa = f$ because $g\varkappa$ is adjoint to t.

(ii) This follows from (i) if we put $f = 1_{F(M)}$ in (i). □

Every profinite module $\varprojlim M_\lambda$ has a topology inherited from $\prod M_\lambda$. This topology does not depend on the system $\{M_\lambda\}$ (prove it). Let $\overline{Mod}$ be the category of profinite topological modules (topologized as above) and continuous homomorphisms.

4.4. Proposition. *Let $f : X \to Y$ be a morphism of spectra. Then $f^* : E^*(Y) \to E^*(X)$ is a morphism of the category $\overline{Mod}$.* □

4.5. Lemma. *$E^*(MU)$ is a projective object of $\overline{Mod}$.*

Proof. Consider the left diagram of (4.6), where σ is epic:

$$
(4.6)\qquad
\begin{array}{ccc}
 & & B \\
 & & \downarrow{\scriptstyle\sigma} \\
E^*(M\mathcal{U}) & \xrightarrow{f} & C
\end{array}
\qquad\qquad
\begin{array}{ccc}
R & \xrightarrow{h} & B \\
\cap\downarrow & & \downarrow{\scriptstyle\sigma} \\
E^*(M\mathcal{U}) & \xrightarrow{f} & C.
\end{array}
$$

Assuming that this is a diagram in $\overline{Mod}$, we must find a continuous homomorphism $g : E^*(M\mathcal{U}) \to B$ with $\sigma g = f$.

It is easy to see that the topology on every $M \in \overline{Mod}$ is such that (every homogeneous component of) M is compact. Hence M admits a unique uniform structure compatible with this topology, see e.g. Bourbaki [2] or Kelley [1]. Furthermore, M is a complete uniform space because M is compact.

Let R be the (discrete) $E^*(S)$-submodule of $E^*(M\mathcal{U})$ generated by finite sums $\sum a_\omega \overline{s}_\omega$. Then R is dense in $E^*(M\mathcal{U})$. Since R is a free $E^*(S)$-module, there exists $h : R \to B$ such that the right diagram of (4.6) commutes. Since R is dense in the complete space $E^*(M\mathcal{U})$ and B is complete, there exists a continuous homomorphism $g : E^*(M\mathcal{U}) \to B$ which extends h. Since $f|R = (\sigma g)|R$ and R is dense in $E^*(S)$, we conclude that $f = \sigma g$. □

4.7. Definition. A sequence (finite or infinite) $\Sigma = \{x_0, x_1, \dots, x_n, \dots\}$, $x_i \in \pi_{d_i}(M\mathcal{U})$, is called *invariant* if $s_\omega(x_i) \in (x_0, \dots, x_{i-1})$ for every i and every $\omega \neq (0)$. Clearly, the invariance implies that $\varphi(x_i) \in (x_0, \dots, x_i)$ for every i and every operation $\varphi \in M\mathcal{U}^*(M\mathcal{U})$.

Recall that $M\mathcal{U}^{\Sigma_k}$ is a (left) $M\mathcal{U}$-quasi-module spectrum.

4.8. Proposition. *Let $\Sigma = \{x_0, \dots, x_n\}, \dim x_0 = 0$, be a proper sequence in $\pi_*(M\mathcal{U})$. Then the $M\mathcal{U}$-quasi-module structure on $M\mathcal{U}^{\Sigma_i}$ can be extended to an $M\mathcal{U}$-module structure on $M\mathcal{U}^{\Sigma_i}$, and this extension is unique. Furthermore, $M\mathcal{U}^{\Sigma_i}$ is the cone of multiplication by $x_i : \Sigma^{d_i} M\mathcal{U}^{\Sigma_{i-1}} \to M\mathcal{U}^{\Sigma_{i-1}}$.*

Proof. By 1.24, $\pi_*(M\mathcal{U}^{\Sigma_k}) = \pi_*(M\mathcal{U})/(x_0, \dots, x_k)$. Hence, the groups $\pi_i(M\mathcal{U}^{\Sigma_k})$ are finite for all i, k. Thus, $M\mathcal{U}^{\Sigma_k}$ is an $M\mathcal{U}$-module spectrum by III.7.8. Since multiplication by $x_i : \pi_*(M\mathcal{U}^{\Sigma_{i-1}}) \to \pi_*(M\mathcal{U}^{\Sigma_{i-1}})$ is monic, $M\mathcal{U}^{\Sigma_k}$ is the cone of x_i by 1.20. □

4.9. Lemma. *Let $\Sigma = \{x_0, \dots, x_n\}, \dim x_0 = 0$, be a proper invariant sequence in $\pi_*(M\mathcal{U})$. If multiplication by $x_i : \Sigma^{d_i} E \to E$ is trivial for every i, then there is an isomorphism of $E^*(M\mathcal{U})$-comodules*

$$E^*(M\mathcal{U}^{\Sigma_n}) \cong E^*(M\mathcal{U}) \otimes \Lambda(q_0, \dots, q_n), \dim q_i = 1 + d_i.$$

Here the coaction

$$\psi : E^*(M\mathcal{U}) \otimes \Lambda(q_0, \dots, q_n) \to E^*(M\mathcal{U}) \widehat{\otimes}_{E^*(S)} E^*(M\mathcal{U}) \otimes \Lambda(q_0, \dots, q_n)$$

has the form $\psi(a \otimes b) = \Delta^E(a) \otimes b$.

Proof. Firstly, we prove that multiplication by

$$x_{i+1} : \Sigma^{d_{i+1}} M\mathcal{U}^{\Sigma_i} \to M\mathcal{U}^{\Sigma_i}$$

induces the zero homomorphism $E^*(M\mathcal{U}^{\Sigma_i}) \to E^*(M\mathcal{U}^{\Sigma_i})$. Consider the commutative diagram

$$\begin{array}{ccccc} E^*(M\mathcal{U}^{\Sigma_i}) \xrightarrow{m^*} & E^*(M\mathcal{U} \wedge M\mathcal{U}^{\Sigma_i}) & \xrightarrow{(x_{i+1}\wedge 1)^*} & E^*(S \wedge M\mathcal{U}^{\Sigma_i}) \\ & a\uparrow\cong & & b\uparrow\cong \\ & M\mathcal{U}^*(M\mathcal{U})\widehat{\otimes} E^*(M\mathcal{U}^{\Sigma_i}) & \xrightarrow{x_{i+1}^*\widehat{\otimes}1} & M\mathcal{U}^*(S)\widehat{\otimes} E^*(M\mathcal{U}^{\Sigma_i}) \end{array}$$

where m^* is induced by the module structure and $\widehat{\otimes}$ means $\widehat{\otimes}_{E^*(S)}$. Now, given $\varphi \in E^*(M\mathcal{U}^{\Sigma_i})$ with $m^*(\varphi) = a\left(\sum_k \varphi_k'\widehat{\otimes}\varphi_k''\right)$, we have

$$b\left((x_{i+1} \wedge 1)^*(m^*\varphi)\right) = b\left((x_{i+1}^* \otimes 1)\left(\sum \varphi_k'\widehat{\otimes}\varphi_k''\right)\right) = b\left(\sum \varphi_k'(x_{i+1})\widehat{\otimes}\varphi_k''\right).$$

However, $\varphi_k'(x_{i+1}) \in (x_0, \dots, x_{i+1})$ because Σ is an invariant sequence. Hence, $b\left((x_{i+1} \wedge 1)^*(m^*\varphi)\right) = 0$. Now, since b is an isomorphism, we conclude that $(x_{i+1}\wedge 1)^*(m^*\varphi) = 0$. But the homomorphism $E^*(M\mathcal{U}^{\Sigma_i}) \to E^*(M\mathcal{U}^{\Sigma_i})$ in question coincides with $(x_{i+1} \wedge 1)^* m^*$.

Now we prove the lemma by induction on n. By 4.8, we have a cofiber sequence $\Sigma^{d_i} M\mathcal{U}^{\Sigma_i} \xrightarrow{x_i} M\mathcal{U}^{\Sigma_i} \xrightarrow{r} M\mathcal{U}^{\Sigma_{i+1}}$. By the above, it yields an exact sequence of $E^*(M\mathcal{U})$-comodules

$$\text{(4.10)} \qquad 0 \to E^*(\Sigma^{1+d_i} M\mathcal{U}^{\Sigma_i}) \to E^*(M\mathcal{U}^{\Sigma_{i+1}}) \to E^*(M\mathcal{U}^{\Sigma_i}) \to 0.$$

By 4.4, this is also a sequence in $\overline{Mod}$. Suppose that we have an isomorphism of $E^*(M\mathcal{U})$-comodules

$$E^*(M\mathcal{U}^{\Sigma_i}) \cong E^*(M\mathcal{U}) \otimes \Lambda(q_0, \dots, q_i) \text{ for } i < n.$$

Then, by 4.5, $E^*(M\mathcal{U}^{\Sigma_i})$ is a projective object of $\overline{Mod}$. So, by 4.4, the sequence (4.10) splits in $\overline{Mod}$, and, therefore, it splits in Mod. Hence, by 4.3(ii), it splits in $\mathcal{C}om$. Thus, we have an isomorphism of $E^*(M\mathcal{U})$-comodules

$$E^*(M\mathcal{U}^{\Sigma_n}) \cong E^*(M\mathcal{U}^{\Sigma_{n-1}}) \otimes \Lambda(q_n) \cong E^*(M\mathcal{U}) \otimes \Lambda(q_0, \dots, q_n).$$

The induction is confirmed. □

4.11. Definition. Let $(M, \psi) \in \mathcal{C}om$. Similarly to II.6.12, we call an element $m \in M$ *simple* if $\psi(m) = u \boxtimes m$, and the submodule of simple elements of M is denoted by $\mathrm{Si}(M)$.

Note that $\mathrm{Si}(E^*(M\mathcal{U})) = E^*(S)$.

4.12. Proposition. *Let (F, m) be an $M\mathcal{U}$-module spectrum. Let $E^*(F)$ be the $E^*(M\mathcal{U})$-comodule with the coaction $m^* : E^*(F) \to E^*(M\mathcal{U})\widehat{\otimes} E^*(F)$*

described after 4.1. *A morphism* $h : F \to E$ *is an* $\mathcal{MU}$*-module morphism iff the element* $h \in E^*(F)$ *is simple.*

Proof. Define $m_E : \mathcal{MU} \wedge E \xrightarrow{u \wedge 1_E} E \wedge E \to E$, where u is the $\mathbb{C}$-marking on E. Because of VII.1.20, the left diagram below induces the right diagram:

$$\begin{array}{ccc} \mathcal{MU} \wedge F & \xrightarrow{m_F} & F \\ {\scriptstyle 1_{\mathcal{MU}} \wedge h}\downarrow & & \downarrow{\scriptstyle h} \\ \mathcal{MU} \wedge E & \xrightarrow{m_E} & E \end{array} \qquad \begin{array}{ccc} E^*(\mathcal{MU})\widehat{\otimes}_{E^*(S)}E^*(F) & \xleftarrow{m_F^*} & E^*(F) \\ {\scriptstyle 1_{E^*(\mathcal{MU})}\widehat{\otimes} h^*}\uparrow & & \uparrow{\scriptstyle h^*} \\ E^*(\mathcal{MU})\widehat{\otimes}_{E^*(S)}E^*(E) & \xleftarrow{m_E^*} & E^*(E). \end{array}$$

We prove that the left diagram commutes iff $m_F^*(h) = u\widehat{\otimes}h$. It is clear that $m_F^*(h) = m_F^* h^*(1_E)$. Hence, the left diagram commutes iff

$$m_F^*(h) = (1_{E^*(\mathcal{MU})}\widehat{\otimes}h^*)m_E^*(1_E),$$

i.e., iff

$$m_F^*(h) = (1_{\mathcal{MU}} \wedge h)^* m_E^*(1_E) = (1_{\mathcal{MU}} \wedge h)^*(u\widehat{\otimes}1_E) = u\widehat{\otimes}h. \qquad \square$$

4.13. Theorem. *Let* $\Sigma = \{x_0, \dots, x_n, \dots\}, \dim x_0 = 0$, *be a proper invariant sequence (finite or not) in* $\pi_*(\mathcal{MU})$. *If multiplication by* $x_i : \Sigma^{d_i}E \to E$ *is trivial for every* i, *then there is an* $\mathcal{MU}$*-module morphism* $\mathcal{MU}^\Sigma \to E$ *such that* $hr \simeq u$ *where* $r : \mathcal{MU} \to \mathcal{MU}^\Sigma$ *is the forgetful morphism.*

Proof. Put $\Sigma_n = \{x_0, \dots, x_n\}$, and let $r_n : \mathcal{MU} \to \mathcal{MU}^{\Sigma_n}$ be the forgetful morphism. By 4.9,

$$\mathrm{Si}(E^*(\mathcal{MU}^{\Sigma_n})) = \mathrm{Si}(E^*(\mathcal{MU})\otimes\Lambda(q_0, \dots, q_n)) = \mathrm{Si}(E^*(\mathcal{MU}))\otimes\Lambda(q_0, \dots, q_n).$$

Now, by 4.12, $u \otimes 1 \in \mathrm{Si}(E^*(\mathcal{MU})) \otimes \Lambda(q_0, \dots, q_n)$ yields an $\mathcal{MU}$-module morphism $h_n : \mathcal{MU}^{\Sigma_n} \to E$.

We prove that $h_n r_n \simeq u$. Following the proof of 4.9, we conclude that there is a commutative diagram

$$\begin{array}{ccc} E^*(\mathcal{MU}^{\Sigma_n}) & \xrightarrow{\cong} & E^*(\mathcal{MU}) \otimes \Lambda(q_0, \dots, q_n) \\ {\scriptstyle r_n^*}\downarrow & & \downarrow{\scriptstyle \overline{r}} \\ E^*(\mathcal{MU}) & = & E^*(\mathcal{MU}) \end{array}$$

where $\overline{r}$ is a morphism of $E^*(\mathcal{MU})$-comodules such that $\overline{r}(q_i) = 0$ and $\overline{r}(a \otimes 1) = a$ for every $a \in E^*(\mathcal{MU})$. Thus, $r_n^*(h_n) = u$, i.e., $h_n r_n \simeq u$.

If the sequence Σ is finite then $\Sigma = \Sigma_n$ and $r = r_n$ for some n, and we can put $h = h_n$. Now, suppose that Σ is infinite. By the definition of h_n, we have $h_n \simeq r_n h_{n+1}$. So, we have the commutative diagram

$$\begin{array}{ccccccc} \cdots \longrightarrow & \mathcal{MU}_*^{\Sigma_n}(-) & \xrightarrow{r_n} & \mathcal{MU}_*^{\Sigma_{n+1}}(-) & \longrightarrow \cdots \\ & h_n\big\downarrow & & \big\downarrow h_{n+1} & \\ \cdots = & E_*(-) & = & E_*(-) & = \cdots \end{array}$$

of morphisms of homology theories. Passing to the direct limit of the top row, we get, by I.2.5, the morphism $h : \mathcal{MU}^{\Sigma}(-) = \varinjlim\{\mathcal{MU}^{\Sigma_n}(-)\} \to E_*(-)$. Now, by III.3.23(ii), h is induced by a morphism $h : \mathcal{MU}^{\Sigma} \to E$ of spectra. The homotopy $hr \simeq u$ can be proved just as the homotopy $h_n r_n \simeq u$ was. □

4.14. Corollary. *Let F be a graded formal group classified by a homomorphism $\rho : \mathcal{MU}^*(S) \to R$. Suppose that ρ is an epimorphism such that* $\operatorname{Ker}\rho = (x_0, \dots, x_n, \dots)$, *where* $\Sigma := \{x_0, \dots, x_n, \dots\}$ *is a proper sequence. Then the formal group F can be realized by a $\mathbb{C}$-marked spectrum. Moreover if* $\dim x_0 = 0$ *and Σ is invariant then this spectrum is unique up to equivalence.*

Proof. Let $\mathscr{C}_T$ be the category described before VII.6.19. It is clear that

$$\left(\mathcal{MU}^{\Sigma}\right)^*(S) \cong R,$$

and so, by VII.1.20, $\left(\mathcal{MU}^{\Sigma}\right)^*(X) \cong \mathcal{MU}^*(X)\widehat{\otimes}_\rho R$ for every $X \in \mathscr{C}_T$. So, $\mathcal{MU}^{\Sigma}$ turns out to be a semiring spectrum. Furthermore, the forgetful morphism $r : \mathcal{MU} \to \mathcal{MU}^{\Sigma}$ turns $\mathcal{MU}^{\Sigma}$ into a $\mathbb{C}$-marked spectrum, and it is clear that the formal group of $(\mathcal{MU}^{\Sigma}, r)$ is F. To prove the uniqueness, consider a $\mathbb{C}$-marked ring spectrum (E, u) having the formal group F. Then, by 4.13, the $\mathbb{C}$-marking u can be decomposed as

$$u : \mathcal{MU} \xrightarrow{r} \mathcal{MU}^{\Sigma} \xrightarrow{h} E,$$

and it is clear that h is an equivalence. □

§5. Realization of Homology Classes by PL Manifolds with Singularities

We have seen in 1.25(a) that every homology class can be realized by a smooth manifold with Sullivan–Baas singularities. But this requires smooth manifolds with a large (in fact, arbitrarily large) number of singularities. It turns out that the situation looks simpler if one uses not smooth but *PL* manifolds. Namely, in this case it suffices to use manifolds with only one Sullivan–Baas singularity, see 5.11 below. This fact is compatible with the following result of Brumfiel [1]: there are many PL manifolds such that their fundamental classes cannot be realized by smooth manifolds, cf. IV.7.38(i).

The results of this section are proved by Rudyak [5,7].

In this section $\mathscr{K}$ is the class of all compact $H\mathbb{Z}$-oriented PL manifolds. Fix an odd prime p. Below $H^*(X)$ denotes $H^*(X;\mathbb{Z}/p)$.

Let $M(p)$ denote the Moore spectrum $M(\mathbb{Z}/p)$, T denote the spectrum $MS\mathcal{PL}^{CP^{p-1}}$, E denote the spectrum $T \wedge M(p)$, and D denote the spectrum $M\mathcal{U} \wedge M(p)$.

5.1. Lemma. (i) *Each of the spectra T, E, D is connected and has finite $\mathbb{Z}$-type. Furthermore, each of the groups $\pi_i(E), \pi_i(D)$ is finite.*

(ii) $\pi_0(T) = \mathbb{Z}$, $\pi_0(E) = \mathbb{Z}/p = \pi_0(D), H^0(E) = \mathbb{Z}/p = H^0(D)$.

Proof. (i) The spectrum $MS\mathcal{PL}$ is connected by IV.5.23(i), and it has finite type by IV.6.4. Now, 1.6 implies that T is connected and has finite $\mathbb{Z}$-type. Furthermore, since $M(p)$ is the cone of a map $S \to S$ of degree p, we have the exact sequence

$$\cdots \to \pi_k(T) \xrightarrow{p} \pi_k(T) \to \pi_k(E) \to \pi_{k-1}(T) \xrightarrow{p} \cdots$$

where p denotes multiplication by p. Hence, E is connected, and $\pi_i(E)$ is a finite p-group. Similarly for D.

(ii) Since $\pi_0(MS\mathcal{PL}) = \mathbb{Z}$, the equality $\pi_0(T) = \mathbb{Z}$ follows from 1.6. Using the exact sequence from (i), we conclude that $\pi_0(E) = \mathbb{Z}/p$. Now, the equality $H^0(E) = \mathbb{Z}/p$ follows from II.7.20. Similarly for D. □

5.2. Lemma. (i) *The spectrum T is a quasi-ring spectrum, and the spectrum $T[p]$ is a commutative quasi-ring spectrum.*

(ii) *If $p > 3$, then E is a commutative ring spectrum. If $p = 3$, then E admits a commutative pairing $E \wedge E \to E$.*

Proof. (i) Note that CP^{p-1} is a complex manifold. Hence, in view of 2.18, the obstructions $\mathfrak{a}(CP^{p-1})$ and $\mathfrak{c}(CP^{p-1})$ belong to

$$\operatorname{Im}(\pi_*(M\mathcal{U}^{CP^{p-1}}) \to \pi_*(T)).$$

Since $\pi_{4p-3}(M\mathcal{U}^{CP^{p-1}}) = 0 = \pi_{6p-3}(M\mathcal{U}^{CP^{p-1}})$, these obstructions are trivial. Hence T is a quasi-ring spectrum. The obstruction $\mathfrak{b}(CP^{p-1})$ to commutativity of its pairing belongs to the group $\operatorname{Im}(\pi_{4p-2}(MSO^{CP^{p-1}}) \to \pi_{4p-2}(T))$ of exponent 4 (because $2\pi_i(MSO) = 0$ for $i \neq 4k$, see IV.6.5 and IV.6.9), and so, by 2.10, $4(xy - (-1)^{|x||y|})\tau_*(yx) = 0$ for every $x, y \in T_*(X)$. Thus, $T[p]$ is a commutative quasi-ring spectrum.

(ii) By 2.14, $M(p)$ is a commutative ring spectrum for $p > 3$. Hence, by (i), the spectrum E admits a quasi-multiplication. By III.7.3(i), this quasi-multiplication can be induced by a multiplication $E \wedge E \to E$. The case $p = 3$ can be considered similarly. □

5.3. Lemma. *Let $a \in H^0(D) = \mathbb{Z}/p$ be a generator. Then $Q_0(a) \neq 0 \neq Q_1(a)$.*

Proof. By 5.1(ii) and II.7.20, $Q_0(a) \neq 0$. The localization $j : M\mathcal{U}^{CP^{p-1}} \to M\mathcal{U}^{CP^{p-1}}[p]$ induces an equivalence

$$j' = j \wedge 1 : D = M\mathcal{U}^{CP^{p-1}} \wedge M(p) \to M\mathcal{U}^{CP^{p-1}}[p] \wedge M(p).$$

Define a morphism

$$f : M\mathcal{U}[p] \xrightarrow{r[p]} M\mathcal{U}^{CP^{p-1}}[p] = M\mathcal{U}^{CP^{p-1}}[p] \wedge S \xrightarrow{1\wedge\iota} M\mathcal{U}^{CP^{p-1}}[p] \wedge M(p) \simeq D,$$

where $r : M\mathcal{U} \to M\mathcal{U}^{CP^{p-1}}$ is the forgetful morphism and the last equivalence is an inverse morphism to j'. Let $\varkappa : BP \to M\mathcal{U}[p]$, $\rho : M\mathcal{U}[p] \to BP$ and $\{v_i\}$ be as in VII.3.19. Let $\varkappa_* : \pi_*(BP) \to \pi_*(M\mathcal{U}[p])$ and $\rho_* : \pi_*(M\mathcal{U}[p]) \to \pi_*(BP)$ be the induced homomorphisms. We have

$$\varkappa_*(v_n) \in I(p, n+1) \subset \pi_*(M\mathcal{U}[p]).$$

Indeed, $v_n = \rho_* x_{p^n-1}$, where $I(p, n+1) = (p, x_{p-1}, \dots, x_{p^n-1})$. Hence $\varkappa_*(v_n) = \varkappa_*\rho_*(x_{p^n-1}) = \Phi(x_{p^n-1})$; but $\Phi(x_{p^n-1}) \in I(p, n+1)$ because $I(p, n+1)$ is an invariant ideal. Since $I(p, 2) = (p, [CP^{p-1}])$, we conclude that $f_*\varkappa_*(v_1) = 0 \in \pi_*(D)$. Hence there exists a morphism $h : C(v_1) \to D$ such that the diagram

$$\begin{array}{ccccc} \Sigma^{2p-2}BP & \xrightarrow{v_1} & BP & \longrightarrow & C(v_1) \\ & & \downarrow{\scriptstyle f\varkappa} & & \downarrow{\scriptstyle h} \\ & & D & = & D \end{array}$$

commutes. Here the first row is a cofiber sequence. We have

$$\pi_i(C(v_1)) = \begin{cases} 0 & \text{if } i < 0, \\ \mathbb{Z}/p & \text{if } i = 0, \\ 0 & \text{if } 0 < i < 2p^2 - 2. \end{cases}$$

Hence the coskeleton $(C(v_1))_{(2p^2-2)}$ is $H\mathbb{Z}[p]$. Hence $Q_1(x) \neq 0$ for a generator $x \in H^0(C(v_1)) = \mathbb{Z}/p$. Since $h^* : H^0(D) \to H^0(C(v_1))$ is an isomorphism, $Q_1(a) \neq 0$. □

5.4. Lemma. *Let $u = u_{\mathbb{C}} \in H^0(M\mathcal{U})$ be the universal Thom class. Then $\mathscr{P}^{\Delta_j}(u) \neq 0$ for every $j = 1, 2, \dots$.*

Proof. Because of the universality of u, it suffices to find a complex vector bundle ξ such that $\mathscr{P}^{\Delta_j}(u_\xi) \neq 0, j = 1, 2, \dots$, where $u_\xi \in H^*(T\xi)$ is the Thom class of ξ. Let η be as in VII.1.3(f). Then, by VII.1.29, we can identify $T\eta$ with CP^∞, and $x := u_\eta \in H^2(CP^\infty)$ generates $H^2(CP^\infty) = \mathbb{Z}$. We prove by induction on j that $\mathscr{P}^{\Delta_j}(x) = x^{p^j}$ for every $j = 1, 2, \dots$. Indeed, $\mathscr{P}^{\Delta_1}(x) = P^1(x) = x^p$. Suppose that $\mathscr{P}^{\Delta_j}(x) = x^{p^j}$. Then

$$\mathscr{P}^{\Delta_{j+1}}(x) = [P^{p^j}, \mathscr{P}^{\Delta_j}](x) = P^{p^j}\mathscr{P}^{\Delta_j}(x) = P^{p^j}(x^{p^j}) = x^{p^{j+1}}. \qquad \square$$

5.5. Theorem. *E is a graded Eilenberg–Mac Lane spectrum.*

Proof. By 5.1(ii), 5.2(ii) and II.7.21, $H^*(E)$ is a connected $\mathscr{A}_p$-coalgebra (it is, probably, non-associative for $p = 3$). Let $v \in H^0(E)$ be the counit of this coalgebra. Define $\nu : \mathscr{A}_p \to H^*(E)$ by setting $\nu(a) = a(v)$. By IV.6.4, E has finite $\mathbb{Z}[p]$-type. Thus, by II.7.24(ii) and II.7.25, is sufficient to prove that $\nu(Q_i) \neq 0$ for $i = 0, 1, \dots$, $\nu(\mathscr{P}^{\Delta_j}) \neq 0$ for $j = 1, 2, \dots$.

Let $r : M\mathcal{SPL} \to T$ be the forgetful morphism. The other forgetful morphism $s : M\mathcal{U} \to M\mathcal{SPL}$ induces a forgetful morphism $\bar{s} : M\mathcal{U}^{CP^{p-1}} \to T$. Consider the morphisms

$$\alpha : D = M\mathcal{U}^{CP^{p-1}} \wedge M(p) \xrightarrow{\bar{s}\wedge 1} T \wedge M(p) = E,$$
$$\beta : M\mathcal{SPL} = M\mathcal{SPL} \wedge S \xrightarrow{r\wedge 1} T \wedge S \xrightarrow{1\wedge\iota} T \wedge M(p) = E,$$
$$\gamma : M\mathcal{U} \xrightarrow{s} M\mathcal{SPL} \xrightarrow{\beta} E.$$

By 5.3, $\alpha^*\nu(Q_i) \neq 0$ for $i = 0, 1$. Hence, $\nu(Q_i) \neq 0$ for $i = 0, 1$. By IV.6.13, $\beta^*\nu(Q_i) \neq 0$ for $i > 1$. So, $\nu(Q_i) \neq 0$ for $i > 1$. By 5.4, $\gamma^*\nu(\mathscr{P}^{\Delta_j}) \neq 0, j = 1, 2, \dots$, because $\gamma^*v = u_{\mathbb{C}}$. Thus, $\nu(\mathscr{P}^{\Delta_j}) \neq 0, j = 1, 2, \dots$. $\square$

5.6. Corollary (the main theorem). *$T[p]$ is a graded Eilenberg–Mac Lane spectrum.*

Proof. This follows from 5.5 and II.7.14 because, in view of II.5.6(ii), $T[p] \wedge M(p) = E$. $\square$

Let $v : T \to H\mathbb{Z}$ be a morphism given by a generator $v \in H^0(T; \mathbb{Z}) = \mathbb{Z}$. It induces certain (Steenrod–Thom) homomorphisms $v^X : T_*(X) \to H_*(X; \mathbb{Z})$ and $v[p]^X : T[p]_*(X) \to H_*(X; \mathbb{Z}[p])$.

5.7. Corollary. *The homomorphism*

$$v[p]^X : T[p]_*(X) \to H_*(X; \mathbb{Z}[p])$$

is an epimorphism for every space X.

Proof. $H\mathbb{Z}[p]$ is a direct summand of the graded Eilenberg–Mac Lane spectrum $T[p]$, because $\pi_0(T[p]) = \mathbb{Z}[p]$. It is clear that $v[p] : T[p] \to H\mathbb{Z}[p]$ gives a generator of the group $H^0(T[p]; \mathbb{Z}[p]) = \mathbb{Z}/p$, and so $v[p]$ is a projection onto a direct summand. $\square$

5.8. Remark. It follows from IV.6.5 and/or IV.7.36 that the Steenrod–Thom homomorphism $MSO[2]_*(X) \to H_*(X; \mathbb{Z}[2])$ is epic for every X.

The Steenrod–Thom homomorphism $v : T_*(X) \to H_*(X; \mathbb{Z})$ can be described as follows. Let M be a closed n-dimensional ($n > 0$) $\mathscr{K}^{CP^{p-1}}$-manifold. Consider a polyhedron K of the form

$$K = \delta M \times C(CP^{p-1}) \cup_\varphi M \tag{5.9}$$

where the inclusion $\delta M \times CP^{p-1} \subset \delta M \times C(CP^{p-1})$ is given by the inclusion of the bottom $CP^{p-1} \subset C(CP^{p-1})$, cf. (1.1). We denote $\delta M \times C(CP^{p-1})$ by A. It is clear that the collapse map $p : K \to K/A$ induces the isomorphism $p_* : H_n(K; \mathbb{Z}) \to H_n(K/A; \mathbb{Z})$. Consider the isomorphism

$$h : H_n(M, \partial M; \mathbb{Z}) \xrightarrow{\cong} H_n(M/\partial M; \mathbb{Z}) = H_n(K/A; \mathbb{Z}) \xleftarrow{p_*} H_n(K; \mathbb{Z})$$

and set $[K] := h[M, \partial M]$.

Given a space X, let $f : M \to X$ be a closed n-dimensional ($n > 0$) singular $\mathscr{K}^{CP^{p-1}}$-manifold in X. Define $g : K \to X$ by setting $g|M = f|M$ and $g(\{a\} \times C(CP^{p-1})) = f(\{a\} \times CP^{p-1})$ for every $a \in \delta M$. According to 1.4, g is well-defined. Finally, define $\mathfrak{t}^X : T_*(X) \to H_*(X; \mathbb{Z})$ by setting $\mathfrak{t}[M, f] = g_*[K]$.

5.10. Lemma. *Fix a generator (one of two) $v \in H^0(T; \mathbb{Z}) = \mathbb{Z}$ such that*

$$v^{\text{pt}} : \mathbb{Z} = T_*(\text{pt}) \to H_*(\text{pt}; \mathbb{Z}) = \mathbb{Z}$$

maps 1 to 1. Then $v^X = \mathfrak{t}^X : T_*(X) \to H_*(X; \mathbb{Z})$ for every space X.

Proof. This can be proved as IV.7.32 was; we leave it to the reader. □

As we remarked in the beginning of this chapter, the homology theory $T_*(X)$ can be considered as a bordism theory based on polyhedra (5.9) as "closed manifolds". Here we have used this implicitly.

5.11. Corollary. *Every homology class $z \in H_n(X)$ can be represented as a sum $\sum_i (g_i)_*([K_i])$, where K_i is as in (5.9) with an arbitrary prime p (depending on i) and $g_i : K_i \to X$ is a map.*

Proof. It follows from 5.7, 5.8 and 5.10 that for every prime p there exists $m_p \in \mathbb{Z}$ such that $(p, m_p) = 1$ and $m_p z$ can be realized as $\sum (g_i)_*([K_i])$ with the given p. Choose a finite set $\{p_1, \dots, p_n\}$ of primes such that $(m_{p_1}, \dots, m_{p_n}) = 1$. (For example, let $m_2 = q_1^{\alpha_1} \cdots q_s^{\alpha_s}$ with different primes q_i. Then we can set $\{p_1, \dots, p_{s+1}\} = \{2, q_1, \dots, q_s\}$.) Then $\sum a_i m_{p_i} = 1$ for certain $a_i \in \mathbb{Z}$, i.e., $\sum a_i m_{p_i} z = z$. Now the result follows from the realizability of $m_{p_i} z$. □

Now we deduce from the above certain algebraic relations for PL manifolds with one CP^{p-1}-singularity.

5.12. Lemma. *$E \simeq MSPL^{\{p,CP^{p-1}\}}$. Furthermore, E admits a structure of a commutative ring spectrum (even for $p = 3$) such that the forgetful morphism $MSPL \to E$ is a ring morphism.*

Proof. Set $B := MSPL^{\{p,CP^{p-1}\}} = T^{\{p\}}$. By IV.4.27(iv), $\pi_i(MSPL) = \pi_i(MSO)$ for $i < 7$. Hence $\pi_2(B) = 0 = \pi_3(B)$. So, by 2.17 and 5.1(i), B admits a commutative and associative quasi-multiplication, compatible with the multiplication in $MSPL$. (Alternatively, one can see that all the obstructions come from MU, and so they belong to trivial groups.) Moreover, by III.7.3, B admits a ring structure because $\pi_i(B)$ are finite. Furthermore, by II.7.7, B is a graded Eilenberg–Mac Lane spectrum because there exists an S-module morphism $H\mathbb{Z}[p] \to T[p] \to B$. (Alternatively, one can copy the proof of 5.5.) Now, because of 1.6 and since $\pi_i(B)$ are finite $\mathbb{Z}/p$-vector spaces,

$$\pi_i(B) \cong \operatorname{Ker}(\pi_{i+1}(T) \xrightarrow{p} \pi_{i+1}(T)) \oplus \operatorname{Coker}(\pi_i(T) \xrightarrow{p} \pi_i(T)) \cong \pi_i(E),$$

and so $B \simeq E$ because B and E are graded Eilenberg–Mac Lane spectra.

Thus, E admits the desired structure of a commutative ring spectrum because B admits it. □

5.13. Proposition. *Let E be equipped with the ring structure in 5.12. Then E admits a $\mathbb{C}$-orientation z such that the formal group of (E, z) is additive.*

Proof. By II.7.30 and 5.5, there is a ring isomorphism $E \simeq H(\pi_*(E))$. In particular, there is a ring morphism $j : H\mathbb{Z}/p \to E$. The formal group of $(H\mathbb{Z}/p, w)$ is additive for any $\mathbb{C}$-orientation $w \in H^2(CP^\infty; \mathbb{Z}/p)$ of $H\mathbb{Z}/p$. Thus, the formal group of (E, j_*w) is additive. □

Let $f(x, y)$ be the universal formal group, i.e., the formal group of the $\mathbb{C}$-oriented spectrum (MU, T), see VII.6.7(e). Set

$$[p]_f(x) = x + \sum_{k>0} a_k x^{k+1}, \quad a_k \in MU^{-2k}(\text{pt}) = \pi_{2k}(MU).$$

Let $\tau : MU \to T[p]$ be the forgetful morphism, $\tau = r[p]s$ in the notation of 5.5.

5.14. Proposition. *$p | \tau_*(a_k)$ for every k.*

Proof. Let $\rho : T[p] \to E$ be the reduction mod p. Since $\rho\tau$ is a ring morphism, $t = (\rho\tau)_*(T) \in \widetilde{E}^2(CP^\infty)$ is a $\mathbb{C}$-orientation of E. Let F be the formal group of (E, t). By VII.6.6, F is equivalent to a formal group of (E, z)

with z as in 5.13. In other words, F is equivalent to the additive formal group. Now, by VII.5.12(i), $[p]_F(x) = \sum(\rho\tau)_*(a_k)x^{k+1}$, and, by VII.5.11(i), $[p]_F(x) = 0$. Thus, $(\rho\tau)_*(a_k) = 0$. Since the sequence

$$\pi_*(T[p]) \xrightarrow{p} \pi_*(T[p]) \xrightarrow{\rho} \pi_*(E).$$

is exact, $p|\tau_*(a_k)$. □

We set $R := \pi_*(T[p])/\operatorname{tors}$, define the homomorphism

$$\overline{\tau} : \pi_*(M\mathcal{U}) \xrightarrow{\tau_*} \pi_*(T[p]) \xrightarrow{\text{quotient}} R$$

and denote $\overline{\tau}[CP^n]$ by $[[CP^n]]$.

5.15. Corollary. *For every n the element $[[CP^n]] \in R$ is divisible by $n+1$.*

Proof. We regard R as a subring of $R \otimes \mathbb{Q}$. Consider the formal group $F = \overline{\tau}_*(f)$ over R. By 5.14, all coefficients of $[p]_F(x) = \sum \overline{\tau}_*(a_k)x^{k+1}$ are divisible by p. Hence, by VII.5.11(ii), all coefficients of the logarithm of F belong to R. By VII.5.12(ii) and VII.6.12, this logarithm has the form

$$x + \sum \frac{[[CP^n]]}{n+1} x^n.$$

Therefore, the element $r_n := \frac{[[CP^n]]}{n+1}$ of $R \otimes \mathbb{Q}$ belongs to R. Finally, $[[CP^n]] = (n+1)r_n$ because R is torsion free. □

Chapter IX. Complex (Co)bordism with Singularities

§1. Brown-Peterson (Co)homology with Singularities

We fix a prime p. Let BP be the corresponding Brown–Peterson spectrum, and let $\varkappa : BP \to M\mathcal{U}[p]$, $\rho : M\mathcal{U}[p] \to BP$ be the pair of morphisms described in VII.3.19(i).

We consider the ring $\pi_*(M\mathcal{U}) = \mathbb{Z}[x_1, \dots, x_n, \dots]$, $\dim x_n = 2n$, and fix a family $\{x_n\}$ of generators. Consider the ordered set $\{i_1 < i_2 < \cdots < i_n < \cdots\}$ of all natural numbers different from $p^k - 1$, $k = 1, 2, \dots$, and set $\Sigma = \{x_{i_1}, \dots, x_{i_n}, \dots\}$. Let $r : M\mathcal{U} \to M\mathcal{U}^\Sigma$ be the forgetful morphism as in VIII.(1.21)

1.1. Theorem. (i) *The spectrum $M\mathcal{U}^\Sigma$ admits a commutative and associative quasi-multiplication such that r is a quasi-ring morphism.*

(ii) *The spectrum $M\mathcal{U}^\Sigma[p]$ admits a ring structure such that the localization $M\mathcal{U}^\Sigma \to M\mathcal{U}^\Sigma[p]$ is a quasi-ring morphism and the morphism*

$$BP \xrightarrow{\varkappa} M\mathcal{U}[p] \xrightarrow{r[p]} M\mathcal{U}^\Sigma[p]$$

is a ring equivalence. In particular, $M\mathcal{U}^\Sigma[p] \simeq BP$ as ring spectra.

Proof. (i) This follows from VIII.2.17.

(ii) The homomorphism $r_* : \pi_*(M\mathcal{U}) \to \pi_*(M\mathcal{U}^\Sigma)$ is just the quotient map $\pi_*(M\mathcal{U}) \to \pi_*(M\mathcal{U})/(x_{i_1}, \dots, x_{i_n}, \dots)$, and so the morphism $r[p]\varkappa : BP \to M\mathcal{U}^\Sigma[p]$ induces an isomorphism of homotopy groups. We equip $M\mathcal{U}^\Sigma[p]$ with a ring structure via the equivalence $r[p]\varkappa$. On the other hand, by (i), localization gives us a quasi-ring structure in $M\mathcal{U}^\Sigma[p]$. Now, the ring structure in $M\mathcal{U}^\Sigma[p]$ is compatible with this quasi-ring structure, because r, and hence $r[p]\varkappa$, is a quasi-ring morphism. □

Thus, the morphism $M\mathcal{U} \xrightarrow{\text{localization}} M\mathcal{U}[p] \xrightarrow{\rho} BP$ can be interpreted as an iterated introduction of singularities with subsequent localization.

We have $\pi_*(BP) = \mathbb{Z}[p][v_1, \dots, v_n, \dots]$, where $v_i := \rho_*(x_{p^n-1})$, $\dim v_n = 2(p^n - 1)$. We also recall the notation $v_0 = p$. Given a subset J of $\mathbb{N} \cup \{0\}$, we consider the set $V := \{ v_j \mid j \in J \}$, and define

$$BP^V := M\mathcal{U}^\Lambda[p], \text{where } \Lambda = \{ x_{p^j-1} \mid j \in J \} \cup \Sigma,\ \Lambda \subset \{p, x_1, \dots, x_n, \dots\}.$$

1.2. Lemma. $H^0(BP^V;\mathbb{Z}/p)=\mathbb{Z}/p$.

Proof. Clearly, BP^V is connected and $\pi_0(BP^V)=\mathbb{Z}$ or $\mathbb{Z}/p$. Hence, by II.4.7(i), $H_0(BP^V)=\mathbb{Z}$ or $\mathbb{Z}/p$. Thus, by II.4.9, $H^0(BP^V;\mathbb{Z}/p)=\mathbb{Z}/p$. □

Let $u\in H^0(BP^V;\mathbb{Z}/p)$ be a generator.

1.3. Theorem (Baas–Madsen [1]). *The homomorphism*

$$\nu:\mathscr{A}_p\to H^*(BP^V;\mathbb{Z}/p),\ \nu(a)=au$$

is epic, and $\operatorname{Ker}\nu=\mathscr{A}_p(Q_j|j\notin J)$. *So,* $H^*(BP^V;\mathbb{Z}/p)\cong\mathscr{A}_p/\mathscr{A}_p(Q_j|j\notin J)$.

Proof. Firstly, we consider the case $p>2$. Let H denote $H\mathbb{Z}/p$ and $H_*(-)$ denote $H_*(-;\mathbb{Z}/p)$. We have the AHSS

$$E^r_{**}=E^r_{**}(J)\Longrightarrow BP^V_*(H),\ E^2_{**}=H_*(H;\pi_*(BP^V)).$$

If $0\in J$, then (cf. VII.3.27)

$$E^2_{**}=H_*(H)\otimes\pi_*(BP^V)=\mathbb{Z}/p\,[\xi_i|i>0]\otimes\Lambda\{\tau_i|i\geq 0\}\otimes\mathbb{Z}/p\,[v_i|i\notin J],$$

where $\operatorname{bideg}\xi_i=(2p^i-2,0),\operatorname{bideg}\tau_i=(2p^i-1,0),\operatorname{bideg}v_i=(0,2p^i-2)$,.
If $0\notin J$, then

$$E^2_{**}=H_*(H;\mathbb{Z})\otimes\pi_*(BP^V)=\mathbb{Z}/p\,[\xi_i|i>0]\otimes\Lambda(\tau_i|i\geq 1)\otimes\mathbb{Z}/p\,[v_i|i\notin J].$$

By VII.3.27, in the case $J=\emptyset$ (i.e., $BP^V=BP$) without loss of generality we can assume that the elements ξ_i are permanent cycles and $d_{2p^i-1}(\tau_i)=v_i$. For arbitrary J, considering the morphism $BP\to BP^V$ and the induced morphism $E^r_{**}(\emptyset)\to E^r_{**}(J)$ of spectral sequences, we conclude that

$$E^\infty_{**}=\mathbb{Z}/p\,[\xi_i|i>0]\otimes\Lambda\{\tau_j|j\in J\}$$

for every J. Hence, $u_*:BP^V_*(H)\to H_*(H)$ is a ring monomorphism, $u_*(\xi_i)=\xi_i$, $u_*(\tau_j)=\tau_j$ for $j\in J$.

Consider the first commutative diagram below, where the horizontal arrows are the switching morphisms. Passing to π_*, we get the second commutative diagram below.

$$\begin{array}{ccc} H\wedge BP^V & \longrightarrow & BP^V\wedge H \\ {\scriptstyle 1\wedge u}\downarrow & & \downarrow{\scriptstyle u\wedge 1} \\ H\wedge H & \longrightarrow & H\wedge H \end{array}\qquad\qquad \begin{array}{ccc} BP^V_*(H) & \longrightarrow & H_*(BP^V) \\ {\scriptstyle u_*}\downarrow & & \downarrow{\scriptstyle H_*(u)} \\ H_*(H) & \xrightarrow{\ \chi\ } & H_*(H) \end{array}.$$

Here χ is the canonical antiautomorphism of the Hopf algebra $\mathscr{A}_p^*=H_*(H)$, see II.6.35. We have $\chi(\xi_i)\equiv\xi_i \bmod \operatorname{Dec}$, $\chi(\tau_i)\equiv\tau_i \bmod \operatorname{Dec}$. Therefore,

$$\operatorname{Im}H_*(u)=\mathbb{Z}/p\,[\xi_i'|i>0]\otimes\Lambda(\tau_j'|j\in J),\ \xi_i'\equiv\xi_i \bmod \operatorname{Dec},\ \tau_j'\equiv\tau_j \bmod \operatorname{Dec}.$$

The elements τ_i of $\mathscr{A}_p^*$ are dual to $Q_i \in \mathscr{A}_p$, and ν is dual to the monomorphism $H_*(u)$. Hence, ν is an epimorphism with $\mathscr{A}_p(Q_j | j \notin J) \subset \operatorname{Ker} \nu$. Thus, for dimensional reasons, $\mathscr{A}_p(Q_j | j \notin J) = \operatorname{Ker} \nu$.

The case $p = 2$ can be considered similarly. If $0 \in J$, then

$$E^2_{**} = \mathbb{Z}/2\,[\zeta_i | i \geq 0] \otimes \mathbb{Z}/2\,[v_j | j \notin J].$$

Here ζ_i^2 plays the role of ξ_{i+1} and ζ_i plays the role of τ_i: ζ_i^2 is a permanent cycle, and $d_{2^i-1}\zeta_i = v_i$ for every $i \geq 1$. If $0 \notin J$, then

$$E^2_{**} = \mathbb{Z}/2\,[\zeta_0^2] \otimes \mathbb{Z}/2\,[\zeta_i | i > 0] \otimes \mathbb{Z}/2\,[v_j | j \notin J],$$

etc. In both cases we have

$$E^\infty_{**} = \mathbb{Z}/2\,[\zeta_i^2 | i \notin J] \otimes \mathbb{Z}/2\,[\zeta_j | j \in J].$$

The further arguments are just the same as for $p > 2$. □

1.4. Corollary. *Let $0 \in J \neq \{0\}$, and let n be the minimal positive number in J. Then Q_n is the first non-trivial Postnikov invariant of BP^V.*

Proof. For simplicity, denote BP^V by X and put $m = 2p^n - 2$. We have $X_{(0)} = H\mathbb{Z}/p = X_{(i)}$ for $i < m$. Consder the Postnikov invariant $\kappa \in H^{m+1}(H\mathbb{Z}/p; \mathbb{Z}/p)$ of X. There is a diagram

$$\begin{array}{ccccc} X_{(m)} & \xrightarrow{p} & H\mathbb{Z}/p & \xrightarrow{\kappa} & \Sigma^{m+1}H\mathbb{Z}/p \\ \| & & \| & & \downarrow f \\ X_{(m)} & \xrightarrow{p} & H\mathbb{Z}/p & \xrightarrow{Q_n} & \Sigma^{m+1}H\mathbb{Z}/p \end{array}$$

where

$$p = p_m : X_{(m)} \to X_{(m-1)} = X_{(0)} = H\mathbb{Z}/p$$

and the top row is a cofiber sequence. We must find an equivalence $f : \Sigma^{m+1}H\mathbb{Z}/p \to \Sigma^{m+1}H\mathbb{Z}/p$ such that $Q_n \simeq f\kappa$.

Let $\tau = \tau_m : X \to X_{(m)}$ be the morphism as in II.4.12. Then $\tau^* : H^i(X_{(m)}; \mathbb{Z}/p) \to H^i(X; \mathbb{Z}/p)$ is an isomorphism for $i \leq m$ and a monomorphism for $i = m + 1$. Since $\tau^* p = u$ and $Q_n u = 0$, the morphism $Q_n p$ is inessential. Hence, there is a morphism $f : \Sigma^{m+1}H\mathbb{Z}/p \to \Sigma^{m+1}H\mathbb{Z}/p$ such that $Q_n = f\kappa$.

We prove that f is an equivalence. Indeed, otherwise f is inessential, and hence so is Q_n. This is a contradiction. □

§2. The Spectra $P(n)$

As in §1, fix any prime p and a system $\{x_i\}$, $\dim x_i = 2i$, of polynomial generators of $\pi_*(MU)$. As in Ch. VII, we denote

$$\pi_*(BP) = \mathbb{Z}[p][v_1, \dots, v_i, \dots], \quad v_i = \rho_*(x_{p^i-1})$$

by Ω and $(v_0, \dots, v_{n-1})$ by I_n. Define a spectrum

$$P(n) := BP^{\{v_0, \dots, v_{n-1}\}}.$$

Furthermore, $P(0) := BP$, $P(\infty) := H\mathbb{Z}/p$.

By VIII.1.6 and/or VIII.(1.18), for every CW-space X we have an exact sequence

$$\begin{aligned} &\cdots \to P(n)_*(X) \xrightarrow{v_n} P(n)_*(X) \xrightarrow{r_n} P(n+1)_*(X) \\ &\xrightarrow{\delta_n} P(n)_*(X) \to \cdots . \end{aligned} \tag{2.1}$$

For general reasons, $P(n)$ is a quasi-module spectrum over $MU[p]$. Hence, it is a quasi-module spectrum over BP via the ring morphism $\varkappa : BP \to MU[p]$. In particular, $\pi_*(P(n))$ is an Ω-module.

2.2. Proposition. (i) *There is an Ω-module isomorphism $\pi_*(P(n)) \cong \Omega/I_n$.*
(ii) *There is an $\mathscr{A}_p$-isomorphism $H^*(P(n);\mathbb{Z}/p) \cong \mathscr{A}_p/\mathscr{A}_p(Q_i|i \geq n)$.*

Proof. (i) This follows from VIII.1.24.
(ii) This follows from 1.3. □

2.3. Remarks. (a) The spectra $P(n)$ were introduced by Morava [1] and considered in detail by Johnson–Wilson [2].

(b) We shall see below (in 2.12) that the spectrum $P(n)$ does not depend (up to equivalence) on the choice of the system $\{x_i\}$.

(c) The finiteness of the groups $\pi_i(P(n)), n > 0$, implies that $P(n)$ has finite $\mathbb{Z}$-type. So, we can and shall assume that $P(n), n > 0$ has finite type.

By VIII.(1.12), the sequence (2.1) is induced by a sequence

$$\cdots \to \Sigma^{2p^n-2}P(n) \xrightarrow{v_n} P(n) \xrightarrow{r_n} P(n+1) \xrightarrow{\delta_n} \Sigma^{2p^n-1}P(n) \xrightarrow{v_n} \cdots \tag{2.4}$$

of spectra. There is also the Bockstein morphism

$$\beta_n := \beta_{v_n} = r_n\delta_n : P(n+1) \to \Sigma^{2p^n-1}P(n+1) \tag{2.5}$$

which induces the Bockstein homomorphisms

$$\begin{aligned} \beta_n &: P(n+1)_i(X) \to P(n+1)_{i-2p^n+1}(X), \\ \beta_n &: P(n+1)^i(X) \to P(n+1)^{i+2p^n-1}(X) \end{aligned}$$

as in VIII.(1.8) and VIII.(1.11).

2.6. Proposition. *The morphisms $v_n, r_n, \delta_n, \beta_n$ are determined uniquely up to homotopy, and the composition of every two adjacent morphisms in* (2.4) *is inessential.*

Proof. By III.5.7, the group $P(k)^0(P(l))$ does not contain phantoms. Furthermore, the spectra $P(n), n > 1$, have finite type, and hence, by III.3.20(iii), the morphisms $r_n, v_n, n > 0$ and $\delta_n, \beta_n, n \geq 0$, are determined uniquely up to homotopy. Moreover, the homotopy uniqueness of v_0 follows from VII.3.2. while the homotopy uniqueness of r_0 follows from III.4.18. Finally, the composition, say,

$$\delta_n r_n : P(n)^i(X) \to P(n)^{i+2p^n-1}(X)$$

is trivial for every finite spectrum X. So, if $n > 0$ then it is trivial for $X = P(n)$ because $P(n)$ has finite type and $P(n)^*(P(n))$ does not contain phantoms. Thus, $\delta_n r_n : P(n) \to \Sigma^{2p^n-1}P(n)$, $n > 0$ is inessential. Moreover, the composition $\delta_0 r_0 : BP \to \Sigma BP$ is inessential by VII.3.2. The compositions $v_n\delta_n$ and $r_n v_n$ can be considered similarly. □

2.7. Theorem. *The tower*

$$BP = P(0) \xrightarrow{r_0} P(1) \to \cdots \to P(n) \xrightarrow{r_n} P(n+1) \to \cdots \to H\mathbb{Z}/p.$$

admits a ring structure, i.e., every spectrum $P(n)$ admits a multiplication $m_n : P(n) \wedge P(n) \to P(n)$ such that every morphism r_n is a ring morphism with respect to these multiplications. Moreover, the multiplication in $P(0)$ coincides with the multiplication in BP. Furthermore, for $p > 2$ the multiplication m_n can be chosen to be commutative for every n, and for $p = 2$ these multiplications can be chosen such that

$$xy + yx = v_n\beta_{n-1}(x)\beta_{n-1}(y).$$

for every $x, y \in P(n)^*(X)$.

Proof. Let $\Sigma = \{x_i | i \neq p^s - 1\}$ be the sequence defined at the beginning of §1. By 1.1(i), the spectrum $V := M\mathcal{U}^\Sigma$ admits a commutative and associative quasi-multiplication. So, by VIII.2.4, the spectrum $P(1) = V^{\{p\}}$ admits a quasi-multiplication, which is associative because $\pi_3(P(1)) = 0$. Moreover, the forgetful morphism $a : V \to P(1)$ is a quasi-ring morphism. Furthermore, every group $\pi_i(P(1))$ is finite, and so, by III.7.3, the quasi-multiplication in $P(1)$ is induced by a multiplication in $P(1)$. Let $BP \simeq V[p]$ be the ring equivalence as in 1.1(ii). Then $r_0 : BP \simeq V[p] \xrightarrow{a} P(1)$ is a quasi-ring morphism of ring spectra, and so, by III.7.5, it is a ring morphism.

Now we construct a ring structure in the tower by induction on n. Suppose that $P(i), i \leq n$, is a ring spectrum and that $r_i : P(i) \to P(i+1)$ is a ring morphism for every $i < n$. Then, by VIII.2.13, $P(n+1)$ admits a quasi-multiplication, which is associative because $\pi_{\mathrm{odd}}(P(n+1)) = 0$. Using arguments as above, one can prove that this quasi-multiplication is induced by a multiplication in $P(n+1)$ and that $r_n : P(n) \to P(n+1)$ is a ring morphism with respect to this multiplication. The induction is confirmed.

If $p > 2$, then the multiplication in $P(n)$ can be chosen to be commutative in view of VIII.2.11, VIII 2.17(ii) and III.7.3(v).

For $p = 2$, the formula $xy + yx = v_n\beta_{n-1}(x)\beta_{n-1}(y)$ follows from VIII.2.17(ii) and VIII.3.11 for a finite CW-space X. Let $\{X_\lambda\}$ be the family of all finite CW-subcomplexes of a CW-complex X. Now, by III.4.17, $P(n)^*(X) = \varprojlim\{P(n)^*(X_\lambda)\}$ because $\pi_i(P(n))$ is a finite group for $n > 0$ and every i, and the result follows. □

In future we always assume that all the spectra $P(n)$ are equipped with the multiplications above.

Given $m \le n$, set $r_n^m = r_{n-1} \circ \cdots \circ r_m : P(m) \to P(n)$ and $\rho_n = r_n^0 : BP \to P(n)$. The pairing

$$P(m) \wedge P(n) \xrightarrow{r_n^m \wedge 1} P(n) \wedge P(n) \xrightarrow{m_n} P(n)$$

turns $P(n)$ into a $P(m)$-module spectrum. In particular, there is a pairing $\mu_n : BP \wedge P(n) \to P(n)$.

Consider the morphism

$$(v_i)_\# : \Sigma^{2p^i-2}P(n) = S^{2p^i-2} \wedge P(n) \xrightarrow{v_i \wedge 1} BP \wedge P(n) \xrightarrow{\mu_n} P(n),$$

where $v_i : S^{2p^i-2} \to BP$ represents the element $v_i \in \pi_*(BP)$.

2.8. Theorem. (i) *For every CW-space X, the homomorphism*

$$(v_n)_\#^X : P(n)_*(X) \to P(n)_*(X)$$

is multiplication by v_n on the $\pi_*(P(n))$-module $P(n)_*(X)$.

(ii) *$P(n+1)$ is a cone of $(v_n)_\#$, and the sequence*

$$\begin{aligned} \cdots \to \Sigma^{2p^n-2}P(n) \xrightarrow{(v_n)_\#} P(n) \xrightarrow{r_n} P(n+1) \\ \xrightarrow{\delta} \Sigma^{2p^n-1}P(n) \xrightarrow{(\Sigma v_n)_\#} \cdots, \end{aligned} \tag{2.9}$$

as well as (2.4), is a long cofiber sequence.

Proof. (i) It is clear that $(v_n)_\#^X$ is multiplication by v_n on the $\pi_*(BP)$-module $P(n)_*(X)$. Since $BP \to P(n)$ is a ring morphism, (i) is proved.

(ii) By (i), $(v_n)_\#$ can play the role of the morphism

$$v_n : \Sigma^{2p^n-2}P(n) \to P(n)$$

in (2.4), i.e., $v_n \simeq (v_n)_\#$. Now, by III.6.7(ii), $P(n+1) \simeq C(v_n) \simeq C((v_n)_\#)$ because every group $\pi_i(P(n))$ is finite for $n > 0$ and every i.

We must prove that there exists a homotopy commutative diagram

$$\begin{array}{ccc}
\Sigma^{2p^n-2}P(n) & = & \Sigma^{2p^n-2}P(n) \\
{\scriptstyle v_n}\downarrow & & \downarrow{\scriptstyle v_n} \\
P(n) & = & P(n) \\
{\scriptstyle \varphi}\downarrow & & \downarrow{\scriptstyle r_n} \\
P(n+1) & \xrightarrow{a_1} & P(n+1) \\
{\scriptstyle \psi}\downarrow & & \downarrow{\scriptstyle \delta} \\
\Sigma^{2p^n-1}P(n) & \xrightarrow{a_2} & \Sigma^{2p^n-1}P(n) \\
{\scriptstyle v_n}\downarrow & & \downarrow{\scriptstyle v_n} \\
\Sigma P(n) & \xrightarrow{a_3} & \Sigma P(n) \\
{\scriptstyle \varphi}\downarrow & & \downarrow{\scriptstyle r_n} \\
\Sigma P(n+1) & \xrightarrow{a_4} & \Sigma P(n+1) \\
\downarrow & & \downarrow \\
\vdots & & \vdots
\end{array},$$

where a_i are equivalences and the left sequence is a long cofiber sequence. By 2.6, the composition of every two adjacent morphisms in right sequence is inessential. Let $H(-)$ denote $H(-;\mathbb{Z}/p)$. Take $u \in H^0(P(n)) = \mathbb{Z}/p$, $u \neq 0$. By 2.2(ii) and II.5.18(ii), a morphism $f : P(n) \to P(n)$ is an equivalence iff $f^*(u) \neq 0$.

An equivalence a_1 was constructed in the proof of III.6.7(ii). Since the morphism $\delta a_1 \varphi$ is inessential, there exists a_2 such that $a_2\psi \simeq \delta a_1$. We prove that $a_2^*(u) \neq 0$. Indeed, if $a_2^*(u) = 0$ then $a_2^* : H^*(P(n)) \to H^*(P(n))$ is zero, and hence $\delta^* : H^*(P(n)) \to H^*(P(n+1))$ is zero. But, by 2.2(ii),

$$Q_{n+1} \in \operatorname{Ker}(r_n^* : H^*(P(n+1)) \to H^*(P(n))) = \operatorname{Im} \delta^*.$$

This is a contradiction. Thus, a_2 is an equivalence.

Furthermore, there exists a_3 such that $a_3 v_n \simeq v_n a_2$. Since a_2 is an equivalence,

$$(a_3)_* : \pi_{2p^n-2}(P(n)) = \pi_{2p^n-1}(\Sigma P(n)) \to \pi_{2p^n-1}(\Sigma P(n)) = \pi_{2p^n-2}(P(n))$$

is an isomorphism. Since $\pi_*(P(n))$ is a $\pi_0(P(n))$-module, we conclude that $(a_3)_* : \pi_0(P(n)) \to \pi_0(P(n))$ is an isomorphism. Hence, $a_3^* : H^0(P(n)) \to H^0(P(n))$ is an isomorphism. Thus, a_3 is an equivalence.

Furthermore, there exists a_4 such that $a_4\varphi \simeq r_n a_3$. Since both homomorphisms

$$(r_n)_*, \varphi_* : \pi_*(P(n)) \to \pi_*(P(n))$$

are epic, $(a_4)_* : \pi_*(P(n+1)) \to \pi_*(P(n+1))$ is an isomorphism, and thus a_4 is an equivalence.

And so on, because of periodicity. □

Let (E, u) be a $\mathbb{C}$-marked ring $\mathbb{Z}[p]$-local spectrum of finite $\mathbb{Z}[p]$-type. Define a ring morphism

$$\text{(2.10)} \qquad \sigma = \sigma_{(E,u)} : BP \xrightarrow{\varkappa} MU[p] \xrightarrow{u[p]} E[p] = E.$$

This morphism σ turns E into a BP-module spectrum.

Let $\sigma_* : \pi_*(BP) \to \pi_*(E)$ be the induced coefficient homomorphism.

2.11. Theorem. *If $\sigma_*(v_i) = 0$ for every $i < m$, then there exists a BP-module morphism $\sigma_m : P(m) \to E$ such that $\sigma = \sigma_m \rho_m$.*

Proof. This is a BP-analog of VIII.4.13. The case $m = 0$ is trivial. So, we assume that $m > 0$. Since $\sigma(v_0) = 0$, we conclude that $p\pi_0(E) = 0$, and hence every group $\pi_i(E)$ is finite.

The multiplication $BP \wedge BP \to BP$ induces a homomorphism

$$\begin{aligned}\Delta : E^*(BP) \to E^*(BP \wedge BP) &= BP^*(BP)\widehat{\otimes}_{BP^*(S)}E^*(BP)\\ &= E^*(BP)\widehat{\otimes}_{E^*(S)}E^*(BP),\end{aligned}$$

where the first equality follows from VII.3.29(v). Now the theorem can be proved just as was VIII.4.13, since $(v_0, \dots, v_{m-1})$ is a proper invariant sequence in $\pi_*(BP)$. □

2.12. Corollary. *The spectrum $P(n)$ does not depend (up to equivalence) on the choice of polynomial generators of $\pi_*(MU)$.*

Proof. If a spectrum $\overline{P}(n)$ is based on some other system of generators, then there is a morphism $\sigma_n : P(n) \to \overline{P}(n)$ which induces an isomorphism of the coefficients. □

2.13. Corollary. *Let $\sigma : BP \to E$ be as in (2.10). If $\sigma_*(v_i) = 0$ for every i then E is a graded Eilenberg–Mac Lane spectrum.*

Proof. Since $\sigma_*(v_i) = 0$, there exists a morphism $H\mathbb{Z}/p \to E$ which preserves the units. Now apply II.7.7. □

2.14. Corollary. *Let (E, u) be as in 2.11. Suppose that the formal group of (E, u) is equivalent to the additive one. Then E is a graded Eilenberg–Mac Lane spectrum.*

Proof. By VII.6.5 and VII.6.22, we can change the orientation of E so that the new $\mathbb{C}$-oriented spectrum has the additive formal group. This new orientation yields a morphism $\varphi : MU \to E$ such that $\varphi_* : \pi_*(MU) \to \pi_*(E)$ classifies the additive formal group. This means that φ_* annihilates all elements of positive dimension in $\pi_*(MU)$. So, $\sigma_*(v_i) = 0$ for every i, where $\sigma := \varphi[p]\varkappa$. □

It would be nice to prove that σ_m from 2.11 can be chosen to be a ring morphism. I can prove this for a commutative spectrum E, see 2.17 below. Notice that there is an additional gain: E does not have to have finite $\mathbb{Z}[p]$ type. We will see that the cases $p = 2$ and $p > 2$ are very different: the spectrum $P(n)$ is commutative for $p > 2$, while every commutative ring spectrum E with $2\pi_*(E) = 0$ is a graded Eilenberg–Mac Lane spectrum, see 5.5 below.

2.15. Lemma. *For every $n \geq 0$ there are the following ring isomorphisms:*
$H_*(P(n); \pi_0(P(n))) \cong \pi_0(P(n))[\xi_1, \dots, \xi_k, \dots] \otimes \Lambda(\tau_0, \dots \tau_{n-1})$ *for* $p > 2$,
$H_*(P(n); \pi_0(P(n))) \cong \pi_0(P(n))[\zeta_0, \dots, \zeta_{n-1}, \zeta_n^2, \dots, \zeta_{n+k}^2, \dots]$ *for* $p = 2$.
Here $\dim \xi_i = 2p^i - 2, \dim \tau_j = 2p^j - 1 = \dim \zeta_j$.

Proof. We consider only the case $p > 2$ and $n > 0$; all the remaining cases can be proved similarly. Set $u = r_\infty^n : P(n) \to H\mathbb{Z}/p$. We proved in 1.3 that

$$H_*(u) : H_*(P(n)) \to H_*(H\mathbb{Z}/p; \mathbb{Z}/p)$$

is a monomorphism and that $\operatorname{Im} H_*(u) = \mathbb{Z}/p\,[\xi_i' | i > 0] \otimes \Lambda(\tau_0', \dots, \tau_n')$. But u is a ring morphism. □

2.16. Lemma. *Let E be a ring spectrum, and let $\sigma : BP \to E$ be a ring morphism. If $\sigma_*(v_i) = 0$ for $i < n$, then the AHSS*

$$E^r_{**}(X) \Longrightarrow E_*(X), \; E^2_{**}(X) = H_*(X; \pi_*(E))$$

collapses for $X = P(n)$ and $X = P(n) \wedge P(n)$.

Proof. We consider only the case $X = P(n), p > 2$; all the remaining cases can be considered similarly. By 2.15,

$$E^2_{**} = H_*(P(n); \pi_*(E)) = \mathbb{Z}[\xi_1, \dots, \xi_k, \dots] \otimes \Lambda(\tau_0, \dots, \tau_{m-1}) \otimes \pi_*(E),$$
$$\operatorname{bideg} \xi_i = (2p^i - 2, 0), \operatorname{bideg} \tau_j = (2p^j - 1, 0).$$

Since E^r_{**} is a spectral sequence of $\pi_*(E)$-algebras, it suffices to prove that $d_r \xi_i = 0 = d_r \tau_j$ for every r, i, j.

Note that σ turns E into a BP-module spectrum.

Firstly, let $n = 0$, i.e., $P(n) = BP$. If $E = BP$ then the AHSS is trivial for dimensional reasons. Thus, the AHSS is trivial for every E and $n = 0$ because E is a BP-module, cf. Adams [8], Lemma 4.2.

Now, let $n > 0$. Consider $\rho_n = r_n^0 : BP \to P(n)$. Since $r_\infty^n \rho_n = r_\infty^0$, we can assume that ρ_n maps the elements ξ for BP to those for $P(n)$. Since ρ_n induces a monomorphism of the E^2-terms, $d_r \xi_i = 0$ for every r, i. The morphism $\sigma_n : P(n) \to E$ in 2.11 induces a morphism $P(n)_{(2p^n-3)} \to E_{(2p^n-3)}$ of coskeletons. But $P(n)_{(2p^m-3)} = H\mathbb{Z}/p$, and hence, by II.7.7, $E_{(2p^n-3)}$ is a graded Eilenberg–Mac Lane spectrum. Thus, non-trivial differentials d_r exist only for $r > 2(p^n - 3)$. But in this case $d_r(\tau_j) = 0$ because $r > \dim \tau_j = 2p^j - 1$ if $j < n$. □

2.17. Theorem. *Let E be a commutative ring spectrum, let $p > 2$, and let $\sigma : BP \to E$ be a ring morphism such that $\sigma(v_i) = 0$ for $i < n$. Then there exists a ring morphism $\sigma_n : P(n) \to E$ with $\sigma = \sigma_n \rho_n$.*

Proof. The triviality of the AHSS from 2.16 yields an isomorphism

$$E_{**}^{\infty}(P(n)) \cong \pi_*(E)[\xi_1, \dots, \xi_k, \dots] \otimes \Lambda(\tau_0, \dots, \tau_{n-1})$$

of $\pi_*(E)$-algebras, i.e., $E_{**}^{\infty}(P(n))$ is a free commutative $\pi_*(E)$-algebra. Since $E_*(P(n))$ is a commutative algebra (the condition $p > 2$ is used here),

$$E_*(P(n)) \cong \pi_*(E)[\xi_1, \dots, \xi_k, \dots] \otimes \Lambda(\tau_0, \dots, \tau_{n-1})$$

as $\pi_*(E)$-algebras. By II.3.45 and 2.16, the evaluation

$$\mathrm{ev}_n = \mathrm{ev}_{P(n)} : E^*(P(n)) \to \mathrm{Hom}_{\pi_*(E)}(E_*(P(n)), \pi_*(E))$$

is an isomorphism. By II.3.46, this evaluation gives us a bijective correspondence between ring morphisms $P(n) \to E$ and homomorphisms $E_*(P(n)) \to \pi_*((E))$ of $\pi_*(E)$-algebras. Now we can get the desired ring morphism σ_n if we put $\mathrm{ev}_n(\sigma_n)(\xi_i) := \mathrm{ev}_0(\sigma)(\xi_i)$, $\mathrm{ev}_n(\sigma_n)(\tau_j) := 0$. □

2.18. Remarks. (a) Theorem 2.17 holds for $p = 2$ because then E is a graded Eilenberg–Mac Lane spectrum, see 5.5 below. Unfortunately, I cannot prove this immediately, without 5.5. Furthermore, Theorem 2.17 is not valid for $p = 2$ and non-commutative E. Namely, by 5.5 and 2.2(ii), $P(1)$ does not admit a commutative multiplication. Let E be the spectrum $P(1)$ with the exotic multiplication $*$ such that $x * y = yx$. Then $\sigma : P(1) \to E$ as in 2.17 must be a ring morphism and induce the identity map of the coefficient rings at the same time. But this is impossible (prove it).

(b) By (a) and 2.17, we can get rid of the finiteness conditions (i.e., E does not have to have finite $\mathbb{Z}[p]$-type) in 2.13 and 2.14 if we require E to be commutative.

(c) The product structures in $P(n)$ were also considered by Johnson–Wilson [2], Shimada–Yagita [1] and Würgler [2]. Theorem 2.17 was proved by Würgler [3].

The following proposition clarifies the Bockstein homomorphism $\beta_n : P(n+1) \to \Sigma^{2p^n-1} P(n+1)$.

2.19. Proposition. *There exists* $\lambda \in \mathbb{Z}/p, \lambda \neq 0$, *such that* $r_\infty^{n+1}\beta_n = \lambda Q_n r_\infty^{n+1}$.

Proof. We set $d := 2p^n - 1$. It suffices to construct a homotopy commutative diagram

$$\begin{array}{ccc} P(n+1) & \xrightarrow{\delta} & \Sigma^d P(n) \\ {\scriptstyle r}\downarrow & & \downarrow{\scriptstyle r'} \\ H & \xrightarrow{\lambda Q_n} & Sigma^d H \end{array}$$

where $H = H\mathbb{Z}/p$ and $r = r_\infty^{n+1}$, $r' = \Sigma^d r_\infty^n$. Consider the diagram

$$\begin{array}{ccccc} P(n) & \xrightarrow{r_n} & P(n+1) & \xrightarrow{\delta} & \Sigma^d P(n) \\ & & {\scriptstyle r}\downarrow & & \downarrow{\scriptstyle u} \\ & & H & \xrightarrow{Q_n} & \Sigma^d H. \end{array}$$

By 2.2(ii), $Q_n r r_n = Q_n r_\infty^n = 0$. Hence, there exists u with $u\delta = Q_n r$. Since

$$H^*(P(n+1); \mathbb{Z}/p) = \mathscr{A}_p / \mathscr{A}_p(Q_i | i > n),$$

we have $u\delta \neq 0$, and so $u \neq 0$. Hence, $u = \mu r'$ with $0 \neq \mu \in \mathbb{Z}/p$. Thus, $r'\delta = \lambda Q_n r$ with $\lambda = \mu^{-1}$. □

§3. Homological Properties of the Spectra $P(n)$

Recall that Ω denotes $\pi_*(BP)$ and that $\pi_*(P(m)) = \Omega/I_m$.

3.1. Lemma. *If* $i < n$ *then the morphism* $(v_i)_\# : \Sigma^{2p^i-2}P(n) \to P(n)$ *is inessential.*

Proof. Since the morphism $S^{2p^i-2} \xrightarrow{v_i} BP \xrightarrow{\rho_n} P(n)$ is inessential, the result follows from II.2.15. □

Recall that $v_i : P(m)_*(P(n)) \to P(m)_*(P(n))$ denotes multiplication by $v_i \in \pi_*(P(m))$ on the $\pi_*(P(m))$-module $P(m)_*(P(n))$.

3.2. Lemma. *If* $i < n$ *then* $v_i : P(m)_*(P(n)) \to P(m)_*(P(n))$ *is zero for every* m. *Thus, the* Ω/I_m*-module* $P(m)_*(P(n))$ *is an* Ω/I_n*-module even for* $m < n$.

Proof. We prove this by induction on i. The assertion is clear for $i = 0$. So, fix some $i > 0, i < n$, and suppose that the homomorphism v_j is zero for every $j < i$. Let $h : \pi_*(BP) \to BP_*(BP)$ be the Hurewicz homomorphism.

By 3.1 and II.3.44, the homomorphism $h(v_i): P(m)_*(P(n)) \to P(m)_*(P(n))$ (the multiplication by $h(v_i)$ on the $BP_*(BP)$-module $P(m)_*(P(n))$) is zero. By VII.3.25(ii), $h(v_i) \equiv v_i \mod I_i BP_*(BP)$, and so, because of the inductive assumption, the homomorphism v_i is zero. The induction is confirmed. □

3.3. Lemma. *There is an isomorphism of* Ω*-modules*

$$BP_*(P(n)) \cong \Omega/I_n[y_1, \dots, y_k, \dots], \dim y_k = 2p^k - 2.$$

Proof. We prove this by induction on n. By VII.3.25(i), this isomorphism holds for $n = 0$. Suppose that the lemma holds for some $n \geq 0$ and prove it for $n + 1$.

Consider the morphism $(v_n)_\# : \Sigma^{2p^n-2} P(n) \to P(n)$. By II.3.44,

$$BP_*((v_n)_\#) = h(v_n) : BP_*(P(n)) \to BP_*(P(n)).$$

Since $P(n + 1)$ is the cone of $(v_n)_\#$, we have the exact sequence

$$\cdots \to BP_*(P(n)) \xrightarrow{h(v_n)} BP_*(P(n)) \xrightarrow{(r_n)_*} BP_*(P(n + 1)) \to \cdots .$$

Because of the inductive assumption, I_n acts trivially on $BP_*(P(n))$. Hence, since $h(v_n) \equiv v_n \mod I_n BP_*(BP)$, this exact sequence has the form

$$\cdots \to BP_*(P(n)) \xrightarrow{v_n} BP_*(P(n)) \xrightarrow{(r_n)_*} BP_*(P(n + 1)) \to \cdots .$$

By the inductive assumption, $BP_*(P(n)) \cong \Omega/I_n[y_1, \dots, y_k, \dots]$. Hence, v_n is monic. Thus, $(r_n)_*$ is epic, and

$$BP_*(P(n + 1)) \cong \Omega/I_{n+1}[y_1, \dots, y_k, \dots].$$ □

3.4. Theorem. *If* $m \leq n$*, then there is an isomorphism of* Ω/I_n*-modules*

$$P(m)_*(P(n)) \cong \Omega/I_n[y_1, \dots, y_k, \dots] \otimes \Lambda(q_1, \dots, q_m), \dim q_k = 2p^k - 1,$$

(where $\Lambda(q_1, \dots, q_m) := \mathbb{Z}$ *for* $m = 0$*). In particular,* $P(m)_*(P(n))$ *is a free* Ω/I_n*-module.*

Proof. We fix $n > 0$ and perform induction on m. If $m = 0$ then we have just 3.3. Suppose that the assertion holds for some $m < n$. Consider the following exact sequence of Ω/I_m-modules:

$$\cdots \to P(m)_*(P(n)) \xrightarrow{v_m} P(m)_*(P(n)) \to P(m + 1)_*(P(n)) \to \cdots .$$

By 3.2, the homomorphism v_m is zero, and so we have the following exact sequence of Ω/I_m-modules:

$$0 \to P(m)_*(P(n)) \to P(m + 1)_*(P(n)) \to P(m)_*(P(n)) \to 0.$$

By 3.2, its terms are Ω/I_n-modules. Since its homomorphisms are homomorphisms of Ω-modules, this sequence is a sequence of Ω/I_n-modules. By the inductive assumption, $P(m)_*(P(n))$ is a free Ω/I_n-module. Hence, this exact sequence splits. Thus,

$$P(m+1)_*(P(n)) \cong P(m)_*(P(n)) \otimes \Lambda(q_m),$$

i.e.,

$$P(m+1)_*(P(n)) \cong \Omega/I_n[y_1, \dots, y_k, \dots] \otimes \Lambda(q_1, \dots, q_{m+1}). \qquad \square$$

3.5. Proposition (cf. Johnson–Wilson [2], Würgler [1]). *If $m > n$ then there is an isomorphism*

$$P(m)^*(P(n)) \cong P(m)^*(BP) \otimes \Lambda(q_0, \dots, q_{n-1}), \ \dim q_i = 2p^i - 1,$$

of $P(m)^*(S)$-modules. Moreover, this isomorphism can be chosen so that

$$\rho_m^* : P(m)^*(P(n)) \to P(m)^*(BP)$$

has the form $\rho_m^*(x \otimes q_i) = 0$ and $\rho_m^*(x \otimes 1) = x$ for every $x \in P(m)^*(BP)$.

Proof. This can be proved as VIII.4.9 was; we leave it to the reader. $\square$

3.6. Corollary. *Let $v_m = v_{m,n} : P(m)^*(P(n)) \to P(m)^*(P(n))$ be multiplication by v_m on the $P(m)^*(S)$-module $(P(m))^*(P(n))$. If $m \geq n$, then the homomorphism $v_{m,n}$ is monic and the homomorphism*

$$(r_m)_* : P(m)^*(P(n)) \to P(m+1)^*(P(n))$$

is epic. In particular, $(r_\infty^m)_* : P(m)^*(P(n)) \to H^*(P(n); \mathbb{Z}/p)$ is epic.

Proof. Firstly, we prove that $v_{m,n}$ is monic. By 3.5 and VII.3.29(iv),

$$\begin{aligned} P(m)^*(P(n)) &\cong P(m)^*(BP) \otimes \Lambda(q_0, \dots, q_{n-1}) \\ &\cong BP^*(BP) \widehat{\otimes}_{BP^*(S)} P(m)^*(S) \otimes \Lambda(q_0, \dots, q_{n-1}) \\ &\cong BP^*(BP) \otimes_{BP^*(S)} P(m)^*(S) \otimes \Lambda(q_0, \dots, q_{n-1}) \end{aligned} \tag{3.7}$$

as $P(m)^*(S)$-modules, and hence $v_{m,n}$ is monic.

Now, the long cofiber sequence (2.9) (for m) induces the exact sequence

$$\cdots \xrightarrow{v_{m,n}} P(m)^*(P(n)) \xrightarrow{(r_m)_*} P(m+1)^*(P(n)) \to P(m)^*(P(n)) \xrightarrow{v_{m,n}} \cdots,$$

where $v_{m,n}$ is monic and so $(r_m)_*$ is epic. $\square$

3.8. Corollary. *If $m \geq n$, then*

$$P(m)^*(P(n)) \cong P(m)^*(S) \otimes^{\text{grad}} H^*(BP; \mathbb{Z}[p]) \otimes \Lambda(q_0, \dots, q_{n-1})$$

as $P(m)^*(S)$-modules. So, there exists a countable family $\{r_\alpha\}, \dim r_\alpha > 0$, in $P(m)^*(P(n))$ such that every $r \in P(m)^*(P(n))$ can be expanded as a countable sum (i.e., a series) $r = \lambda_0 + \sum \lambda_\alpha r_\alpha$ with $\lambda_0, \lambda_\alpha \in P(m)^*(S)$.

Proof. By VII.(4.2), $BP^*(BP) \cong BP^*(S) \otimes^{\text{grad}} H^*(BP; \mathbb{Z}[p])$ as abelian groups. So, the required isomorphism follows from (3.7). Now, we choose a $\mathbb{Z}[p]$-basis $\{b_\alpha\}$ of the $\mathbb{Z}[p]$-module

$$(\oplus_{i>0} H^i(BP; \mathbb{Z}[p])) \otimes \Lambda(q_0, \dots, q_{n-1})$$

and define $\{r_\alpha\}$ to be the family which corresponds to $\{1 \otimes^{\text{grad}} b_\alpha\}$ under the isomorphism. □

Following II.3.47, we see that the $P(m)^*(S)$-algebra $P(m)^*(P(m))$ acts on every group $P(m)^*(X)$. We define

$$\nu : P(m)^*(P(m)) \to P(m)^*(S), \quad \nu(\varphi) = \varphi(1),$$

where $1 \in P(m)^*(S)$, and set $R := \operatorname{Ker} \nu$. Clearly, $P(m)^*(P(m)) = P(m)^*(S) \oplus R$.

3.9. Proposition. *Choose $\{r_\alpha\}$ as in 3.8 with $m = n$. Then every $r \in R$ has the form $r = \sum \lambda_\alpha r_\alpha$.*

Proof. Indeed, if $r = \lambda_0 + \sum \lambda_\alpha r_\alpha$ then $r(1) = \lambda_0$. Thus, $\lambda_0 = 0$. □

An additional information about the algebra $P(m)^*(P(m))$ can be found in Johnson–Wilson [1], Würgler [1], Yagita [2].

3.10. Proposition. *If X is a finite CW-space (or a finite spectrum), then $P(n)_*(X)$ and $P(n)^*(X)$ are coherent, and hence finitely generated, modules over the coefficient ring.*

Proof. Because of the duality arguments, it suffices to consider $P(n)_*(X)$ only. For $n = 0$ this is proved in VII.3.29(i). If $n > 0$ then $\pi_*(P(n))$ is a polynomial ring over a field. Hence, by VII.1.13(i), it is coherent. Now the proposition follows from VII.1.14. □

3.11. Proposition. *Let X be a spectrum bounded below. Suppose that every group $H_i(X; \mathbb{Z}[p])$ is a finitely generated $\mathbb{Z}[p]$-module. Then $P(n)_*(X)$ is a free module over the coefficient ring iff the homomorphism $(r^n_\infty)^X : P(n)_*(X) \to H_*(X; \mathbb{Z}/p)$ is epic.*

Proof. Firstly, let $n = 0$. If $BP_*(X)$ is a free $\pi_*(BP)$-module then, by VII.3.29(ii), the homomorphism $BP_*(X) \to H_*(X; \mathbb{Z}[p])$ is epic and $H_*(X; \mathbb{Z}[p])$ is a free $\mathbb{Z}[p]$-module. Hence the mod p reduction $H_*(X; \mathbb{Z}[p]) \to H_*(X; \mathbb{Z}/p)$ is epic. Hence, $(r^n_\infty)^X$ is epic.

Conversely, if $(r_\infty^n)^X$ is epic, then the mod p reduction $H_*(X;\mathbb{Z}[p]) \to H_*(X;\mathbb{Z}/p)$ is epic. Hence, the $\mathbb{Z}[p]$-module $H_*(X;\mathbb{Z}[p])$ is free since it is finitely generated. Thus, by VII.3.29(ii), $BP_*(X)$ is a free $\pi_*(BP)$-module.

Now we consider $n > 0$. Throughout the proof $H_*(X)$ denotes $H_*(X;\mathbb{Z}/p)$.

Suppose $P(n)_*(X)$ is a free $\pi_*(P(n))$-module. Then $v_n : P(n)_*(X) \to P(n)_*(X)$ is monic. Hence, the exact sequence (2.1) converts into an exact sequence

$$0 \to P(n)_*(X) \xrightarrow{v_n} P(n)_*(X) \xrightarrow{r_n^X} P(n+1)_*(X) \to 0,$$

and, therefore,

$$P(n+1)_*(X) \cong P(n)_*(X) \otimes_{\pi_*(P(n))} \pi_*(P(n+1)).$$

Thus, $P(n+1)_*(X)$ is a free $\pi_*(P(n+1))$-module and

$$r_n^X : P(n)_*(X) \to P(n+1)_*(X)$$

is epic. An iteration of these arguments gives us a sequence of epimorphisms

$$P(n)_*(X) \to P(n+1)_*(X) \to \cdots \to P(n+k)_*(X) \xrightarrow{r_{n+k}^X} \cdots .$$

Thus, $(r_\infty^n)^X : P(n)_*(X) \to H_*(X)$ is epic, because $H_*(X) = \varinjlim_N P(N)_*(X)$.

Conversely, suppose that $(r_\infty^n)^X$ is epic. Consider the AHSS

$$E_{p,q}^s \Longrightarrow P(n)_*(X), E_{p,q}^2 = H_p(X;\pi_q(P(n))).$$

Since $(r_\infty^n)_*$ is epic, all differentials $d_s : E_{m,0}^s \to E_{m-s,s-1}^s$ are trivial. Thus, all differentials $d_s : E_{m,r}^s \to E_{m-s,r+s-1}^s$ are trivial because this spectral sequence is a spectral sequence of $\pi_*(P(n))$-modules. Hence, $E^2 = E^\infty$, and we can identify E_{**}^∞ with $P(n)_*(X)$. Let $\{x_j\}$ be a basis of the $\mathbb{Z}/p$-vector space $H_*(X)$, let $\{a_i\}$ be a basis of the $\mathbb{Z}/p$-vector space $\pi_*(P(n))$, and let y_j be the image of x_j in $E_{*,0}^\infty$. Then $\{a_i y_j\}$ is a basis of the $\mathbb{Z}/p$-vector space $P(n)_*(X)$. Define additive homomorphisms

$$\varphi : H_*(X) \to P(n)_*(X), \quad \varphi(x_j) = y_j$$

and

$$\psi : \pi_*(P(n)) \otimes H_*(X) \to P(n)_*(X), \quad \psi(a \otimes x) = a\varphi(x).$$

It is easy to see that ψ is a $\pi_*(P(n))$-module homomorphism. Moreover, ψ is epic because $\sum a_i y_j = \psi(\sum a_i \otimes x_j)$. Hence, ψ is an isomorphism because the isomorphic groups $P(n)_i(X)$ and $\sum_k \pi_k(P(n)) \otimes H_{i-k}(X)$ are finite. Thus, $P(n)_*(X)$ is isomorphic to a free $\pi_*(P(n))$-module $\pi_*(P(n)) \otimes H_*(X)$. □

3.12. Proposition. *Let X be an arbitrary spectrum, and let $n > 0$. Then the following conditions* (i) *and* (ii) *are equivalent:*

(i) *The homomorphism* $(r^n_\infty)^X : P(n)_*(X) \to H_*(X;\mathbb{Z}/p)$ *is epic;*

(ii) *For every* s, *the homomorphism*

$$(r^n_\infty)^{X^{(s)}} : P(n)_*(X^{(s)}) \to H_*(X^{(s)};\mathbb{Z}/p)$$

is epic.

Furthermore, if X *has finite type then the following conditions* (iii) *and* (iv) *are equivalent:*

(iii) *The homomorphism* $(r^n_\infty)_X : P(n)^*(X) \to H^*(X;\mathbb{Z}/p)$ *is epic;*

(iv) *For every* s, *the homomorphism*

$$(r^n_\infty)_{X^{(s)}} : P(n)^*(X^{(s)}) \to H^*(X^{(s)};\mathbb{Z}/p)$$

is epic.

Proof. For simplicity, denote $P(n)_*(X)$ by $P_*(X)$ and $H_*(X;\mathbb{Z}/p)$ by $H_*(X)$.

We prove that (i) $\Rightarrow$ (ii). The cofiber sequence $X^{(k)} \subset X^{(k+1)} \to \vee S^{k+1}$ yields the following commutative diagram with exact rows:

$$\begin{array}{ccccccc}
P_{d+1}(\vee S^{k+1}) & \longrightarrow & P_d(X^{(k)}) & \longrightarrow & P_d(X^{(k+1)}) & \longrightarrow & P_d(\vee S^{k+1}) \\
\downarrow{\scriptstyle a_1} & & \downarrow{\scriptstyle a} & & \downarrow{\scriptstyle b} & & \downarrow{\scriptstyle a_2} \\
H_{d+1}(\vee S^{k+1}) & \longrightarrow & H_d(X^{(k)}) & \longrightarrow & H_d(X^{(k+1)}) & \longrightarrow & H_d(\vee S^{k+1}),
\end{array}$$

where the vertical homomorphisms are induced by r^n_∞. Here $a_i, i = 1, 2$, is epic. Clearly, a is epic if b is. (Indeed, if a_2 is monic then this follows from the Five Lemma. If a_2 is not monic then $P_d(S^{k+1}) \neq 0$, and so $d \geq k+1$. Hence $H_d(X^{(k)}) = 0$.) Thus, if $P_d(X^{(s)}) \to H_d(X^{(s)})$ is not epic for some s then $P_d(X^{(r)}) \to H_d(X^{(r)})$ is not epic for every $r \geq s$, and so $P_d(X) \to H_d(X)$ is not epic because $P_d(X^{(r)}) = P_d(X)$, $H_d(X^{(r)}) = H_d(X)$ for $r > d+1$.

We prove that (ii) $\Rightarrow$ (i). Consider the commutative diagram

$$\begin{array}{ccccccc}
\cdots & \longrightarrow & P_*(X^{(s)}) & \longrightarrow & P_*(X^{(s+1)}) & \longrightarrow & \cdots \\
 & & \downarrow{\scriptstyle (r^n_\infty)^{X^{(s)}}} & & \downarrow & & \\
\cdots & \longrightarrow & H_*(X^{(s)}) & \longrightarrow & H_*(X^{(s+1)}) & \longrightarrow & \cdots .
\end{array}$$

If all the homomorphisms $(r^n_\infty)^{X^{(s)}} : P_*(X^{(s)}) \to H_*(X^{(s)})$ are epic, then so is the homomorphism

$$P_*(X) = \varinjlim P_*(X^{(s)}) \to \varinjlim H_*(X^{(s)}) = H_*(X).$$

We prove that (iii) $\Rightarrow$ (iv). Fix any $d \in \mathbb{Z}$. If $P^d(X) \to H^d(X)$ is epic then the homomorphism

$$P^d(X) \to P^d(X^{(k)}) \to H^d(X^{(k)})$$

is epic for every $k \geq d+1$. So, it remains to prove that $r^n_\infty : P^d(X^{(k)}) \to H^d(X^{(k)})$ is epic for $k \leq d$.

If $r < d$ then $P^d(S^{r+1}) = 0 = P^{d+1}(S^{r+1})$, $H^d(S^{r+1}) = 0 = H^{d+1}(S^{r+1})$. Hence, if $r \leq d-1$ then the cofiber sequence $X^{(r)} \subset X^{(r+1)} \to \vee S^{r+1}$ yields the following commutative diagram with exact rows:

$$\begin{array}{ccccccc} 0 & \longleftarrow & P^d(X^{(r)}) & \longleftarrow & P^d(X^{(r+1)}) & \longleftarrow & 0 \\ & & \big\downarrow{\scriptstyle a} & & \big\downarrow{\scriptstyle b} & & \\ 0 & \longleftarrow & H^d(X^{(r)}) & \longleftarrow & H^d(X^{(r+1)}) & \longleftarrow & 0. \end{array}$$

Hence, b is epic provided that a is epic and $r \leq d-1$. Since X is bounded below, there is an integer number N such that $H^d(X^{(r)}) = 0$ for every $r \leq N$, and so a is epic for every $r \leq N$. Now, we perform an obvious induction and prove that $P^d(X^{(k)}) \to H^d(X^{(k)})$ is epic for $k \leq d$ (do it).

We prove that (iv) $\Rightarrow$ (iii). Consider the commutative diagram

$$\begin{array}{ccccccc} \cdots & \longleftarrow & P^*(X^{(s)}) & \longleftarrow & P^*(X^{(s+1)}) & \longleftarrow & \cdots \\ & & \big\downarrow & & \big\downarrow & & \\ \cdots & \longleftarrow & H^*(X^{(s)}) & \longleftarrow & H^*(X^{(s+1)}) & \longleftarrow & \cdots. \end{array}$$

Because of III.4.18 and finiteness of $P^i(X^{(s)})$, the homomorphism $P^*(X) \to \varprojlim P^*(X^{(s)})$ is an isomorphism. Furthermore, by III.2.17,

$$\varprojlim\{(r^\infty_n)_{X^{(s)}}\} : \varprojlim P^*(X^{(s)}) \to \varprojlim H^*(X^{(s)})$$

is an epimorphism because the groups $P^i(X^{(s)})$ and $H^i(X^{(s)})$ are finite. Thus,

$$(r^\infty_n)_X : P^*(X) \to \varprojlim P^*(X^{(s)}) \to \varprojlim H^*(X^{(s)}) = H^*(X)$$

is an epimorphism. □

3.13. Corollary. *The homomorphism*

$$r^n_\infty : P(n)^*(P(n)^{(s)}) \to H^*(P(n)^{(s)}; \mathbb{Z}/p)$$

is epic for every $n > 0$ and every s.

Proof. This follows from 3.6 and 3.12. □

3.14. Theorem (cf. Yosimura [1]). *For every $n \geq 0$ and every k-connected spectrum E, there exists a morphism $f : W \to E$ such that W is a k-connected spectrum, $P(n)_*(W)$ is a free $\pi_*(P(n))$-module and*

$$f_* : P(n)_*(W) \to P(n)_*(E)$$

is an epimorphism.

Proof. The case $n = 0$ was done in VII.3.29(iii), so we assume that $n > 0$. Fix s, and let Y be a spectrum dual to to $P(n)^{(s)}$. By 3.12, the homomorphism $r^n_\infty : P(n)^*(P(n)^{(s)}) \to H^*(P(n)^{(s)}; \mathbb{Z}/p)$ is epic. Hence, the homomorphism $P(n)_*(Y) \to H_*(Y; \mathbb{Z}/p)$ is epic. Thus, by 3.11, $P(n)_*(Y)$ is a free $\pi_*(P(n))$-module. Now we can complete the proof following that of VII.1.17. □

Now we give an analog of the Filtration Theorem VII.4.18 for the spectra $P(m)$. Consider the diagram

$$BP^*(BP) \xrightarrow{(\rho_m)_*} P(m)^*(BP) \xleftarrow{\rho_m^*} P(m)^*(P(m)).$$

By 3.6, $(\rho_m)_*$ is epic, and ρ_m^* is epic because of 3.5.

The pairing $\mu : BP \wedge P(m) \to P(m)$ gives us a homomorphism

$$\Delta : P(m)^*(P(m)) \xrightarrow{\mu^*} P(m)^*(BP \wedge P(m)) \xrightarrow{\cong} BP^*(BP) \widehat{\otimes}_\Omega P(m)^*(P(m)).$$

Clearly,

$$\Delta(\varphi) = \theta \widehat{\otimes} 1 + 1 \widehat{\otimes} \varphi + \sum \theta_i \widehat{\otimes} \varphi_i,$$

where $(\rho_m)_*(\theta) = \rho_m^*(\varphi)$. Furthermore, we can and shall assume that $\varphi_i \in R$.

Consider the following category $\mathscr{L}(m)$. Objects of this category are coherent graded Ω/I_m-modules (and hence Ω-modules) L equipped with a $P(m)^*(P(m))$-action such that:

1. $\dim \varphi(x) = \dim x - \dim \varphi$ for every $\varphi \in P(m)^*(P(m)), x \in L$.
2. If $\Delta(\varphi) = \theta \widehat{\otimes} 1 + 1 \widehat{\otimes} \varphi + \sum \theta_i \widehat{\otimes} \varphi_i$, then

$$\varphi(ax) = \theta(a)x + a\varphi(x) + \sum \theta_i(a)\varphi_i(x)$$

for every $a \in \Omega, x \in L$.

Note that the term $\sum \theta_i(a)\varphi_i(x)$ is well-defined since $\theta_i(a)\varphi_i(x) \neq 0$ for all but finitely many i's. Also, $P(m)_*(X) \in \mathscr{L}(m)$ for every finite spectrum X

Morphisms of $\mathscr{L}(m)$ are $P(m)^*(P(m))$-equivariant Ω-module morphisms.

3.15. Theorem (Yagita [1], Yosimura [1]). *Every object L of $\mathscr{L}(m)$ admits a finite filtration in $\mathscr{L}(m)$*

$$0 = L_0 \subset L_1 \subset \cdots \subset L_r = L,$$

such that L_i/L_{i-1} is stably isomorphic in $\mathscr{L}$ to Ω/I_{n_i}. In particular, this holds for $L = P(m)_*(X)$ where X is a finite spectrum.

Proof. Firstly, note the following. Given $a \in \Omega$, let $\overline{a}$ denote the image of a under the epimorphism $\Omega \to \Omega/I_m$. If $\theta \in BP^*(BP)$ and $\varphi \in P(m)^*(P(m))$ are such that $(\rho_m)_*(\theta) = \rho_m^*(\varphi)$ then $\overline{\theta(a)} = \varphi(\overline{a})$ for every $a \in \Omega$. In particular, if I is a $BP^*(BP)$-invariant ideal of Ω then $I/(I \cap I_m)$ is a $P(m)^*(P(m))$-invariant ideal of Ω/I_m.

We prove the theorem by induction on the number of Ω-generators of L. If L has just one generator x, then L is stably isomorphic in $\mathscr{L}$ to Ω/J where $J = \operatorname{Ann} x$. Indeed, it is clear that the homomorphism $f : \Omega \to L, f(a) = ax$, yields a stable isomorphism $L \approx \Omega/J$ of Ω-modules, and it remains to prove that f yields a stable isomorphism in $\mathscr{L}$. So, we must prove that $\varphi(ax) = \theta(a)x$ for every $\varphi \in P(m)^*(P(m))$ where

$$\Delta(\varphi) = \theta\widehat{\otimes}1 + 1\widehat{\otimes}\varphi + \sum \theta_i\widehat{\otimes}\varphi_i, \ (\rho_m)_*(\theta) = \rho_m^*(\varphi).$$

Note that, by 3.9, $r(x) = 0$ for every $r \in R$ since $\dim r_\alpha > 0$ and x is an element of least dimension in L. Hence,

$$\varphi(ax) = \theta(a)x + a\varphi(x) + \sum \theta_i(a)\varphi_i(x) = \theta(a)x$$

since $\varphi_i \in R$. Hence, L and Ω/J are stably isomorphic in $\mathscr{L}$. So, it suffices to prove that Ω/J admits a desired filtration.

Clearly, $I_m \subset J$, and J is a coherent ideal of Ω. We prove that J is $BP^*(BP)$-invariant. Consider the operations $s_\omega \in BP^*(BP)$ (in fact, $\rho s_\omega \varkappa$) defined in VII.(4.1). We must prove that $s_\omega(a) \in J$ for every $a \in J$ and every $\omega \neq (0)$. Choose any $\varphi \in P(m)^*(P(m))$ such that $(\rho_m)_*(s_\omega) = \rho_m^*(\varphi)$. (Such φ exists since both homomorphisms $(\rho_m)_*$ and ρ_m^* are epic.) So,

$$\Delta(\varphi) = s_\omega\widehat{\otimes}1 + 1\widehat{\otimes}\varphi + \sum \theta_i\widehat{\otimes}\varphi_i.$$

Furthermore, $\varphi \in R$ since $\omega \neq (0)$. Now,

$$0 = \varphi(ax) = s_\omega(a)x + a\varphi(x) + \sum \theta_i(a)\varphi_i(x) = s_\omega(a)x$$

since $\varphi, \varphi_i \in R$. So, $s_\omega(a) \in J$, i.e., $J = \operatorname{Ann} x$ is a $BP^*(BP)$-invariant ideal of Ω.

Hence, by VII.4.18, there is a filtration

$$J = J_0 \subset J_1 \subset \cdots \subset J_n = \Omega$$

such that $J_k/J_{k-1} \approx \Omega/I_{k_i}$ in $\mathscr{L}$ and each J_k, $k < n$, is a $BP^*(BP)$-invariant ideal of Ω. So, by the above, J_k/I_m is a $P(m)^*(P(m))$-invariant ideal of Ω/I_m. Thus, the above filtration yields the filtration

$$J/I_m = J_0/I_m \subset J_1/I_m \subset \cdots \subset J_n/I_m = \Omega/I_m$$

in $\mathscr{L}$. Thus, $\{J_k/J\}$ is a desired filtration $\{J_k/I_m\}$ of Ω/J.

Assume that the theorem holds for every $N \in \mathscr{L}$ such that the Ω-module N can be generated by n elements. Consider any $L \in \mathscr{L}$ which admits a family $\{x_1, \dots, x_{n+1}\}$ of Ω-generators. We suppose that $\dim x_1 \leq \dim x_i$ for every $i = 2, \dots, n+1$. Consider the submodule Ωx_1 of L. Since L is coherent, Ωx_1 is coherent by definition. Since x_1 is an element of least dimension, we conclude that $\Omega x_1 \in \mathscr{L}$ (i.e., Ωx_1 is a $P(m)^*(P(m))$-invariant submodule of

L). So, by the above, Ωx_1 admits a desired filtration. On the other hand, by inductive assumption, $L/\Omega x_1$ admits a desired filtration. Thus, L itself admits a desired filtration. □

§4. The Exactness Theorem

The results of this section are stimulated by the following Conner–Floyd Theorem. The spectrum K of complex K-theory admits a canonical $\mathbb{C}$-orientation (see, e.g., Stong [3], Ch. 9), and so there is a ring morphism (Thom–Dold class) $u : M\mathcal{U} \to K$. Furthermore,

$$u_* : \pi_*(M\mathcal{U}) \to \pi_*(K) = \mathbb{Z}[t, t^{-1}], \quad \dim t = 2,$$

coincides with $Td : \pi_*(M\mathcal{U}) \to \mathbb{Z}[t, t^{-1}], Td[M] = T(M)t^{(\dim M)/2}$, where $[M]$ is the bordism class and $T(M)$ is the Todd genus of a stably almost complex manifold M (i.e., $T(M) = \langle T(\tau M), [M]_H \rangle$, where τM is the tangent bundle of M). The homomorphism Td turns $\mathbb{Z}[t, t^{-1}]$ into a $\pi_*(M\mathcal{U})$-module ${}_{Td}\mathbb{Z}[t, t^{-1}]$, and so for every X we have a homomorphism

$$\overline{u} : M\mathcal{U}_*(X) \otimes_{Td} \mathbb{Z}[t, t^{-1}] \to K_*(X).$$

Conner–Floyd [1] proved that $\overline{u}$ is an isomorphism for every spectrum X. (In fact, they proved that the corresponding cohomological map is an isomorphism for every finite CW-space X. Furthermore, to be precise, they considered another orientation, which gives the genus $\mathscr{T}(M) = (-1)^{(\dim M)/2} Td(M)$, cf. V.3.5.) Note that the proof would be simpler if we knew a priori that $M\mathcal{U}_*(-) \otimes_{Td} \mathbb{Z}[t, t^{-1}]$ is a homology theory (on $\mathscr{S}$): clearly, $\overline{u}$ is an isomorphism for $X = S$, and one can apply II.3.19(iii). So, it makes sense to describe all homomorphisms $\rho : \pi_*(M\mathcal{U}) \to \mathbb{Z}[t, t^{-1}]$ such that $M\mathcal{U}_*(-) \otimes_\rho \mathbb{Z}[t, t^{-1}]$ is a homology theory. This question was considered in Rudyak [1], §3. (There are some gaps in this paper, but the answer to this question is correct.)

More generally, it makes sense to describe all $\pi_*(M\mathcal{U})$-modules M such that the functor $M\mathcal{U}_*(-) \otimes_{\pi_*(M\mathcal{U})} M$ is a homology theory. In order to solve this problem it suffices to solve the corresponding local problem, i.e., to describe all $\pi_*(BP)$-modules M such that $BP_*(-) \otimes_{\pi_*(BP)} M$ is a homology theory. Furthermore, one can settle a similar problem for $P(m)$ instead of BP.

Recall that Ω denotes $\pi_*(BP)$ and that $\pi_*(P(m)) = \Omega/I_m$.

The following theorem solves this problem.

4.1. Theorem. *Let M be an Ω/I_m-module. The following conditions are equivalent:*

(i) *The functor* $P(m)_*(-) \otimes_{\Omega/I_m} M$ *is a homology theory on* $\mathscr{S}$ *(i.e., it satisfies the exactness axiom);*

(ii) *For every* $n \geq m$, $\mathrm{Tor}_1^{\Omega/I_m}(\Omega/I_n, M) = 0$;

(iii) *For every* $n \geq m$, *the homomorphism*

$$v_n : M/I_nM \to M/I_nM$$

is monic (i.e., $\{v_m, \dots, v_{m+k}, \dots\}$ *is an* M*-proper sequence).*

Thus, each of the conditions (ii), (iii) is necessary and sufficient for the exactness of the functor $P(m)_*(X, A) \otimes_{\Omega/I_m} M$. The sufficiency of the conditions (ii), (iii) for the exactness is a purely algebraic fact, while in order to prove necessity we use certain topological information. For these reasons we shall prove necessity and sufficiency separately, see 4.3–4.7 below. But first we prove the equivalence of (ii) and (iii).

4.2. Proposition. *The conditions* (ii) *and* (iii) *of* 4.1 *are equivalent.*

Proof. Applying the functor $\otimes_{\Omega/I_m} M$ to the exact sequence

$$0 \to \Omega/I_n \xrightarrow{v_n} \Omega/I_n \to \Omega/I_{n+1} \to 0,$$

we obtain an exact sequence

$$\mathrm{Tor}_1^{\Omega/I_m}(\Omega/I_n, M) \to \mathrm{Tor}_1^{\Omega/I_m}(\Omega/I_{n+1}, M) \to M/I_nM \xrightarrow{v_n} M/I_nM.$$

Thus, (ii) implies (iii). Conversely, suppose (iii) holds; we prove (ii) by induction. Clearly, $\mathrm{Tor}_1^{\Omega/I_m}(\Omega/I_m, M) = 0$. Assuming that $\mathrm{Tor}_1^{\Omega/I_m}(\Omega/I_n, M) = 0$ for some $n \geq m$, we conclude that $\mathrm{Tor}_1^{\Omega/I_m}(\Omega/I_{n+1}, M) = 0$ because of the exactness of the sequence above. The induction is confirmed. □

Let $\mathscr{L}(m)$ be the category described in §3.

4.3. Lemma. *Let* M *be an* Ω/I_m*-module such that* $\mathrm{Tor}_1^{\Omega/I_m}(\Omega/I_n, M) = 0$ *for every* $n \geq m$. *Then for every exact sequence* $L' \to L \to L''$ *in* $\mathscr{L}(m)$ *the induced sequence*

$$L' \otimes_{\Omega/I_m} M \to L \otimes_{\Omega/I_m} M \to L'' \otimes_{\Omega/I_m} M$$

is exact.

Proof. Because of 3.15, every object $V \in \mathscr{L}(m)$ admits a filtration

$$0 = V_0 \subset \cdots \subset V_r = V$$

such that $V_i/V_{i-1} \approx \Omega/I_{n_i}$. Thus, the condition $\mathrm{Tor}_1^{\Omega/I_m}(\Omega/I_n, M) = 0$ implies that $\mathrm{Tor}_1^{\Omega/I_m}(V, M) = 0$ for every $V \in \mathscr{L}(m)$. Since the kernel and cokernel of every morphism of $\mathscr{L}(m)$ belong to $\mathscr{L}(m)$, the equality $\operatorname{Tor}_1^{\Omega/I_m}(V, M) = 0$ for all $V \in \mathscr{L}(m)$ implies the exactness of $\otimes_{\Omega/I_m} M$ on $\mathscr{L}(m)$, see e.g. Bourbaki [3]. □

4.4. Corollary (the sufficiency). *Let M be an Ω/I_m-module such that $\operatorname{Tor}_1^{\Omega/I_m}(\Omega/I_n, M) = 0$ for every $n \geq m$. Then $P(m)_*(-) \otimes_{\Omega/I_m} M$ is a homology theory on $\mathscr{S}$.*

Proof. One must check the exactness axiom for $P(m)_*(-) \otimes_{\Omega/I_m} M$. By 4.3, it holds for finite spectra, since $P(m)_*(X) \in \mathscr{L}(m)$ for every finite spectrum X. Given an arbitrary spectrum X, let $\{X_\lambda\}$ be the family of all finite subspectra of X. Recall that $\varinjlim$ commutes with the tensor product, see e.g. Bourbaki [1]. Now,

$$\begin{aligned} P(m)_*(X) \otimes_{\Omega/I_m} M &= \left(\varinjlim\{P(m)_*(X_\lambda)\}\right) \otimes_{\Omega/I_m} M \\ &= \varinjlim\{P(m)_*(X_\lambda) \otimes_{\Omega/I_m} M\}, \end{aligned}$$

and the result follows because, by I.2.7, $\varinjlim$ preserves exactness. □

4.5. Proposition. *If the functor $P(m)_*(-) \otimes_{\Omega/I_m} M$ is a homology theory on $\mathscr{S}$, then $\operatorname{Tor}_1^{\Omega/I_m}(P(m)_*(E), M) = 0$ for every spectrum E bounded below.*

Proof. Let $f : W \to E$ be as in 3.14. Then the cofiber sequence

$$W \xrightarrow{f} E \to Cf$$

yields an exact sequence

$$0 \to P(m)_*(Cf) \to P(m)_*(W) \to P(m)_*(E) \to 0. \tag{4.6}$$

Applying the functor $\otimes_{\Omega/I_m} M$ to this sequence, and using that the functor $P(m)_*(-) \otimes_{\Omega/I_m} M$ is a homology theory, we obtain an exact sequence

$$\begin{aligned} 0 \to P(m)_*(Cf) \otimes_{\Omega/I_m} M &\to P(m)_*(W) \otimes_{\Omega/I_m} M \\ &\to P(m)_*(E) \otimes_{\Omega/I_m} M \to 0. \end{aligned}$$

On the other hand, there is an exact sequence

$$\begin{aligned} 0 = \operatorname{Tor}_1^{\Omega/I_m}(P(m)_*(W), M) &\to \operatorname{Tor}_1^{\Omega/I_m}(P(m)_*(E), M) \to \\ &\to P(m)_*(Cf) \otimes_{\Omega/I_m} M \\ &\to P(m)_*(W) \otimes_{\Omega/I_m} M \to P(m)_*(E) \otimes_{\Omega/I_m} M \to 0. \end{aligned}$$

Thus, $\operatorname{Tor}_1^{\Omega/I_m}(P(m)_*(E), M) = 0$. □

4.7. Corollary (the necessity). *If the functor $P(m)_*(-) \otimes_{\Omega/I_m} M$ is a homology theory on $\mathscr{S}$, then* $\mathrm{Tor}_1^{\Omega/I_m}(\Omega/I_n, M) = 0$ *for every $n \geq m$.*

Proof. Choose $n > m$. Then, by 4.5,

$$\mathrm{Tor}_1^{\Omega/I_m}(P(m)_*(P(n)), M) = 0. \tag{4.8}$$

But, by 3.4, $P(m)_*(P(n))$ is a free Ω/I_n-module. Thus,

$$\mathrm{Tor}_1^{\Omega/I_m}(\Omega/I_n, M) = 0. \qquad \square$$

Thus, we have proved 4.1.

Now we prove the exactness theorem for MU. Let $\{x_n\}$, $\dim x_n = 2n$, be a system of free polynomial generators of $\pi_*(MU)$ such as in VII.1.9(v). In other words, for every prime p all the characteristic numbers of x_{p^n-1} are divisible by p, i.e., $I(p,n) = (p, x_{p-1}, \dots, x_{p^n-1})$. (For instance, one can choose x_{p^n-1} to be the coefficient of t^{p^n} in the formal power series $[p]_f(t)$, where f is the (universal) formal group of complex cobordism, see VII.6.11.)

4.9. Theorem. *Let M be a $\pi_*(MU)$-module. The following conditions are equivalent:*

(i) *The functor $MU^*(-) \otimes_{\pi_*(MU)} M$ is a homology theory on $\mathscr{S}$;*

(ii) *For every prime p and every natural number n the homomorphisms $p : M \to M$ and $x_{p^n-1} : M/I(p,n)M \to M/I(p,n)M$ are monic;*

(iii) *For every prime p and every natural number n,*

$$\mathrm{Tor}_1^{\pi_*(MU)}(\pi_*(MU)/I(p,n), M) = 0.$$

Proof. This can be proved just as was 4.1. The equivalence of (ii) and (iii) can be proved as was 4.2. In order to prove the sufficiency one must use the Filtration Theorem VII.4.23. In order to prove the necessity one needs to prove that $MU_*(P(n))$ is a free $\pi_*(MU)/I(p,n)$-module, $n > 0$. But

$$MU_*(P(n)) = BP_*(P(n)) \otimes_{\pi_*(BP)} \pi_*(MU[p])$$

for every $n > 0$, and $\Omega/I_n \otimes_\Omega \pi_*(MU[p]) \cong \pi_*(MU)/I(p,n)$. $\square$

Now we prove the Conner–Floyd Theorem mentioned above.

4.10. Theorem. *For every spectrum X the homomorphism*

$$\overline{u} : MU_*(X) \otimes_{Td} \mathbb{Z}[t, t^{-1}] \to K_*(X)$$

is an isomorphism.

Proof. Since $\overline{u}$ is an isomorphism for $X = S$, it suffices to prove that

$$MU_*(-) \otimes_{Td} \mathbb{Z}[t, t^{-1}]$$

is a homology theory. One can take $x_{p-1} = [CP^{p-1}]$ in 4.9(ii). Clearly,

$$p : \mathbb{Z}[t, t^{-1}] \to \mathbb{Z}[t, t^{-1}]$$

is monic. Furthermore, $Td(CP^{p-1}) = t^{p-1}$, and so

$$x_{p-1} : \mathbb{Z}/p\,[t, t^{-1}] \to \mathbb{Z}/p\,[t, t^{-1}]$$

coincides with multiplication by t^{p-1}, and so it is monic. Finally,

$${}_{Td}\mathbb{Z}[t, t^{-1}]/(p, CP^{p-1}) = 0,$$

and therefore multiplication by $x_{p^n-1}, n > 1$, is trivially monic. Thus, $MU_*(-) \otimes_{Td} \mathbb{Z}[t, t^{-1}]$ is a homology theory. □

Dualizing 4.1, we get the following result.

4.11. Theorem. *Let M be an Ω/I_m-module. The following conditions are equivalent:*

(i) *The functor $P(m)^*(-) \otimes_{\Omega/I_m} M$ is a cohomology theory on $\mathscr{S}_{\mathrm{f}}$ (i.e., it satisfies the exactness axiom);*

(ii) *For every $n \geq m$, $\mathrm{Tor}_1^{\Omega/I_m}(\Omega/I_n, M) = 0$;*

(iii) *For every $n \geq m$, the homomorphism*

$$v_n : M/I_nM \to M/I_nM$$

is monic, (i.e., the sequence $\{v_m, \dots, v_{m+k}, \dots,\}$ is M-proper).

Proof. Clearly, $P(m)^*(-) \otimes_{\Omega/I_m} M$ is a cohomology theory on $\mathscr{S}_{\mathrm{f}}$ iff $P(m)_*(-) \otimes_{\Omega/I_m} M$ is a homology theory on $\mathscr{S}_{\mathrm{f}}$, i.e., iff $P(m)_*(-) \otimes_{\Omega/I_m} M$ is a homology theory on $\mathscr{S}$, cf. II.3.20(iii). □

4.12. Lemma. *Let R be a commutative ring. Then for every exact sequence $A \to B \to C$ of R-modules the sequence*

$$A \otimes_R R[x^{-1}] \to B \otimes_R R[x^{-1}] \to C \otimes_R R[x^{-1}]$$

is exact. In other words, $R[x^{-1}]$ is a flat R-module.

Proof. This is a special case of the well-known theorem stating that localization preserves the exactness, see Bourbaki [3], Ch II, §2, n° 4, Th. 1. □

Let k denote connective k-theory.

4.13. Proposition. *Let (E, u) be a $\mathbb{C}$-marked spectrum with*

$$\pi_*(E) \cong \mathbb{Z}[t], \quad \dim t = 2,$$

and let $u_* : \pi_*(MU) \to \pi_*(E)$ *be the coefficient homomorphism induced by* u. *If* $u_*(x) = Td(x)$ *for every* $x \in \pi_*(MU)$, *then* $E \simeq k$.

Proof. By 4.12, $F^*(-) := E^*(-) \otimes_{\mathbb{Z}[t]} \mathbb{Z}[t, t^{-1}]$ is a cohomology theory on $\mathscr{S}_{\mathrm{f}}$. (In fact, it is a non-additive cohomology theory on $\mathscr{S}$.) By III.3.20(iii), the morphism

$$q : E^*(X) \to F^*(X), \quad x \mapsto x \otimes 1$$

of cohomology theories on $\mathscr{S}_{\mathrm{f}}$ is induced by a morphism $q : E \to F$ of spectra, and it is clear that q is a connective covering. We set $f := qu : MU \to F$. Then $f_* = Td : \pi_*(MU) \to \pi_*(F)$, and so f induces a natural transformation

$$\{v^X : MU^*(X) \otimes_{Td} \mathbb{Z}[s, s^{-1}] \to F^*(X), \quad X \in \mathscr{S}_{\mathrm{f}}\}.$$

By 4.10, the family $\{v^X\}$ turns out to be a morphism $v : K^*(-) \to F^*(-)$ of cohomology theories on $\mathscr{S}_{\mathrm{f}}$. Since this morphism is an isomorphism for $X = S$, it is for every $X \in \mathscr{S}_{\mathrm{f}}$. In other words, we get an isomorphism $v : K^*(-) \to F^*(-)$ of cohomology theories on $\mathscr{S}_{\mathrm{f}}$. By III.3.20(iii), this isomorphism v is induced by an equivalence $K \to F$ of spectra. Passing to connective coverings, we get an equivalence $k \to E$. □

Again, consider the homomorphism

$$Td : \pi_*(MU) = \mathbb{Z}[x_1, \dots, x_n, \dots] \to \mathbb{Z}[t], \quad Td(x_i) = T(x_i)t^i.$$

We set $y_i := x_i - T(x_i)[CP^1]^i$. Since $T(CP^1) = 1, Td(y_i) = 0$. It is easy to see that $\operatorname{Ker} Td = (y_1, \dots, y_i, \dots,)$. Set $\Sigma = \{y_1, \dots, y_i, \dots,\}$.

4.14. Corollary. $MU^\Sigma \simeq k$.

Proof. By VIII.2.17, MU^Σ is a commutative ring spectrum. It is clear that the forgetful morphism $r : MU \to MU^\Sigma$ gives a $\mathbb{C}$-orientation of MU^Σ. Finally, $\pi_*(MU^\Sigma) = \mathbb{Z}[t]$, and r induces Td on the coefficient rings. □

4.15. Remark. For an arbitrary system $\{x_i\}$ the spectrum $MU^{\{x_i | i>1\}}$ can be different from k. For example, it is not the spectrum k if $x_{p-1} = [CP^{p-1}] - [CP^1]^{p-1}$ for some prime p, cf. Rudyak [2].

Given a prime p, set $\ell := BP^{\{v_i | i>1\}}$, $\ell^*(\mathrm{pt}) = \mathbb{Z}[p][v_1]$.

4.16. Proposition (cf. Adams [6]). $k[p] \simeq \vee_{i=0}^{p-1} \Sigma^{2i} \ell$.

Proof. Let $\varkappa : BP \to MU[p]$ be the inclusion of the direct summand. Define a homomorphism $\sigma : BP^*(\mathrm{pt}) \to \mathbb{Z}[p][v, v^{-1}]$, $\dim v = 2 - 2p$ by setting

$$\sigma(v_i) := T(\varkappa_* v_i) v^{\frac{p^i - 1}{p-1}},$$

where $T : MU[p]^*(\mathrm{pt}) \to \mathbb{Z}[p]$ is the Todd genus. Consider the ring homomorphism

$$a : \mathbb{Z}[p][v, v^{-1}] \to \mathbb{Z}[p][s, s^{-1}], \quad a(v) = s^{p-1}, \dim s = -2.$$

Define

$$h : BP^*(X) \otimes_\sigma \mathbb{Z}[p][v, v^{-1}] \to MU[p]^*(X) \otimes_{Td} \mathbb{Z}[p][s, s^{-1}],$$
$$h(x \otimes y) = \varkappa(x) \otimes a(y).$$

Set $L^*(X) := \ell^*(X) \otimes_{\ell^*(\mathrm{pt})} \ell^*(\mathrm{pt})[v_1^{-1}]$. The forgetful morphism

$$BP^*(-) \to \ell^*(-)$$

induces a morphism

$$\varphi : BP^*(-) \otimes_\sigma \mathbb{Z}[p][v, v^{-1}] \to L^*(-)$$

of cohomology theories on $\mathscr{S}_{\mathrm{f}}$. Here $BP^*(-) \otimes_\sigma \mathbb{Z}[p][v, v^{-1}]$ is a cohomology theory by 4.11, while $L^*(-)$ is a cohomology theory by 4.12. Since

$$\varphi^S : BP^*(S) \otimes_\sigma \mathbb{Z}[p][v, v^{-1}] \to L^*(S)$$

is an isomorphism, we conclude that φ is an isomorphism of cohomology theories on $\mathscr{S}_{\mathrm{f}}$. Thus, h can be treated a morphism $h : L^*(-) \to K[p]^*(-)$ of cohomology theories on $\mathscr{S}_{\mathrm{f}}$. By III.3.20(iii), h is induced by a morphism $h : L \to K[p]$ of spectra. We define a morphism

$$h_n : \Sigma^{2n} L \xrightarrow{\Sigma^{2n} h} \Sigma^{2n} K[p] \xrightarrow{s^n} K[p]$$

and consider a morphism

$$f : \bigvee_{i=0}^{p-1} \Sigma^{2i} L \to K[p]$$

which gives us the element $h_0 \oplus \cdots \oplus h_{p-1} \in K[p]^0(\bigvee_{i=0}^{p-1} \Sigma^{2i} L)$. Clearly, f is an equivalence. Passing to connective coverings, we obtain an equivalence $\bigvee_{i=0}^{p-1} \Sigma^{2i} \ell \to k[p]$. □

4.17. Corollary. *$BU[p] \simeq X_1 \times \cdots \times X_{p-1}$, where $\pi_k(X_i) = \mathbb{Z}[p]$ for $k \equiv 2i \bmod (p-1)$ and $\pi_k(X_i) = 0$ otherwise. Furthermore, if p is an odd prime then* $BSO[p] \simeq \prod_{i=1}^{\frac{p-1}{2}} X_{2i}$

Proof. The splitting in 4.16 gives us a splitting

$$\Omega^\infty k[p] \simeq \prod_{i=0}^{p-1} \Omega^\infty(\Sigma^{2i} l).$$

Now put $X_i := \Omega^\infty(\Sigma^{2i}l)$. To prove the last assertion, consider the realification $R: BU \to BO$. By IV.4.27(iii), $R_*: \pi_{4i}(BU[p]) \to \pi_{4i}(BO[p])$ is an isomorphism, and the result follows. □

4.18. Remarks. Corollary 4.4 (together with 4.2) was published by Landweber [5] for $m = 0$, cf. also Rudyak [2]. Yagita [1] and Yosimura [1] proved it for $m > 0$. Corollary 4.7 was proved by Rudyak [4].

§5. Commutative Ring Spectra of Characteristic 2

The main result of this section is Theorem 5.5 below. Here we assume that $p = 2$, the spectra $P(n)$ are considered under this assumption, H denotes $H\mathbb{Z}/2$, and $\mathscr{A}$ denotes $\mathscr{A}_2$. We also denote $r_\infty^n : P(n) \to H$ just by r. In other words, r is the canonical projection in the Postnikov tower of $P(n)$, i.e., $r = \tau_0$ in the notation of II.4.12.

5.1. Lemma. *Let F be a commutative ring spectrum with $2\pi_0(F) = 0$. If $\pi_i(F) = 0$ for $i > 2^{n+1} - 2$, and if $F_{(2^{n+1}-3)}$ is a graded Eilenberg–Mac Lane spectrum, then there exists a ring morphism $\varphi: P(n) \to F$.*

Furthermore, if $\varphi_*(v_n) = 0 \in \pi_*(F)$ then F is a graded Eilenberg–Mac Lane spectrum.

Proof. Consider the AHSS for $F_*(P(n))$. By 2.15, we have a ring isomorphism

$$\begin{aligned} E^2_{**} &= H_*(P(n); \pi_*(F)) \cong H_*(P(n)) \otimes \pi_*(F) \\ &\cong \pi_*(F) \otimes \mathbb{Z}/2\,[\zeta_0, \dots, \zeta_{n-1}, \zeta_n^2, \dots, \zeta_{n+k}^2, \dots]. \end{aligned}$$

By 2.16, this spectral sequence is trivial. Thus, E^∞_{**} is a free $\pi_*(F)$-module with the basis $\{\zeta_0^{i_0} \cdots \zeta_{n-1}^{i_{n-1}} \zeta_n^{2j_0} \cdots \zeta_{n+k}^{2j_k}\}$, $k \geq 0$.

Consider the homomorphism

$$\pi_*(F)[x_1, \dots, x_n, \dots] = F_*(BP) \xrightarrow{(\rho_n)_*} F_*(P(n))$$

and set $\alpha_i = (\rho_n)_*(x_i)$ for $i \geq n$. Clearly, α_i corresponds to $\zeta_i^2 \in E^\infty_{**}$, $i \geq n$.

Since $\overline{F} := F_{(2^{n+1}-3)}$ is a graded Eilenberg–Mac Lane spectrum,

$$\overline{F}_*(P(n)) = \pi_*(\overline{F})[y_0, \dots, y_{n-1}, y_n^2, \dots], \dim y_i = 2^i - 1,$$

see 2.15. Let $\tau = \tau_{2^{n+1}-3}: F \to \overline{F}$ be the projection onto the Postnikov stage as in II.4.12. Then the induced homomorphism $\tau_*: F_*(P(n)) \to \overline{F}_*(P(n))$ is an isomorphism up to dimension $2^{n+1} - 4$. Set $\beta_i = \tau_*^{-1}(y_i), i = 0, \dots, n-1$. Clearly, $\beta_i\beta_j = \beta_j\beta_i$ because $y_iy_j = y_jy_i$. Furthermore, β_i corresponds to $\zeta_i \in E^\infty_{**}, 0 \leq i < n$.

Since $\alpha_i\alpha_j = \alpha_j\alpha_i$ and $\beta_i\beta_j = \beta_j\beta_i$, $F_*(P(n))$ is a free $\pi_*(F)$-module with the basis $\{\beta^I\alpha^J := \beta_0^{i_0}\cdots\beta_{n-1}^{i_{n-1}}\alpha_n^{j_0}\cdots\alpha_{n+k}^{j_k}\cdots\}$, where each i_s is a non-negative integer and $j_0,\ldots,j_k,\ldots$ is a sequence of non-negative integers such that $j_k=0$ for all but finitely many k. Consider the $\pi_*(F)$-submodule M of $F_*(P(n))$ generated by $\beta^I\alpha^J$ with $\beta^I\alpha^J\neq 1$. We prove that M is an ideal in the ring $F_*(P(n))$. Since $\pi_*(F)$ commutes with all elements of $F_*(P(n))$, it suffices to prove that $\beta^I\alpha^J\beta^{I'}\alpha^{J'}$ belongs to M provided $\beta^I\alpha^J\beta^{I'}\alpha^{J'}\neq 1$. If $J=0$, this is valid because $\beta_i\beta_j=\beta_j\beta_i$. If $J\neq 0$, then $\dim\alpha^J > 2^{n+1}-1$, and so $\beta^I\alpha^J\beta^{I'}\alpha^{J'}$ belongs to M for dimensional reasons (i.e., because $\pi_i(F)=0$ for $i>2^{n+1}-2$).

Now we can construct a ring morphism $\varphi : P(n)\to F$ using II.3.46. (By 2.16, we are able to apply II.3.46). Namely, set $\operatorname{ev}(\varphi)(M)=0$, $\operatorname{ev}(\varphi)(x)=x$ for every $x\in\pi_*(F)$, where

$$\operatorname{ev}: F^*(P(n))\to \operatorname{Hom}_{\pi_*(F)}(F_*(P(n)),\pi_*(F))$$

is the evaluation.

Now, suppose that $\varphi_*(v_n)=0$. Then, by II.2.15, the morphism

$$\Sigma^{2p^n-2}P(n)\xrightarrow{v_n}P(n)\xrightarrow{\varphi}F$$

is inessential. Hence, there is a morphism $\psi : P(n+1)\to F$ such that $\varphi\simeq\psi r_n$. Consider a cofiber sequence

$$X\to P(n+1)\xrightarrow{r_\infty^{n+1}}H.$$

Since $\pi_i(X)=0$ for $i<2n+1-2$, we conclude, by II.4.1(iv), that $[X,F]=0$. So, there is a morphism $f: H\to F$ with $fr_\infty^{n+1}=\psi$, and thus, by II.7.7, F is a graded Eilenberg–Mac Lane spectrum. □

5.2. Lemma. *Let X be a finite CW-space, and let F be a connected commutative ring spectrum. Suppose that the AHSS*

(5.3) $$E_r^{**}(X)\Longrightarrow F^*(X),\ E_2^{**}(X)=H^*(X;F^0(\mathrm{pt}))$$

is trivial. Let $\tau=\tau_0: F\to H(F^0(\mathrm{pt}))$ be as in II.4.12, *let $z\in F^*(X)$ be such that $\tau_*(z)\neq 0$, and let $a\in F^0(\mathrm{pt})$ be such that*

$$a\otimes\tau_*(z)\neq 0\in E_2^{**}(X).$$

Then $az\neq 0$.

Proof. Recall that there is a canonical ring homomorphism $h: F^k(X)\to \oplus E_\infty^{i,k-i}(X)$. Since $a\otimes\tau(z)\neq 0\in E_2^{**}=E_\infty^{**}$, we conclude that

$$h(az)=h(a)h(z)=a\otimes\tau(z)\neq 0,$$

and thus $az\neq 0$. □

5.4. Lemma. *Let F be a connected spectrum with $2\pi_0(F) = 0$. If $\pi_i(F) = 0$ for $i > 2k - 2$, and if $F_{(2k-3)}$ is a graded Eilenberg–Mac Lane spectrum, then the AHSS* (5.3) *for $X = RP^k \times RP^k$ is trivial.*

Proof. Since $F_{(2k-3)}$ is a graded Eilenberg–Mac Lane spectrum, only the triviality of $d_{2k-1}^{0,0}$ and $d_{2k-1}^{0,1}$ needs to be proved. Firstly, consider

$$d_{2k-1}^{0,1} : H^1(RP^k \times RP^k; F^0(\mathrm{pt})) \to H^{2k}(RP^k \times RP^k; F^{2-2k}(\mathrm{pt})).$$

Note that

$$d_{2k-1}^{0,1} : H^1(-; F^0(\mathrm{pt})) \to H^{2k}(-; F^{2-2k}(\mathrm{pt}))$$

is a natural transformation, and so it can be treated as a cohomology operation. Furthermore, $F^0(\mathrm{pt})$ is a $\mathbb{Z}/2$-vector space, $F^0(\mathrm{pt}) = \oplus \mathbb{Z}/2$, and so $H^1(Y; F^0(\mathrm{pt})) = \oplus H^1(Y)$ for every finite CW-space Y, see II.1.16(ii). Hence, we have a natural in Y splitting

$$\begin{aligned}\mathrm{Hom}(H^1(Y; F^0(\mathrm{pt})), H^{2k}(Y; F^i(\mathrm{pt}))) &= \mathrm{Hom}(\oplus H^1(Y), H^{2k}(Y; F^i(\mathrm{pt}))) \\ &= \prod \mathrm{Hom}(H^1(Y), H^{2k}(Y; F^i(\mathrm{pt}))).\end{aligned}$$

So, it suffices to prove that

$$\theta : H^1(RP^k \times RP^k) \to H^{2k}(RP^k \times RP^k; F^{2-2k}(\mathrm{pt}))$$

is zero for every cohomology operation $\theta : H^1(-) \to H^{2k}(-; F^{2-2k}(\mathrm{pt}))$. Let $a \in H^1(RP^k) = \mathbb{Z}/2$ be the generator. We set $a_i := p_i^* a$ where

$$p_i : RP^k \times RP^k \to RP^k$$

is the projection on the i-th factor, $i = 1, 2$. Since the group

$$H^1(RP^k \times RP^k) = \mathbb{Z}/2 \oplus \mathbb{Z}/2$$

is generated by a_1 and a_2, it suffices to prove that $\theta : H^1(RP^k) \to H^{2k}(RP^k; F^{2-2k}(\mathrm{pt}))$ is zero. But $H^{2k}(RP^k; F^{2-2k}(\mathrm{pt})) = 0$. Hence, $\theta = 0$, and thus $d_{2k-1}^{0,1} = 0$

Finally, $d_{2k-1}^{0,0} = 0$ because $\theta(y) = 0$ for every $\theta \in \mathscr{A}$, $\dim \theta > 0$, and every $y \in H^0(Y)$. □

5.5. Theorem. *Let E be a commutative ring spectrum with $2\pi_*(E) = 0$. Then E is a graded Eilenberg–Mac Lane spectrum. Furthermore, there is a ring equivalence $E \simeq H(\pi_*(E))$.*

Proof. Firstly, we assume that E is connected. By II.7.3(iii), it suffices to prove that every coskeleton $E_{(i)}$ of E is a graded Eilenberg–Mac Lane spectrum. We prove this by induction. Clearly, this holds for $i = 0$. Suppose that $E_{(i)}$ is a graded Eilenberg–Mac Lane spectrum and consider n such that $2^n - 2 \le i < 2^{n+1} - 2$. We set $F = E_{(i+1)}$ and prove that F is a graded

Eilenberg–Mac Lane spectrum. Consider a ring morphism $\varphi : P(n) \to F$ as in 5.1. By 5.1, it suffices to prove that $\varphi_*(v_n) = 0 \in \pi_*(F)$.

If $i \le 2^{n+1} - 3$ then $\pi_{2^{n+1}-2}(F) = 0$, and hence $\varphi_*(v_n) = 0$. So, assume that $i = 2^{n+1} - 3$. Let $a \in H^1(RP^{2^n}) = \mathbb{Z}/2$ be the generator. In the AHSS

$$E_r^{**} \Longrightarrow P(n)^*(RP^{2^n}), E_2^{**} = H^*(RP^{2^n}) \otimes P(n)^*(\mathrm{pt})$$

the element $a \otimes 1 \in E_2^{1,0}$ survives for dimensional reasons. Hence, there exists $t \in P(n)^1(RP^{2^n})$ with $r_*(t) = a$.

By 2.19, $r_*(\beta_{n-1}(t)) = Q_{n-1}r_*(t) = Q_{n-1}(a) = a^{2^n}$.

Let $p_k : RP^{2^n} \times RP^{2^n} \to RP^{2^n}$ be the projection on the k-th factor. We set $t_k := p_k^* t, a_k := p_k^* a, k = 1, 2$.

Let $\tau : F \to H(F^0(\mathrm{pt}))$ be as in 5.2. By II.7.1, there is a morphism $f : H \to H(F^0(\mathrm{pt}))$ such that $f_* : \mathbb{Z}/2 = \pi_0(H) \to \pi_0(H(F^0(\mathrm{pt}))) = F^0(\mathrm{pt})$ is the unit of the $\mathbb{Z}/2$-algebra $F^0(\mathrm{pt})$. We have the homotopy commutative diagram

$$\begin{array}{ccc} P(n) & \xrightarrow{\varphi} & F \\ {\scriptstyle r}\downarrow & & \downarrow{\scriptstyle \tau} \\ H & \xrightarrow{f} & H(F^0(\mathrm{pt})). \end{array}$$

Since f is the inclusion of a direct summand, we conclude that

$$f_* : H^*(RP^{2^n} \times RP^{2^n}) \to H^*(RP^{2^n} \times RP^{2^n}; F^0(\mathrm{pt}))$$

is monic. Hence, $f_*(a_1^{2^n} a_2^{2^n}) \ne 0$, and so

$$\tau_*\varphi_*(\beta_{n-1}(t_1)\beta_{n-1}(t_2)) = f_*(r_*(\beta_{n-1}(t_1)\beta_{n-1}(t_2))) = f_*(a_1^{2^n} a_2^{2^n}) \ne 0.$$

Now, if $\varphi_*(v_n) \ne 0$ then $\varphi_*(v_n) \otimes \tau_*\varphi_*(\beta_{n-1}(t_1)\beta_{n-1}(t_2)) \ne 0 \in E_2^{**}$, and so, by 5.2,

$$\varphi_*(v_n)\varphi_*(\beta_{n-1}(t_1)\beta_{n-1}(t_2)) \ne 0.$$

On the other hand, $\varphi_*(t_1t_2 + t_2t_1) = \varphi_*(t_1)\varphi_*(t_2) + \varphi_*(t_2)\varphi_*(t_1) = 0$ because F is commutative. By 2.7,

$$t_1t_2 + t_2t_1 = v_n\beta_{n-1}(t_1)\beta_{n-1}(t_2).$$

Hence,

$$0 = \varphi_*(t_1t_2 + t_2t_1) = \varphi_*(v_n)\varphi_*(\beta_{n-1}(t_1)\beta_{n-1}(t_2)).$$

This is a contradiction. The induction is confirmed. Thus, E is a graded Eilenberg–Mac Lane spectrum.

If E is not connected, consider a connective covering $p : \widetilde{E} \to E$. Then $\widetilde{E}$ is a graded Eilenberg–Mac Lane spectrum by the above and II.4.28(i,iv). Thus, by II.7.8, E is a graded Eilenberg–Mac Lane spectrum.

Finally, a ring equivalence $E \simeq H(\pi_*(E))$ follows from II.7.30. □

5.6. Remarks. (a) Theorem 5.5 was proved by Pazhitnov–Rudyak [2], Würgler [5], and Hopkins–Mahowald (unpublished).

(b) It follows from 1.3 that $M\mathcal{U} \wedge M(\mathbb{Z}/2)$, as well as $K \wedge M(\mathbb{Z}/2)$, is not a graded Eilenberg–Mac Lane spectrum, cf. also Stong [4]. Thus, because of 5.5, none of these spectra admits a commutative multiplication.

(c) See Remark VIII.3.12(b).

Now we give some applications of 5.5.

5.7. Corollary (cf. IV.6.2). *The spectrum $M\mathcal{V}$ is a graded Eilenberg–Mac Lane spectrum, $M\mathcal{V} \simeq \vee \Sigma^d H\mathbb{Z}/2$.* □

Let $f : S^9 \to BO$ be a generator of $\pi_9(BO) = \mathbb{Z}/2$, and let $g : S^3 \to \Omega^6 BO$ be the adjoint map. We regard the map

$$h : \Omega^2 S^3 \xrightarrow{\Omega^2 g} \Omega^8 BO \simeq BO \times \mathbb{Z} \xrightarrow{\text{proj}} BO$$

as a stable vector bundle ξ.

5.8. Theorem (Mahowald [1]). *$T\xi \simeq H\mathbb{Z}/2$. In particular, $H\mathbb{Z}/2$ is a Thom spectrum.*

Proof. Since h is a double loop map, $T\xi$ is a commutative ring spectrum. Since

$$h_* : \pi_1(\Omega^2 S^3) \to \pi_1(BO)$$

is epic, ξ is non-orientable, and so $\pi_0(T\xi) = \mathbb{Z}/2$. Hence, $T\xi$ is a spectrum of characteristic 2. Hence, by 5.5, $T\xi \simeq H\mathbb{Z}/2 \vee E$ for some E, and hence

$$\dim H^k(T\xi) \geq \dim \mathscr{A}_k,$$

where dim is the dimension of $\mathbb{Z}/2$-vector spaces and $\mathscr{A}_k = H^k(H)$. The theorem will be proved once we have proved that the inequality above is an equality.

Using Kudo–Araki–Dyer–Lashof operations, one can prove the ring isomorphism

$$H_*(\Omega^2 S^3) \cong \mathbb{Z}/2\,[x_i|, \dim x_i = 2^i - 1],$$

see e.g. Cohen–Lada–May [1]. Hence, $\dim H_k(\Omega^2 S^3) = \dim \mathscr{A}_k^* = \dim \mathscr{A}_k$, see II.6.25. Now,

$$\dim H^k(T\xi) = \dim H_k(T\xi) = \dim H_k(\Omega^2 S^3) = \dim \mathscr{A}_k. \qquad \square$$

5.9. Example (cf. Bullett [1]). Let Σ_n be the symmetric group of degree n, and let $(k_1, \dots, k_n)$ be a typical permutation. We define the monomorphisms

$$i_n : \Sigma_n \to \Sigma_{n+1}, i_n(k_1, \dots, k_n) = (k_1, \dots, k_n, n+1)$$

and let Σ be the direct limit of the sequence

$$\{\cdots \to \Sigma_n \xrightarrow{i_n} \Sigma_{n+1} \to \cdots\}.$$

In other words, Σ is the group of "infinite permutations" $(k_1, \dots, k_n, \dots)$ where $k_n = n$ for all but a finite number of n's. We define a homomorphism $m : \Sigma \times \Sigma \to \Sigma$ by setting

$$m((j_1, \dots, j_n, \dots), (k_1, \dots, k_n, \dots)) = (j_1, k_1, \dots, j_n, k_n, \dots).$$

Then m induces a map

$$\mu : B\Sigma \times B\Sigma = B(\Sigma \times \Sigma) \xrightarrow{Bm} B\Sigma,$$

and one can prove that μ is homotopy commutative and associative.[18]

Finally, we define homomorphisms $t_n : \Sigma_n \to \mathcal{O}_n$ by setting $t_n(\sigma)(\mathbf{e}_i) = \mathbf{e}_{\sigma(i)}$ where $\sigma \in \Sigma_n$ and $\{\mathbf{e}_1, \dots, \mathbf{e}_n\}$ is the standard basis of $\mathbb{R}^n$.

Clearly, $B\Sigma$ is (homotopy equivalent to) the telescope of the sequence

$$\{\cdots \to B\Sigma_n \xrightarrow{Bi_n} B\Sigma_{n+1} \to \cdots\}.$$

Hence, the family $\{Bt_n : B\Sigma_n \to B\mathcal{O}_n\}$ yields a map

$$\varphi := \varinjlim\{Bt_n\} : B\Sigma \to B\mathcal{O}.$$

Moreover, one can prove that the diagram

$$\begin{array}{ccc} B\Sigma \times B\Sigma & \xrightarrow{\varphi \times \varphi} & B\mathcal{O} \times B\mathcal{O} \\ {\scriptstyle\mu}\downarrow & & \downarrow{\scriptstyle\mu^{\mathcal{O}}} \\ B\Sigma & \xrightarrow{\varphi} & B\mathcal{O} \end{array}$$

commutes up to homotopy.

The map $\varphi : B\Sigma \to B\mathcal{O}$ yields a Thom spectrum $M\Sigma := T(B\Sigma, \varphi)$.

5.10. Theorem. *$M\Sigma$ is a commutative ring spectrum of characteristic 2, and so $M\Sigma \simeq H(\pi_*(M\Sigma)) \simeq \vee\Sigma^d H$.*

Proof. Following IV.5.22, one can prove $M\Sigma$ is a commutative ring spectrum (the unit is given by the root). Furthermore, the stable vector bundle $\varphi : B\Sigma \to B\mathcal{O}$ is not orientable since $\varphi_* : \pi_1(B\Sigma) \to \pi_1(B\mathcal{O})$ is an epimorphism. Hence, by IV.5.23(i), $\pi_0(M\Sigma) = \mathbb{Z}/2$, and thus $M\Sigma$ is a spectrum of characteristic 2. □

[18]Notice that $B\Sigma$ is *not* an H-space (there is no unit) since $\pi_1(B\Sigma)$ is not an abelian group.

5.11. Example. Let β_n be the Artin braid group, see e.g. Birman [1]. The presentation of β_n is

$$\langle \sigma_1, \dots, \sigma_{n-1} \mid \sigma_i\sigma_j = \sigma_j\sigma_i \text{ if } |i-j| > 1 \text{ and } \sigma_i\sigma_{i+1}\sigma_i = \sigma_{i+1}\sigma_i\sigma_{i+1}\rangle.$$

We define the monomorphism $i_n : \beta_n \to \beta_{n+1}$, $i(\sigma_k) = \sigma_k$, and let β be the direct limit of the sequence

$$\{\cdots \to \beta_n \xrightarrow{i_n} \beta_{n+1} \to \cdots\}.$$

Furthermore, we define a homomorphism $m : \beta \times \beta \to \beta$ by setting

$$m((\sigma_1, \dots, \sigma_k, \dots), (\sigma_1', \dots, \sigma_k', \dots)) = (\sigma_1, \sigma_1', \dots, \sigma_k, \sigma_k', \dots).$$

Finally, we define a homomorphism $u_n : \beta_n \to \Sigma_n$ where the permutation $u_n(\sigma_i)$ interchanges i and $i+1$ and does not move other symbols. Then the family $\{u_n\}$ yields a homomorphism $u : \beta \to \Sigma$ and hence the map $Bu : B\beta \to B\Sigma$. So, we get a stable vector bundle

$$B\beta \xrightarrow{Bu} B\Sigma \xrightarrow{B\varphi} BO$$

with φ as in 5.9, and we set $M\beta := T(B\beta, \varphi \circ Bu)$.

5.12. Theorem (F. Cohen [1]). *$M\beta \simeq H$. In particular, H is a Thom spectrum.*

Proof. As in 5.10, we can prove that $M\beta \simeq \Sigma^d H$, i.e., $M\beta \simeq H \vee E$. So, as in 5.8, it suffices to prove the existence of an additive isomorphism $H^*(M\beta) \cong \mathscr{A}$. This has been done in Fuks [1]. $\square$

5.13. Remark. Comparing 5.8 and 5.12, we see that it makes sense to compare the spaces $\Omega^2 S^3$ and $B\beta$. In fact, there is a map $f : B\beta \to \Omega^2 S^3$ which induces an isomorphism in integral homology, see F. Cohen [1]. So, f is a Quillenization (a plus-construction in terms of Adams [9]). Hence, Tf induces an isomorphism in $\mathbb{Z}/2$-homology, and one can prove that Tf is an equivalence.

§6. The Spectra $BP\langle n\rangle$ and Homological Dimension

Let $\Omega\langle n\rangle$ denote the subring $\mathbb{Z}[p][v_1, \dots, v_n]$ of $\Omega = \mathbb{Z}[p][v_1, \dots, v_n, \dots]$. Clearly, the inclusion $\Omega\langle n\rangle \subset \Omega$ turns Ω into an $\Omega\langle n\rangle$-module. On the other hand, there is the quotient ring homomorphism

$$\Omega \to \Omega\langle n\rangle, \quad v_k \mapsto \begin{cases} v_k & \text{if } k \le n, \\ 0 & \text{if } k > n. \end{cases}$$

This homomorphism turns $\Omega\langle n\rangle$ into an Ω-module.

Given an $\Omega\langle n\rangle$-module M, consider the homomorphism

$$l : M \to M[v_k^{-1}] := M \otimes_{\Omega\langle n\rangle} \Omega\langle n\rangle[v_k^{-1}], a \mapsto a \otimes 1.$$

6.1. Lemma. *The $\Omega\langle n\rangle$-module $\Omega\langle n\rangle[v_k^{-1}]$ is flat. In particular, the functor $M \mapsto M[v_k^{-1}]$ is exact on the category of $\Omega\langle n\rangle$-modules.*

Proof. This follows from 4.12. □

6.2. Definition. We say that an $\Omega\langle n\rangle$-module M is v_k *torsion free* if the homomorphism $v_k : M \to M$ (multiplication by v_k) is monic. Clearly, it holds iff the homomorphism $l : M \to M[v_k^{-1}]$ is monic.

6.3. Lemma. *Let R be a commutative ring such that every projective R-module is free. Let A be a graded commutative connected R-algebra. Then every graded projective A-module is free provided that it is bounded below. In particular, every graded projective $\Omega\langle n\rangle$-module is free provided that it is bounded below.*

Proof. Let $\varepsilon : A \to R$ be the augmentation. As in II.6.5, set $GM = M/\bar{A}M = M \otimes_\varepsilon R$.

Consider a projective A-module P. Then there exists P' such that $P \oplus P'$ is a free A-module. It is easy to see that $G(P \oplus P') = GP \oplus GP'$ is a free R-module. Hence, GP is a projective R-module, and hence it is a free R-module by our assumption.

Set $F := A \otimes_R GP$ and equip F with an A-module structure by setting $a(x \otimes y) := (ax) \otimes y, a, x \in A, y \in GP$. Let $\{e_i | i \in I\}$ be a free R-basis of GP. Then $\{f_i := 1 \otimes e_i\}$ is a free A-basis of F. Choose elements $p_i \in P$ such that the canonical epimorphism $P \to GP$ maps p_i to e_i.

Now suppose that P is bounded below. Define a homomorphism of A-modules $\varphi : F \to P$ by setting $\varphi(f_i) = p_i$. It is clear that $G\varphi : GF \to GP$ is an epimorphism, and so, by II.6.6(i), $\varphi : F \to P$ is an epimorphism. Since P is a projective A-module, φ splits by some monomorphism $\psi : P \to F$, $\varphi\psi = 1_P$. Since $G\psi$ is an isomorphism, ψ is an epimorphism. Hence, ψ is an isomorphism. Thus, P is a free A-module. □

6.4. Definition. Let M be a module over a commutative ring R. We say that M has *homological dimension* $\leq n$ if it admits a projective resolution of length n, i.e., if there is an exact sequence of R-modules

$$0 \to P_n \to \cdots \to P_1 \to P_0 \to M \to 0$$

where every P_i is projective. In this case we write $\text{hom. dim}_R M \leq n$. We say that M has homological dimension n, and write $\text{hom. dim}_R M = n$, if $\text{hom. dim}_R M \leq n$ while it is false that $\text{hom. dim}_R M \leq n-1$.

6.5. Proposition. *Let R be a commutative ring.*

(i) hom. $\dim_R M \le n$ *iff* $\operatorname{Ext}_R^{n+1}(B,M)=0$ *for every R-module B.*

(ii) *Let $0 \to N \to P \to M \to 0$ be a short exact sequence of R-modules such that P is projective. Then*

$$\operatorname{Ext}_R^{i+1}(M,B) \cong \operatorname{Ext}_R^{i}(N,B) \text{ and } \operatorname{Tor}^R_{i+1}(M,B) \cong \operatorname{Tor}^R_{i}(N,B)$$

for every $i > 0$ and every R-module B.

(iii) *Let $0 \to N \to P \to M \to 0$ be a short exact sequence of R-modules such that P is projective. If* hom. $\dim_R M > 0$ *then* hom. $\dim_R M =$ hom. $\dim_R N + 1$.

Proof. (i) See Mac Lane [1], Theorem VII.1.1.

(ii) See Mac Lane [1], Theorems III.3.1 and V.8.4.

(iii) This follows from (i) and (ii). □

6.6. Proposition. *Let F be a free Ω-module. Let N be a finitely generated submodule of F. Then* hom. $\dim_\Omega N < \infty$.

Proof. Let $S = \{s_i\}$ be a free Ω-basis of F. Define $F\langle n\rangle$ to be the free $\Omega\langle n\rangle$-module generated by S. We regard $F\langle n\rangle$ as the subgroup of F consisting of linear combinations $\sum a_i s_i$, $a_i \in \Omega\langle n\rangle \subset \Omega$. Define a homomorphism of $\Omega\langle n\rangle$-modules

$$\psi_n : F\langle n\rangle \otimes_{\Omega\langle n\rangle} \Omega \to F, \quad \psi\left(\left(\sum a_i s_i\right) \otimes b\right) := \sum b a_i s_i,\ b \in \Omega.$$

It is clear that ψ_n is an isomorphism.

Let T be a finite set of generators of N. Choose n so large that $T \subset F\langle n\rangle$. Set $N\langle n\rangle := N \cap F\langle n\rangle$, and let $i_n : N\langle n\rangle \to F\langle n\rangle$ be the inclusion. Since Ω is a free $\Omega\langle n\rangle$-module, the homomorphism $i_n' := i_n \otimes 1 : N\langle n\rangle \otimes_{\Omega\langle n\rangle} \Omega \to F\langle n\rangle \otimes_{\Omega\langle n\rangle} \Omega$ is monic. Let φ_n be the unique homomorphism such that the following diagram commutes:

$$\begin{array}{ccc} N\langle n\rangle \otimes_{\Omega\langle n\rangle} \Omega & \xrightarrow{\varphi_n} & N \\ {\scriptstyle i_n'}\downarrow & & \downarrow \\ F\langle n\rangle \otimes_{\Omega\langle n\rangle} \Omega & \xrightarrow{\psi_n} & F. \end{array}$$

Then φ_n becomes a homomorphism of Ω-modules if we equip $N\langle n\rangle \otimes_{\Omega\langle n\rangle} \Omega$ with the following Ω-module structure:

$$a(x \otimes b) := x \otimes (ab), \quad a \in \Omega, x \in N, b \in \Omega\langle n\rangle \subset \Omega,.$$

Now, φ_n is epic since $T \subset F\langle n\rangle$, and φ_n is monic since $\psi_n i_n'$ is monic. Hence, φ_n is an isomorphism.

By the Hilbert Syzygy Theorem (see Mac Lane [1], VII.6.4), there exists an exact sequence of $\Omega\langle n\rangle$-modules

$$0 \to F_{n+1} \to F_n \to \cdots \to F_1 \to F_0 \to N\langle n\rangle \to 0$$

where every F_i is a free $\Omega\langle n\rangle$-module. Since Ω is a free $\Omega\langle n\rangle$-module, the sequence

$$0 \to F_{n+1} \otimes_{\Omega\langle n\rangle} \Omega \to F_n \otimes_{\Omega\langle n\rangle} \Omega \to \cdots \to F_1 \otimes_{\Omega\langle n\rangle} \Omega$$
$$\to F_0 \otimes_{\Omega\langle n\rangle} \Omega \to N \otimes_{\Omega\langle n\rangle} \Omega \to 0$$

is exact. Equipping each term with the Ω-module structure as above, we conclude that every Ω-module $F_i \otimes_{\Omega\langle n\rangle} \Omega$ is free, and hence projective. Since $N \cong N\langle n\rangle \otimes_{\Omega\langle n\rangle} \Omega$, we conclude that hom. $\dim_\Omega N \le n < \infty$. □

6.7. Corollary. *If M is a coherent Ω-module then* hom. $\dim_\Omega M < \infty$. *In particular,* hom. $\dim_\Omega BP_*(X) < \infty$ *for every finite spectrum X.*

Proof. Since M is finitely generated, there is an exact sequence

$$0 \to N \to F \to M \to 0$$

where F is a free finitely generated Ω-module. Note that, by VII.1.13(i), Ω is a coherent ring. Hence, by VII.1.13(ii), F is a coherent Ω-module. So, again by VII.1.13(ii), the Ω-module N is coherent, and hence finitely generated. Hence, by 6.6, hom. $\dim_\Omega N < \infty$, and thus, by 6.5(iii), hom. $\dim_\Omega M < \infty$. □

Consider the spectrum

$$BP\langle n\rangle := BP^{\{v_{n+1},\ldots,v_{n+m},\ldots\}}.$$

Because of VIII.1.24, $\pi_*(BP\langle n\rangle) \cong \Omega\langle n\rangle$. Notice that $BP\langle 0\rangle = H\mathbb{Z}[p]$ and $BP\langle -1\rangle = H\mathbb{Z}/p$. By VIII.1.6, for every $X \in \mathscr{C}$ there is an exact sequence

$$\text{(6.8)} \qquad \cdots \to BP\langle n\rangle_*(X) \xrightarrow{v_n} BP\langle n\rangle_*(X) \xrightarrow{r_n} BP\langle n-1\rangle_*(X) \xrightarrow{\delta_n} \cdots .$$

Furthermore, we have the obvious forgetful morphism $\rho_n : BP_*(X) \to BP\langle n\rangle_*(X)$ of homology theories.

The morphisms r_n of homology theories form a tower

$$\text{(6.9)} \qquad \cdots \to BP\langle n+1\rangle_*(-) \xrightarrow{r_{n+1}} BP\langle n\rangle_*(-) \xrightarrow{r_n} \cdots .$$

6.10. Theorem. *The tower* (6.9) *admits a quasi-ring structure, i.e., every homology theory $BP\langle n\rangle_*(X)$ admits a commutative and associative quasi-multiplication such that every morphism r_n is a quasi-ring morphism with respect to these quasi-multiplications. Moreover, all the morphisms ρ_n are quasi-ring morphisms.*

Proof. This follows from VIII.2.17. □

6.11. Remark. In view of III.4.22, we can and shall assume that the homomorphisms $r_n : BP\langle n\rangle_*(X) \to BP\langle n-1\rangle_*(X)$ and $\rho_n : BP_*(X) \to BP\langle n\rangle_*(X)$ are defined when X is a *spectrum.*

6.12. Lemma. *If $0 \le k \le n$, then the following holds for every spectrum X:*
(i) *The homomorphism $\rho_n[v_k^{-1}] : BP_*(X)[v_k^{-1}] \to BP\langle n\rangle_*(X)[v_k^{-1}]$ is epic;*
(ii) *The homomorphism $v_n : BP\langle n\rangle_*(X)[v_k^{-1}] \to BP\langle n\rangle_*(X)[v_k^{-1}]$ is monic.*

Proof. (i) For every spectrum X, consider the homomorphism

$$\tau : BP_*(X) \otimes_\Omega \Omega\langle n\rangle\,[v_k^{-1}] \to BP\langle n\rangle_*(X)[v_k^{-1}],\ \tau(a \otimes b) = \rho_n(a)b.$$

By 4.1, $BP_*(-)\otimes_\Omega\Omega\langle n\rangle\,[v_k^{-1}]$ is a homology theory. By 6.1, $BP\langle n\rangle_*(-)[v_k^{-1}]$ is a homology theory. Hence, τ is a morphism of homology theories. Since it is an isomorphism for $X = S$, it is for every spectrum X. But the homomorphism

$$BP_*(X)[v_k^{-1}] = BP_*(X) \otimes_\Omega \Omega[v_k^{-1}] \to BP_*(X) \otimes_\Omega \Omega\langle n\rangle[v_k^{-1}]$$

is evidently epic, and the result follows.
(ii) This is clear if $k = n$. So, assume that $k < n$. Because of the exactness of (6.8), it suffices to prove that

$$r_n[v_k^{-1}] : BP\langle n\rangle_*(X)[v_k^{-1}] \to\ BP\langle n-1\rangle_*(X)[v_k^{-1}]$$

is epic. But this follows from (i). □

6.13. Corollary. *Let X be a spectrum, and let $k < n$. If $BP\langle n\rangle_*(X)$ is v_k torsion free, then $BP\langle n\rangle_*(X)$ is v_{k+1} torsion free.*

Proof. Consider the commutative diagram

$$\begin{array}{ccc} BP\langle n\rangle_*(X) & \xrightarrow{v_{k+1}} & BP\langle n\rangle_*(X) \\ {\scriptstyle l}\big\downarrow & & \big\downarrow{\scriptstyle l} \\ BP\langle n\rangle_*(X)[v_k^{-1}] & \xrightarrow{v_{k+1}} & BP\langle n\rangle_*(X)[v_k^{-1}]. \end{array}$$

Since $BP\langle n\rangle_*(X)$ is v_k torsion free, l is monic. By 6.12(ii), the bottom arrow is monic. Hence, the top arrow is monic. □

6.14. Lemma. *Let X be a spectrum bounded below, and let $k \le n$ where $n \ge 0$. Suppose that $BP\langle n\rangle_*(X)$ is v_k torsion free. Then:*
(i) *$BP\langle n+1\rangle_*(X)$ is v_k torsion free.*
(ii) *$\rho_n : BP_*(X) \to BP\langle n\rangle_*(X)$ is epic.*

Proof. (i) Consider the following diagram with exact columns. Here the left column is (6.8), and the right column is exact by 6.1 and 6.12(ii).

$$\begin{array}{ccc}
BP\langle n\rangle_{i+2p^{n+1}-1}(X) & \xrightarrow{l_1} & 0 \\
\downarrow & & \downarrow \\
BP\langle n+1\rangle_i(X) & \xrightarrow{l_2} & BP\langle n+1\rangle_i(X)[v_k^{-1}] \\
{\scriptstyle v_{n+1}}\downarrow & & \downarrow{\scriptstyle v_{n+1}} \\
BP\langle n+1\rangle_{i+2p^{n+1}-2}(X) & \xrightarrow{l_3} & BP\langle n+1\rangle_{i+2p^{n+1}-2}(X)[v_k^{-1}] \\
\downarrow & & \downarrow \\
BP\langle n\rangle_{i+2p^{n+1}-2}(X) & \xrightarrow{l_4} & BP\langle n\rangle_{i+2p^{n+1}-2}(X)[v_k^{-1}].
\end{array}$$

We prove that l_2 and l_3 are monic by induction on i. Note that l_4 is monic because $BP\langle n\rangle_*(X)$ is v_k torsion free. Since X is bounded below, there exists m such that $BP\langle n+1\rangle_i(X) = 0$ for $i < m$. Hence, l_2 is monic for every $i < m$. Hence, by the Five Lemma I.2.1, l_3 is monic for every $i < m + 2p^{n+1} - 2$. Hence, l_2 is monic for every $i < m + 2p^{n+1} - 2$. Hence, l_3 is monic for every $i < m + 2(2p^{n+1} - 2)$, and so on.

(ii) By (i), $BP\langle n+1\rangle_*(X)$ is v_k torsion free. Hence, by 6.13, it is v_{n+1} torsion free. Therefore,

$$r_{n+1} = r_{n+1}^X : BP\langle n+1\rangle_*(X) \to BP\langle n\rangle_*(X)$$

is epic. Similarly, $r_{n+2} : BP\langle n+2\rangle_*(X) \to BP\langle n+1\rangle_*(X)$ is epic. And so on. Finally, X is bounded below, and so, by II.4.5(ii), for every i there exists $m = m(i)$ such that $\rho_m : BP_i(X) \to BP\langle m\rangle_i(X)$ is epic. □

6.15. Corollary. *For every $n \geq -1$ and every spectrum X bounded below, the homomorphism $\rho_n = \rho_n^X : BP_*(X) \to BP\langle n\rangle_*(X)$ is epic iff $BP\langle n+1\rangle_*(X)$ is v_{n+1} torsion free.*

Proof. If ρ_n is epic then $r_{n+1} : BP\langle n+1\rangle_*(X) \to BP\langle n\rangle_*(X)$ is epic. Hence, because of the exactness of (6.8), $BP\langle n+1\rangle_*(X)$ is v_{n+1} torsion free. Conversely, suppose that $BP\langle n+1\rangle_*(X)$ is v_{n+1} torsion free. Then, by 6.14(i), $BP\langle n+2\rangle_*(X)$ is v_{n+1} torsion free. Hence, by 6.13, $BP\langle n+2\rangle_*(X)$ is v_{n+2} torsion free. And so on. Now we can prove that ρ_n is epic just as in 6.14(ii). □

We define $\widetilde{\rho}_n : BP_*(X) \otimes_\Omega \Omega\langle n\rangle \to BP\langle n\rangle_*(X)$, $\widetilde{\rho}_n(a \otimes b) = \rho_n(a)b$.

6.16. Corollary. *Let W be a spectrum bounded below. Suppose that the group $H_i(W; \mathbb{Z}[p])$ is a free $\mathbb{Z}[p]$-module for every i. Then, for every $n \geq -1$,*

$$\rho_n : BP_*(W) \to BP\langle n\rangle_*(W)$$

is an epimorphism and $\widetilde{\rho}_n : BP_*(W) \otimes_\Omega \Omega\langle n\rangle \to BP\langle n\rangle_*(W)$ is an isomorphism.

Proof. Firstly, consider $n \geq 0$. By 6.14(i), $BP\langle n\rangle_*(W)$ is v_0 torsion free because $BP\langle 0\rangle_*(W) = H_*(W)$ is v_0 torsion free. Hence, by 6.14(ii), ρ_n is epic. So $\widetilde{\rho}_n$ is epic for every $n \geq -1$. By II.7.13,

$$\widetilde{\rho}_n \otimes 1 : BP_*(Y) \otimes_\Omega \Omega\langle n\rangle \otimes \mathbb{Q} \to BP\langle n\rangle_*(Y) \otimes \mathbb{Q}$$

is an isomorphism for every Y. So, we will have proved that

$$\widetilde{\rho}_n : BP_*(W) \otimes_\Omega \Omega\langle n\rangle \to BP\langle n\rangle_*(W)$$

is monic if we prove that $BP_*(W) \otimes_\Omega \Omega\langle n\rangle$ is torsion free. But this follows since, by VII.3.29(ii), $BP_*(W)$ is a free Ω-module. Thus, $\widetilde{\rho}_n$ is an isomorphism for every $n \geq 0$.

Finally, if $n = -1$ then

$$\begin{aligned}BP\langle -1\rangle_*(W) = H_*(W;\mathbb{Z}/p) &\cong H_*(W,\mathbb{Z}[p]) \otimes \mathbb{Z}/p = BP\langle 0\rangle_*(W) \otimes \mathbb{Z}/p\\ &\cong BP_*(W) \otimes_\Omega \mathbb{Z}/p,\end{aligned}$$

and it is clear that this isomorphism is given by $\widetilde{\rho}_{-1}$. □

6.17. Lemma. *Let M be a graded Ω-module bounded below. If $k > -1$ and* $\mathrm{Tor}_1^\Omega(M, \Omega\langle k\rangle) = 0$ *then* $Tor_1^\Omega(M, \Omega\langle k+1\rangle) = 0$.

Proof. Consider the exact sequence

$$0 \to \Omega\langle k+1\rangle \xrightarrow{v_{k+1}} \Omega\langle k+1\rangle \to \Omega\langle k\rangle \to 0.$$

It yields the exact sequence of graded Ω-modules

$$\cdots \to \mathrm{Tor}_1^\Omega(M, \Omega\langle k+1\rangle) \xrightarrow{v_{k+1}} \mathrm{Tor}_1^\Omega(M, \Omega\langle k+1\rangle) \to \mathrm{Tor}_1^\Omega(M, \Omega\langle k\rangle) = 0.$$

But v_{k+1} increases the degree while $\mathrm{Tor}_1^\Omega(M, \Omega\langle k+1\rangle)$ is bounded below. □

6.18. Lemma. *Let X be a spectrum bounded below, and let $f : W \to X$ be as in* VII.3.29(iii)*. Consider a cofiber sequence*

$$A \xrightarrow{g} W \xrightarrow{f} X.$$

Then $\mathrm{Tor}_{i+1}^\Omega(BP_*(X), K) = \mathrm{Tor}_i^\Omega(BP_*(A), K)$ *for every $i > 1$ and every Ω-module K. Furthermore,*

$$\mathrm{hom.\ dim}_\Omega\, BP_*(X) = \mathrm{hom.\ dim}_\Omega\, BP_*(A) + 1$$

provided hom. $\dim_\Omega BP_*(X) > 0$.

Proof. The cofiber sequence $A \xrightarrow{g} W \xrightarrow{f} X$ induces an exact sequence

$$0 \to BP_*(A) \xrightarrow{g_*} BP_*(W) \xrightarrow{f_*} BP_*(X) \to 0$$

where g_* and f_* are the induced homomorphisms and $BP_*(W)$ is a free Ω-module. Now the result follows from 6.5(ii,iii). □

6.19. Notation. Given a prime p, we denote by $\mathscr{P} = \mathscr{P}(p)$ the class of spectra X such that X is bounded below and every group $H_i(X; \mathbb{Z}[p])$ is a finitely generated $\mathbb{Z}[p]$-module. For example, $\mathscr{P}$ contains all $\mathbb{Z}[p]$-local spectra of finite $\mathbb{Z}[p]$-type and all spectra of finite $\mathbb{Z}$-type.

6.20. Remark. If $X \in \mathscr{P}$ then there exists a cofiber sequence

$$A \xrightarrow{g} W \xrightarrow{f} X$$

as in 6.18 with $A, W \in \mathscr{P}$. This follows immediately from VII.3.29(iii).

6.21. Theorem (Johnson–Wilson [1]). *Fix any $n \geq -1$. For every spectrum $X \in \mathscr{P}$ the following conditions are equivalent:*

(i) hom. $\dim_\Omega BP_*(X) \leq n+1$;

(ii) *The homomorphism $\rho_n : BP_*(X) \to BP\langle n\rangle_*(X)$ is epic;*

(iii) *The homomorphism $\widetilde{\rho}_n : BP_*(X) \otimes_\Omega \Omega\langle n\rangle \to BP\langle n\rangle_*(X)$ is an isomorphism;*

(iv) $\operatorname{Tor}_1^\Omega(BP_*(X), \Omega\langle n\rangle) = 0$;

(v) $\operatorname{Tor}_i^\Omega(BP_*(X), \Omega\langle n\rangle) = 0$ *for every $i > 0$.*

Proof. We prove that (i) $\iff$ (ii) $\iff$ (iii) $\Rightarrow$ (iv) $\Rightarrow$ (i) $\Rightarrow$ (v) $\Rightarrow$ (iii).

We prove that (i) $\iff$ (ii) by induction on n. By 3.11, $\rho_{-1} : BP_*(X) \to H_*(X; \mathbb{Z}/p)$ is epic iff $BP_*(X)$ is a free Ω-module. By 6.3, this holds, in turn, iff $BP_*(X)$ is a projective Ω-module, i.e., iff hom. $\dim_\Omega BP_*(X) = 0$. Assume that (i)$\iff$ (ii) for $n-1$. The cofiber sequence $A \xrightarrow{g} W \xrightarrow{f} X$ as in 6.18 yields the following commutative diagram with exact rows:

$$\begin{array}{ccccccccc}
0 & \longrightarrow & BP_*(A) & \xrightarrow{g_*} & BP_*(W) & \xrightarrow{f_*} & BP_*(X) & \longrightarrow & 0 \\
 & & \downarrow & & \downarrow{\scriptstyle \rho_n^W} & & \downarrow{\scriptstyle \rho_n} & & \\
\cdots & \longrightarrow & BP\langle n\rangle_*(A) & \xrightarrow{g_\bullet} & BP\langle n\rangle_*(W) & \xrightarrow{f_\bullet} & BP\langle n\rangle_*(X) & \longrightarrow & \cdots .
\end{array}$$

Recall that $BP_*(W)$ is a free $\pi_*(BP)$-module. Hence, in view of VII.3.29(ii), $H_*(W; \mathbb{Z}[p])$ is a free $\mathbb{Z}[p]$-module, and so, by 6.16, ρ_n^W is an epimorphism. Furthermore, again by 6.16, $BP_*(W) \otimes_\Omega \Omega\langle n\rangle \cong BP\langle n\rangle_*(W)$, and so $BP\langle n\rangle_*(W)$ is v_k torsion free for every $k \leq n$. By 6.20, we can assume that $A \in \mathscr{P}$. Now:

ρ_n is epic $\iff$ $f_\bullet$ is epic $\iff$ $g_\bullet$ is monic $\iff$ $BP\langle n\rangle_*(A)$ is v_n torsion free
$\iff$ $\rho_{n-1} : BP_*(A) \to BP\langle n-1\rangle_*(A)$ is epic (by 6.15)
$\iff$ hom. $\dim_\Omega BP_*(A) \leq n$
$\iff$ hom. $\dim_\Omega BP_*(X) \leq n+1$ (by 6.5(iii)).

It is clear that (iii)$\Rightarrow$ (ii). We prove that (ii) $\Rightarrow$ (iii). If ρ_{-1} is epic, then, by 3.11, $BP_*(X)$ is a free Ω-module. Hence, by VII.3.29(ii) and 6.16, $\widetilde{\rho}_{-1}$ is an isomorphism. Thus, assume that $n \geq 0$.

Consider any X such that $\rho_n : BP_*(X) \to BP\langle n\rangle_*(X)$ is epic. The diagram above yields the following diagram with exact rows, where $\rho'' = \widetilde{\rho}_n^W$ is an isomorphism and $\widetilde{\rho}_n$ is an epimorphism:

(6.22)

$$\begin{array}{ccccc} BP_*(A)\otimes_\Omega \Omega\langle n\rangle & \xrightarrow{\widetilde{g}_*} & BP_*(W)\otimes_\Omega \Omega\langle n\rangle & \xrightarrow{\widetilde{f}_*} & BP_*(X)\otimes_\Omega \Omega\langle n\rangle \longrightarrow 0 \\ \rho'\big\downarrow & & \rho''\big\downarrow\cong & & \big\downarrow\widetilde{\rho}_n \\ BP\langle n\rangle_*(A) & \xrightarrow{g_\bullet} & BP\langle n\rangle_*(W) & \xrightarrow{f_\bullet} & BP\langle n\rangle_*(X). \end{array}$$

If (ii) holds then $f_\bullet$ is epic and $g_\bullet$ is monic. Since (i) $\Longleftrightarrow$ (ii), we conclude that

$$\text{hom. dim}_\Omega BP_*(A) \leq \text{hom. dim}_\Omega BP_*(X) - 1 \leq n.$$

Again, since (i) $\Longleftrightarrow$ (ii), ρ' is epic. Now diagram chasing (the Five Lemma) shows that $\widetilde{\rho}_n$ is monic.

We prove that (iii) $\Rightarrow$ (iv). If (iii) holds then $g_\bullet$ in (6.22) is monic, and hence ρ' is an isomorphism. So, $\operatorname{Tor}_1^\Omega(BP_*(X), \Omega\langle n\rangle) = \operatorname{Ker}\widetilde{g}_* = \operatorname{Ker} g_\bullet = 0$.

We prove that (iv) $\Rightarrow$ (i). We have $\operatorname{Tor}_1^\Omega(BP_*(X), \Omega\langle n\rangle) = 0$. Suppose

$$\text{hom. dim}_\Omega BP_*(X) = m + 2 > n + 1.$$

Consider the diagram (6.22) with m instead of n. By 6.18,

$$\text{hom. dim}_\Omega BP_*(A) \leq m + 1,$$

and so ρ' is an isomorphism since (i) $\Rightarrow$ (iii). Furthermore, by 6.17,

$$\operatorname{Tor}_1^\Omega(BP_*(X), \Omega\langle m\rangle) = 0,$$

and so $\widetilde{g}_*$ is monic, and so $g_\bullet$ is monic, and so $f_\bullet$ is epic. Hence, $\widetilde{\rho}_m$ is an epimorphism, and therefore $\text{hom. dim}_\Omega BP_*(X) \leq m + 1$. This is a contradiction.

We prove that (i) $\Rightarrow$ (v). Consider the following claim $\mathfrak{A}_m$:
Let $Y \in \mathscr{P}$. Suppose that $\text{hom. dim}_\Omega BP_*(Y) \leq m$. *Then*

$$\operatorname{Tor}_i^\Omega(BP_*(Y), \Omega\langle n\rangle) = 0$$

for every $i > 0$.

It suffices to prove $\mathfrak{A}_{n+1}$. We prove by induction that $\mathfrak{A}_m, m \leq n+1$, is valid. It is clear that $\mathfrak{A}_0$ is valid. We assume that $\mathfrak{A}_{k-1}, 1 \leq k \leq n+1$, is valid and prove $\mathfrak{A}_k$. Consider any spectrum $X \in \mathscr{P}$ with $\text{hom. dim}_\Omega BP_*(X) \leq k$. Then $\operatorname{Tor}_1^\Omega(BP_*(X), \Omega\langle n\rangle) = 0$ (because (i) $\Rightarrow$(iv)). Consider a cofiber sequence $A \xrightarrow{g} W \xrightarrow{f} X$ as in 6.18. By 6.20, we can assume that $A \in$

$\mathscr{P}$. By 6.18, $\text{hom. dim}_\Omega BP_*(A) \leq k-1$, and so, by the inductive assumption, $\text{Tor}_i^\Omega(BP_*(A), \Omega\langle n\rangle) = 0$ for every $i > 0$. Hence, by 6.18, $\text{Tor}_i^\Omega(BP_*(X), \Omega\langle n\rangle) = 0$ for every $i > 1$, and thus $\text{Tor}_i^\Omega(BP_*(X), \Omega\langle n\rangle) = 0$ for every $i > 0$.

Finally, we prove that (v) $\Rightarrow$ (iii). Consider the following claim $\mathfrak{B}_m$: *Let $Y \in \mathscr{P}$. Suppose that* $\text{hom. dim}_\Omega BP_*(Y) \leq m$. *If*

$$\text{Tor}_i^\Omega(BP_*(Y), \Omega\langle n\rangle) = 0$$

for every $i > 0$, then $\widetilde{\rho}_n : BP_*(Y) \otimes_\Omega \Omega\langle n\rangle \to BP\langle n\rangle_*(Y)$ is epic.

We prove $\mathfrak{B}_m$ by induction. By 6.16, $\mathfrak{B}_0$ is valid. We assume that $\mathfrak{B}_{k-1}, 1 \leq k$, is valid and prove $\mathfrak{B}_k$. Let $X \in \mathscr{P}$. Supose that

$$\text{hom. dim}_\Omega BP_*(X) \leq k$$

and that $\text{Tor}_i^\Omega(BP_*(X), \Omega\langle n\rangle) = 0$ for every $i > 0$. By 6.18,

$$\text{Tor}_i^\Omega(BP_*(A), \Omega\langle n\rangle) = 0 \text{ for every } i > 0,$$

where A is as in 6.18. By 6.20, we can assume that $A \in \mathscr{P}$. Since

$$\text{hom. dim}_\Omega BP_*(A) \leq k-1,$$

the homomorphism ρ' in (6.22) is an isomorphism by the inductive assumption. Finally, $\text{Tor}_1^\Omega(BP_*(X), \Omega\langle n\rangle) = 0$, and so $\widetilde{g}_*$ is a monomorphism. Thus, $\widetilde{\rho}_n : BP_*(X) \otimes_\Omega \Omega\langle n\rangle \to BP\langle n\rangle_*(X)$ is an isomorphism. □

6.23. Theorem. (i) *Given a spectrum $X \in \mathscr{P}$, suppose that $v_n x = 0$ for some $x \in BP_*(X), x \neq 0$. Then* $\text{hom. dim}_\Omega BP_*(X) > n$.

(ii) *For every prime p and every natural number n, there exists a finite spectrum X with* $\text{hom. dim}_\Omega BP_*(X) = n$.

Proof. (i) Since $BP_*(X)$ is not v_n torsion free, $BP\langle N\rangle_*(X)$ is not v_n torsion free for N large enough, see II.4.5(i). Hence, by 6.14(i), $BP\langle n\rangle_*(X)$ is not v_n torsion free. Hence, by 6.15, $\rho_{n-1} : BP_*(X) \to BP\langle n\rangle_*(X)$ is not epic. Thus, by 6.21, $\text{hom. dim}_\Omega BP_*(X) > n$.

(ii) If $\text{hom. dim}_\Omega BP_*(X) > 0$ then

$$\text{hom. dim}_\Omega BP_*(A) = \text{hom. dim}_\Omega BP_*(X) - 1$$

for A as in 6.18. Hence, it suffices to prove that for every n there exists a finite spectrum X with $\text{hom. dim}_\Omega BP_*(X) > n$. Fix a natural number n. By 3.3, $v_n a = 0$ for every $a \in BP_0(P(n)) = \mathbb{Z}/p$. Hence, $v_n a = 0$ for every $a \in BP_0(X)$, where $X = P(n)^{(2p^n-1)}$. Thus, by (i), $\text{hom. dim}_\Omega BP_*(X) > n$. □

6.24. Remarks. Conner–Smith [1] demonstrated that homological algebra over $\pi_*(MU)$ has a geometrical meaning. In particular, $\text{hom. dim}_{\pi_*(MU)}(X)$

turns out to be a useful geometric invariant of X. The proofs of 6.3 and 6.6 are taken from this paper. We remark that 6.6 was proved by Adams [6], Lect. 5, Th. 2, p. 114. Besides, Conner–Smith [1] remarked that Adams suggested the idea of the proof of 6.3. Conner–Smith [1] proved the MU-analog of 6.21 for $n = 0$ and, partially, for $n = 1$ (considering k instead of $MU\langle 1\rangle$). All other results of this section were proved by Johnson–Wilson [1]. Our proof of 6.12 is taken from Landweber [5].

(b) Following Baas [1], fix a system $\{x_i\}$ of free polynomial generators of $\pi_*(MU)$ and let

$$MU\langle n\rangle := MU^{\{x_{n+1},\dots x_{n+k}\dots\}}.$$

Conner–Smith [1] proved that $u^H : MU^*(X) \to H_*(X)$ is epic iff

$$\text{hom. dim}_{\pi_*(MU)}(X) \leq 1,$$

and Conner–Smith [1], Johnson–Smith [1] proved that $u^k : MU_*(X) \to k_*(X)$ is epic iff hom. $\dim_{\pi_*(MU)}(X) \leq 2$. Since $H = MU\langle 0\rangle$ and $k = MU\langle 1\rangle$ (the last one for a suitable system $\{x_i\}$), one can conjecture that the forgetful homomorphism $MU_*(X) \to MU\langle n\rangle_*(X)$ is epic iff hom. $\dim_{\pi_*(MU)}(X) \leq n+1$. This conjecture is wrong, see Johnson–Wilson [1], Rudyak [2]. However, Theorem 6.21 demonstrates that a local version of this conjecture is valid.

(c) I think (conjecture) that there is a spectral sequence

$$E^r_{**}(X) \Longrightarrow BP_*(X), \quad E^2_{**}(X) = BP\langle n\rangle \otimes_{\omega\langle n\rangle} \Omega.$$

For $n = 0$ it is just the AHSS, for n=1 such a spectral is constructed by Johnson [1]

Yosimura [1] generalized Theorem 6.21 as follows. Consider the spectrum $BP[m,n] := BP^{\Sigma}$, where $\Sigma = \{v_0, \dots, v_{m-1}, v_{n+1}, \dots, v_{n+k}, \dots\}$, $m-1 \leq n$. So, $BP[m,\infty] = P(m)$ and $BP[0,n] = BP\langle n\rangle$. Furthermore, $BP[n,n-1] = H\mathbb{Z}/p$. The forgetful morphism $\rho_m : P(m) \to BP[m,n]$ can be defined in an obvious way. Set $\Omega[m,n] := \pi_*(BP[m,n])$.

6.25. Theorem. *Fix any m,n with $0 \leq m \leq n$. For every spectrum $X \in \mathscr{P}$ the following conditions are equivalent:*

(i) hom. $\dim_{\Omega/I_m} P(m)_*(X) \leq n - m + 1$;

(ii) *The homomorphism* $\rho_n : P(m)_*(X) \to BP[m,n]_*(X)$ *is epic;*

(iii) *The homomorphism* $\widetilde{\rho}_n : P(m)_*(X) \otimes_{\Omega/I_m} \Omega[m,n] \to BP[m,n]_*(X)$ *is an isomorphism;*

(iv) $\mathrm{Tor}_1^{\Omega/I_m}(P(m)_*(X), \Omega[m,n]) = 0$;

(v) $\mathrm{Tor}_i^{\Omega/I_m}(P(m)_*(X), \Omega[m,n]) = 0$ *for every* $i > 0$. □

This theorem generalizes 6.21 (namely, 6.21 is its special case with $m = 0$), and it can be proved similarly to 6.21.

§7. Morava K-Theories

Fix a prime p.

7.1. Definition. Given a natural number n, consider the ring $\mathbb{Z}/p[x]$, $\dim x = 2(1-p^n)$. A *connected Morava k-theory* $k(n)$ is any ring $\mathbb{C}$-marked cohomology theory on $\mathscr{S}$(i.e., a cohomology theory represented by a ring $\mathbb{C}$-marked spectrum) with the following properties:

1. $k(n)^*(S) = \mathbb{Z}/p[x]$;
2. The formal group of $k(n)$ has height n.

We require the ring spectrum $k(n)$ to be commutative (i.e., $\mathbb{C}$-oriented) if $p > 2$. Furthermore, we set $k(0) := H\mathbb{Z}[p]$, $k(\infty) := H\mathbb{Z}/p$.

7.2. Definition. Given a natural number n, consider the ring $\mathbb{Z}/p[x, x^{-1}]$ where $\dim x = 2(1-p^n)$. A *periodic Morava K-theory* $K(n)$ is any ring $\mathbb{C}$-marked cohomology theory on $\mathscr{S}$ with the following properties:

1. $K(n)^*(S) = \mathbb{Z}/p[x, x^{-1}]$;
2. The formal group of $K(n)$ has height n.

We require the ring spectrum $K(n)$ to be commutative (i.e., $\mathbb{C}$-oriented) if $p > 2$. Furthermore, we set $K(0) := H\mathbb{Q}$, $K(\infty) := H\mathbb{Z}/p$.

We now give examples (i.e., prove the existence) of Morava theories. Consider any family

$$R = \{v_1, \dots, v_n, \dots\}, \quad \dim v_i = 2(p^i - 1),$$

of free polynomial generators of $\pi_*(BP)$. Set $k^R(n) := BP^V$, where $V = \{p, v_1, \dots, v_{n-1}, v_{n+1}, \dots\}$.

7.3. Proposition. *$k^R(n)$ can be equipped with a $\mathbb{C}$-marking which turns it into a Morava k-theory $k(n)$.*

Proof. By VIII.2.17, $k^R(n)$ admits an admissible associative multiplication which can be chosen to be commutative for $p > 2$. Therefore, the morphism $MU \xrightarrow{\rho} BP \xrightarrow{r} k^R(n)$ is a ring $\mathbb{C}$-marking. Let

$$\rho_S : MU[p]^*(S) \to BP^*(S), \; r_S : BP^*(S) \to k^R(n)^*(S)$$

be the coefficient homomorphisms, and let α_n be the coefficient of x^{p^n} in $[p]_f(x)$. We have $I_n = (p, v_1, \dots, v_{n-1})$, and so, by VII.6.15(ii), $\rho(\alpha_n) \equiv b_n v_n \bmod I_n, b_n \in \mathbb{Z}/p, b_n \neq 0$. Hence, $r_S \rho_S(\alpha_i) = 0$ for $i < n$ and $r_S \rho_S(\alpha_n) \neq 0$. Thus, $\mathrm{ht}((r_S)_*(\rho_S)_* f) = n$. □

As usual, we denote by $\{X_\lambda\}$ the family of all finite subspectra of a spectrum X.

7.4. Proposition. *Let $k(n)$ be a connected Morava k-theory, $k(n)^*(S) = \mathbb{Z}/p[x], \dim x = 2(1-p^n), n \in \mathbb{N}$. Then*

$$E^*(X) := \varprojlim\{k(n)^*(X_\lambda) \otimes_{\mathbb{Z}/p[x]} \mathbb{Z}/p[x, x^{-1}]\}, X \in \mathscr{S},$$

is an additive cohomology theory on $\mathscr{S}$ which is a periodic Morava K-theory $K(n)^*(X)$.

Proof. By 6.1, $k(n)^*(X) \otimes_{\mathbb{Z}/p[x]} \mathbb{Z}/p[x, x^{-1}]$ is a cohomology theory on $\mathscr{S}_{\mathrm{f}}$. Hence, by III.4.17, the functor

$$E^*(X) := \varprojlim\{k(n)^*(X_\lambda) \otimes_{\mathbb{Z}/p[x]} \mathbb{Z}/p[x, x^{-1}]\}$$

is a cohomology theory on $\mathscr{S}$. The additivity follows from III.1.21. □

7.5. Proposition. *If $n > 0$, then a connective covering of any periodic Morava K-theory $K(n)$ is a connected Morava k-theory.*

Proof. This follows from II.4.28 because MU is a connected ring spectrum. □

Consider the diagram $BP \xrightarrow{\varkappa} MU[p] \xrightarrow{\rho} BP$ with $\rho\varkappa = 1_{BP}$. Set $f_p := l_*(f)$, where $l : MU^*(S) \to MU^*(S)[p] = MU[p]^*(S)$ is the $\mathbb{Z}[p]$-localization and f is the universal formal group over $MU^*(S)$. Let $\varkappa_S : BP^*(S) \to MU[p]^*(S)$ and $\rho_S : MU[p]^*(S) \to BP^*(S)$ be the coefficient homomorphisms. Set $f^{BP} := (\rho_S)_* f_p$.

7.6. Lemma. (i) *The formal groups $(\varkappa_S)_* f^{BP}$ and f_p are equivalent.*

(ii) *Let F be a graded formal group over a graded $\mathbb{Z}[p]$-algebra R. Then there exists a graded formal group G over R such that $F \simeq G$ and $G = \varphi_* f^{BP}$ for some $\varphi : BP^*(S) \to R$.*

Proof. (i) Note that $(\varkappa_S)_* f^{BP}$ is the formal group of the $\mathbb{C}$-oriented spectrum $(MU[p], ((\varkappa\rho)_{CP^\infty})_*(T))$, where

$$(\varkappa\rho)_{CP^\infty} : MU[p]^*(CP^\infty) \to MU[p]^*(CP^\infty)$$

is the induced homomorphism. Now the assertion follows from VII.6.5.

(ii) We have $F = h_* f_p$ for some $h : MU[p]^*(S) \to R$. Set $\varphi := h\varkappa_S : BP^*(S) \to R$ and $G := \varphi_* f^{BP}$. Now

$$G = \varphi_* f^{BP} = h_*(\varkappa_S)_* f^{BP} \simeq h_* f_p = F. \qquad \square$$

7.7. Lemma. *Let $\varphi : BP \to R$ be a ring homomorphism such that* $\mathrm{ht}(\varphi_* f^{BP}) = n$. *Then $\varphi(v_k) = 0$ for $k < n$ and $\varphi(v_n) \neq 0$.*

Proof. Let $\overline{\alpha}_k$ be the coefficient of x^{p^k} in $[p]_{f^{BP}}(x)$. Since $\mathrm{ht}(\varphi_* f^{BP}) = n$, we conclude that $\varphi(\overline{\alpha}_k) = 0$ for $k < n$ and $\varphi(\overline{\alpha}_n) \neq 0$. It follows from VII.6.15(ii) that $\overline{\alpha}_k \equiv b_k v_k \bmod I_k, b_k \in \mathbb{Z}/p, b_k \neq 0$. Thus, $\varphi(v_k) = 0$ for $k < n$ and $\varphi(v_n) \neq 0$. □

7.8. Theorem. *Let $\mathbb{F}$ be a field of characteristic $p > 0$, and let x be an indeterminate, $\deg x = 2(1-p^n)$. Let R be one of the rings $\mathbb{F}[x], \mathbb{F}[x, x^{-1}]$. Let F be a graded formal group over R with $\operatorname{ht}(F) = n$. Then there exists a ring $\mathbb{C}$-marked spectrum (E, u) whose formal group is F. Moreover, E can be chosen to be commutative if $p > 2$.*

Proof. We consider the cases $R = \mathbb{F}[x, x^{-1}]$ and $R = \mathbb{F}[x]$ separately.

1. $R = \mathbb{F}[x, x^{-1}]$. Let G be any formal group equivalent to F. If we construct a $\mathbb{C}$-marked ring spectrum whose formal group is G, then, by VII.6.22, the same spectrum with another $\mathbb{C}$-marking realizes the formal group F. So, it suffices to construct a spectrum whose formal group G is equivalent to F.

By 7.6(ii), there is a formal group G such that $G \simeq F$ and $G = \varphi_* f^{BP}$ for some $\varphi : BP^*(S) \to \mathbb{F}[x, x^{-1}]$. Since $\operatorname{ht}(F) = n = \operatorname{ht}(G)$, $\varphi(v_i) = 0$ for $i < n$ and $\varphi(v_n) = ax, a \in \mathbb{F}, a \neq 0$. Hence, there exists a ring homomorphism $\psi : P(n)^*(S) \to \mathbb{F}[x, x^{-1}]$ such that the following diagram commutes:

$$\begin{array}{ccc} BP^*(S) & \xrightarrow{\varphi} & \mathbb{F}[x, x^{-1}] \\ {\scriptstyle r_*}\downarrow & & \| \\ P(n)^*(S) & \xrightarrow{\psi} & \mathbb{F}[x, x^{-1}]. \end{array}$$

This homomorphism ψ turns $\mathbb{F}[x, x^{-1}]$ into a $P(n)^*(S)$-module ${}_\psi\mathbb{F}[x, x^{-1}]$. Moreover, $P(n)^*(-) \otimes_\psi \mathbb{F}[x, x^{-1}]$ is a cohomology theory on $\mathscr{S}_{\mathrm{f}}$. Indeed, we prove that $\mathbb{F}[x, x^{-1}]$ satisfies the condition 4.11(iii). Multiplication by v_n is an isomorphism because $\psi(v_n) = ax, a \neq 0$. Furthermore, $\mathbb{F}[x, x^{-1}]/(x) = 0$, and hence multiplication by v_m, $m > n$ is monic for trivial reasons.

By III.3.20(i), the cohomology theory $P(n)^*(-) \otimes_\psi \mathbb{F}[x, x^{-1}]$ is represented by a spectrum E. Furthermore, by III.4.17,

$$E^*(X) := \varprojlim \{P(n)^*(X_\lambda) \otimes_\psi \mathbb{F}[x, x^{-1}]\}$$

for every $X \in \mathscr{S}$. So, E turns out to be a ring spectrum, and

$$\tau : MU[p] \xrightarrow{\rho} BP \xrightarrow{r} P(n) \to E$$

is a ring $\mathbb{C}$-marking of E. Clearly, G is the formal group of (E, τ), i.e., G is realized.

2. $R = \mathbb{F}[x]$. Let $i : \mathbb{F}[x] \subset \mathbb{F}[x, x^{-1}]$ be the inclusion, and let F be a formal group over R with $\operatorname{ht}(F) = n$. By the above, there is a ring $\mathbb{C}$-marked spectrum $(\mathscr{E}, \tau)$ whose formal group is i_*F. Let $q : E \to \mathscr{E}$ be a connective covering. Then, by II.4.16, there is a morphism $t : MU \to E$ with $qt = \tau$. Clearly, F is the formal group of (E, t). □

7.9. Corollary. (i) *Let F be a graded formal group over*

$$\mathbb{Z}/p[x, x^{-1}], \quad \dim x = 2(1-p^n),$$

with $\mathrm{ht}(F) = n$. *Then there exists a periodic Morava* K*-theory* $K(n) = K_F(n)$ *whose formal group is* F.

(ii) *Let* F *be a graded formal group over* $\mathbb{Z}/p[x], \dim x = 2(1-p^n)$, *with* $\mathrm{ht}(F) = n$. *Then there exists a connected Morava* k*-theory* $k(n) = k_F(n)$ *whose formal group is* F. □

7.10. Exercise. Prove 7.8 by constructing $k_F(n)$ as $k^R(n)$ for suitable R, see 7.3, and $K_F(n)$ as in 7.4.

7.11. Theorem. *Let* $\mathbb{F}$ *be a field of characteristic* $p > 2$, *and let* x *be an indeterminate,* $\deg x = 2(1-p^n)$. *Let* R *be one of the rings* $\mathbb{F}[x], \mathbb{F}[x, x^{-1}]$, *and let* F *be a formal group over* R *with* $\mathrm{ht}(F) = n$. *Let* (E, t) *and* (E', t') *be two* $\mathbb{C}$*-oriented spectra such that* F *is the formal group of* (E, t) *as well as of* (E', t'). *Then* E *and* E' *are equivalent ring spectra.*

Proof. 1. Let $R = \mathbb{F}[x, x^{-1}]$. Let $u : MU \to E$ be a ring morphism which gives and is given by the formal group F, see VII.(2.7). Consider the ring morphism

$$\sigma : BP \xrightarrow{\varkappa} MU[p] \xrightarrow{u[p]} E.$$

By 7.6(i), $(\varkappa_S)_* f^{BP} \simeq f_p$. Hence, $(\sigma_S)_* f^{BP} \simeq F$, where $\sigma_S : BP^*(S) \to E^*(S) = R$ is the coefficient homomorphism. Since $\mathrm{ht}((\sigma_S)_* f^{BP}) = \mathrm{ht}(F) = n$, $\sigma_S(v_i) = 0$ for $i < n$. By 2.17, there is a ring morphism $\sigma_n : P(n) \to E$ with $\sigma = \sigma_n \rho_n$. Note that $\psi := (\sigma_n)_S : P(n)^*(S) \to E^*(S) = R$ depends only on the formal group F. Furthermore, by 4.11, $P(n)^*(-) \otimes_\psi R$ is a cohomology theory on $\mathscr{S}_{\mathrm{f}}$.

We define

$$\alpha_Y : P(n)^*(Y) \otimes_\psi R \to E^*(Y), \; \alpha_Y(a \otimes b) := \psi(a)b.$$

It is clear that the family $\{\alpha_Y\}$ gives a ring morphism of cohomology theories on $\mathscr{S}_{\mathrm{f}}$. By III.4.16(ii), $E^*(X) = \varprojlim E^*(X_\lambda)$ for every $X \in \mathscr{S}$. We define

$$\beta_X : \varprojlim \{P(n)^*(X_\lambda) \otimes_\psi R\} \xrightarrow{\varprojlim \alpha_{X_\lambda}} \varprojlim E^*(X_\lambda) = E^*(X).$$

By III.2.17 and III.1.21, $\varprojlim \{P(n)^*(X_\lambda) \otimes_\psi R\}$ is an additive cohomology theory on $\mathscr{S}$. Hence, the family $\{\beta_X\}$ gives us a morphism of cohomology theories on $\mathscr{S}$, and, by II.3.19(iii), it is an isomorphism. Since the cohomology theory on the left hand side depends only on ψ, E and E' are equivalent ring spectra.

2. Let $R = \mathbb{F}[x]$. We have E and E' with $\pi_*(E) = \mathbb{F}[x] = \pi_*(E')$. Set $\mathscr{E}^*(X) := \varprojlim \{E^*(X_\lambda) \otimes_{F[x]} F[x, x^{-1}]\}$. Then $\mathscr{E}^*(X)$ is an additive cohomology theory on $\mathscr{S}$, and E is a connective covering of $\mathscr{E}$. Since $\mathscr{E} \simeq \mathscr{E}'$ as ring spectra, we conclude that $E \simeq E'$ as ring spectra. □

7.12. Corollary. *Two Morava theories with p odd prime are ring equivalent iff their formal groups are equivalent.*

Proof. Let $k(n), k'(n)$ have equivalent formal groups F, G respectively. Then, by VII.6.22, $k'(n)$ admits a $\mathbb{C}$-orientation t such that F is the formal group of $(k'(n), t)$. Hence, by 7.11, $k(n)$ and $k'(n)$ are equivalent ring spectra. Similarly for $K(n)$. □

What about $p = 2$ and 7.11? Consider any spectrum $l = k_F(n)$. There is a multiplication $\mu : l \wedge l \to l$. Consider the multiplication $\mu' : l \wedge l \xrightarrow{\tau} l \wedge l \to l$, where τ switches the factors. By 5.5, the multiplication μ is not commutative, and hence $\mu \neq \mu'$. Moreover, there is no morphism $f : l \to l$ which induces a ring isomorphism $(l, \mu) \to (l, \mu')$. Thus, there are at least two ring spectra with the same formal group F. (There is a reason that there are just two $\mathbb{C}$-marked spectra with the same F, but I do not have any rigorous proof.) However, the following uniqueness result holds.

7.13. Theorem. *Let $\mathbb{F}$ be a finite field of characteristic $p > 0$, and let x be an indeterminate, $\deg x = 2(1 - p^n)$. Let R be one of the rings $\mathbb{F}[x], \mathbb{F}[x, x^{-1}]$, and let F be a formal group over R with $\mathrm{ht}(F) = n$. Let (E, u) and (E', u') be two $\mathbb{C}$-marked ring spectra such that F is the formal group of (E, u) as well as of (E', u'). Then E and E' are equivalent BP-module spectra.*

Proof. The proof is similar to that of 7.11. Consider the ring morphism $\sigma : BP \xrightarrow{\varkappa} MU[p] \xrightarrow{u[p]} E$. By 2.11, there is a BP-module (but, possibly, not ring) morphism

$$\sigma_n : P(n) \to E \tag{7.14}$$

with $\sigma = \sigma_n \rho_n$. As in 7.11, it induces a BP-module morphism

$$\varprojlim \{P(n)^*(X_\lambda) \otimes_{(\sigma_n)_*} R\} \to E^*(X)$$

of cohomology theories on $\mathscr{S}$. Now the proof can be completed similarly to the proof of 7.11. □

7.15. Theorem. *The homotopy type of a spectrum $k(n)$, as well as $K(n)$, is uniquely determined by the number n (and prime p, of course).*

Proof. We prove this only for $k(n)$; the proof for $K(n)$ is similar. Consider any Morava k-theories $k = k_F(n), l = k_G(n)$ with some formal groups F, G over $\mathbb{Z}/p[x]$ where $\deg x = 2(1 - p^n)$ and $\mathrm{ht}(F) = n = \mathrm{ht}(G)$. Let $\mathbb{F}$ be the algebraic closure of $\mathbb{Z}/p$. We set

$$K^*(X) := \varprojlim \{k^*(X_\lambda) \otimes_{\mathbb{Z}/p} \mathbb{F}\}, L^*(X) := \varprojlim \{l^*(X_\lambda) \otimes_{\mathbb{Z}/p} \mathbb{F}\},$$

where $\{X_\lambda\}$ is the family of all finite subspectra of X. Then $K^*(-), L^*(-)$ are additive cohomology theories on $\mathscr{S}$, and hence they are represented by spectra K, L respectively. Let $i : \mathbb{Z}/p \to \mathbb{F}$ be the inclusion homomorphism, $i(1) = 1$, and let $\varepsilon : \mathbb{F} \to \mathbb{Z}/p$ be a $\mathbb{Z}/p$-linear map of $\mathbb{Z}/p$-vector spaces such that $\varepsilon i = 1_{\mathbb{Z}/p}$. By VII.5.15, i_*F and i_*G are isomorphic formal groups.

Firstly, let $p > 2$. Then, by VII.6.6 and 7.11, there is an equivalence $h : K \simeq L$. Now, for every finite spectrum Y we have a natural homomorphism

$$k^*(Y) \xrightarrow{i_*} K^*(Y) \xrightarrow{h_Y} L^*(Y) \xrightarrow{\varepsilon_*} l^*(Y),$$

where i_* has the form

$$k^*(Y) = k^*(Y) \otimes_{\mathbb{Z}/p} \mathbb{Z}/p \xrightarrow{1 \otimes i} k^*(Y) \otimes_{\mathbb{Z}/p} \mathbb{F} = K^*(Y) \tag{7.16}$$

and ε_* has the form

$$L^*(Y) = l^*(Y) \otimes_{\mathbb{Z}/p} \mathbb{F} \xrightarrow{1 \otimes \varepsilon} l^*(Y) \otimes_{\mathbb{Z}/p} \mathbb{Z}/p = l^*(Y).$$

By III.4.16(ii), $E^*(X) = \varprojlim\{E^*(X_\lambda)\}$ for $E = k, K, L, l$. Hence, passing to $\varprojlim$, we get a morphism

$$k^*(X) \xrightarrow{i_*} K^*(X) \xrightarrow{h_Y} L^*(X) \xrightarrow{\varepsilon_*} l^*(X)$$

of cohomology theories on $\mathscr{S}$. This morphism is an isomorphism because it induces an isomorphism for $X = S$.

For $p = 2$ the proof needs a modification. We want to apply 7.13 instead of 7.11, but the field $\mathbb{F}$ is not finite. However, why do we need $\mathbb{F}$ to be finite? The only reason is that we are able to use 2.11 in the proof of 7.13. So, the theorem will be proved once we have constructed a morphism

$$\sigma_n = \sigma_n^E : P(n) \to E$$

as in (7.14) for $E = K, L$.

The morphism $i_* : k^*(-) \to K^*(-)$ of cohomology theories in (7.16) is represented by a morphism $i_* : k \to K$ of spectra. By 2.11, there is a BP-module morphism $\sigma_n^k : P(n) \to k$ as in 7.14. We define the required morphism σ_n^K to be the composition

$$\sigma_n^K : P(n) \xrightarrow{\sigma_n^k} k \xrightarrow{i_*} K.$$

Similarly for L. □

7.17. Corollary. *$H^*(k(n); \mathbb{Z}/p) = \mathscr{A}_p/\mathscr{A}_p Q_n$. Thus, the first non-trivial Postnikov invariant of $k(n)$ is Q_n.*

Proof. By 1.3, $H^*(k^S(n); \mathbb{Z}/p) = \mathscr{A}_p/\mathscr{A}_p Q_n$ where $k^S(n)$ is as in 7.3. But, by 7.15, $k(n) \simeq k^S(n)$. The last assertion follows from 1.4. □

7.18. Corollary. *Let p be a prime.*

(i) *Let (E, u) be a $\mathbb{C}$-marked ring spectrum such that*

$$E^*(S) = \mathbb{Z}/p\,[y, y^{-1}], \quad \dim y = 2s < 0.$$

Suppose the formal group of (E, u) has height $n \le \infty$. Then E splits into a wedge of suspensions of periodic Morava K-theories $K(n)$, i.e.,

$$E \simeq \vee_i \Sigma^{d_i} K(n).$$

(ii) *Let (E, u) be a $\mathbb{C}$-marked ring spectrum such that*

$$E^*(S) = \mathbb{Z}/p\,[y], \quad \dim y = 2s < 0.$$

Suppose the formal group of E has height $n \le \infty$. Then E splits into a wedge of suspensions of connected Morava k-theories $k(n)$, i.e.,

$$E \simeq \vee_i \Sigma^{d_i} k(n).$$

Proof. (i) Let F be the formal group of (E, u). If $\mathrm{ht}(F) = n = \infty$, then, by VII.5.11(i), F is additive. Hence, by 2.14, E is a graded Eilenberg–Mac Lane spectrum, and the result holds. So, we assume that $n < \infty$.

We have $K(n)^*(S) = \mathbb{Z}/p\,[x, x^{-1}]$ where $\dim x = 2(1-p^n)$. Since $\mathrm{ht}(F) = n$, we conclude that $s|(p^n - 1)$. We set $l := (p^n - 1)/s$ and define the ring homomorphism

$$h : \mathbb{Z}/p\,[x, x^{-1}] \to \mathbb{Z}/p\,[y, y^{-1}]$$

by setting $h(x) := y^l$.

Consider the ring morphism

$$\sigma : BP \xrightarrow{\varkappa} M\mathcal{U}[p] \xrightarrow{u[p]} E.$$

By 2.11, we can construct a BP-module morphism $\sigma_n : P(n) \to E$ such that $\sigma = \sigma_n \rho_n$. It is clear that the homomorphism

$$(\sigma_n)_S : P(n)^*(S) \to E^*(S)$$

can be decomposed as

$$P(n)^*(S) \xrightarrow{\psi} \mathbb{Z}/p\,[x, x^{-1}] \xrightarrow{h} \mathbb{Z}/p\,[y, y^{-1}] = E^*(S).$$

Given $Y \in \mathscr{S}_{\mathrm{f}}$, we have an isomorphism

$$\alpha_Y : P(n)^*(Y) \otimes_\psi \mathbb{Z}/p\,[x, x^{-1}] \cong K(n)^*(Y),$$

cf. the proof of 7.11. Now, we define a homomorphism

$$\rho_Y : K(n)^*(Y) \cong P(n)^*(Y) \otimes_\psi \mathbb{Z}/p\,[x, x^{-1}] \to E^*(Y),\ \rho(a \otimes b) = \sigma_n(a)h(b).$$

It is clear that the family $\{\rho_Y\}$ gives us a morphism of BP-module cohomology theories on $\mathscr{S}_{\mathrm{f}}$. Given $X \in \mathscr{S}$, define

$$\tau_X : K(n)^*(X) = \varprojlim\{K(n)^*(X_\lambda)\} \xrightarrow{\varprojlim \rho_{X_\lambda}} \varprojlim\{E^*(X_\lambda)\} = E^*(X).$$

Here the equalities hold because of III.4.17. Clearly, the family $\{\tau_X\}$ yields a morphism of additive BP-module cohomology theories on $\mathscr{S}$.

The homomorphism h turns $\mathbb{Z}/p[y, y^{-1}]$ into the $\mathbb{Z}/p[x, x^{-1}]$-module ${}_h\mathbb{Z}/p[y, y^{-1}]$, and hence $E^*(X)$ becomes a natural $\mathbb{Z}/p[x, x^{-1}]$-module for every $X \in \mathscr{S}$. Since τ_X is a homomorphism of $\mathbb{Z}/p[x, x^{-1}]$-modules, we have a well-defined homomorphism

$$\varphi_X : K(n)^*(X) \otimes_h \mathbb{Z}/p[y, y^{-1}] \to E^*(X), \; \varphi_X(a \otimes b) := \tau_X(a)b.$$

Now, the family $\{\varphi_X\}$ is an isomorphism of cohomology theories on $\mathscr{S}$, and thus

$$E \simeq \bigvee_{i=0}^{l-1} \Sigma^{is} K(n).$$

(ii) This follows from (i), because $(\Sigma^r K(n))|0 = \Sigma^r k(n)$ for $0 \leq r < 2(p^n - 1)$. □

7.19. Corollary. *Let $p > 2$.*

(i) *Let E be a commutative ring spectrum with*

$$E^*(S) = \mathbb{Z}/p[y, y^{-1}], \quad \dim y = 2s < 0.$$

Then there is $n \leq \infty$ such that $E \simeq \vee_i \Sigma^{d_i} K(n)$.

(ii) *Let E be a commutative ring spectrum with*

$$E^*(S) = \mathbb{Z}/p[y], \quad \dim y = 2s < 0.$$

Then there is $n \leq \infty$ such that $E \simeq \vee_i \Sigma^{d_i} k(n)$.

Proof. Considering the AHSS for $E^*(CP^\infty)$, we conclude that E is a $\mathbb{C}$-orientable spectrum. □

Corollary 7.19 describes the homotopy type of E, but the number n in 7.19 is described in terms of some extra structure on E (namely, $\mathbb{C}$-orientation). So, it makes sense to describe this n as an explicit homotopy invariant of E. This is done in the following corollary.

7.20. Corollary. *Let E be as in 7.19(ii). Suppose that E is not a graded Eilenberg–Mac Lane spectrum. Then the first nontrivial Postnikov invariant of E has dimension $2p^n - 1$ for some n such that $s|(p^n - 1)$, and*

$$E \simeq \bigvee_{i=0}^{\frac{p^n-1}{s}-1} \Sigma^{is} k(n).$$

Proof. By 7.19(ii), we have the splitting as above with some n. Hence, by 7.17, the first nontrivial Postnikov invariant of E has dimension $2p^n - 1$. $\square$

7.21. Corollary. *Let ℓ be the direct summand of $k[p]$ described before* 4.16. *Then $\ell \wedge M(\mathbb{Z}/p) \simeq k(1)$.*

Proof. By VIII.2.17, $\ell = BP^{\{v_i | i>1\}}$ can be equipped with a multiplication such that there is a ring morphism $h : BP \to \ell$. So, $\ell \wedge M(\mathbb{Z}/p)$ becomes a ring $\mathbb{C}$-marked spectrum via the $\mathbb{C}$-marking

$$MU \xrightarrow{l} MU[p] \xrightarrow{\rho} BP \xrightarrow{h} \ell = \ell \wedge S \to \ell \wedge M(\mathbb{Z}/p).$$

It is clear that $\pi_*(\ell \wedge M(\mathbb{Z}/p)) = \mathbb{Z}/p\,[v_1]$ and that the formal group of $\ell \wedge M(\mathbb{Z}/p)$ has height 1, and hence, by 7.18(ii), $\ell \wedge M(\mathbb{Z}/p) \simeq k(1)$. $\square$

The Postnikov tower of $k(n)$ can be constructed as that of k was, see VI.2.6. It is easy to see that

$$(\Sigma^{2-2p^n} k(n))|0 = k(n). \tag{7.22}$$

We set

$$k^r(n) := \text{ the cone of } \{x^{r+1} : \Sigma^{2r(p^n-1)} k(n) \to k(n)\}.$$

There is the cofiber sequence (cf. VI.2.3)

$$k^r(n) \xrightarrow{p_r} k^{r-1}(n) \xrightarrow{\sigma_r} \Sigma^{2r(p^n-1)+1} H\mathbb{Z}/p.$$

7.23. Theorem. *The tower*

$$\begin{array}{ccccc}
 & & \vdots & & \\
 & & \downarrow & & \\
 & & k^{r+1}(n) & & \\
 & & \downarrow & & \\
k(n) & \xrightarrow{\tau_{2r(p^n-1)}} & k^r(n) & & \\
 & & \downarrow{\scriptstyle p_r} & & \\
 & & k^{r-1}(n) & \xrightarrow{\sigma_r} & \Sigma^{2r(p^n-1)+1} H\mathbb{Z}/p \\
 & & \downarrow & & \\
 & & \vdots & & \\
 & & \downarrow & & \\
H\mathbb{Z}/p & = & k^0(n) & \xrightarrow{\sigma_1} & \Sigma^{2p^n-1} H\mathbb{Z}/p
\end{array}$$

is the Postnikov tower of $k(n)$. *Here*

(i) $\sigma_1 = Q_n$.

(ii) $(\Omega^\infty \sigma_r)|2(p^n-1) \neq 0$ *for all* $r > 1$.

(iii) $Q_n \sigma_{n-1} = 0$, *and* σ_n *is associated with this relation.*

Proof. (i) This follows from 7.17.

(ii) By (7.22) and (i), we have

$$\Omega^\infty(\sigma_2)|2(p^n-1) = Q_n : K(\mathbb{Z}/p, 2p^n-2) \to K(\mathbb{Z}/p, 4p^n-1).$$

This map is essential, and the assertion about $\sigma_r, r > 2$, follows from (7.22).

(iii) This can be proved similarly to VI.2.6(iii). □

Now we consider the $k(n)$-orientability problem. (By V.1.16 and 7.5, it is equivalent to the $K(n)$-orientabilty problem.) Namely, following Ch. V, §5, for every $k(n)$ we introduce the corresponding characteristic classes $\varkappa_r$ and consider the realizability problem for these classes.

7.24. Theorem (cf. Rudyak–Khokhlow [1]). *Let* $p = 2$, *and let* n *be an arbitrary natural number. Then every characteristic class* $\varkappa_r, r \geq 1$, *can be realized by vector bundles, i.e., there exists a vector bundle* ξ *which is* $k^{r-1}(n)$-*orientable and is not* $k^r(n)$-*orientable.*

Proof. By V.5.1 or V.5.6, in view of 7.23(ii) it suffices to prove that $\varkappa_1(\gamma_0) \neq 0$. But here γ_0 is the universal bundle $\gamma^{\mathcal{O}}$ over $B\mathcal{O}$, and

$$\varkappa_1(\gamma_0) = \varphi^{-1} Q_n(u) \in H^{2^{n+1}-1}(B\mathcal{O}; \mathbb{Z}/2)$$

where $u \in H^0(M\mathcal{O}; \mathbb{Z}/2)$ is the Thom class and $\varphi : H^0(B\mathcal{O}) \to H^0(M\mathcal{O})$ is the Thom isomorphism. By IV.6.2, $H^*(M\mathcal{O}; \mathbb{Z}/2)$ is a free $\mathscr{A}_2$-module. Hence, $\varkappa_1(\gamma_0) \neq 0$. (In fact, it is easy to see that $\varkappa_1(\gamma_0)$ is just the universal class $s_{2^{n+1}-1} \in H^*(B\mathcal{O}; \mathbb{Z}/2)$, i.e., the primitive class which is represented as $\sum t_i^{2^{n+1}-1}$ in terms of Wu generators t_i.) □

If $p > 2$ then, by V.4.9, $k(n)$-orientability implies $H\mathbb{Z}$-orientability, i.e., $B(\mathcal{V}, k^0(n)) = BS\mathcal{V}$ and $\gamma_0^{\mathcal{V}} = \gamma^{S\mathcal{V}}$. So, only $S\mathcal{V}$-objects are able to realize the classes $\varkappa_n$.

7.25. Proposition. *If* $p > 2$ *then every* $S\mathcal{O}$-*bundle is* $k(n)$-*orientable for every* n. *In other words, no* $k(n)$-*characteristic class* $\varkappa_r$ *with* $p > 2$ *can be realized by vector bundles.*

Proof. Firstly, the $\mathbb{C}$-marking $u : M\mathcal{U} \to k(n)$ gives a $k(n)$-orientation of the universal stable complex vector bundle. By IV.6.9 and IV.4.29(ii) (together with IV.5.23(ii)) $\pi_i(MS\mathcal{O}[p])$ and $H_i(MS\mathcal{O}[p])$ are free finitely generated $\mathbb{Z}[p]$-modules for every i. Hence, by VII.3.21, there is a projection $MS\mathcal{O}[p] \to BP$ which preserves the units. Thus, the morphism

$$MSO \xrightarrow{\iota} MSO[p] \to BP \xrightarrow{\varkappa} MU[p] \xrightarrow{u[p]} k(n)$$

gives a $k(n)$-orientation of $\mathcal{SO}$-bundles. □

7.26. Theorem. (i) *For $p > 2$ every $\mathcal{STOP}$-bundle is $k(1)$-orientable. Thus, in this case no class $\varkappa_r$ can be realized by $\mathcal{TOP}$-bundles.*

(ii) *If $p > 2$ and $n > 1$ then every $k(n)$-characteristic class $\varkappa_r, r \geq 1$, can be realized by $\mathcal{PL}$-bundles.*

(iii) *For every prime p and natural number n all the $k(n)$-characteristic classes $\varkappa_r, r \geq 1$, can be realized by spherical fibrations.*

Proof. (i) As we remarked above, every $\mathcal{TOP}$-bundle is $k\,[1/2]$-orientable, see the text before VI.3.4. Therefore, it is $k[p]$-orientable. Hence, every $\mathcal{STOP}$-bundle is ℓ-orientable, where ℓ is the direct summand of $k\,[p]$ in 4.16. Thus, by 7.21, every $\mathcal{STOP}$-bundle is $k(1)$-orientable.

(ii) Let $u \in H^0(M\mathcal{SPL}; \mathbb{Z}/p)$ denote the Thom class. Because of 7.23(i), $\varkappa_1(\gamma_0^{\mathcal{PL}}) = \varphi^{-1}Q_n u$ where $\varphi : H^0(B\mathcal{SPL}) \to H^0(M\mathcal{SPL})$ is the Thom isomorphism. Furthermore, by IV.6.13, $Q_n u \neq 0$ since $n > 1$. Hence, $\varkappa_1(\gamma_0^{\mathcal{PL}}) \neq 0$. Thus, by V.5.1 (or V.5.6) and 7.23(ii), all the classes $\varkappa_r, r \geq 1$, can be realized by $\mathcal{PL}$-bundles.

(iii) By 7.24, it suffices to consider the case $p > 2$ only. By IV.6.10, $Q_n(u) \neq 0$, and so $\varkappa_1(\gamma_0^{\mathcal{G}})$. Now, the theorem follows from V.5.1 (or V.5.6) and 7.23(ii). □

Résumé on $k(n)$-orientability

Recall the hierarchy $BO \to B\mathcal{PL} \to B\mathcal{TOP} \to B\mathcal{G}$.

Theorem. (i) *For $p = 2$ and arbitrary n all the $k(n)$-characteristic classes $\varkappa_r, r \geq 1$, can be realized by vector bundles.*

(ii) *For $p > 2$ and arbitrary n every $\mathcal{SO}$-bundle is $k(n)$-orientable.*

(iii) *For $p > 2$ every $\mathcal{STOP}$-bundle (and hence $\mathcal{SPL}$-bundle) is $k(1)$-orientable.*

(iv) *For $p > 2$ and $n > 1$ all the $k(n)$-characteristic classes $\varkappa_r, r \geq 1$, can be realized by $\mathcal{SPL}$-bundles.*

(v) *For every p and every n all the $k(n)$-characteristic classes $\varkappa_r, r \geq 1$, can be realized by spherical fibrations.*

We interpret $\mathscr{A}_p^*$ as $H_*(H\mathbb{Z}/p; \mathbb{Z}/p)$. Let $(Q_n)_* : \mathscr{A}_p^* \to \mathscr{A}_p^*$ be the homomorphism defined before II.6.36. We used the following proposition in II.4.8 and II.4.31.

7.27. Proposition. (i) $H_*(K(n); \mathbb{Z}/p) = 0 = H_*(K(n))$.

(ii) $H_*(K(n)_{(0)};\mathbb{Z}/p) \simeq \mathscr{A}_p^*/\mathscr{A}_p^*(Q_n)_*$.

Proof. (i) Consider the AHSS

$$E^r_{**} \Rightarrow K(n)_*(H\mathbb{Z}/p),\ E^2_{**} = H_*(H\mathbb{Z}/p;\pi_*(K(n))).$$

By 7.17, $d_{2p^n-1} = (Q_n)_*$ and $d_i = 0$ for $i < 2p^n - 1$. It is well known (and easy to see) that $\operatorname{Ker} Q_n = \operatorname{Im} Q_n$ in $\mathscr{A}_p$. Hence, $E^{2p^n}_{**} = 0$, and so $H_*(K(n);\mathbb{Z}/p) = 0$.

Now, $H_*(K(n);\mathbb{Z}/p) = H_*(K(n) \wedge M(\mathbb{Z}/p))$. Hence, we have an exact sequence

$$H_*(K(n)) \xrightarrow{p} H_*(K(n)) \to H_*(K(n);\mathbb{Z}/p)),$$

and $pH_*(K(n)) = 0$ since $p1_{K(n)} \simeq 0$. Thus, $H_*(K(n)) \subset H_*(K(n);\mathbb{Z}/p) = 0$.

(ii) Consider the AHSS

$$E^r_{**} \Rightarrow (K(n)_{(0)})_*(H\mathbb{Z}/p),\ E^2_{**} = H_*(H\mathbb{Z}/p;\pi_*(K(n)_{(0)})).$$

Again, $d_i = 0$ for $i < 2p^n - 1$, $d_{2p^n-1} = (Q_n)_*$. Hence, $E^{2p^n}_{m,*} = 0$ for $m \neq 0$, $E^{2p^n}_{0,*} = \mathscr{A}^*/\mathscr{A}^*(Q_n)_*$. □

7.28. Remarks. (a) It was Morava [1] who introduced periodic $K(n)$-theories in the form of $\mathbb{Z}/2$-graded cohomology theories. He called them extraordinary K-theories, and 7.21 clarifies this term. Morava introduced $K(n)$ in a very sophisticated (interesting, but still mysterious) way. Namely, he considered the spectrum (in the sense of algebraic geometry) of the ring $\pi_*(M\mathcal{U})$ and an action on it of a certain group scheme Γ, and theories $K(n)$ arise as some invariants (orbit types) of this action. In fact, $\Gamma = \operatorname{Spec} \mathbb{S}$, and the action above is induced by the $\mathbb{S}$-action on $\pi_*(M\mathcal{U})$. The purely topological description of Morava K-theories was done in Johnson–Wilson [2].

(b) Classification Theorems 7.8–7.11 were proved by Pazhitnov–Rudyak (see Khokhlov–Pazhitnov–Rudyak [1]), Würgler [4], and Hopkins (unpublished). Würgler [4] proved 7.13. Theorem 7.15 was proved by Würgler [4], Pazhitnov [1] (for the connective case with $p > 2$) and, partially, Hopkins (unpublished). The Postnikov tower of $k(n)$ was constructed by Madsen–Margolis (unpublished) and Pazhitnov–Rudyak [1].

The usefulness of Morava K- and k-theories is based on the following. On the one hand, they are rather close to classical (co)homology: they have a simple coefficient ring and simple Postnikov tower. So, they are quite manageable. On the other hand, considering all Morava K- or k-theories (for all p, n) of a certain space, we can get a lot of information on the stable homotopy types of this space. In fact, Morava K-theories give a good global picture of stable homotopy type of finite CW-spaces. I am not able to discuss it here and refer the reader to the monograph of Ravenel [3], but I want to formulate here the remarkable result of Devinatz–Hopkins–Smith [1]. Recall that an endomorphism $h : A \to A$ of an abelian group is called nilpotent if some iteration h^n of it is zero. Similarly, given a spectrum X and a morphism $f : \Sigma^d X \to X$, we say that f is nilpotent if some finite iteration

$$\Sigma^d X \xrightarrow{f} X \xrightarrow{\Sigma^{-d} f} \Sigma^{-d} X \xrightarrow{\Sigma^{-2d} f} \cdots$$

is inessential.

7.29. Theorem. *Let X be a finite spectrum, and let $f : \Sigma^d X \to X$ be a morphism. Then the following three conditions are equivalent:*
(i) *The morphism f is nilpotent;*
(ii) *The homomorphism $f_* : MU_*(\Sigma^d X) \to MU_*(X)$ is nilpotent;*
(iii) *The homomorphism $f_* : K(n)_*(\Sigma^d X) \to K(n)_*(X)$ is nilpotent for every prime p and every n, $0 \le n \le \infty$.* □

Let us see that this theorem implies the following well-known result of Nishida [1]: *Every morphism $f : S^n \to S^0, n > 0$, of sphere spectra is nilpotent.* Indeed, the element $[f] \in \pi_n(S^0)$ has finite order, while $MU_*(S)$ is torsion free. Therefore $f_* : MU_*(S^n) \to MU_*(S^0)$ is the zero (and so a nilpotent) homomorphism.

References

Adams, J.F.

1. An example in homotopy theory. Proc. Cambridge Phil. Soc. **53**, 922–923 (1957)
2. On Chern characters and the structure of the unitary group. Proc. Cambridge Phil. Soc. **57**, 189–199 (1961)
3. Vector fields on spheres. Ann. of Math **75**, 603–632 (1962)
4. On the groups $J(X)$, I–IV. Topology **2**, 181–195 (1963); **3**, 137–171, 193–222 (1964); **5**, 21–71 (1966)
5. Stable homotopy theory. Lecture Notes in Mathematics **3**, Springer, Berlin Heidelberg New York 1964
6. Lectures on generalized cohomology. Lecture Notes in Mathematics **99**, Springer, Berlin Heidelberg New York 1969, pp. 1–138
7. A variant of E.H. Brown's representability theorem. Topology **10**, 185–198 (1971)
8. Stable homotopy and generalised cohomology. The Univ. of Chicago Press, Chicago 1974
9. Infinite loop spaces. Ann. Math. Studies **90**, Princeton Univ. Press, Princeton, New Jersey 1978

Adams, J.F., Priddy, S.

1. Uniqueness of BSO. Proc. Cambridge Phil. Soc. **80**, 475–509 (1976)

Adams, J.F., Walker, G.

1. An example in homotopy theory. Proc. Cambridge Phil. Soc. **60**, 699–700 (1964)

Allaud, G.

1. On the classification of fiber spaces. Math. Z. **92**, 110–125 (1966)
2. Concerning universal fibrations and a theorem of E. Fadell. Duke Math. J. **37**, 213–224 (1970)

Anderson, D.

1. There are no phantom cohomology operations in K-theory. Pacific J. Math. **107** (2), 279–306 (1983)

Araki, S.

1 Typical formal groups in complex cobordism and K-theory. Kinokuniya Bookstore Co, Ltd, Tokyo 1973

Araki, S., Yosimura, Z.

1. A spectral sequence associated with a cohomology theory of infinite CW-complexes. Osaka J. Math. **9**, 351–365 (1972)

Araki, S., Toda, H.

1. Multiplicative structure in mod q cohomology theories. I – Osaka J. Math. **2** (1), 71–115 (1965), II – Osaka J. Math. **3** (1), 81–120 (1966)

Atiyah, M.F.

1. Bordism and cobordism. Proc. Cambridge Phil. Soc **57**, 200–208 (1961)
2. Characters and cohomology of finite groups. Publ. Math. IHES **9**, 247–288 (1961)
3. Thom complexes. Proc. London Math. Soc. **11** (3), 291–310 (1961)
4. K-Theory. Benjamin, New York 1967

Atiyah, M.F., Bott, R., Shapiro, A.

1. Clifford modules. Topology **3**, suppl. 1, 3–38 (1964)

Averbuch, B.G. (Авербух, Б.Г.)

1. Алгебраическая структура групп внутренних гомологий. Доклады АН СССР, **125**, 11–14 (1959) The intrinsic homology groups]

Baas, N.A.

1. On bordism theories of manifolds with singularities. Math. Scand. **33**, 285–298 (1973)
2. On formal group laws and singularities in complex cobordism. Math.Scand. **33**, 303-313 (1973)

Baas, N., Madsen, I.

1. On the realization of certain modules over the Steenrod algebra. Math. Scand. **31**, 220–224 (1972)

Barratt, M.G.

1. Track groups, I, II. Proc. London Math. Soc. **5**, 71–106, 285–329 (1955)

Birman, J.

1. Braids, links, and mapping class groups. Ann. Math. Studies **82**, Princeton Univ. Press, Princeton, New Jersey 1974

Boardman, J.

1. Graded Eilenberg–Mac Lane spectra. Amer. J. Math. **105** (5), 979–1010 (1980)
2. Splitting of MU and certain other spectra. Lecture Notes in Mathematics **658**, Springer, Berlin Heidelberg New York 1978, pp. 27–79

Becker, J.T.

1. Extensions of cohomology theories. Illinois J. Math, **14**, pp. 551–584 (1970)

Boardman, J.M., Vogt, R.M.

1. Homotopy invariant algebraic structures on topological spaces. Lecture Notes in Mathematics **347**, Springer, Berlin Heidelberg New York 1973

Booth, P., Heath, P., Piccinini, R.

1. Characterizing universal fibrations. Lecture Notes in Mathematics **673**, Springer, Berlin Heidelberg New York 1978, pp. 165–185

Borel, A.

1. Sur la cohomologie des espaces fibrés principaux et des espaces homogènes de groups de Lie compact. Ann. of Math. **57**, 115–207 (1953)

Bott, R.

1. The stable homotopy of the classical groups. Ann. of Math. **70**, 313–337 (1959)

Botvinnik, B.

1. Manifolds with singularities and the Adams–Novikov Spectral Sequence. London Mathematical Society, Lecture Note Series **170**, Cambridge Univ. Press, Cambridge 1992.

Bourbaki, N.

1. Algèbre I, Ch. 1–3. Hermann, Paris 1970 [English translation: Algebra I, Ch. 1–3 Springer, Berlin Heidelberg New York Tokyo 1989]
2. Topologie Générale, Ch.1,2. Hermann, Paris 1971 [English translation: General topology, Chapters 1–4, Springer, Berlin Heidelberg New York London Tokyo 1989]
3. Algèbre commutative, Ch.1–7. Masson, Paris 1985 [English translation: Commutative Algebra, Ch.1–7, Springer, Berlin Heidelberg New York London Tokyo 1989]

Bousfield, A.K., Kan, D.M.

1. Homotopy limits, Completion and Localization. Lecture Notes in Mathematics **304**, Springer, Berlin Heidelberg New York 1972

Bröcker, T., tom Dieck, T.

1. Kobordismentheorie. Lecture Notes in Mathematics **178**, Springer, Berlin Heidelberg New York 1970.

Browder, W.

1. Homotopy type of differentiable manifolds. Proceedings of the Aarhus Symposium, Aarhus Univ. 1962, 42–46
2. The Kervaire invariant of framed manifolds and its generalization. Ann. of Math. **90**, 157–186 (1968)
3. Surgery on simply connected manifolds. Springer, Berlin Heidelberg New York 1972

Brown, E.H.

1. Cohomology theories. Ann. of Math. **75**, 467–484 (1962); correction, Ann. of Math. **78**, 201 (1963)

Brown, E.H., Peterson, F.P.

1. A spectrum whose cohomology is the algebra of reduced p^{th}-powers. Topology **5**, 39–52 (1964)
2. Relations between characteristic classes I. Topology **3**, 149–154 (1966)

Brumfiel, G.

1. Realizing homology classes by PL-manifolds. Lecture Notes in Mathematics **644**, Springer, Berlin Heidelberg New York 1978, pp. 43–60

Brumfiel, G., Madsen, I., Milgram, J.

1. PL characteristic classes and cobordism. Ann. of Math. **97**, 82–159 (1973)

Buhštaber, V.M. (Бухштабер, В.М.)

1. Характер Чженя–Дольда в кобордизмах, I. Матем. Сборник **83(125)** (4), 575–595 (1970) [English translation: The Chern–Dold character in cobordisms, I. Math. USSR Sbornik **12** (4), 573–594 (1970)]
2. Новые методы в теории кобордизмов. An Addendunm to the Russian translation of Stong [3], Мир, Москва 1972
3. Характеристические классы в кобордизмах и топологические приложения теории однозначных и двузначных формальных групп. Современные проблемы математики **10**, 5–178 (1978) [English translation: Characteristic classes in cobordism and topological applications of single-valued and two-valued formal groups. J. Soviet Math. **11** (6), (1979)]

Buhštaber, V.M., Miščenko, A.S., Novikov, S.P. (Бухштабер, В.М., Мищенко, А.С, Новиков, С.П.)

1. Формальные группы и их роль в аппарате алгебраической топологии. Успехи Матем. Наук **26** (2), 131-154 (1971) [English translation: Formal groups and their role in the apparatus of algebraic topology. Russ. Math. Surveys **26** (6), 63–90 (1971)]

Bullett, S.

1. Permutations and braids in cobordism theory, Proc. of the London Math. Soc. **38**, Ser. 3, 517–531 (1979)

Cartan, H.

1. Séminaire H. Cartan. Paris 1959.

Cartan, H., Eilenberg, S.

1. Homological algebra, Princeton Univ. Press, Princeton,1956

Cerf, J.

1. La nullité du groupe Γ_4. Sem. H. Cartan, **8,9,10,20,21**, Paris, 1962–63

Clapp, M., Puppe, D.

1. Invariants of Lusternik–Schnirelmann type and the topology of critical sets. Trans. Amer. Math. Soc. **288** (2), 603–620 (1986)

Cohen, F.R.

1. Braid orientations and bundles with flat connections. Inventiones Math. **46**, 99-110 (1978)

Cohen, F.R., Lada, T.J., May, J.P.

1. The homology of iterated loop spaces. Lecture Notes in Mathematics **533**, Springer, Berlin Heidelberg New York 1976

Cohen, J.M.

1. The Hurewicz homomorphism on MU. Invent. Math. **10**, 177–186 (1970)

Conner, P.

1. Differentiable periodic maps (second edition). Lecture Notes in Mathematics **738**. Springer, Berlin Heidelberg New York 1979

Conner, P., Floyd, E.

1. The relation of cobordism to K-theories. Lecture Notes in Mathematics **28**. Springer, Berlin Heidelberg New York 1966

Conner, P., Smith, L.

1. On complex bordism of finite complexes. Publ. Math. IHES **37**, 117–221 (1969)

Devinatz, E., Hopkins, M.J., Smith, J.H.

1. Nilpotence and stable homotopy theory. Ann. of Math. **128**, 207–242 (1988)

Dold, A.

1. Relations between ordinary and extraordinary cohomology. Notes. Aarhus Colloquium in Algebraic Topology, Aarhus 1962
2. Partitions of unity in the theory of fibrations. Ann. of Math. **78**, 223–255 (1963)
3. Halbexakte Homotopiefunktoren. Lecture Notes in Mathematics **12**, Springer, Berlin Heidelberg New York 1966
4. Chern classes in general cohomology. Inst. Nazionale di Alta Matematica, Symposia Mathematica **V**, 385–410 (1970)
5. Lectures on algebraic topology. Springer, Berlin Heidelberg New York 1972
6. Geometric Cobordism and the Fixed point Transfer. Lecture Notes in Mathematics **673**, Springer, Berlin Heidelberg New York 1978, pp. 33–87

Dold, A., Lashof, R.

1. Principal quasifibrations and fibre homotopy equivalence of bundles. Illinois J. Math. **3**, 285–305 (1959)

Dold, A., Puppe, D.

1. Двойственность, след и трансфер. Труды Матем. Инст. им. В.А. Стеклова **154**, 81–97 (1983) [English translation: Duality, trace and transfer. Proceedings of the Steklov Inst. of Math. **4**, 85–103 (1984)]

Dold, A., Thom, R.

1. Quasifaserungen und unendlische symmetrische Produkte. Ann. of Math. **67**, 239–281 (1958)

Donaldson, S.

1. An application of gauge theory to the topology of 4-manifolds. J. Diff. Geom. **18**, 279–315 (1983)

Dowker, C.H.

1. Topology of metric complexes. Amer. J. Math. **74**, 555–577 (1952)

tom Dieck, T., Kamps, K., Puppe, D.

1. Homotopientheorie. Lecture Notes in Mathematics **157**, Springer, Berlin Heidelberg New York 1970

tom Dieck, T.

1. Operation in Kobordismen-Theorien. Math. Z., **107**, 380–401 (1967)
2. Topologie, Walter de Gruyter, Berlin New York 1991

Dubrovin, B.A., Novikov, S.P., Fomenko, A.T. (Дубровин, Б.А., Новиков, С.П., Фоменко, А.Т.)

1. Современная геометрия, II. Наука, Москва 1979 [English translation: Modern geometry, II. Springer, Berlin Heidelberg New York 1985.]

Dyer, E.

1. Cohomology theories. Benjamin, New York 1969

Dyer, E., Lashof, R.

1. Homology of iterated loop spaces. Amer. J. Math. **84**, 53-88 (1962)

Eilenberg, S.

1. Problems in topology. Ann. of Math. **50**, 246-260 (1949)

Eilenberg, S., Steenrod, N.

1. Foundations of algebraic topology. Princeton Mathematical Series **15**, Princeton Univ. Press, Princeton, New Jersey 1952

Elmendorf, A.D., Kriz, I., Mandell, M.A., May, J.P.

1. Rings, Modules, and Algebras in Stable Homotopy Theory. Mathematical Surveys and Monographs **47**, AMS, Providence, Rhode Island 1997.

Fomenko, A.T., Fuchs, D.B., Gutenmacher, V.L. (Фоменко, А.Т., Фукс, Д.Б., Гутенмахер, В.Л.)

1. Гомотопическая топология. Изд-во МГУ, Москва 1972 [English translation: Homotopic topology. Akadémiai Kiadó, Budapest 1986]

Fox, R.

1. Homotopy type and deformation retracts. Ann. of Math. **44**, 40–50 (1943)

Freedman, M.

1. The topology of four-dimensional manifolds. J. Diff. Geom. **17**, 157–454 (1982)

Fritsch, R., Piccinini, R.A.

1. Cellular structures in topology. Cambridge Univ. Press, Cambridge, 1990

Fröhlich, A.

1. Formal Groups. Lecture Notes in Mathematics **74**, Springer, Berlin Heidelberg New York 1968

Fuks, D.B. (Фукс, Д.Б.)

1. Когомологии группы кос mod 2. Функциональный анализ и его приложения **4** (2), 62–75 (1970) [English translation: Cohomology of the braid group mod 2. [19] Functional analysis and its applications, **4** (2), 143–151 (1970)]

Fuks, D.B., Rokhlin, V.A. (Фукс, Д.Б., Рохлин, В.А.)

1. Начальный курс топологии. Наука, Москва 1977 [English translation: Beginners Course in Topology. Springer, Berlin Heidelberg New York Tokyo 1984]

Godement, R.

1. Topologie algébrique et theorie des faisceaux. Paris, Hermann 1958.

Gottlieb, D.H.

1. The total space of universal fibration. Pacific J. Math. **46** (2), 415–417 (1973)

Gray, B.

1. Homotopy theory. Academic Press, New York London San Francisco 1975

Gromov, M., Lawson, H.B.

1. The classification of simply connected manifolds of positive scalar curvature. Ann. of Math. **111**, 423–434 (1980)

Hattory, A.

1. Integral characteristic numbers for weakly almost complex manifolds. Topology **5**, 259-280 (1966)

Hazewinkel, M.

1. Formal groups and applications. Academic Press, New York London San Francisco 1978

Hegenbarth, F.

1. Secondary cohomology operations applied to Thom class. Lecture Notes in Mathematics **778**, Springer, Berlin Heidelberg New York 1980, pp. 442–455

Heller, A.

1. On the representability of homotopy functors. J. London Math. Soc. **23** (2), 551-562 (1981)

---

[19] The title of the English translation is " Cohomology of the group COS mod 2. The Russian word кос was just transliterated, and so the words "группа кос" turned into "group COS" instead of "braid group".

Hilton, P., Mislin, G., Roitberg, J.

1. Localization of Nilpotent Groups and Spaces. North Holland Publishing Company, Amsterdam 1975

Hilton, P., Wiley, S.

1. Homology theory. Cambridge Univ. Press, Cambridge 1965

Hirsch, M.

1. Immersions of manifolds. Trans. Amer. Math. Soc., **93**, 242–276 (1959)

Hirsch, M., Mazur, B.

1. Smoothing of piecewise linear manifolds. Ann. Math. Studies **80**, Princeton Univ. Press, Princeton, New Jersey 1974

Hirzebruch, F.

1. Topological methods in algebraic geometry (third edition, translated). Springer, Berlin Heidelberg New York 1966

Hopf, H.

1. Uber die Topologie der Gruppen-Mannigfaltigkeiten und ihrer Verallgemeinnerungen. Ann. of Math. **42**, 22–52 (1941)

Honda, T.

1. Formal groups and zeta-function. Osaka J. Math. **5**, 199–219 (1968)

Hopkins, M.J.

1. Global methods in homotopy theory. Homotopy theory. Proceedings Symp. Durham/England 1985. London Mathematical Society, Lecture Note Series **117**, Cambridge Univ. Press, Cambridge 1987, pp. 73–96

Hopkins, M.J., Smith, J.H.

1. Nilpotence and stable homotopy theory II. (to appear in Ann. of Math.)

Hu, S.T.

1. Homotopy theory. Pure and Applied Mathematics VIII. Academic Press, New York 1959

Husemoller, D.

1. Fibre bundles (second edition). Springer, Berlin Heidelberg New York 1972 1975

James, I.M., Whitehead, J.H.K.

1. Homology with zero coefficients. Quart. J. Math. **9**, 317–320 (1958)

Jensen, C.U.

1. Les Foncteurs Dérivés de lim et leurs Application en Théorie des Modules. Lecture Notes in Mathematics **254**, Springer, Berlin Heidelberg New York 1972

Johnson, D.C.

1. A Stong–Hattory spectral sequence. Trans. Amer. Math. Soc., **179**, 211-225 (1973)

Johnson, D.C., Smith, L.

1. On the relation of complex cobordism to connective K-theory. Proc. Cambridge Phil. Soc., **70**, 19–22 (1971)

Johnson, D.C., Wilson, W.S.

1. Projective dimension and Brown–Peterson homology. Topology **12**, 327–353 (1973)
2. BP-operations and Morava's extraordinary K-theories. Math. Z. **144** (1), 55–75 (1975)

Kamps, K.

1. Kan-Bedingungen und abstrakte Homotopientheorie. Math. Z. **124**, 215–236 (1972)

Karoubi, M.

1. K-Theory. Springer, Berlin Heidelberg New York 1978

Kelley, J.

1. General Topology (reprint edition). Springer, Berlin Heidelberg New York 1975

Kervaire, M.A., Milnor, J.W.

1. Groups of homotopy spheres I. Ann. of Math. **77**, 504-537 (1963)

Kirby, R., Siebenmann, L.

1. Foundational Essays on Topological Manifolds, Smoothing and Triangulations. Ann of Math. Studies **88**, Princeton Univ. Press, Princeton, New Jersey 1977

Kister, J.

1. Microbundles and fibre bundles. Ann. Math. **80**, 190–199 (1964)

Khokhlov, A.V., Pazitnov, A.V., Rudyak, Yu.B.

1. On the homotopy properties of Morava extraordinary k-theories. Lecture Notes in Mathematics **1060**, Springer, Berlin Heidelberg New York 1984, pp. 151–166

Knapp, K.-H., Ossa, E.

1. KO-codegree and real line bundles. Lecture Notes in Mathematics **1418**, Springer, Berlin Heidelberg New York 1990, pp. 130–139

Kudo, T., Araki, S.

1. On $H_*(\Omega^n S^n; \mathbb{Z}_2)$. Proc. Japan Acad. Sci.**32**, 333–335 (1956)
2. Topology of H_n-spaces and H_n-squaring operations. Mem. Fac. Sci. Kyusyu. Univ., Ser. A 10, 85–120 (1956)

Kuiper, N., Lashof, R.

1. Microbundles and bundles, I. Elementary theory. Invent. Math. **1**, 1–17 (1966)

Landweber, P.S.

1. Cobordism operations and Hopf algebras. Trans. Amer. Math. Soc. **129**, 94–110 (1967)

2. Complex bordism of classifying spaces. Proc. Amer. Math. Soc. **27**, 175–179 (1971)
3. Annihilator ideals and primitive elements in complex bordism. Illinois J. Math. **17**, 273–284 (1973)
4. Associated prime ideals and Hopf algebras. J. Pure and Applied Algebra **3**, 43–58 (1973)
5. Homological properties of comodules over $MU_*(MU)$ and $BP_*(BP)$. Amer. J. Math., **98**, 591–610 (1976)

Lang, S.

1. Algebra. Addison–Wesley, Reading, Massachusets 1984

Lashof, R.

1. Poincaré duality and cobordism. Trans. Amer. Math. Soc. **189**, 257–277 (1963)

Lashof, R., Rothenberg, M.

1. Microbudles and smoothing. Topology **3**, 357–388 (1965)

Lazard, M.

1. Lois de groupes et analyseurs. Ann Écoles norm. Sup., **72**, 299–400 (1955)
2. Sur les groupes de Lie formels á un paramètre. Bull. Soc. Math. France, **83**, 251–274 (1955)

Lewis, L.G.

1. When is the natural map $X \to \Omega\Sigma X$ a cofibration? Trans. Amer. Math. Soc. **273**, 147–155 (1982)

Lewis, L.G., May, J.P., Steinberger, M.

1. Equivariant Stable Homotopy Theory. Lecture Notes in Mathematics **1213**, Springer, Berlin Heidelberg London Paris New York Tokyo 1986

Lima, E.L.

1. The Spanier–Whitehead duality in new categories. Summa Brasilias Math. **4**, 9-48 (1959)
2. Stable Postnikov invariants and their duals. Summa Brasilias Math. **4**, 193–281 (1960)

Mac Lane, S.

1. Natural associativity and commutativity. Rice University Studies **49** (4), Houston 1963
2. Homology (third edition). Springer, Berlin Heidelberg New York 1975

Madsen, I.

1. Higher torsion in SG and BSG. Math. Z. **143**, 55–80 (1975)

Madsen, I., Milgram, R.

1. The classifying spaces for surgery and cobordism of manifolds. Ann. of Math. Studies **92**, Princeton Univ. Press, Princeton, New Jersey 1979.

Mahowald, M.

1. A new family in ${}_2\pi_*^S$. Topology **16**, 249–256 (1977)

Mahowald, M., Ray, N.

1. A note on the Thom isomorphism. Proc. Amer. Math. Soc. **82**, 307–308 (1981)

Margolis, H.R.

1. Spectra and the Steenrod algebra. North Holland, Amsterdam New York Toronto 1983

Massey, W.

1. On the Stiefel–Whitney classes of a manifold. Amer. J. Math. **82**, 92–102 (1960)

Matsumoto, Y.

1. Topological t-regularity and Rohlin's theorem. J. Fac. Sci. Univ. Tokyo, Sect I, **18**, 97-108 (1971)

Matveev, A.G. (Матвеев, А.Г.)

1. Представимость полуточных функторов на категории пунктированных клеточных пространств. Матем. Заметки **39** (6), 869–876 (1986) [English translation: Representability of semiexact functors on the category of pointed cellular spaces. Math. Notes **39**, 471–474 (1986)]

May, J.P.

1. The geometry of iterated loop spaces. Lecture Notes in Mathematics **271**, Springer, Berlin Heidelberg New York 1972
2. Classifying spaces and fibrations. Memoirs of the American Mathematical Society., issue 1, no 155, AMS, Providence, Rhode Island 1975.
3. (with contributions by N.Ray, F.Quinn and J. Tørnehave) E_∞-ring spaces and E_∞-ring spectra. Lecture Notes in Mathematics **577**, Springer, Berlin Heidelberg New York 1977.
4. Infinite loop spaces theory. Bull. Amer. Math. Soc. **83** (4), 456–494 (1977)

McCord, M.C.

1. Classifying spaces and infinite symmetric products. Trans. Amer. Math. Soc. **146**, 273–298 (1969)

McGibbon, C.A.

1. Phantoms Maps. Handbook of Algebraic Topology, ed. by I.M.James, Elsevier, Amsterdam Lausanne NewYork Oxford Shannon Tokyo 1995, pp. 1209–1257

Milgram, R.

1. The bar construction and abelian H-spaces. Ill. J. Math. **11**, 242–250 (1967)
2. The structure over the Steenrod algebra of some 2-stage Postnikov system. Quart. J. Math. Oxford **20**, 161–169 (1969)

Miller, H.R., Ravenel, D.C.

1. Morava stabilizer algebras and the localization of Novikov's E_2-term. Duke Math. J. **44** (2), 433–447 (1977)

Milnor, J.W.

1. Construction of universal bundles II. Ann. of Math. **63**, 430–436 (1956)
2. The Steenrod algebra and its dual. Ann. of Math. **67**, 150–171 (1958)
3. On spaces having the homotopy type of a CW-complex. Trans. Amer. Math. Soc. **90**, 272–280 (1959)
4. On the cobordism ring Ω^* and a complex analogue, Part I. Amer. J.Math. **82**, 505–521 (1960)
5. On axiomatic homology theory. Pacific J. Math. **12**, 337–341 (1962)
6. Morse theory. Ann of Math. Studies **51**, Princeton Univ. Press, Princeton, New Jersey 1963
7. Microbundles, I. Topology **3**, Suppl. 1, 53–80 (1964)
8. Lectures on the h-cobordism theorem. Princeton Univ. Press, Princeton, New Jersey 1965

Milnor, J.W., Moore, J.C.

1. On the structure of Hopf algebras. Ann. of Math. **81**, 211–264 (1965)

Milnor, J.W., Spanier, E.H.

1. Two remarks on fiber homotopy type. Pacif. J. Math. **10**, 585–590 (1960)

Milnor, J.W., Stasheff, J.D.

1. Characteristic classes. Ann. of Math. Studies **76**, Princeton Univ. Press, Princeton, New Jersey 1974

Mironov, O.K. (Миронов, О.К.)

1. Существование мультипликативных структур в теориях кобордизмов с особенностями. Известия АН СССР **39** (5), 1056–1092 (1975) [English translation: Existence of multiplicative structures in the theory of cobordism with singularities. Math. USSR Izvestia **9** (5), 1007–1034 (1975)]
2. Умножения в теориях кобордизмов с особенностями и операции Стинрода–Дика. Известия АН СССР **42** (4), 789–806 (1978) [English translation: Multiplications in the theory of cobordism with singularities and Steenrod–tom Dieck operations.Math. USSR Izvestia **13** (1), 89–107 (1979)]

Miyazaki, H.

1. Paracompactness of CW-complexes. Tohoku. Math. J. **4**, 309–313 (1952)

Moore, J., Smith, L.

1. Hopf Algebras and Multiplicative Fibrations, I. Amer. J. Math. **90**, 753–778 (1968)

Morava, J.

1. Extraordinary K-theories. Preprint of Columbia University, 1969.
2. Noetherian localization of categories of cobordism comodules. Ann. of Math. **121**, 1–39 (1985)

Mosher, R.E., Tangora, M.C.

1. Cohomology operations and applications in homotopy theory. Harper and Row, New York 1968

Munkres, J.

1. Differential isotopies of the 2-sphere. Michigan Math. J. **7**, 193–197 (1960)
2. Topology. Prentice Hall, Inc., Englewood Cliffs, New Jersey, 1965
3. Elementary Differential Topology, Ann. of Math. Studies **54**, Princeton Univ. Press, Princeton, New Jersey 1966

Nishida, G.

1. Cohomology operations in iterated loop spaces. Proc. Japan. Acad. **44**, 104–109 (1968)

Novikov, S.P. (Новиков, С.П.)

1. О некоторых задачах топологии многообразий, связанных с теорией пространств Тома. Доклады АН СССР **132** (5), 1031–1034 (1960) [English translation: Some problems in the topology of manifolds connected with the theory of Thom spaces. Soviet Math. Dokl. **1**, 717–720 (1960)]
2. О диффеоморфизме односвязных многообразий. Доклады АН СССР **143** (5), 1046–1049 (1962) [English translation: Diffeomorphism of simply connected manifolds. Soviet Math. Dokl. **3**, 540–543 (1962)]
3. Гомотопически еквивалентные гладкие многообразия, I. Известия АН СССР **28**, 365–474 (1964) [English translation: Homotopy equivalent smooth manifolds, I. AMS Translation **48**, 271–396 (1965)]
4. Методы алгебраическои топологии с точки зрения теории кобордизмов. Известия АН СССР **31** (4) 855–951 (1967) [English translation: The methods of algebraic topology from the viewpoint of cobordism theory. Math. USSR Izvestija **1** (4), 827–913 (1967)]
5. Pontrjagin classes, the Fundamental Group and Some Problems of Stable Algebra, 147–155, Essays on Topology and Related Topics, Memoires dédiés à Georges de Rham, 1970

Ossa, E.

1. Topologie, Vieweg, Braunschweig Wiesbaden, 1992

Pazhitnov, A.V. (Пажитнов, А.В.)

1. Теоремы единственности для обобщенных теории когомологий. Известия АН СССР **47** (3), 518–543 (1983) [English translation: Uniqueness theorems for generalized cohomology theories. Math. USSR Izvestia **22**, 483–506 (1984)]

Pazhitnov, A.V., Rudyak, Yu.B. (Пажитнов, А.В., Рудяк, Ю.Б.)

1. О гомотопическом типе высших K-теорий Моравы, Препринт Новосибирского Университета, Новосибирск 1981[On the homotopy type of the higher Morava K-theories]
2. О коммутативных кольтцевых спектрах характеристики 2. Матем. Сборник **124** (4), 486–494 (1984) [English translation: On commutative ring spectra of characteristic 2. Math. USSR Sbornik **52** (2), 471–479 (1985)]

Palais, R.

1. Seminar on the Atiyah–Singer Index theorem. Princeton Univ. Press, Princeton, New Jersey 1965

Peterson, F.P.

1. Some results on PL cobordism. J. Math. Kyoto Univ. **9**, 189–194 (1969)
2. Twisted cohomology operations and exotic characteristic classes. Advances in Math. **4**, 81–90 (1970)

Peterson, F.P., Toda, H.

1. On the structure of $H^*(BSF;\mathbb{Z}_p)$. Journal of Math. of Kyoto Univ. **7** (2), 113–121 (1967)

Poincaré, H.

1. Analysis Situs. Journale de l'Ecole Polytechnique **1**, 1–121, (1895)

Pontrjagin, L.S. (Понтрягин, Л.С.)

1. Характеристичецкие циклы на дифференцируемых многообразиях. Матем. сб., **21(63)**, 233–284, (1947) [English translation: Characteristic cycles on differentiable manifolds. AMS translations, ser. 1, **32**, AMS, Providence, RI, 1959]
2. Гладкие многообразия и их применения в теории гомотопий. Труды Матем. Инст. им. В.А. Стеклова, **45**, 1-139, Moscow, 1955 [English translation: Smooth manifolds and their applications in homotopy theory. AMS translations, ser. 2, **11**, 1–114, AMS, Providence, RI, 1959]

Postnikov, M.M. (Постников, М.М.)

1. Локализация топологических пространств. Успехи. Матем. Наук **32** (6), 117–181 (1967) [English translation: Localization of topological spaces. Russ. Math. Surveys **32** (6), 121–184 (1977)]
2. Лекции по алгебраической топологии. Теория гомотопий клеточных пространств. Наука, Москва 1985[Lectures on algebraic topology. Homotopy theory of cellular spaces]

Prieto, C.

1. The relative spectral sequence of Leray-Serre for fibration pairs, Monografias del Instituto de Matematicas, no. 8, UNAM, Mexico 1979

Puppe, D.

1. Homotopiemenge und ihre induzierten Abbildungen, I. Math. Z., **69**, 299-344 (1958)
2. On the stable homotopy category. Topology and its Applications. Budva 1972 (Beograd 1973), pp. 200–212

Quillen, D.

1. On the formal group laws of unoriented and complex cobordism theories. Bull. Amer. Math. Soc. **75**, 1293–1298 (1969)
2. Elementary proofs of some results of cobordism theory using Steenrod operations. Advances in Math. **7**, 29–56 (1971)

Ravenel, D.

1. A definition of exotic characteristic classes of spherical fibrations. Comm. Math. Helv. **47**, 421–436 (1972)
2. Complex Cobordism and Stable Homotopy Groups. Academic Press, Orlando, Florida 1986.
3. Nilpotence and Periodicity in Stable Homotopy Theory. Ann. Math. Studies **128**, Princeton Univ. Press, Princeton, New Jersey 1992

Rokhlin, V.A. (Рохлин, В.А.)

1. Трехмерное многообразие — граница четырехмерного. Докл. АН СССР **81**, 355–357, (1951) [A three-dimensional manifold is a boundary of a four-dimensional one.]

Roos, J. E.

1. Sur le dérivés de $\varprojlim$. Applications. C.R. Acad. Sci. Paris **252**, 3702–3704 (1961)
2. Bidualité et structure des foncteurs dérives de $\varprojlim$ dans la categorie des modules sur un anneau regulier. C.R. Acad. Sci. Paris **254**, 120–1722 (1962)

Rourke, C., Sanderson, B.

1. Introduction to piecewise-linear topology. Springer, Berlin Heidelberg New York 1972

Rudyak, Yu. B. (Рудяк, Ю.Б.)

1. Формальные группы и бордизмы с особенностями. Матем. Сборник **96** (4), 523–542 (1975) [English translation: Formal groups and bordism with singularities. Math. USSR Sbornik **25** (4), 487–506 (1975)]
2. Теоремы типа Коннера–Флойда в теории унитарных кобордизмов. Изв. АН СССР **42** (3), 654–666 (1978) [English translation: Theorems of Conner–Floyd type in the theory of unitary cobordism. Math. USSR Izvestia **12** (3), 605–616 (1978)]
3. Об изоморфизме Тома–Дольда для неориентируемых расслоений. Доклады АН СССР **255** (6), 1323–1325 (1980) [English translation: On the Thom–Dold isomorphism for non-orientable vector bundles. Soviet Math. Doklady **22** (3), 842–844 (1980)]
4. Теоремы о точности в теориях когомологий MU, BP и $P(n)$. Матем. Заметки **40** (1), 115–126 (1986) [English translation: Exactness theorems for the cohomology theories MU, BP and $P(n)$). Math. Notes **40**, no. 1-2, 562–569 (1986)]
5. О реализации гомологических классов PL многообразиями с особенностями. Матем. Заметки **41** (5), 741–749 (1987) [English translation: On the realization of homology classes by PL manifolds with singularities. Math. Notes **41**, no. 5-6, 417–422 (1987)]
6. Об ориентируемости сферических, топологических и кусочно ли-нейных расслоений в комплексной K-теории. Доклады АН СССР **298** (6), 1338–1341 (1988) [English translation: On the orientability of spherical fibrations and topological and piecewise linear bundles in complex K-theory. Soviet Math. Doklady **37** (1), 283–286 (1988)]

7. Некоторые соотношения в кольце кусочно линейных кобордизмов. Матем. Заметки **46** (6), 75–84 (1989) [English translation: Some relations in the piecewise linear cobordism ring. Math. Notes **46**, no. 5-6, 939–945 (1989)]
8. Ориентируемость расслоений: теория препятствий и приложения к K-теории. Матем. Сборник **180** (11), 1561–1582 (1989) [English translation: Orientability of bundles and fibrations: obstruction theory and applications to K-theory. Math. USSR Sbornik **68** (2), 429–451 (1991)]
9. On the orientability of bundles and fibrations: the obstruction theory and applications to K, KO and Morava K-theories. Algebraic Topology Oaxtepec 1991. Contemp. Math. **146**, AMS, Providence, Rhode Island 1993, pp. 384–405

Rudyak, Yu.B, Khokhlov, A.V. (Рудяк, Ю.Б., Хохлов, А.В.)

1. Высшие когомологические операции на классах Тома. Вестник Московск. Университета, Мат., Мех. 6, 26–31 (1984) [English translation: Higher cohomology operations on Thom classes. Moscow Univ. Math. Bull, **39** (6), 36–43 (1984)]

Scharlemann, M.

1. Transversality theories in dimension 4. Invent. Math. **33**, 1-14, (1976)

Schön, R.

1. The Brownian classification of fiber spaces. Archiv der Math. **39** (4), 359–365 (1982)

Serre, J-P.

1. Homologie singulière des espaces fibrés. Ann. of Math. **54**, 425–505 (1951)
2. Cohomologie modulo 2 des complexes d'Eilenberg– Mac Lane, Comm. Math. Helv. **27**, 198–231 (1953)
3. Groupes d'homotopie et classes de groupes abéliens. Ann. of Math. **58**, 258–294 (1953)

Shimada, N., Yagita, N.

1. Multiplication in the complex bordism theory with singularities. Publ. Res. Inst. Math. Sci. **12** (1), 251—293 (1976/1977)

Siebenmann, L., Guillou, L., Hahl, H.

1. Les voisinages ouverts réguliers. Annales Ecole Normal Sup. **6**, 253–293 (1973)

Smale, S.

1. Classification of immersions of spheres to Euclidean space. Ann. of Math. **69**, 327–344 (1959)
2. Diffeomorphisms of 2-sphere. Proc. Amer. Math. Soc. **10**, 621–626 (1959)
3. On the Structure of Manifolds. Amer. J. Math. **84**, 287–399 (1962)

Smith, L.

1. On the finite generation of $\Omega_*^U(X)$. Journal of Math. and Mech. **12**, 10, 1017–1023 (1969)

Spanier, E.H.

1. Function spaces and duality. Ann. of Math. **70**, 338–378 (1959)
2. Algebraic Topology. McGraw-Hill, New York 1966.

Spanier, E.H., Whitehead, J.H.C.

1. Duality in homotopy theory. Mathematika, **2**, 56–80 (1955)

Stasheff, J.

1. A classification theorem for fiber spaces. Topology **2**, 239–246 (1963)

Steenrod, N.E.

1. The topology of fibre bundles. Princeton Mathematical Series **14**, Princeton, Princeton Univ. Press, Princeton, New Jersey 1951
2. A convenient category of topological spaces. Michgan Math. J. **14**, 133–152 (1967)

Steenrod, N.E., Epstein, D.B.A.

1. Cohomology operations. Annals of Mathematics Studies **50**, Princeton, Princeton Univ. Press 1962

Stolz, S.

1. Simply connected manifolds of positive scalar curvature. Ann. of Math. **136**, 511–540 (1992)

Stong, R.E.

1. Determination of $H^*(BO(k,...,\infty);\mathbb{Z}/2)$ and $H^*(BU(k,...,\infty);\mathbb{Z}/2)$. Trans. Amer. Math. Soc. **107**, 526–544 (1963)
2. Relations among characteristic numbers, I, II. Topology **4** (1965), 267–281 (1965); **5**, 133-148 (1966)
3. Notes on cobordism theory. Mathematical notes. Princeton Univ. Press, Princeton, New Jersey 1968
4. Orientability mod 2. J. London Math. Soc. (2) **14**, 296–302 (1976)

Strøm, A.

1. Note on cofibrations I. Math. Scand. **19**, 1–14 (1966)
2. Note on cofibrations II. Math. Scand. **22**, 130–142 (1968)

Sullivan, D.

1. On the Hauptvermutung for manifolds. Bull. Amer. Math. Soc. **73**, 598–600 (1967)
2. Geometric topology. Localization, Periodicity and Galois symmetry. The M.I.T. Press, Cambridge, Massachusets 1970

Switzer, R.W.

1. Algebraic Topology – Homotopy and Homology. Springer, Berlin Heidelberg New York 1975

Thom, R.

1. Espaces fibrés en sphères et carrés de Steenrod. Ann. Sci. Ecole Norm. Sup., **69** (3), 109-181, (1952)
2. Quelques propriétés globales des variétés differentiables. Comm. Math. Helv. **28**, 17–86 (1954)

Toda, H.

1. Composition methods in homotopy groups of spheres. Ann. of Math. Studies **49**, Princeton Univ. Press, Princeton, New Jersey 1962

Tsuchia, A.

1. Characteristic classes for PL-microbundles. Nagoya Math. J. **43**, 169–198 (1971)
2. Characteristic classes for spherical fiber spaces. Nagoya Math. J. **43**, 1–39 (1971)

Vershinin, V.V.

1. Cobordisms and Spectral Sequence. Translation of the Mathematical Monographs **130**, AMS, Providence, Rhode Island 1993.

Vick, J.

1. Homology theory. Springer, Berlin Heidelberg New York London Tokyo, 1994

Vogt, R.

1. Convenient Categories of Topological Spaces for Homotopy Theory. Arch. der Math. **22**, 545–555 (1971)

Wall, C.T.C.

1. Determination of the cobordism ring. Ann. of Math. **7**, 292–311 (1960)

Whitehead , G.W.

1. Generalized homology theory. Trans. Amer. Math. Soc. **102**, 227-283 (1962)
2. Homotopy theory. The M.I.T. Press, Cambridge, Mass. 1966

Wilkerson, C.

1. Classification of spaces of the same n-type for all n. Proc. Amer. Math. Soc. **60**, 279–285 (1976)

Williamson, R.

1. Cobordism of combinatorial manifolds. Ann. of Math. **83**, 1–33 (1966)

Wilson, W.S.

1. Brown–Peterson homology: an introduction and sampler. Regional Conference series in Mathematics **48**, Amer. Math. Soc., Providence, Rhode Island 1982.

Würgler, U.

1. Cobordism Theories of Unitary Manifolds with Singularities and Formal Group Laws. Math. Z., **150**, 239–260 (1976)
2. On the product in a family of cohomology theories associated to the invariant prime ideals of $\pi_*(BP)$. Comm. Math. Helv. **52**, 457–481 (1977)
3. A splitting for certain cohomology theories associated to BP. Manuscripta Math. **29**, 93–111 (1979)

4. On a class of 2-periodic cohomology theories. Math. Ann. **267**, 251–269 (1984)
5. Commutative ring-spectra of characteristic 2. Comm. Math. Helv. **61**, 33–45 (1986)

Yagita, N.

1. The exact functor theorem for BP_*/I_n-theory. Proc. Japan. Acad. **52**, 1–3 (1976)
2. On the algebraic structure of cobordism operations with singularities. J. London Math. Soc. **16**, 131–141 (1977)

Yosimura, Z.

1. Projective dimension of Brown–Peterson homology with modulo $(p, v_1, \ldots, v_{n-1})$. Osaka J. Math. **13**, 289–309 (1976)

List of Notations

This symbol list contains only non-standard notations; so, such symbols as $\mathbb{N}, \mathbb{Z}, \mathbb{Q}, \mathbb{R}, \mathbb{C}$, Hom, $\otimes$, Ext, $\mathcal{O}$, $\mathcal{PL}$, $\mathcal{TOP}$, $\mathcal{G}, \mathcal{U}$, $\mathcal{TOP}/\mathcal{PL}$, CP^n, CP^∞, etc, are not included in the list. Recall that we use the uniform symbol $\mathcal{V}$ for $\mathcal{O}$, $\mathcal{PL}$, $\mathcal{TOP}$, $\mathcal{G}$. So, when we write that $M\mathcal{V}_n$ is the Thom space of the universal $\mathcal{V}_n$-object this means, say, that $M\mathcal{O}_n$ is the Thom space of the universal n-dimensional vector bundle and $M\mathcal{G}_n$ is the Thom space of the universal S^{n-1}-fibration. Here we keep the convention used for the Subject Index.

Categories, Classes, Special Sets and Groups

Usually we mention only objects of categories when morphisms are clear (or can be found in the correponding subsection).

$\mathscr{A}_p$	the Steenrod algebra	II.6.25
$\mathscr{A}\mathscr{G}$	the category of abelian groups and homomorphisms	I,§2
$\mathfrak{C}$	a Serre class of abelian groups	II.4.21
$\mathscr{C}$	the category of CW-spaces	I.3.40 ff.
$\mathscr{C}^\bullet$	the category of pointed CW-spaces	I.3.40 ff.
$\mathscr{C}_{\mathrm{con}}$	the category of connected CW-spaces	I.3.40 ff.
$\mathscr{C}^\bullet_{\mathrm{con}}$	the category of connected CW-spaces	I.3.40 ff.
$\mathscr{C}_{\mathrm{f}}$	the category of finite CW-spaces	I.3.40 ff.
$\mathscr{C}^\bullet_{\mathrm{f}}$	the category of pointed finite CW-spaces	I.3.40 ff.
$\mathscr{C}_{\mathrm{fd}}$	the category of finite dimensional CW-spaces	I.3.40 ff.
$\mathscr{C}^\bullet_{\mathrm{fd}}$	the category of pointed finite dimensional CW-spaces	I.3.40 ff.
$\mathscr{C}^2$	the category of CW-pairs	II,§3
$\mathscr{C}^2_{\mathrm{f}}$	the category of finite CW-pairs	II,§3
$\mathscr{C}^2_{\mathrm{fd}}$	the category of finite dimensional CW-pairs	II,§3
$\mathscr{E}ns$	the category of sets	I,§1
$\mathscr{E}ns^\bullet$	the category of pointed sets	I,§1
$\mathscr{H}\mathscr{C}$	the homotopy category of the category $\mathscr{C}$	I.3.40 ff.
$\mathscr{H}\mathscr{C}^\bullet$	the homotopy category of $\mathscr{C}^\bullet$	I.3.40 ff.

$\mathscr{H}(F)$	the monoid of self-equivalences of a space F	IV.1.68 ff.
$\mathscr{H}\mathscr{S}$	the homotopy category of $\mathscr{S}$	II.1.9 ff.
$\mathscr{L}(m)$	the category of coherent $\pi_*(P(m))$-modules equipped with a $P(m)^*(P(m))$-action	IX.3.14 ff.
$\mathscr{M}$	the category of coherent $\pi_*(BP)$-modules equipped with a $BP^*(BP)$-action	VII.4.11 ff.
Ω	the ring $\pi_*(BP) = \mathbb{Z}[p]\,[v_0, \dots, v_n, \dots], \dim v_i = 2p^i - 2$	VII,§4
$\mathscr{P}$	the class of spectra bounded below and whose $\mathbb{Z}[p]$-homology are finitely generated over $\mathbb{Z}[p]$	IX.6.19
$\mathbb{S}$	the subalgebra of $M\mathcal{U}^*(M\mathcal{U})$ generated by s_ω	VII.3.1
$\mathscr{S}$	the category of spectra	II.1.3
$\mathscr{S}_{\mathrm{f}}$	the category of finite spectra	II.3.9 ff.
$\mathscr{S}_{\mathrm{fd}}$	the category of finite dimensional spectra	II.3.9 ff.
$\mathscr{S}_{\mathrm{s}}$	the category of suspension spectra	II.3.9 ff.
$\mathscr{S}_{\mathrm{sfd}}$	the category of spectra of the form $\Sigma^n\Sigma^\infty X$, $n \in \mathbb{Z}$, $X \in \mathscr{C}^\bullet_{\mathrm{fd}}$	II.3.9 ff.
$\mathscr{W}$	the category of weak Hausdorff compactly generated spaces	I.3.2 ff.
$\mathscr{W}^\bullet$	the category of pointed weak Hausdorff compactly generated spaces	I.3.2 ff.
$\mathbb{Z}/m$	the cyclic group of order m	I,§2
$\mathbb{Z}[p]$	the subring of $\mathbb{Q}$ consisting of fractions m/n with $(m,n) = 1 = (n,p)$, p prime	II,§5
$\mathbb{Z}[1/p]$	the subring of $\mathbb{Q}$ consisting of fractions m/p^k, p prime	II.5.16 ff.

Some Special Spaces, Spectra and Bundles

$B\mathcal{F}$	the classifying space for stable $\mathcal{F}$-objects	IV.4.8 ff.
$B\mathcal{F}_n$	the classifying space for $\mathcal{F}_n$-objects	IV.4.8 ff.
$BM\mathcal{V}$	the homotopy fiber of $w_2 : BS\mathcal{V} \to K(\mathbb{Z}/2, 2)$	VI.4.6 ff.
BP	the Brown–Peterson spectrum	VII.3.20
$BP\langle n\rangle$	the spectrum BP with singularities $v_i, i > n$	IX.6.7 ff.
$BR\mathcal{V}$	the homotopy fiber of $\delta w_2 : BS\mathcal{V} \to K(\mathbb{Z}, 3)$	VI.3.4 ff.
$B\mathcal{U}$	the classifying space for stable complex vector bundles	IV.4.16
$B\mathcal{U}_n$	the classifying space for n-dimensional complex vector bundles	
$B\mathcal{V}$	the classifying space for stable $\mathcal{V}$-objects	IV.4.4 ff.
$B(\mathcal{V}, E)$	the classifying space for E-oriented stable $\mathcal{V}$-objects	V.1.17
$B\mathcal{V}_n$	the classifying space for $\mathcal{V}_n$-objects	IV.4.2
$B(\mathcal{V}_n, E)$	the classifying space for E-oriented $\mathcal{V}_n$-objects	V.1.11
$\gamma_{\mathbb{C}}$	the universal stable complex vector bundle	IV.4.17 ff.

$\gamma_{\mathbb{C}}^n$	the universal n-dimensional complex vector bundle	IV.4.16
D^n	the n-dimensional disk	I.3.11 ff.
$\gamma_{\mathcal{V}}$	the universal stable $\mathcal{V}$-object	IV.4.5
$\gamma_{\mathcal{V}}^n$	the universal $\mathcal{V}_n$-object	IV.4.2
$\gamma_n^{\mathcal{V}}$	the universal E_n-oriented stable $\mathcal{V}$-object	V.4.10 ff.
η	the complex line bundle over CP^∞	VII.1.3
$\mathcal{F}_n$	the monoid of pointed self-equivalences $S^n \to S^n$	IV.4.15 ff.
$H(G)$	the graded Eilenberg–Mac Lane spectrum for a graded group G	II.3.32(d)
$H(\pi)$	the Eilenberg–Mac Lane spectrum for a group π	II.3.24
k	the spectrum of the connected complex k-theory	II.3.32(f)
K	the spectrum of the complex K-theory	II.3.32(e)
k^r	the $2r$-coskeleton of k	VI.2.1 ff.
$k\mathcal{O}$	the spectrum of the connected real k-theory	II.3.32(g)
$K\mathcal{O}$	the spectrum of the real k-theory	II.3.32(g)
$k(n)$	the spectrum of the connected Morava k-theory	IX.7.1
$K(n)$	the spectrum of the periodic Morava K-theory	IX.7.2
$K(\pi, n)$	the Eilenberg–Mac Lane space	II.3.24
$M(A)$	the Moore spectrum for an abelian group A	II.4.32
$M\mathcal{U}$	the Thom spectrum of the universal stable complex vector bundle	IV.7.31
$M\mathcal{U}_n$	the Thom space of the universal n-dimensional complex vector bundle	IV.7.31
$M\mathcal{V}$	the Thom spectrum of the universal stable $\mathcal{V}$-object	IV.5.12
$M\mathcal{V}_n$	the Thom space of the universal $\mathcal{V}_n$-object	IV.5.2
$M(\mathcal{V}, E)$	the Thom spectrum of the universal E-oriented stable $\mathcal{V}$-object	V.1.17 ff.
$P(n)$	the spectrum BP with singularities $v_i, i < n$	IX,§2
pt	the one-point space	I.3.1 ff.
$\mathbb{R}_+^{N+n}$	the Euclidean half-space	IV.7.8 ff.
S	the sphere spectrum	II.1.1 ff.
S^n	the n-dimensional sphere	I.3.11 ff.
θ_B^F	the standard trivial F-bundle over B	IV.1.11
θ_B^n	the standard trivial $\mathcal{V}_n$-object over B	IV.1.11

Some Constructions For a Space (Spectrum) A

A_Λ	the Λ-localization of A	II.5.2, II.5.12
$A^{(n)}$	the n-skeleton of A	I.1.1
$A_{(n)}$	the n-coskeleton of A	II.4.15
$A\|n$	the $(n-1)$-killing space (spectrum) for A	II.4.14, IV.1.39
$A^\perp$	the dual to A spectrum	II.2.3
$A[0]$	the $\mathbb{Q}$-localization of A	II.5.16 ff.
$A[p]$	the $\mathbb{Z}[p]$-localization of A	II.5.16 ff.
$A[1/p]$	the $\mathbb{Z}[1/p]$-localization of A	II.5.16 ff.
ΩA	the loop space of a space A	I.3.9
$\Omega^\infty A$	the infinite delooping of a spectrum A	II.3.27
$\Omega^n_k S^n$	the component of all maps of degree k of $\Omega^n S^n$	IV.4.22 ff.
$\pi_k(A)$	the i-th homotopy group of a spectrum (space) A	II.1.9 ff., II.3.32(a)
$\Pi_k(A)$	the i-th stable homotopy group of a space A	II.1.9 ff., II.3.32(a)
SA	the suspension over a space A	I.3.18, I.3.23
$S^n A$	the iterated suspension over a space A	I.3.18, I.3.23
ΣA	the suspension over a spectrum A	I.1.1
$\Sigma^\infty A$	the suspension spectrum over a space A	I.1.1
τA	the telescope of the spectrum A	II.1.23

Some Constructions for a Bundle (Fibration) ζ

$\operatorname{bs}\zeta$	the base of ζ	IV.1.1
$\chi(\zeta)$	the Euler class of ζ	V.1.24
$\mathfrak{F}_\varphi$	a morphism $\zeta \to (\operatorname{bs}\varphi)^*\eta$, where $\varphi : \zeta \to \eta$	IV.1.9(ii)
$\mathfrak{I}_{f,\zeta}$	a morphism $f^*\zeta \to \zeta$	IV.1.8
p_ζ	the projection in ζ	IV.1.1
$\hat{p}_\zeta$	the morphism $\zeta \to 1_{\operatorname{bs}\zeta}$	IV.1.1
$\operatorname{Prin}\zeta$	the principal bundle for ζ	IV.1.68 ff.
$\operatorname{Prin}^\bullet\zeta$	the principal bundle for sectioned ζ	IV.1.74 ff.
$\operatorname{proj}_\zeta$	the projection in ζ	IV.1.1
s_ζ	the section of the sectioned bundle ζ	IV.1.1
$\hat{s}_\zeta$	the morphism $1_{\operatorname{bs}\zeta} \to \zeta$	IV.1.1
$\operatorname{Sec}\zeta$	the set of all sections of a bundle ζ	IV.1.2
$[\operatorname{Sec}\zeta]$	the set of homotopy classes of all sections of a bundle ζ	IV.1.2
$\operatorname{ts}\zeta$	the total space of ζ	IV.1.1
$T\zeta$	the Thom space of ζ	IV.5.1, IV.5.2
$\mathfrak{z}$	the zero section of a Thom space of ζ	IV.5.4
ζ_{st}	the stabilization of ζ	IV.4.8(c)

Miscellaneous Symbols

$\mathfrak{a}(P)$	the obstruction to the existence of a quasi-multiplication	VIII.2.2 ff.
B_F	the classifying space for F-fibrations	IV.1.58
$B_{(F,*)}$	the classifying space for $(F,*)$-fibrations	IV.1.74 ff.
$\mathfrak{b}(P)$	the obstruction to the commutativity of a quasi-multiplication	VIII.2.9 ff.
$B_\bullet(Y,M,X)$	something	IV.1.63
Cf	the cone of a map (morphism) f	I.3.16, II.1.7
c^N	the Browder–Novikov map	IV.7.15(a)
$\mathfrak{c}(P)$	the obstruction to the associativity of a quasi-multiplication	VIII.2.9ff.
Cyl	the mapping cylinder over a space	IV.1.4
DCyl	the double mapping cylinder over a space	IV.1.4
Dec	the ideal of decomposable elements	II.6.5
γ^F	the universal F-fibration	IV.1.58
e_n	a map $(CP^\infty)^n \to BU_n$	VII.1.3(g)
ht	height of the formal group	VII.5.14
hom. dim	homological dimension	IX.6.4
j_Λ	the Λ-localization	II.5.2, II.5.12
j_n	usually, a map (morphism) $?_n \to ?_\infty$, for example:	
	$j_n^{\mathcal{V}} : B\mathcal{V}_n \to B\mathcal{V}$	IV.(4.7)
	$j_n^{\mathcal{U}} : B\mathcal{U}_n \to B\mathcal{U}$	IV.4.25
	$j_n := l_\infty^n : CP^n \to CP^\infty$	VII.1.3(a)
κ_n	the Postnikov invariant	II.4.19
l_n^m	the inclusion $CP^m \subset CP^n$	VII.1.3(a)
$\mathrm{Lift}_p\, f$	the set of p-liftings of f	IV.1.2
$[\mathrm{Lift}_p\, f]$	the set of homotopy classes of p-liftings of f	IV.1.2
Mf	the mapping cylinder of f	I.3.16
∇	the coaddition of a spectrum	II.1.16 ff.
ν	the normal bundle of a manifold	IV.7.12
P_n^{2r}	the Steenrod–tom Dieck operation	VII.7.7
$\mathrm{Pr}\, C$	primitive elements of a coalgebra C	II.6.21
QA	indecomposable elements of an algebra A	II.6.5
Q_i	a primitive element of the Steenrod algebra	II.(6.26) ff.
Q^i	the Kudo–Araki–Dyer–Lashof operation	VI,§1
$\mathscr{R}$	the Roos construction	III 2.8 ff.
r_n	usually, a map (morphism) $?_n \to ?_{n+1}$, for example:	
	$r_n : B\mathcal{V}_n \to B\mathcal{V}_{n+1}$	IV.4.3 ff.
	$r_n^{\mathcal{U}} : B\mathcal{U}_n \to B\mathcal{U}_{n+1}$	IV.4.16
	$r_n : L^{\Sigma_n} \to L^{\Sigma_{n+1}}$	VIII.2.16
	$r_n : P(n) \to P(n+1)$	IX.2.7

r_n^m	usually, a map (morphism) $?_m \to ?_n$, for example:	
	$r_n^m : B\mathcal{U}_m \to B\mathcal{U}_n$	VII.1.3(b)
	$r_n^m : P(m) \to P(n)$	IX.2.7 ff..
$\mathrm{Si}\, V$	simple elements of a comodule V	II.6.12
$\mathfrak{s}$	the suspension isomorphism	II.3.4
$\widehat{\mathfrak{s}}$	the suspension isomorphism	II.3.10
$\widehat{\otimes}$	the profinitely completed tensor product	III.1.15
$\otimes^{\mathrm{grad}}$	the completed graded tensor product	III.1.15
$\mathscr{T}$	the uniform symbol for DIFF, PL, TOP	IV,§7
$T\mathfrak{X}$	the telescope of a filtration $\mathfrak{X}$	I.3.19, I.3.23
$\mathcal{V}$	the uniform symbol for $\mathcal{O}$, $\mathcal{PL}$, $\mathcal{TOP}$, $\mathcal{G}$	IV.4.2
$\mathcal{V}^{\mathscr{T}}$	a uniform symbol	IV,§7
$[X, Y]$	the set of homotopy classes of maps $X \to Y$	I.3.12
$[X, Y]^{\bullet}$	the set of pointed homotopy classes of pointed maps $X \to Y$	I.3.22
$\times_B$	the total space of the induced bundle	IV.1.10
$\oplus$	the Whitney sum	IV.4.5 ff.
$*$	the join	I.3.18, IV.1.4
$\vee$	the wedge	I.3.20
$\vee^h$	the homotopy wedge	I.3.34, IV.1.4
$\wedge$	the smash product	I.3.20, II.2.1
$\wedge^h$	the homotopy smash product	I.3.36, IV.1.4
$\cap$	the cap-multiplication	II.3.44 ff.
$\cup$	the cup-multiplication	II.3.44 ff.
$\sqcup$	the disjoint union	I,§1
$\cong$	homeomorphism, isomorphism	I,§1
$\simeq$	homotopy, homotopy equivalence	I.3.12
$\simeq$	equivalence of formal groups	VII.5.1
$\simeq_B$	homotopy over a space B, homotopy equivalence over B	IV.1.5
$\simeq^{\mathrm{bun}}$	the bundle homotopy	IV.1.5
$\simeq^{CW}$	CW-equivalence	I.3.42
$\langle \varphi_\lambda \mid \varinjlim \rangle$	the universal map from the direct limit	I.2.5 ff.
$\{ f_\lambda \mid \varprojlim \}$	the universal map to the inverse limit	III.1.7

Subject Index

The notation, say, **spectrum II.1.1** means that a notion of a spectrum is defined in II.1.1. The notation, say, **wedge of spectra II.1.16 ff.** means that a notion of a wedge of spectra is defined not in II.1.16 but just after II.1.16 (and before II.1.17). Finally, the notation, say, **manifold IV, §7** means that the notion of manifold is discussed in the very beginning of §7 of Chapter IV, before IV.7.1. We arrange that all letters go after the space (blank) while the signs - and – go after all letters: e.g. **Thom–Dold isomorphism** goes after **Thom space**, ***F*-fibration** goes after **formal group** and **profinitely completed...** goes after **profinite module**.

Printing: Druckhaus Beltz, Hemsbach
Binding: Buchbinderei Schäffer, Grünstadt